Robot Manipulators

Edited by
Marco Ceccarelli

Robot Manipulators

Edited by Marco Ceccarelli

This is a reprint of the book published in 2008 by Intech

Republished by InTech

Janeza Trdine 9, 51000 Rijeka, Croatia

Technical Editor Teodora Smiljanic

Cover Designer Dino Smrekar

Additional hard copies can be obtained from orders@intechopen.com

Robot Manipulators, Edited by Marco Ceccarelli
p. cm.
978-953-7619-06-0

For EU product safety concerns:
IN TECH d.o.o., Prolaz Marije Krucifikse Kozulić 3, 51000 Rijeka, Croatia,
info@intechopen.com or visit our website at intechopen.com.

Preface

Robots can be considered as the most advanced automatic systems and robotics, as a technique and scientific discipline, can be considered as the evolution of automation with interdisciplinary integration with other technological fields.

A robot can be defined as a system which is able to perform several manipulative tasks with objects, tools, and even its extremity (end-effector) with the capability of being reprogrammed for several types of operations. There is an integration of mechanical and control counterparts, but it even includes additional equipment and components, concerned with sensorial capabilities and artificial intelligence. Therefore, the simultaneous operation and design integration of all the above-mentioned systems will provide a robotic system

The State-of-Art of Robotics already refers to solutions of early 70s as very obsolete designs, although they are not yet worthy to be considered as pertaining to History. The fact that the progress in the field of Robotics has grown very quickly has given a shorten of the time period for the events being historical, although in many cases they cannot be yet accepted as pertaining to the History of Science and Technology.

It is well known that the word "robot" was coined by Karel Capek in 1921 for a theatre play dealing with cybernetic workers, who/which replace humans in heavy work. Indeed, even in today life-time robots are intended with a wide meaning that includes any system that can operate autonomously for given class of tasks. Sometimes intelligent capability is included as a fundamental property of a robot, as shown in many fiction works and movies, although many current robots, mostly in industrial applications, have only flexible programming and are very far to be intelligent machines.

From technical viewpoint a unique definition of robot has taken time for being universally accepted.

In 1988 the International Standard Organization gives: "An industrial robot is an automatic, servo-controlled, freely programmable, multipurpose manipulator, with several axes, for the handling of work pieces, tools or special devices. Variably programmed operations make possible the execution of a multiplicity of tasks".

However, still in 1991 for example, the IFToMM International Federation for the Promotion and Mechanism and Machine Science (formerly International Federation for the Theory of Machines and Mechanisms) gives its own definitions: Robot as "Mechanical system under automatic control that performs operations such as handling and locomotion"; and Manipulator as "Device for gripping and controlled movements of objects".

Even roboticists use their own definition for robots to emphasize some peculiarities, as for example from IEEE Community: "a robot is a machine constructed as an assemblage of joined links so that they can be articulated into desired positions by a programmable controller and precision actuators to perform a variety of tasks".

However, different meanings for robots are still persistent from nation to nation, from technical field to technical field, from application to application.

Nevertheless, a robot or robotic system can be recognized when it has the three main characteristics: mechanical versatility, reprogramming capacity, and intelligent capability.

Summarizing briefly the concepts, one can understand the above-mentioned terms as follows; mechanical versatility refers to the ability of the mechanical design to perform several different manipulative tasks; reprogramming capacity concerns with the possibility to update the operation timing and sequence, even for different tasks, by means of software programming; intelligent capability refers to the skill of a robot to recognize its owns state and neighbour environment by using sensors and human-like reasoning, even to update automatically the operation.

Therefore, a robot can be considered as a complex system that is composed of several systems and devices to give:

- mechanical capabilities (motion and force);
- sensorial capabilities (similar to human beings and/or specific others);
- intellectual capabilities (for control, decision, and memory).

In this book we have grouped contributions in 28 chapters from several authors all around the world on the several aspects and challenges of research and applications of robots with the aim to show the recent advances and problems that still need to be considered for future improvements of robot success in worldwide frames. Each chapter addresses a specific area of modeling, design, and application of robots but with an eye to give an integrated view of what make a robot a unique modern system for many different uses and future potential applications.

Main attention has been focused on design issues as thought challenging for improving capabilities and further possibilities of robots for new and old applications, as seen from today technologies and research programs. Thus, great attention has been addressed to control aspects that are strongly evolving also as function of the improvements in robot modeling, sensors, servo-power systems, and informatics. But even other aspects are considered as of fundamental challenge both in design and use of robots with improved performance and capabilities, like for example kinematic design, dynamics, vision integration.

Maybe some aspects have received not a proper attention or discussion as an indication of the fecundity that Robotics can still express for a future benefit of Society improvement both in term of labor environment and productivity, but also for a better quality of life even in other fields than working places.

Thus, I believe that a reader will take advantage of the chapters in this edited book with further satisfaction and motivation for her or his work in professional applications as well as in research activity.

I thank the authors who have contributed with very interesting chapters in several subjects, covering the many fields of Robotics. I thank the editor I-Tech Education and Publishing KG in Wien and its Scientific Manager prof Aleksandar Lazinica for having supported this editorial initiative and having offered a very kind editorial support to all the authors in elaborating and delivering the chapters in proper format in time.

Editor

Marco Ceccarelli

LARM: Laboratory of Robotics and Mechatronics, DiMSAT - University of Cassino
Via Di Biasio 43, 03043 Cassino (Fr), Italy

Contents

1

Experimental Results on Variable Structure Control for an Uncertain Robot Model

K. Bouyoucef[1], K. Khorasani[1] and M. Hamerlain[2]
[1]Department of Electrical and Computer Engineering, Concordia University,
[2]Centre de Développement des Technologies Avancées (CDTA)
[1]Canada, [2]Algeria

1. Introduction

To reduce computational complexity and the necessity of utilizing highly nonlinear and strongly coupled dynamical models in designing robot manipulator controllers, one of the solutions is to employ robust control techniques that do not require an exact knowledge of the system. Among these control techniques, the sliding mode variable structure control (SM-VSC) is one that has been successfully applied to systems with uncertainties and strong coupling effects.

The sliding mode principle is basically to drive the nonlinear plant operating point along or nearby the vicinity of the specified and user-chosen hyperplane where it 'slides' until it reaches the origin, by means of certain high-frequency switching control law. Once the system reaches the hyperplane, its order is reduced since it depends only on the hyperplane dynamics.

The existence of the sliding mode in a manifold is due to the discontinuous nature of the variable structure control which is switching between two distinctively different system structures. Such a system is characterized by an excellent performance, which includes insensitivity to parameter variations and a complete rejection of disturbances. However, since this switching could not be practically implemented with an infinite frequency as required for the ideal sliding mode, the discontinuity generates a chattering in the control, which may unfortunately excite high-frequency dynamics that are neglected in the model and thus might damage the actual physical system.

In view of the above, the SM-VSC was restricted in practical applications until progresses in the electronics area and particularly in the switching devices in the nineteen seventies. Since then, the SM-VSC has reemerged with several advances for alleviating the undesirable chatter phenomenon. Among the main ideas is the approach based on the equivalent control component which is added to the discontinuous component (Utkin, 1992; Hamerlain et al, 1997). In fact, depending on the model parameters, the equivalent control corresponds to the SM existence condition. Second, the approach studied in (Slotine, 1986) consists of the allocation of a boundary layer around the switching hyperplane in which the discontinuous control is replaced by a continuous one. In (Harashima et al, 1986; Belhocine et al, 1998), the gain of the discontinuous component is replaced by a linear function of errors. In (Furuta et

al, 1989), the authors propose a technique in which the sliding mode is replaced by a sliding sector.
Most recent approaches consider that the discontinuity occurs at the highest derivatives of the control input rather than the control itself. These techniques can be classified as a higher order sliding mode approaches in which the state equation is differentiated to produce a differential equation with the derivative of the control input (Levant & Alelishvili, 2007; Bartolini et al, 1998). Among them, a particular approach that is introduced in (Fliess, 1990) and investigated in (Sira-Ramirez, 1993; Bouyoucef et al, 2006) uses differential algebraic mathematical tools. Indeed, by using the differential primitive element theorem in case of nonlinear systems and the differential cyclic element theorem in case of linear systems, this technique transforms the system dynamics into a new state space representation where the derivatives of the control inputs are involved in the generalization of the system representation. By invoking successive integrations to recover the actual control the chattering of the so-called Generalized Variable Structure (GVS) control is filtered out.
In this paper, we present through extensive simulations and experimentations the results on performance improvements of two GVS algorithms as compared to a classical variable structure (CVS) control approach. Used as a benchmark to the GVS controllers, the CVS is based on the equivalent control method. The CVS design methodology is based on the differential geometry whereas the GVS algorithms are designed using the differential algebraic tools. Once the common state representation of the system with the derivatives of the control input is obtained, the first GVS algorithm is designed by solving the well-known sliding condition equation while the second GVS algorithm is derived on the basis of what is denoted as the hypersurface convergence equation.
The remainder of this chapter is organized as follows. After identifying experimentally the robot axes in Section 2, the procedure for designing SM-VSC algorithms is studied in Section 3. In order to evaluate the chattering alleviation and performance improvement, simulations and experimentations are performed in Section 4. Finally, conclusions are stated in Section 5.

2. Identification of robot manipulator axes

In this study, Generalized Variable Structure (GVSC) control techniques are implemented on the Robot Manipulator (RP41) as illustrated in Fig. 1-a. From the schematic that is depicted in Fig. 1-b, one can observe that the RP41 is a SCARA robot with four degrees of freedom. The three first joints (J1, J2, and J3) are rotoide while the fourth one (T) is prismatic. To each robot axis, one assigns a controller that uses only a measured angular signal that is generated by a shaft encoder via a 12 bit Analog/Digital converter. As far as control is concerned, it is digitized from 0 to 4096. As illustrated in Table 1, this interval corresponds to an analog input of the converter spanning from – 5 to + 5 Volts. In order to activate the DC drive of each robot joint, these low voltages are amplified by a power board to the range of -24 to +24 Volts.
In virtue of the robustness properties, uncertain linear models of the robot are obtained for the design of the SM-VS controllers. This section briefly presents the experimental identification of the three robot axes resulting in a suitable second order linear model for each manipulator axis.

(a) The SCARA Robot RP41

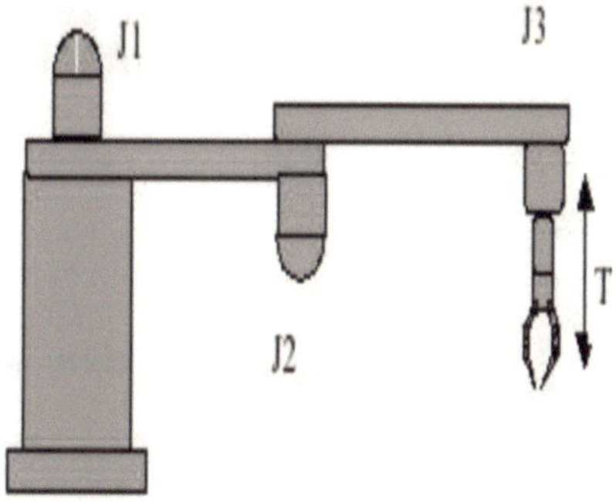

(b) Schematic of the SCARA RP41 mechanism

Figure 1. The SCARA Robot Manipulator (RP 41), *(Centre de Développement des Technologies Avancées, Algiers)*

Digital controller output	D/A Converter output [volts]	Robot DC motors input [volts]
0	+5	+24
2048	0	0
4096	-5	-24

Table 1. Digital and analog control ranges

For further explanations on the identification of the arm axes, the reader can refer to our previous investigations (Youssef et al, 1998). The complete Lagrange formalism-based dynamic model of the considered SCARA robot has been experimentally studied in (Bouyoucef et al, 1998), in which the model parameters are identified and then validated by using computed torque control algorithm. The well-known motion dynamics of the three joints manipulator is described by equation (1)

$$M(q)\,.\,\ddot{q} + h(q\,,\dot{q}) + g(q) = u \tag{1}$$

Where $q, \dot{q}, \ddot{q} \in R^3$ are the vectors of angular position, velocity and acceleration, respectively, $M(.) \in R^{3x3}$ is the symmetric positive definite inertia matrix, $h(.) \in R^{3x3}$ is the coefficient matrix of the centripetal and Coriolis torques, $g(.) \in R^3$ is the vector of the gravitational torques, and $u(.) \in R^3$ is the vector of torques applied to the joints of the manipulator.

As developed in (Youssef et al, 1998), considering the diagonal elements preponderance of the non singular matrix $M(q)$, and replacing $h(q,\dot{q})$ and $g(q)$ by $C(q,\dot{q})\dot{q}$ and $G(q)q$, respectively, $M^{-1}(q)\,C(q,\dot{q})$, $M^{-1}(q)\,G(q)$ and $M^{-1}(q)$ by A_1, A_0 and B, respectively, equation (1) can be written as follows:

$$\ddot{q} + A_1\dot{q} + A_0 q = B\,u \tag{2}$$

where each of the diagonal matrices A_0, A_1 and B contains the dynamic parameters of the three robot axes for the angle, rate and control variables, respectively. On the basis of the plant input/output data, the parametric identification principle consisting of the estimation of the model parameters according to a priori user-chosen structure was performed. Adopting the ARX (Auto regressive with exogenous input) model, and using the Matlab software, the off-line identification generated the robot parameters according to model (2), which are illustrated in Table 1.

$A_0 = diag\,[-5.4 \quad -2.41 \quad -117]$
$A_1 = diag\,[560.7 \quad 200 \quad 413.5]$
$B = diag\,[0.5 \quad 0.65 \quad 7.5]$

Table 1. The identification of the robot parameters corresponding to model (2)

Note that in compliance with model (2), the obtained parameters correspond to the robot model that is used in the CVS control approach, which constitutes in this study as the benchmark to our proposed GVS control approaches. In order to implement GVS approaches, model (2) is not suitable since it doesn't exhibit the derivatives of the control, however, model (3) that contains the zeros dynamics is utilized instead, namely

$$\ddot{q} + A_1\dot{q} + A_0 q = B_1\,\dot{u} + B_0 u \tag{3}$$

Using the same identification procedure as before, the parameters for model (3) are now given in Table 2.

$A_0 = diag\,[-2.4 \quad -2.965 \quad -13.258]$
$A_1 = diag\,[201 \quad 200.92 \quad 213.92]$
$B_0 = diag\,[0.65 \quad 0.67 \quad 0.61]$
$B_1 = diag\,[0.041 \quad 0.004 \quad 0.006]$

Table 2. The identification of the robot parameters corresponding to model (3)

3. Sliding mode-based variable structure control strategy

The CVS control has been used for a number of years. The switching occurs on the control variable, and this is discussed in the next subsection in the context of differential geometry and constitutes in this study as the benchmark to GVS control approaches. Recently the GVS scheme was introduced in (Fliess, 1990) where the switching occurs on the highest derivative of the control input. The GVS analysis and design are studied in the context of the differential algebra. In subsection 3.2, we design two GVS control approaches, the first GVS approach is designed by solving the well-known sliding condition equation, while the second GVS approach is derived on the basis of what is denoted as the hypersurface convergence equation.

3.1 Classical variable structure control in the differential geometry context

Consider the nonlinear dynamical system in which the time variable is not explicitly indicated, that is

$$\frac{dx}{dt} = f(x) + g(x)U \tag{4}$$

where $x \in X$ is an open set of R^n, $f(x) = [f_1, f_2, \cdots, f_n]^T$ and $g(x) = [g_1, g_2, \cdots, g_n]^T$ are vector fields defined on R^n with $g(x) \neq 0 \ \forall x \in X$, and the control is defined so that $U : R^n \to R$.

Assume a hypersurface $S = \{x \in R^n : S(x) = 0\}$ is denoted as the 'sliding surface' on which discontinuous control functions of the type

$$U = \begin{cases} U^+(x) & if \quad S(x) > 0 \\ U^-(x) & if \quad S(x) < 0 \end{cases} \tag{5}$$

make the surface attractive to the representative point of the system such that it slides until the equilibrium point is reached. This behavior occurs whenever the well-known sliding condition $S\dot{S} < 0$ is satisfied (Utkin, 1992).

Using the directional derivative $L_h \sigma$, this condition can be represented as

$$\begin{cases} \lim\limits_{S \to 0^+} L_{f+gU^+} S < 0 \\ \lim\limits_{S \to 0^-} L_{f+gU^-} S > 0 \end{cases} \tag{6}$$

or by using the gradient ∇ of S and the scalar product $< .,. >$ as

$$\begin{cases} \lim\limits_{S \to 0^+} \left\langle \nabla S, f + g \cdot U^+ \right\rangle < 0 \\ \lim\limits_{S \to 0^-} \left\langle \nabla S, f + g \cdot U^- \right\rangle > 0 \end{cases} \tag{7}$$

A geometric illustration of this behavior is shown in Figure 2, in which the switching of the vector fields occurs on the hypersurface $S(x) = 0$.

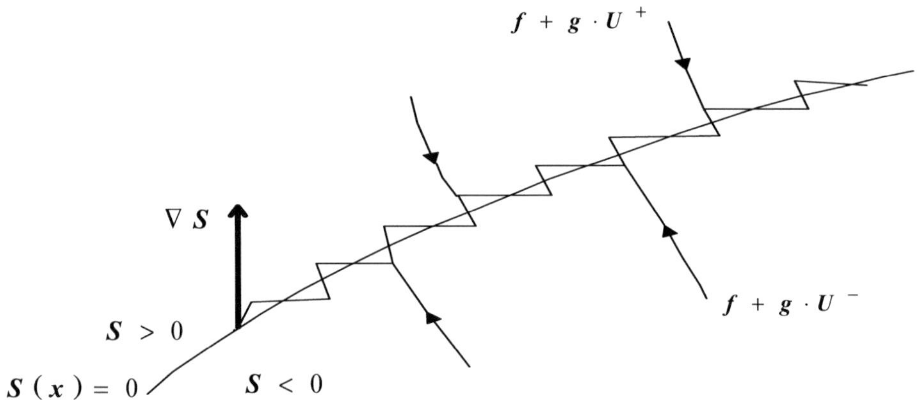

Figure 2. The geometric illustration of the sliding surface and switching of the vector fields on the hypersurface $S(x)=0$

Depending on the system state with respect to the surface, the control is selected such that the vector fields converge to the surface. Specifically, using the equivalent control method, the classical variable structure control law can be finally expressed as a sum of two components as follows,

$$U = U_{eq} + \delta U \tag{8}$$

where the equivalent control component U_{eq} is derived for the ideal sliding mode so that the previously defined hypersurface is a local invariant manifold. Therefore, if $S(x)=0$,

$$L_{f+g \cdot U_{eq}} S = \left\langle \nabla S, f + g \cdot U_{eq} \right\rangle = 0 \text{, then}$$

$$U_{eq} = -\frac{\langle \nabla S, f \rangle}{\langle \nabla S, g \rangle} = -\frac{L_f S}{L_g S} \tag{9}$$

whereas the second component corresponds to the discontinuous control so that $\delta U = -M sign(S)$ where the gain M should be chosen to be greater than the perturbation signal amplitude. A typical control U (dotted line), and its components U_{eq} (solid line), and δU (dashed line) are illustrated in Fig. 3. It can be seen that the state of the discontinuous component δU changes from continuous and positive to discontinuous with variable sign. This change coincides to the first crossing of the surface $S(x)=0$. In compliance with the equivalent control component that is always positive, it also switches since it is derived by using the derivative of the surface but with a small amplitude. The switching of the equivalent control component occurs one iteration later than the discontinuous component switching. The control U that is always postive corresponds to the geometric sum of both U_{eq} and δU .

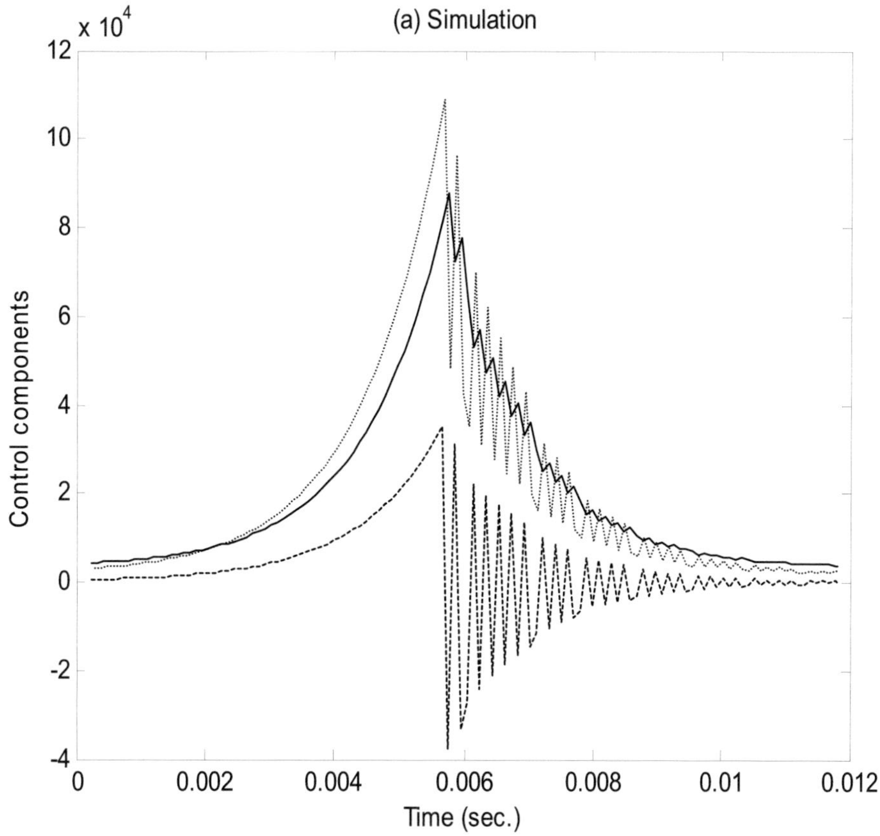

Figure 3. The control and its equivalent and discontinuous components (U dotted line, U_{eq} solid line, and δU dashed line)

3.2 Generalized variable structure control

In the context of differential algebra, and under the existence conditions of the differential primitive element for nonlinear systems, or cyclic element in the case of linear systems, the elimination of the state in the original Kalman state representation leads to the pair Generalized Control Canonical Form (GCCF) and Generalized Observable Canonical Form (GOCF). By associating for example the output equation $y = h(x)$ to the given state equation (4), the elimination of x in both state and output equations leads to the following differential equation:

$$\varsigma(y, \dot{y}, \cdots, y^{(d-1)}, y^{(d)}, u, \dot{u}, \cdots, u^{(\alpha)}) = 0 \tag{10}$$

where $\alpha = d - r$ is a strictly positive integer related to the relative degree r of the output function y with respect to the scalar input u. The integer d is defined such that the rank condition (11) should be satisfied

$$rank\frac{\delta(h,h,\cdots,h^{(d-1)})}{\delta x} = rank\frac{\delta(h,h,\cdots,h^{(d-1)},h^{(d)})}{\delta x} \tag{11}$$

Defining a new variable $\eta_i = y^{i-1}$ where $i=1,\cdots,\ d$ and assuming that the Jacobian $\dfrac{\delta(\xi)}{\delta(y)^{(d)}}$ is locally nonsingular, the transformation from the implicit input-output representation to the locally explicit GOCF given by (12) may be accomplished as follows:

$$(GOCF)\begin{cases} \dot{\eta}_i = \eta_{i-1};\quad (i=1,2,\cdots,n-1) \\ \dot{\eta}_n = \varsigma(\eta,u,\dot{u},\cdots,u^{(\alpha)}) \\ y=\eta_1 \end{cases} \tag{12}$$

For tracking control purposes, one considers a reference trajectory $W_R(t)=[y_R(t),\dot{y}_R(t),...,y_R^{(n-1)}(t)]^T$ and defines the tracking error vector $e(t)=[e_1,e_2,...,e_n]^T = \eta(t)-W_R(t)$ such that

$$\begin{cases} e_1 = y - y_R \\ e_i = e_1^{(i-1)} = \eta_i - y_R^{(i-1)},\quad i=2,\cdots,n \end{cases} \tag{13}$$

The system (12) is now rewritten in the error coordinates as follows:

$$\begin{cases} \dot{e}_i = e_{i+1},\quad (i=1,\cdots,n-1) \\ \dot{e}_n = \varsigma(W_R(t)+e(t),u,\dot{u},...,u^{(\alpha)}) - y_R^{(n)}(t) \\ e_1 = y - y_R \end{cases} \tag{14}$$

It is worth mentioning that the state representation of the system obtained with the derivatives of the control input constitutes the core of the GVS control design procedure. Using the common state representation (14), several GVS control approaches can be studied. Three approaches are considered for deriving the GVS control algorithm, namely

a. by solving the well-known sliding condition $S\dot{S}<0$ after substitution of (14) (the reader can refer to (Belhocine et al, 1998) for more details and experimental results about this approach) ;
b. by using what is denoted as the hypersurface convergence equation (15),

$$\frac{dS}{dt} + \mu\, S = -\mu\,\Omega sign(S) \tag{15}$$

where μ and Ω are positive design parameters, and $sign\,(S)$ is the signum function, and

c. by considering the following feedback control that is introduced in (Messager, 1992).

$$\varsigma\left(\eta_1,\ldots,\eta_n,u,\dot{u},\ldots,u^{(\alpha)}\right)=\sum_{i=1}^{n}\bar{a}_i\eta_i+\sum_{i=1}^{\alpha}\bar{b}_i v_i \quad (16)$$

where $\eta=\eta_1,\ldots,\eta_n$, $v=v_1,\cdots,v_\alpha$ is the new input, and $\bar{a}_i$, $\bar{b}_i$ are chosen according to the stability of the resulting system.
In this study, a linear hypersurface can be defined in the error state space $e(t)$ as

$$S(t)=e_n+\sum_{i=1}^{n-1}s_i e_i \quad (17)$$

The representative operating point of the control system, whose structure switches on the highest derivative of the input as shown in Fig. 4, slides on the defined surface $S(t)=0$ until the origin, whenever the existence condition $S\dot{S}<0$ holds.

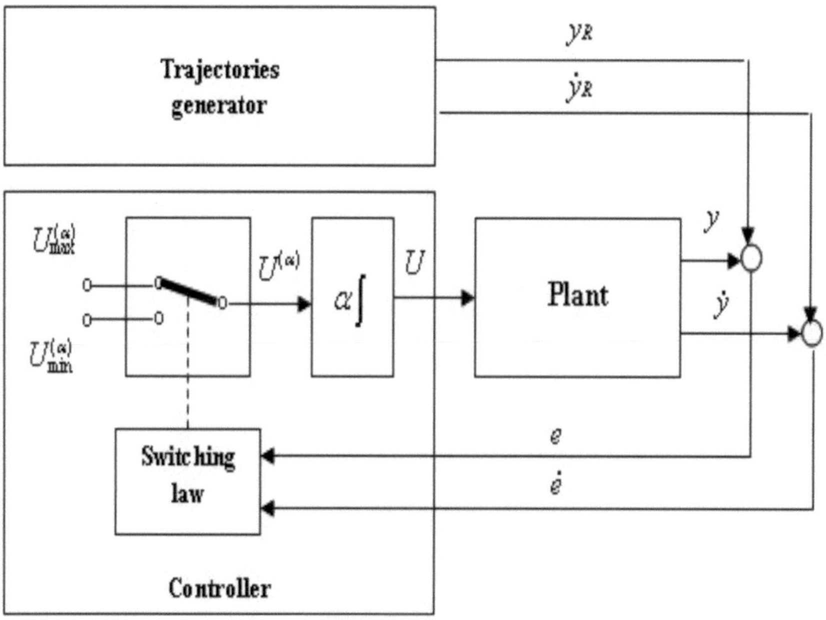

Figure 4. The switching principle in the GVS control system

3.3 The generalized variable structure control controller design

Through its robustness property, the GVS is insensitive to the interactions between manipulator axes which can be regarded as disturbances. Therefore, for the considered MIMO robotic system, the controller can be designed as a SISO system, in addition, the control design does not require an accurate model. As mentioned earlier, in order to design the aforementioned GVS control laws, one starts by deriving the common state

representation of the system with the derivatives of the control input by using the second-order linear model (3) that was obtained in robot axes identification section.
However, let us rewrite for each arm axis the second-order linear model (3) where q, $\dot{q}$, and $\ddot{q}$ are the joint angles, angular rates, and accelerations, respectively,

$$\ddot{q}_i + a_{1i}\dot{q}_i + a_{0i}q_i = b_{0i}u_i + b_{1i}\dot{u}_i; \quad (i=1,\cdots,3) \tag{18}$$

Let us now introduce a new set of state variables that are given according to

$$\begin{cases} \eta_{1i} = q_i \\ \eta_{2i} = \dot{q}_i = \dot{\eta}_{1i} \\ \dot{\eta}_{2i} = \ddot{q}_i \end{cases} \tag{19}$$

Therefore, by substituting the new variables (19) into (18), the GOCF representation is obtained as follows:

$$(GOCF)\begin{cases} \dot{\eta}_{1i} = \eta_{2i} \\ \dot{\eta}_{2i} = -a_{0i}\eta_{1i} - a_{1i}\eta_{2i} + b_{0i}u_i + b_{1i}\dot{u}_i \\ q_i = \eta_{1i} \end{cases} \tag{20}$$

Now to consider the control system design in the error state space, let us define the error variables as shown in (21):

$$\begin{cases} e_{1i} = \eta_{1i} - \eta_{iref} \\ \dot{e}_{1i} = e_{2i} = \dot{\eta}_{1i} - \dot{\eta}_{iref} = \eta_{2i} - \dot{\eta}_{iref} \\ \dot{e}_{2i} = \dot{\eta}_{2i} - \ddot{\eta}_{iref} \end{cases} \tag{21}$$

where η_{iref} is the angular reference corresponding to the robot axis i. By substituting these into the GOCF model (20), we get

$$(GOCF)\begin{cases} \dot{e}_{1i} = e_{2i} \\ \dot{e}_{2i} = -a_{0i}e_{1i} - a_{1i}e_{2i} + b_{0i}u_i + b_{1i}\dot{u}_i - a_{0i}\eta_{iref} - a_{1i}\dot{\eta}_{iref} - \ddot{\eta}_{iref} \\ q_i = e_{1i} + \eta_{iref} \end{cases} \tag{22}$$

Following the GOCF representation that is obtained in the error state space and is given by (22), the derivation of two GVS control approaches are performed for the same linear switching surface that is given by (17). In fact, the objective is to ensure robustness of the closed-loop system against uncertainties, unmodeled dynamics and external perturbations. This will be the case provided that for the above system (22), a sliding mode exists in some neighborhood of the switching surface (23) with intersections that are defined in the same space of which function (23) is characterized, that is

$$S_i(t) = s_{1i}e_{1i} + \dot{e}_{1i} = s_{1i}e_{1i} + e_{2i}; \quad (i=1,\cdots,3) \tag{23}$$

with the derivatives as given by

$$\dot{S}_i(t) = s_{1i}\dot{e}_{1i} + \dot{e}_{2i} = s_{1i}e_{2i} + \dot{e}_{2i}; \quad (i = 1, \cdots, 3) \tag{24}$$

where s_{1i} designates positive parameters corresponding to each robot axis.
Let us designate the first GVS control approach as GVS1 which is derived by solving the sliding existence condition $S\dot{S} < 0$ in which the GOCF model (22) is substituted, that is

$$\begin{cases} \dot{u}_i = b_{1i}^{-1}[-b_{0i}u_i + a_{0i}e_{1i} + (a_{1i} - s_{1i})e_{2i} \\ + a_{0i}\eta_{iref} + a_{1i}\dot{\eta}_{iref} + \ddot{\eta}_{iref} - M_i sign(S_i)] \end{cases} \tag{25}$$

The second GVS control approach that is designated as GVS2 is derived by imposing that the defined surfaces are solutions to the differential equations given by (26):

$$\frac{dS_i}{dt} + \mu_i S_i = -\mu_i \Omega_i sign(S_i); \quad (i = 1, \cdots, 3) \tag{26}$$

where the positive design parameters μ and Ω are chosen such that the dynamics described by the differential equation (27) is asymptotically stable. The latter can be written in an explicit form by substituting (23) and (24) into (26), that is

$$\dot{e}_{2i} = -s_{1i}e_{2i} - \mu_i(\Omega_i sign(S_i) + S_i); \quad (i = 1, \cdots, 3) \tag{27}$$

Now, let us consider (27) as a feedback control to model (22), that is

$$\begin{cases} \dot{e}_{2i} = -a_{0i}e_{1i} - a_{1i}e_{2i} + b_{0i}u_i + b_{1i}\dot{u}_i - a_{0i}\eta_{iref} - a_{1i}\dot{\eta}_{iref} - \ddot{\eta}_{iref} \\ \quad = -(s_{1i} + \mu_i)e_{2i} - \mu_i s_{1i}e_{1i} - \mu_i \Omega_i sign(S_i) \\ \quad = \sum_{j=1}^{2} \bar{a}_{ji}e_{ji} + v_i \end{cases} \tag{28}$$

One can observe that the above is the feedback control given by (16), where $\sum_{j=1}^{2} \bar{a}_{ji}e_{ji} + v_i$ is the resulting system corresponding to each axis, and the coefficients $\bar{a}_{ji}$ are chosen such that stability of the closed-loop system is ensured. Consequently, the above feedback control leads to the control law specified below:

$$\begin{cases} \dot{u}_i = b_{1i}^{-1}[b_{0i}u_i + (a_{0i} - \mu_i s_{1i})e_{1i} + (a_{1i} - s_{1i} - \mu_i)e_{2i} \\ \quad + a_{0i}\eta_{iref} + a_{1i}\dot{\eta}_{iref} + \ddot{\eta}_{iref} - \mu_i \Omega_i sign(S_i)] \end{cases} \tag{29}$$

Governed by integration of (29), the signal u constitutes the control variable that is sent to the plant and represents the advantages of the chattering alleviation scheme in comparison to the approach given by (16). This is possible since the differential equation permits one to better adjust the convergence of $S_i(t)$.

4. Simulations and experimental results

The simulations and experimentations presented in this section are conducted to demonstrate the effectiveness of the results obtained on the SCARA robot manipulator as illustrated in Fig. 1-a by using the two generalized variable control approaches designated as GVS1 and GVS2 in the previous section. These two GVS approaches are compared to the CVS control approach in terms of performance improvement and chattering alleviation capability.
Simulations are first performed on the model of the robot axis 3, for a step input, which show the benefits of the GVS control approaches and particularly the GVS2 in terms of performance and chattering alleviation by comparison with the CVS control approach. Subsequently, a set of experimentations are conducted on the three axes of the robot to confirm for a step input the simulation results as well as the robustness of the proposed algorithm. Note that in the comparative study only the results corresponding to the rotoide axis 3 are presented since its dynamics is faster than those of the two other axes 2 and 3 and the end effector that it supports.
Furthermore, the experimental results corresponding to the three axes that are obtained in the tracking mode are also presented.
In the simulation presented below, (a) the sampling time is set to $1ms$, (b) the control parameters corresponding to the CVS are designed to be $s_1 = 80$, $M = 5.10^3$, and those of the GVS1 are set so that $s_1 = 100$, $M = 2.10^5$; whereas those of the GVS2, they are designed to be $s_1 = 150$, $\mu = 1000$, and $\Omega = 50$; and (c) the reference angles are set to $q_{ref} = 2.1\, rad$. In the simulation results presented in Fig. 5 (a), one can observe that corresponding to step responses of the CVS (solid blue line), the GVS1 (dotted red line) and the GVS2 (dashed green line) on one hand the axis angle converges to its reference by using any control approach, and on the other hand, the system performance is improved, particularly in terms of the system response time when the GVS2 is used. The last three rows of Fig. 5 are dedicated to the control and the phase plane characteristics of the CVS, GVS1 and GVS2, respectively. The control characteristics are shown in the left hand side column while the characteristics of the phase plane are illustrated in the right hand side column. As far as the control characteristics (b), (d), and (f) are concerned the alleviation of the chattering phenomenon is readily validated when the GVS control approaches are used. This alleviation is more significant by using the GVS2 approach. However, the price that is paid for this improvement is in the increase of the control effort. From the plots in Fig. 5(c), (e), (g), one can observe that the behaviour of the system consists of first being attracted to the surface and then sliding on the surface with a slope s_{1i} linearly towards the origin. Furthermore, note that by increasing the design parameter s_{1i}, one improves the performance of the control system.

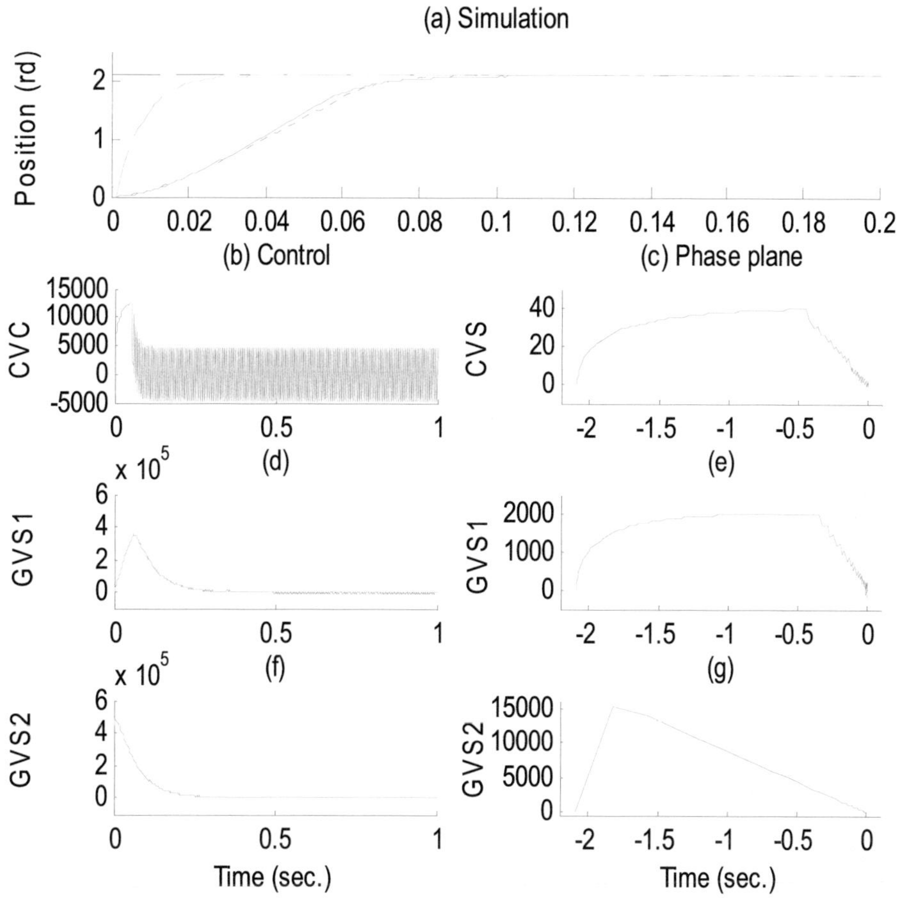

Figure 5. Comparative simulations of the GVS1, GVS2, and CVS controllers on the robot axis model

After designing the CVS and GVS controllers on the basis of a three axes model parameters discussed in the identification section, the experimental implementation results of the three robot axes are illustrated in Figs. 6-10.

In the experimentations presented below, (a) the sampling time is set to $5ms$, (b) the best controller parameters are: $s_{1i} = [20\ \ 80\ \ 40]$ and $M_i = [65\ \ 10^3\ \ 200]$ for the CVS, and $s_{1i} = [80\ \ 60\ \ 5]$ and $M_i = [1500\ \ 500\ \ 25]$ for the GVS1 whereas for the GVS2 they are set to $s_{1i} = [120\ \ 60\ \ 5]$, $\mu_i = [60\ \ 40\ \ 80]$ and $\Omega_i = [20\ \ 10\ \ 0.1]$, with $i = 1, \cdots, 3$ corresponding to the robot axes, and (c) the three axes initial angles are set to $q_{initial} = [0.92\ \ 0.92\ \ 0.52]$ radians and the final angles are set to $q_{final} = [2.09\ \ 2.09\ \ 2.09]$ radians as illustrated in Table 4.

The first experiment corresponding to Figs. 6 and 7 is conducted to confirm the simulation results that are illustrated in Fig. 5. These results are obtained without a payload and

external disturbances using step reference inputs and the initial and final angles as stated earlier. Figure 6(a) shows that the fastest step response is obtained by using the GVS2 approach and where the GVS1 approach step response is faster than that of the CVS approach. In addition, Fig. 6(b), (d), (f), that correspond to the control characteristics of the CVS, GVS1 and GVS2, respectively, demonstrate the chattering alleviation of the GVS control approaches. In order to show this improvement, the three control characteristics in Fig. 6 are zoomed in the steady state (i.e., 1s ≤ time ≤ 3s) and are depicted in Fig. 7. Indeed, one can clearly observe that the GVS1 control approach illustrated by graph (b) alleviates the chattering better than the CVS control that is given by graph (a). However, our best results are obtained with the GVS2 control approach as shown in graph (c).
The diminishing of the chattering is also visible through the smooth plots in Fig. 6(e) and (g) corresponding to the phase plane of GVS1 and GVS2 in comparison to the CVS that is shown in Fig. 6(c). From the above comparative results, it can be seen that the GVS2 control approach enjoys the best capabilities for chattering alleviation and performance improvement in comparison to the CVS and GVS1 control approaches.

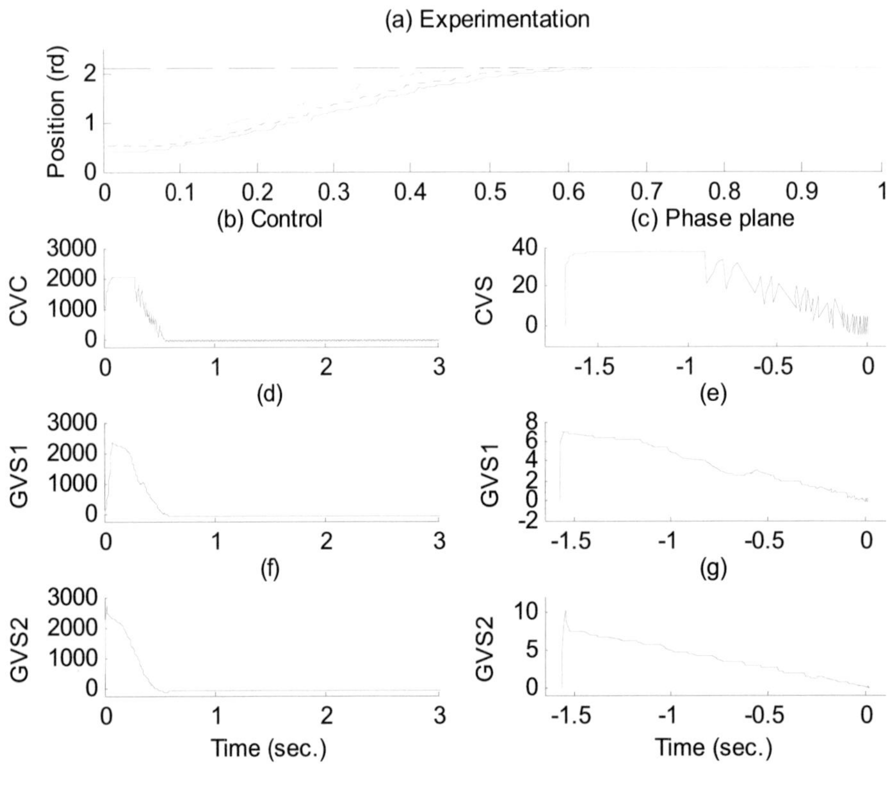

Figure 6. Comparative experimental results of the GVS1, GVS2, and CVS controllers for the robot axis model

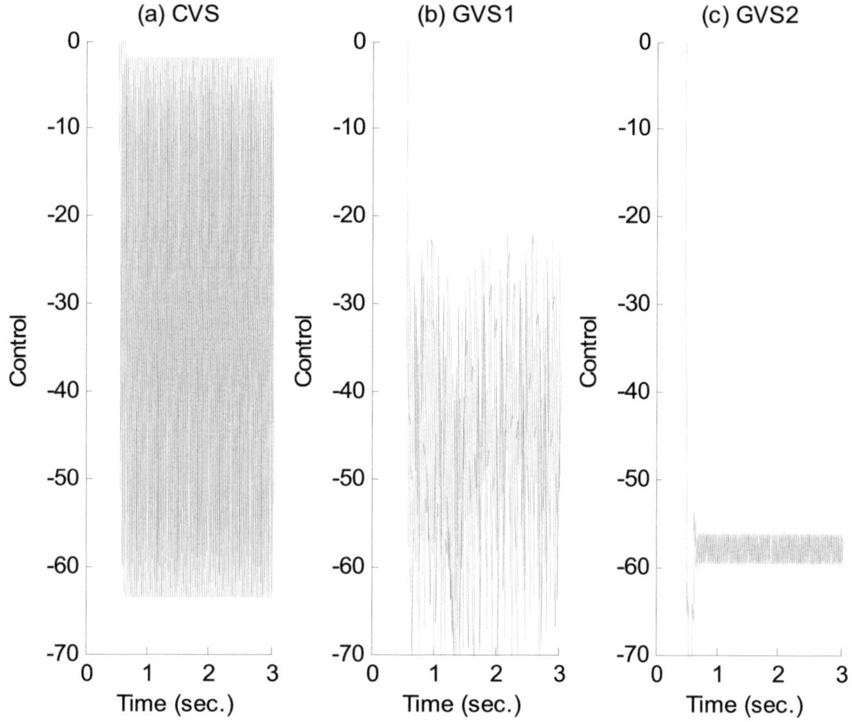

Figure 7. The zoomed graphs of the control characteristics shown in Fig. 6(b), (d), and (f) in the steady state

The second experimentation is operated in the tracking mode on our SCARA robot in order to show the robustness properties of our proposed GVS2 controller to an external disturbance and parameter variations caused by a 900 g payload. As shown in the following algorithm, depending on the difference absolute value between the final and the initial axis angles, the reference trajectory (Belhocine et al, 1998) applied to each robot axis has a trapezoidal or triangular profile, whose cruising velocity and acceleration amplitude are presented in Table 4. Note that, the initial and final configurations of the reference trajectory given in Table 4 correspond to the same initial and final angles that are used in the first experimentation conducted in the regulation mode.

Note also that for this experimentation, the sampling period for the new reference trajectory is $35ms$ whereas the regulation sampling time is kept at $5ms$, as in the first experimentation.

Axis	$q_i(rd)$	$q_f(rd)$	V (rd/s)	A(rd/s^2)
1,2	0.92	2.09	0.35	0.35
3	0.52	2.09	3.00	3.00

Table 4. The reference trajectory settings

Algorithm generating the reference trajectories:

If $\left|q_f - q_i\right| \geq \frac{V^2}{A}$

then use the trapezoidal profile

$$t_{acc} = \frac{V}{A}$$

$$t_{vit} = \left(\frac{1}{V}\right)\left|q_f - q_i\right| - t_{acc}$$

$$t_{tot} = t_{vit} + 2t_{acc}$$

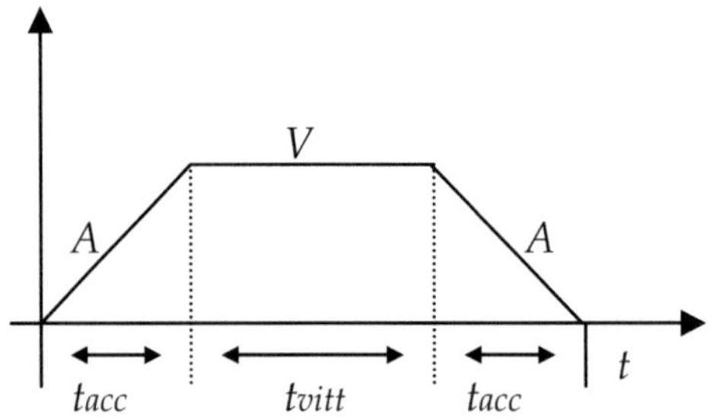

Otherwise use the triangular profile

$$t_{acc} = \frac{V}{A}$$

$$t_{vit} = 0$$

$$V = \left(\left|q_f - q_i\right|.A\right)^{\frac{1}{2}}$$

$$t_{tot} = 2t_{acc}$$

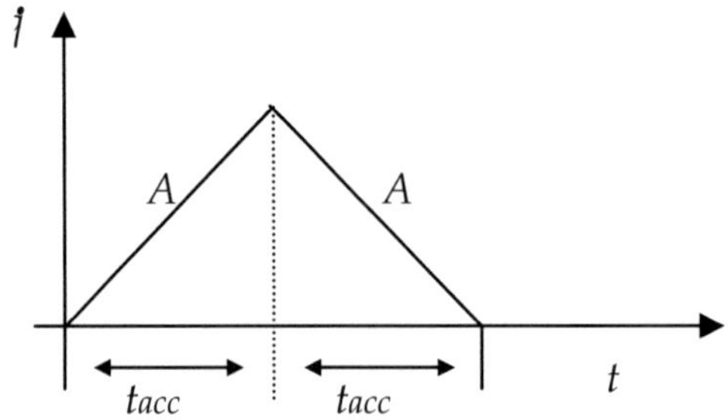

where q_i and q_f are the initial and final angles of each axis, respectively, V, A are the cruising velocity and acceleration amplitude, and $t_{acc}, t_{vitt}, t_{tot}$ are the accelaration, constant velocity and total times. Moreover, in compliance with Figs. 8-10 corresponding to the results obtained on axes 1-3, respectively, by using our proposed GVS2 in the tracking mode, one can see through the graphs (a) and (b) of each figure how the robot angles and rates follow the reference trajectories even when a strong disturbance and payload are added. It can also be noted that the spikes present in the velocity characteristics are due to the fact that the velocity is not measured but computed. In addition, one can observe that the control characteristics depicted by graph (c) of each figure is chattering-free.

5. Conclusion

Motivated by the VSC robustness properties, uncertain linear models of a robot are obtained for design of VS-based controllers. In this study, we presented through extensive simulation and experimentations results on chattering alleviation and performance improvements for two GVS algorithms and compared them to a classical variable structure (CVS) control approach. The GVS controllers are based on the equivalent control method. The CVS design methodology is based on the differential geometry whereas the GVS controllers are designed by using the differential algebraic tools. The results obtained from implementation of the above controllers can be summarized as follows: a) the GVS controllers do indeed filter out the control chattering characteristics and improve the system performance when compared to the CVS approach; b) the filtering and performance improvements are more clearly evident by using the GVS algorithm that is obtained with the hypersurface convergence equation; and c) in the tracking mode, in addition to the above improvements, the GVS control also enjoys insensitivity to parameter variations and disturbance rejection properties.

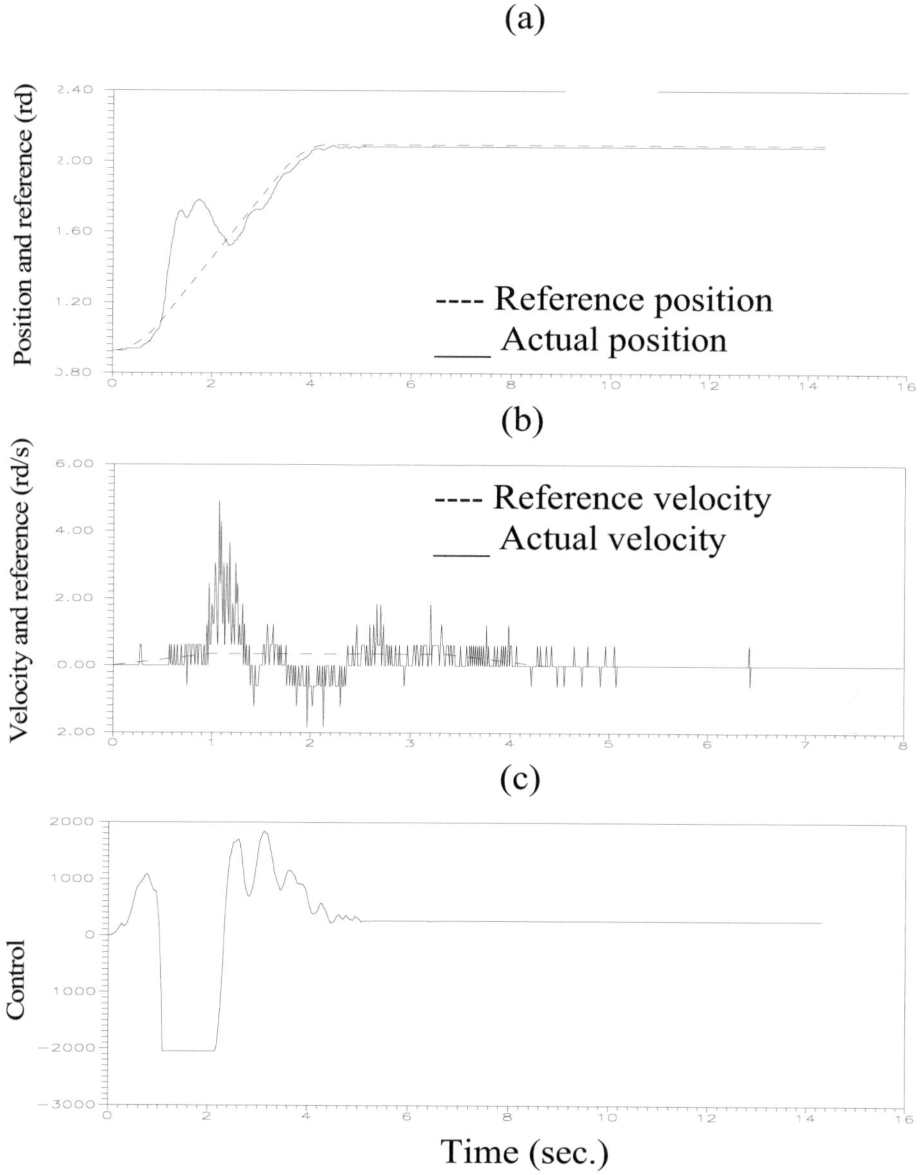

Figure 8. Experimental results obtained in the tracking mode using our proposed GVS2 controller on axis 1 in presence of 900g payload and external disturbance

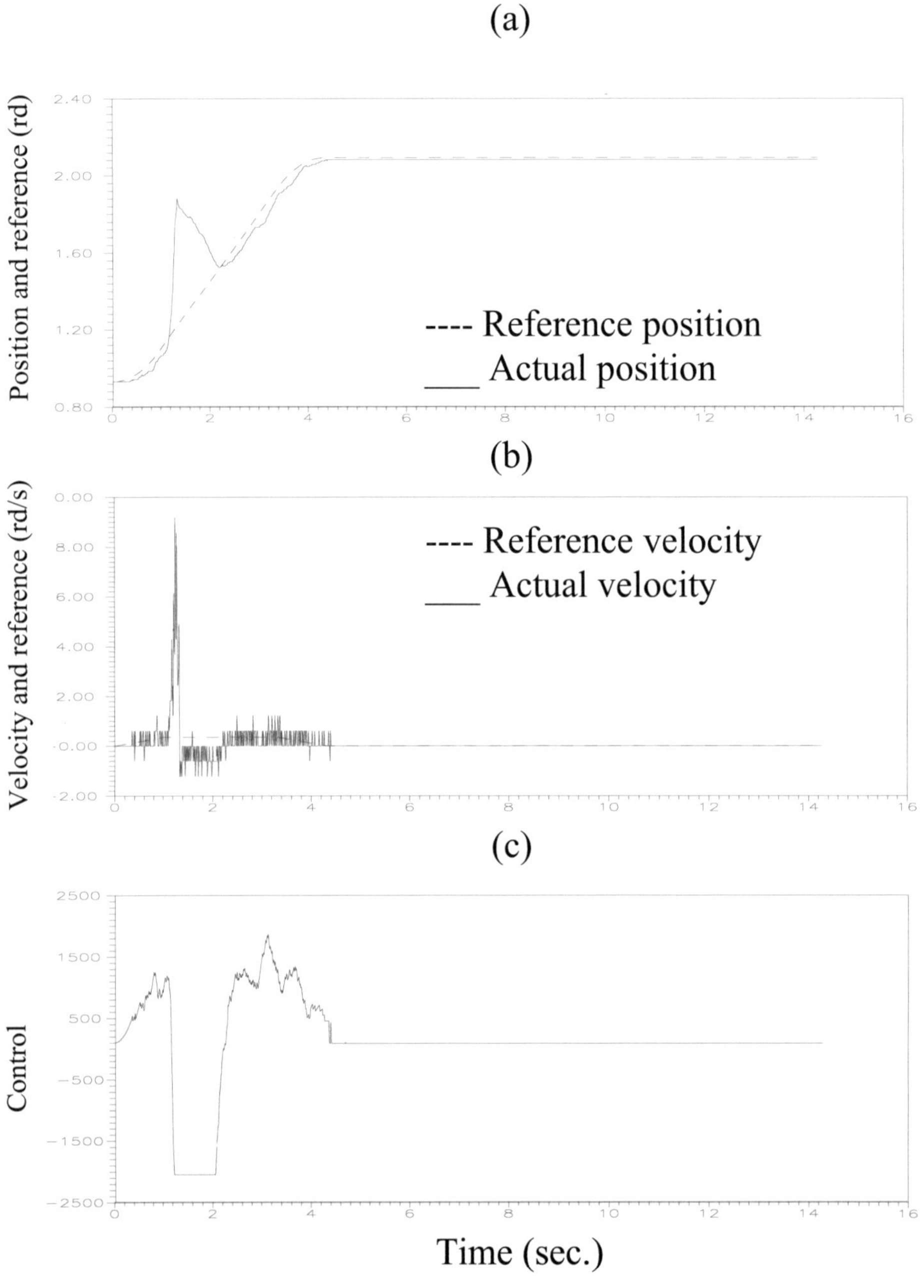

Figure 9. Experimental results obtained in the tracking mode using our proposed GVS2 controller on axis 2 in presence of 900g payload and external disturbance

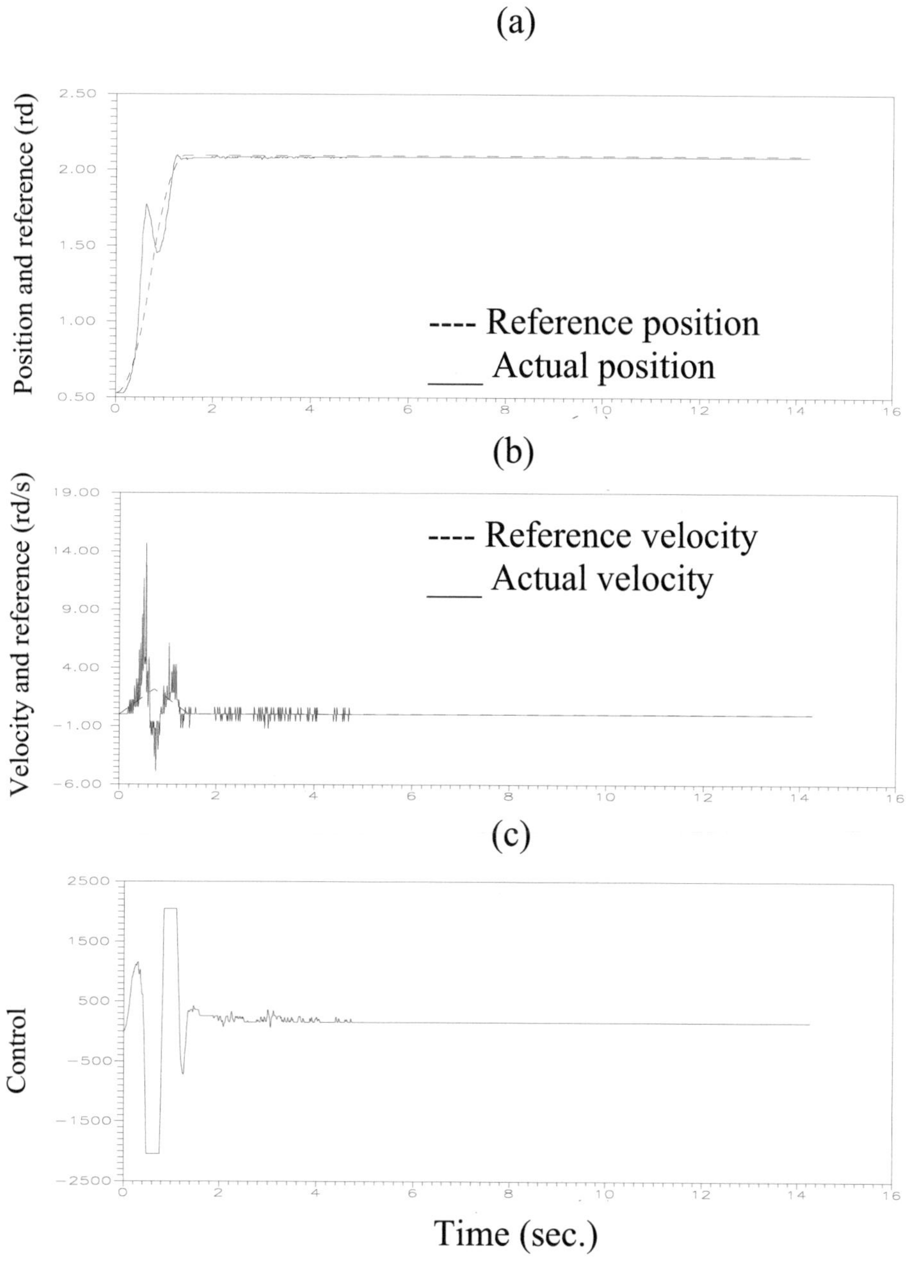

Figure 10. Experimental results obtained in the tracking mode using our proposed GVS2 controller on axis 3 in presence of 900g payload and external disturbance

7. References

Bartolini, G.; Ferrara, A. & Usai, E. (1998). Chattering avoidance by second order sliding mode control, *IEEE Transaction on Automatic Control*, Vol. 43, No. 2, Feb. 1998, pp. 241-246, ISSN: 0018-9286.

Belhocine, M.; Hamerlain, M. & Bouyoucef, K. (1998). Tracking trajectory for robot using Variable Structure Control, *Proceeding of the 4th ECPD, International Conference on Advanced Robotics, Intelligent Automation and Active Systems*, pp. 207-212, 24-26 August 1998, Moscow.

Bouyoucef, K.; Kadri, M. & Bouzouia, B. (1998). Identification expérimentale de la dynamique du robot manipulateur RP41, *1er Colloque National sur la Productique*, CNP98, May 1998, Tizi Ouzou.

Bouyoucef, K.; Khorasani, K. & Hamerlain, M. (2006). Chattering-alleviated Generalized Variable Structure Control for Experimental Robot Manipulators, *Proceeding of the IEEE Thirty-Eighth Southeastern Symposium on System Theory*, SSST '06, pp. 152 - 156, March 2006, Cookeville, TN, ISBN: 0-7803-9457-7.

Fliess, M.(1990). Generalized Controller Canonical Forms for linear and nonlinear dynamic, *IEEE Transactions AC*, Vol. 35, September 1990, pp. 994-1001, ISSN: 0018-9286.

Furuta, K.; Kosuge, K. & Kobayshi, K. (1989). VSS-type self-tuning Control of direct drive motor in *Proceedings of IEEE/IECON'89*, pp. 281-286, November 1989, Philadelphia, PA, 281-286.

Hamerlain M.; Belhocine, M. & Bouyoucef, K. (1997). Sliding Mode Control for a Robot SCARA, *Proceeding of IFAC/ACE'97*, pp153-157, 14-16 July 1997, Istambul, Turkey.

Harashima, E, Hashimoto H. & Maruyama K. (1986). Practical robust control of robot arm using variable structure systems, *Proceeding of IEEE, International Conference on Robotics and automation*, pp. 532-538, 1986, San Francisco, USA.

Levant, A. & Alelishvili L.(2007). Integral High-Order Sliding Modes, *IEEE Transaction on Automatic control*, Vol. 5, No. 7, July 2007, pp. 1278 - 1282.

Messager, E. (1992). Sur la stabilisation discontinue des systèmes, *PhD thesis in Sciences Automatiques*, Orsay, Paris, France, 1992.

Nouri, A. S.; Hamerlain, M.; Mira, C. & Lopez, P. (1993). Variable structure model reference adaptive control using only input and output measurements for two real one-link manipulators, *System, Man, and Cybernetics*, pp. 465-472, October 1993, ISBN: 0-7803-0911-1.

Sira-Ramirez, H. (1993). A Dynamical Variable Structure Control Strategy, in Asymptotic Output Tracking Problems, *IEEE Transactions on Automatic Control*, Vol. 38, No. 4, April 1993, pp. 615-620, ISSN: 0018-9286.

Slotine, J.J. E. & Coetsee, J.A. (1986). Adaptive sliding controller synthesis for nonlinear systems, *International Journal of Control*, Vol. 43, No. 6, Nov. 1986, pp. 1631-1651, ISSN:0018-9464.

Utkin, V. I. (1992). Sliding mode in control and optimization, Springer-Verlag, Berlin ISBN: 0387535160

Youssef, T.; Bouyoucef, K. & Hamerlain, M. (1998). Reducing the chattering using the generalized variable structure control applied to a manipulator arm, IEE, UKACC, *International conference on CONTROL'98*, pp. 1204-1211, ISBN: 0-85296-708-X, September 1998, University of Wales Swansea.

2

Unit Quaternions: A Mathematical Tool for Modeling, Path Planning and Control of Robot Manipulators

Ricardo Campa and Karla Camarillo
Instituto Tecnológico de la Laguna
Mexico

1. Introduction

Robot manipulators are thought of as a set of one or more kinematic chains, composed by rigid bodies (links) and articulated joints, required to provide a desired motion to the manipulator's end-effector. But even though this motion is driven by control signals applied directly to the joint actuators, the desired task is usually specified in terms of the pose (i.e., the position and orientation) of the end-effector. This leads to consider two ways of describing the configuration of the manipulator, at any time (Rooney & Tanev, 2003): via a set of joint variables or pose variables. We call these configuration spaces the *joint space* and the *pose space*, respectively.

But independently of the configuration space employed, the following three aspects are of interest when designing and working with robot manipulators:

- Modeling: The knowledge of all the physical parameters of the robot, and the relations among them. Mathematical (kinematic and dynamic) models should be extracted from the physical laws ruling the robot's motion. Kinematics is important, since it relates joint and pose coordinates, or their time derivatives. Dynamics, on the other hand, takes into account the masses and forces that produce a given motion.
- Task planning: The process of specifying the different tasks for the robot, either in pose or joint coordinates. This may involve from the design and application of simple time trajectories along precomputed paths (this is called *trajectory planning*), to complex computational algorithms taking real-time decisions during the execution of a task.
- Control: The elements that allow to ensure the accomplishment of the specified tasks in spite of perturbances or unmodeled dynamics. According to the type of variables used in the control loop, we can have joint space controllers or pose space controllers. Robot control systems can be implemented either at a low level (e.g. electronic controllers in the servo-motor drives) or via sophisticated high-level programs in a computer.

Fig. 1 shows how these aspects are related to conform a robot motion control system. By *motion control* we refer to the control of a robotic mechanism which is intended to track a desired time-varying trajectory, without taking into account the constraints given by the environment, i.e., as if moving in free space. In such a case the desired task (a time function along a desired path) is generated by a trajectory planner, either in joint or pose variables. The motion controller can thus be designed either in joint or pose space, respectively.

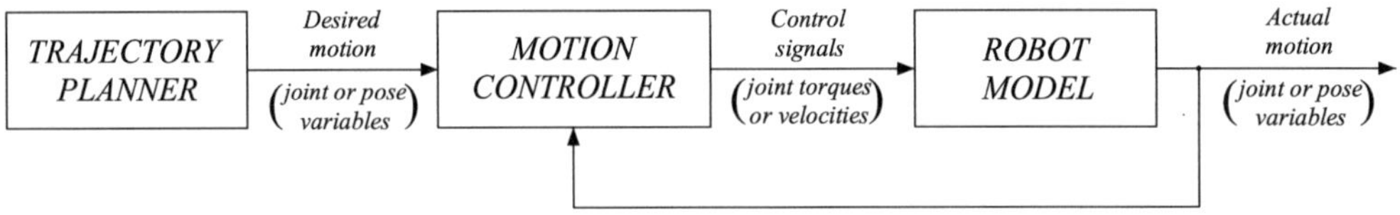

Figure 1. General scheme of a robot motion control system

Most of industrial robot manipulators are driven by *brushless DC* (*BLDC*) servoactuators. A BLDC servo system is composed by a permanent-magnet synchronous motor, and an electronic drive which produces the required power signals (Krause et al., 2002). An important feature of BLDC drives is that they contain several nested control loops, and they can be configured so as to accept as input for their inner controllers either the motor's desired position, velocity or torque. According to this, a typical robot motion controller (such as the one shown in Fig. 1) which is expected to include a position tracking control loop, should provide either velocity or torque reference signals to each of the joint actuators' inner loops. Thus we have kinematic and dynamic motion controllers, respectively.
On the other hand, it is a well-known fact that the number of degrees of freedom required to completely define the pose of a rigid body in free space is six, being three for position and three for orientation. In order to do so, we need to attach a coordinate frame to the body and then set the relations between this moving frame and a fixed one (see Fig. 2). Position is well described by a position vector $p \in \mathbb{R}^3$, but in the case of orientation there is not a generalized method to describe it, and this is mainly due to the fact that orientation constitutes a three-dimensional manifold, M^3, which is not a vector space, but a *Lie group*.

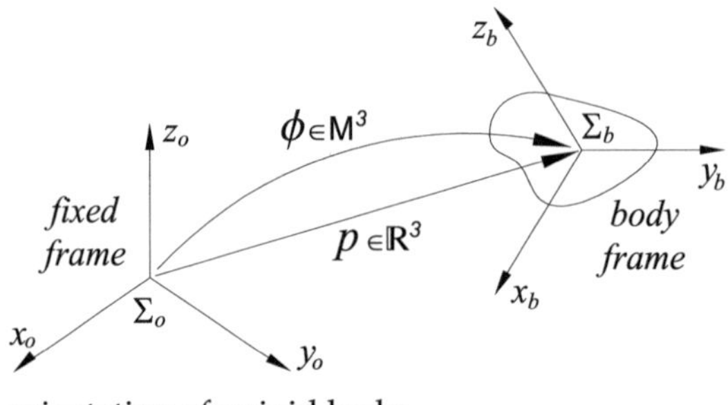

Figure 2. Position and orientation of a rigid body

Minimal representations of orientation are defined by three parameters, e.g., Euler angles. But in spite of their popularity, Euler angles suffer the drawbacks of representation singularities and inconsistency with the task geometry (Natale, 2003). There are other nonminimal parameterizations of orientation which use a set of $3+k$ parameters, related by k holonomic constraints, in order to keep the required three degrees of freedom. Common examples are the rotation matrices, the angle-axis pair and the Euler parameters. For a list of these and other nonminimal parameterizations of orientation see, e.g., (Spring, 1986).
This work focuses on Euler parameters for describing orientation. They are a set of four parameters with a unit norm constraint, so that they can be considered as points lying on the surface of the unit hypersphere $\mathrm{S}^3 \subset \mathbb{R}^4$. Therefore, Euler parameters are *unit quaternions*, a subset of the quaternion numbers, first described by Sir W. R. Hamilton in 1843, as an extension to complex numbers (Hamilton, 1844).

The use of Euler parameters in robotics has increased in the latter years. They are an alternative to rotation matrices for defining the kinematic relations among the robot's joint variables and the end-effector's pose. The advantage of using unit quaternions over rotation matrices, however, appears not only in the computational aspect (e.g., reduction of floating-point operations, and thus, of processing time) but also in the definition of a proper orientation error for control purposes.
The term *task space control* (Natale, 2003) has been coined for referring to the case of pose control when the orientation is described by unit quaternions, in opposition to the traditional *operational space control*, which employs Euler angles (Khatib, 1987). Several works have been done using this approach (see, e.g., Lin, 1995; Caccavale et al., 1999; Natale, 2003; Xian et al., 2004; Campa et al., 2006).
The aim of this work is to show how unit quaternions can be applied for modeling, path planning and control of robot manipulators. After recalling the main properties of quaternion algebra in Section 2, we review some quaternion-based algorithms that allow to obtain the kinematics and dynamics of serial manipulators (Section 3). Section 4 deals with path planning, and describes two algorithms for generating curves in the orientation manifold. Section 5 is devoted to task-space control. It starts with the specification of the pose error and the control objective, when orientation is given by unit quaternions. Then, two common pose controllers (the *resolved motion rate controller* and the *resolved acceleration controller*) are analyzed. Even though these controllers have been amply studied in literature, the Lyapunov stability analysis presented here is novel since it uses a quaternion-based space-state approach. Concluding remarks are given in Section 6.

Throughout this paper we use the notation $\lambda_m\{A\}$ and $\lambda_M\{A\}$ to indicate the smallest and largest eigenvalues, respectively, of a symmetric positive definite matrix $A(x)$ for all $x \in \mathbb{R}^n$. The Euclidean norm of vector x is defined as $\|x\| = \sqrt{x^T x}$. Also, for a given matrix A, the spectral norm $\| A \|$ is defined as $\| A \| = \sqrt{\lambda_M\{A^T A\}}$.

2. Euler Parameters

For the purpose of this work, an n-dimensional manifold $\mathrm{M}^n \subseteq \mathbb{R}^m$, with $n \leq m$, can be easily understood as a subset of the Euclidean space $\mathbb{R}^m$ containing all the points whose coordinates $x_1, x_2, ..., x_m$ satisfy $r = m - n$ holonomic constraint equations of the form $\gamma_i(x_1, x_2, ..., x_m) = 0$ ($i = 1, 2, ..., r$). As a typical example of an n-dimensional manifold we have the generalized unit (hyper)sphere $\mathrm{S}^n \subset \mathbb{R}^{n+1}$, defined as

$$\mathrm{S}^n = \{x \in \mathbb{R}^{n+1} : \| x \| = 1\}.$$

It is well-known that the configuration space of a rigid-body's orientation is a tree-dimensional manifold, $\mathrm{M}^3 \subset \mathbb{R}^m$, where m is the number of parameters employed to describe the orientation. Thus we have minimal ($m = 3$) and non-minimal ($m > 3$) parameterizations of the orientation, being the most common the following:

1. Euler angles, $\mathrm{M}^3 \equiv \mathbb{R}^3$.
2. Rotation matrices, $\mathrm{M}^3 \equiv SO(3) \subset \mathbb{R}^{3\times 3} (\equiv \mathbb{R}^9)$.

3. Angle/axis pair, $M^3 \equiv \mathbb{R} \times S^2 \subset \mathbb{R}^4$.
4. Euler parameters, $M^3 \equiv S^3 \subset \mathbb{R}^4$.

Given the fixed and body coordinate frames (Fig. 2), Euler angles indicate the sequence of rotations around one of the frame's axes required to make them coincide with those of the other frame. There are several conventions of Euler angles (depending of the sequence of axes to rotate), but all of them have inherently the problem of singular points, i.e., orientations for which the set of Euler angles is not uniquely defined (Kuipers, 1999).

Euler also showed that the relative orientation between two frames with a common origin is equivalent to a single rotation of a minimal angle around an axis passing through the origin. This is known as the Euler's rotation theorem, and implies that other way of describing orientation is by specifying the minimal angle $\theta \in \mathbb{R}$ and a unit vector $u \in S^2 \subset \mathbb{R}^3$ indicating the axis of rotation. This parameterization, known as the *angle/axis pair*, is non-minimal since we require four parameters and one holonomic constraint, i.e., the unit norm condition of u.

The angle/axis parameterization, however, has some drawbacks that make difficult its application. First, it is easy to see that the pairs $\begin{bmatrix} \theta & u^T \end{bmatrix}^T$ and $\begin{bmatrix} -\theta & -u^T \end{bmatrix}^T \in \mathbb{R} \times S^2$ both give the same orientation (in other words, $\mathbb{R} \times S^2$ is a *double cover* of the orientation manifold). But the main problem is that the angle/axis pair has a singularity when $\theta = 0$, since in such case u is not uniquely defined.

A most proper way of describing the orientation manifold is by means of an orthogonal rotation matrix $R \in SO(3)$, where the *special orthogonal group*, defined as

$$SO(3) = \{R \in \mathbb{R}^{3\times 3} : R^T R = I, \det(R) = 1\}$$

is a 3-dimensional manifold requiring nine parameters, that is, $M^3 \equiv SO(3) \subset \mathbb{R}^{3\times 3} \equiv \mathbb{R}^9$ (it is easy to show that the matrix equation $R^T R = I$ is in fact a set of six holonomic constraints, and $\det(R) = 1$ is one of the two possible solutions). Furthermore, being SO(3) a multiplicative matrix group, it means that the composition of two rotation matrices is also a rotation matrix, or

$$R_1, R_2 \in SO(3) \Rightarrow R_1 R_2 \in SO(3).$$

A group which is also a manifold is called a *Lie group*.

Rotation matrices are perhaps the most extended method for describing orientation, mainly due to the convenience of matrix algebra operations, and because they do not have the problem of singular points, but rotation matrices are rarely used in control tasks due to the difficulty of extracting from them a suitable vector orientation error.

Quaternion numbers (whose set is named $\mathbb{H}$ after Hamilton) are considered the immediate extension to complex numbers. In the same way that complex numbers can be seen as arrays of two real numbers (i.e. $\mathbb{C} \equiv \mathbb{R}^2$) with all basic arithmetic operations defined, quaternions become arrays of four real numbers ($\mathbb{H} \equiv \mathbb{R}^4$). It is useful however to group the last three components of a quaternion as a vector in $\mathbb{R}^3$. Thus $\begin{bmatrix} a & b & c & d \end{bmatrix}^T \in \mathbb{H}$, and $\begin{bmatrix} a & v^T \end{bmatrix}^T \in \mathbb{H}$, with $v \in \mathbb{R}^3$, are both valid representations of quaternions (a is called the scalar part, and v

the vector part, of the quaternion). Given two quaternions $z_1, z_2 \in \mathbb{H}$, where $z_1 = [a_1 \quad v_1^T]^T$ and $z_2 = [a_2 \quad v_2^T]^T$, the following quaternion operations are defined:

- Conjugation:

$$z_1^* = \begin{bmatrix} a_1 \\ -v_1 \end{bmatrix} \in \mathbb{H}$$

- Inversion:

$$z_1^{-1} = \frac{z_1^*}{\| z_1 \|^2} \in \mathbb{H}$$

where the quaternion Euclidean norm is defined as usual $\| z_1 \| = \sqrt{z_1^T z_1} = \sqrt{a_1^2 + v_1^T v_1}$.

- Addition/subtraction:

$$z_1 \pm z_2 = \begin{bmatrix} a_1 \pm a_2 \\ v_1 \pm v_2 \end{bmatrix} \in \mathbb{H}$$

- Multiplication:

$$z_1 \otimes z_2 = \begin{bmatrix} a_1 a_2 - v_1^T v_2 \\ a_1 v_2 + a_2 v_1 + S(v_1) v_2 \end{bmatrix} \in \mathbb{H} \tag{1}$$

where $S(\cdot)$ is a matrix operator such that, for all $x = [x_1 \quad x_2 \quad x_3]^T \in \mathbb{R}^3$ produces a skew-symmetric matrix, given by:

$$S(x) = \begin{bmatrix} 0 & -x_3 & x_2 \\ x_3 & 0 & -x_1 \\ -x_2 & x_1 & 0 \end{bmatrix}.$$

Skew-symmetric matrices have several useful properties, such as the following (for all $x, y, z \in \mathbb{R}^3$):

$$S(x)^T = S(-x) = -S(x), \tag{2}$$

$$S(x)(y+z) = S(x)y + S(x)z, \tag{3}$$

$$S(x)y = -S(y)x, \tag{4}$$

$$S(x)x = 0, \tag{5}$$

$$y^T S(x) y = 0, \tag{6}$$

$$x^T S(y) z = z^T S(x) y, \tag{7}$$

$$S(x)S(y) = yx^T - x^T y I. \tag{8}$$

Note that, by (4), quaternion multiplication is not commutative, that is, $z_1 \otimes z_2 \neq z_2 \otimes z_1$.

- Division: Due to the non–commutativity of the quaternion multiplication, dividing z_1 by z_2 can be either

$$z_1 \otimes z_2^{-1} = \frac{z_1 \otimes z_2^*}{\| z_2 \|^2}, \text{ or } z_2^{-1} \otimes z_1 = \frac{z_2^* \otimes z_1}{\| z_2 \|^2}.$$

Euler parameters are another way of describing the orientation of a rigid body. They are four real parameters, here named η, ε_1, ε_2, $\varepsilon_3 \in \mathbb{R}$, subject to a unit norm constraint, so that they are points lying on the surface of the unit hypersphere $\mathrm{S}^3 \subset \mathbb{R}^4$. Euler parameters are also *unit quaternions*. Let $\xi = \begin{bmatrix} \eta & \varepsilon^T \end{bmatrix}^T \in \mathrm{S}^3$, with $\varepsilon = \begin{bmatrix} \varepsilon_1 & \varepsilon_2 & \varepsilon_3 \end{bmatrix}^T \in \mathbb{R}^3$, be a unit quaternion. Then the unit norm constraint can be written as

$$\| \xi \| = \eta^2 + \varepsilon^T \varepsilon = 1. \tag{9}$$

An important property of unit quaternions is that they form a multiplicative group (in fact, a Lie group) using the quaternion multiplication defined in (1). More formally:

$$\begin{bmatrix} \eta_1 \\ \varepsilon_1 \end{bmatrix}, \begin{bmatrix} \eta_2 \\ \varepsilon_2 \end{bmatrix} \in \mathrm{S}^3 \Rightarrow \begin{bmatrix} \eta_1 \\ \varepsilon_1 \end{bmatrix} \otimes \begin{bmatrix} \eta_2 \\ \varepsilon_2 \end{bmatrix} = \begin{bmatrix} \eta_1\eta_2 - \varepsilon_1^T \varepsilon_2 \\ \eta_1 \varepsilon_2 + \eta_2 \varepsilon_1 + S(\varepsilon_1)\varepsilon_2 \end{bmatrix} \in \mathrm{S}^3. \tag{10}$$

The proof of (10) relies in showing that the multiplication result has unit norm, by using (9) and some properties of the skew–symmetric operator.
Given the angle/axis pair for a given orientation, the Euler parameters can be easily obtained, since (Sciavicco & Siciliano, 2000):

$$\eta = \cos\left(\frac{\theta}{2}\right), \quad \varepsilon = \sin\left(\frac{\theta}{2}\right)u. \tag{11}$$

And it is worth noticing that the Euler parameters solve the singularity of the angle/axis pair when $\theta = 0$. If a rotation matrix $R = \{r_{ij}\} \in \mathrm{SO}(3)$ is given, then the corresponding Euler parameters can be extracted using (Craig, 2005):

$$\begin{bmatrix} \eta \\ \varepsilon_1 \\ \varepsilon_2 \\ \varepsilon_3 \end{bmatrix} = \frac{1}{2\sqrt{r_{11} + r_{22} + r_{33} + 1}} \begin{bmatrix} r_{11} + r_{22} + r_{33} + 1 \\ r_{32} - r_{23} \\ r_{13} - r_{31} \\ r_{21} - r_{12} \end{bmatrix}. \tag{12}$$

Conversely, given the Euler parameters $\begin{bmatrix} \eta & \varepsilon^T \end{bmatrix}^T \in \mathrm{S}^3$ for a given orientation, the corresponding rotation matrix $R \in \mathrm{SO}(3)$ can be found as (Sciavicco & Siciliano, 2000):

$$R(\eta, \varepsilon) = (\eta^2 - \varepsilon^T \varepsilon) I + 2\eta S(\varepsilon) + 2\varepsilon\varepsilon^T. \tag{13}$$

However, note in (13) that $R(\eta,\varepsilon) = R(-\eta,-\varepsilon)$, meaning that S^3 is a double cover of $\mathrm{SO}(3)$. That is to say that every given orientation $\phi \in \mathrm{M}^3$ corresponds to one and only one rotation matrix, but maps into two unit quaternions, representing antipodes in the unit hypersphere (Gallier, 2000):

$$R \in \mathrm{SO}(3) \quad \Leftrightarrow \quad \pm\begin{bmatrix} \eta \\ \varepsilon \end{bmatrix} \in \mathrm{S}^3. \tag{14}$$

Also from (13) we can verify that $R(\eta,-\varepsilon) = R(\eta,\varepsilon)^T$ and, as $\begin{bmatrix} \eta & -\varepsilon^T \end{bmatrix}^T \in \mathrm{S}^3$ is the conjugate (or inverse) of $\begin{bmatrix} \eta & \varepsilon^T \end{bmatrix}^T \in \mathrm{S}^3$, then we have that

$$R^T \in \mathrm{SO}(3) \quad \Leftrightarrow \quad \pm\begin{bmatrix} \eta \\ \varepsilon \end{bmatrix}^* = \pm\begin{bmatrix} \eta \\ -\varepsilon \end{bmatrix} \in \mathrm{S}^3.$$

Another important fact is that quaternion multiplication is equivalent to the rotation matrix multiplication. Thus, if $\pm\xi_1, \pm\xi_2 \in \mathrm{S}^3$ are the Euler parameters corresponding to the rotation matrices $R_1, R_2 \in \mathrm{SO}(3)$, respectively, then:

$$R_1 R_2 \quad \Leftrightarrow \quad \pm\xi_1 \otimes \xi_2. \tag{15}$$

Moreover, the premultiplication of a vector $v \in \mathbb{R}^3$ by a rotation matrix $R \in \mathrm{SO}(3)$ produces a transformed (rotated) vector $w = Rv \in \mathbb{R}^3$. The same transformation can be performed with quaternion multiplication using $\overline{w} = \xi \otimes \overline{v} \otimes \xi^*$, where $\xi \in \mathrm{S}^3$ is the unit quaternion corresponding to R and quaternions $\overline{v}$, $\overline{w} \in \mathbb{H}$ are formed adding a null scalar part to the corresponding vectors ($\overline{v} = \begin{bmatrix} 0 & v^T \end{bmatrix}^T$, $\overline{w} = \begin{bmatrix} 0 & w^T \end{bmatrix}^T$). In sum

$$Rv \in \mathbb{R}^3 \quad \Leftrightarrow \quad \xi_1 \otimes \overline{v} \otimes \xi_1^* \in \mathbb{R}^4. \tag{16}$$

In the case of a time-varying orientation, we need to establish a relation between the time derivatives of Euler parameters $\dot{\xi} = \begin{bmatrix} \dot{\eta} & \dot{\varepsilon}^T \end{bmatrix}^T \in \mathbb{R}^4$ and the angular velocity of the rigid body $\omega \in \mathbb{R}^3$. That relation is given by the so-called *quaternion propagation rule* (Sciavicco & Siciliano, 2000):

$$\dot{\xi} = \frac{1}{2} E(\xi)\omega. \tag{17}$$

where

$$E(\xi) = E(\eta,\varepsilon) = \begin{bmatrix} -\varepsilon^T \\ \eta I - S(\varepsilon) \end{bmatrix} \in \mathbb{R}^{4\times 3}. \tag{18}$$

By using properties (2), (8) and the unit norm constraint (9) it can be shown that $E(\xi)^T E(\xi) = I$, so that ω can be resolved from (17) as

$$\omega = 2E(\xi)^T \dot{\xi} \tag{19}$$

3. Robot Kinematics and Dynamics

Serial robot manipulators have an *open kinematic chain*, i.e., there is only one path from one end (the base) to the other (the end-effector) of the robotic chain (see Fig. 3). A serial manipulator with n joints, either translational (prismatic) or rotational, has n *degrees of freedom* (dof). This section deals only with the modeling of serial manipulators.

As explained in (Craig, 2005), any serial robot manipulator can be described by using four parameters for each joint, that is, a total of $4n$ parameters for a n-dof manipulator. Denavit and Hartenberg (Denavit & Hartenberg, 1955) proposed a general method to establish those parameters in a systematic way, by defining $n+1$ coordinate frames, one per each link, including the base. The four Denavit-Hartenberg (or simply D-H) parameters for the i-th joint can then be extracted from the relation between frame $i-1$ and frame i, as follows:

a_i : Distance from axis z_{i-1} to z_i, along axis x_i.

α_i : Angle between axes z_{i-1} and z_i, about axis x_i.

d_i : Distance from axis x_{i-1} to x_i, along axis z_{i-1}.

θ_i : Angle between axes x_{i-1} and x_i, about axis z_{i-1}.

Note that there are three constant D-H parameters for each joint; the other one (θ_i for rotational joints, d_i for translational joints) describes the motion produced by such i-th joint. To follow the common notation, let q_i be the variable D-H parameter corresponding to the i-th joint. Then the so-called joint configuration space is specified by the vector

$$q = \begin{bmatrix} q_1 & q_2 & \cdots & q_n \end{bmatrix}^T \in \mathbb{R}^n,$$

while the pose (position and orientation) of the end-effector frame with respect to the base frame (see Fig. 3), denoted here by x, is given by a point in the 6-dimensional pose configuration space $\mathbb{R}^3 \times \mathrm{M}^3$, using whichever of the representations for $\mathrm{M}^3 \subset \mathbb{R}^m$. That is

$$x = \begin{bmatrix} p \\ \phi \end{bmatrix} \in \mathbb{R}^3 \times \mathrm{M}^3 \subset \mathbb{R}^{3+m}.$$

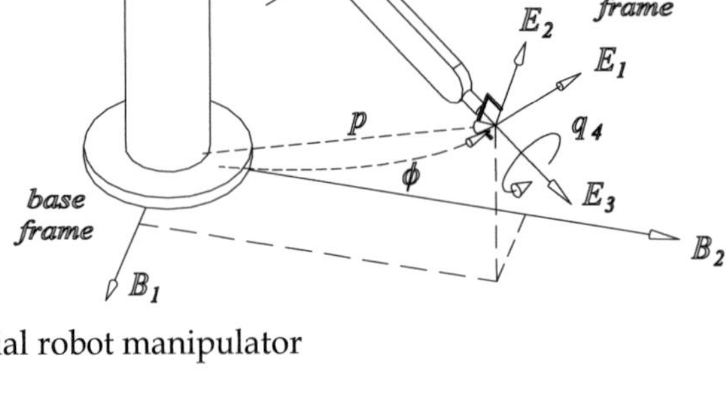

Figure 3. A 4-dof serial robot manipulator

3.1 Direct Kinematics

The direct kinematics problem consists on expressing the end-effector pose x in terms of q. That is, to find a function (called the *kinematics function*) $h:\mathbb{R}^n \rightarrow \mathbb{R}^3 \times \mathrm{M}^3$ such that

$$x = h(q) = \begin{bmatrix} p(q) \\ \phi(q) \end{bmatrix}. \tag{20}$$

Traditionally, direct kinematics is solved from the D-H parameters using *homogeneous matrices* (Denavit & Hartenberg, 1955). A homogeneous matrix combines a rotation matrix and a position vector in an extended 4×4 matrix. Thus, given $p \in \mathbb{R}^3$ and $R \in \mathrm{SO}(3)$, the corresponding homogeneous matrix T is

$$T = \begin{bmatrix} R & p \\ 0^T & 1 \end{bmatrix} \in \mathrm{SE}(3) \subset \mathbb{R}^{4\times4}. \tag{21}$$

$\mathrm{SE}(3)$ is known as the *special Euclidean group* and contains all homogeneous matrices of the form (21). It is a group under the standard matrix multiplication, meaning that the product of two homogeneous matrices $T_1, T_2 \in \mathrm{SE}(3)$ is given by

$$T_2T_1 = \begin{bmatrix} R_2 & p_2 \\ 0^T & 1 \end{bmatrix}\begin{bmatrix} R_1 & p_1 \\ 0^T & 1 \end{bmatrix} = \begin{bmatrix} R_2R_1 & R_2p_1 + p_2 \\ 0^T & 1 \end{bmatrix} \in \mathrm{SE}(3). \tag{22}$$

In terms of the D-H parameters, the relative position vector and rotation matrix of the i-th frame with respect to the previous one are (Sciavicco & Siciliano, 2000):

$$^{i-1}R_i = \begin{bmatrix} C_{\theta_i} & -S_{\theta_i}C_{\alpha_i} & S_{\theta_i}S_{\alpha_i} \\ S_{\theta_i} & C_{\theta_i}C_{\alpha_i} & -C_{\theta_i}S_{\alpha_i} \\ 0 & S_{\alpha_i} & C_{\alpha_i} \end{bmatrix} \in \mathrm{SO}(3), \qquad ^{i-1}p_i = \begin{bmatrix} a_iC_{\theta_i} \\ a_iS_{\theta_i} \\ d_i \end{bmatrix} \in \mathbb{R}^3. \tag{23}$$

where C_x, S_x stand for $\cos(x)$, $\sin(x)$, respectively.

The original method proposed by (Denavit & Hartenberg, 1955) uses the expressions in (23) to form relative homogeneous matrices $^{i-1}T_i$ (each depending on q_i), and then the composition rule (22) to compute the homogeneous matrix of the end-effector with respect to the base frame, $T(q) \in \mathrm{SE}(3)$:

$$T(q) = \begin{bmatrix} R(q) & p(q) \\ 0^T & 1 \end{bmatrix} = {}^0T_1\,{}^1T_2 \cdots {}^{n-1}T_n \in \mathrm{SE}(3).$$

This leads to the following general expressions for the pose of the end-effector, in terms of $p(q)$ and $R(q)$

$$R(q) = {}^0R_1\,{}^1R_2 \cdots {}^{n-1}R_n \tag{24}$$

$$p(q) = {}^0R_1\,{}^1R_2 \cdots {}^{n-2}R_{n-1}\,{}^{n-1}p_n + {}^0R_1\,{}^1R_2 \cdots {}^{n-3}R_{n-2}\,{}^{n-2}p_{n-1} + \ldots + {}^2R_1\,{}^1p_2 + {}^0p_1 \tag{25}$$

An alternative method for computing the direct kinematics of serial manipulators using quaternions, instead of homogeneous matrices, was proposed recently by (Campa et al.,

2006). The method is inspired in one found in (Barrientos et al. 1997), but it is more general, and gives explicit expressions for the position $p(q) \in \mathbb{R}^3$ and orientation $\xi(q) \in \mathrm{S}^3 \subset \mathbb{R}^4$ in terms of the original D-H parameters.
Let us start from the expressions for ${}^{i-1}R_i$ and ${}^{i-1}p_i$ in (23). We can use (12) and some trigonometric identities to obtain the corresponding quaternion ${}^{i-1}\xi_i$ from ${}^{i-1}R_i$; also, we can extend ${}^{i-1}p_i$ to form the quaternion ${}^{i-1}\overline{p}_i$. The results are

$$
{}^{i-1}\xi_i = \begin{bmatrix} \cos\left(\frac{\theta_i}{2}\right)\cos\left(\frac{\alpha_i}{2}\right) \\ \cos\left(\frac{\theta_i}{2}\right)\sin\left(\frac{\alpha_i}{2}\right) \\ \sin\left(\frac{\theta_i}{2}\right)\sin\left(\frac{\alpha_i}{2}\right) \\ \sin\left(\frac{\theta_i}{2}\right)\cos\left(\frac{\alpha_i}{2}\right) \end{bmatrix} \in \mathrm{S}^3, \qquad {}^{i-1}\overline{p}_i = \begin{bmatrix} 0 \\ a_i\cos(\theta_i) \\ a_i\sin(\theta_i) \\ d_i \end{bmatrix} \in \mathbb{H}
$$

Then, we can apply properties (15) and (16) iteratively to obtain expressions equivalent to (24) and (25):

$$
\begin{aligned}
\xi(q) &= {}^0\xi_1 \otimes {}^1\xi_2 \otimes \cdots \otimes {}^{n-1}\xi_n, \\
\overline{p}(q) &= ({}^0\xi_1 \otimes {}^1\xi_2 \otimes \cdots \otimes {}^{n-2}\xi_{n-1}) \otimes {}^{n-1}\overline{p}_n \otimes ({}^0\xi_1 \otimes {}^1\xi_2 \otimes \cdots \otimes {}^{n-2}\xi_{n-1})^* + \\
&\quad ({}^0\xi_1 \otimes {}^1\xi_2 \otimes \cdots \otimes {}^{n-3}\xi_{n-2}) \otimes {}^{n-2}\overline{p}_{n-1} \otimes ({}^0\xi_1 \otimes {}^1\xi_2 \otimes \cdots \otimes {}^{n-3}\xi_{n-2})^* + \ldots + {}^0\xi_1 \otimes {}^1\overline{p}_2\, {}^0\xi_1^* + {}^0\overline{p}_1.
\end{aligned}
$$

$\xi(q)$ is the unit quaternion (Euler parameters) expressing the orientation of the end-effector with respect to the base frame. The position vector $p(q)$ is the vector part of the quaternion $\overline{p}(q)$.
An advantage of the proposed method is that the position and orientation parts of the pose are processed separately, using only quaternion multiplications, which are computationally more efficient than homogeneous matrix multiplications. Besides, the result for the orientation part is given directly as a unit quaternion, which is a requirement in task space controllers (see Section 5).
The *inverse kinematics problem*, on the other hand, consists on finding explicit expressions for computing the joint coordinates, given the pose coordinates, that is equivalent to find the inverse funcion of h in (20), mapping $\mathbb{R}^3\times\mathrm{M}^3$ to $\mathbb{R}^n$:

$$
q = h^{-1}(x). \tag{26}
$$

The inverse kinematics, however, is much more complex than direct kinematics, and is not derived in this work.

3.2 Differential Kinematics

Differential kinematics gives the relationship between the joint velocities and the corresponding end-effector's linear and angular velocities (Sciavicco & Siciliano, 2000).

Thus, if $\dot{q} = \frac{d}{dt}q \in \mathbb{R}^n$ is the vector of joint velocities, $v \in \mathbb{R}^3$ is the linear velocity, and $\omega \in \mathbb{R}^3$ is the angular velocity of the end–effector, then the *differential kinematics equation* is given by

$$\begin{bmatrix} v \\ \omega \end{bmatrix} = \begin{bmatrix} J_p(q) \\ J_o(q) \end{bmatrix} \dot{q} = J(q)\dot{q} \tag{27}$$

where matrix $J(q) \in \mathbb{R}^{6 \times n}$ is the *geometric Jacobian* of the manipulator, which depends on its configuration. $J_p(q)$, $J_o(q) \in \mathbb{R}^{3 \times n}$ are the components of the geometric Jacobian producing the translational and rotational parts of the motion.
Now consider the time derivative of the generic *kinematics function* (20):

$$\dot{x} = \frac{\partial h(q)}{\partial q}\dot{q} = \begin{bmatrix} \dfrac{\partial p(q)}{\partial q} \\ \dfrac{\partial \phi(q)}{\partial q} \end{bmatrix} \dot{q} = J_A(q)\dot{q}. \tag{28}$$

Matrix $J_A(q) \in \mathbb{R}^{(3+m) \times n}$ is known as the *analytic Jacobian* of the robot, and it relates the joint velocities with the time derivatives of the pose coordinates (or *pose velocities*).
In the particular case of using Euler parameters for the orientation, $\phi(q) = \xi(q) \in S^3 \subset \mathbb{R}^4$, and $J_A(q) = J_{A_\xi}(q) \in \mathbb{R}^{7 \times n}$, i.e.

$$\dot{x} = \begin{bmatrix} \dot{p} \\ \dot{\xi} \end{bmatrix} = J_{A_\xi}(q)\dot{q}. \tag{29}$$

The linear velocity v of the end–effector is simply the time derivative of the position vector p (i.e. $v = \dot{p}$); instead, the angular velocity ω is related with $\dot{\xi}$ by (19). So, we have

$$\begin{bmatrix} v \\ \omega \end{bmatrix} = J_{R_\xi}(q) \begin{bmatrix} \dot{p} \\ \dot{\xi} \end{bmatrix} \tag{30}$$

where the *representation Jacobian*, for the case of using Euler parameters, is defined as

$$J_{R_\xi}(q) = \begin{bmatrix} I & 0 \\ 0 & 2E(\xi(q))^T \end{bmatrix} \in \mathbb{R}^{6 \times 7} \tag{31}$$

with matrix $E(\xi)$ as in (18). Also notice, by combining (27), (29), and (30), that

$$J(q) = J_{R_\xi}(q) J_{A_\xi}(q)$$

Whichever the Jacobian matrix be (geometric, analytic, representation), it is often required to find its inverse mapping, although this is possible only if such a matrix has full rank. In the case of a matrix $J \in \mathbb{R}^{n \times m}$, with $n > m$, then there exists a *left pseudoinverse* matrix of J, $J^\dagger \in \mathbb{R}^{m \times n}$, defined as

$$J^\dagger = \left(J^T J\right)^{-1} J^T$$

such that $J^{\dagger}J = I \in \mathbb{R}^{m\times m}$. On the contrary, if $n < m$, then there exist a *right pseudoinverse* matrix of J, ${}^{\dagger}J \in \mathbb{R}^{m\times n}$, defined as

$$ {}^{\dagger}J = J^{T}\left(JJ^{T}\right)^{-1} $$

such that $J\,{}^{\dagger}J = I \in \mathbb{R}^{n\times n}$. It is easy to show that in the case $n = m$ (square matrix) then $J^{\dagger} = {}^{\dagger}J = J^{-1}$.

3.3 Lagrangian Dynamics

Robot kinematics studies the relations between joint and pose variables (and their time derivatives) during the motion of the robot's end-effector. These relations, however, are established using only a geometric point of view. The effect of mechanical forces (either external or produced during the motion itself) acting on the manipulator is studied by *robot dynamics*.

As pointed out by (Sciavicco & Siciliano, 2000), the derivation of the dynamic model of a manipulator plays an important role for simulation, analysis of manipulator structures, and design of control algorithms. There are two general methods for derivation of the dynamic equations of motion of a manipulator: one method is based on the *Lagrange formulation* and is conceptually simple and systematic; the other method is based on the *Newton-Euler formulation* and allows obtaining the model in a recursive form, so that it is computationally more efficient.

In this section we consider the application of the Lagrangian formulation for obtaining the robot dynamics in the case of using Euler parameters for describing the orientation. This approach is not new, since it has been employed for modeling the dynamics of rigid bodies (see, e.g., Shivarama & Fahrenthold, 2004; Lu & Ren, 2006, and references therein).

An important fact is that Lagrange formulation allows to obtain the dynamic model independently of the coordinate system employed to describe the motion of the robot. Let us consider a serial robot manipulator with n degrees of freedom. Then we can choose any set of $m = n + k$ coordinates, say $\rho \in \mathbb{R}^{m}$, with $k \geq 0$ holonomic constraints of the form $\gamma_i(\rho) = 0$ ($i = 1, 2, \ldots, k$), to obtain the robot's dynamic model. If $k = 0$ we talk of independent or *generalized coordinates*, if $k \geq 1$ we have dependent or *constrained coordinates*.

The next step is to express the total (kinetic and potential) energy of the mechanical system in terms of the chosen coordinates. Let $\mathcal{K}(\rho, \dot{\rho})$ be the kinetic energy, and $\mathcal{U}(\rho)$ the potential energy of the system; then the *Lagrangian function* $\mathcal{L}(\rho, \dot{\rho})$ is defined as

$$ \mathcal{L}(\rho, \dot{\rho}) = \mathcal{K}(\rho, \dot{\rho}) - \mathcal{U}(\rho) $$

The Lagrange's dynamic equations of motion, for the general case (with constraints), are expressed by

$$ \frac{d}{dt}\left\{\frac{\partial \mathcal{L}(\rho, \dot{\rho})}{\partial \dot{\rho}}\right\} - \frac{\partial \mathcal{L}(\rho, \dot{\rho})}{\partial \rho} = \tau_{\rho} + \left(\frac{\partial \gamma(\rho)}{\partial \rho}\right)^{T} \lambda \tag{32} $$

where $\tau_{\rho} \in \mathbb{R}^{m}$ is the vector of (generalized or constrained) forces associated with each of the coordinates ρ, vector $\gamma(\rho)$ is defined as

$$\gamma(\rho)=\begin{bmatrix}\gamma_1(\rho)\\ \gamma_2(\rho)\\ \vdots\\ \gamma_k(\rho)\end{bmatrix}\in\mathbb{R}^k, \quad \text{so that} \quad \frac{\partial\gamma(\rho)}{\partial\rho}\in\mathbb{R}^{k\times m}$$

and $\lambda\in\mathbb{R}^k$ is the vector of k *Lagrange multipliers*, which are employed to reduce the dimension of the system, producing only n independent equations in terms of a new set $r\in\mathbb{R}^n$ of generalized coordinates. In sum, a general constrained system such as (32), with $\rho\in\mathbb{R}^{n+k}$, can be transformed in an unconstrained system of the form

$$\frac{d}{dt}\left\{\frac{\partial\mathcal{L}(r,\dot{r})}{\partial\dot{r}}\right\}-\frac{\partial\mathcal{L}(r,\dot{r})}{\partial r}=\tau_r \tag{33}$$

Moreover, as explained by (Spong et al., 2006), equation (33) can be rewritten as:

$$M_r(r)\ddot{r}+C_r(r,\dot{r})\dot{r}+g_r(r)=\tau_r$$

where $M_r(r)\in\mathbb{R}^{n\times n}$ is a symmetric positive definite matrix known as the inertia matrix, $C_r(r,\dot{r})\in\mathbb{R}^{n\times n}$ is the matrix of centripetal and Coriolis forces, and $g_r(r)\in\mathbb{R}^n$ is the vector of forces due to gravity. These matrices satisfy the following useful properties (Kelly et al., 2005):

- Inertia matrix is directly related to kinetic energy by

$$\mathcal{K}(r,\dot{r})=\frac{1}{2}\dot{r}^TM_r(r)\dot{r} \tag{34}$$

- For a given robot, matrix $C_r(r,\dot{r})$ is not unique, but it can always be chosen so that $\dot{M}_r(r)-2C_r(r,\dot{r})$ be skew-symmetric, that is

$$x^T\left[\dot{M}_r(r)-2C_r(r,\dot{r})\right]x=0,\quad \forall x,r,\dot{r}\in\mathbb{R}^n \tag{35}$$

- The vector of gravity forces is the gradient of the potential energy, that is

$$g_r(r)=\frac{\partial\mathcal{U}(r)}{\partial r} \tag{36}$$

Now consider the total mechanical energy of the robot $\mathcal{E}(r,\dot{r})=\mathcal{K}(r,\dot{r})+\mathcal{U}(r)$. Using properties (34)-(36), it can be shown that the following relation stands

$$\dot{\mathcal{E}}(r,\dot{r})=\dot{r}^T\tau_r. \tag{37}$$

And, as the rate of change of the total energy in a system is independent of the generalized coordinates, we can use (37) as a way of convert generalized forces between two different generalized coordinate systems.

For an n -dof robot manipulator, it is common to express the dynamic model in terms of the joint coordinates, i.e. $r \equiv q \in \mathbb{R}^n$, this leads us to the well-known equation (Kelly et al., 2005):

$$M(q)\ddot{q} + C(q,\dot{q})\dot{q} + g(q) = \tau, \quad (q,\dot{q},\ddot{q},\tau \in \mathbb{R}^n). \tag{38}$$

Subindexes have been omitted in this case for simplicity; τ is the vector of joint forces (and torques) applied to each of the robot joints.
Now consider the problem of modeling the robot dynamics of a non-redundant ($n \leq 6$) manipulator, in terms of the pose coordinates of the end-effector, given by $x = \begin{bmatrix} p^T & \xi^T \end{bmatrix}^T \in \mathbb{R}^3 \times S^3 \subset \mathbb{R}^7$. Notice that in this case we have holonomic constraints. If $n = 6$ then the only constraint is given by (9), so that:

$$\gamma(x) = \eta^2 + \varepsilon^T \varepsilon - 1;$$

If $n < 6$ then the pose coordinates require $6 - n$ additional constraints to form the vector $\gamma(x) \in \mathbb{R}^{7-n}$ The equations of motion would be

$$M_x(x)\ddot{x} + C_x(x,\dot{x})\dot{x} + g_x(x) = \tau_x + \frac{\partial \gamma(x)}{\partial x}^T \lambda, \quad (x,\dot{x},\ddot{x},\tau_x \in \mathbb{R}^7), \tag{39}$$

where matrices $M_x(x)$, $C_x(x,x)$, $g_x(x)$ can be computed from $M(q)$, $C(q,q)$, $g(q)$ in (38) using a procedure similar to the one presented in (Khatib, 1987) for the case of operational space. We get

$$M_x(x) = {}^{\dagger}(J_{A_\xi}(q)^T) M(q) J_{A_\xi}(q)^{\dagger}\Big|_{q=h^{-1}(x)}$$

$$C_x(x,\dot{x}) = {}^{\dagger}(J_{A_\xi}(q)^T) C(q, J_{A_\xi}(q)^{\dagger}\dot{x}) J_{A_\xi}(q)^{\dagger} + {}^{\dagger}(J_{A_\xi}(q)^T) M(q) \dot{J}_{A_\xi}(q)^{\dagger}\Big|_{q=h^{-1}(x)}$$

$$g_x(x) = {}^{\dagger}(J_{A_\xi}(q)^T) g(q)\Big|_{q=h^{-1}(x)}$$

where $J_{A_\xi}(q)$ is the analytic Jacobian in (29). To obtain the previous equations we are assuming, according to (37), that

$$\dot{q}^T \tau = \dot{x}^T \tau_x = \dot{q}^T J_{A_\xi}(q)^T \tau_x$$

so that

$$\tau_x = {}^{\dagger}(J_{A_\xi}(q)^T)\tau\Big|_{q=h^{-1}(x)}.$$

Equation (39) expresses the dynamics of a robot manipulator in terms of the pose coordinates (p, ξ) of the end-effector. In fact, these are a set of seven equations which can be reduced to n, by using the $7 - n$ Lagrange multipliers and choosing a subset of n

generalized coordinates taken from x. This procedure, however, can be far complex, and sometimes unsolvable in analytic form.

In theory, we can use (32) to find a minimal system of equations in terms of any set of generalized coordinates. For example, we could choose the scalar part of the Euler parameters in each joint, η_i, ($i = 1,2,...,n$) as a possible set of coordinates. To this end, it is useful to recall that the kinetic energy of a free-moving rigid body is a function of the linear and angular velocities of its center of mass, v and ω, respectively, and is given by

$$\mathcal{K}(v,\omega) = \frac{1}{2} m v^T v + \frac{1}{2} \omega^T H \omega, \tag{40}$$

where m is the mass, and H is the matrix of moments of inertia (with respect to the center of mass) of the rigid body. The potential energy depends only on the coordinates of the center of mass, p, and is given by

$$\mathcal{U}(p) = m g^T p,$$

where g is the vector of gravity acceleration. Now, using (30), (31), we can rewrite (40) as

$$\mathcal{K}(\dot{p},\xi,\dot{\xi}) = \frac{1}{2} m \dot{p}^T \dot{p} + 2\dot{\xi}^T E(\xi) H E(\xi)^T \dot{\xi} \tag{41}$$

Considering this, we propose the following procedure to compute the dynamic model of robot manipulators in terms of any combination of pose variables in task space:

1. Using (41), find the kinetic and potential energy of each of the manipulator's links, in terms of the pose coordinates, as if they were independent rigid bodies. That is, for link i, compute $\mathcal{K}_i(\dot{p}_i,\xi_i,\dot{\xi}_i)$ and $\mathcal{U}_i(p_i)$.
2. Still considering each link as an independent rigid body, compute the total Lagrangian function of the n -dof manipulator as

$$\mathcal{L}(p,\xi,\dot{p},\dot{\xi}) = \sum_{i=1}^{n} \left[\mathcal{K}_i(\dot{p}_i,\xi_i,\dot{\xi}_i) - \mathcal{U}_i(p_i) \right]$$

where $\rho \in \mathbb{R}^{7n}$ is a vector containing the pose coordinates of each link, i.e.

$$\rho = \begin{bmatrix} x_1 \\ x_2 \\ \vdots \\ x_n \end{bmatrix}, \quad \text{with } x_i = \begin{bmatrix} p_i \\ \xi_i \end{bmatrix}$$

3. Now use (9), for each joint, and the knowledge of the geometric relations between links to find $6n$ holonomic constraints, in order to form vector $\gamma(\rho) \in \mathbb{R}^{6n}$.
4. Write the constrained equations of motion for the robot, i. e. (32).
5. Finally, solve for the $6n$ Lagrange multipliers, and get de reduced (generalized) system in terms of the chosen n generalized coordinates.

This procedure, even if not simple (specially for robots with $n>3$) can be useful when requiring explicit equations for the robot dynamics in other than joint coordinates.

4. Path Planning

The minimal requirement for a manipulator is the capability to move from an initial posture to a final assigned posture. The transition should be characterized by motion laws requiring the actuators to exert joint forces which do not violate the saturation limits and do not excite typically unmodeled resonant modes of the structure. It is then necessary to devise planning algorithms that generate suitably *smooth* trajectories (Sciavicco & Siciliano, 2000).
To avoid confusion, a *path* denotes simply a locus of points either in joint or pose space; it is a pure geometric description of motion. A *trajectory*, on the other hand, is a path on which a time law is specified (e.g., in terms of velocities and accelerations at each point of the path). This section shows a couple of methods for designing paths in the task space, i.e., when using Euler parameters for describing orientation.
Generally, the specification of a path for a given motion task is done in pose space. It is much easier for the robot's operator to define a desired position and orientation of the end-effector than to compute the required joint angles for such configuration. A time-varying position $p(t)$ is also easy to define, but the same is not valid for orientation, due to the difficulty of visualizing the orientation coordinates.
Of the four orientation parameterizations mentioned in Section 2, perhaps the more intuitive is the angle/axis pair. To reach a new orientation from the actual one, employing a minimal displacement, we only need to find the required axis of rotation $\tilde{u}\in S^2$ and the corresponding angle to rotate $\tilde{\theta}\in\mathbb{R}$. Given the Euler parameters for the initial orientation, $\begin{bmatrix}\eta_o & \varepsilon_o^T\end{bmatrix}^T\in S^3$, and the desired orientation, $\begin{bmatrix}\eta_d & \varepsilon_d^T\end{bmatrix}^T\in S^3$, it can be shown, using (11) and quaternion operations, that (Campa & de la Torre, 2006):

$$\tilde{\theta}=2\arccos\left(\eta_o\eta_d+\varepsilon_o^T\varepsilon_d\right),\quad \tilde{u}=\frac{\eta_o\varepsilon_d-\eta_d\varepsilon_o+S(\varepsilon_o)\varepsilon_d}{\|\eta_o\varepsilon_d-\eta_d\varepsilon_o+S(\varepsilon_o)\varepsilon_d\|} \tag{42}$$

We can use expressions in (42) to define the minimal path between the initial and final orientations. A velocity profile can be employed for $\tilde{\theta}$ in order to generate a suitable $\theta(t)$. However, during a motion along such a path, it is important that $\tilde{u}$ remains fixed. Also notice that when the desired orientation coincides with the initial (i.e., $\eta_o=\eta_d$, $\varepsilon_o=\varepsilon_d$), then $\tilde{\theta}=0$ and $\tilde{u}$ is not well defined.
Now consider the possibility of finding a path between a given set of points in the task space using interpolation curves. The simplest way of doing so is by means of linear interpolation. Let the initial pose be given by $\begin{bmatrix}p_o^T & \xi_o^T\end{bmatrix}^T$ and the final pose by $\begin{bmatrix}p_d^T & \xi_d^T\end{bmatrix}^T$, and assume, for simplicity, that the time required for the motion is normalized, that is, we need to find a trajectory $\begin{bmatrix}p(t)^T & \xi(t)^T\end{bmatrix}^T$, such that

$$\begin{bmatrix}p(0)\\ \xi(0)\end{bmatrix}=\begin{bmatrix}p_o\\ \xi_o\end{bmatrix}\in\mathbb{R}^3\times S^3,\ \text{ and }\ \begin{bmatrix}p(1)\\ \xi(1)\end{bmatrix}=\begin{bmatrix}p_d\\ \xi_d\end{bmatrix}\in\mathbb{R}^3\times S^3$$

For the position part of the motion, the linear interpolation is simply given by

$$p(t) = (1-t)\,p_o + t\,p_d$$

but for the orientation part we can not apply the same formula, because, as unit quaternions do not form a vector space, then such $\xi(t)$ would not be a unit quaternion anymore. Shoemake (1985) solved this by defining a *spherical linear interpolation*, or simply *slerp*, which is given by the following expression

$$\xi(t) = \frac{\sin([1-t]\Omega)\xi_o + \sin(t\Omega)\xi_d}{\sin(\Omega)} \tag{43}$$

where $\Omega = \cos^{-1}(\xi_o^T \xi_d)$.

It is possible to show (Dam et al., 1998) that such $\xi(t)$ in (43) satisfies that $\xi(t) \in S^3$, for all $0 \leq t \leq 1$. (Dam et al., 1998) is also a good reference for understanding other types of interpolation using quaternions, including the so called *spherical spline quaternion interpolation* or *squad,* which produces smooth curves joining a series of points in the orientation manifold.

Even though these algorithms have been used in the fields of computer graphics and animation for years, as far as the authors' knowledge, there are few works using them for motion planning in robotics.

5. Task-space Control

As pointed out by (Sciavicco & Siciliano, 2000), the joint space control is sufficient in those cases where it is required motion control in free space; however, for the most common applications of interaction control (robot interacting with the environment) it is better to use the so-called *task space* control (Natale,2003). Thus, in the later years, more research has been done in this kind of control schemes.

Euler parameters have been employed in the later years for the control of generic rigid bodies, including spaceships and underwater vehicles. The use of Euler parameters in robot control begins with Yuan (1988) who applied them to a particular version of the *resolved-acceleration control* (Luh et al., 1980). More recently, several works have been published on this topic (see, e.g., Lin, 1995; Caccavale et al., 1999; Natale, 2003; Xian et al., 2004; Campa et al., 2006, and references therein).

The so-called *hierarchical control* of manipulators is a technique that solves the problem of motion control in two steps (Kelly & Moreno-Valenzuela, 2005): first, a task space kinematic control is used to compute the desired joint velocities from the desired pose trajectories; then, those desired velocities become the inputs for an inner joint velocity control loop (see Fig. 4). This two-loop control scheme was first proposed by Aicardi et al. (1995) in order to solve the pose control problem using a particular set of variables to describe the end-effector orientation.

The term *kinematic controller* (Caccavale et al., 1999) refers to that kind of controller whose output signal is a joint velocity and is the first stage in a hierarchical scheme. On the other hand *dynamic controller* is employed here for those controllers also including a position controller but producing joint torque signals.

In this section we analyze two common task-space control schemes: the *resolved-motion rate control* (*RMRC*), which is a simple kinematic controller, and the *resolved-acceleration control* (*RAC*), a dynamic controller. These controllers are well-known in literature, however, the stability analysis presented here is done employing Euler parameters as space-state variables for describing the closed-loop control system.

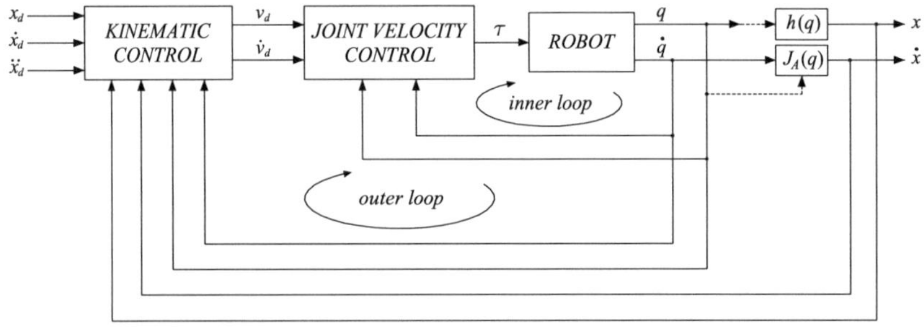

Figure 4. Pose-space hierarchical control

5.1 Pose Errror and Control Objective

For the analysis forthcoming let us consider that the motion of the manipulator's end-effector is described through its pose, given by $x(q)=\left[p(q)^T \quad \xi(q)^T\right]^T \in \mathbb{R}^3 \times S^3$, and its linear and angular velocities referred to the base frame, $\dot{p}(q) \in \mathbb{R}^3$ and $\omega(q) \in \mathbb{R}^3$, respectively. These terms can be computed from the measured joint variables ($q, \dot{q} \in \mathbb{R}^n$) using the direct kinematics function (20) and the differential kinematics equation (27).

Also, let the desired end-effector trajectory be specified by the desired pose $x_d(t)=\left[p_d(t)^T \quad \xi_d(t)^T\right]^T \in \mathbb{R}^3 \times S^3$, where $\xi_d(t)=\left[\eta_d(t) \quad \varepsilon_d(t)^T\right]^T \in S^3$, and the desired linear and angular velocities referred to the base frame, $\dot{p}_d(t), \omega_d(t) \in \mathbb{R}^3$. The time derivative of the desired Euler parameters is related to the desired angular velocity through an expression similar to (17), i.e.

$$\dot{\xi}_d = \frac{1}{2} E(\xi_d)\omega_d. \tag{44}$$

with matrix operator $E(\cdot)$ defined in (18). The desired pose x_d can be considered as describing the position and orientation of a desired frame with respect to the base frame, as shown in Fig. 5.

The position error vector is simply the difference between the desired and actual positions the robot's end-effector, that is $\tilde{p} = p_d - p$; the linear and angular velocity errors are computed in a similar way

$$\dot{\tilde{p}} = \dot{p}_d - \dot{p}, \qquad \tilde{\omega} = \omega_d - \omega. \tag{45}$$

However, as unit quaternions do not form a vector space, they cannot be subtracted to form the orientation error; instead, we should use the properties of the quaternion group algebra. Let $\tilde{\xi} = \left[\tilde{\eta} \quad \tilde{\varepsilon}^T\right]^T$ be the Euler parameters of the orientation error, then (Lin, 1995):

$$\tilde{\xi} = \xi_d \otimes \xi^* = \begin{bmatrix} \eta_d \\ \varepsilon_d \end{bmatrix} \otimes \begin{bmatrix} \eta \\ -\varepsilon \end{bmatrix}$$

and applying (1) we get (Sciavicco & Siciliano, 2000):

$$\tilde{\xi} = \begin{bmatrix} \tilde{\eta} \\ \tilde{\varepsilon} \end{bmatrix} = \begin{bmatrix} \eta\eta_d + \varepsilon^T \varepsilon_d \\ \eta\varepsilon_d - \eta_d \varepsilon + S(\varepsilon)\varepsilon_d \end{bmatrix} \in S^3. \tag{46}$$

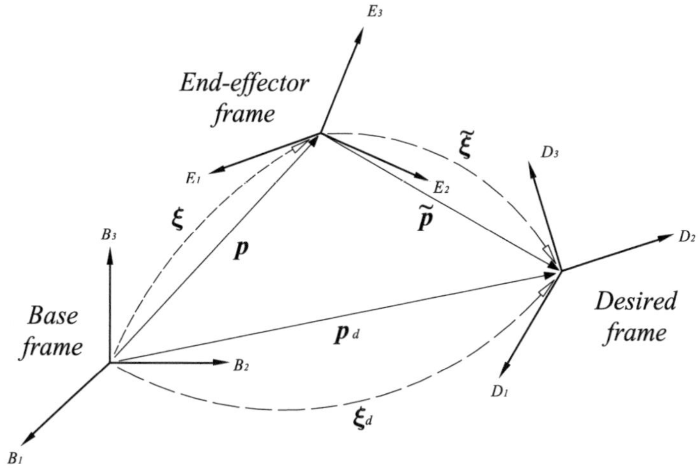

Figure 5. Actual and desired coordinate frames for defining the pose error

Taking the time derivative of (46), considering (17), (44) and properties (3), (4), and (8), it can be shown that (Fjellstad, 1994):

$$\begin{bmatrix} \dot{\tilde{\eta}} \\ \dot{\tilde{\varepsilon}} \end{bmatrix} = \frac{1}{2} \begin{bmatrix} -\tilde{\varepsilon}^T \\ \tilde{\eta} I + S(\tilde{\varepsilon}) \end{bmatrix} \tilde{\omega} - \begin{bmatrix} 0 \\ S(\tilde{\varepsilon}) \end{bmatrix} \omega_d, \tag{47}$$

where $\tilde{\omega}$ is defined as in (45).

Notice that, considering the properties of unit quaternions, when the end-effector orientation equals the desired orientation, i.e. $\xi = \xi_d$, the orientation error becomes $\tilde{\xi} = \begin{bmatrix} 1 & 0^T \end{bmatrix}^T \in S^3$. If $\xi = -\xi_d$, then $\tilde{\xi} = \begin{bmatrix} -1 & 0^T \end{bmatrix}^T \in S^3$, however, by (14) this represents the *same orientation*.

From the previous analysis we can establish the *pose control objective* as

$$\lim_{t\to\infty} \begin{bmatrix} \tilde{p}(t) \\ |\tilde{\eta}(t)| \\ \tilde{\varepsilon}(t) \end{bmatrix} = \begin{bmatrix} 0 \\ 1 \\ 0 \end{bmatrix}; \quad \tilde{p} \in \mathbb{R}^3, \begin{bmatrix} \tilde{\eta} \\ \tilde{\varepsilon} \end{bmatrix} \in S^3, \tag{48}$$

or, in words, that the end–effector position and orientation converge to the desired position and orientation.

5.2 A Hierarchical Controller

Let us consider a control scheme such as the one in Fig. 4. As the robot model we take the joint–space dynamics given in (38).

The RMRC is a kinematic controller first proposed by Whitney (1969) to handle the problem of motion control in operational space via joint velocities (rates). Let $x, x_d \in \mathbb{R}^6$ be the actual and desired operational coordinates, respectively, so that $\tilde{x} = x_d - x$ is the pose error in operational space. The RMRC controller is given by

$$v_d = J_A(q)^{\dagger}\left[\dot{x}_d + K_x\tilde{x}\right] \tag{49}$$

where $J_A(q)^{\dagger} \in \mathbb{R}^{n\times 6}$ is the left pseudoinverse of the analytic Jacobian in (28), $K_x \in \mathbb{R}^{n\times n}$ is a positive definite matrix, and $v_d \in \mathbb{R}^n$ is the vector of (resolved) desired joint velocities. In fact, in the ideal case where $x = x_d$, then $\tilde{x} = 0$ and we have the inverse analytic Jacobian mapping. In the case of using Euler parameters, then $x = \begin{bmatrix} p^T & \xi^T \end{bmatrix}^T$, $x_d = \begin{bmatrix} p_d^T & \xi_d^T \end{bmatrix}^T$, and we can use (Campa et al., 2006):

$$v_d = J(q)^{\dagger}\begin{bmatrix} \dot{p}_d + K_p\tilde{p} \\ \omega_d + K_o\tilde{\varepsilon} \end{bmatrix} \tag{50}$$

which has a similar structure than (49) but now we employ the geometric Jacobian instead of the analytic one, and the orientation error is taken as $\tilde{\varepsilon}$ (the vector part of the unit quaternion $\tilde{\xi}$), which becomes zero when the actual orientation coincides with the desired one. $K_p, K_o \in \mathbb{R}^{3\times 3}$ are positive definite gain matrices for the position and orientation parts, respectively.

In order to apply the proposed control law it is necessary to assume that the manipulator does not pass through singular configurations —where $J(q)$ losses rank— during the execution of the task. Also it is convenient to assume that the submatrices of $J(q)$, $J_p(q)$ and $J_o(q)$, are bounded, i.e., there exist positive scalars k_{J_p} and k_{J_o} such that, for all $q \in \mathbb{R}^n$ we have

$$\| J_p(q) \| \leq k_{J_p}, \quad \text{and} \quad \| J_o(q) \| \leq k_{J_o}. \tag{51}$$

Finally, for the joint velocity controller in Fig. 4 we use the same structure proposed in (Aicardi et al., 1995):

$$\tau = M(q)[\dot{v}_d + K_v\tilde{v}] + C(q,\dot{q})\dot{q} + g(q) \tag{52}$$

where $K_v \in \mathbb{R}^{n\times n}$ is a positive definite gain matrix, $\tilde{v} \in \mathbb{R}^n$ is the joint velocity error, given by

$$\tilde{v} = v_d - \dot{q}, \tag{53}$$

and $\dot{v}_d$ can be precomputed from (50) as

$$\dot{v}_d = \dot{J}(q)^{\dagger}\begin{bmatrix} \dot{p}_d + K_p\tilde{p} \\ \omega_d + K_o\tilde{\varepsilon} \end{bmatrix} + J(q)^{\dagger}\begin{bmatrix} \ddot{p}_d + K_p\dot{\tilde{p}} \\ \dot{\omega}_d + K_o\left[\frac{1}{2}[\tilde{\eta}I + S(\tilde{\varepsilon})]\tilde{\omega} - S(\tilde{\varepsilon})\omega_d\right] \end{bmatrix} \tag{54}$$

where we have employed (47). Fig. 6 shows in a detailed block diagram how this controller is implemented.

Substituting the control law (52) in the robot dynamics (38), and considering (53) we get $\dot{\tilde{v}}+K_v\tilde{v}=0$. On the other hand, replacing (50) in (53), premultiplying by $J(q)$, and considering (27), and (45) we get two decoupled equations

$$\dot{\tilde{p}}+K_p\tilde{p}=J_p(q)\tilde{v} \tag{55}$$

$$\tilde{\omega}+K_o\tilde{\varepsilon}=J_o(q)\tilde{v} \tag{56}$$

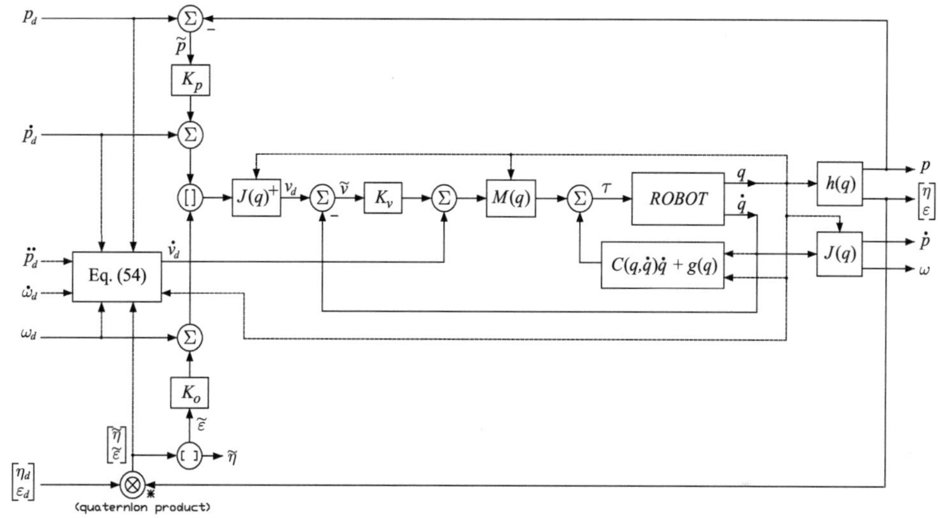

Figure 6. Hierarchical control: RMRC + inverse-dynamics velocity controller

The closed-loop equation of the whole controller can be obtained taking $\tilde{p}$, $\tilde{\eta}$ and $\tilde{\varepsilon}$ as states for the outer (task space) loop and v for the inner loop. The state equation becomes

$$\frac{d}{dt}\begin{bmatrix}\tilde{p}\\ \tilde{\eta}\\ \tilde{\varepsilon}\\ \tilde{v}\end{bmatrix}=\begin{bmatrix}-K_p\tilde{p}+J_p(q)\tilde{v}\\ \frac{1}{2}\tilde{\varepsilon}^T[K_o\tilde{\varepsilon}-J_o(q)\tilde{v}]\\ -\frac{1}{2}[\tilde{\eta}I+S(\tilde{\varepsilon})][K_o\tilde{\varepsilon}-J_o(q)\tilde{v}]-S(\tilde{\varepsilon})\omega_d\\ -K_v\tilde{v}\end{bmatrix}, \tag{57}$$

where (56) has been used to substitute $\tilde{\omega}$ in (47). The domain of the state space, D_{rr}, is defined as follows

$$D_{rr}=\mathbb{R}^3\times\mathrm{S}^3\times\mathbb{R}^n=\left\{\begin{bmatrix}\tilde{p}\\ \tilde{\eta}\\ \tilde{\varepsilon}\\ \tilde{v}\end{bmatrix}\in\mathbb{R}^{7+n}:\tilde{\eta}^2+\tilde{\varepsilon}^T\tilde{\varepsilon}-1=0\right\}.$$

System (57) is non-autonomous due to the term $\omega_d(t)$, and it is easy to see that it has two equilibria :

$$\begin{bmatrix} \tilde{p} \\ \tilde{\eta} \\ \tilde{\varepsilon} \\ \tilde{\nu} \end{bmatrix} = \begin{bmatrix} 0 \\ 1 \\ 0 \\ 0 \end{bmatrix} = E_1 \in D_{rr}, \quad \text{and} \quad \begin{bmatrix} \tilde{p} \\ \tilde{\eta} \\ \tilde{\varepsilon} \\ \tilde{\nu} \end{bmatrix} = \begin{bmatrix} 0 \\ -1 \\ 0 \\ 0 \end{bmatrix} = E_2 \in D_{rr}.$$

Notice that both equilibria represent the case when the end-effector's pose has reached the desired pose, and the joint velocity error, $\tilde{\nu}$, is null. However, E_1 is an asymptotically stable equilibrium while E_2 is an unstable equilibrium.

The Lyapunov stability analysis presented below for equilibrium E_1 of system (57) was taken from (Campa et al., 2006). More details can be found in (Campa, 2005).

Let us consider the function:

$$V(\tilde{p},\tilde{\eta},\tilde{\varepsilon},\tilde{\nu}) = \alpha\left[\tilde{\varepsilon}^T\tilde{\varepsilon} + (\tilde{\eta}-1)^2 + \frac{1}{2}\tilde{p}^T\tilde{p}\right] + \frac{1}{2}\tilde{\nu}^T\tilde{\nu} \tag{58}$$

with α such that

$$0 < \alpha < \frac{4\lambda_m\{K_p\}\lambda_m\{K_o\}\lambda_m\{K_v\}}{2\lambda_m\{K_p\}k_{J_o}^2 + \lambda_m\{K_o\}k_{J_p}^2}, \tag{59}$$

and k_{Jp}, k_{Jo} satisfying (51).

Notice that $\alpha > 0$ is enough to ensure that $V(\tilde{p},\tilde{\eta},\tilde{\varepsilon},\tilde{\nu})$ is positive definite in D_{rr}, around the equilibrium E_1, that is to say that $V(0,1,0,0)=0$ and there is a neighborhood $B \subset D_{rr}$ around E_1 such that $V(\tilde{p},\tilde{\eta},\tilde{\varepsilon},\tilde{\nu}) > 0$, for all $\begin{bmatrix}\tilde{p}^T & \tilde{\eta} & \tilde{\varepsilon}^T & \tilde{\nu}^T\end{bmatrix}^T \neq \begin{bmatrix}0^T & 1 & 0^T & 0^T\end{bmatrix}^T \in B$.

Now, taking the time derivative of (58) along the trajectories of system (57) we get:

$$\dot{V}(\tilde{p},\tilde{\eta},\tilde{\varepsilon},\tilde{\nu}) = -\alpha\left[\tilde{p}^T K_p \tilde{p} + \varepsilon^T K_o \tilde{\varepsilon} - \tilde{p}^T J_p(q)\tilde{\nu} - \tilde{\varepsilon}^T J_o(q)\tilde{\nu}\right] - \nu^T K_v \tilde{\nu}$$

and it can be shown that it is bounded by

$$\dot{V}(\tilde{p},\tilde{\eta},\tilde{\varepsilon},\tilde{\nu}) \leq -\begin{bmatrix} \|\tilde{p}\| \\ \|\tilde{\varepsilon}\| \\ \|\tilde{\nu}\| \end{bmatrix}^T Q \begin{bmatrix} \|\tilde{p}\| \\ \|\tilde{\varepsilon}\| \\ \|\tilde{\nu}\| \end{bmatrix} - \frac{1}{2}\alpha\lambda_m\{K_o\}\left(1-\tilde{\eta}^2\right).$$

where

$$Q = \begin{bmatrix} \alpha\lambda_m\{K_p\} & 0 & -\frac{1}{2}\alpha k_{Jp} \\ 0 & \frac{1}{2}\alpha\lambda_m\{K_o\} & -\frac{1}{2}\alpha k_{Jo} \\ -\frac{1}{2}\alpha k_{Jp} & -\frac{1}{2}\alpha k_{Jo} & \lambda_m\{K_v\} \end{bmatrix}$$

If α satisfies condition (59), then Q is positive definite matrix and $\dot{V}(\tilde{p},\tilde{\eta},\tilde{\varepsilon},\tilde{v})$ is negative definite function in D_{rr}, around E_1, that is, $\dot{V}(0,1,0,0)=0$ and there is a neighborhood $B \subset D_{rr}$ around E_1 such that $\dot{V}(\tilde{p},\tilde{\eta},\tilde{\varepsilon},\tilde{v})<0$, for all $\begin{bmatrix}\tilde{p}^T & \tilde{\eta} & \tilde{\varepsilon}^T & \tilde{v}^T\end{bmatrix} \neq \begin{bmatrix}0^T & 1 & 0^T & 0^T\end{bmatrix}^T \in B$. Thus, according to classical Lyapunov theory (see e.g. (Khalil, 2002)), we can conclude that the equilibrium E_1 is asymptotically stable in D_{rr}. This implies that, starting from an initial condition near E_1 in D_{rr}, the control objective (48) is fulfilled.

5.3 Resolved Acceleration Controller

The resolved acceleration control (RAC) is a well-known technique proposed by Luh et al. (1980) to solve the problem of pose space control of robot arms. It is based on the inverse dynamics methodology (Spong et al., 2006), and assumes that the Jacobian is full rank, in order to get a closed-loop system which is *almost* linear. The original RAC scheme used a parameterization of orientation known as the *Euler rotation,* but some other parameterizations have also been studied later (see Caccavale et al., 1998). The number of RAC closed-loop equilibria and their stability properties depend on the parameterization used for describing orientation.
The general expression for the RAC control law when using Euler parameters is given by

$$\tau = M(q)J^{\dagger}(q)\left[\begin{bmatrix}\ddot{p}_d + K_{V_p}\dot{\tilde{p}} + K_{P_p}\tilde{p} \\ \dot{\omega}_d + K_{V_o}\tilde{\omega} + K_{P_o}\tilde{\varepsilon}\end{bmatrix} - \dot{J}(q)\dot{q}\right] + C(q,\dot{q})\dot{q} + g(q) \tag{60}$$

where $M(q)$, $C(q,\dot{q})$, $g(q)$ are elements of the joint dynamics (38); $J^{\dagger}(q)$ denotes the left pseudo-inverse of $J(q)$, and K_{P_p}, K_{V_p}, K_{P_o} are symmetric positive definite matrices, $K_{V_o}=k_v I$ is diagonal with $k_v>0$. As in the RMRC controller in the previous subsection, we use $\tilde{\varepsilon}$ as an orientation error vector. Fig. 7 shows how to implement RAC's control law.
Substituting the control law (60) into the robot dynamics (38), considering that $J(q)$ is full rank, and the time derivative of (27) is given by

$$\begin{bmatrix}\ddot{p} \\ \dot{\omega}\end{bmatrix} = J(q)\ddot{q} + \dot{J}(q)\dot{q},$$

we get the (almost linear) system:

$$\begin{bmatrix}\ddot{\tilde{p}} + K_{V_p}\dot{\tilde{p}} + K_{P_p}\tilde{p} \\ \dot{\tilde{\omega}} + K_{V_o}\tilde{\omega} + K_{P_o}\tilde{\varepsilon}\end{bmatrix} = 0 \in \mathbb{R}^6. \tag{61}$$

To obtain the full state-space representation of the RAC closed-loop system, we use (61) and include dynamics (47), to get

$$\frac{d}{dt}\begin{bmatrix}\tilde{p}\\ \dot{\tilde{p}}\\ \tilde{\eta}\\ \tilde{\varepsilon}\\ \tilde{\omega}\end{bmatrix}=\begin{bmatrix}\dot{\tilde{p}}\\ -K_{V_p}\dot{\tilde{p}}-K_{P_p}\tilde{p}\\ -\frac{1}{2}\tilde{\varepsilon}^T\tilde{\omega}\\ \frac{1}{2}[\tilde{\eta}I+S(\tilde{\varepsilon})]\tilde{\omega}-S(\tilde{\varepsilon})\omega_d\\ -K_{V_o}\tilde{\omega}-K_{P_o}\tilde{\varepsilon}\end{bmatrix} \tag{62}$$

where now the domain of the state-space is

$$D_{ra}=\mathbb{R}^6\times\mathrm{S}^3\times\mathbb{R}^3=\left\{\begin{bmatrix}\tilde{p}\\ \dot{\tilde{p}}\\ \tilde{\eta}\\ \dot{\tilde{\varepsilon}}\\ \tilde{\omega}\end{bmatrix}\in\mathbb{R}^{13}:\tilde{\eta}^2+\tilde{\varepsilon}^T\tilde{\varepsilon}-1=0\right\}.$$

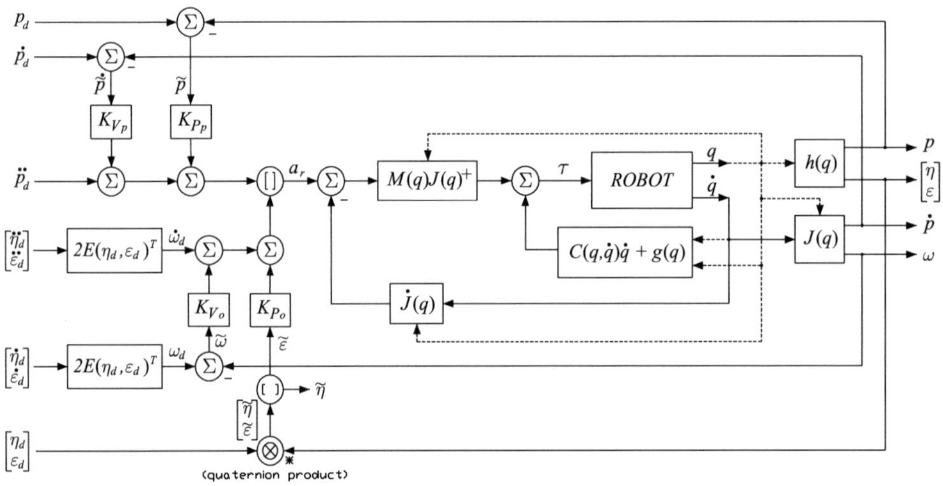

Figure 7. Resolved acceleration controller with Euler parameters

The closed-loop system (62) is nonautonomous, because of the presence of $\omega_d(t)$, and it has two equilibria:

$$\begin{bmatrix}\tilde{p}\\ \dot{\tilde{p}}\\ \tilde{\eta}\\ \tilde{\varepsilon}\\ \tilde{\omega}\end{bmatrix}=\begin{bmatrix}0\\0\\1\\0\\0\end{bmatrix}=E_1\in D_{ra},\quad\text{and}\quad\begin{bmatrix}\tilde{p}\\ \dot{\tilde{p}}\\ \tilde{\eta}\\ \tilde{\varepsilon}\\ \tilde{\omega}\end{bmatrix}=\begin{bmatrix}0\\0\\-1\\0\\0\end{bmatrix}=E_2\in D_{ra}.$$

As shown in (Campa, 2005) these two equilibria have different stability properties (E_1 is asymptotically stable, E_2 is unstable). Notwithstanding, they represent the same pose.

In order to prove asymptotic stability of the equilibrium E_1, let us consider the following Lyapunov function (Campa, 2005):

$$V(\tilde{p},\dot{\tilde{p}},\tilde{\eta},\tilde{\varepsilon},\tilde{\omega}) = V_p(\tilde{p},\dot{\tilde{p}}) + V_o(\tilde{\eta},\tilde{\varepsilon},\tilde{\omega})$$

where $V_p(\tilde{p},\dot{\tilde{p}})$ and $V_o(\tilde{\eta},\tilde{\varepsilon},\tilde{\omega})$ stand for the position and orientation parts, given by

$$V_p(\tilde{p},\dot{\tilde{p}}) = \frac{1}{2}\dot{\tilde{p}}^T\dot{\tilde{p}} + \frac{1}{2}\tilde{p}^T\left[K_{P_p} + \alpha K_{V_p}\right]\tilde{p} + \alpha\tilde{p}^T\dot{\tilde{p}},$$

$$V_o(\tilde{\eta},\tilde{\varepsilon},\tilde{\omega}) = (\tilde{\eta}-1)^2 + \tilde{\varepsilon}^T\tilde{\varepsilon} + \frac{1}{2}\tilde{\omega}^T K_{P_o}^{-1}\tilde{\omega} + \beta\tilde{\varepsilon}^T\tilde{\omega},$$

and α, β, satisfy

$$0 < \alpha < \lambda_m\{K_{V_p}\} \tag{63}$$

$$0 < \beta < \min\left\{\sqrt{2\lambda_m\left\{K_{P_o}^{-1}\right\}}, \frac{2k_v\lambda_m\{K_{P_o}\}\lambda_m\{K_{P_o}^{-1}\}}{\lambda_m\{K_{P_o}\} + \left[k_v + \|\omega_d\|_M\right]^2}\right\}. \tag{64}$$

where $\|\omega_d\|_M$ stands for the supremum of the norm $\|\omega_d\|$.
Let us notice that the position part of system is a second-order linear system, and it is easily proved that this part, considering (63), is asymptotically stable (see, e.g., the proof of the *computed-torque* controller in (Kelly et al., 2005)). Regarding the orientation part, it should be noticed that

$$V_o(\tilde{\eta},\tilde{\varepsilon},\tilde{\omega}) \geq (\tilde{\eta}-1)^2 + \frac{1}{2}\begin{bmatrix} \|\tilde{\varepsilon}\| \\ \|\tilde{\omega}\| \end{bmatrix}^T \begin{bmatrix} 2 & -\beta \\ -\beta & \lambda_m\{K_{P_o}^{-1}\} \end{bmatrix} \begin{bmatrix} \|\tilde{\varepsilon}\| \\ \|\tilde{\omega}\| \end{bmatrix}.$$

Thus, given (64), we have that $V_o(\tilde{\eta},\tilde{\varepsilon},\tilde{\omega})$ is a positive definite function in $S^3\times\mathbb{R}^3$, around E_1, that is to say that $V_o(1,0,0) = 0$, and there is a neighborhood $B \subset S^3\times\mathbb{R}^3 \subset D_{ra}$, around E_1, such that $V_o(\tilde{\eta},\tilde{\varepsilon},\tilde{\omega}) > 0$, for all $\begin{bmatrix}\tilde{\eta} & \tilde{\varepsilon}^T & \tilde{\omega}^T\end{bmatrix}^T \neq \begin{bmatrix}1 & 0^T & 0^T\end{bmatrix}^T \in B$.
The time derivative of $V_o(\tilde{\eta},\tilde{\varepsilon},\tilde{\omega})$ along the trajectories of system (62) results in

$$\dot{V}_o(\tilde{\eta},\tilde{\varepsilon},\tilde{\omega}) = -\beta\tilde{\varepsilon}^T K_{P_o}\tilde{\varepsilon} - \tilde{\omega}^T\left[K_{P_o}^{-1}K_{V_o} - \frac{1}{2}\beta\tilde{\eta}I\right]\tilde{\omega} - \beta\tilde{\varepsilon}^T\left[K_{V_o} + S(\omega_d)\right]\tilde{\omega},$$

where properties (4) to (7) on $S(\tilde{\varepsilon})$ have been evoked. Hence, we have

$$\dot{V}_o(\tilde{\eta},\tilde{\varepsilon},\tilde{\omega}) \leq -\frac{1}{2}\begin{bmatrix} \|\tilde{\varepsilon}\| \\ \|\tilde{\omega}\| \end{bmatrix}^T Q \begin{bmatrix} \|\tilde{\varepsilon}\| \\ \|\tilde{\omega}\| \end{bmatrix} - \frac{1}{2}\beta\lambda_m\{K_{P_o}\}\left(1-\tilde{\eta}^2\right),$$

where

$$Q=\begin{bmatrix} \beta\lambda_m\{K_{P_o}\} & -\beta\left[k_v+\|\omega_d\|_{\rm M}\right] \\ -\beta\left[k_v+\|\omega_d\|_{\rm M}\right] & 2k_v\lambda_m\{K_{P_o}^{-1}\}-\beta \end{bmatrix},$$

the property $\|S(x)\|=\|x\|$ has ben considered, and the fact $\|\varepsilon\|^2=\left(1-\eta^2\right)$ was employed to get the last term.
A condition for $\dot{V}_o(\tilde{\eta},\tilde{\varepsilon},\tilde{\omega})$ to be negative definite in $\mathrm{S}^3\times\mathbb{R}^3$ is that $Q>0$, which is true if β satisfies (5). Thus, $\dot{V}_o(\tilde{\eta},\tilde{\varepsilon},\tilde{\omega})$ is negative definite in $\mathrm{S}^3\times\mathbb{R}^3$, around E_1, that is to say that $\dot{V}_o(1,0,0)=0$, and there is a neighborhood $B\subset\mathrm{S}^3\times\mathbb{R}^3\subset D_{ra}$, around E_1, such that $\dot{V}_o(\tilde{\eta},\tilde{\varepsilon},\tilde{\omega})<0$, for all $\begin{bmatrix}\tilde{\eta} & \tilde{\varepsilon}^T & \tilde{\omega}^T\end{bmatrix}^T\neq\begin{bmatrix}1 & 0^T & 0^T\end{bmatrix}^T\in B$.
Thus, we can conclude that the equilibrium E_1 is asymptotically stable in D_{ra}. This implies that, starting from an initial condition near E_1 in D_{ra}, the control objective (48) is fulfilled.

6. Conclusion

Among the different parameterizations of the orientation manifold, Euler parameters are of interest because they are unit quaternions, and form a Lie group. This chapter was intended to recall some of the properties of the quaternion algebra, and to show the usage of Euler parameters for modeling, path planning and control of robot manipulators.
For many years, the Denavit–Hartenberg parameters have been employed for obtaining the direct kinematic model of robotic mechanisms. Instead of using homogeneous matrices for handling the D-H parameters, we proposed an algorithm that uses the quaternion multiplication as basis, thus being computationally more efficient than matrix multiplication. In addition, we sketched a general methodology for obtaining the dynamic model of serial manipulators when Euler parameters are employed. The so–called spherical linear interpolation (or slerp) is a method for interpolating curves in the orientation manifold. We reviewed such an algorithm, and proposed its application for motion planning in robotics. Finally, we studied the stability of two task space controllers that employ unit quaternions as state variables; such analysis allows us to conclude that both controllers are suitable for motion control tasks in pose space.
As a conclusion, even though Euler parameters are still not very known, it is our believe that they constitute a useful tool, not only in the robotics field, but wherever the description of objects in 3-D space is required. More research should be done in these topics.

7. Acknowledgement

This work was partially supported by DGEST and CONACYT (grant 60230), Mexico.

8. References

Aicardi, M.; Caiti, A.; Cannata, G. & Casalino, G. (1995). Stability and robustness analysis of a two layered hierarchical architecture for the closed loop control of robots in the operational space, *Proceedings of the IEEE International Conference on Robotics and Automation*, pp. 2771-2778, Nagoya, Japan, May 1995.

Barrientos, A.; Peñín, L. F.; Balaguer, C. & Aracil, R. (1997). *Fundamentos de Robótica*, McGraw-Hill, Madrid.

Caccavale, F.; Natale, C.; Siciliano, B. & Villani, L. (1998) Resolved-acceleration control of robot manipulators: A critical review with experiments. *Robotica*, Vol. 16, pp. 565-573.

Caccavale, F.; Siciliano, B. & Villani, L. (1999). The role of Euler parameters in robot control, *Asian Journal of Control*, Vol. 1, pp. 25-34.

Campa, R. (2005). *Control de Robots Manipuladores en Espacio de Tarea*. Doctoral thesis (in Spanish), CICESE Research Center, Ensenada, Mexico, March 2005.

Campa, R. & de la Torre, H. (2006). On the representations of orientation for pose control tasks in robotics. *Proceedings of the VIII Mexican Congress on Robotics* Mexico, D.F., October 2006.

Campa, R.; Camarillo, K. & Arias, L. (2006). Kinematic modeling and control of robot manipulators via unit quaternions: Application to a spherical wrist. *Proceedings of the 45th IEEE Conference on Decision and Control*, San Diego, CA, December 2006.

Craig, J. J. (2005) *Introduction to Robotics: Mechanics and Control*, Pearson Prentice-Hall, USA, 2005.

Dam, E. B.; Koch, M. & Lillholm, M. (1998) *Quaternions, Interpolation and Animation*. Technical report DIKU-TR-98/5, University of Copenhagen, Denmark.

Denavit, J. & Hartenberg, E. (1955) A kinematic notation for lower-pair mechanisms based on matrices. *Journal of Applied Mechanics*, Vol. 77, pp. 215-221.

Fjellstad, O. E. (1994). "Control of unmanned underwater vehicles in six degrees of freedom: a quaternion feedback approach", Dr.ing. thesis, Norwegian Institute of Technology, Norway.

Gallier, J. (2000). *Geometric Methods and Applications for Computer Science and Engineering*. Springer-Verlag, New York.

Hamilton, W. R. (1844). On a new species of imaginary quantities connected with a theory of quaternions. *Proceedings of the Royal Irish Academy*. Vol. 2, pp. 424-434. (http://www.maths.tcd.ie/pub/HistMath/People/Hamilton/Quatern1).

Kelly, R. & Moreno-Valenzuela, J. (2005). Manipulator Motion Control in Operational Space Using Joint Velocity Inner Loop, *Automatica*, Vol. 41, pp. 1423-1432.

Kelly, R.; Santibáñez, V. & Loría, A. (2005) *Control of Robot Manipulators in Joint Space*, Springer-Verlag, London.

Khalil, H. K. (2002) *Nonlinear Systems*, Prentice-Hall, New Jersey..

Khatib, O. (1987). A unified approach for motion and force control of robot manipulators: The operational space formulation. *IEEE Journal of Robotics and Automation*. Vol. 3, pp. 43-53.

Krause, P. C.; Wasynczuk, O. & Sudhoff S. D. (2002). *Analysis of Electric Machinery and Drive Systems*. Wiley-Interscience.

Kuipers, J. B. (1999). *Quaternions and rotation sequences*. Princeton University Press. Princeton, NJ.

Lin, S. K. (1995) Robot control in Cartesian space, In: *Progress in Robotics and Intelligent Systems*, Zobrit, G. W. & Ho, C. Y. (Ed.), pp. 85-124, Ablex, New Jersey.

Lu, Y. J. & Ren G. X. (2006) A symplectic algorithm for dynamics of rigid-body. *Applied Mathematics and Dynamics*, Vol. 27, No. 1, pp. 51-57.

Luh, J. Y. S., Walker M. W., Paul R. P. C. (1980). Resolved-acceleration control of mechanical manipulators. *IEEE Transactions on Automatic Control*. Vol. 25, pp. 486–474.

Natale, C. (2003). *Interaction Control of Robot Manipulators: Six degrees-of-freedom Tasks*, Springer, Germany.

Rooney, J. & Tanev, T. K. (2003). Contortion and formation structures in the mappings between robotic jointspaces and workspaces. *Journal of Robotic Systems*, Vol. 20, No. 7, pp. 341-353.

Sciavicco, L. & Siciliano, B. (2000). *Modelling and Control of Robot Manipulators*, Springer-Verlag, London.

Shivarama, R. & Fahrenthold, E. P. (2004) Hamilton's equation with Euler parameters for rigid body dynamics modeling *Journal of Dynamics Systems, Measurement and Control* Vol. 126, pp. 124-130.

Shoemake, K. (1985) Animating rotation with quaternion curves, *ACM SIGGRAPH Computer Graphics* Vol. 19, No. 3, pp. 245-254.

Spong, M. W.; Hutchinson, S. & Vidyasagar, M. (2006) *Robot Modeling and Control*, John Wiley and Sons, USA, 2006.

Spring, K. W. (1986). Euler parameters and the use of quaternion algebra in the manipulation of finite rotations: A review. *Mechanism and Machine Theory*, Vol. 21, No. 5, pp. 365-373.

Whitney, D. E. (1969). Resolved motion rate control of manipulators and human prostheses. *IEEE Transactions on Man-Machine Systems*, Vol. 10, no. 2, pp. 47-53.

Xian, B.; de Queiroz, M. S.; Dawson, D. & Walker, I. (2004). Task-space tracking control of robot manipulators via quaternion feedback. *IEEE Transactions on Robotics and Automation*. Vol. 20, pp. 160–167.

Yuan, J. S. C. (1988). Closed-loop manipulator control using quaternion feedback. *IEEE Journal of Robotics and Automation*. Vol. 4, pp. 434–440.

3

Kinematic Design of Manipulators

Marco Ceccarelli and Erika Ottaviano
LARM: Laboratory of Robotics and Mechatronics, DiMSAT - University of Cassino Italy

1. Introduction

Robots can be considered as the most advanced automatic systems and robotics, as a technique and scientific discipline, can be considered as the evolution of automation with interdisciplinary integration with other technological fields.
A robot can be defined as a system which is able to perform several manipulative tasks with objects, tools, and even its extremity (end-effector) with the capability of being re-programmed for several types of operations. There is an integration of mechanical and control counterparts, but it even includes additional equipment and components, concerned with sensorial capabilities and artificial intelligence. Therefore, the simultaneous operation and design integration of all the above-mentioned systems will provide a robotic system, as illustrated in Fig. 1, (Ceccarelli 2004).
In fact, more than in automatic systems, robots can be characterized as having simultaneously mechanical and re-programming capabilities. The mechanical capability is concerned with versatile characteristics in manipulative tasks due to the mechanical counterparts, and re-programming capabilities concerned with flexible characteristics in control abilities due to the electric-electronics-informatics counterparts. Therefore, a robot can be considered as a complex system that is composed of several systems and devices to give:

- mechanical capabilities (motion and force);
- sensorial capabilities (similar to human beings and/or specific others);
- intellectual capabilities (for control, decision, and memory).

Initially, industrial robots were developed in order to facilitate industrial processes by substituting human operators in dangerous and repetitive operations, and in unhealthy environments. Today, additional needs motivate further use of robots, even from pure technical viewpoints, such as productivity increase and product quality improvements. Thus, the first robots have been evolved to complex systems with additional capabilities.
Nevertheless, referring to Fig. 1, an industrial robot can be thought of as composed of:

- a mechanical system or manipulator arm (mechanical structure), whose purpose consists of performing manipulative operation and/or interactions with the environment;
- sensorial equipment (internal and external sensors) that is inside or outside the mechanical system, and whose aim is to obtain information on the robot state and scenario, which is in the robot area;

- a control unit (controller), which provides elaboration of the information from the sensorial equipment for the regulation of the overall systems and gives the actuation signals for the robot operation and execution of desired tasks;
- a power unit, which provides the required energy for the system and its suitable transformation in nature and magnitude as required for the robot components;
- computer facilities, which are required to enlarge the computation capability of the control unit and even to provide the capability of artificial intelligence.

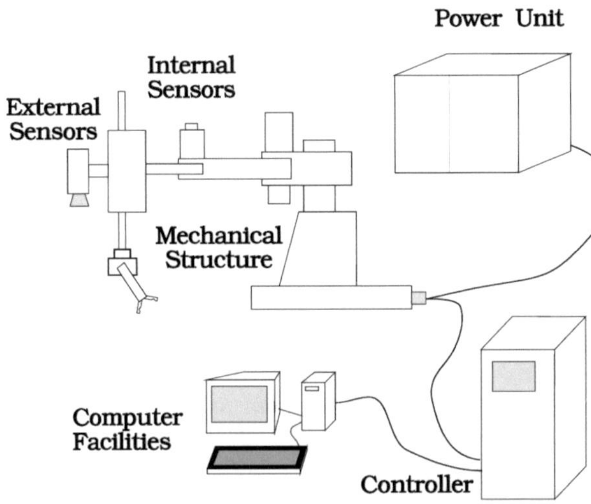

Figure 1. Components of an industrial robot

Thus, the above-mentioned combination of sub-systems gives the three fundamental simultaneous attitudes to a robot, i.e. mechanical action, data elaboration, and re-programmability.

Consequently, the fundamental capability of robotic systems can be recognized in:

- mechanical versatility;
- re-programmability.

Mechanical versatility of a robot can be understood as the capability to perform a variety of tasks because of the kinematic and mechanical design of its manipulator arm.

Re-programmability of a robot can be understood as the flexibility to perform a variety of task operations because of the capability of its controller and computer facilities.

These basic performances give a relevant flexibility for the execution of several different tasks in a similar or better way than human arms. In fact, nowadays robots are well-established equipment in industrial automation since they substitute human operators in operations and situations.

The mechanical capability of a robot is due to the mechanical sub-system that generally is identified and denominated as the 'manipulator', since its aim is the manipulative task.

The term manipulation refers to several operations, which include:

- grasping and releasing of objects;
- interaction with the environment and/or with objects not related with the robot;
- movement and transportation of objects and/or robot extremity.

Consequently, the mechanical sub-system gives mechanical versatility to a robot through kinematic and dynamic capability during its operation. Manipulators can be classified according to the kinematic chain of their architectures as:

- serial manipulators, when they can be modeled as open kinematic chains in which the links are jointed successively by binary joints;
- parallel manipulators, when they can be modeled as closed kinematic chains in which the links are jointed to each other so that polygonal loops can be determined.

In addition, the kinematic chains of manipulators can be planar or spatial depending on which space they operate. Most industrial robotic manipulators are of the serial type, although recently parallel manipulators have aroused great interest and are even applied in industrial applications.

In general, in order to perform similar manipulative tasks as human operators, a manipulator is composed of the following mechanical sub-systems:

- an arm, which is devoted to performing large movements, mainly as translations;
- a wrist, whose aim is to orientate the extremity;
- an end-effector, which is the manipulator extremity that interacts with the environment.

Several different architectures have been designed for each of the above-mentioned manipulator sub-systems as a function of required specific capabilities and characteristics of specific mechanical designs. It is worthy of note that although the mechanical design of a manipulator is based on common mechanical components, such as all kinds of transmissions, the peculiarity of a robot design and operation requires advanced design of those components in terms of materials, dimensions, and designs because of the need for extreme lightness, compactness, and reliability.

The sensing capability of a robot is obtained by using sensors suitable for knowing the status of the robot itself and surrounding environment. The sensors for robot status are of fundamental importance since they allow the regulation of the operation of the manipulator. Therefore, they are usually installed on the manipulator itself with the aim of monitoring basic characteristics of manipulations, such as position, velocity, and force. Additionally, an industrial robot can be equipped with specific and/or advanced sensors, which give human-like or better sensing capability. Therefore, a great variety of sensors can be used, to which the reader is suggested to refer to in specific literature.

The control unit is of fundamental importance since it gives capability for autonomous and intelligent operation to the robot and it performs the following aims:

- regulation of the manipulator motion as a function of current and desired values of main kinematic and dynamic variables by means of suitable computations and programming;
- acquisition and elaboration of sensor signals from the manipulator and surrounding environment;
- capability of computation and memory, which is needed for the above-mentioned purposes and robot re-programmability.

In particular, an intelligence capability has been added to some robotic systems concerned mainly with decision capability and memory of past experiences by using the means and techniques of expert systems and artificial intelligence. Nevertheless, most of the current industrial robots have no intelligent capability since the control unit properly operates for the given tasks within industrial environments. Nowadays industrial robots are usually equipped with minicomputers, since the evolution of low-cost PCs has determined the wide use of PCs in robotics so that sequencers, which are going to be restricted to PLC units only, will be used mainly in rigid automation or low-flexible systems.

Generally, the term manipulator refers specifically to the arm design, but it can also include the wrist when attention is addressed to the overall manipulation characteristics of a robot.
A kinematic study of robots deals with the determination of configuration and motion of manipulators by looking at the geometry during the motion, but without considering the actions that generate or limit the manipulator motion. Therefore, a kinematic study makes it possible to determine and design the motion characteristics of a manipulator but independently from the mechanical design details and actuator's capability.
This aim requires the determination of a model that can be deduced by abstraction from the mechanical design of a manipulator and by stressing the fundamental kinematic parameters.
The mobility of a manipulator is due to the degrees of freedom (d.o.f.s) of the joints in the kinematic chain of the manipulator, when the links are assumed to be rigid bodies.
A kinematic chain can be of open architecture, when referring to serial connected manipulators, or closed architecture, when referring to parallel manipulators, as in the examples shown in Fig. 2.

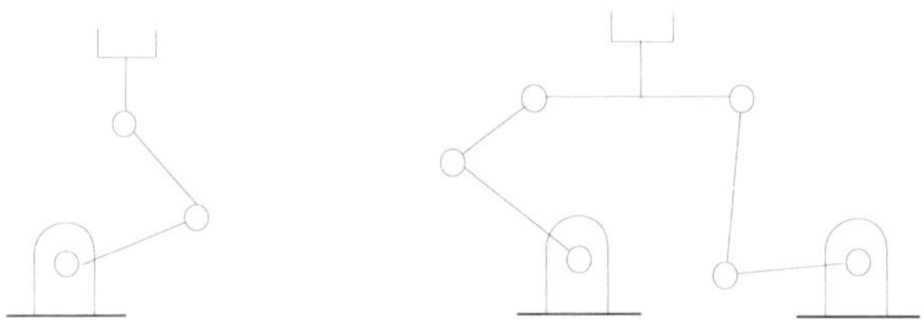

a) b)

Figure 2. Planar examples of kinematic chains of manipulators: a) serial chain as open type; b) parallel chain as closed type

a) b)

Figure 3. Schemes for joints in robots: a) revolute joint; b) prismatic joint

Of course, it is also possible to design mixed chains for so-called hybrid manipulators.
Regarding the joints, although there are several designs both from theoretical and practical viewpoints, usually the joint types in robots are related to prismatic and revolute pairs with one degree of freedom. They can be modeled as shown in Fig. 3.
However, most of the manipulators are designed by using revolute joints, which have the advantage of simple design, long durability, and easy operation and maintenance. But the revolute joints also allow a kinematic chain and then a mechanical design with small size, since a manipulator does not need a large frame link and additionally its structure can be of small size in a work-cell.
In addition, it is possible to also obtain operation of other kinematic pairs with revolute joints only, when they are assembled in a proper way and sequence. For example, three revolute joints can obtain a spherical joint and depending on the assembling sequence they may give different practical spherical joints.
In general the multidisciplinarity aspects of structure and operation of robots will require a

complex design procedure with a mechatronic approach of integration of all constraints and requirements of the different natures of the robot components. In Fig.4 a general scheme is reported as referring to a procedure, which is based on step by step design approach for the different aspects but by considering and integrating them from each other. Nevertheless it is stressed the fundamentals of the design of the manipulator structure which will affect and will be affected from the other components of a robot. Indeed, each component will affect the design and operation of other part of a robot when a design and operation is conceived with full exploit of the capability of each component. The design of manipulator can be considered as a starting point of an iterative process in which each aspect will contribute and will affect the previous and next solution to a mechatronic integrated solution of the robot system. Similarly important are the characteristics and requirements of the task and application to which the robot is devoted. Thus an so-called optimal design of a robot will be achieved only after a reiteration of design process both for the components and the whole systems, by looking at each component separately and integrated approach. Thus, even the design of the manipulator can be considered at the same time as starting and final point of the design process.

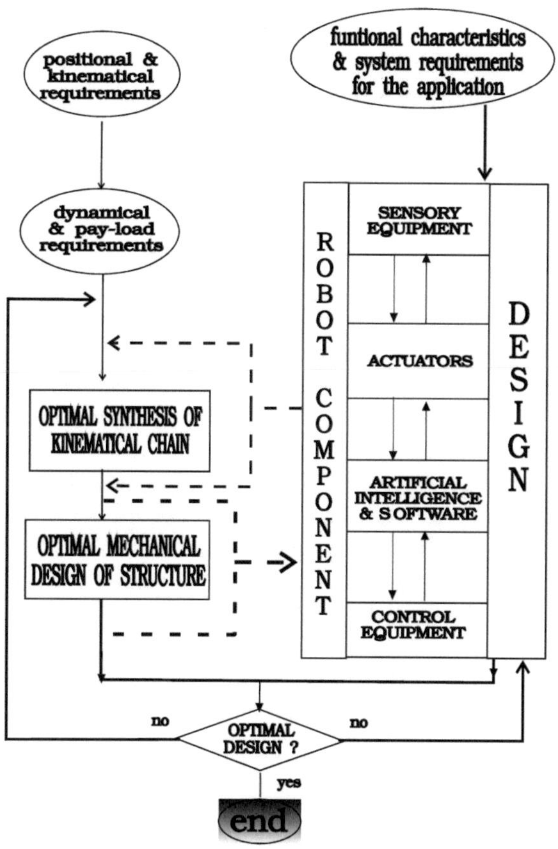

Figure 4. A general scheme for a design procedure of robots

Kinematic design of manipulators refers to the determination of the dimensional parameters of a kinematic chain, i.e. link lengths and link angles. Once the kinematic architecture of a manipulator is sized by means of a kinematic design, a manipulator can be completely defined by means of a mechanical design that specifies all the sizes and details for a physical construction. Indeed, kinematic design is a fundamental step in a design procedure of any mechanical system and its accuracy will affect strongly the basic properties of a mechanical systems. In the case of manipulators the kinematic design is of a particular importance since the manipulator tasks can be performed when the kinematic architecture has been properly conceived or chosen and specifically synthesized (i.e. kinematic design).

Several approaches have been formulated for the kinematic design of mechanisms and many of them have been specialized for robotic manipulators. General procedures and specific algorithms both for general kinematic architectures and specific designs of manipulators have been proposed in a very rich literature. A limited list of references is reported with the aim to give to the readers basic sources and suggestions for further reading on the topic.

In this chapter a survey of current issues is presented by using basic concepts and formulations in order to emphasize on problem formulation and computational efforts. Indeed, a great attention is still addressed to kinematic design of manipulators by robot designers and researchers mainly with the aim to improve computational efficiency, generality and optimality of the algorithms, even with respect to new and new requirements for robotic manipulations. In addition, theoretical and numerical works are usually validated by the same investigators through tests with prototypes and experimental activity on performance characteristics.

This survey has overviewed the currently available procedures for kinematic design of manipulators that can be grouped in three main approaches, namely extension of Synthesis of Mechanisms (for example: Precision Point Techniques, Workspace Design, Inversion algorithms, Optimization formulation), application of Screw Theory, application of 3D Kinematics/Geometry (for example: Lie Group Theory, Dual Numbers, Quaternions, Grassmann Geometry).

A kinematic design procedure is aimed to obtain closed-form formulation and/or numerical algorithms, which can be used not only for design purposes but even to investigate effects of design parameters on design characteristics and operation performance of manipulators.

Usually, there is a distinction between open-chain serial manipulators and closed-chain parallel manipulators. This distinction is also considered as a constraint for the kinematic design of manipulators and in fact different procedures and formulation have been proposed to take into account the peculiar differences in their kinematic design. Nevertheless, recently, attempts have been made to formulate a unique view for kinematic design both of serial and parallel manipulators, mainly with an approach using optimization problems.

Future challenges in the field of robot design can be recognized mainly in the aspects for computational efficiency and in conceiving new manipulator architectures with a fully insight of design degeneracy both of the kinematic possibilities and proposed numerical algorithms.

2. The design problem

The manipulator architecture of a robot is composed of an arm mostly for translation movements, a wrist for orientation movement, and an end-effector for interaction with the environment and/or external objects, as shown in Fig. 1. Generally, the term manipulator refers specifically to the arm design, but it can also include the wrist when attention is addressed to the overall manipulation characteristics of a robot.

A kinematic study of robots deals with the determination of configuration and motion of manipulators by looking at the geometry during the motion, but without considering the actions that generate or limit the manipulator motion. Therefore, a kinematic study makes possible to determine and design the motion characteristics of a manipulator but independently from the mechanical design details and actuator capability.

A kinematic chain can be of open architecture, when referring to serial connected manipulators, or closed architecture, when referring to parallel manipulators, as in the example in Fig. 5b).

The kinematic model of a manipulator can be obtained in the form of a kinematic chain or mechanism by using schemes for joints and rigid links through essential dimensional sizes for connections between two joints. The mobility of a manipulator is due to the degrees of freedom (d.o.f.s) of the joints in the kinematic chain, when the links are assumed to be rigid bodies. In order to determine the geometrical sizes and kinematic parameters of open-chain general manipulators, one can usually refer to a scheme like that in Fig. 5a) by using a H–D notation, in agreement with a procedure that was proposed by Hartenberg and Denavit in 1955.

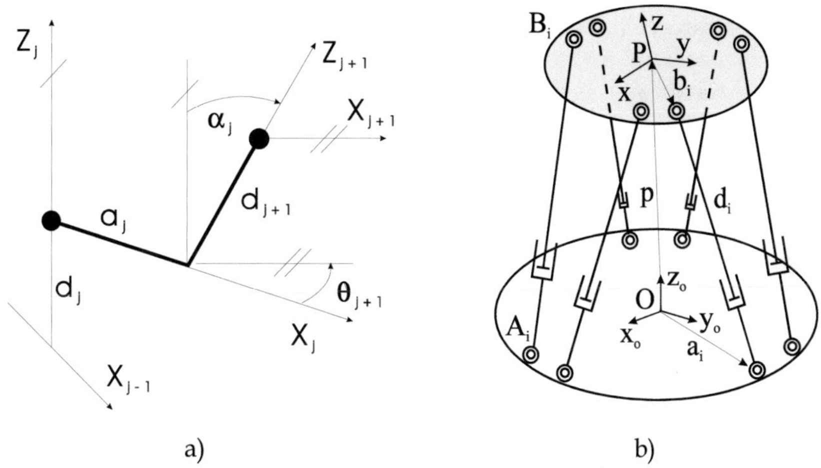

Figure 5. A kinematic scheme for manipulator link parameters: a) according to H-D notation; b) for parallel architectures

This scheme gives the minimum number of parameters that are needed to describe the geometry of a link between two joints, but also indicates the joint variables. The joints in Fig. 5a) are indicated as big black points in order to stress attention to the link geometry and H–D parameters. In particular, referring to Fig. 5a) for j-link, the j-frame $X_jY_jZ_j$ is assumed as fixed to j-link, with the Z_j axis coinciding with the joint axis, with the X_j axis lying on the common normal between Z_j and Z_{j+1} and pointing to Z_{j+1}.

The kinematic parameters of a manipulator can be defined according to the H–D notation in Fig. 5a) as:

- a_j, link length that is measured as the distance between the Z_j and Z_{j+1} axes along X_j;
- αj, twist angle that is measured as the angle between the Z_j and Z_{j+1} axes about X_j;
- d_{j+1}, link offset that is measured as the distance between the X_j and X_{j+1} axes along Z_{j+1};

- θ_{j+1}, joint angle that is measured as the angle between the X_j and X_{j+1} axes about Z_{j+1}

When a joint can be modelled as a rotation pair, the angle θ_{j+1} is the corresponding kinematic variable. When a joint is a prismatic pair, the distance d_{j+1} is the corresponding kinematic variable. Other H-D parameters can be considered as dimensional parameters of the links.

The H-D notation is very useful for the formulation of the position problems of manipulators through the so-called transformation matrix by using matrix algebra.

The position problem of manipulators, both with serial and parallel architectures, consists of determining the position and orientation of the end-effector as a function of the manipulator configuration that is given by the link position that is defined by the joint variables.

In general, the position problem can be considered from different viewpoints depending on the unknowns that one can solve in the following formulations:

- Kinematic Direct Problem in which the dimensions of a manipulator are given through the dimensional H-D parameters of the links but the position and orientation of the end-effector are determined as a function of the values of the joint variables;
- Kinematic Inverse Problem in which the position and orientation of the end-effector of a given manipulator are given, and the configuration of the manipulator chain is determined by computing the values of the joint variables.

A third kinematic problem can be formulated as:

- Kinematic Indirect Problem (properly 'Kinematic Design Problem') in which a certain number of positions and orientations of the end-effector are given but the type of manipulator chain and its dimensions are the unknowns of the problem.

  Although general concepts are common both for serial and parallel manipulators, peculiarities must be considered for parallel architectures chains.

In parallel manipulators one can consider as generalized coordinates the position coordinates of the center point P of the moving platform with respect to a fixed frame (X_o Y_o Z_o), Fig. 5b), and the direction is described by Euler angles defining the orientation of the moving platform with respect to a fixed frame. A matrix R defines the orthogonal 3 × 3 rotation matrix defined by the Euler angles, which describes the orientation of the frame attached to the moving platform with respect to the fixed frame, Fig. 5b). Let Ai and Bi be the attachment points at the base and moving platform, respectively, and di the leg lengths. Let $\mathbf{a}_i$ and $\mathbf{b}_i$ be the position vectors of points A_i and B_i in the fixed and moving coordinate frames, respectively. Thus, for parallel manipulators the Inverse Kinematics Problem can be solved by using

$$A_iB_i = \mathbf{p} + R\mathbf{b}_i - \mathbf{a}_i \tag{1}$$

to extract the joint variables from leg lengths. The length of the i-th leg can be obtained by taking the dot product of vector A_iB_i with itself, for i:1,...,6 in the form

$$\mathbf{d}_i^2 = [\mathbf{p} + R\mathbf{b}_i - \mathbf{a}_i]^t [\mathbf{p} + R\mathbf{b}_i - \mathbf{a}_i] \tag{2}$$

The Direct Kinematics Problem describes the mapping from the joint coordinates to the generalized coordinates. The problem for parallel manipulators is quite difficult since it involves the solution of a system of nonlinear coupled algebraic equations (1), and has many solutions that refer to assembly modes. For a general case of Gough-Stewart Platform with planar base and platform, the Direct Kinematics Problem may have up to 40 solutions. A 20-th degree polynomial has been derived leading to 40 mutually symmetric assembly modes.

3. Algorithms for kinematic design of manipulators

Synthesis deals with reverse problems of Analysis. Thus, Synthesis of mechanisms and manipulators deals with design of the kinematic chain as function of manipulative tasks.
Characteristic manipulative tasks of manipulators concern with manipulation of objects as movement and orientation of grasped objects or end-effector itself during a suitably programmed motion of a manipulator. But manipulation includes also other aspects of functional and operation characteristics, and nowadays mechatronic approaches are also used to consider those other aspects in fully integrated approaches.
Design calculation of kinematic chain of mechanisms and manipulators is usually attached through three problems, namely type synthesis, number synthesis, and dimensional synthesis. Number synthesis concerns with the determination of the number of links and joints in the chain, which are useful or necessary to obtain a desired mobility and manipulation capability of a manipulator mechanism. Similarly, type synthesis concerns with the determination of the structure of the kinematic chain, i.e. the type of joints and kinematic architecture, that are useful or necessary to obtain a desired mobility and manipulation capability of a manipulator mechanism. Finally, dimensional synthesis (i.e. kinematic design) concerns with the calculation of the link sizes and range mobility of joints that are useful or necessary to obtain a desired mobility and manipulation capability of a manipulator mechanism.
Type synthesis and number synthesis are related to morphologies of manipulator architectures and today they are approached with designer's own experience or through complex design procedures that most of the time can be understood as data bases in informatics expert systems.
The traditional design activity on manipulators is still recognized in the problem of the dimensional design of a manipulator when its kinematic architecture is given. This is the problem that is surveyed in this paper.
The manipulative tasks that are used as design data and constraints, are related mainly to kinematic features such as workspace, path planning, Static accuracy; but other aspects can be also considered within the Mechanics of robots, such as singularities, stiffness behaviour, dynamic response. One aspect of relevant significance for manipulator design is the workspace analysis that is often used as design means yet, beside as a criteria for evaluating the quality of designed solutions. Positioning and orientation capability can be evaluated by computing position and orientation workspaces that give the reachable regions by the manipulator extremity or end-effector as function of the mobility range of the manipulator joints. Position workspace refers to reachable points by a reference point on manipulator extremity, and orientation workspace describes the angles that can be swept by reference axes on manipulator extremity. Once the workspace points (both in position and orientation) are determined, one can use them to perform an evaluation of workspace characteristics and a feasibility evaluation of kinematic design solutions. In particular, a cross-section area can be determined by selecting from the computed workspace points those that lay on a cross-section plane under examination. Thus, the shape can be illustrated by the result in a plot form. The computation of the value of a cross-section area can be obtained by using a grid evaluation or an algebraic formula.

By referring to the scheme of Fig. 6a) for a grid evaluation, one can calculate the area measure A as a sum of the scanning resolution rectangles over the scanned area as

$$A = \sum_{i=1}^{I} \sum_{j=1}^{J} \left({}_{A}P_{ij} \, \Delta i \, \Delta j \right) \tag{3}$$

by using the ${}_{A}P_{ij}$ entries of a binary matrix that are related to the cross-section plane for A.

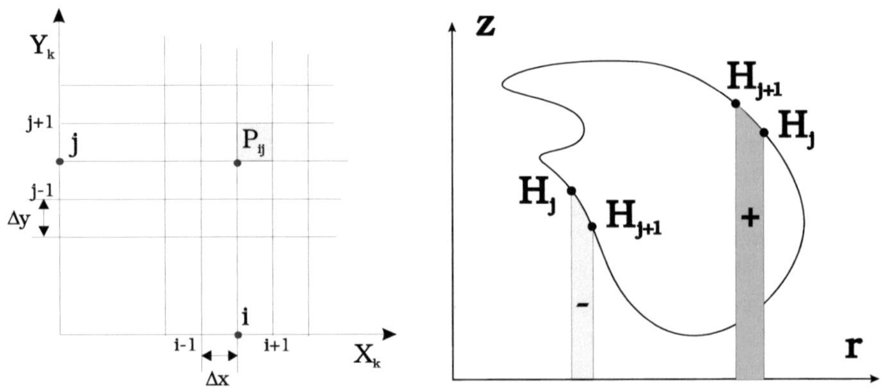

a) b)

Figure 6. General schemes for an evaluation of manipulator workspace: a) through binary representation; b) through geometric properties for algebraic formulation

Alternatively, one can use the workspace points of the boundary contour of a cross-section area that can be determined from an algebraic formulation or using the entries of the binary matrix. Thus, referring to the scheme of Fig. 6b) and by assuming as computed the coordinates of the cross-section contour points as an ordinate set (r_j, z_j) of the contour points H_j with j=1, …, N, the area measure A can be computed as

$$A = \sum_{j=1}^{N} \left(z_{1,j+1} + z_{1,j} \right) \left(r_{1,j} - r_{1,j+1} \right) \tag{4}$$

By extending the above-mentioned procedures, the workspace volume V can be computed by using the grid scanning procedure in a general form as

$$V = \sum_{i=1}^{I} \sum_{j=1}^{J} \sum_{k=1}^{K} \left[P_{ijk} \, \Delta i \, \Delta j \, \Delta k \right] \tag{5}$$

in which P_{ijk} is the entry of a binary representation in a 3D grid.

When the workspace volume is a solid of revolution, by using the boundary contour points through the Pappus-Guldinus Theorem the workspace volume V can be computed within the binary mapping procedure, Fig.6, but yet in the form as

$$V = 2\pi \sum_{i=1}^{I} \sum_{j=1}^{J} \left[P_{ij} \, \Delta i \, \Delta j \left(i \, \Delta i + \frac{\Delta i}{2} \right) \right] \tag{6}$$

or within the algebraic formulation in the form

$$V=\frac{\pi}{2}\sum_{j=1}^{N}\left(z_{1,j+1}+z_{1,j}\right)\left(r_{1,j}^{2}-r_{1,j+1}^{2}\right) \tag{7}$$

Therefore, it is evident that the formula of Eq. (6) has a general application, while Eqs. (6) and (7) are restricted to serial open-chain manipulators with revolute joints. Those approaches and formulation can be proposed and used for a numerical evaluation of workspace characteristics of parallel manipulators too.
Similarly, hole and void regions, as unreachable regions, can be numerically evaluated by using the formulas of Eqs (3) to (7) to obtain the value of their cross-sections and volumes, once they have been preliminarily determined.
Orientation workspace can be similarly evaluated by considering the angles in a Cartesian frame representation.
A design problem for manipulators can be formulated as a set of equations, which give the position and orientation of a manipulator in term of its extremity (such as workspace formulation) together with additional expressions for required performance in term of suitable criteria evaluations.

3.1 Synthesis procedures for mechanisms

Since manipulators can be treated as spatial mechanisms, the traditional techniques for mechanism design can be used once suitable adaptations are formulated to consider the peculiarity of the open chain architecture.
Two ways can be approached as referring to general model for closure equations: elaboration of closure equations for the open polygon either by adding a fictitious link with its joints either by using the coordinates of the manipulator extremity, (Duffy 1980), as shown in the illustrative example of Fig.7.
In any case, traditional techniques for mechanisms are used by considering the manipulator extremity/end-effector as a coupler link whose kinematics is the purpose of the formulation. Thus, Direct and Inverse Kinematics can be formulated and Synthesis problems can be attached by using Precision Points as those points (i.e. poses in general) at which the pose and/or other performances are prescribed as to be reached and/or fulfilled exactly.

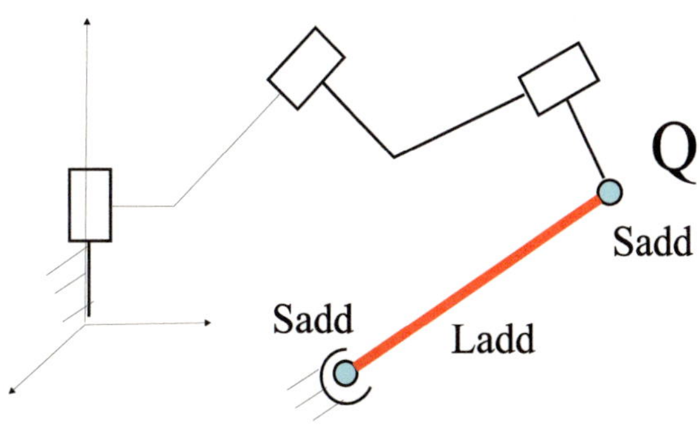

Figure 7. Closing kinematic chain of a3R manipulator by adding a fictious link and spherical joint or by looking at coordinates of point Q

This can be expressed in a general form as

$$F_i = F(X_i) \tag{8}$$

in which F_i is the performance evaluation at i-th precision pose whose coordinates X_i are function of the mechanism configuration that can be obtained by solving closure equations through any traditional methods for mechanism analysis.

Thus, design requirements and design equations can be formulated for the Precision Points, whose maximum number for a mathematical defined problem can be determined by the number of variables. But, the pose accuracy and path planning as well as the performance value away from the Precision Points will determine errors in the behaviour of the manipulator motion, whose evaluation can be still computed by using the design equations in proper procedures for optimization purposes.

Precision Points techniques for mechanisms have been developed for path positions, but for manipulator design the concept has been extended to performance criteria such as workspace boundary points and singularities. Thus, new specific algorithms have been developed for manipulator design by using approaches from Mechanism Design but with specific formulation of the peculiar manipulative tasks. Approaches such as Newton-Raphson numerical techniques, dyad elimination, Graph Theory modeling, mobility analysis, Instantaneous kinematic invariants have been developed for manipulator architectures as extension of those basic properties of planar mechanisms that have been investigated and used for design purposes since the second half of 19-th century. Of course, the complexity of 3D architectures have requested development of new more efficient calculation means, such as a suitable use of Matrix Algebra, 3D Geometry considerations, and Screw Theory formulation.

3.2 Application of 3D geometry and Screw Theory

Three dimensional Geometry of spatial manipulators has required and requires specific consideration and investigation on the 3D characteristics of a general motion. Thus, different mathematizations can be used by taking into account of generality of 3D motion.

Dual numbers and quaternions have been introduced in the last decades to study Mechanism Design and they are specifically applied to study 3D properties of rigid body motion in manipulator architectures.

The structure of mathematical properties of rigid body motion has been also addressed for developing or applying new Algebra theories for analysis and design purposes of spatial mechanisms and manipulators. Recently Lie Group Theory and Grassman Geometry have been adapted and successfully applied to develop new calculation means for designing new solutions and characterizing manipulator design in general frames.

A group ***G*** is a non-empty set endowed with a closed product operation in the set satisfying some definition conditions. A subset ***H*** of elements of a group ***G*** is called a subgroup of ***G*** if the subset ***H*** constitutes a group which has common group operation with the group ***G***. Furthermore, a group ***G*** is called a Lie group if ***G*** is an analytic manifold and the mapping ***G***x***G*** to ***G*** is analytic. The set {***D***} of rigid body motions or displacements is a 6-dimensional Lie group of transformations, which acts on the points of the 3-dimensional Euclidean affine space. The Lie subgroups of {***D***} play a key role in the mobility analysis and synthesis of mechanisms. Therefore using the mathematics of this algebra is possible to describe general features in a synthetic form that allows also fairly easy investigation of new particular conditions.

For example in Fig.8, (Lee and Hervè 2004), a hybrid spherical-spherical spatial 7R mechanism is a combination of two trivial spherical chains. Both chains are the spherical four-revolute chains **A-B-C-D** and **G-F-E-D** with the apexes O_1 and O_2 respectively. The mechanical bond $\{L(4,7)\}$ between links 4 and 7 as the intersection set of two subsets $\{G_1\}$ and $\{G_2\}$ is given by

$$\{G_1\}=\{R(O_1,\mathbf{u}_{Z1})\}\{R(O_2,\mathbf{u}_{Z2})\}\{R(O_1,\mathbf{u}_{Z3})\}\{R(O_2,\mathbf{u}_{Z4})\}\{G_2\}=\{R(O_2,\mathbf{u}_{Z7})\}\{R(O_2,\mathbf{u}_{Z6})\}\{R(O_1,\mathbf{u}_{Z5}) \quad (9)$$

where a mechanical bond is a mechanical connection between rigid bodies and it can be described by a mathematical bond, i.e. connected subset of the displacement group.

Hence, the relative motion between links 4 and 7 is depicted by

$$\{L(4,7)\}= \{G_1\} \cap \{G_2\} = \quad (10)$$

$$= \{R(O_1,\mathbf{u}_{Z1})\}\{R(O_2,\mathbf{u}_{Z2})\}\{R(O_1,\mathbf{u}_{Z3})\}\{R(O_2,\mathbf{u}_{Z4})\} \cap \{R(O_2,\mathbf{u}_{Z7})\} \{R(O_2,\mathbf{u}_{Z6})\} \{R(O_1,\mathbf{u}_{Z5})\}$$

In general, $\{R(O_1,\mathbf{u}_{Z1})\}$ $\{R(O_2,\mathbf{u}_{Z2})\}\{R(O_1,\mathbf{u}_{Z3})\}$ $\{R(O_2,\mathbf{u}_{Z4})\}$ $\{R(O_1,\mathbf{u}_{Z5})\}$ $\{R(O_2,\mathbf{u}_{Z6})\}$ is a 6-dimensional kinematic bond and generates the displacement group $\{D\}$. Therefore, $\{R(O_1,\mathbf{u}_{Z1})\}$ $\{R(O_2,\mathbf{u}_{Z2})\}$ $\{R(O_1,\mathbf{u}_{Z3})\}$ $\{R(O_2,\mathbf{u}_{Z4})\}$ $\{R(O_1,\mathbf{u}_{Z5})\}$ $\{R(O_2,\mathbf{u}_{Z6})\}$ $\cap\{R(O_2,\mathbf{u}_{Z7})\}$ = $\{D\} \cap \{R(O_2,\mathbf{u}_{Z7})\}$ = $\{R(O_2,\mathbf{u}_{Z7})\}$. This yields that the **A-G-B-F-C-E-D** 7R chain has one dof when all kinematic pairs move and consequently $\{L(4,7)\}$ includes a 1-dimensional manifold denoted by $\{L(1/D)(4,7)\}$. If all the pairs move and joint axes do not intersect again, any possible mobility characterized by this geometric condition stops occurring and we have $\{L(4,7)\} \supseteq \{L(1/D)(4,7)\}$. Summarizing, the kinematic chain works like a general spatial 7R chain whose general mobility is with three dofs, but with the above.-mentioned condition is constrained to one dof, since it acts like a spherical four-revolute **A-B-C-D** chain with one dof, or a spherical four-revolute **G-F-E-D** chain with one dof.

Grassman Geometry and further developments have been used to describe the Line Geometry that can be associated with spatial motion. Plucker coordinates and suitable algebra of vectors are used in Grassman Geometry to generalize properties of motion of a line that can be fixed on any link of a manipulator, but mainly on its extremity.

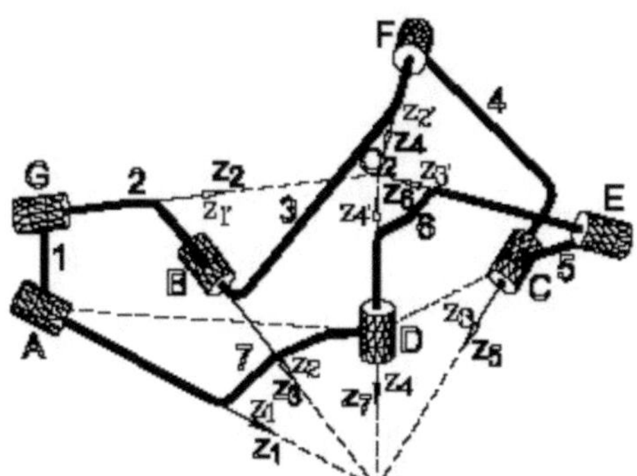

Figure 8. Hybrid spherical-spherical discontinuously movable 7R mechanism, (Lee and Hervè 2004)

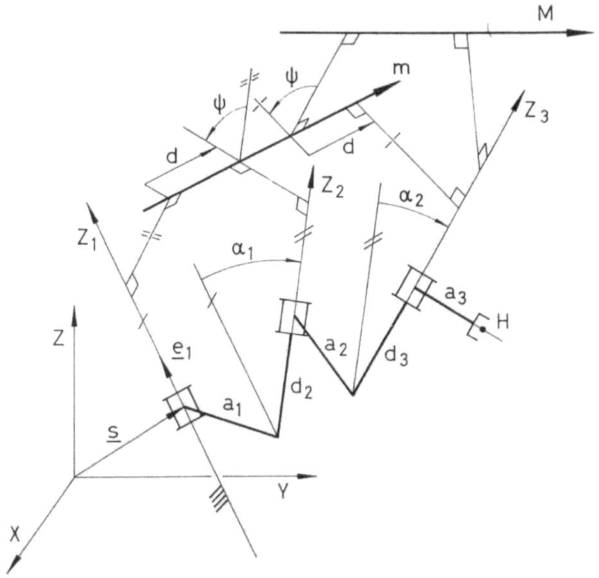

Figure 9. A scheme of screw triangle for 3R manipulator design

Screw Theory was developed to investigate the general motion of rigid bodies in its form of helicoidal (screw) motion in 3D space. A screw entity was defined to describe the motion and to perform computation still through vector approaches.

A unit screw is a quantity associated with a line in the three-dimensional space and a scalar called pitch, which can be represented by a 6 x 1 vector $\$ = [\,\mathbf{s}\,,\mathbf{r} \times \mathbf{s} + \lambda\mathbf{s}\,]T$ where **s** is a unit vector pointing along the direction of the screw axis, **r** is the position vector of any point on the screw axis with respect to a reference frame and λ is the pitch of the screw. A screw of intensity ρ is represented by $S = \rho\ \$$. When a screw is used to describe the motion state of a rigid body, it is often called a twist, represented by a 6 x 1 vector as $\$ = [\boldsymbol{\omega}, \mathbf{v}\,]\ T$, where $\boldsymbol{\omega}$ represents the instant angular velocity and **v** represents the linear velocity of a point O which belongs to the body and is coincident with the origin of the coordinate system.

Screw Theory has been applied to manipulator design by using suitable models of manipulator chains, both with serial and parallel architectures, in which the joint mobility is represented by corresponding screws, (Davidson and Hunt 2005).

Thus, screw systems describe the motion capability of manipulator chains and therefore they can be used still with a Precision Point approach to formulate design equations and characteristics of the architectures. In Fig.9 an illustrative example is reported as based on the fundamental so-called Screw Triangle model for efficient computational purposes, even to deduce closed-form design expressions.

3.3 Optimization problem design

The duality between serial and parallel manipulators is not anymore understood as a competition between the two kinematic architectures. The intrinsic characteristics of each architecture make each architecture as devoted to some manipulative tasks more than an alternative to the counterpart. The complementarities of operation performance of serial and parallel manipulators make them as a complete solution set for manipulative operations.

The differences but complementarities in their performance have given the possibility in the past to treat them separately, mainly for design purposes. In the last two decades several analysis results and design procedures have been proposed in a very rich literature with the aim to characterize and design separately the two manipulator architectures.
Manipulators are said useful to substitute/help human beings in manipulative operations and therefore their basic characteristics are usually referred and compared to human manipulation performance aspects. A well-trained person is usually characterized for manipulation purpose mainly in terms of positioning skill, arm mobility, arm power, movement velocity, and fatigue limits. Similarly, robotic manipulators are designed and selected for manipulative tasks by looking mainly to workspace volume, payload capacity, velocity performance, and stiffness. Therefore, it is quite reasonable to consider those aspects as fundamental criteria for manipulator design. But generally since they can give contradictory results in design algorithms, a formulation as multi-objective optimization problem can be convenient in order to consider them simultaneously. Thus, an optimum design of manipulators can be formulated as

$$\min \mathbf{F(X)} = \min \left[f_1(\mathbf{X}), f_2(\mathbf{X}), \ldots, f_N(\mathbf{X}) \right]^T \tag{11}$$

subjected to

$$\mathbf{G}(\mathbf{X}) < \mathbf{0} \tag{12}$$

$$\mathbf{H}(\mathbf{X}) = \mathbf{0} \tag{13}$$

where T is the transpose operator; **X** is the vector of design variables; **F(X)** is the vector of objective functions f_i that express the optimality criteria, **G(X)** is the vector of constraint functions that describes limiting conditions, and **H(X)** is the vector of constraint functions that describes design prescriptions.
There is a number of alternative methods to solve numerically a multi-objective optimization problem. In particular, in the example of Fig. 10 the proposed multi-objective optimization design problem has been solved by considering the min-max technique of the Matlab Optimization Toolbox that makes use of a scalar function of the vector function **F (X)** to minimize the worst case values among the objective function components f_i.
The problem for achieving optimal results from the formulated multi-objective optimization problem consists mainly in two aspects, namely to choose a proper numerical solving technique and to formulate the optimality criteria with computational efficiency.
Indeed, the solving technique can be selected among the many available ones, even in commercial software packages, by looking at a proper fit and/or possible adjustments to the formulated problem in terms of number of unknowns, non-linearity type, and involved computations for the optimality criteria and constraints. On the other hand, the formulation and computations for the optimality criteria and design constraints can be deduced and performed by looking also at the peculiarity of the numerical solving technique.
Those two aspects can be very helpful in achieving an optimal design procedure that can give solutions with no great computational efforts and with possibility of engineering interpretation and guide.
Since the formulated design problem is intrinsically high no-linear, the solution can be obtained when the numerical evolution of the tentative solutions due to the iterative process converges to a solution that can be considered optimal within the explored range. Therefore

a solution can be considered an optimal design but as a local optimum in general terms. This last remark makes clear once more the influence of suitable formulation with computational efficiency for the involved criteria and constraints in order to have a design procedure, which is significant from engineering viewpoint and numerically efficient.

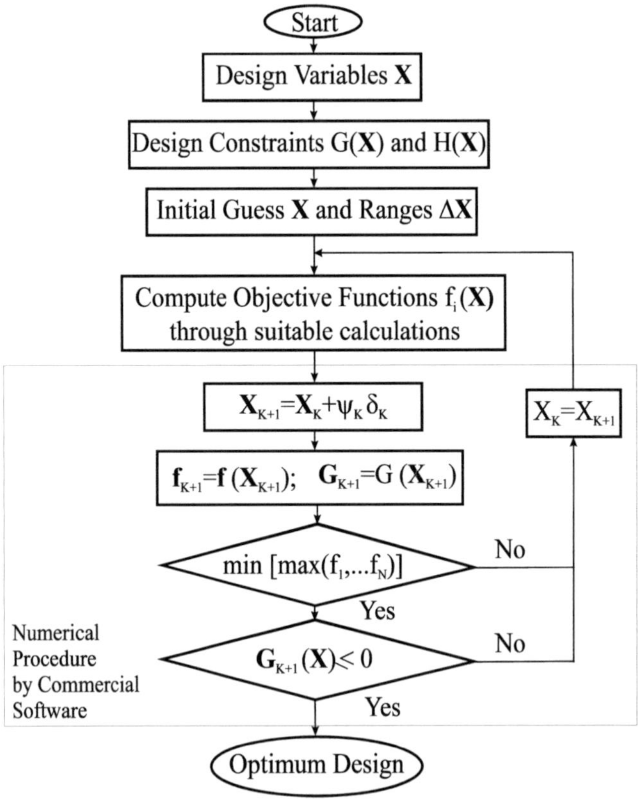

Figure 10. A general scheme for optimum design procedure by using multi-objective optimization problem solvable by commercial software

4. Experimental validation of manipulators

Engineering approach for kinematic design is completed by experimental activity for validation of theories and numerical algorithms and for validation and evaluation of prototypes and their performance as last design phase. Experimental activity can be carried out at several levels depending on the aims and development sequence:

- by checking mechanical design and assembly problems for manipulators and test-beds;
- by looking at operation characteristics of tasks and manipulator architectures;
- by simulating manipulators both in terms of kinematic capability and dynamic actions;
- by validating prototype performance in term of evaluation of errors from expected behavior.

Construction activity is aimed to check the feasibility of practical implementation of designed manipulators. Assembly possibilities are investigated also by looking at alternative

components. The need to obtain quickly a validation of the prototypes as well as of novel architectures has developed techniques of rapid prototyping that facilitate this activity both in term of cost and time. Test-beds are developed by using or adjusting specific prototypes or specific manipulator architectures. Once a physical system is available, it can be used both to characterize performance of built prototypes and to further investigate on operation characteristics for optimality criteria and validation purposes. At this stage a prototype can be used as a test-bed or even can be evolved to a test-bed for future studies. This activity can be carried out as an experimental simulation of built prototypes both for functionality and feasibility in novel applications. From mechanical engineering viewpoint, experimental activity is understood as carried out with built systems with considerable experiments for verifying operation efficiency and mechanical design feasibility. Recently experimental activity is understood even only through numerical simulations by using sophisticated simulation codes (like for example ADAMS).
The above mentioned activity can be also considered as completing or being preliminary to a rigorous experimental validation, which is carried out through evaluation of performance and task operation both in qualitative and quantitative terms by using previously developed experimental procedures.

5. Experiences at LARM in Cassino

As an example of the above-mentioned aspects illustrative cases of study are reported from the activity of LARM: Laboratory of Robotics and Mechatronics in Cassino in Figs. 11-19.
Since the beginning of 1990s at LARM in Cassino, a research line has been dedicated to the development of analysis formulation and experimental activity for manipulator design and performance characterization. More details and further references can be found in the LARM webpage http://webuser.unicas.it/weblarm/larmindex.htm.
Workspace has been analyzed to characterize its manifold and to formulate efficient evaluation algorithms. Scanning procedure and algebraic formulation for workspace boundary have been proposed. Results can be obtained likewise in the illustrative examples in Fig. 11.

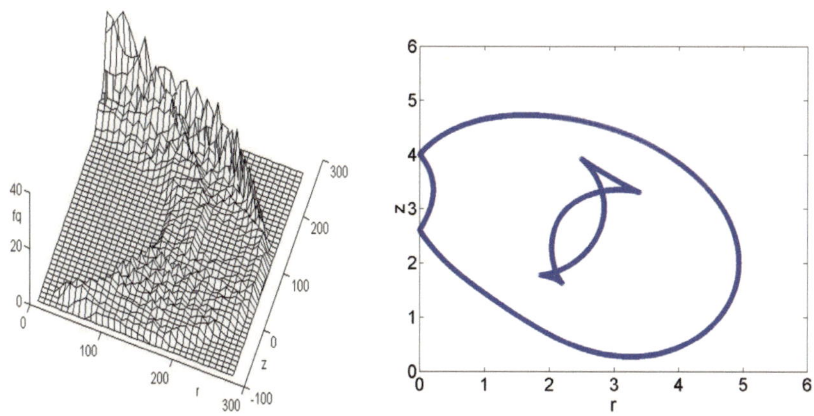

a) b)

Figure 11. Illustrative examples of results of workspace determination through : a) binary representation in scanning procedure; b) algebraic formulation of workspace boundary

A design algorithm has been proposed as an inversion of the algebraic formulation to give all possible solutions like for the reported case of 3R manipulator in Fig.12.
Further study has been carried out to characterize the geometry of ring (internal) voids as outlined in Fig.13.
A workspace characterization has been completed by looking at design constraints for solvable workspace in the form of the so-called Feasible Workspace Regions. The case of 2R manipulators has been formulated and general topology has been determined for design purposes, as reported in Fig. 14.
Singularity analysis and stiffness evaluation have been approached to obtain formulation and procedure that are useful also for experimental identification, operation validation, and performance testing. Singularity analysis has been approached by using arguments of Descriptive Geometry to represent singularity conditions for parallel manipulators through suitable formulation of Jacobians via Cayley-Grassman determinates or domain analysis. Figure 15 shows examples how using tetrahedron geometry in 3-2-1- parallel manipulators has determined straightforward the shown singular configurations.

no.	a_1	α_1	d_2	α_2	d_3	a_3	a_2
1	1.00	30.00	1.97	35.96	1.93	5.90	1.47
2	1.00	30.00	1.97	-35.96	1.93	5.90	13.27
3	1.00	30.00	1.97	75.35	6.17	3.58	0.89
4	1.00	30.00	1.97	-75.35	6.17	3.58	8.06
5	1.00	-30.00	-1.97	30.51	6.38	6.82	13.54
6	1.00	-30.00	-1.97	-30.51	6.38	6.82	-0.10
7	1.00	-30.00	-1.97	51.25	8.79	4.44	7.38
8	1.00	-30.00	-1.97	-51.25	8.79	4.44	1.49
9	1.80	64.00	1.00	12.27	6.12	32.20	28.25
10	1.80	64.00	1.00	-12.27	6.12	32.20	36.14
11	1.80	64.00	1.00	--	--	--	--
12	1.80	64.00	1.00	--	--	--	--
13	1.80	-64.00	-1.00	27.30	8.97	14.92	15.46
14	1.80	-64.00	-1.00	-27.30	8.97	14.92	14.38
15	1.80	-64.00	-1.00	--	--	--	--
16	1.80	-64.00	-1.00	--	--	--	--

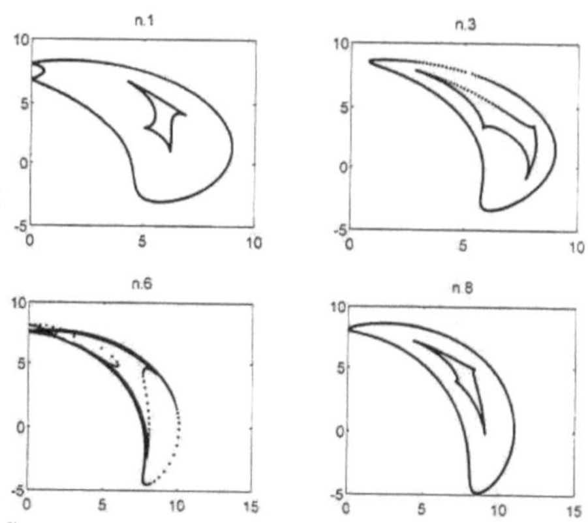

Figure 12. Design solutions for 3R manipulators by inverting algebraic formulation for workspace boundary when boundary points are given: a) all possible solutions; b) feasible workspace designs

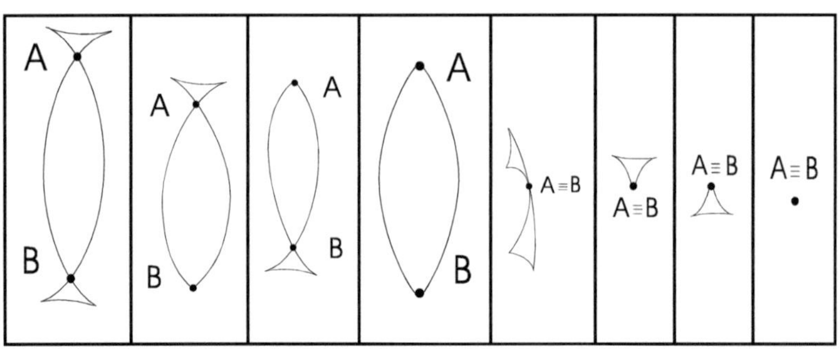

Figure 13. Manifolds for ring void of 3R manipulators

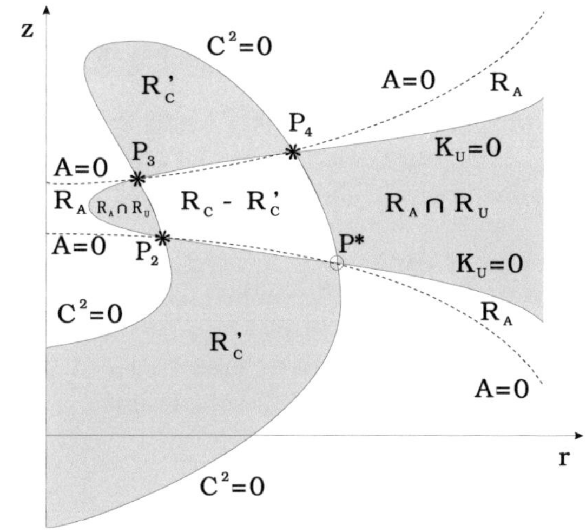

Figure 14. General geometry of Feasible Workspace Regions for 2R manipulators depicted as grey area

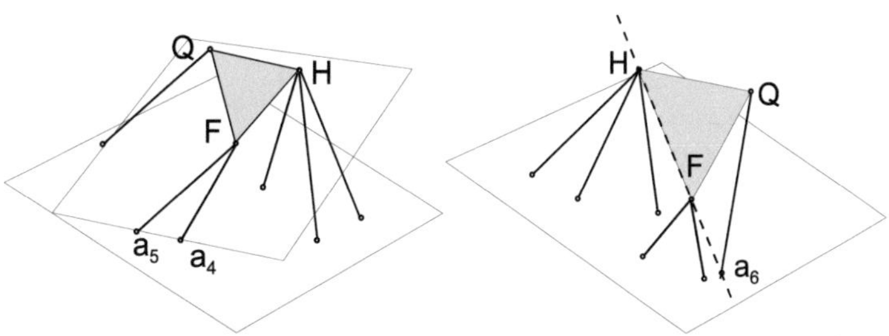

Figure 15. Determination of singularity configuration of a wire 3-2-1 parallel manipulator by looking at the descriptive geometry of the manipulator architecture

Recently, optimal design procedures have been formulated and experienced by using multi-criteria optimization problem when Precision Points equations have been combined with suitable numerical evaluation of performances. An attempt has been proposed to obtain a unique design procedure both for serial and parallel manipulators through the objective formulation

$$
\begin{aligned}
f_1(\mathbf{X}) &= \left| 1 - \left(V_{pos} / V_{pos}' \right) \right| \\
f_2(\mathbf{X}) &= \left| 1 - \left(V_{or} / V_{or}' \right) \right| \\
f_3(\mathbf{X}) &= -\frac{\min(\det J)}{(\det J_0)} \\
f_4(\mathbf{X}) &= 1 - \left(|\Delta \mathbf{U}_d| / |\Delta \mathbf{U}_g| \right) \\
f_5(\mathbf{X}) &= 1 - \left(|\Delta \mathbf{Y}_d| / |\Delta \mathbf{Y}_g| \right)
\end{aligned}
\tag{14}
$$

where V_{pos} and V_{or} values correspond to computed position and orientation workspace volume V and prime values describe prescribed data; J is the manipulator Jacobian with respect to a prescribed one J_o; $\Delta\mathbf{U}_d$ and $\Delta\mathbf{U}_g$ are compliant displacements along X, Y, and Z-axes, $\Delta\mathbf{Y}_d$ and $\Delta\mathbf{Y}_g$ are compliant rotations about φ, θ and ψ; d and g stand for design and given values, respectively. Illustrative example results are reported in Figs.16 and 17 as referring to a PUMA-like manipulator and a CAPAMAN (Cassino Parallel Manipulator) design.

Experimental activity has been particularly focused on construction and functionality validation of prototypes of parallel manipulators that have been developed at LARM under the acronym CAPAMAN (Cassino Parallel Manipulator). Figures 18 and 19 shows examples of experimental layouts and results that have been obtained for characterizing design performance and application feasibility of CAPAMAN design.

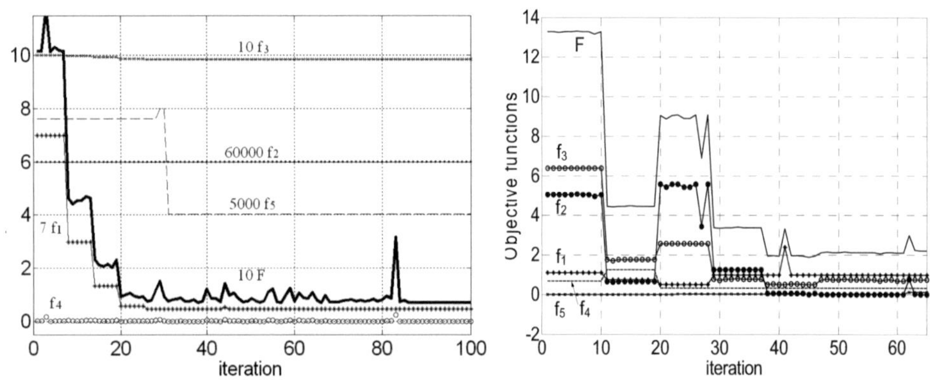

a) b)

Figure 16. Evolution of the function F and its components versus number of iterations in an optimal design procedure: a) for a PUMA-like robot; b) a CAPAMAN design in Fig.15a). (position workspace volume as f_1; orientation workspace volume as f_2; singularity condition as f_3; compliant translations and rotations as f_4 and f_5)

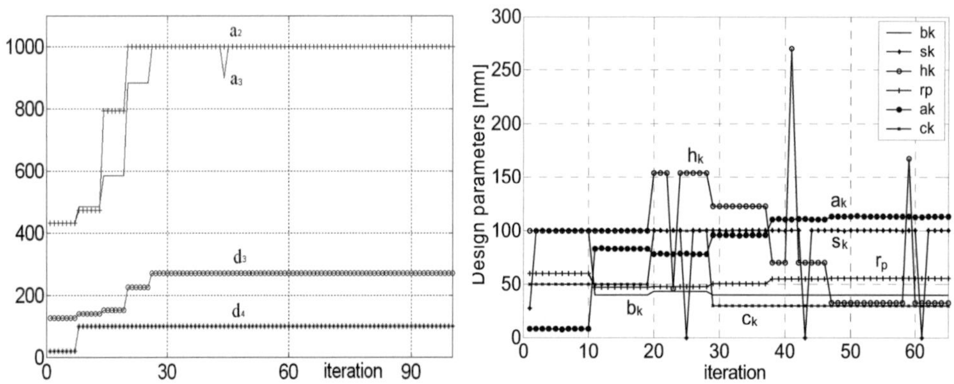

a) b)

Figure 17. Evolution of design parameters versus number of iterations for: a) PUMA-like robot in Fig.16a); CAPAMAN in Fig.16b)

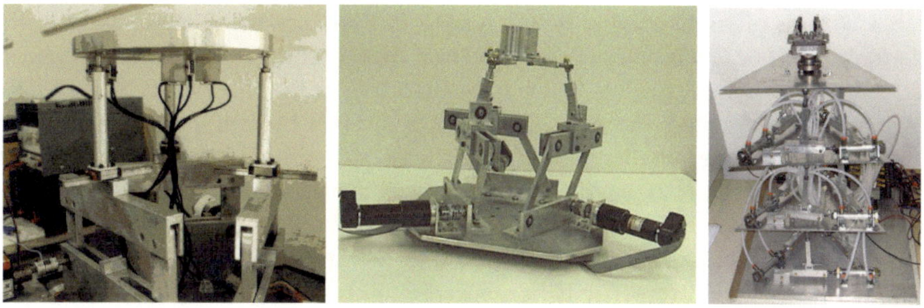

a) b) c)

Figure 18. Built prototypes of different versions of CAPAMAN: a) basic architecture; b) 2-nd version; c) 3-rd version in multi-module assembly

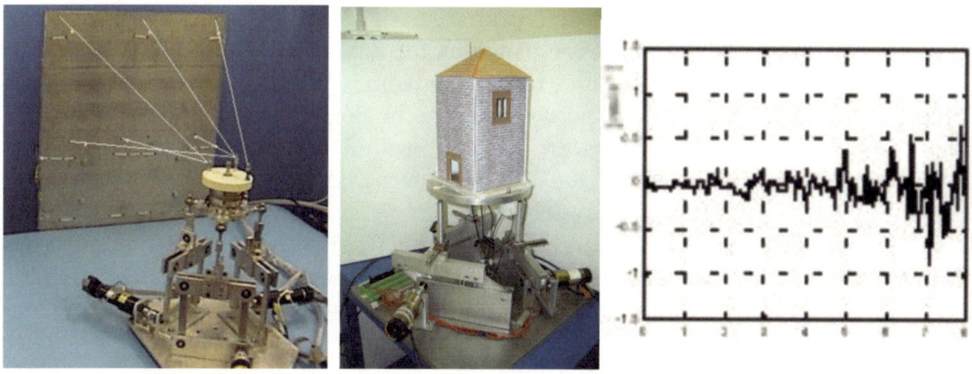

a) b) c)

Figure 19. Examples of validation tests for numerical evaluation of CAPAMAN: a) in experimental determination of workspace volume and compliant response; b) in an application as earthquake simulator; c) results of numerical evaluation of acceleration errors in simulating an happened earthquake

6. Future challenges

The topic of kinematic design of manipulators, both for robots and multi-body systems, addresses and will address yet attention for research and practical purposes in order to achieve better design solutions but even more efficient computational design algorithms. An additional aspect that cannot be considered of secondary importance, can be advised in the necessity of updating design procedures and algorithms for implementation in modern current means from Informatics Technology (hardware and software) that is still evolving very fast.

Thus, future challenges for the development of the field of kinematic design of manipulators and multi-body systems at large, can be recognized, beside the investigation for new design solutions, in:

- more exhaustive design procedures, even including mechatronic approaches;
- updated implementation of traditional and new theories of Kinematics into new Informatics frames.

Research activity is often directed to new solutions but because the reached highs in the field mainly from theoretical viewpoints, manipulator design still needs a wide application in practical engineering. This requires better understanding of the theories at level of practicing engineers and user-oriented formulation of theories, even by using experimental activity. Thus, the above-mentioned challenges can be included in a unique frame, which is oriented to a transfer of research results to practical applications of design solutions and procedures.
Mechatronic approaches are needed to achieve better practical design solutions by taking into account the construction complexity and integration of current solutions and by considering that future systems will be overwhelmed by many sub-systems of different natures other than mechanical counterpart. Although the mechanical aspects of manipulation will be always fundamental because of the mechanical nature of manipulative tasks, the design and operation of manipulators and multi-body systems at large will be more and more influenced by the design and operation of the other sub-systems for sensors, control, artificial intelligence, and programming through a multidisciplinary approach/integration. This aspect is completed by the fact that the Informatics Technology provides day by day new potentialities both in software and hardware for computational purposes but even for technical supports of other technologies. This pushes to re-elaborate design procedures and algorithms in suitable formulation and logics that can be used/adapted for implementation in the evolving Informatics.
Additional efforts are requested by system users and practitioner engineers to operate with calculation means (codes and procedures in commercial software packages) that are more and more efficient in term of computation time and computational results (numerical accuracy and generality of solutions) as well as more and more user-oriented design formulation in term of understand ability of design process and its theory. This is a great challenge: since while more exhaustive algorithms and new procedures (with mechatronic approaches) are requested, nevertheless the success of future developments of the field strongly depends on the capability of the researchers of expressing the research result that will be more and more specialist (and sophisticated) products, in a language (both for calculation and explanatory purposes) that should not need a very sophisticate expertise.

7. Conclusion

Since the beginning of Robotics the complexity of the kinematic design of manipulators has been solved with a variety of approaches that are based on Theory of Mechanisms, Screw Theory, or Kinematics Geometry. Algorithms and design procedures have evolved and still address research attention with the aim to improve the computational efficiency and generality of formulation in order to obtain all possible solutions for a given manipulation problem, even by taking into account other features in a mechatronic approach. Theoretical and numerical approaches can be successfully completed by experimental activity, which is still needed for performance characterization and feasibility tests of prototypes and design algorithms

8. References

The reference list is limited to main works for further reading and to author's main experiences. Citation of references has not included in the text since the subjects refer to a very reach literature that has not included for space limits.

Angeles, J. (1997). *Fundamentals of Robotic Mechanical Systems*, Springer-Verlag, NewYork

Angeles, J. (2002). The Robust Design of Parallel Manipulators, *Proceedings of the 1rst Int. Colloquium Coll. Research Center 562*, pp. 9-30, Braunschweig, 2002.

Bhattacharya, S.; Hatwal, H. & Ghosh, A. (1995). On the Optimum Design of Stewart Platform Type Parallel Manipulators, *Journal of Robotica*, Vol. 13, pp. 133-140.

Carbone G., Ottaviano E., Ceccarelli M. (2007), An Optimum Design Procedure for Both Serial and Parallel Manipulators, *Proceedings of the Institution of Mechanical Engineers IMechE Part C: Journal of Mechanical Engineering Science*. Vol. 221, No.7, pp.829-843.

Ceccarelli, M. (1996). A Formulation for the Workspace Boundary of General N-Revolute Manipulators, *Mechanism and Machine Theory*, Vol. 31, pp. 637-646.

Ceccarelli, M. (2002). Designing Two-Revolute Manipulators for Prescribed Feasible Workspace Regions, *ASME Journal of Mechanical Design*, Vol. 124, pp.427-434.

Ceccarelli, M. (2002). An Optimum Design of Parallel Manipulators: Formulation and Experimental Validation", *Proceedings of the 1rst Int. Coll. Collaborative Research Center 562*, Braunschweig, pp. 47-63.

Ceccarelli, M. (2004). *Fundamentals of Mechanics of Robotic Manipulation*, Kluwer/Springer.

Ceccarelli M.. & Lanni C. (2004), Multi-Objective Optimum Design of General 3R Manipulators for Prescribed Workspace Limits, *Mechanism and Machine Theory*, Vol. 39, N.2, pp.119-132.

Ceccarelli, M. & Carbone, G. (2005). Numerical and experimental analysis of the stiffness performance of parallel manipulators, *Proceedings of the 2nd Intern.l Coll. Collaborative Research Centre 562*, Braunschweig, pp.21-35.

Ceccarelli, M.; Carbone, G. & Ottaviano, E. (2005). An Optimization Problem Approach for Designing both Serial and Parallel Manipulators, *Proceedings of MUSME 2005, Int. Symposium on Multibody Systems and Mechatronics*, Uberlandia, Paper No. 3-MUSME05.

Davidson J. & Hunt K. (2005), *Robot and Screw Theory*, Oxford University Press.

Duffy, J. (1980). *Analysis of Mechanisms and Robot Manipulators*, Arnold, London.

Duffy, J. (1996). *Statics and Kinematics with Applications to Robotics*, Cambridge University Press.

Dukkipati, R. V. (2001). *Spatial Mechanisms*, CRC Press.

Erdman, A. G. (Editor) (1993), *Modern Kinematics*, Wisley.

Freudenstein, F. & Primrose, E.J.F. (1984). On the Analysis and Synthesis of the Workspace of a Three-Link, Turning-Pair Connected Robot Arm, *ASME Journal of Mechanisms, Transmissions and Automation in Design*, Vol.106, pp. 365-370.

Gosselin, C. M. (1992). The Optimum Design of Robotic Manipulators Using Dexterity Indices, *Robotics and Autonomous Systems*, Vol. 9, pp. 213-226.

Gupta, K. C. & Roth, B. (1982). Design Considerations for Manipulator Workspace, *ASME Journal of Mechanical Design*, Vol. 104, pp. 704-71.

Hao, F. & Merlet, J.-P. (2005). Multi-Criteria Optimal Design of Parallel Manipulators Based on Interval Analysis, *Mechanism and Machine Theory*, Vol. 40, pp. 157-171.

Lee, C. & Hervè, J.M. (2004). Synthesis of Two Kinds of Discontinuously Movable Spatial 7R Mechanisms through the Group Algebraic Structure of Displacement Set, *Proceedings of the 11th World Congress in Mechanism and Machine Science*, Tianjin, paper CK49.

Manoochehri, S. & Seireg, A.A. (1990). A Computer-Based Methodology for the Form Synthesis and Optimal Design of Robot Manipulators, *Journal of Mechanical Design*, Vol. 112, pp. 501-508.

Merlet, J-P. (2005). *Parallel Robots*, Kluwer/Springer, Dordrecht.

Ottaviano E., Ceccarelli M. (2006), An Application of a 3-DOF Parallel Manipulator for Earthquake Simulations, *IEEE Transactions on Mechatronics*, Vol. 11, No. 2, pp. 240-246.

Ottaviano E., Ceccarelli M. (2007), Numerical and experimental characterization of singularities of a six-wire parallel architecture, *International Journal ROBOTICA*, Vol.25, pp.315-324.

Ottaviano E., Husty M., Ceccarelli M. (2006), Identification of the Workspace Boundary of a General 3-R Manipulator, *ASME Journal of Mechanical Design*, Vol.128, No.1, pp.236-242.

Ottaviano E., Ceccarelli M., Husty M. (2007), Workspace Topologies of Industrial 3R Manipulators, *International Journal of Advanced Robotic Systems*, Vol.4, No.3, pp.355-364.

Paden, B. & Sastry, S. (1988). Optimal Kinematic Design of 6R Manipulators, *The Int. Journal of Robotics Research*, Vol. 7, No. 2, pp. 43-61.

Park, F.C. (1995). Optimal Robot Design and Differential Geometry, *Transaction ASME, Special 50th Anniversary Design Issue*, Vol. 117, pp. 87-92.

Roth, B. (1967). On the Screw Axes and Other Special Lines Associated with Spatial Displacements of a Rigid Body, *ASME Jnl of Engineering for Industry*, pp. 102-110.

Schraft, R.D. & Wanner, M.C. (1985). Determination of Important Design Parameters for Industrial Robots from the Application Point if View: Survey Paper, *Proceedings of ROMANSY '84*, Kogan Page, London, pp. 423-429.

Selig, J.M. (2000). *Geometrical Foundations of Robotics*, Worlds Scientific, London.

Seering, W.P. & Scheinman, V. (1985). Mechanical Design of an Industrial Robot, *Ch.4, Handbook of Industrial Robotics*, Wiley, New York, pp. 29-43.

Sobh, T.M. & Toundykov, D.Y. (2004). Kinematic synthesis of robotic manipulators from task descriptions - optimizing the tasks at hand, *IEEE Robotics & Automation Magazine*, Vol. 11, No. 2, pp. 78-85.

Takeda, Y. & Funabashi, H. (1999). Kinematic Synthesis of In-Parallel Actuated Mechanisms Based on the Global Isotropy Index, *Journal of Robotics and Mechatronics*, Vol. 11, No. 5, pp. 404-410.

Tsai, L.W. (1999). Robot Analysis: *The Mechanics of Serial and Parallel Manipulators*, Wiley, New York.

Uicker, J.J.; Pennock, G.R. & Shigley, J.E. (2003). *Theory of Machines and Mechanisms*, Oxford University Press.

Vanderplaats, G. (1984). *Numerical Optimization Techniques for Engineers Design*, McGraw-Hill.

4

Gentle Robotic Handling Using Acceleration Compensation

Suei Jen Chen
Universität Karlsruhe (TH)
Germany

1. Introduction

Nowadays, robot manipulators are being widely used in the industrial and manufacturing sector, because of their high precision, speed, endurance and reliability. In the automotive industry, their typical applications include welding, painting, assembly of body, pick and place, etc. Normally, the main task is restricted to simple operations like moving a tool in different working locations. In the sector of logistics, they have to fulfill material handling and storage tasks like palletizing, depalletizing and transportation of goods. Without doubt, the main goal is to maximize the productivity. For this purpose, mostly the robotic manipulators are programmed to drive as fast as possible. But, reaching a time-optimal path does not mean the attainment of a perfect task performance. Since important aspects such as quality and safety should not be overlooked. Usually, most of the handling goods require a special and delicate treatment, due to the object's characteristic and complexity. Otherwise, quality degradations or damages to the transferring object may occur. In the process of material handling, it is common to encounter different kind of problems because of too high accelerated motions. For example: unexpected separation of the transferring object from its grasping tool, the spill-over of fluid materials in highly accelerated liquid containers, such as the transfer of hot molten steel or glass in a casting process, or dangerous collisions caused by swinging suspended objects in harbors or in construction areas, etc.
To overcome the mentioned problems, new efficient approaches based on the so-called *Acceleration Compensation Principle* will be introduced in this chapter. The main idea is to modify the reference motion trajectory, in order to compensate undesirable disturbances. The position and the orientation of the robot's end-effector are adapted in such a way, that undesired acceleration side effects can be compensated and minimized. In addition, a derivation from the first proposed technique is introduced to compensate undesirable swinging oscillations. *Gentle robotic handling* and *time optimized movements* are considered as two main research objectives in the context of this work. Additionally, economic and complexity factor should be also taken into account. Beyond the achievement of optimal cycle time and the avoidance of collisions, the new algorithms are independent of the object's physical aspects and are evaluated with serial robot manipulators, as main part of the experimental test-bed, in order to verify the feasibility and effectiveness of the proposed approaches.

Since the problems concerning to robotic handling issue could be rather broad, this chapter is limited to analyze and to evaluate three particular problem-scenarios, which are commonly encountered in most industrial applications. Depending on the nature of the handling operation, a fast movement may induce particular collateral effects, such as:

a. **Presence of excessive shear forces:** undesired sliding effects or loss of transferring objects from the grasping tool, as consequence of too large lateral accelerations.
b. **Undesired oscillations in liquid containers:** abrupt liquid sloshing and spill-over of liquid materials in highly accelerated open containers, such as molten steel, glass, etc.
c. **Swing oscillations in suspended loads:** dangerous swing balancing effects in suspended transferring objects, which could induce unexpected collisions.

Based on this classification and before going into the details, a brief overview about the existing methodologies concerned with general robotic handling problems will be introduced in the next section.

2. State of the Art

So far, many manufacturers still apply conventional methods to deal with the problems described before. One common and inefficient solution is the reduction of motion accelerations, until the object can be safely transported to the programmed destination. This leads to tedious trial-and-error teaching procedures, and implies also an enormous increase in the cycle time. A further solution is the implementation of active feedback control. In such systems, the controlled parameters are monitored through sensors. Unfortunately, the feasibility of installing sensor devices into the controlling system depends mainly on the properties of the transferring object itself and also on the system infrastructure. Furthermore, most of the sensor systems require extra calibrations and maintenances, which increase enormously the cost and system complexity.

Based on the proposed classification-criteria, important literature surveys dealing with these three problem-scenarios can be seen as follows:

I. *Presence of large shear forces:* large shear forces can induce undesired sliding effects due to highly accelerated transfer, e.g., transfer of cartons containing fresh fruit or bakery products in the food manufacturing. Here, it is very important that the goods are handled carefully, to maintain the quality of each individual item. The worse scenario is the unwanted losses of goods during the motion. This happens when the resulting forces applied on the handled object have exceeded the maximum permissible. This could cause irremediable damages to the object, the carrying device or/and its surrounding area. A well known example is the transfer of fragile glass cathode ray tubes (CRT) using vacuum suction gripper in the manufacturing of TV.

   Dealing with the shear force minimizing, several investigations proposed feedback control system using different kinds of sensor technologies. In [Kolluru et al., 2000], a reconfigurable robotic gripper system was developed to ensure a reliable manipulation of deformable limp material. Later, [Tsourveloudi et al., 2000] incorporated a fuzzy logic control system to regulate the air flow and to ensure safe handling. In this approach, an adequate amount of contact pressure at the gripping points during the entire manipulation process was determined. Other authors dealt with this problem using mobile transportation systems, where the acceleration of the mobile platform is actively compensated to stabilize the transported object, e.g. the well-known self-balancing mobile robots [Segway, 2004], [Crasser et al., 2002] and [Lauwers et al., 2005].

Active Acceleration Compensation technique was firstly proposed by [Graf & Dillmann, 1997]. In this system, a Stewart Platform was mounted on top of a mobile platform. By tilting this platform, it was possible to compensate the acceleration of the mobile platform in such a way that no shear forces were applied to the transported object. A similar work was established by [Dang et al., 2001, 2002, 2004]. In this case, a parallel platform manipulator was utilized and it tried to emulate a virtual pendulum, with the aim to compensate actively for disturbances in acceleration input. The undesired shear forces were compensated by imitating the behaviour of a pendulum instead of tilting the object.

II. *Undesirable oscillations produced by objects with dynamical behaviors:* objects with dynamical constraints may suffer oscillation problems during high-speed motions. In this study, relying on the nature of the handling object, the oscillations can be roughly classified into two groups:

a. The first category deals with undesirable sloshing (fluid oscillations) produced by containers with fluid contents. High-speed transfer may cause the spillover of the fluid content and leading to possible contaminations. A popular example is in the casting process, where the transportation of open containers filled with hot molten steel or glass should be executed in high speed and with high positioning accuracy, in order to avoid any undesired cooling of molten material. Hence, it is crucial that such motions are accomplished within a minimum cycle-time and with utmost delicacy, to prevent disturbances produced by undesired forces or oscillations.

   Interesting works to be mentioned are the approaches from [Feddema et al., 1996, 1997], which used two command shaping techniques for controlling the surface of a liquid in an open container, as it was being carried by a robot arm. In this work, oscillation and damping of the liquids were predicted using the Boundary Element Method (BEM). Numerous studies dealing with automatic pouring systems in casting industries have been realized. In [Yano & Terashima, 2001], [Noda et al., 2002, 2004], [Kaneko et al., 2003], the Hybrid Shape Approach was adapted to design an advanced control system for automatic pouring processes. The method introduced by Yano, comprised a suitable nominal model and determined an appropriate reference trajectory, in order to construct a high-speed and robust liquid transfer system, reducing in this way undesired endpoint residual vibrations. The behaviour of sloshing in the liquid container was approximated by a pendulum-type model. A H-Infinity feedback control system was applied.

b. The second category comprises swinging effects caused by suspended objects. These problems are widely encountered in many sectors of industry. These include: operations of cranes on construction site, warehouse, harbor, shipboard handling systems, etc., where the factor 'safety' plays a crucial role.

   Several investigations were based on open-loop control technique, in which the concept of optimal control was taken into account [Mita & Kanai, 1979], [Blazicevic & Novakovic, 1996] and [Sakawa & Shindo, 1982]. Most of them used bang-coast-bang control, consisting of a sequence of constant acceleration pulses in conjunction with zero acceleration periods. For the control of swing-free boom cranes, Blazicevic adopted a set of velocity basis functions to minimize swing oscillations. The dynamic model of the crane was proposed and transformed into state space for the purpose of examining the dynamic behaviour of a rotary crane, which combined speed profiles were employed for the software simulation. In 1957, [Smith, 1957] introduced for first time, a technique which generated a "shaped" input trajectory to the system, with the goal to suppress oscillation effects. The main idea of a command shaper was to excite the system with a sequence of impulses during the entire trajectory, to cancel the perturbing oscillations. This method was later extended by many other researchers [Starr, 1985], [Singer & Seering, 1990], [Parker et al., 1995] and [Singhose et al., 1997].

As observed, most of the existing studies found in the literature have been proposing either complex modeling of the system or required the help of external sensors for the performance of their proposed controlling methods, at the expense of large efforts, time and money.

In this work, we propose open-loop control methods based on the adaptation of the tool position and orientation. The main aim is to establish a gentle handling of objects without compromising cycle time. With this focusing on the background, the new approaches lead to new robot optimal trajectories, which reduce large lateral accelerations and avoid undesired oscillations acting on the transferring objects, for moving them fast and in a gentle way.

3. Basic Concept. Acceleration Compensation Principle

3.1 Introduction

Several issues arisen during the handling tasks due to undesired large acceleration effects have been presented in the previous section. Additionally, important scientific-research works dealing with these problems, together with their distinguishing characteristics, were briefly described. As a relevant observation, the majority of these methods require either additional sensors or an accurate model of the system, which consequently imply major computational effort and algorithm complexity. As we observe, problems with material handling may occur in different ways, in which each issue require a careful analysis and an individual solution. The main contribution of this investigation is to provide generalized solutions, which are simple, time-efficient and feasible to solve handling problems.

Considering the nonlinearity of the system and the conservation of product's quality without overlooking the safety and the cost reducing, Acceleration Compensation Principle (ACP) is considered as an appropriate solution, offering great efficiency and robustness. Based fundamentally on the approach proposed by [Graf & Dillmann, 1999, 2001], our system employs mainly robot manipulators as transferring devices, which move and tilt their carrying tool in such a way, that no undesired accelerations effects could exist, to affect the current state of the transporting object.

3.2 The "Waiter-Tray" Model

This idea was born by observing humans carrying objects (those which need special care and attention) rapidly from one location to another. A good example is a waiter walking inside a restaurant holding up with his hand a tray full of plates and glasses, without throwing away the transported elements. Probably, without knowing it, the waiter tries to incline the tray in such a way that unwanted accelerations and forces acting on the carried objects are avoided. The main approach presented in this work has a similar mechanism: while the waiter is orienting his hand to tilt the tray in an appropriate manner, the orientation of the robot's end-effector is adapted as well to compensate for undesired acceleration effects [Fig. 1].

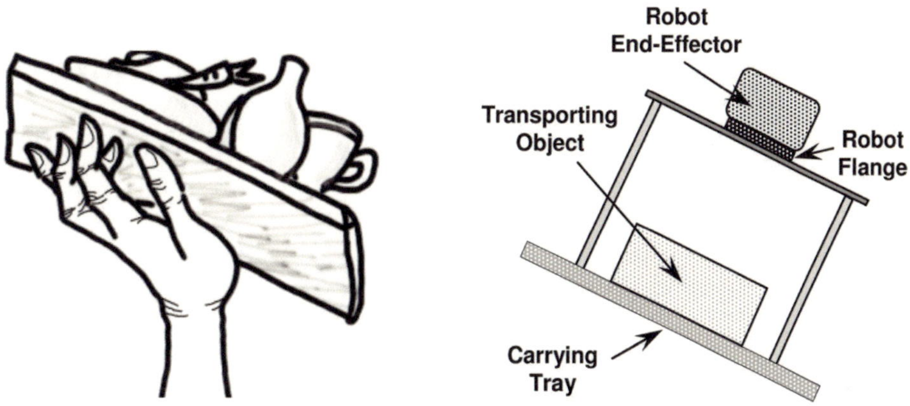

Figure 1. Robot arm imitating a human hand to carry objects

3.3 The Test-Environment

According to the problem-scenarios described before, three test-beds consisting of a robot manipulator with different kind of tools, which functionally replicate the transfer system for testing, are used to verify the effectiveness of the proposed approaches [Fig.2].

4. Problem-Scenario I: Reduction of Shear Forces

4.1 Introduction

Commonly, the handling robots employ vacuum suction grippers as gripping tool to manipulate and to move goods from one place to another. Although vacuum suction grippers have a lot of advantages, there is still one significant drawback: due to the phenomenon of large lateral accelerations, the highly accelerated load could fall off the gripper. To solve this problem, the orientation of the robot's end-effector is adapted in such a way, that the undesired large shear forces, between the contact surface of the grasping tool and its carrying object, can be minimized until the end of the motion.

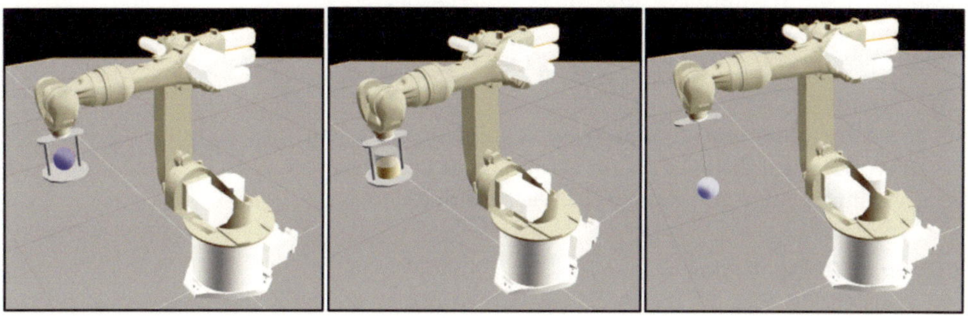

Figure 2. Three problem-scenarios in evaluation. Minimizing of shear forces (a), liquid sloshing (b) and sway oscillations (c)

4.2. Model Description

With the need to describe the problem in question, and to observe the effects created by large shear forces, a simplified model using a tray is proposed. Fig. 3 shows a simplified free-body diagram with the principal forces involved in such a system. F_{sh}, indicates the shear force. F_f is the friction force, it resists to any force that tries to move an object along a surface and it acts oppositely to the shear force, thus $F_f = - F_{sh}$, as long as the object is not slipping. F_N is the normal force that the tray exerts on the transporting object, it remains perpendicular to the tray's surface, g is the gravitational acceleration and thus, mg represents the object's weight.

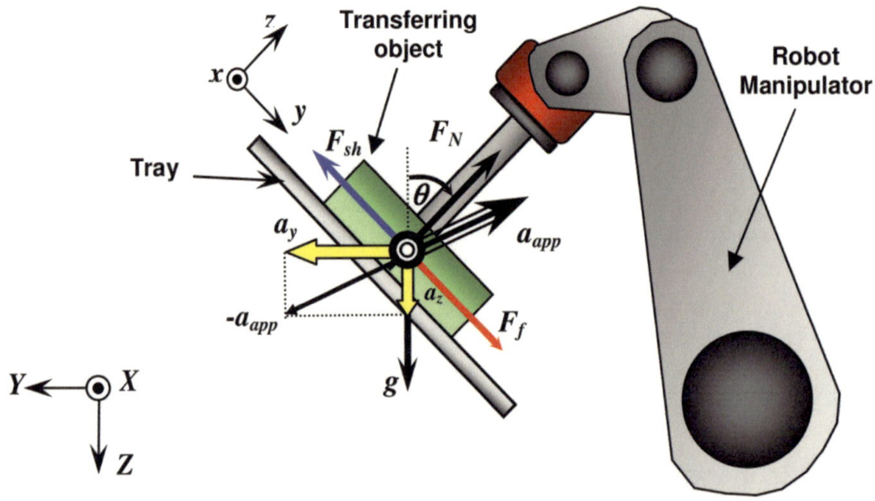

Figure 3. Free body diagram. Robot manipulator with the transferring object

a_{app} represents the total acceleration applied by the end-effector, it points in the direction of the motion. a_x, a_y and a_z are the inertial accelerations opposed to any acceleration acting on the object in x-, y- and z-direction respectively, due to horizontal/vertical motion of the end-effector (relative to the **X-Y-Z** world coordinate system), x, y and z establish the reference frame system in the tray. θ symbolizes the tilting angle of the TCP(Tool Center Point).

The fundamental requisite to guarantee that the carried object remains on spot in the carrying tray is, that the magnitude of the shear force must be less than or equal to the magnitude of the maximum friction force:

$$\left|F_{sh}\right| \leq \left|F_{f\,\max}\right|. \tag{1}$$

This is a very important condition, which has to be taken into account. The proposed method works when this inequation is fulfilled. It is also important to notice that in an ideal case, the optimal value of the shear force is considered as zero.

4.3 Optimal Tilting Angles

Now, considering that all forces acting on the object are referring to the tray reference coordinate system. The forces along the contact surface of the object in y-direction (x-direction can be treated in the same way) is defined as:

$$F_{sh_y} = m(g\sin(\theta_y) + a_z \sin(\theta_y) - a_y \cos(\theta_y)) . \tag{2}$$

This yields to the definition of the shear force. To achieve the condition (1), a change in the orientation of the TCP is needed. This guarantees that the shear force F_{sh} is minimized in every time-instance. Hence, deriving from (2) and assuming the "ideal case", where the shear force is zero, the mass m can be eliminated. This means that the role of the mass for the new approach is irrelevant and this gives:

$$(g + a_z)\sin(\theta_y) - a_y \cos(\theta_y) = 0 . \tag{3}$$

Thus, the value of θ_y can be computed and as well in an analogue way for θ_x. These are the *optimal tilting angles* due to y- and x-horizontal movement respectively. Note that the tilting angles are functions of each time-instance t:

$$\theta_x(t) = \tan^{-1}\left(\frac{a_x(t)}{(g + a_z(t))}\right), \tag{4}$$

$$\theta_y(t) = \tan^{-1}\left(\frac{a_y(t)}{(g + a_z(t))}\right). \tag{5}$$

During the motion, if the robot controller adapts its tool orientation according to (4) and (5), the load is guaranteed to be kept on spot in the carrying tray.

4.4 Lateral Acceleration

Once that the definition of shear force has been introduced, the lateral acceleration a_{lat} can be determined according to (2) and it is expressed as follows

$$a_{lat_y} = F_{sh_y} / m = g\sin(\theta_y) + a_z \sin(\theta_y) - a_y \cos(\theta_y) , \tag{6}$$

a_{lat_y} is the lateral acceleration, consequence of the remaining shear force. Let's consider the model described in Fig. 3. In case of no existence of tilting movement (that is $\theta_y = 0$), then the lateral acceleration is equivalent to $a_{lat_y} = -a_y$ until the end of motion. This means, when no compensation is executed, then the lateral acceleration acting on the transfer object has the same magnitude but opposite to the original applied acceleration. In this situation, if the friction is not large enough, the object will easily fall down from the carrying tray.

4.5 Motion Feasibility

As seen in (4) and (5), a fast increasing of acceleration will lead to fast changes in the tilting angles. Because modern robots are highly dynamic machines, the acceleration is ramped up very fast to the maximum (trapezoidal motion profile for the velocity) leading to high jerks. This fast change of orientation can not be achieved with a standard commercial robot, because of severe dynamic performance limits, such as maximum motor torque and maximum gear load. In order to avoid too large jerk effects and therefore to achieve suitable motions, one possible solution is the modification of shape of velocity and acceleration profile using filtering algorithms, with the aim to reach smoother motions during the periods of acceleration and deceleration [Chen et al., 2006].

4.6 Simulation and Experimental Results

After analyzing theoretically the lateral acceleration, a linear motion is first put into examination using software simulation, which its results and performance are later evaluated through a virtual robot controller, in order to verify its feasibility before putting the motion into the test environment with real robot manipulators. Here, the results are also verified though the measurements using an axis accelerometer, a development kit ADSL202 from ANALOG DEVICES with a measurement range of ±2 g.

A. Test Setup

To simplify the simulation analysis, a simple linear motion along *Y*-axis is evaluated. The trajectory begins at the starting position **A** = [0.370, 0.300, 0.390] and ends at position **B** = [0.370, -0.300, 0.390][m]. The positions are expressed in Cartesian coordinate system [Fig. 4]. Every interpolation cycle t_{ipo}, is 0.012 s. The maximum acceleration for continuous motions is set to 4.6 m/s^2 and maximum velocity to 2 m/s.

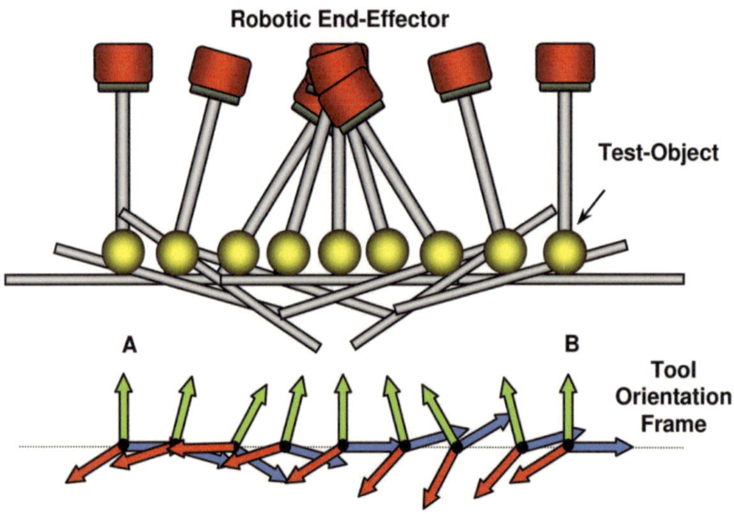

Figure 4. Trajectory of robot TCP with compensated tilting angles

B. Results

Using simulation software, the above motion is tested with diverse filter lengths L= 9, 15 and 30 [Fig. 5]. In this particular case, the reference acceleration requires a total of 0.816 s to complete the programmed movement. The best compensated motion is when the filter length L = 9, with a filtered lateral acceleration (maximum absolute) of 2.1 m/s^2 and a cycle time of 1.032 s. When L < 9, the adaptation of the TCP-orientation can not be accomplished due to the robot dynamic limits. Thus L = 9 is the best suitable filter length. The measurement results can be observed in [Fig. 6].

It is worth to mention that there is a difference between the simulated and the measured curves, due to the fact that the sensor signals themselves have to be filtered to be useful (filter length of sensor = 20 ti_{po}, where ti_{po} represents the robot controller's interpolation cycle time). Taking this fact into account, the experimental results confirm satisfactorily the simulation results.

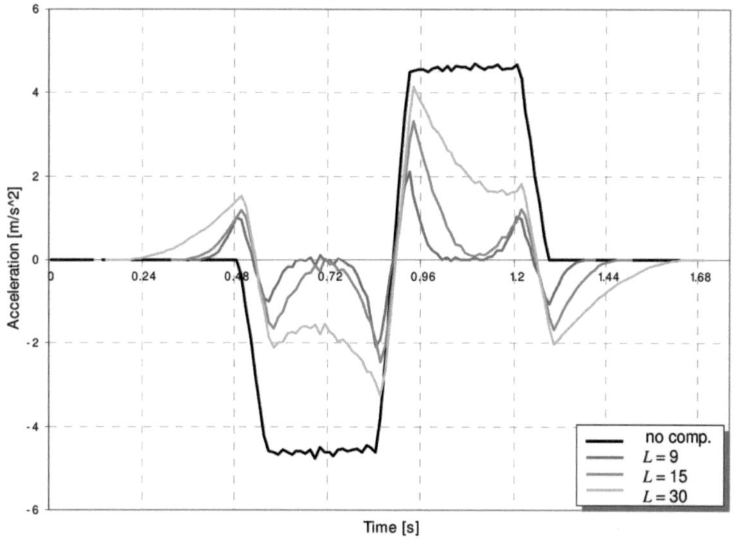

Figure 5. Lateral acceleration with different filter lengths in the software simulation

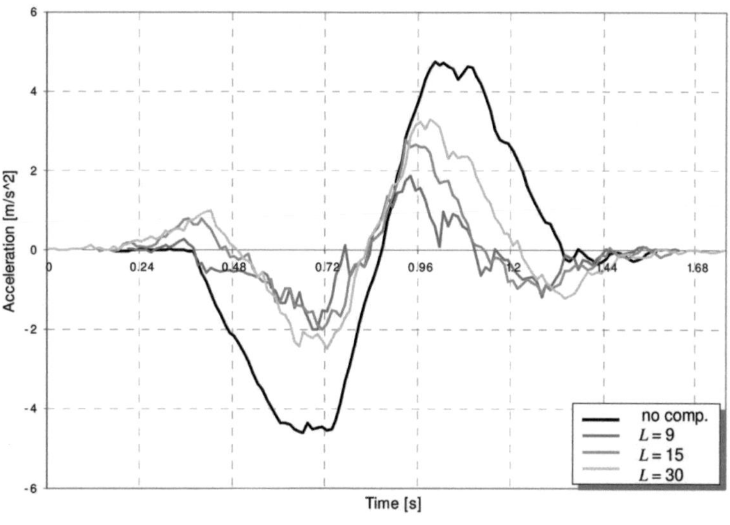

Figure 6. Measurements of lateral acceleration with different filter lengths

5. Problem-Scenario II: Sloshing Suppression

5.1 Introduction

In the last decades, the topic of suppressing the sloshing or liquid vibration has become a great interest to many sectors, especially in the areas of industry and research. The prevention and avoidance of overflow caused by sloshing is fundamental for many industrial applications, which include liquid container transfer systems. In such systems, like the transportation of molten metal from a furnace [Yano & Terashima, 2001], [Hamaguchi et al., 1994], [Feddema et al., 1997], automatic pouring system in the casting industries [Terashima & Yano, 2001], [Linsay, 1983], [Lerner & Laukhin, 2000] or transfer operations with fluid products in the food industry, etc. Here, the sloshing suppression is essential, in order to prevent any negative effect on the product's quality and as well to avoid any possible contaminations.

5.2 Acceleration Compensation "against" Sloshing?

Now, a difficult and even more challenging handling problem is introduced. Compared to the manipulation of rigid objects as described in the preceding section, the fluid behaviour is much more difficult to be modelled and controlled, due to its nonlinear dynamical characteristics. Here, any disturbance may cause its deformation and hence, the producing of instabilities to the entire system.

Theoretically, as long as there is no relative motion between the liquid and the container, then the liquid is assumed as a rigid-body. In this case, ACP may be also suitable and applicable to solve the sloshing problems. As main idea, this method operates basically in maintaining the normal of the liquid surface to be opposite to the accelerations of the entire system, until the completion of the transportation [Fig. 7]. In this way, undesired sloshing effects could be enormously reduced, which assures that the liquid can stay safely inside of its container during the entire motion process.

(a)

(b)

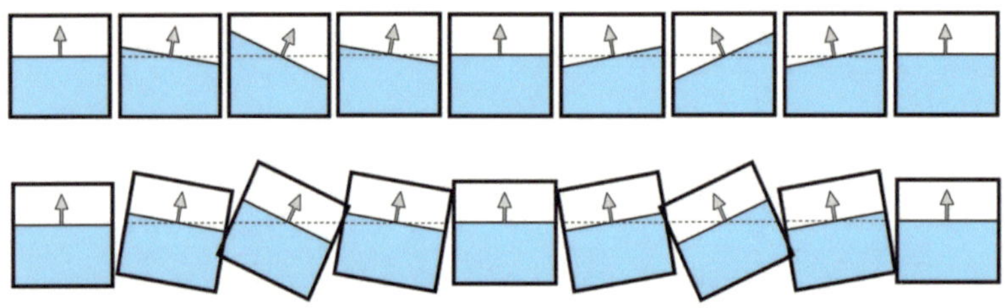

Figure 7. Trajectory of a rectangular liquid container without (a) and with (b) compensated tilting angles

Next, with the aim to validate the proposed theory, important fluid dynamic fundamentals will be introduced, in order to deduce the optimal tilting angles required to compensate the sloshing effects inside an accelerated liquid container.

5.3 Model Description

First, an open container of liquid is considered. It translates along a straight path with a constant acceleration $\vec{a}$ as illustrated in Fig. 8. Since $a_x = 0$, the pressure gradient in the x-direction is zero ($\partial p / \partial x = 0$) [Chen et al., 2006]. In the y- and z-directions, it yields

$$-\frac{\partial p}{\partial y} = \rho a_y$$

$$\frac{\partial p}{\partial z} = -\rho (g + a_z) \qquad (7)$$

Now, the variation in pressure between two closely spaced points located at y and z is considered. Thus, $y+dy$ and $z+dz$ can be expressed as

$$dp = \frac{\partial p}{\partial y}dy + \frac{\partial p}{\partial z}dz \,. \tag{8}$$

Using the results obtained from Eq. (7) and (8),

$$dp = -\rho a_y\, dy - \rho(g + a_z)dz \,. \tag{9}$$

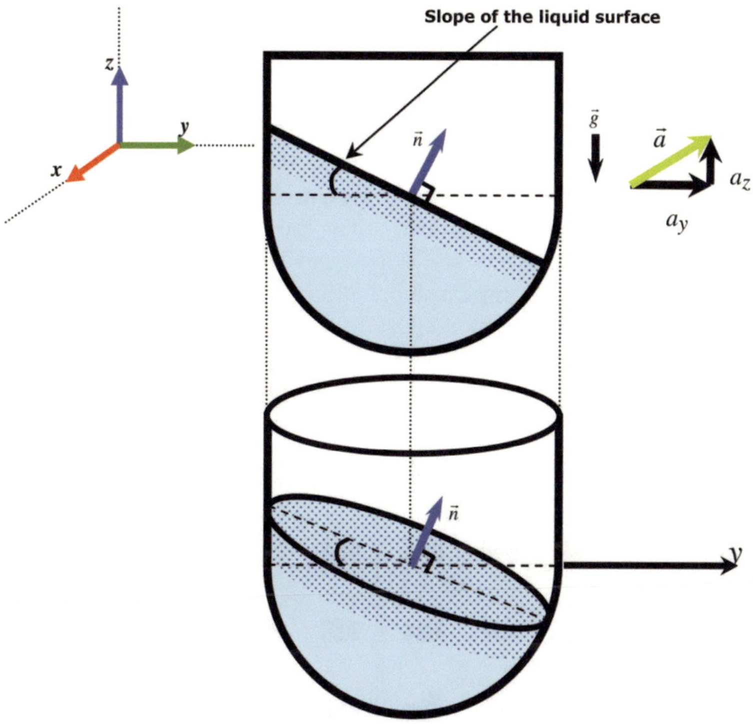

Figure 8. Free body diagram. An accelerating fluid container without compensation

5.4 Optimal Tilting Angles

Now, if the pressure is considered constant, then $dp = 0$. According to Eq. (9), it follows that the slope of the liquid surface is given by the relationship

$$\frac{dz}{dy} = -\frac{a_y}{(g + a_z)}, \tag{10}$$

where dz/dy is equivalent to $tan(\theta_y)$. Therefore, it gives

$$\theta_y = \tan^{-1}\left(\frac{a_y}{(g + a_z)}\right), \tag{11}$$

the same conclusion as obtained in Eq. (4).

5.5 Simulation and Experimental Results

For the verification, a test-bed consisting of a KUKA KR16 industrial manipulator with 6-DOF (Degrees Of Freedom) and a metal tray as carrying tool has been used. A camera was adopted as sensor apparatus to observe the liquid behaviour in 2D, and it has been attached directly on the experimental tray, opposing to the glass-recipient.

A. Test Setup

For the evaluation, two motions are carried out: one motion without, and the other one, with acceleration compensation. As test motion, a linear motion along the Y-axis is used. It starts at the position **A** = [0.930 0.800 1.012] and ends at the position **B** = [0.930 -0.800 1.012].

B. Results

As test object, a transparent rectangular glass-recipient containing water is used. The static liquid level is 40 mm. The maximum acceleration for continuous motions is set to 4.3 m/s^2. The sequences of the filtered images, obtained from the experimentation-videos, verify that the sloshing effects can be diminished significantly after applying the compensation algorithm [Fig. 9]. For such compensated movement, the adopted filter length is $L = 9$.

Here, some residual oscillations still exist slightly during and at the end of the motion, as a consequence of the phase-delay generated after the filtering [Chen et al., 2007]. This oscillation effect is caused because of the differences between the original and the filtered motion. In the case of the non-compensated motion, the maximum deviation of the peak elevation is approximately 37.5 mm respect to the corresponding static level [Fig. 10]. Contrary to this, the compensated motion has only a maximum deviation around 5.7 mm at its fluid surface. This represents a reduction of approximately 84.7 %.

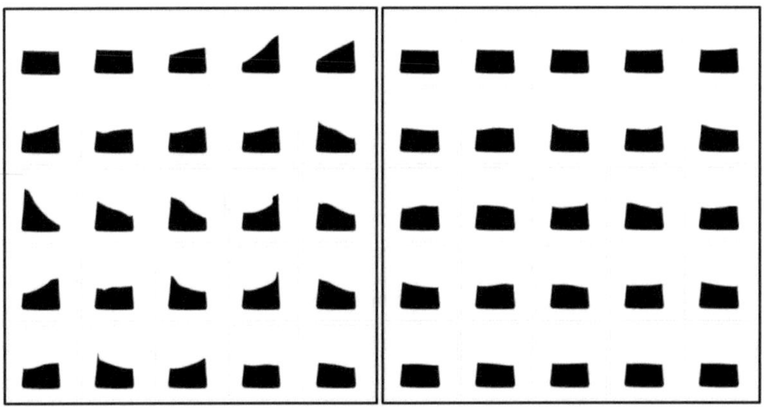

(a) (b)

Figure 9. Part of image sequences from a linear movement without (a) and with (b) acceleration compensation. The sequence order follows from left to right and from top to bottom

Notice that in the compensated motion, the liquid surface reestablishes its resting state faster than in the non-compensated case. The maximum amplitude of those oscillations is below 6 mm, which assures that the liquid can stay safely inside of its container during the entire motion process.

This approach is very efficient, since it neither requires any complex fluid modelling, nor the help of an external sophisticated sensing system, or the feedback of any vibration information.

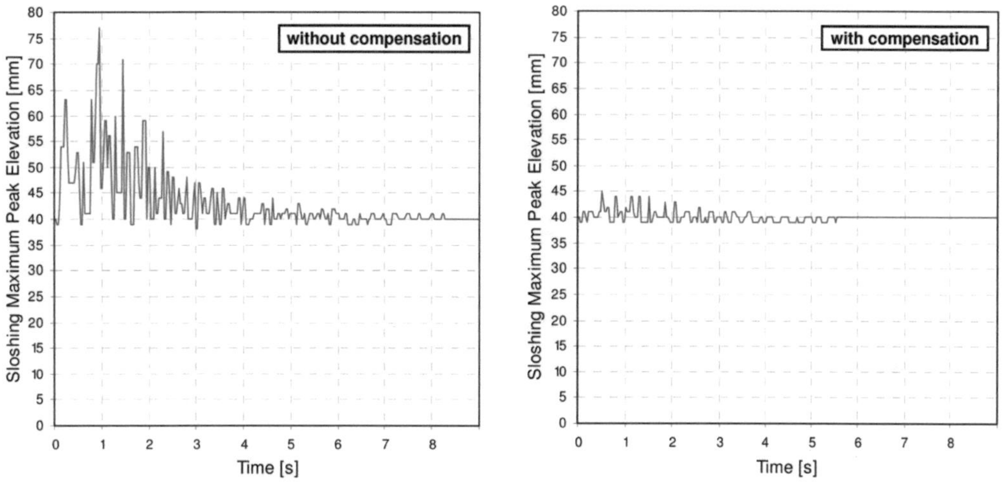

Figure 10. Non-compensation versus compensation. Results obtained from the sensor-camera images

6. Problem-Scenario III: Minimization of Swing Oscillation

6.1 Introduction

As we mentioned before, high-speed transportation of loads using rope may cause undesired swing oscillations. Next, with the purpose to solve this problem in a simple way, a new open-loop technique is proposed. It derives immediately from the ACP and consists mainly on modifying the reference-trajectory (which is comprised by a set of suspension points) in such a way, that any undesirable swing effects arisen during the motion execution can be suppressed without the help of any external sensor system.

6.2 The New Compensation Strategy

This new technique is as a result of the model proposed in Section 4. In this former method, the orientation of robot's end-effector (optimal tilting angles) was remodified according to the compensation algorithm, in order to minimize and compensate undesired acceleration side effects.

On the basis of this principle, our new methodology uses mainly the computed optimal tilting angles to calculate a new compensated trajectory, composed by a new set of suspension points. This means, based on the ACP, the reference input trajectory is modified in such a way, that the resulting movements from the robot's TCP permit the minimizing of undesired swing effects caused by large motion accelerations. Referring to the model described in [Fig. 3], a robot manipulator with a carrying tool is reconsidered [Fig. 11] and it performs a linear horizontal motion from position **A** to **B**.

The optimal tilting angles θ_x and θ_y are defined as indicated in Eq. (4) and (5). As shown in Fig. 11, each computed optimal tilting angle is comprised between the axis of the tool and the vertical plane. Additionally, the turning point is referred to the programmed TCP. In this case, the TCP locates at the center of the imaginary tool.

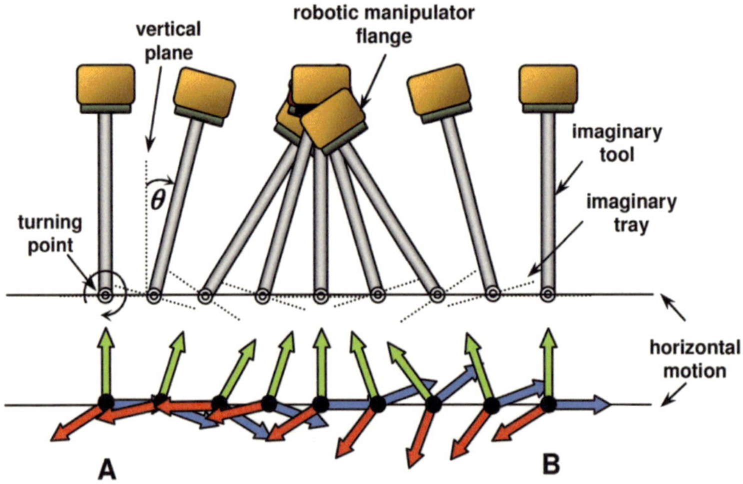

Figure 11. Former acceleration compensation model proposed in section 4

6.3 A New Set of Suspension Points. Model Description

A pendulum model is proposed [Chen et al., 2007]. An object with mass m is connected to a robot manipulator through a rigid rope with invariant length l [Fig. 12].

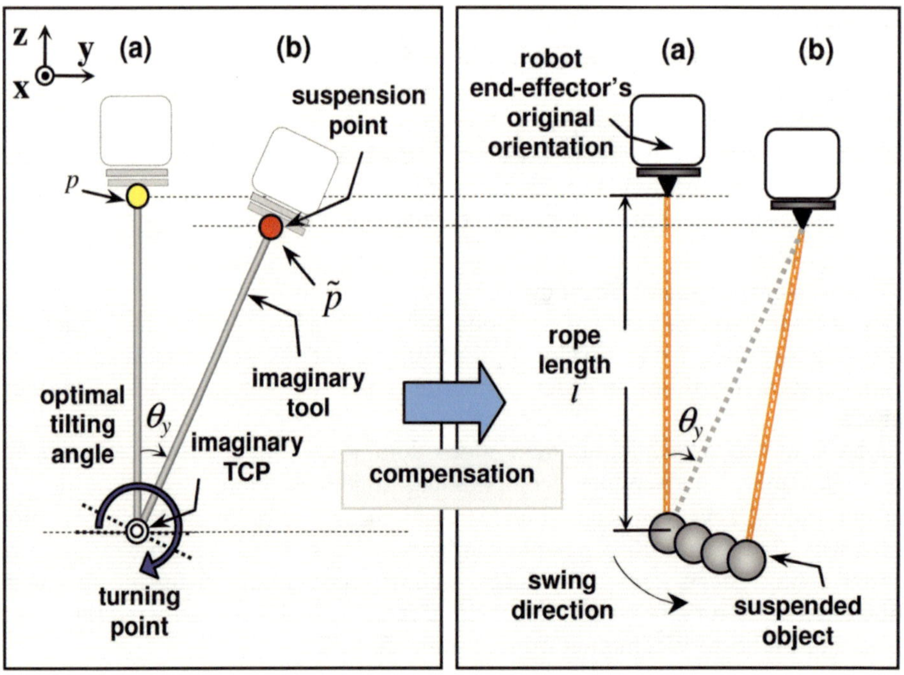

Figure 12. Left: the "imaginary tool" rotates its "imaginary TCP" according to the optimal tilting angle θ_y. Right: new compensated trajectory with the suspended object

The suspension point is attached directly to the robot flange, and the friction due to the bending of the rope is neglected. To compute the new compensated trajectory, the existence of an "imaginary tool" with the same length l as the rigid rope is assumed [Fig. 12(left)]. At its lower endpoint, where theoretically locates the suspended object, an "imaginary TCP" (the turning point) is considered. Now, the robot end-effector moves horizontally to the right [Fig. 12 (left) from (a) to (b)], with the initial conditions $v(0)=0$ and $a(0)=0$. p is a Cartesian position from the reference input trajectory, v and a are respectively the rate of displacement (velocity) and rate of velocity variation with respect to time (acceleration).
In order to compute the new compensated trajectory, which is composed by a set of new suspension points of the pendulum, the set of $\tilde{p}$ [Fig. 12 (left)] is calculated with the help of a rotation matrix (Eq. 12). Together with the information of the rope length l, a rotation in the turning point is performed, according to θ_x and θ_y. To specify the orientation, the convention roll, pitch and yaw angles are used. The rotation around axis x is represented by θ_x (roll), rotation about y by θ_y (pitch), and around axis z by θ_z (yaw).

$$R_{rpy}(\theta_x,\theta_y,\theta_z) = Rot(Z,\theta_z)Rot(Y,\theta_y)Rot(X,\theta_x)$$

$$= \begin{bmatrix} c_z & -s_z & 0 \\ s_z & c_z & 0 \\ 0 & 0 & 1 \end{bmatrix} \begin{bmatrix} c_y & 0 & s_y \\ 0 & 1 & 0 \\ -s_y & 0 & c_y \end{bmatrix} \begin{bmatrix} 1 & 0 & 0 \\ 0 & c_x & -s_x \\ 0 & s_x & c_x \end{bmatrix},$$

$$= \begin{bmatrix} c_z c_y & c_z s_y s_x - s_z c_x & c_z s_y c_x + s_z s_x \\ s_z c_y & s_z s_y s_x + c_z c_x & s_z s_y c_x - c_z s_x \\ -s_y & c_y s_x & c_y c_x \end{bmatrix} \quad (12)$$

where $c_x = \cos\theta_x$, $c_y = \cos\theta_y$, $c_z = \cos\theta_z$, $s_x = \sin\theta_x$, $s_y = \sin\theta_y$ and $s_z = \sin\theta_z$. Now using this rotation matrix, the new suspended point $\tilde{p}$ is computed. Considering that there is no rotation around Z-axis, then $\theta_z = 0$. The position $\tilde{p}$ is defined as

$$\begin{bmatrix} \tilde{p}_x \\ \tilde{p}_y \\ \tilde{p}_z \end{bmatrix} = \begin{bmatrix} p_x \\ p_y \\ p_z - l \end{bmatrix} + \begin{bmatrix} c_y & s_y s_x & s_y c_x \\ 0 & c_x & -s_x \\ -s_y & c_y s_x & c_y c_x \end{bmatrix} \begin{bmatrix} 0 \\ 0 \\ l \end{bmatrix} = \begin{bmatrix} p_x + l s_y c_x \\ p_y - l s_x \\ p_z + l c_y c_x - l \end{bmatrix}, \quad (13)$$

then, the new suspension point is given by

$$\begin{aligned} \tilde{p}_x &= p_x + l \sin\theta_y \cos\theta_x \\ \tilde{p}_y &= p_y - l \sin\theta_x \\ \tilde{p}_z &= p_z + l \cos\theta_y \cos\theta_x - l \end{aligned} \quad . \quad (14)$$

This is the new position for the suspension point $\tilde{p}$ at the time instant i. The set of "old" suspension points are shifted in such a way, that possible rest swing oscillations induced by a high-speed motion can be reduced enormously by applying the acceleration compensation method.

6.4 Compensation vs. Non-Compensation

To evaluate the shape of the resulting trajectory established by the new suspension points, two examples are illustrated in Fig. 13.

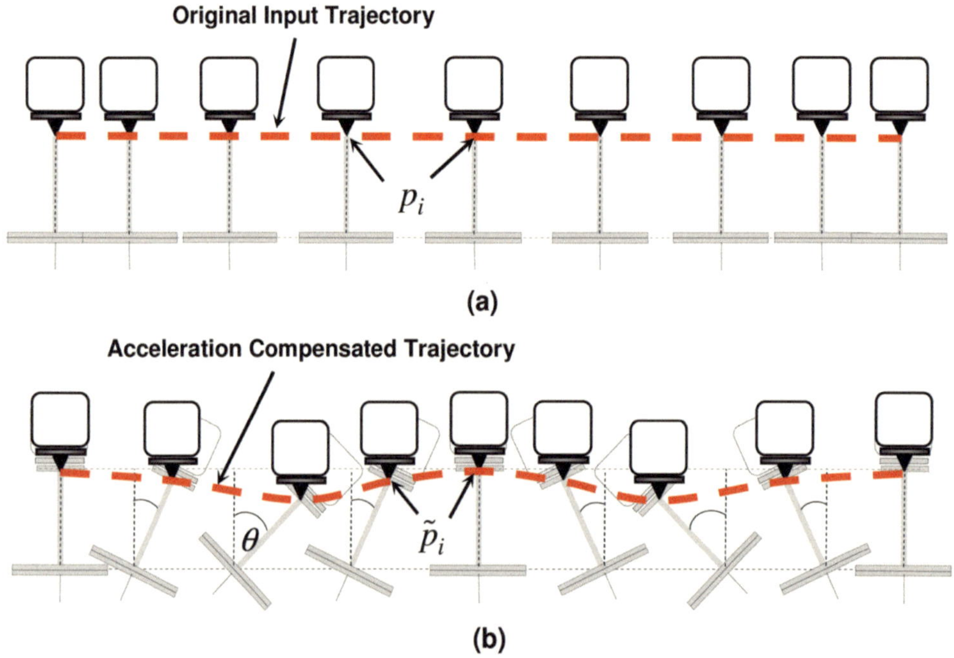

Figure 13 (a) Trajectory established by the original suspension points, (b) New trajectory applying acceleration compensation

The original trajectory without applying acceleration compensation technique remains a straight horizontal trajectory. On the other hand, the location of each suspension point from a compensated motion, involves not only motions in horizontal-plane, but also vertical movements to eliminate the oscillation effects. Two motions with different lengths are shown in Fig. 14. As observed, the short motion in a reduced time term requires "more movements" to compensate the oscillations.

6.5 Software Simulation

Since the swing effect relies principally on the driving accelerations, it is interesting to analyze its behaviour before and after the compensation algorithm.

For simplicity, a test-motion translates in horizontal Y-direction. The utilized rigid rope has a length of 0.340 m. Since the reference test motion is a linear continuous-path (CP), the maximum velocity is set to 2 m/s. The resulting velocity and acceleration profiles before and after applying the compensation algorithm are illustrated in Fig. 15 respectively. All illustrations on the left side of each figure belong to the reference motion without acceleration compensation and on the right side is the compensated motion. The resulting swing angles of both motions are shown in Fig. 16.

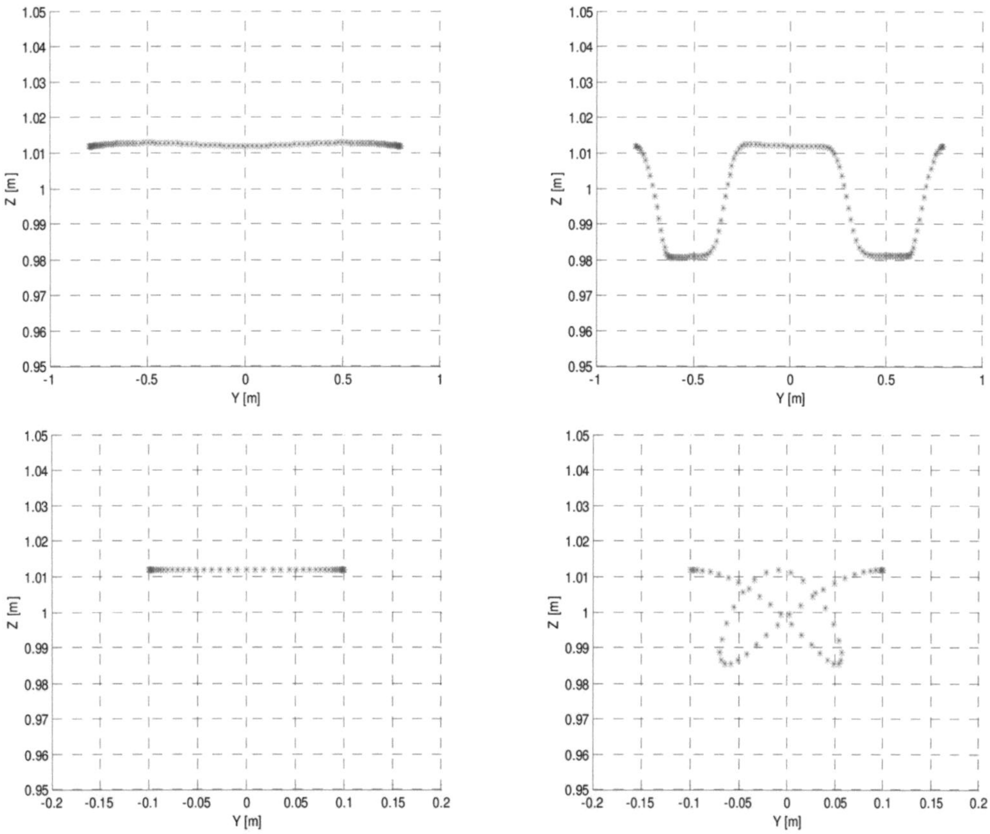

Figure 14. Two test-trajectories. Left: non-compensated motion. Right: compensated motion

6.6 Acceleration Excitation

Now, we proceed to evaluate the excitation effects of acceleration in the compensation of sways. Recalling the Fig. 15, in order to compensate the swing effects induced by the driving accelerations, additional acceleration excitations are applied [Fig. 15 (right)] at the changing phases from the reference acceleration profile [Fig. 15 (left)].

As described in section 2, the work from [Mita & Kanai, 1997] reveal that it is possible to obtain a swing-free motion at the end of the desired trajectory by constraining the velocity profile. This means that an optimal time swing-free motion could be achieved using a sequence of constant acceleration pulses to cancel undesirable swing effects in transportation of suspended objects. The similar principle and resulting effects can be observed with the results obtained in this study.

However, a proper modification of the reference motion using the ACP is more efficient and convenient, because of the simplicity of its computing algorithm and great facility to be implemented into the robot system. Furthermore, there is no need of pre-designing the velocity profile, where the knowledge of different phase-time instances are necessarily to be known in advance.

<table>
<tr><th colspan="2">Test Motion</th></tr>
<tr><th>Without Acceleration Compensation</th><th>With Acceleration Compensation</th></tr>
<tr><td>Velocity</td><td>Veloctiy</td></tr>
<tr><td>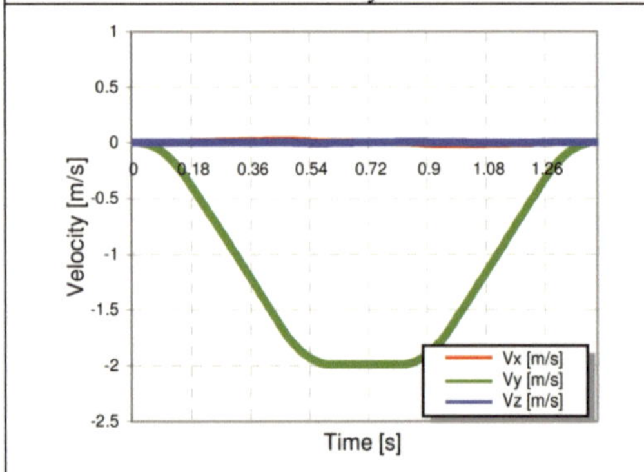
</td><td>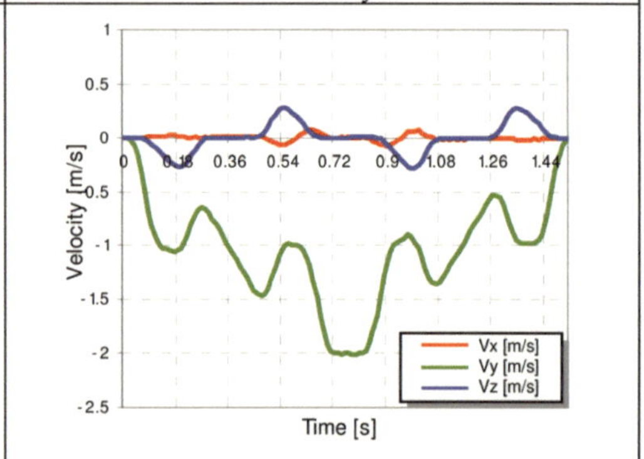
</td></tr>
<tr><td>Acceleraion</td><td>Acceleration</td></tr>
<tr><td>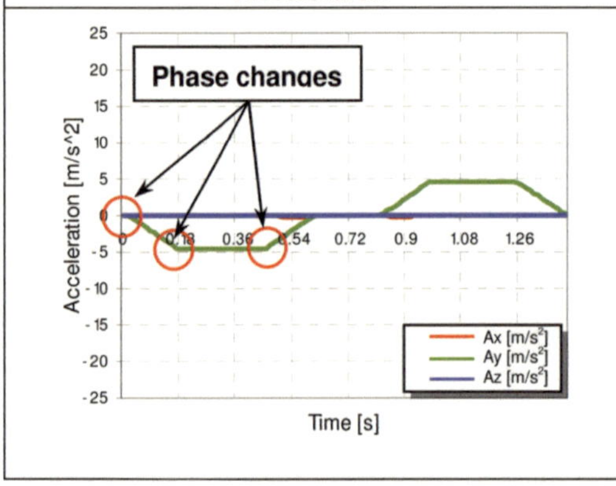
</td><td>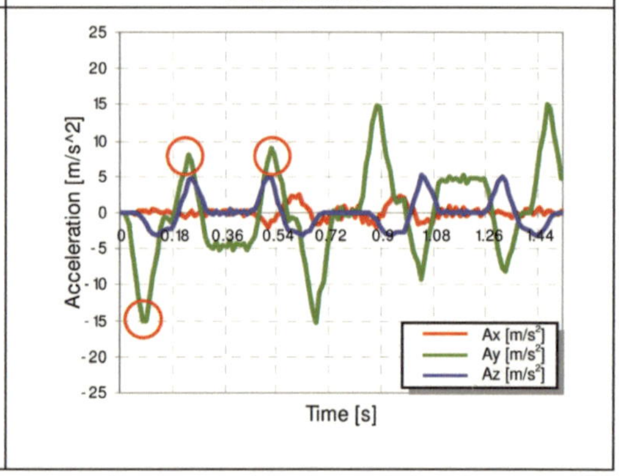
</td></tr>
</table>

Figure 15. Velocity and acceleration profiles from the test motion. Non-compensated motion vs. compensated motion

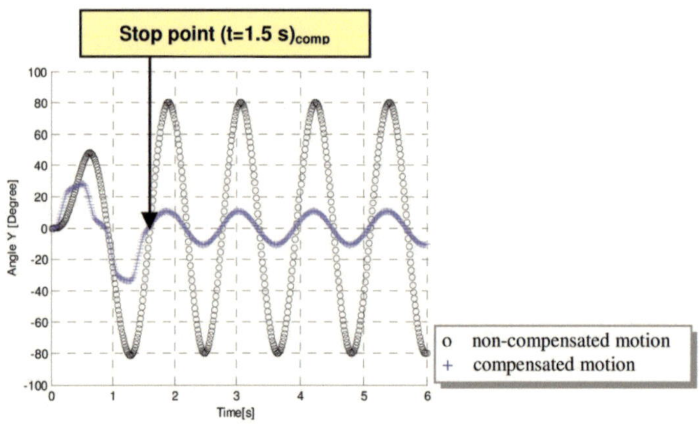

Figure 16. Test motion. Swing angles in indirection. Non-compensation vs. compensation

6.7 Integrating Physics Engine into the Simulation Environment

Before the execution of the compensated motion with the real robot system, open-source physics engine 'Bullet' has been incorporated into our software simulation system, in order to observe in advance the nonlinear behaviours of those test-objects with dynamical characteristics, to avoid in this way, the causing of any possible hazard [Fig. 17].

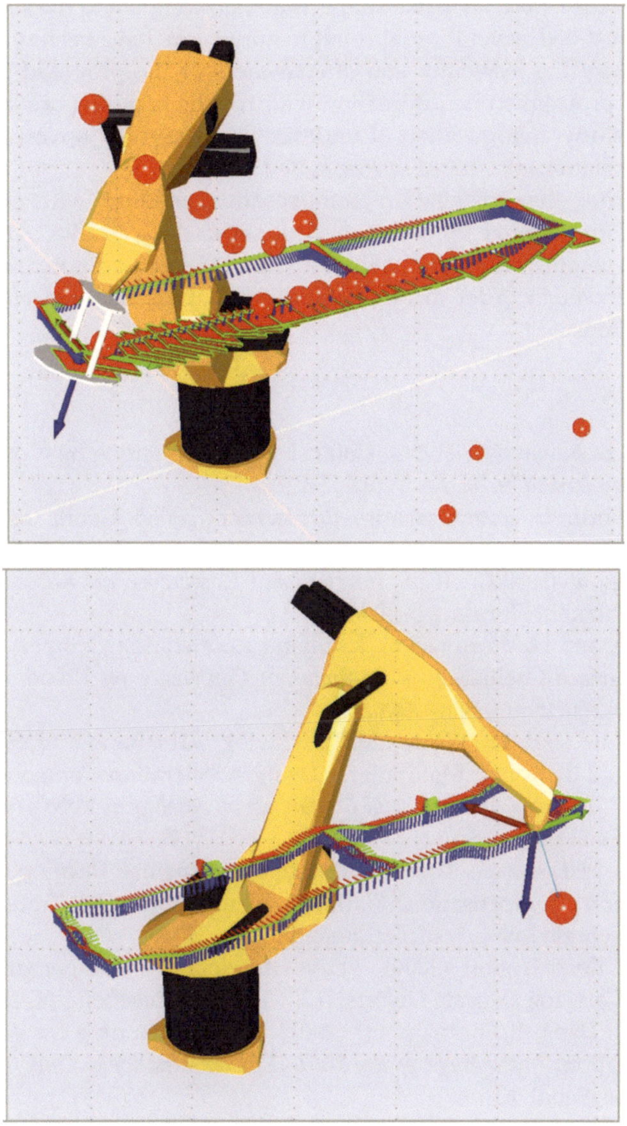

Figure 17. Software simulation using physics engine 'Bullet'

7. Conclusion

The acceleration compensation principle (ACP) proved to be beneficial and powerful for compensating side effects caused by acceleration, commonly produced on handling objects with dynamical constraints. In order to deal with different kinds of handling problems caused by highly accelerated motions, new efficient solutions based on this principle have been established. Gentle robotic handling and time optimized movements are considered as two main research objectives. Beyond the achievement of optimal cycle time and the avoidance of undesirable collisions, the proposed algorithms are independent of the physical aspects of the object, such like shape, material, weight and the size. As main part of the experimental test-bed, several serial robot manipulators have been utilized with diverse test objects, to verify the feasibility and effectiveness of the proposed approaches. Three different kinds of problem-scenarios dealing with robotic handling are investigated within the scope of this study: minimization of undesirable large shear forces, fluid sloshing and residual swing oscillations produced in suspended objects.

As final conclusion, the acceleration compensation method with its simplicity and robustness, has demonstrated its potential with satisfactory results. As future work, the incorporation of a compensation on the fly will be suggested as a potential improvement of the proposed methods, in order to make them more applicable and efficient for the real industrial applications.

8. References

N. Blazicevic and B. Novakovic (1996). Control of a rotary crane by a combined profile of speeds, *Strojarstvo*, v. 38, no. 1, pp. 13-21.

S. J. Chen, H. Worn, U. Zimmermann, R Bischoff (2006). Gentle Robotic Handling - Adaptation of Gripper-Orientation To Minimize Undesired Shear Forces - *Proceedings of the 2006 IEEE International Conference on Robotics and Automation (ICRA)*, Orlando, Florida, pp. 309-314.

S. J. Chen, B. Hein, and H. Worn (2006). Applying Acceleration Compensation to Minimize Liquid Surface Oscillation, - *Australasian Conference on Robotics and Automation ,* December, Auckland, New Zealand.

S. J. Chen, B. Hein, and H. Worn (2007). Swing Attenuation of Suspended Objects Transported by Robot Manipulator Using Acceleration Compensation, *Proceedings of the 2007 IEEE/RSJ International Conference on Intelligent Robots and Systems (IROS)*, October, San Diego, USA.

A.X. Dang (2002). *PHD Thesis*: Theoretical and Experimental Development of an Active Acceleration Compensation Platform Manipulator for Transport of Delicate Objects, Georgia Institute of Technology.

A.X. Dang and I. Ebert-Uphoff (2004), Active Acceleration Compensation for Transport Vehicles Carrying Delicate Objects, *IEEE Trans, on Robotics*, Vol. 20, No. 5.

M.W. Decker, A.X. Dang, I. Ebert-Uphoff (2001). Motion Planning for Active Acceleration Compensation, *Proceedings of the 2001 IEEE International Conference on Robotics & Automation*, Seoul, Korea.

J. Feddema, C. Dohrmann, G. Parker, R. Robinett, V. Romero and D. Schmitt (1996). Robotically Controlled Slosh-Free Motion of an open Container of Liquid, *IEEE International Conf. On Robotics and Automation*, Minneapolis, Minnesota.

J. Feddema, C. Dohrmann, G. Parker, R. Robinett, V. Romero and D. Schmitt (1997). A Comparison of Maneuver Optimization and Input Shaping Filters for Robotically Controlled Slosh-Free Motion of an Open Container of Liquid, *Proceedings of the American Control Conference*, Albuquerque, New Mexico.

J. T. Feddema et al.(1997), Control for slosh-free motion of an open container, *IEEE Contr. Syst. Mag.*, vol. 17, no. 1, pp. 29-36, Feb.

R. Graf, R. Dillmann (1997). Active acceleration compensation using a Stewart-platform on a mobile robot, *Proceedings of the 2nd Euromicro Workshop on Advanced Mobile Robots,* EUROBOT '97, Brescia, Italy, pp. 59-64.

R. Graf, R. Dillmann (1999). Acceleration Compensation Using a Stewart-Platform on a Mobile Robot, *Proceedings of the 3rd European Workshop on Advanced Mobile Robots,* EUROBOT '99, Zurich, Switzerland.

R. Graf (2001). *PHD Thesis*: Aktive Beschleunigungskompensation mittels einer Stewart-Plattform auf einem mobilen Roboter, Universitat Karlsruhe. ISBN: 3-935363-37-0.

F. Crasser, A. D'Arrigo, S. Colombi, and A. Rufer (2002). Joe: A mobile, inverted pendulum, *IEEE Trans. Ind. Electron.*, vol. 49, no. 1, pp. 107-114.

M. Hamaguchi, K. Terashima, and H. Nomura (1994). Optimal control of liquid container transfer for several performance specifications, */. Adv. Automat. Technol,* vol. 6, pp. 353-360.

M. Kaneko, Y. Sugimoto, K. Yano (2003). Supervisory Control of Pouring Process by Tilting-Type Automatic Pouring Robot, *Proceedings of the 2003 lEEE/RSf International Conference on Intelligent Robots and Systems,* Las Vegas, Nevada.

R. Kolluru, K.P. Valavanis, S.A. Smith and N. Tsourveloudis (2000). Design Fundamentals of a Reconfigurable Robotic Gripper System, *IEEE Transactions on Systems, Man, and Cybernetics-Part A: Systems and Humans,* Vol. 30, No.2.

T. Lauwers, G. Kantor and R. Hollis (2005). One is Enough!, *12th Int'l Symp. On Robotics Research.* San Francisco.

Y. Lerner and N. Laukhin (2000). Development trends in pouring technology, *Foundry Trade }.,* vol. 11, pp. 16-21.

W. Linsay (1983). Automatic pouring and metal distribution system, *Foundry Trade }.,* pp. 151-165.

T. Mita, and T. Kanai (1979). Optimal Control of the Crane System Using the Maximum Speed of the Trolley, *Trans. Soc. Instrument. Control Eng.* (Japan), 15, pp. 833-838.

Y. Noda, K. Yano, K. Terashima (2002). Tracking Control with Sloshing-Suppression of Self-Transfer Ladle to Continuous-Type Mold Line in Automatic Pouring System, *Proc. of SICE Annual Conference (2002),* Osaka.

Y. Noda, K. Yano, S. Horihata and K. Terashima (2004). Sloshing suppression control during liquid container transfer involving dynamic tilting using wigner distribution analysis, *43th IEEE Conference on Decision and Control,* Atlantis, Bahamas.

G. G. Parker, B. Petterson, C. R. Dohrmann, and R. D. Robinett III (1995).Vibration Suppression of Fixed-Time Jib Crane Maneuvers, Sandia National Laboratories report SAND95-0139C.

Y. Sakawa and Y. Shindo (1982). Optimal control of container cranes, *Automatica,* v. 18, no. 3, pp. 257-266.

Segway Human Transporter (2004). [Online]. Available: http://www.segway.com

N. C. Singer and W. P. Seering (1990). Preshaping Command Inputs to Reduce System Vibration, *ASME Journal of Dynamic Systems,* Measurement,and Control.

W. E. Singhose, L. J. Porter, and W. P. Seering (1997). Input Shaped Control of a Planar Gantry Crane with Hoisting, American Control Conference, Albuquerque, NM.

O. J. M. Smith (1957). Posicast Control of Damped Oscillatory Systems, *Proc. of the IRE,* pp. 1249-1255.

G. P. Starr (1985). Swing-Free Transport of Suspended Objects with a Path-Controlled Robot Manipulator, ASME *Journal of Dynamic Systems,* Measurement, and Control, v. 107.

K. Terashima and K. Yano (2001). Sloshing analysis and suppression control of tilting-type automatic pouring machine, *IF AC J. Contr. Eng. Practice,* vol. 9, no. 6, pp. 607-620.

N.C. Tsourveloudis, R. Kolluru, K.P. Valavanis and D. Gracanin (2000). Suction Control of a Robotic Gripper: A Neuro-Fuzzy Approach, *Journal of Intelligent and Robotic System* 27: 215-235.

K. Yano and K. Terashima (2001). Robust Liquid Container Transfer Control for Complete Sloshing Suppression, *IEEE Transactions, on Control Systems Technology,* Vol. 9,No. 3.

5

Calibration of Robot Reference Frames for Enhanced Robot Positioning Accuracy

Frank Shaopeng Cheng
Central Michigan University
United States

1. Introduction

Industrial robot manipulators are important components of most automated manufacturing systems. Their design and applications rely on modeling, analyzing, and programming the robot tool-center-point (TCP) positions with the best accuracy. Industrial practice shows that creating accurate robot TCP positions for robot applications such as welding, material handling, and inspection is a critical task that can be very time-consuming depending on the complexity of robot operations. Many factors may affect the accuracy of created robot TCP positions. Among them, variations of robot geometric parameters such as robot link dimensions and joint orientations represent the major cause of overall robot positioning errors. This is because the robot kinematic model uses robot geometric parameters to determine robot TCP position and corresponding joint values in the robot system. In addition, positioning variations of the robots and their end-effectors also affect the accuracy of robot TCP positions in a robot work environment.

Model-based robot calibration is an integrated solution that has been developed and applied to improve robot positioning accuracy through software rather than changing the mechanical structure or design of the robot itself. The calibration technology involves four steps: modeling the robot mechanism, measuring strategically planned robot TCP positions, identifying true robot frame parameters, and compensating existing robot TCP positions for the best accuracy. Today, commercial robot calibration systems play an increasingly important role in industrial robot applications because they are able to minimize the risk of having to manually recreate required robot TCP positions for robot programs after the robots, end-effectors, and fixtures are slightly changed in robot workcells. Due to the significant reduction of robot production downtime, this practice is extremely beneficial to robot applications that may involve a rather large number of robot TCP positions.

This chapter provides readers with methods of calibrating the positions of robot reference frames for enhancing robot positioning accuracy in industrial robot applications. It is organized in the following sections: Section 2 introduces basic concepts and methods used in modeling static positions of an industrial robot. This includes robot reference frames, joint parameters, frame transformations, and robot kinematics. Section 3 discusses methods and techniques for identifying the true parameters of robot reference frames. Section 4 presents applications of robot calibration methods in transferring existing robot programs among "identical" robot workcells. Section 5 describes calibration procedures for conducting true robotic simulation and offline programming. Finally, conclusions are given in Section 6.

2. Robot Reference Frames and Frame Transformations

An industrial robot is a multiple degrees-of-freedom (DOF) manipulator built with a series of mechanical joints. The static positions of the robot are represented by Cartesian reference frames and frame transformations as shown in Fig 1. The robot world reference frame, R (x, y, z), is a fixed one in the robot world space and the robot default tool-center-point (TCP) frame, Def_TCP (n, o, a), is a moving one representing the robot end-point. The 4 × 4 homogeneous transformation matrix in Eq. (1) mathematically represents the position of default TCP frame Def_TCP relative to robot base frame R:

$$ {}^{R}T_{Def\_TCP} = \begin{bmatrix} n_x & o_x & a_x & p_x \\ n_y & o_y & a_y & p_y \\ n_z & o_z & a_z & p_z \\ 0 & 0 & 0 & 1 \end{bmatrix}, \qquad (1) $$

where the coordinates of point vector p represent TCP frame location and the coordinates of three unit directional vectors [n, o, a] of the frame represent TCP frame orientation.
The arrow from robot base frame R to default TCP frame Def_TCP graphically represents transformation matrix ${}^{R}T_{Def\_TCP}$ as shown in Fig. 1. The inverse of ${}^{R}T_{Def\_TCP}$ denoted as $({}^{R}T_{Def\_TCP})^{-1}$ represents the position of robot base frame R measured relative to default TCP frame Def_TCP denoted as ${}^{Def\_TCP}T_{R}$. Generally, the definition of a frame transformation matrix or its inverse as described with the examples of ${}^{R}T_{Def\_TCP}$ and $({}^{R}T_{Def\_TCP})^{-1}$ can be applied to any available reference frame in the robot system.
A robot joint consists of an input link and an output link. The relative position between the input and output links defines the joint position. It is obvious that different robot joint positions result in different robot TCP frame positions. Mathematically, robot kinematics called robot kinematic model describes the geometric motion relationship between a given robot TCP frame position expressed in Eq. (1) and corresponding robot joint positions. Specifically, robot forward kinematics will enable the robot system to determine where the TCP frame position will be if all the joint positions are known. Robot inverse kinematics will enable the robot system to calculate what each joint position must be if the robot TCP frame is desired to be at a particular position.
Developing robot kinematics equations starts with the use of Cartesian reference frames for representing the relative position between two successive robot links as shown in Fig. 1. The Denavit-Hartenberg (D-H) representation is an effective way of systematically modeling the link positions of robot joints, regardless of their sequences or complexities (Denavit & Hartenberg, 1955). It allows the robot designer to assign the link frames of a robot joint with the following rules as shown in Fig. 2. The z-axis of the input-link frame is always along the joint axis with arbitrary directions. The x-axis of the output-link frame must be perpendicular and intersecting to the z-axis of the input-link frame of the joint. The y-axis of a link frame follows the right hand rule of the Cartesian frame. For an n-joint robot, frame 0 on the first link (i.e., link 0) serves as robot base frame R and frame n on the last link (i.e., link n) serves as robot default TCP frame Def_TCP. The origins of frames R and Def_TCP

are freely located. Generally, an industrial robot system defines the origin of the robot default TCP frame, Def_TCP, at the center of the robot wrist mounting plate.

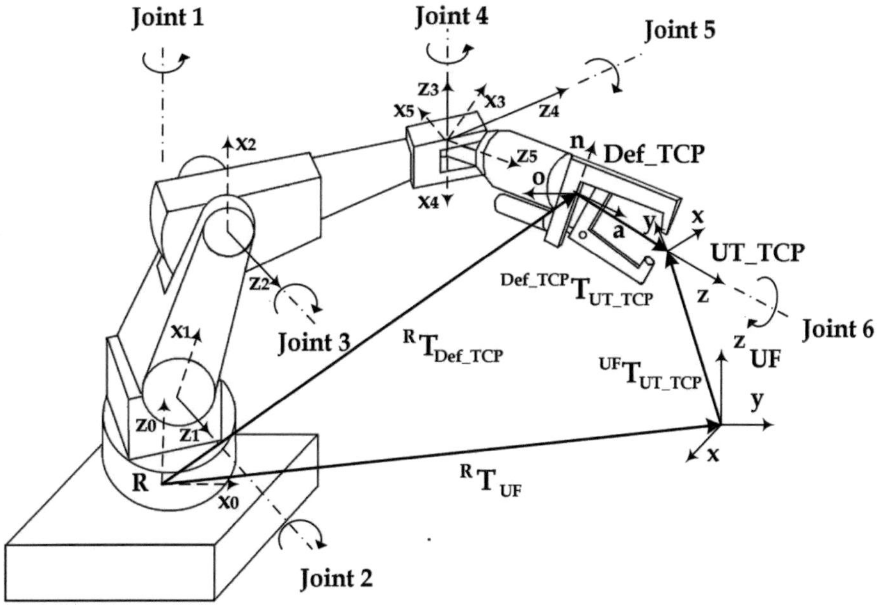

Figure 1. Robot reference frames and frame transformations

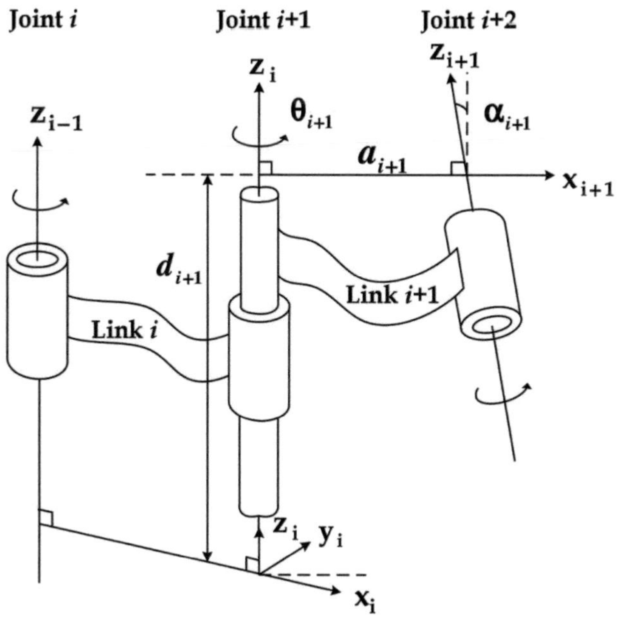

Figure 2. D-H link frames and joint parameters

After each joint link has an assigned D-H frame, link frame transformation ${}^{i}T_{i+1}$ defines the relative position between two successive link frames i and i+1, where i = 0,1,..., (n -1).

Transformation ${}^{i}T_{i+1}$ is called A_{i+1} matrix in the D-H representation and determined by Eq. (2):

$$A_{i+1} = Rot(z,\theta_{i+1}) \times Trans(0,0,d_{i+1}) \times Trans(a_{i+1},0,0) \times Rot(x,\alpha_{i+1})$$
$$= \begin{bmatrix} \cos\theta_{i+1} & -\sin\theta_{i+1}\cos\alpha_{i+1} & \sin\theta_{i+1}\sin\alpha_{i+1} & a_{i+1}\cos\theta_{i+1} \\ \sin\theta_{i+1} & \cos\theta_{i+1}\cos\alpha_{i+1} & -\cos\theta_{i+1}\sin\alpha_{i+1} & a_{i+1}\sin\theta_{i+1} \\ 0 & \sin\alpha_{i+1} & \cos\alpha_{i+1} & d_{i+1} \\ 0 & 0 & 0 & 1 \end{bmatrix}, \qquad (2)$$

where θ_{i+1}, d_{i+1}, a_{i+1}, α_{i+1} are the four standard D-H parameters defined by link frames i and i+1 as shown in Fig. 2.

Parameter θ_{i+1} is the rotation angle about z_i-axis for making x_i and x_{i+1} axes parallel. For a revolute joint, θ is the joint position variable representing the relative rotary displacement of the output link to the input link. Parameter d_{i+1} is the distance along z_i-axis for making x_i and x_{i+1} axes collinear. For a prismatic joint, d is the joint position variable representing the relative linear displacement of the output link to the input link. Parameter a_{i+1} is the distance along x_{i+1}-axis for making the origins of link frames i and i+1 coincident. It represents the length of the output link in the case of two successive parallel joint axes. Parameter α_{i+1} is the rotation angle about x_{i+1}-axis for making z_i and z_{i+1} axes collinear. In the case of two successive parallel joint axes, the rotation angle about y-axis represents another joint parameter called twisting angle "β"; however, the standard D-H representation does not consider it.

With link frame transformation ${}^{i}T_{i+1}$ denoted as A_{i+1} matrix in Eq. (2), the transformation between robot default TCP frame Def_TCP (i.e., link frame n) and robot base frame R (i.e., link frame 0) is expressed as:

$${}^{R}T_{Def\_TCP} = {}^{0}T_1 \times {}^{1}T_2 \times {}^{2}T_3 \ldots \times {}^{n-1}T_n = A_1 \times A_2 \times A_3 \ldots \times A_n. \qquad (3)$$

Eq. (3) is the total chain transformation of the robot that the robot designer uses to derive the robot forward and inverse kinematics equations. Comparing to the time-consuming mathematical derivation of the robot inverse kinematics equations, formulating the robot forward kinematics equations is simple and direct by making each of the TCP frame coordinates in Eq. (1) equal to the corresponding matrix element in Eq. (3). Clearly, the known D-H parameters in the robot forward kinematics equations determine the coordinates of ${}^{R}T_{Def\_TCP}$ in Eq. (1).

The industrial robot controller implements the derived robot kinematics equations in the robot system for determining the actual robot TCP positions. For example, the robot forward kinematics equations in the robot system allow the robot programmer to create the TCP positions by using the "online robot teaching method." To do it, the programmer uses the robot teach pendent to jog the robot joints for a particular TCP position within the robot work volume. The incremental encoders on the joints measure joint positions when the robot joints are moving, and the robot forward kinematics equations calculate the corresponding coordinates of ${}^{R}T_{Def\_TCP}$ in Eq. (1) based on the encoder measurements. After the programmer records a TCP position with the teach pendant, the robot system saves both the coordinates of ${}^{R}T_{Def\_TCP}$ and the corresponding measured joint positions.

Besides the robot link frames, the robot programmer can also define different robot user frame UF relative to robot base frame R and different robot user tool TCP frame UT_TCP relative to the default TCP frame. They are denoted as transformations ${}^{R}T_{UF}$ and ${}^{Def\_TCP}T_{UT\_TCP}$ as shown in Fig. 1, respectively. After a user frame and a user tool frame are established and set active in the robot system, Eq. (4) determines the coordinates of the actual robot TCP position, ${}^{UF}T_{UT\_TCP}$, as shown in Fig. 1:

$$ {}^{UF}T_{UT\_TCP} = ({}^{R}T_{UF})^{-1} \times {}^{R}T_{Def\_TCP} \times {}^{Def\_TCP}T_{UT\_TCP} . \quad (4) $$

3. Calibration of Robot Reference Frames

Eq. (4) shows that an industrial robot is an open-loop, multiple degrees-of-freedom (DOF) mechanism modeled by a series of frame transformations. If any frame transformation in Eq. (4) is different from the corresponding one used by the actual robot system, the actual robot TCP positions are to be different from those calculated by the robot kinematics equations in the robot system, which results in robot TCP position errors. This section introduces robot calibration methods and techniques for enhancing robot positioning accuracy.

3.1 Calibration of Robot Link Frames

The nominal values of the D-H parameters used in Eq. (3) are often different from the corresponding ones of the actual robot system, causing robot TCP position errors of an existing or new industrial robot from about 5 mm to 15 mm (Motta, et al., 2001). The variations of robot D-H parameters are the result of both the geometric effects such as link tolerance and link twist and the non-geometric effects including joint compliance, link flexibility, gravity loading, joint backlash, gear train harmonics, controller steady state error, and temperature variation (Roth et al., 1987 and Greenway, 2000). For a well-built robot operated within its load and speed limitations, the geometric effects represent approximately 90 percent of the overall robot TCP position errors (Bernhardt, 1997 and Schroer, 1993). In this sense, every industrial robot has its own "true geometric parameters" and the calculated robot TCP positions are not accurate if the robot kinematic model does not use the true robot parameters.

One important concern about the robot TCP position error is the variations of the "zero" position values of the robot joint variables. As mentioned in Section 2, most industrial robot systems use incremental encoders to measure robot joint positions. Because an incremental encoder is not able to measure the absolute joint position without a defined "zero" joint position value, the robot programmer must calibrate the "zero" values of the robot encoders by performing the robot "mastering" procedure. During the "mastering" process, the programmer manually jogs each robot joint to a position within the joint travel limit and then set the corresponding joint encoder value to "zero" in the robot system. As long as the defined "zero" values of the joints and the corresponding joint encoders remain unchanged, the recorded robot TCP positions are accurate via the actual encoder measurements. But the programmer must calibrate the robot system each time when any of the actual encoders looses the pre-defined "zero" value in the robot system. This may occur due to either replacing the broken joint motor and encoder, or accidentally losing the battery support to all actual robot encoders after the robot power-off. However, performing the robot "mastering" procedure usually results in different "zero" joint positions for the joint encoders. Consequently, all pre-recorded robot TCP positions are no longer accurate for the originally programmed robot operations simply because for a given robot joint position (e.g., 30 degrees) the actual position of the joint output link can be completely different depending on the actual defined "zero" joint position in the system. To reuse the pr-recorded robot TCP positions in the robot operations after losing the robot "zero" joint positions, the robot programmer needs to precisely recover these values by identifying the true robot parameters.

Researchers have investigated the problems and methods of calibrating the robot parameters for many years. Most works have been focused on the kinematic model-based calibration that is considered a global calibration method of enhancing the robot positioning accuracy across the robot work volume. Because of the physical explicit significance of the D-H model and its widespread use in the industrial robot control software, identification methods have been developed to directly obtain the true robot D-H parameters and the additional twist angle "β" in the case of consecutive parallel joint axes (Stone, 1987, Abderrahim & Whittaker, 2000). To better understand the process of robot parameter identification, assume that $p = [p_1^T \ldots p_n^T]^T$ is the parameter vector for an n-joint robot, and p_{i+1} is the link parameter vector for joint i+1. Then, the actual link frame transformation ${}^iA_{i+1}$ in Eq. (2) can be expressed as (Driels & Pathre, 1990):

$$ {}^iA_{i+1} = {}^iT_{i+1} + \Delta T_{i+1} \qquad \text{and} \qquad \Delta T_{i+1} = \Delta T_{i+1}(\Delta p_{i+1}), \tag{5} $$

where Δp_{i+1} is the link parameter error vector for joint i+1. Thus, the exact actual robot chain transformation in Eq. (3) is:

$$ {}^0\hat{A}_n = {}^0\hat{T}_n + \Delta\hat{T} \qquad \text{and} \qquad \Delta\hat{T} = \Delta\hat{T}(q, \Delta p), \tag{6} $$

where $\Delta p = [\Delta p_1^T\ \Delta p_2^T \ldots \Delta p_n^T]^T$ is the robot parameter error vector and $q = [q_1, q_2, \ldots q_n]^T$ is the vector of robot joint position variables. Studies show that $\Delta\hat{T}$ in Eq. (6) is a non-linear function of robot parameter error vector Δp. For identifying the true robot parameters, an external measuring system must be used to measure a number of strategically planned robot TCP positions. Then, an identification algorithm computes parameter values $p^* = p + \Delta p$ that result in an optimal fit between the actually measured TCP positions and those computed by the robot kinematic model in the robot control system.

Practically, the robot calibration process consists of four steps. The first step is to teach and record a sufficient number of robot TCP positions within the robot work volume so that the individual joint is able to move as much as possible for exciting the joint parameters. The second step is to "physically" measure the recorded robot TCP positions with an appropriate external measurement system. The measuring methods and techniques used by the measuring systems include laser interferometry, stereo vision, and mechanical "string pull" devices (Greenway, 2000). The third step is to determine the relevant actual robot parameters through a specific mathematical solution such as the standard non-linear least squares optimization with the Levenberg-Marquardt algorithm (Dennis & Schnabel, 1983). The final step is to compensate the existing robot kinematic model in the robot control system with the identified robot D-H parameters. However, due to the difficulties in modifying the kinematic parameters in the actual robot controller directly, compensations are made to the corresponding joint positions of all pre-recorded robot TCP positions by solving the robot inverse kinematics equations with the identified D-H parameters.

Among the available commercial robot calibration systems, the DynaCal Robot Cell Calibration System developed by Dynalog Inc. is able to identify and use the true robot D-H parameters including the additional twist angle "β" in the case of consecutive parallel axes to compensate the robot TCP positions used in any industrial robot programs. As shown in Fig. 3, the DynaCal measurement device defines its own measurement frame through a precise base adaptor mounted at an alignment point. The device uses a high resolution, low inertia optical encoder to constantly measure the extension of the cable that is connected to the robot TCP frame through a DynaCal TCP adaptor, and sends the encoder measurements to the Window-based DynaCal software for determining the relative position between the robot TCP frame and the measurement device frame. The DynaCal software allows the programmer to calibrate the robot parameters, as well as the positions of the end-effector and fixture in the robot workcell, and perform the compensation to the robot TCP positions used in the industrial robot programs. The software also supports other measurement devices including Leica Smart Laser Interferometer system, Metronor Inspection system, Optotrack camera, and Perceptron camera, etc. Prior to the DynaCal robot calibration, the robot programmer needs to conduct the calibration experiment in which a developed robot calibration program moves the robot TCP frame to a set of pre-taught robot calibration points. Depending on the required accuracy, at least 30 calibration points are required. It is also important to select robot calibration points that are able to move each robot joint as much as possible in order to "excite" its calibration parameters. This is achieved by not only moving the robot TCP frame along the x-, y-, and z-axes of robot base frame R but also changing the robot orientation around the robot TCP frame, as well as robot configurations (e.g., Flip vs. No Flip). During the calibration experiment, the DynaCal measurement device measures the corresponding robot TCP positions and sends measurements to the DynaCal software. The programmer also needs to specify the following items required by the DynaCal software for conducting the robot calibration:

- Select the robot calibration program (in ASCII format) used in the robot calibration experiment from the Robot File field.
- Select the measurement file (.MSR) from the Measurement File field that contains the corresponding robot TCP positions measured by the DynaCal system.
- Select the robot parameter file (.PRM) from the Parameter File field that contains the relevant kinematic, geometric, and payload information for the robot to be calibrated. When a robot is to be calibrated for the first time, the only selection will be the robot default nominal parameter file provided by the robot manufacturer. After that, a previous actual robot parameter file can be selected.
- Specify the user desired standard deviation for the calibration in the Acceptance Angular Standard Deviation field. During the identification process, the DynaCal software will compare the identified standard deviation of the joint offset values to this value. By default, the DynaCal system automatically sets this value according to the needed accuracy of the robot application.
- Select the various D-H based robot parameters to be calibrated from the Robot Calibration Parameters field. By default, the DynaCal software highlights a set of parameters according to the robot model and the robot application type.
- Select the calibration mode from the Nominal/Previous Calibration field. In the "Normal" mode, the DynaCal software sets the robot calibration parameters to the nominal parameters provided by the default parameter file (.PRM) from the robot manufacturer. This is the normal setting for performing a standard robot calibration, for example, when identifying the default set of the robot parameters. By selecting the

"Previous" mode instead, the robot calibration parameters are set to their previously calibrated values stored in the selected parameter file (.PRM). This is convenient, for example, when re-calibrating a sub-set of the robot parameters (e.g. a specific joint offset) while leaving all other parameters to their previously calibrated values. This re-calibration process would typically allow the use of less robot calibration points than that used for a standard robot calibration.

After the robot calibration, a new robot parameter file appears in the Parameter File field that contains the identified true robot parameters. The DynaCal software uses the "now-identified" actual robot parameters to modify the corresponding joint positions of all robot TCP positions used in any robot program during the DynaCal compensation process.

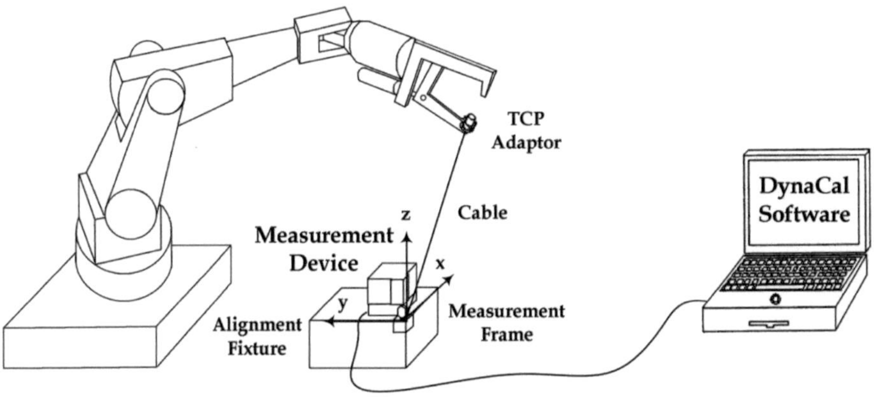

Figure 3. The DynaCal robot cell calibration system

3.2 Calibration of a Robot User Tool Frame

A robot user tool frame (UT) plays an important role in robot programming as it not only defines the actual tool-tip position of the robot end-effector but also addresses its variations during robot operations. The damage or replacement of the robot end-effector would certainly change the pre-defined tool-tip position used in recording the robot TCP positions. Tool-tip position variations also occur when the robot wrist changes its orientation by rotating about the x, y, and z axes of the default TCP frame at a robot TCP location. This type of tool-tip position variation can be a serious problem if the robot programs use the robot TCP positions generated by an external system such as the robot simulation software or vision system with any given TCP frame orientations.

To eliminate or minimize the TCP position errors caused by the tool-tip position variations mentioned above, it is always necessary for the robot programmer to establish and calibrate a UT frame at the desired tool-tip reference point of the robot end-effector used in industrial robot applications. The task can be done by using either the robot system or an external robot calibration system such as the DynaCal system. In the case of using the robot system, the programmer teaches six robot TCP positions relative to both the desired tool-tip reference point of the end-effector and the reference point on any tool-reachable surface as shown in Fig. 4. The programmer records the first position when the desired tool-tip reference point touches the surface reference point. To record the second position, the programmer rotates the robot end-effector about the x-axis of the default TCP frame for at least 180 degrees, and then moves the tool-tip reference point back to the surface reference point. Similarly, the programmer teaches the third position by rotating the robot end-effector about the y-axis of the default TCP frame. The other three positions are recorded after the programmer moves each axis of the default TCP frame along the corresponding axis of robot base frame R for at least one-foot distance relative to the surface reference point, respectively. The robot system uses these six recorded robot TCP positions to calculate the actual UT frame position at the tool-tip reference point and saves its value in the robot system. The UT calibration procedure also allows the established UT frame at the tool-tip position to stay at any given TCP location regardless of the orientation change of robot end-effector as shown in Fig. 4. Technically, the programmer needs to re-calibrate the UT frame each time the robot end-effector is changed, for example, after an unexpected collision. Some robot manufacturers have developed the "online robot UT frame adjustment" option for their robot systems in order to reduce the robot production downtime. The robot programmer can also simultaneously obtain an accurate UT frame position at the desired tool-tip point on the robot end-effector during the DynaCal robot calibration. The procedure is called the "DynaCal end-effector calibration." To achieve it, the programmer needs to specify at least three non-collinear measurement points on the robot end-effector and input their locations relative to the desired tool-tip point in the DynaCal system.

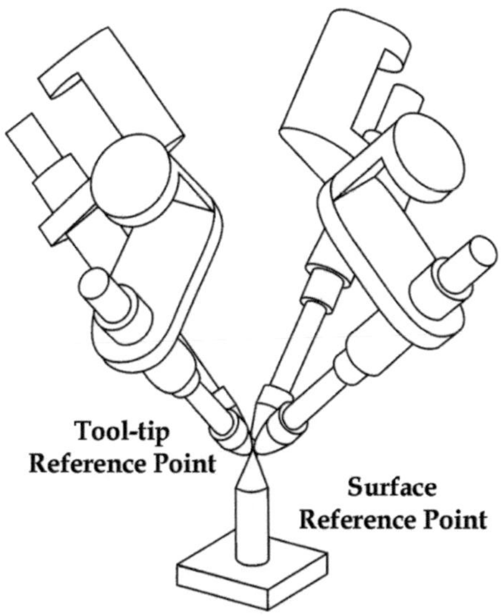

Figure 4. Calibration of a robot user tool frame with the robot

During the DynaCal robot calibration, the DynaCal TCP adaptor should be mounted at each of the three measurement points in between the measurements of the robot calibration TCP points. For example, with a total of 30 robot calibration points, the first 10 calibration points are measured with the TCP adaptor at the first measurement point, the next 10 calibration points are measured with the TCP adaptor at the second measurement point, etc. However, when only UT frame location (i.e. x, y, z,) needs to be calibrated, one measurement point on the end-effector suffices and choosing the measurement point at the desired tool-tip point further simplifies the process because its location relative to the desired tool-tip point is then simply zero. After the calibration, the DynaCal system uses the values of the measurement points to calculate the actual UT frame position at the desired tool-tip reference point and saves the value in the parameter file (.PRM).

3.3 Calibration of Robot User Frame

Like robot base frame R, an active robot user frame, UF, in the robot system defines the robot TCP positions as shown in Eq (4). However, unlike the robot base frame that is inside the robot body, robot user frame UF can be established relative to robot base frame R at any position within the robot work volume. The main application of robot user frame UF is to identify the position changes of robot TCP frame and robot base frame in a robot workcell.

Fig. 5 shows a workpiece held by a rigid mechanism called "fixture" in a robot workcell. Often, the programmer establishes a robot user frame, UF, on the fixture and uses it in recording the required robot TCP positions for the workpiece (e.g., ${}^{UF}T_{TCP1}$ in Fig. 5). To do so, the programmer uses the robot and a special cylindrical-type pointer to record at least three non-collinear TCP positions in robot base frame R as shown in Fig. 6. These three reference points define the origin, x-axis, and x-y plane of the user frame, respectively. As long as the workpiece and its position remain the same, the developed robot program can accurately move the robot TCP frame to any of the pre-taught robot TCP positions (e.g., ${}^{UF}T_{TCP1}$). However, if the workpiece is in a different position in the workcell, the existing robot program is not able to perform the programmed robot operations to the workpiece because there are no corresponding recorded robot TCP positions (e.g., ${}^{UF}T_{TCP1'}$ in Fig. 5) for the robot program to use. Instead of having to manually teach each of the new required robot TCP positions (e.g., ${}^{UF}T_{TCP1'}$, etc.), the programmer may identify the position change or "offset" of the user frame on the fixture denoted as transformation ${}^{UF}T_{UF}$ in Fig. 5. With identified transformation ${}^{UF}T_{UF'}$, Eq. (7) and Eq. (8) are able to change all existing robot TCP positions (e.g., ${}^{UF}T_{TCP1}$) into their corresponding ones (e.g., ${}^{UF}T_{TCP1'}$) for the workpiece in the new position:

$$ {}^{UF}T_{TCP1'} = {}^{UF}T_{UF'} \times {}^{UF'}T_{TCP1'}, \tag{7} $$

$$ {}^{UF'}T_{TCP1'} = {}^{UF}T_{TCP1}. \tag{8} $$

Fig. 7 shows another application in which robot user frame UF serves as a common calibration fixture frame for measuring the relative position between two robot base frames R and R'. The programmer can establish and calibrate the position of the fixture frame by using the DynaCal system rather than the robot system as mentioned earlier. Prior to that, the programmer has to perform the DynaCal end-effector calibration as introduced in Section 3.2. During the DynaCal fixture calibration, the programmer needs to mount the DynaCal measurement device at three (or four) non-collinear alignment points on the fixture as shown in Fig. 7. The DynaCal measurement device measures each alignment point relative to robot base frame R (or R') through the DynaCal cable and TCP adaptor connected to the end-effector. The DynaCal software uses the measurements to determine the transformation between fixture frame Fix (or Fix') and robot base frame R (or R'), denoted as ${}^{R}T_{Fix}$ (or ${}^{R'}T_{Fix}$) in Fig. 7. After the fixture calibration, the DynaCal system calculates the relative position between two robot base frames R and R', denoted as ${}^{R}T_{R'}$, in Fig. 7, and saves the result in the alignment file (.ALN).

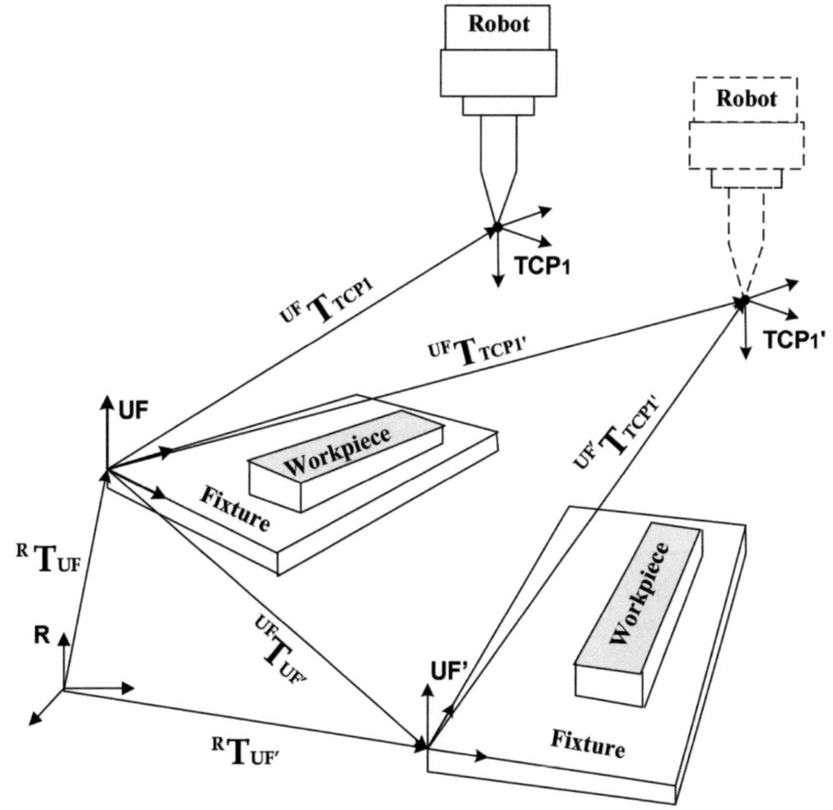

Figure 5. Application of a robot user frame

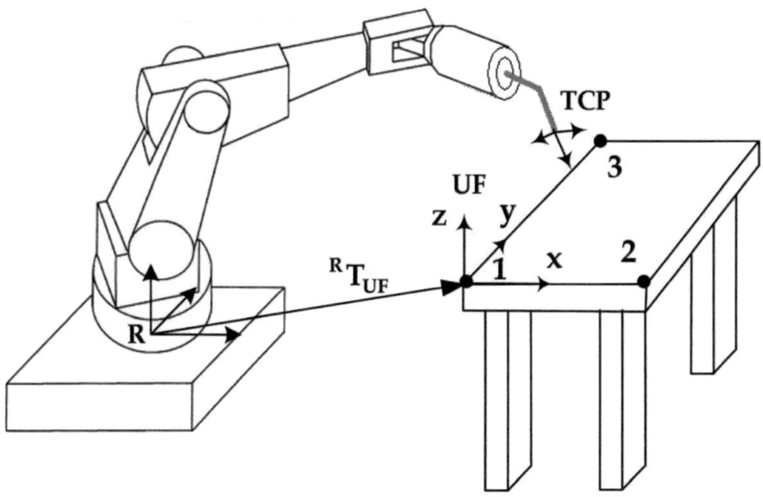

Figure 6. Calibrating a robot user frame with the robot

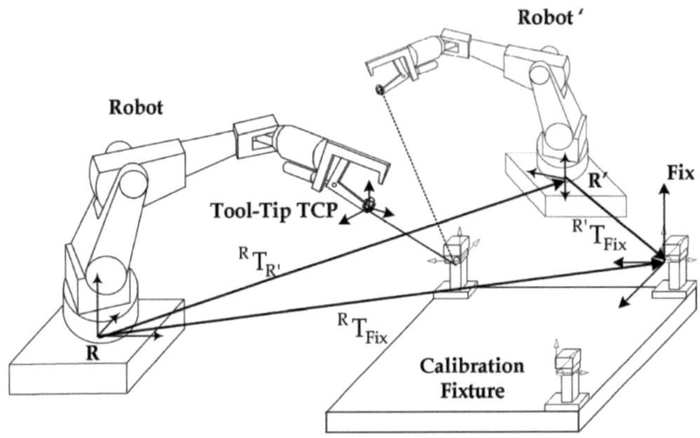

Figure 7. Calibration of a fixture frame with the DynaCal system

4. Calibration of Existing Robot TCP Positions for Programming Identical Robot Workcells

Creating the accurate robot TCP positions for industrial robot programs is an important task in robot applications. It can be very time-consuming depending on the type and complexity of the robot operations. For example, it takes about 400 hours to teach the required robot points for a typical wielding line with 30 robots and 40 welding spots per robot (Bernhardt, 1997). The programmer often expects to quickly and reliably transfer or download robot production programs to the corresponding robots in the "identical" robot workcells for execution. These applications called the "clone" or "move" of robot production programs can greatly reduce robot program developing time and robot downtime. However, because of the actual manufacturing tolerances and relative positioning variations of the robots, end-effectors, and fixtures in "identical" robot cells, it is not feasible to copy or transfer existing robot TCP positions and programs to "identical" robot cells directly.

To conduct the "clone" or "move" applications successfully, the robot programmer has to identify the dimension differences between two "identical" robot cells by applying the methods and techniques of robot cell calibration as introduced in Section 3. Fig. 8 shows two identical robot workcells where the two robots, Robot and Robot', and their corresponding peripheral components are very "close" in terms of their corresponding frame transformations. The application starts with calibrating the corresponding robots and robot UT frames in the robot cells as discussed in Sections 3.1 and 3.2. In the case of using the DynaCal system, a "clone" (or "move") application consists of a Master Project and the corresponding Clone (or Move) Project. Each project contains specific files for the calibration. For calibrating the robot and its UT frame position in the original robot cell, the programmer needs to select three files from the Master Project. They are the robot calibration program(s) in the Robot File field, the corresponding DynaCal measurement file(s) in the Measurement File field, and the default/previous robot parameter file in the Parameter File (.PRM) field. For calibrating the counterparts in the "identical" robot cell, the DynaCal system automatically copies the corresponding robot calibration program(s) and parameter file(s) from the Master Project to the Clone (or Move) Project since the corresponding robots and robot end-effectors in both robot cells are very "close." The programmer may also use the same robot calibration program(s) to generate the corresponding DynaCal measurement file(s) in the Clone (or Move) Project. After the calibration, a new parameter file appears in the Clone (or Move) Project that contains the identified parameters of the robot and robot UT frame position in the "identical" robot cell. During the compensation process, the DynaCal system utilizes the "now-identified" parameters to compensate the joint positions of the TCP positions used by the actual robot (e.g., Robot′) existed in the "identical" robot cell. As a result, actual frame transformations ${}^{R}T_{UT\_TCP}$ and ${}^{R'}T_{UT\_TCP}$ of the two robots as shown in Fig. 8 are exactly the same for any of the TCP positions used by the actual robot (e.g., Robot) in the original robot cell. The programmer also identifies the dimension differences of the two robot base frames relative to the corresponding robot workpieces in the two robot cells. To this end, the programmer first calibrates a common fixture frame (i.e., a robot user frame) on the corresponding robot workpiece in each robot cell by using the method as introduced in Section 3.3. It is important that the relative position between the workpiece and the three (or four) measurement points on the fixture remains exactly the same in the two "identical" robot cells. This condition can be easily realized by embedding these required measurement points in the design of the workpiece and its fixture. In the case of conducting the DynaCal fixture calibration, the DynaCal system automatically selects a default alignment file (.ALN) for the Master Project, and copies it to the Clone (or Move) Project since they are very "close." During the fixture calibration, the DynaCal system identifies the positions of fixture frames Fix and Fix′ relative to robot base frames R and R′, respectively. They are denoted as ${}^{R}T_{Fix}$ and ${}^{R'}T_{Fix'}$ in Fig. 8.

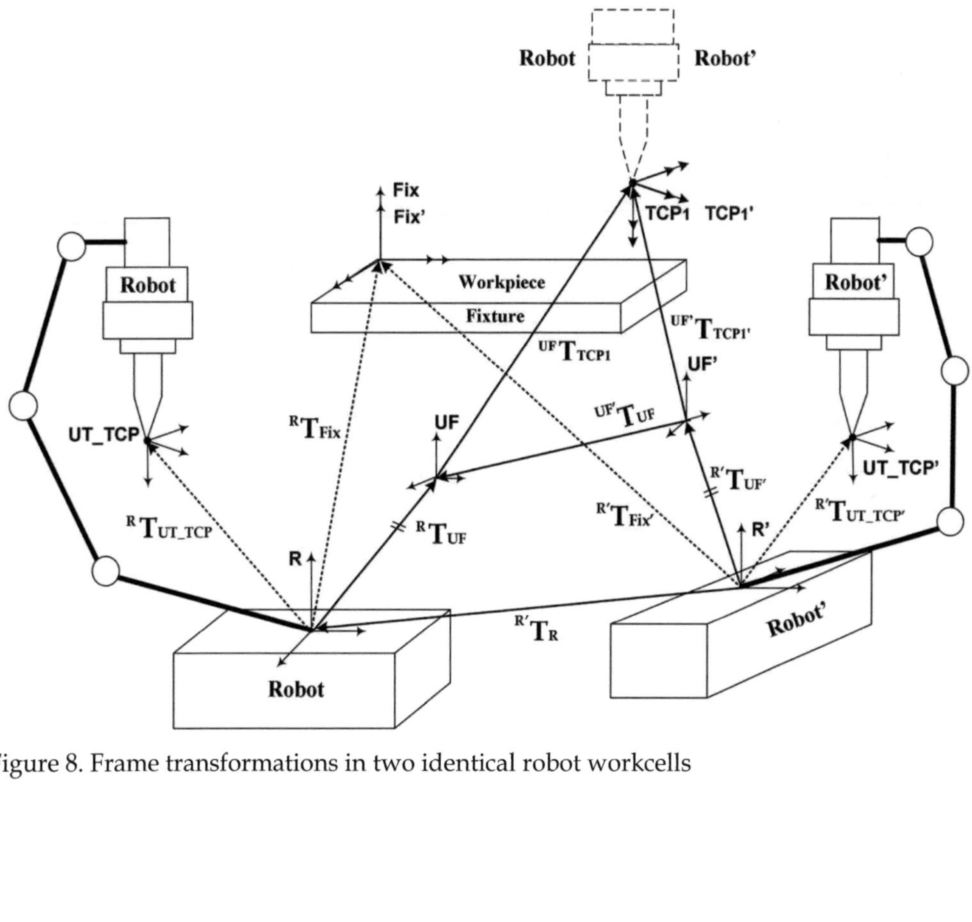

Figure 8. Frame transformations in two identical robot workcells

After identifying the dimension differences of the two "identical" robot workcells, the programmer can start the compensation process by changing the robot TCP positions used in the original robot cell into the corresponding ones for the "identical" robot cell. The following frame transformation equations show the method. Given that the coincidence of fixture frames Fix and Fix′ as shown in Fig. 8 represents the common calibration fixture used in the two "identical" robot cells, then, Eq. (9) calculates the transformation between two robot base frames R′ and R as:

$$^{R'}T_R = {}^{R'}T_{Fix'} \times ({}^{R}T_{Fix})^{-1}. \tag{9}$$

In the case of fixture calibration in the "identical" robot cell, the DynaCal system calculates $^{R'}T_R$ in Eq. (9) and saves the value in a new alignment file in the Clone (or Move) Project.
It is also possible to make transformation $^{R'}T_{UF'}$ equal to transformation $^{R}T_{UF}$ as shown in Eq. (10):

$$^{R'}T_{UF'} = {}^{R}T_{UF}, \tag{10}$$

where user frames UF and UF′ are used in recording the robot TCP positions (e.g., TCP1 and TCP1′ in Fig. 8) in the two "identical" robot cells, respectively. With the results from Eq. (9) and Eq. (10), Eq. (11) calculates the relative position between two user frames UF′ and UF as:

$$^{UF'}T_{UF} = ({}^{R'}T_{UF'})^{-1} \times {}^{R'}T_R \times {}^{R}T_{UF} \tag{11}$$

After $^{UF'}T_{UF}$ is identified, Eq. (12) is able to change any existing robot TCP position (e.g., $^{UF}T_{TCP1}$) for the robot workpiece in the original robot cell into the new one (e.g., $^{UF'}T_{TCP1'}$) for the corresponding robot workpiece in the "identical" robot cell:

$$^{UF'}T_{TCP1'} = {}^{UF'}T_{UF} \times {}^{UF}T_{TCP1}. \tag{12}$$

In the DynaCal system, a successful "clone' (or "move") application generates a destination robot program file (in ASCII format) in the Robot File field of the Clone (or Move) Project that contains all modified robot TCP positions. The programmer then converts the file into an executable robot program with the file translation service provided by the robot manufacturer, and executes it in the actual robot controller with the expected accuracy of the robot TCP positions. For example, in the "clone" application conducted for two "identical" FANUC robot cells (Cheng, 2007), the DynaCal system successfully generated the destination FANUC Teach Pendant (TP) program after the robot cell calibration and compensation. The programmer used the FANUC PC File Service software to convert the ASCII robot file into the executable FANUC TP program and download it to the actual FANUC M6i robot in the "identical" cell for execution. As a result, the transferred robot program allowed the FANUC robot to complete the same operations to the workpiece in the "identical" robot cell within the robot TCP position accuracy of 0.5 mm.

5. Calibration of Simulated Robot TCP Positions for Robot Offline Programming

Offline robot programming is, by definition, the technique of generating reliable, efficient, and executable robot TCP positions and programs without using a real robot. One approach is to use the commercial robot simulation software such as IGRIP developed by Delmia

Corporation (Cheng, 2003). This new method represents several advantages over the conventional "online robot programming method" in terms of improved robot workcell design and reduced robot downtime. However, there are inevitable differences between the simulated robot workcell and the corresponding actual one due to the manufacturing tolerance and the poisoning variations of the robot workcell components. Among them, the positioning differences of robot "zero" joint positions, robot base frames and UT frames cause 80 percent of the inaccuracy in a typical robot simulation workcell. Therefore, it is not feasible to download the robot TCP positions and programs created in the robot simulation workcell to the actual robot controller for execution directly.

Calibrating the robotic and non-robotic device models in the simulation workcell makes it possible to conduct the true simulation-based offline robot programming. The method uses the calibration functions provided by the robot simulation software (e.g., IGRIP) and a number of actually measured robot TCP positions in the real robot cell to modify the nominal values of the robot simulation workcell including robot parameters, and positions of robot base frame, UT frame, and fixture frame. The successful application of the technology allows robot TCP positions and programs to be created in the nominal simulation workcell, verified in the calibrated simulation workcell, and finally downloaded to the actual robot controllers for execution.

5.1 Calibrating Robot Device Model in Robot Simulation Workcell

The purpose of calibrating a robot device model in the simulation workcell is to infer the best fit of the robot kinematic parameters and UT frame position based on a number of measured robot TCP positions. The IGRIP simulation software uses non-linear least squares model fitting with the Levenberg-Marquardt method to obtain the parameter identification solution.

The calibration procedure consists of three steps. First, in the simulation workcell the programmer selects both the robot device model to be calibrated and the device model that represents the actual external coordinate measurement system (CMS) such as the DynaCal measurement device in Fig. 9. The position of the CMS device model in the simulation workcell represents a starting guess of the true CMS position in the actual robot workcell. Second, the programmer develops a robot calibration program that moves the actual default robot TCP frame to a number of pre-taught robot calibration TCP positions. The actual external CMS such as the DynaCal measuring device also measures each of these robot TCP positions. Finally, the programmer uploads the two sets of robot calibration TCP positions into the simulation workcell and saves them as Tag Points in two Tag Paths. The first Tag Path contains the coordinates of the default TCP frame positions at the corresponding robot calibration TCP positions used by the robot calibration program. They are uploaded from the actual robot controller and attached to the base of the robot device model in the simulation workcell. The second Tag Path contains the coordinates of the corresponding robot calibration TCP positions measured by the external CMS. They are uploaded from the actual CMS and attached to the corresponding CMS device model in the simulation workcell.

In IGRIP software, the robot calibration algorithm works by minimizing the mean square position error at the Tag Points of the measured Paths. After the robot calibration, the simulation software adjusts the nominal parameters of the robot device model (i.e. "zero" joint positions and UT frame position) in the simulation workcell for the best fit of the corresponding real identified parameters.

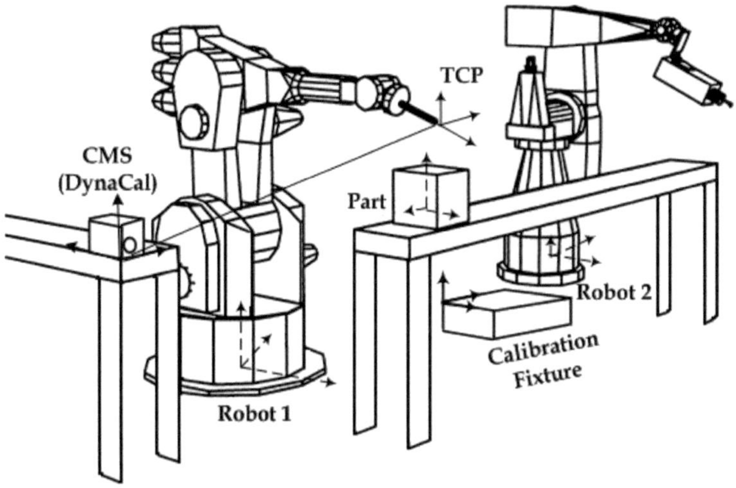

Figure 9. A robot simulation workcell

5.2 Calibrating the Position of Robot Device Model in Simulation Workcell

The robot position calibration is to infer the best fit of the robot base frame position relative to a calibration fixture frame in the workcell as shown in Fig. 9. The base frame of the calibration fixture device model represents a pristine position in the simulation workcell and the robot device model is adjusted with respect to it after the calibration. This calibration function is very important in that a real robot workcell may have more than one robot and the relative position between two different robots needs to be precisely identified. In addition, as long as the calibration fixture remains at no change, any one robot may be moved and recalibrated without having to calibrate the rest of the robots in the workcell. The calibration procedure requires the programmer to develop a real robot program that moves the robot TCP frame to a number of robot TCP positions recorded in robot base frame R. These robot TCP positions are uploaded from the real robot controller into the simulation workcell and saved as Tag points in a Tag Path attached to the base frame of the robot device model. The programmer also calibrates a robot user frame (i.e. fixture frame) on the real calibration fixture by using either the robot system or an external measuring system as introduced in Section 3.3. In the case of using the robot system, the programmer creates another robot program that moves the robot tool-tip TCP frame to the same robot TCP positions recorded in the established user frame (i.e. fixture frame). These robot TCP positions are then uploaded into the simulation workcell and saved as Tag Points in another Tag Path attached to the calibration fixture device model. In the case of using an external CMS such as the DynaCal measurement device, the corresponding measured robot TCP positions in the second Tag Path must be expressed in the calibration fixture frame, which can be determined by the transformation between the calibration fixture frame and the CMS base frame.

In IGRIP software, the calibration algorithm works by minimizing the mean square position and/or orientation errors at the Tag Points of the measured Paths. After the calibration, the simulation software adjusts the base frame position of the robot device model relative to the base frame of the calibration fixture device model, which results in the best fit of the robot base frame position.

5.3 Calibrating Non-Robotic Device Model in Simulation Workcell

This type of calibration is usually the last step in the overall simulation workcell calibration. It intends to adjust the position of a non-kinematic model such as a workpiece or a table relative to robot base frame R and UT frame in the robot simulation workcell in order to match the counterpart in the real robot workcell. The method uses measurements from the actual robot workcell. After the calibration, the programs developed in the robot simulation workcell will contain the corrected robot TCP positions that can be directly downloaded to the actual robot workcell.

The calibration procedure requires the programmer to create three non-collinear feature Tag points on the workpiece device model in the robot simulation workcell and save them in the first Tag Path attached to the workpiece device model. Then, the programmer develops a real robot program that moves the tool-tip TCP frame to the corresponding three points on the real workpiece. These robot TCP positions are then uploaded from the actual robot controller and saved as Tag Points in the second Tag Path attached to the workpiece device model. During the calibration, the simulation software moves the workpiece device model in simulation workcell by matching up the first set of three Tag Points with the second set. Other methods such as the six-point method and the Least Squares method allow the programmer to use more robot points on the workpiece for conducting the same calibration.

6. Conclusion

This chapter discussed the importance and methods of conducting robot workcell calibration for enhancing the accuracy of the robot TCP positions in industrial robot applications. It shows that the robot frame transformations define the robot geometric parameters such as joint position variables, link dimensions, and joint offsets in an industrial robot system. The D-H representation allows the robot designer to model the robot motion geometry with the four standard D-H parameters. The robot kinematics equations use the robot geometric parameters to calculate the robot TCP positions for industrial robot programs. The discussion also shows that because of the geometric effects, the true geometric parameters of an industrial robot or a robot workcell are often different from the corresponding ones used by the robot kinematic model, or "identical" robots and robot workcells, which results in robot TCP position errors.

Throughout the chapter, it has been emphasized that the model-based robot calibration methods are able to minimize the robot TCP position errors through identifying the true geometric parameters of a robot and robot workcell based on the measurements of strategically planned robot TCP positions and the mathematical solutions of non-linear least squares optimization. The discussions have provided readers with the methods of calibrating the robot, end-effector, and fixture in the robot workcell. The integrated calibration solution shows that the robot programmer is able to use the calibration functions provided by the robot systems and commercial products such as the DynaCal Robot Cell Calibration System and IGRIP robotic simulation software to conduct true offline robot programming for "identical" and simulated robot cells with enhanced accuracy of robot TCP positions. The successful applications of the robot cell calibration techniques bring great benefits to industrial robot applications in terms of reduced robot program developing time and robot production downtime.

7. References

Abderrahim, M. & Whittaker, A. R. (2000). Kinematic Model Identification of Industrial Manipulators, *Robotics and Computer-Integrated Manufacturing,* Vol. 16, No. 1, February 2000, pp 1-8.

Bernhardt, R. (1997). Approaches for Commissioning Time Reduction, *Industrial Robot,* Vol. 24, No. 1, pp. 62-71.

Cheng, S. F. (2007). The Method of Recovering TCP Positions in Industrial Robot Production Programs, *Proceedings of 2007 IEEE International Conference on Mechatronics and Automation,* August 2007, pp. 805-810.

Cheng, S. F. (2003). The Simulation Approach for Designing Robotic Workcells, *Journal of Engineering Technology,* Vol. 20, No. 2, Fall 2003, pp. 42-48.

Denavit, J. & Hartenberg, R. S. (1955). Kinematic Modelling for Robot Calibration, *Trans. ASME Journal of Applied Mechanics,* Vol. 22, June 1955, pp. 215-221.

Dennis, J. E. & Schnabel, R. B. (1983). *Numerical Methods for Unconstrained Optimization and Nonlinear Equations,* New Jersey: Prentice-Hall.

Driels, M. R. & Pathre, U. S. (1990). Significance of Observation Strategy on the Design of Robot Calibration Experiments, *Journal of Robotic Systems,* Vol. 7, No. 2, pp. 197-223.

Greenway, B. (2000). Robot Accuracy, *Industrial Robot,* Vol. 27, No. 4,2000, pp 257-265.

Mooring, B., Roth, Z. and Driels, M. (1991). *Fundamentals of manipulator calibration,* Wiley, New York.

Motta, J. M. S. T., Carvalho, G. C. and McMaster, R. S. (2001). Robot Calibration Using a 3-D Vision-Based Measurement System with a Single Camera, *Robotics and Computer-Integrated Manufacturing,* Ed. Elsevier Science, U.K., Vol. 17, No. 6, pp. 457-467.

Roth, Z. S., Mooring, B. W. and Ravani, B. (1987). An Overview of Robot Calibration, *IEEE Journal of Robotics and Automation,* RA-3, No. 3, pp. 377-85.

Schroer, K. (1993). Theory of Kinematic Modeling and Numerical Procedures for Robot Calibration, *Robot Calibration,* Chapman & Hall, London.

Stone H. W. (1987). *Kinematic modeling, identification, and control of robotic manipulators.* Kluwer Academic Publishers.

6

Control of Robotic Systems Undergoing a Non-Contact to Contact Transition

S. Bhasin, K. Dupree and W. E. Dixon
Department of Mechanical and Aerospace Engineering, University of Florida
USA

1. Introduction

The problem of controlling a robot during a non-contact to contact transition has been a historically challenging problem that is practically motivated by applications that require a robotic system to interact with its environment. The control challenge is due, in part, to the impact effects that result in possible high stress, rapid dissipation of energy, and fast acceleration and deceleration (Tornambe, 1999). Over the last two decades, results have focused on control designs that can be applied during the non-contact to contact transition. One main theme in these results is the desire to prescribe, reduce, or control the interaction forces during or after the robot impact with the environment, because large interaction forces can damage both the robot and/or the environment or lead to degraded performance or instabilities.

In the past, researchers have used different techniques to design controllers for the contact transition task. (Hogan, 1985) treated the environment as mechanical impedance and used force feedback in an impedance control technique to stabilize the robot undergoing contact transition. (Yousef-Toumi & Gutz, 1989) used integral force compensation with velocity feedback to improve transient impact response. In recent years, two main approaches have been exploited to accommodate for the non-contact to contact transition. The first approach is to exploit kinematic redundancy of the manipulator to reduce the impact force (Walker, 1990; Gertz et al., 1991; Walker, 1994). The second mainstream approach is to exploit a discontinuous controller that switches based on the different phases of the dynamics (i.e., non-contact, robot impact transition, and in-contact coupled manipulator and environment) as in (Chiu et al., 1995; Pagilla & Yu, 2001; Lee et al., 2003). Typically, these discontinuous controllers consist of a velocity controller in the pre-contact phase that switches to a force controller during the contact phase. Motivation exists to explore alternative methods because kinematic redundancy is not always possible, and discontinuous controllers require infinite control frequency (i.e., exhibit chattering). Also, force measurements can be noisy and may lead to degraded performance. The focus of this chapter is control design and stability analysis for systems with dynamics that transition from non-contact to contact conditions. As a specific example, the development will focus on a two link planar robotic arm that transitions from free motion to contact with an unactuated mass-spring system. The objective is to control a robot from a non-contact initial condition to a desired (contact) position so that a stiff mass-spring system is regulated to a desired compressed state, while limiting the impact force. The mass-spring system models a stiff environment for the robot, and the robot/mass-spring system collision is modeled as a linear differentiable impact, with the impact force being proportional to the mass deformation. The presence of parametric uncertainty in the robot/mass-spring system and the impact model is accounted for by using a Lyapunov based adaptive backstepping technique. The feedback elements of the controller are contained inside of hyperbolic tangent functions as a means to limit the impact force (Liang et al., 2007). A two-stage stability analysis (contact and non-contact) is included to prove that the continuous controller achieves the control objective despite parametric uncertainty throughout the system, and without the use of acceleration or force measurements.

In addition to impact with stiff environments, applications such as robot interaction with human tissue in clinical procedures and rehabilitative tasks, cell manipulation, finger manipulation, etc. (Jezernik & Morari, 2002; Li et al., 1989; Okamura et al., 2000) provide practical motivation to study robotic interaction with viscoelastic materials. (Hunt & Crossley, 1975) proposed a compliant contact model, which not only included both the stiffness and damping terms, but also eliminated the discontinuous impact forces at initial contact and separation, thus making it more suitable for robotic contact with soft environments. The model has found acceptance in the scientific community (Gilardi & Sharf, 2002; Marhefka & Orin; 1999; Lankarani & Nikravesh, 1990; Diolaiti et al., 2004), because it has been shown (Gilardi & Sharf, 2002) to better represent the physical nature of the energy transfer process at contact. In addition to the controller design using a linear spring contact model, this chapter also describes how the more general Hunt-Crossley model can be used to design a controller (Bhasin et al., 2008) for viscoelastic contact. A Neural Network (NN) feedforward term is inserted into the controller (Bhasin et al., 2008) to estimate the environment uncertainties which are not linear in the parameters. Similar to the control design in the first part of the chapter, the control objective is achieved despite uncertainties in the robot and the impact model, and without the use of acceleration or force measurements. However, the NN adaptive element enables adaptation for a broader class of uncertainty due to viscoelastic impact.

2. Control of Robotic Contact with a stiff environment

Robot applications in the industry like assembling, material handling, painting etc. have motivated the study of robot interaction with a stiff environment. We consider an academic problem of a two link robot undergoing a non-contact to contact transition with a stiff mass-spring system (Fig. 1). The contact dynamics of the stiff environment can be modeled using a simple linear spring, wherein the contact force is directly proportional to the local deformation at impact. Besides regulating the mass to a desired final position, another objective is to limit the impact force. The feedback elements for the controller are contained inside of hyperbolic tangent functions as a means to limit the impact forces resulting from large initial conditions as the robot transitions from a non-contact to a contact state. Although saturating the feedback error is an intuitive solution, several new technical challenges arise in the stability analysis. The main challenge is that the use of saturated feedback does not allow some coupling terms to be canceled in the stability analysis, resulting in the need to develop state dependent upper bounds that result in semi-global stability.

2.1 Dynamic Model

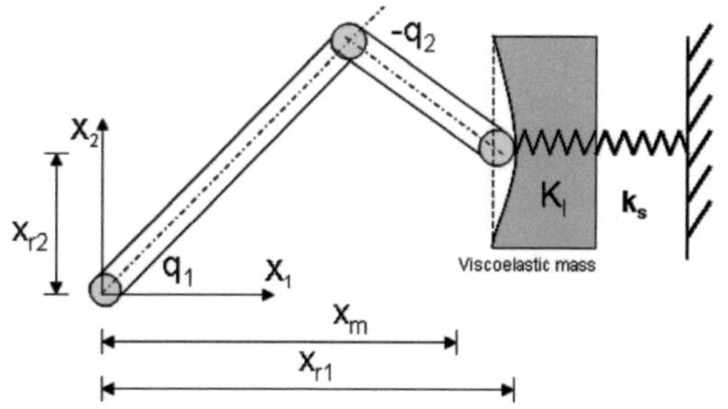

Figure 1 Robot contact with a Mass-Spring System (MSR), modeled as a linear spring

The subsequent development is motivated by the academic problem illustrated in Fig. 1. The dynamic model for the two-link planar robot depicted in Fig. 1 can be expressed in the joint-space as

$$M(q)\ddot{q} + C(q,\dot{q})\dot{q} + h(q) = \tau, \tag{1}$$

where $q(t)$, $\dot{q}(t)$, $\ddot{q}(t) \in \mathbb{R}^2$ represent the angular position, velocity, and acceleration of the robot links, respectively, $M(q) \in \mathbb{R}^{2\times 2}$ represents the uncertain inertia matrix, $C(q,\dot{q}) \in \mathbb{R}^{2\times 2}$ represents the uncertain Centripetal-Coriolis effects, $h(q) \triangleq [h_1(q),\ h_2(q)]^T \in \mathbb{R}^2$ represents uncertain conservative forces (e.g., gravity), and $\tau(t) \in \mathbb{R}^2$ represents the torque control inputs. The Euclidean position of the end-point of the second robot link is denoted by $x_r(t) \triangleq [x_{r1}(t),\ x_{r2}(t)]^T \in \mathbb{R}^2$, which can be related to the joint-space through the following kinematic relationship:

$$\dot{x}_r = J(q)\dot{q}, \tag{2}$$

where $J(q) \in \mathbb{R}^{2\times 2}$ denotes the manipulator Jacobian. The unforced dynamics of the mass-spring system in Fig. 1 are

$$m\ddot{x}_m + k_s(x_m - x_0) = 0, \tag{3}$$

where $x_m(t)$ and $\ddot{x}_m(t) \in \mathbb{R}$ represent the displacement and acceleration of the unknown mass $m \in \mathbb{R}$, $x_0 \in \mathbb{R}$ represents the initial undisturbed position of the mass, and $k_s \in \mathbb{R}$ represents the unknown stiffness of the spring.

After pre-multiplying the robot dynamics by the inverse of the Jacobian transpose and utilizing (2), the dynamics in (1) and (3) can be rewritten as (Dupree et al., 2006 a; Dupree et al., 2006 b)

$$\bar{M}(x_r)\ddot{x}_r + \bar{C}(x_r,\dot{x}_r)\dot{x}_r + \bar{h}(x_r) + \begin{bmatrix} F_m \\ 0 \end{bmatrix} = F, \tag{4}$$

$$m\ddot{x}_m + k_s(x_m - x_0) = F_m, \tag{5}$$

where $F(t) \triangleq J^{-T}(q)\tau(t) \in \mathbb{R}^2$ denotes the manipulator force. In (4) and (5), $F_m(x_r, x_m) \in \mathbb{R}$ denotes the impact force acting on the mass that occurs when $x_{r1}(t) \geq x_m(t)$ (Fig. 1) that is assumed to have the following form (Tornambe, 1999; Indri & Tornambe, 2004)

$$F_m = K_l \Lambda(x_{r1} - x_m), \tag{6}$$

where $K_l \in \mathbb{R}$ represents an unknown positive stiffness constant, and $\Lambda(x_r, x_m) \in \mathbb{R}$ is defined as

$$\Lambda = \begin{cases} 1 & x_{r1} \geq x_m \\ 0 & x_{r1} < x_m. \end{cases} \tag{7}$$

The dynamic model in (4) exhibits the following properties that will be utilized in the subsequent analysis.

Property 1: The inertia matrix $\bar{M}(x_r)$ is symmetric, positive definite, and can be lower and upper bounded as

$$a_1 \|\xi\|^2 \leq \xi^T \bar{M} \xi \leq a_2 \|\xi\|^2 \quad \forall \xi \in \mathbb{R}^2, \tag{8}$$

where a_1, $a_2 \in \mathbb{R}$ are positive constants.

Property 2: The following skew-symmetric relationship is satisfied

$$\xi^T (\frac{1}{2} \dot{\bar{M}}(x_r) - \bar{C}(x_r, \dot{x}_r)) \xi = 0 \quad \forall \xi \in \mathbb{R}^2. \tag{9}$$

Property 3: The robot dynamics given in (4) can be linearly parameterized as

$$Y(x_r, x_m, \dot{x}_r, \ddot{x}_r)\theta = \bar{M}(x_r)\ddot{x}_r + \bar{C}(x_r, \dot{x}_r)\dot{x}_r + \bar{h}(x_r) + \begin{bmatrix} F_m \\ 0 \end{bmatrix}, \tag{10}$$

where $\theta \in \mathbb{R}^p$ contains the constant unknown system parameters, and $Y(x_r, x_m, \dot{x}_r, \ddot{x}_r) \in \mathbb{R}^{2 \times p}$ denotes the known regression matrix .

Property 4: The following inequalities are valid for all $\xi = [\xi_1, ..., \xi_n]^T \in \mathbb{R}^n$ (Dixon et al., 1999)

$$\|\xi\| \geq \|Tanh(\xi)\|$$

$$\|\xi\| + 1 \geq \frac{\|\xi\|}{\tanh(\|\xi\|)} \tag{11}$$

$$\|\xi\|^2 \geq \sum_{i=1}^{n} \ln(\cosh(\xi_i)) \geq \ln(\cosh(\|\xi\|)) \tag{12}$$

$$\xi^T Tanh(\xi) \geq Tanh^T(\xi) Tanh(\xi) = \|Tanh(\xi)\|^2 \geq \tanh^2(\|\xi\|). \tag{13}$$

Remark 1: To aid the subsequent control design and analysis, we define the vector $Tanh(\cdot) \in \mathbb{R}^n$ as follows

$$Tanh(\delta) = [\tanh(\delta_1), ..., \tanh(\delta_n)]^T, \tag{14}$$

where $\delta = [\delta_1, ..., \delta_n]^T \in \mathbb{R}^n$.

The following assumptions will be utilized in the subsequent control development.

Assumption 1: We assume that $x_{r1}(t)$ and $x_m(t)$ can be bounded as

$$\zeta_{x_r} \leq x_{r1}(t) \quad x_m(t) \leq \zeta_{x_m}, \tag{15}$$

where $\zeta_{x_r} \in \mathbb{R}$ is a known constant that is determined by the minimum coordinate of the robot along the X_1 axis, and $\zeta_{x_m} \in \mathbb{R}$ is a known positive constant. The lower bound assumption for $x_{r1}(t)$ is based on the geometry of the robot, and the upper bound assumption for $x_m(t)$ is based on the physical fact that the mass is attached by the spring to some object, and the mass will not be able to move past that object.

Assumption 2: We assume that the mass of the mass-spring system can be upper and lower bounded as

$$\underline{m} \leq m \leq \overline{m}, \tag{16}$$

where $\underline{m}$, $\overline{m} \in \mathbb{R}$ denote known positive bounding constants. The unknown stiffness constants K_I and k_s are also assumed to be bounded as

$$\underline{K_I} \leq K_I \leq \overline{K_I}, \tag{17}$$

$$\underline{k_s} \leq k_s \leq \overline{k_s}, \tag{18}$$

where $\underline{K_I}$, $\overline{K_I}$, $\underline{k_s}$, $\overline{k_s} \in \mathbb{R}$ denote known positive bounding constants.

Assumption 3: The subsequent development is based on the assumption that $q(t), \dot{q}(t)$, $x_m(t)$, and $\dot{x}_m(t)$ are measurable, and that $x_r(t)$ and $\dot{x}_r(t)$ can be obtained from $q(t)$ and $\dot{q}(t)$.

Remark 2: During the subsequent control development, we assume that the minimum singular value of $J(q)$ is greater than a known, small positive constant $\delta > 0$, such that $\max\{\|J^{-1}(q)\|\}$ is known a priori, and hence, all kinematic singularities are always avoided.

2.2 Control Development

The subsequent control design is based on integrator backstepping methods. A desired trajectory is designed for the robot (i.e., a virtual control input) to ensure the robot impacts the mass and regulates it to a desired position. A force controller is developed to ensure that the robot tracks the desired trajectory despite the transition from free motion to an impact collision and despite parametric uncertainties throughout the MSR system.

2.2.1 Control Objective

The control objective is to regulate the states of the mass-spring system via a two-link planar robot that transitions from non-contact to contact with the mass-spring through an impact collision. An additional objective is to limit the impact force to prevent damage to the robot or the environment (i.e., the mass-spring system). An error signal, denoted by $e(t) \in \mathbb{R}^3$, is defined to quantify this objective as

$$e = \begin{bmatrix} e_m & e_r^T \end{bmatrix}^T,$$

where $e_r(t) \triangleq [e_{r1}(t), e_{r2}(t)]^T \in \mathbb{R}^2$ and $e_m(t) \in \mathbb{R}$ denote the errors for the end-point of the second link of the robot and mass-spring system (Fig. 1), respectively, and are defined as

$$e_r = x_{rd} - x_r \quad e_m = x_{md} - x_m. \tag{19}$$

In (19), $x_{md} \in \mathbb{R}$ denotes the constant known desired position of the mass, and $x_{rd}(t) \triangleq [x_{rd1}(t), x_{rd2}]^T \in \mathbb{R}^2$ denotes the desired position of the end-point of the second link of the robot. To facilitate the subsequent control design and stability analysis, filtered tracking errors, denoted by $\eta_m(t) \in \mathbb{R}$ and $r_r(t) \in \mathbb{R}^2$, are defined as (Dixon et al., 2000)

$$\begin{aligned} \eta_m &= \dot{e}_m + \alpha_1 \tanh(e_m) + \alpha_2 \tanh(e_f) \\ r_r &= \dot{e}_r + \alpha e_r, \end{aligned} \tag{20}$$

where $\alpha,\ \alpha_1,\ \alpha_2 \in \mathbb{R}$ are positive, constant gains, and $e_f(t) \in \mathbb{R}$ is an auxiliary filter variable designed as (Dixon et al., 2000)

$$\dot{e}_f = -\alpha_3 \tanh(e_f) + \alpha_2 \tanh(e_m) - k_1 \cosh^2(e_f)\eta_m, \tag{21}$$

where $k_1 \in \mathbb{R}$ is a positive constant control gain, and $\alpha_3 \in \mathbb{R}$ is a positive constant filter gain. The filtered tracking error $r_r(t)$ is introduced to reduce the terms in the Lyapunov analysis (i.e., $r_r(t)$ can be used in lieu of including both $e_r(t)$ and $\dot{e}_r(t)$ in the stability analysis). The filtered tracking error $\eta_m(t)$ and the auxiliary signal $e_f(t)$ are introduced to eliminate a dependence on acceleration in the subsequently designed force controller (Dixon et al., 2003).

2.2.2 Closed-Loop Error System

By taking the time derivative of $m\eta_m(t)$ and utilizing (5), (6), (19), and (20), the following open-loop error system can be obtained:

$$m\dot{\eta}_m = Y_d\theta_d - K_I\Lambda(x_{r1} - x_m) + \alpha_2 m\cosh^{-2}(e_f)\dot{e}_f + \alpha_1 m\cosh^{-2}(e_m)\dot{e}_m. \tag{22}$$

In (22), $Y_d(x_m) = (x_m - x_o)$ and $\theta_d = k_s$. To facilitate the subsequent analysis, the following notation is introduced (Dixon et al., 2000):

$$Y_d\theta_d = Y_d K_I K_I^{-1}\theta_d = Y_{dk}\theta_{dk} = K_I(x_m - x_o)\left[\frac{k_s}{K_I}\right]. \tag{23}$$

After using (20) and (21), the expression in (22) can be rewritten as

$$m\dot{\eta}_m = Y_d\theta_d + K_I(x_{rd1} - \Lambda x_{r1}) + K_I\Lambda x_m - K_I x_{rd1} + \chi - \alpha_2 m k_1 \eta_m, \tag{24}$$

where $\chi(e_m,\ e_f,\ \eta_m,\ t) \in \mathbb{R}$ is an auxiliary term defined as

$$\begin{aligned} \chi = {} & \alpha_1 m\cosh^{-2}(e_m)(\eta_m - \alpha_1 \tanh(e_m)) - \alpha_1\alpha_2 m\cosh^{-2}(e_m)\tanh(e_f) \\ & + \alpha_2 m\cosh^{-2}(e_f)\big(\alpha_2 \tanh(e_m) - \alpha_3 \tanh(e_f)\big). \end{aligned} \tag{25}$$

The auxiliary expression, $\chi(e_m,\ e_f,\ \eta_m,\ t)$ defined in (25) can be upper bounded as

$$|\chi| \leq \zeta_1 \|z\|, \tag{26}$$

where $\zeta_1 \in \mathbb{R}$ is a positive bounding constant, and $z(t) \in \mathbb{R}^3$ is defined as

$$z = \begin{bmatrix} \eta_m & \tanh(e_m) & \tanh(e_f) \end{bmatrix}^T. \tag{27}$$

Based on (24) and the subsequent stability analysis, the desired robot link position is designed as

$$\begin{aligned} x_{rd1} &= Y_d\hat{\theta}_{dk} + x_m + k_2 \tanh(e_m) - k_1 k_2 \cosh^2(e_f)\tanh(e_f) \\ x_{rd2} &= \varepsilon. \end{aligned} \tag{28}$$

In (28), $\varepsilon \in \mathbb{R}$ is an appropriate positive constant (i.e., ε is selected so the robot will impact the mass-spring system in the vertical direction), $k_2 \in \mathbb{R}$ is a positive constant control gain, and the control gain $k_1 \in \mathbb{R}$ is defined as

$$k_1 = \frac{1}{\underline{m}}\left(3 + k_{n1}\zeta_1^2\right), \tag{29}$$

where $k_{n1} \in \mathbb{R}$ is a positive constant nonlinear damping gain. The parameter estimate $\hat{\theta}_{dk}(t) \in \mathbb{R}$ in (28) is generated by the adaptive update law

$$\dot{\hat{\theta}}_{dk} = proj(\Gamma Y_d \eta_m). \tag{30}$$

In (30), $\Gamma \in \mathbb{R}$ is a positive constant, and $proj(\cdot)$ denotes a sufficiently smooth projection algorithm (Cai et al., 2006) utilized to guarantee that $\hat{\theta}_{dk}(t)$ can be bounded as

$$\underline{\theta}_{dk} \leq \hat{\theta}_{dk} \leq \overline{\theta}_{dk}, \tag{31}$$

where $\underline{\theta}_{dk}, \overline{\theta}_{dk} \in \mathbb{R}$ denote known, constant lower and upper bounds for $\theta_{dk}(t)$, respectively. After substituting (28) into (24), the closed-loop error system for $\eta_m(t)$ can be obtained as

$$\begin{aligned} m\dot{\eta}_m &= K_I\left(x_{rd1} - \Lambda x_{r1}\right) + K_I\left(\Lambda x_m - x_m\right) + K_I k_1 k_2 \cosh^2(e_f)\tanh(e_f) + Y_{dk}\tilde{\theta}_{dk} \\ &\quad - K_I k_2 \tanh(e_m) + \chi - \alpha_2 m k_1 \eta_m. \end{aligned} \tag{32}$$

In (32), the parameter estimation error $\tilde{\theta}_{dk}(t) \in \mathbb{R}$ is defined as

$$\tilde{\theta}_{dk} = \theta_{dk} - \hat{\theta}_{dk}.$$

The open-loop robot error system can be obtained by taking the time derivative of $r_r(t)$ and premultiplying by the robot inertia matrix as

$$\overline{M}\dot{r}_r = Y_r\theta_r - \overline{C}r_r - F, \tag{33}$$

where (4), (19), and (20) were utilized, and

$$Y_r\theta_r = \bar{M}\ddot{x}_{rd} + \alpha\bar{M}\dot{e}_r + \bar{h} + \bar{C}\dot{x}_{rd} + \alpha\bar{C}x_{rd} + \begin{bmatrix} K_I\Lambda(x_{r1} - x_m) \\ 0 \end{bmatrix} - \alpha\bar{C}x_r, \tag{34}$$

where $Y_r(x_r, \dot{x}_r, x_m, \dot{x}_m, e_f, \eta_m, t) \in \mathbb{R}^{2\times P}$ denotes a known regression matrix, and $\theta_r \in \mathbb{R}^P$ denotes an unknown constant parameter vector. By making substitutions from the dynamic model and the previous error systems, $\ddot{x}_{rd}(t)$ can be expressed without a dependence on acceleration terms (Liang et al., 2007). Based on (33) and the subsequent stability analysis, the robot force control input is designed as

$$F = Y_r\hat{\theta}_r + Tanh(e_r) + k_3 Tanh(r_r), \tag{35}$$

where $k_3 \in \mathbb{R}$ is a positive constant control gain, and $\hat{\theta}_r(t) \in \mathbb{R}^P$ is an estimate for θ_r generated by the following adaptive update law

$$\dot{\hat{\theta}}_r = proj(\Gamma_r Y_r^T r_r). \tag{36}$$

In (36), $\Gamma_r \in \mathbb{R}^{P\times P}$ is a positive definite, constant, diagonal, adaptation gain matrix, and *proj(·)* denotes a projection algorithm utilized to guarantee that the *i-th* element of $\hat{\theta}_r(t)$ can be bounded as

$$\underline{\theta}_{ri} \le \hat{\theta}_{ri} \le \bar{\theta}_{ri}, \tag{37}$$

where $\underline{\theta}_{ri}$, $\bar{\theta}_{ri} \in \mathbb{R}$ denote known, constant lower and upper bounds for each element of $\theta_r(t)$, respectively. The closed-loop error system for $r_r(t)$ can be obtained after substituting (35) into (33) as

$$\bar{M}\dot{r}_r = Y_r\tilde{\theta}_r - \bar{C}r_r - Tanh(e_r) - k_3 Tanh(r_r). \tag{38}$$

In (38), the parameter estimation error $\tilde{\theta}_r(t) \in \mathbb{R}^P$ is defined as

$$\tilde{\theta}_r = \theta_r - \hat{\theta}_r. \tag{39}$$

2.3 Stability Analysis

Theorem: The controller given by (28), (30), (35), and (36) ensures semi-global asymptotic regulation of the MSR system in the sense that

$$|e_m(t)| \to 0 \quad \|e_r(t)\| \to 0 \;\; as \;\; t \to \infty$$

provided the control gains are chosen sufficiently large (Liang et al., 2007).

Proof: Let $V(t) \in \mathbb{R}$ denote the following non-negative, radially unbounded function (i.e., a Lyapunov function candidate):

$$V = \frac{1}{2}r_r^T\bar{M}r_r + \frac{1}{2}\tilde{\theta}_r^T\Gamma_r^{-1}\tilde{\theta}_r + \frac{1}{2}\tilde{\theta}_{dk}^T K_I \Gamma^{-1}\tilde{\theta}_{dk} + k_2 K_I\left[\ln(\cosh(e_m)) + \ln\left(\cosh\left(e_f\right)\right)\right]$$
$$+ e_r^T e_r + \frac{1}{2}m\eta_m^2 + \ln(\cosh(e_{r1})) + \ln(\cosh(e_{r2})). \tag{40}$$

After using (9), (13), (20), (21), (29), (30), (32), (36), and (38), the time derivative of (40) can be determined as

$$\begin{aligned}\dot{V} \leq & -k_3 \tanh^2(\|r_r\|) - \alpha_1 k_2 K_I \tanh^2(e_m) - \alpha_3 k_2 K_I \tanh^2(e_f) + 2e_r^T r_r - 2\alpha e_r^T e_r - 3\alpha_2 \eta_m^2 \\ & -k_{n1}\zeta_1^2 \alpha_2 \eta_m^2 - \alpha \tanh^2(\|e_r\|) + \eta_m \left[K_I (x_{rd1} - \Lambda x_{r1}) + K_I (\Lambda x_m - x_m) + \chi\right].\end{aligned} \quad (41)$$

The expression in (41) will now be examined under two different scenarios.

Case 1-Non-contact: For this case, the systems are not in contact ($\Lambda = 0$) and (41) can be rewritten as

$$\begin{aligned}\dot{V} \leq & -k_3 \tanh^2(\|r_r\|) - \alpha_1 k_2 K_I \tanh^2(e_m) - \alpha_3 k_2 K_I \tanh^2(e_f) + 2e_r^T r_r - 2\alpha e_r^T e_r \\ & -3\alpha_2 \eta_m^2 - k_{n1}\zeta_1^2 \alpha_2 \eta_m^2 - \alpha \tanh^2(\|e_r\|) + \eta_m [K_I x_{rd1} - K_I x_m + \chi].\end{aligned} \quad (42)$$

Based on the development in (Liang et al., 2007), the above expression can be reduced to

$$\dot{V} \leq -\lambda \|y\|^2 + \varepsilon_x, \quad (43)$$

where λ is a known constant, and $y(t) \in \mathbb{R}^5$ is defined as

$$y = \begin{bmatrix} z^T & \tanh(\|e_r\|) & \tanh(\|r_r\|) \end{bmatrix}^T. \quad (44)$$

Based on (40) and (43), if $\lambda \|y(t)\|^2 > \varepsilon_x$, then Barbalat's Lemma can be used to conclude that $\dot{V}(t) \to 0$ since $V(t)$ is lower bounded, $\dot{V}(t)$ is negative semi-definite, and $\dot{V}(t)$ can be shown to be uniformly continuous (UC). As $\dot{V}(t) \to 0$, eventually $\lambda \|y(t)\|^2 \leq \varepsilon_x$. While $\lambda \|y(t)\|^2 > \varepsilon_x$, (40) and (43) can be used to conclude that $V(t) \in \mathbf{L}_\infty$. When $\lambda \|y(t)\|^2 \leq \varepsilon_x$, the definitions (27) and (44) can be used to conclude that $e_m(t)$, $e_f(t)$, $e_r(t)$, $r_r(t)$, $\eta_m(t) \in \mathbf{L}_\infty$.
Since $\tilde{\theta}_r(t)$ and $\tilde{\theta}_{dk}(t) \in \mathbf{L}_\infty$ from the use of a projection algorithm, the previous facts can be used to conclude that $V(t) \in \mathbf{L}_\infty$. Signal chasing arguments can be used to prove the remaining closed-loop signals are also bounded during the non-contact case provided the control gains are chosen sufficiently large.
If the initial conditions for $V(0)$ are inside the region defined by ε_x, then $V(t)$ can grow larger until $\lambda \|y(t)\|^2 \leq \varepsilon_x$. Therefore, further development is required to determine how large $V(t)$ can grow. Sufficient gain conditions are developed in (Liang et al., 2007), which guarantee a semi-global stability result and ensure that the manipulator makes contact with the mass-spring system.

Case 2-Contact: For the case when the dynamic systems collide ($\Lambda = 1$) and the two dynamic systems become coupled, then (41) can be rewritten as

$$\begin{aligned}\dot{V} \leq & -k_3 \tanh^2(\|r_r\|) - \alpha_1 k_2 \underline{K_I} \tanh^2(e_m) - \alpha_3 k_2 \underline{K_I} \tanh^2(e_f) - 3\alpha_2 \eta_m^2 - \alpha \tanh^2(\|e_r\|) \\ & -\left[\alpha \|e_r\|^2 - \overline{K_I} |\eta_m| \|e_r\|\right] - \left[k_{n1}\zeta_1^2 \alpha_2 \eta_m^2 - \zeta_1 \|z\| |\eta_m|\right] - [\alpha \|e_r\|^2 - 2\|e_r\| \|r_r\|],\end{aligned}$$

where (17) was substituted for K_I and (26) was substituted for $\chi(e_m, e_f, \eta_m, t)$. Completing the square on the three bracketed terms yields

$$\dot{V} \leq -\beta\|z\|^2 - \alpha\tanh^2(\|e_r\|) - \left(\alpha_2 - \frac{\overline{K_I}}{4\alpha}\right)\eta_m^2 - \left(k_3 - \frac{\|r_r\|^2}{\alpha\tanh^2(\|r_r\|)}\right)\tanh^2(\|r_r\|). \tag{45}$$

Because (40) is non-negative, as long as the gains are picked sufficiently large (Liang et al., 2007), (45) is negative semi-definite, and $r_r(t)$, $\tilde{\theta}_r(t)$, $\tilde{\theta}_{dk}(t)$, $e_r(t)$, $e_m(t)$, $e_f(t)$ and $\eta_m(t) \in \mathbf{L}_\infty$. Based on the development in (Liang et al., 2007), Barbalat's Lemma can be used to conclude that $\tanh(e_m(t))$, $\tanh(\|r_r(t)\|)$, $|\eta_m(t)| \to 0$ as $t \to \infty$, which also implies $|e_m(t)|$, $\|r_r(t)\| \to 0$ as $t \to \infty$. Based on the fact that $\|r_r(t)\| \to 0$ as $t \to \infty$, standard linear analysis methods (see Lemma A.15 of (Dixon et al., 2003)) can then be used to prove that $\|e_r(t)\| \to 0$ as $t \to \infty$. Signal chasing arguments can be used to show that all signals remain bounded.

2.4 Experimental Results

The testbed depicted in Fig. 2 and Fig. 3 was developed for experimental demonstration of the proposed controller. The testbed is composed of a mass-spring system and a two-link planar robot. The body of the mass-spring system includes a U-shaped aluminum plate (item (8) in Fig. 2) mounted on an undercarriage with porous carbon air bearings which enables the undercarriage to glide on an air cushion over a glass covered aluminum rail. A steel core spring (item (1) in Fig. 2) connects the U-shaped aluminum plate to an aluminum frame, and a linear variable displacement transducer (LVDT) (item (2) in Fig. 2) is used to measure the position of the undercarriage assembly. The impact surface consists of an aluminum plate connected to the undercarriage assembly through a stiff spring mechanism (item (7) in Fig. 2). A capacitance probe (item (3) in Fig. 2) is used to measure the deflection of the stiff spring mechanism. The two-link planar robot (items (4-6) in Fig. 2) is made of two aluminum links, mounted on 240.0 [Nm] (base link) and 20.0 [Nm] (second link) direct-drive switched reluctance motors. The motor resolvers provide rotor position measurements with a resolution of 614400 pulses/revolution, and a standard backwards difference algorithm is used to numerically determine angular velocity from the encoder readings. A Pentium 2.8 GHz PC operating under QNX hosts the control algorithm, which was implemented via a custom graphical user-interface to facilitate real-time graphing, data logging, and the ability to adjust control gains without recompiling the program. Data acquisition and control implementation were performed at a frequency of 2.0 kHz using the ServoToGo I/O board. The initial conditions for the robot coordinates and the mass-spring position were (in [m])

$$[x_{r1}(0) \quad x_{r2}(0) \quad x_m(0)] = [0.060 \quad 0.436 \quad 0.206]. \tag{46}$$

The initial velocity of the robot and mass-spring were zero, and the desired mass-spring position was (in [m])

$$x_{md} = 0.236. \tag{47}$$

That is, the tip of the second link of the robot was initially 176 [mm] from the desired setpoint and 146 [mm] from x_0 along the X_1 -axis (Fig. 2). Once the initial impact occurs, the robot is required to depress the spring (item (1) in Fig. 2) to move the mass 30 [mm] along the X_1 -axis. The control and adaptation gains were selected as in (Liang et al., 2007). The mass-spring and robot errors (i.e., $e(t)$) are shown in Fig. 4. The peak steady-state position error of the end-point of the second link of the robot along the X_1-axis (i.e., $|e_{r1}(t)|$) and along the X_2 -axis (i.e., $|e_{r2}(t)|$) are 0.216 [mm] and 0.737 [mm], respectively. The peak steady-state position error of the mass (i.e., $|e_m(t)|$) is 2.56 [μ m]. The relatively large $|e_{r1}(t)|$ is due to the mismatch between the estimate value $\hat{\theta}_{dk}(t)$ and the real value $\theta_{dk}(t)$ in $e_{r1}(t)$. The relatively large $|e_{r2}(t)|$ is due to the inability of the model to capture friction and slipping effects on the contact surface. In this experiment, a significant friction is present along the X_2 -axis between the robot tip and the contact surface due to a large normal spring force applied along the X_1-axis. The input control torques (i.e., $J^T(q)F(t)$) are shown in Fig. 5. The resulting desired trajectory along the X_1-axis (i.e., $x_{rd1}(t)$) is depicted in Fig. 6, and the desired trajectory along the X_2 -axis was chosen as x_{rd2} = 0.358 [m]. Fig. 7 depicts the value of $\hat{\theta}_{dk}(t)$ and Figs. 8-10 depict the values of $\hat{\theta}_r(t)$. The order of the curves in the plots is based on the relative scale of the parameter estimates rather than numerical order in $\hat{\theta}_r(t)$.

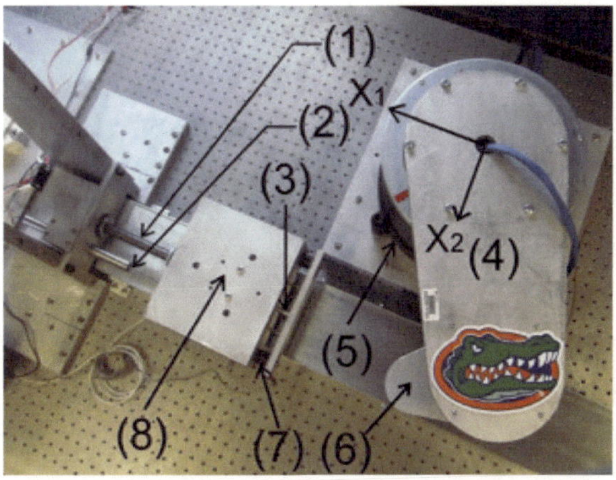

Figure 2. Top view of the experimental testbed including: (1) spring, (2) LVDT, (3) capacitance probe, (4) link1, (5) motor1, (6) link2, (7) stiff spring mechanism, (8) mass

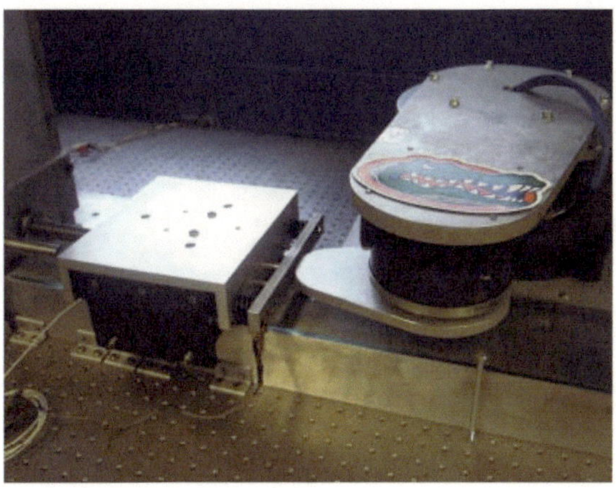

Figure 3. Side view of the experimental testbed

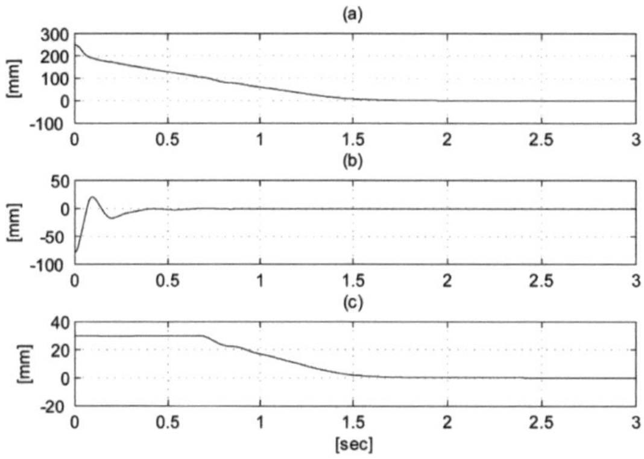

Figure 4. The mass-spring and robot errors $e(t)$. Plot (a) indicates the position error of the robot tip along the X_1-axis (i.e., $e_{r1}(t)$), (b) indicates the position error of the robot tip along the X_2-axis (i.e., $e_{r2}(t)$), and (c) indicates the position error of the mass-spring (i.e., $e_m(t)$)

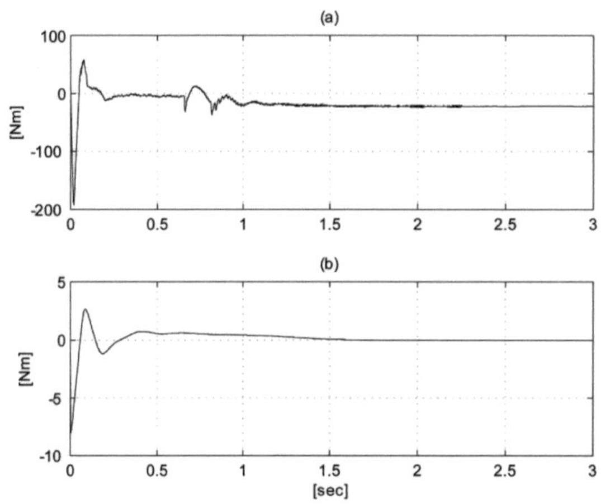

Figure 5. Applied control torques $J^T(q)F(t)$, for the (a) base motor and (b) second link motor

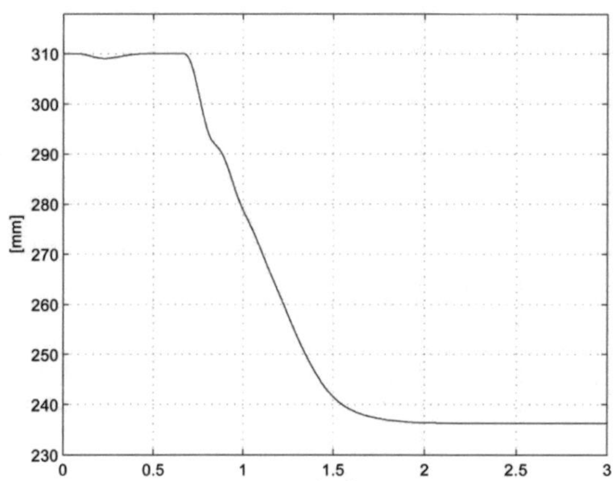

Figure 6. Computed desired robot trajectory, $x_{rd1}(t)$

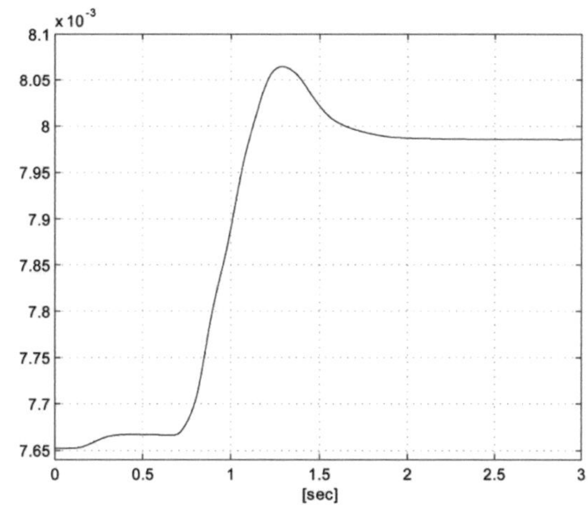

Figure 7. Unitless parameter estimate $\hat{\theta}_{dk}(t)$ introduced in (28)

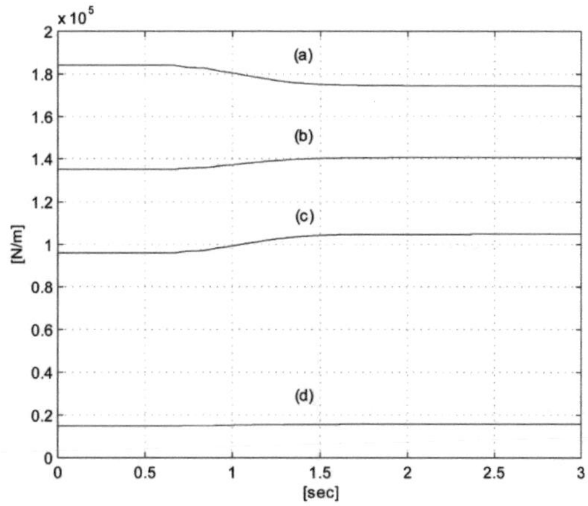

Figure 8. Estimate for the unknown constant parameter vector $\hat{\theta}_r(t)$. (a) $\hat{\theta}_{r10}(t)$ (b) $\hat{\theta}_{r4}(t)$, (c) $\hat{\theta}_{r1}(t)$, and (d) $\hat{\theta}_{r7}(t)$

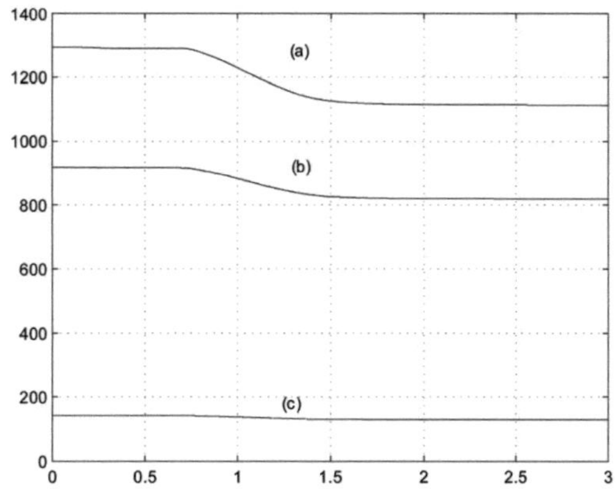

Figure 9. Estimate for the unknown constant parameter vector $\hat{\theta}_r(t)$. (a) $\hat{\theta}_{r5}(t)$ (b) $\hat{\theta}_{r2}(t)$, (c) $\hat{\theta}_{r8}(t)$

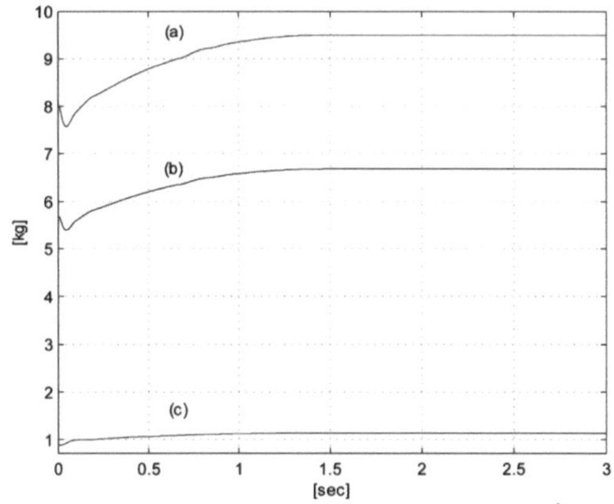

Figure 10. Estimate for the unknown constant parameter vector $\hat{\theta}_r(t)$. (a) $\hat{\theta}_{r6}(t)$ (b) $\hat{\theta}_{r3}(t)$, (c) $\hat{\theta}_{r9}(t)$

3. Control of Robotic Contact with a Viscoelastic Environment

The study of robot interaction with a viscoelastic environment is motivated by the increasing applications involving human-robot interaction. Since viscoelastic materials exhibit damping, the linear spring model used for stiff environments, in the previous section, would be inadequate to accurately represent the physical phenomena during contact. A nonlinear Hunt-Crossley model used in this section includes both stiffness and damping terms to account for the energy dissipation at contact. The differences in the contact model result in differences in the control development/stability analysis with regard to the controller in the previous section for stiff environments. The control structure in this section includes a desired robot velocity as a virtual control input to the unactuated viscoelastic mass spring system, coupled with a force controller to ensure that the actual robot position tracks the desired position. A NN feedforward term is used in the controller to estimate the parametric and non-parametric uncertainties.

3.1 Dynamic Model

The dynamic model for a rigid two-link revolute robot interacting with a compliant viscoelastic environment (Fig. 11) is given as

$$\overline{M}(x_r)\ddot{x}_r + \overline{C}(x_r, \dot{x}_r)\dot{x}_r + \overline{h}(x_r) + \begin{bmatrix} F_m \\ 0 \end{bmatrix} = F \tag{48}$$

$$m\ddot{x}_m + k_s(x_m - x_0) = F_m. \tag{49}$$

The interaction force $F_m(x_{r1}, \dot{x}_{r1}, x_m, \dot{x}_m)$ between the robot and the viscoelastic mass is modeled as

$$F_m = \Lambda F_v, \tag{50}$$

where $\Lambda(x_{r1}, x_m) \in \mathbb{R}$ is defined in (7) as a function which switches at impact, and $F_v(x_{r1}, \dot{x}_{r1}, x_m, \dot{x}_m) \in \mathbb{R}$ denotes the Hunt-Crossley force defined as (Hunt & Crossley, 1975)

$$F_v \triangleq \lambda\delta^n + b\dot{\delta}\delta^n. \tag{51}$$

In (51), $\lambda \in \mathbb{R}$ is the unknown contact stiffness of the viscoelastic mass, $b \in \mathbb{R}$ is the unknown impact damping coefficient, $\delta(x_r, x_m) \in \mathbb{R}$ denotes the local deformation of the material and is defined as

$$\delta \triangleq x_{r1} - x_m. \tag{52}$$

Also, in (51), $\dot{\delta}(t)$ is the relative velocity of the contacting bodies, and n is the unknown Hertzian compliance coefficient which depends on the surface geometry of contact. The model in (50) is a continuous contact force-based model wherein the contact forces increase from zero upon impact and return to zero upon separation. Also, the energy dissipation during impact is a function of the damping constant which can be related to the impact velocity and the coefficient of restitution (Hunt & Crossley, 1975), thus making the model more consistent with the physics of contact. The contact is considered to be direct-central and quasi-static (i.e., all the stresses are transmitted at the time of contact and sliding and friction effects during contact are negligible) where plastic deformation effects are assumed to be negligible. In addition to the properties in (8) and (9), the following property will be utilized in the subsequent control development.

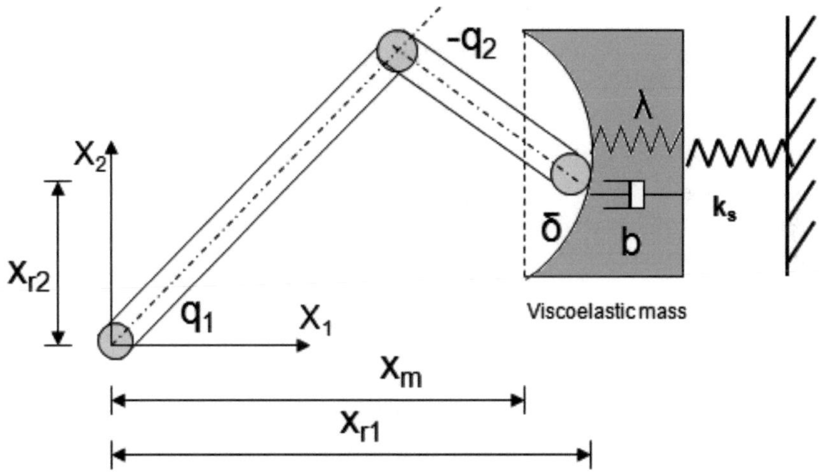

Figure 11. Robot contact with a viscoelastic mass

Property 5: The expression for the interaction force $F_m(x_r, \dot{x}_r, x_m, \dot{x}_m)$ in (50) can be written, using (7) and (51), as

$$F_m = \begin{cases} 0 & \delta < 0 \\ \lambda\delta^n + b\dot{\delta}\delta^n & \delta \geq 0. \end{cases} \tag{53}$$

Based on the fact that

$$\lim_{\delta \to 0^-} F_m = 0 \quad \lim_{\delta \to 0^+} F_m = 0, \tag{54}$$

the interaction force $F_m(t)$ is continuous.

3.2 Neural Network Control

In the subsequent control development, the desired robot velocity is designed as a virtual control input to the unactuated viscoelastic mass. The desired velocity is designed to ensure that the robot impacts and then regulates the mass to a desired position. A force controller is developed to ensure that the robot tracks the desired trajectory despite the non-contact to contact transition and parametric uncertainties in the robot and the viscoelastic mass-spring system. The viscoelastic model requires that the backstepping error be developed in terms of the desired robot velocity. To overcome this limitation, a Neural Network (NN) feedforward term is used in the controller to estimate the environmental uncertainties which are not linear in the parameters (such as the Hertzian compliance coefficient in (51)). In addition to the assumptions made in (15) and (16), the following assumptions are made

Assumption 4: The local deformation of the viscoelastic material during contact, $\delta(x_r, x_m)$, defined in (52), is assumed to be upper bounded, and hence δ^n can be upper bounded as

$$\delta^n \leq \mu, \tag{55}$$

where $\mu \in \mathbb{R}$ is a positive bounding constant.

Assumption 5: The damping constant, b, in (51), is assumed to be upper bounded as

$$b \leq \bar{b}, \tag{56}$$

where $\bar{b} \in \mathbb{R}$ denotes a known positive bounding constant.

3.2.1 Control Objective

The control objective of regulating the position of a viscoelastic mass attached to a spring via a robotic contact is quantified as in (19). To facilitate the subsequent control design and stability analysis, filtered tracking errors for the robot and the mass-spring, denoted by $r_r(t) \in \mathbb{R}^2$ and $r_m(t) \in \mathbb{R}$ respectively, are redefined as

$$\begin{aligned} r_r &\triangleq \dot{e}_r + \alpha e_r \\ r_m &\triangleq \dot{e}_m + \gamma_1 e_m + \gamma_2 e_f, \end{aligned} \tag{57}$$

where $\alpha \in \mathbb{R}^{2\times 2}$ is a positive, diagonal, constant gain matrix γ_1, $\gamma_2 \in \mathbb{R}$ are positive constant gains, and $e_f(t) \in \mathbb{R}$ is an auxiliary filter variable designed

$$\dot{e}_f = -\gamma_3 e_f + \gamma_2 e_m - k_1 r_m, \tag{58}$$

where k_1, $\gamma_3 \in \mathbb{R}$ are positive constant control gains.

3.2.2 NN Feedforward Estimation

NN-based estimation methods are well suited for control systems where the dynamic model contains uncertainties as in (48), (49) and (51). The universal approximation property of the Neural Network lends itself nicely to control system design. Multilayer Neural Networks have been shown to be capable of approximating generic nonlinear continuous functions. Let S be a compact simply connected set of $\mathbb{R}^{N_1+1}$. With map $f : S \to \mathbb{R}^n$, define $C^n(S)$ as the space where f is continuous. There exist weights and thresholds such that some function $f(x) \in C^n(S)$ can be represented by a three-layer NN as (Lewis, 1999; Lewis et al., 2002)

$$f(x) = W^T \sigma\left(V^T x\right) + \varepsilon(x), \tag{59}$$

for some given input $x(t) \in \mathbb{R}^{N_1+1}$. In (59), $V \in \mathbb{R}^{(N_1+1)\times N_2}$ and $W \in \mathbb{R}^{(N_2+1)\times n}$ are bounded constant ideal weight matrices for the first-to-second and second-to-third layers respectively, where N_1 is the number of neurons in the input layer, N_2 is the number of neurons in the hidden layer, and n is the number of neurons in the third layer. The activation function in (59) is denoted by $\sigma(\cdot) \in \mathbb{R}^{N_2+1}$, and $\varepsilon(x) \in \mathbb{R}^n$ is the functional reconstruction error. Note that, augmenting the input vector $x(t)$ and activation function $\sigma(\cdot)$ by 1 allows the thresholds as the first columns of the weight matrices (Lewis, 1999; Lewis et al., 2002). Thus, any tuning of W and V then includes tuning of thresholds as well. The computing power of the NN comes from the fact that the activation function $\sigma(\cdot)$ is nonlinear and the weights W and V can be modified or tuned through some learning procedure (Lewis et al., 2002). Based on (59), the typical three-layer NN approximation for $f(x)$ is given as (Lewis, 1999; Lewis et al., 2002)

$$\hat{f}(x) \triangleq \hat{W}^T \sigma(\hat{V}^T x), \tag{60}$$

where $\hat{V}(t) \in \mathbb{R}^{(N_1+1)\times N_2}$ and $\hat{W}(t) \in \mathbb{R}^{(N_2+1)\times n}$ are subsequently designed estimates of the ideal weight matrices. The estimate mismatch for the ideal weight matrices, denoted by $\tilde{V}(t) \in \mathbb{R}^{(N_1+1)\times N_2}$ and $\tilde{W}(t) \in \mathbb{R}^{(N_2+1)\times n}$, are defined as

$$\tilde{V} \triangleq V - \hat{V}, \quad \tilde{W} \triangleq W - \hat{W},$$

and the mismatch for the hidden-layer output error for a given $x(t)$, denoted by $\tilde{\sigma}(x) \in \mathbb{R}^{N_2+1}$, is defined as

$$\tilde{\sigma} \triangleq \sigma - \hat{\sigma} = \sigma(V^T x) - \sigma(\hat{V}^T x). \tag{61}$$

The NN estimate has several properties that facilitate the subsequent development. These properties are described as follows.

Property 6: *(Taylor Series Approximation)* The Taylor series expansion for $\sigma\left(V^T x\right)$ for a given x may be written as (Lewis, 1999; Lewis et al., 2002)

$$\sigma(V^T x) = \sigma(\hat{V}^T x) + \sigma'(\hat{V}^T x)\tilde{V}^T x + O(\tilde{V}^T x)^2, \tag{62}$$

where $\sigma'(\hat{V}^T x) \equiv d\sigma\left(V^T x\right)/d\left(V^T x\right)|_{V^T x = \hat{V}^T x}$, and $O(\hat{V}^T x)^2$ denotes the higher order terms. After substituting (62) into (61) the following expression can be obtained:

$$\tilde{\sigma} = \hat{\sigma}' \tilde{V}^T x + O(\tilde{V}^T x)^2, \tag{63}$$

where $\hat{\sigma}' \triangleq \sigma'(\hat{V}^T x)$.

Property 7: *(Boundedness of the Ideal Weights)* The ideal weights are assumed to exist and be bounded by known positive values so that

$$\|V\|_F^2 \le \bar{V}_B \tag{64}$$

$$\|W\|_F^2 \le \bar{W}_B, \tag{65}$$

where $\|\cdot\|_F$ is the Frobenius norm of a matrix, and $tr(\cdot)$ weights, $\hat{W}(t)$ and $\hat{V}(t)$, can be bounded using the projection algorithm as in (Patre et al., 2008).

Property 9: *(Boundedness of activation function* σ and $\dot{\sigma}$) The typical choice of activation function is the sigmoid function

$$\sigma(z) = \frac{1}{1+e^{\kappa z}}, \tag{66}$$

where

$$\|\sigma\| < 1 \text{ and } \|\dot{\sigma}\| \le \sigma_n,$$

where $\sigma_n \in \mathbb{R}$ is a known positive constant.

Property 10: (Boundedness of functional reconstruction error $\varepsilon(x)$) On a given compact set S, the net reconstruction error $\varepsilon(x)$ is bounded as

$$\|\varepsilon(x)\| \le \varepsilon_n,$$

where $\varepsilon_n \in \mathbb{R}$ is a known positive constant.

Property 11: *(Property of trace)* If $A \in \mathbb{R}^{n\times m}$ and $B \in \mathbb{R}^{m\times n}$, then

$$tr(AB) = tr(BA).$$

3.2.3 Closed-Loop Error System

The open-loop error system for the mass can be obtained by multiplying (57) by m and then taking its time derivative as

$$m\dot{r}_m = k_s(x_m - x_0) - \Lambda\lambda\delta^n - \Lambda b\dot{\delta}\delta^n + \chi - m\gamma_1^2 e_m - m\gamma_2 k_1 r_m, \tag{67}$$

where $\chi(e_m, r_m, e_f, t) \in \mathbb{R}$ is an auxiliary term defined as

$$\chi \triangleq m\gamma_1 r_m + m\gamma_2^2 e_m - (m\gamma_1\gamma_2 + m\gamma_2\gamma_3)e_f. \tag{68}$$

The auxiliary expression $\chi(e_m, r_m, e_f, t) \in \mathbb{R}$ defined in (68) can be upper bounded as

$$|\chi| \le \zeta_2 \|z\|, \tag{69}$$

where $\zeta_2 \in \mathbb{R}$ is a known positive constant, and $z(t) \in \mathbb{R}^3$ is defined as

$$z \triangleq \begin{bmatrix} r_m & e_m & e_f \end{bmatrix}. \tag{70}$$

The expression in (67) can be written as

$$m\dot{r}_m = f_1 - \Lambda b\dot{\delta}\delta^n + \chi - m\gamma_1^2 e_m - m\gamma_2 k_1 r_m, \tag{71}$$

where the function $f_1(t) \in \mathbb{R}$ containing the uncertain spring and damping constants, is defined as

$$f_1 \triangleq k_s(x_m - x_0) - \Lambda\lambda\delta^n. \tag{72}$$

The auxiliary function in (72) can be represented by a three layer NN as

$$f_1 = W_1^T \sigma_1\left(V_1^T x_1\right) + \varepsilon_m\left(x_1\right), \tag{73}$$

where the NN input $x_1(t) \in \mathbb{R}^3$ is defined as $x_1 \triangleq [1,\, x_m,\, \Lambda\delta]^T$, $W_1 \in \mathbb{R}^{(N_{m2}+1)\times 1}$, and $V_1 \in \mathbb{R}^{3\times N_{m2}}$ are ideal NN weights, and $N_{m2} \in \mathbb{R}$ denotes the number of hidden layer neurons of the NN. Since the open loop error expression for the mass in (71) does not have an actual control input, a virtual control input, $\dot{x}_{rd1}(t)$, is introduced by adding and subtracting $(1-\Lambda b\delta^n)\dot{x}_{rd1}$ to (71) as

$$m\dot{r}_m = f_1 - \Lambda b\dot{\delta}\delta^n + \chi - m\gamma_1^2 e_m - m\gamma_2 k_1 r_m + (1-\Lambda b\delta^n)\dot{x}_{rd1} - (1-\Lambda b\delta^n)\dot{x}_{rd1}. \tag{74}$$

To facilitate the subsequent backstepping-based design, the virtual control input to the unactuated mass-spring system, $\dot{x}_{rd1}(t)$ is designed as

$$\dot{x}_{rd1} = \hat{f}_1. \tag{75}$$

Also, $x_{rd2} = \rho$ where $\rho \in \mathbb{R}$ is an appropriate positive constant, selected so the robot will impact the mass-spring system. In (75), $\hat{f}_1(t) \in \mathbb{R}$ is the estimate for $f_1(t)$ and is defined as

$$\hat{f}_1 \triangleq \hat{W}_1^T \sigma_1(\hat{V}_1^T x_1), \tag{76}$$

where $\hat{W}_1(t) \in \mathbb{R}^{(N_{m2}+1)\times 1}$ and $\hat{V}_1(t) \in \mathbb{R}^{3\times N_{m2}}$ are the estimates of the ideal weights, which are updated based on the subsequent stability analysis as

$$\dot{\hat{W}}_1 = \Gamma_{w1}\hat{\sigma}_1 r_m - \Gamma_{w1}\hat{\sigma}_1'\hat{V}_1^T x_1 r_m \qquad \dot{\hat{V}}_1 = \Gamma_{v1} x_1 \hat{W}_1^T \hat{\sigma}_1' r_m \tag{77}$$

where $\Gamma_{w1} \in \mathbb{R}^{(N_{m2}+1)\times(N_{m2}+1)}$, $\Gamma_{v1} \in \mathbb{R}^{3\times 3}$ are constant, positive definite, symmetric gain matrices. The estimates for the NN weights in (77) are generated on-line (there is no off-line learning phase). The closed loop error system for the mass can be developed by substituting (75) into (74) and using (19) as

$$m\dot{r}_m = f_1 - \hat{f}_1 + \Lambda b\delta^n \dot{e}_{r1} - \Lambda b\delta^n \dot{e}_m + \chi - m\gamma_1^2 e_m - m\gamma_2 k_1 r_m + (1-\Lambda b\delta^n)\dot{x}_{rd1}. \tag{78}$$

Using (73) and (76), the expression in (78) can be written as

$$\begin{aligned} m\dot{r}_m = {} & W_1^T\sigma_1\left(V_1^T x_1\right) - \hat{W}_1^T \sigma_1(\hat{V}_1^T x_1) + \varepsilon_m\left(x_1\right) + \Lambda b\delta^n \dot{e}_{r1} - \Lambda b\delta^n \dot{e}_m + \chi - m\gamma_1^2 e_m \\ & - m\gamma_2 k_1 r_m + (1-\Lambda b\delta^n)\dot{x}_{rd1}. \end{aligned} \tag{79}$$

Adding and subtracting $W_1^T\hat{\sigma}_1 + \hat{W}_1^T\tilde{\sigma}_1$ to (79), and then using the Taylor series approximation in (63), the following expression for the closed loop mass error system can be obtained

$$m\dot{r}_m = \tilde{W}_1^T\hat{\sigma}_1 - \tilde{W}_1^T\hat{\sigma}_1'\hat{V}_1^T x_1 + \hat{W}_1^T\hat{\sigma}_1'\tilde{V}_1^T x_1 + w_1 - m\gamma_1^2 e_m - m\gamma_2 k_1 r_m, \tag{80}$$

where the notations $\hat{\sigma}_1$ and $\tilde{\sigma}_1$ were introduced in (61), and $w_1(t)\in\mathbb{R}$ is defined as

$$w_1 = \tilde{W}_1^T\hat{\sigma}_1' V_1^T x_1 + W_1^T O(\tilde{V}_1^T x_1)^2 + \varepsilon_m(x_1) + \Lambda b\delta^n \dot{e}_{r1} - \Lambda b\delta^n \dot{e}_m + \chi + (1-\Lambda b\delta^n)\dot{x}_{rd1}.$$

It can be shown from Property 7, Property 8, and (Lewis et al., 1996) that $w_1(t)$ can be bounded as

$$|w_1| \le c_{m1} + c_{m2}\|z\| + c_{m3}\|e_r\| + c_{m4}\|r_r\|, \tag{81}$$

where $c_{mi}\in\mathbb{R}$, (i = 1,2,...,4) are computable known positive constants. The open-loop robot error system can be obtained by taking the time derivative of $r_r(t)$), premultiplying by the robot inertia matrix $\overline{M}(x_r)$, and utilizing (19), (48), and (57) as

$$\overline{M}\dot{r}_r = f_2 - \overline{C}r_r - F, \tag{82}$$

where the function $f_2(t)\in\mathbb{R}^2$, contains the uncertain robot and Hunt-Crossley model parameters, and is defined as

$$f_2 \triangleq \overline{M}\ddot{x}_{rd} + \alpha\overline{M}\dot{e}_r + \overline{h} + \overline{C}\dot{x}_{rd} + \alpha\overline{C}x_{rd} + \begin{bmatrix}\Lambda(\lambda\delta^n + b\dot{\delta}\delta^n)\\ 0\end{bmatrix} - \alpha\overline{C}x_r.$$

By representing the function $f_2(t)$ by a NN, the expression in (82) can be written as

$$\overline{M}\dot{r}_r = W_2^T\sigma_2\left(V_2^T x_2\right) + \varepsilon_r(x_2) - \overline{C}r_r - F, \tag{83}$$

where the NN input $x_2(t)\in\mathbb{R}^{13}$ is defined as $x_2(t)\triangleq[1,\ \Lambda\delta,\ x_r^T,\ x_{rd}^T,\ \dot{\delta},\ \dot{x}_r^T,\ \dot{x}_{rd}^T,\ \ddot{x}_{rd}^T]^T$, $W_2\in\mathbb{R}^{(N_{r2}+1)\times 2}$ and $V_2\in\mathbb{R}^{13\times N_{r2}}$ are ideal NN weights and $N_{r2}\in\mathbb{R}$ denotes the number of hidden layer neurons of the NN. An expression for $\ddot{x}_{rd1}(t)$ can be developed to illustrate that the second derivative of the desired trajectory is continuous and does not require acceleration measurements. Based on (83) and the subsequent stability analysis, the robot force control input is designed as

$$F = \hat{W}_2^T\sigma_2(\hat{V}_2^T x_2) + k_2 r_r + e_r, \tag{84}$$

where $k_2\in\mathbb{R}$ is a constant positive control gain, and $\hat{W}_2(t)\in\mathbb{R}^{(N_{r2}+1)\times 2}$ and $\hat{V}_2(t)\in\mathbb{R}^{13\times N_{r2}}$ are the estimates of the ideal weights, which are designed based on the subsequent stability analysis as

$$\dot{\hat{W}}_2 = \Gamma_{w2}\hat{\sigma}_2 r_r^T - \Gamma_{w2}\hat{\sigma}_2'\hat{V}_2^T x_2 r_r^T \qquad \dot{\hat{V}}_2 = \Gamma_{v2}x_2 r_r^T\hat{W}_2^T\hat{\sigma}_2' \tag{85}$$

where $\Gamma_{w2}\in\mathbb{R}^{(N_{r2}+1)\times(N_{r2}+1)}$, $\Gamma_{v2}\in\mathbb{R}^{13\times 13}$ are constant, positive definite, symmetric gain matrices. Substituting (84) into (83) and following a similar approach as in the mass error system in (78)-(80), the closed loop error system for the robot is obtained as

$$\overline{M}\dot{r}_r = \tilde{W}_2^T\hat{\sigma}_2 - \tilde{W}_2^T\hat{\sigma}_2'\hat{V}_2^T x_2 + \hat{W}_2^T\hat{\sigma}_2'\tilde{V}_2^T x_2 + w_2 - \overline{C}r_r - k_2 r_r - e_r, \tag{86}$$

where $w_2(t) \in \mathbb{R}^2$ is defined as

$$w_2 = \tilde{W}_2^T \hat{\sigma}_2' V_2^T x_2 + W_2^T O(\tilde{V}_2^T x_2)^2 + \varepsilon_r(x_2). \tag{87}$$

It can be shown from Property 7, Property 8 and (Lewis et al., 1996) that $w_2(t)$ can be bounded as

$$\|w_2\| \le c_{r1} + c_{r2}\|z\| + c_{r3}\|e_r\| + c_{r4}\|r_r\|, \tag{88}$$

where $c_{ri} \in \mathbb{R}$, $(i = 1,2,\ldots,4)$ are computable known positive constants.

3.2.4 Stability Analysis

Theorem: The controller given by (75), (77), (84), and (85) ensures uniformly ultimately bounded regulation of the MSR system in the sense that

$$|e_m(t)|,\ \|e_r(t)\| \to \varepsilon_0 \exp(-\varepsilon_1 t) + \varepsilon_2 \tag{89}$$

provided the control gains are chosen sufficiently large (Bhasin et al., 2008).

Proof: Let $V(t) \in \mathbb{R}$ denote a non-negative, radially unbounded function (i.e., a Lyapunov function candidate) defined as

$$V = \frac{1}{2} r_r^T \overline{M} r_r + \frac{1}{2} e_r^T e_r + \frac{1}{2} tr(\tilde{W}_1^T \Gamma_{w1}^{-1} \tilde{W}_1) + \frac{1}{2} tr(\tilde{V}_1^T \Gamma_{v1}^{-1} \tilde{V}_1) + \frac{1}{2} tr(\tilde{W}_2^T \Gamma_{w2}^{-1} \tilde{W}_2) + \frac{1}{2} tr(\tilde{V}_2^T \Gamma_{v2}^{-1} \tilde{V}_2)$$
$$+ \frac{1}{2} m r_m^2 + \frac{1}{2} \gamma_1^2 m e_f^2 + \frac{1}{2} \gamma_1^2 m e_m^2. \tag{90}$$

It follows directly from the bounds given in (8), Property 8, (64) and (65), that $V(t)$ can be upper and lower bounded as

$$\lambda_3 \|y\|^2 \le V(t) \le \lambda_4 \|y\|^2 + \theta, \tag{91}$$

where λ_3, λ_4, $\theta \in \mathbb{R}$ are known positive bounding constants, and $y(t) \in \mathbb{R}^7$ is defined as

$$y = \begin{bmatrix} r_r^T & e_r^T & z^T \end{bmatrix}^T. \tag{92}$$

The time derivative of $V(t)$ in (90) can be upper bounded (Bhasin et al., 2008) as

$$\dot{V}(t) \le -\frac{\beta}{\lambda_4} V(t) + \varepsilon_x, \tag{93}$$

where ε_x, $\beta \in \mathbb{R}$ are positive constants which can be adjusted through the control gains (Bhasin et al., 2008). Provided the gains are chosen sufficiently large (Bhasin et al., 2008), the definitions in (70) and (92), and the expressions in (90) and (93) can be used to prove that $r_r(t)$, $e_r(t)$, $r_m(t)$, $e_m(t)$, $e_f(t) \in \mathcal{L}_\infty$. In a similar approach to the one developed in the first section, it can be shown that all other signals remain bounded and the controller given by (75), (77), (84), and (85) is implementable.

4. Conclusion

In this chapter, we consider a two link planar robotic system that transitions from free motion to contact with an unactuated mass-spring system. In the first half of the chapter, an adaptive nonlinear Lyapunov-based controller with bounded torque input amplitudes is designed for robotic contact with a stiff environment. The feedback elements for the controller are contained inside of hyperbolic tangent functions as a means to limit the impact forces resulting from large initial conditions as the robot transitions from non-contact to contact. The continuous controller in (35) yields semi-global asymptotic regulation of the spring-mass and robot links. Experimental results are provided to illustrate the successful performance of the controller. In the second half of the chapter, a Neural Network controller is designed for a robotic system interacting with an uncertain Hunt-Crossley viscoelastic environment. This result extends our previous result in this area to include a more general contact model, which not only accounts for stiffness but also damping at contact. The use of NN-based estimation in (Bhasin et al., 2008) provides a method to adapt for uncertainties in the robot and impact model.

5. References

S. Bhasin, K. Dupree, P. M. Patre, and W. E. Dixon (2008), Neural Network Control of a Robot Interacting with an Uncertain Hunt-Crossley Viscoelastic Environment, *ASME Dynamic Systems and Control Conference,* Ann Arbor, Michigan, to appear.

Z. Cai, M.S. de Queiroz, and D.M. Dawson (2006), A Sufficiently Smooth Projection Operator, *IEEE Transactions on Automatic Control,* Vol. 51, No. 1, pp. 135-139.

D. Chiu and S. Lee (1995), Robust Jump Impact Controller for Manipulators, *Proceedings of the IEEE/RSJ International Conference on Intelligent Robots and Systems,* Pittsburgh, Pennsylvania, pp. 299-304.

N Diolaiti, C Melchiorri, S Stramigioli (2004), Contact impedance estimation for robotic systems, *Proceedings of IEEE/RSJ International Conference on Intelligent Robots and Systems,* Italy, pp. 2538-2543.

W. E. Dixon, M. S. de Queiroz, D. M. Dawson, and F. Zhang (1999), Tracking Control of Robot Manipulators with Bounded Torque Inputs, *Robotica,* Vol. 17, pp. 121-129.

W. E. Dixon, E. Zergeroglu, D. M. Dawson, and M. W. Hannan (2000), Global Adaptive Partial State Feedback Tracking Control of Rigid-Link Flexible-Joint Robots, *Robotica,* Vol. 18. No 3. pp. 325-336.

W. E. Dixon, A. Behal, D. M. Dawson, and S. Nagarkatti (2003), *Nonlinear Control of Engineering Systems: A Lyapunov-Based Approach,* Birkhauser, ISBN 081764265X, Boston.

K. Dupree, C. Liang, G. Hu and W. E. Dixon (2006) a, Lyapunov-Based Control of a Robot and Mass-Spring System Undergoing an Impact Collision, *International Journal of Robotics and Automation,* to appear; see also *Proceedings of the IEEE American Control Conference,* Minneapolis, MN, pp. 3241-3246.

K. Dupree, C. Liang, G. Hu and W. E. Dixon (2006) b, Global Adaptive Lyapunov-Based Control of a Robot and Mass-Spring System Undergoing an Impact-Collision, IEEE Transactions on Systems, Man and Cybernetics, to appear; see also *Proceedings of the IEEE Conference on Decision and Controls,* San Deigo, California, pp. 2039-2044.

M. W. Gertz, J. Kim, and P. K. Khosla (1991), Exploiting Redundancy to Reduce Impact Force, *IEEE/RSJ International Workshop on Intelligent Robots and Systems IROS,* Osaka, Japan, pp. 179-184.

G. Gilardi and I. Sharf (2002), Literature survey of contact dynamics modelling, *Mechanism and Machine Theory,* Volume 37, Issue 10, Pages 1213-1239.

N. Hogan (1985), Impedance control: An approach to manipulation: Parts I, II, and III, /. *Dynamic Sys. Measurement Control* 107:1-24.

K.H. Hunt and F.R.E. Crossley (1975), Coefficient of restitution interpreted as damping in vibroimpact, *Journal of Applied Mechanics* 42, Series E, pp. 440 – 445.

M. Indri and A. Tornambe (2004), Control of Under-Actuated Mechanical Systems Subject to Smooth Impacts, *Proceedings of the IEEE Conference on Decision and Control,* Atlantis, Paradise Island, Bahamas, pp. 1228-1233.

S. Jezernik, M Morari (2002), Controlling the human-robot interaction for robotic rehabilitation of locomotion, *7th International Workshop on Advanced Motion Control.*

H.M. Lankarani and P.E. Nikravesh (1990), A contact force model with hysteresis damping for impact analysis of multi-body systems, *Journal of Mechanical Design* 112, pp. 369--376.

E. Lee, J. Park, K. A. Loparo, C. B. Schrader, and P. H. Chang (2003), Bang-Bang Impact Control Using Hybrid Impedance Time-Delay Control, *IEEE/ASME Transactions on Mechatronics,* Vol. 8, No. 2, pp. 272-277.

F. L. Lewis, A. Yesildirek, and K. Liu (1996), Multilayer neural-net robot controller: structure and stability proofs, *IEEE Transactions. Neural Networks.*

F. L. Lewis (1999), Nonlinear Network Structures for Feedback Control, *Asian Journal of Control,* Vol. 1, No. 4, pp. 205-228.

F. L. Lewis, J. Campos, and R. Selmic (2002), *Neuro-Fuzzy Control of Industrial Systems with Actuator Nonlinearities,* SIAM, PA.

Z Li, P Hsu, S Sastry (1989), Grasping and Coordinated Manipulation by a Multifingered Robot Hand, *The International Journal of Robotics Research,* Vol. 8, No. 4,33-50.

C. Liang, S. Bhasin, K. Dupree and W. E. Dixon (2007), An Impact Force Limiting Adaptive Controller for a Robotic System Undergoing a Non-Contact to Contact Transition, IEEE Transactions on Control Systems Technology, submitted; see also *Proceedings of the IEEE Conference on Decision and Controls,* Louisiana, pp. 3555-3560.

D.W. Marhefka and D.E. Orin (1999), A compliant contact model with nonlinear damping for simulation of robotic systems, *IEEE Transactions on Systems, Man, and Cybernetics-Part A: Systems and Humans,* pp. 566 – 572.

A.M. Okamura, N. Smaby, M.R. Cutkosky (2000), An overview of dexterous manipulation, *IEEE International Conference on Robotics and Automation.*

P. R. Pagilla and B. Yu (2001), A Stable Transition Controller for Constrained Robots, *IEEE/ASME Transactions on Mechatronics,* Vol. 6, No. 1, pp. 65-74.

P. M. Patre, W. MacKunis, C. Makkar, W. E. Dixon (2008), Asymptotic Tracking for Uncertain Dynamic Systems via a Multilayer NN Feedforward and RISE Feedback Control Structure, *IEEE Transactions on Control Systems Technology,* Vol. 16, No. 2, pp. 373-379.

A. Tornambe (1999), Modeling and Control of Impact in Mechanical Systems: Theory and Experimental Results, *IEEE Transactions on Automatic Control,* Vol. 44, No. 2, pp. 294-309.

I. D. Walker (1990), The Use of Kinematic Redundancy in Reducing Impact and Contact Effects in Manipulation, *Proceedings of the IEEE International Conference on Robotics and Automation,* Cincinnati, OH, pp. 434-439.

I. D. Walker (1994), Impact Configurations and Measures for Kinematically Redundant and Multiple Armed Robot Systems, *IEEE Transactions on Robotics and Automation,* Vol. 10, No. 5, pp 346-351.

K. Youcef-Toumi and D. A. Guts (1989), Impact and Force Control, *Proceedings of the IEEE International Conference on Robotics and Automation,* AZ, pp. 410-416.

7

Motion Control of a Robot Manipulator in Free Space Based on Model Predictive Control

Vincent Duchaine, Samuel Bouchard and Clément Gosselin
Université Laval
Canada

1. Introduction

The majority of existing industrial manipulators are controlled using PD controllers. This type of basically linear control does not represent an optimal solution for the motion control of robots in free space because robots exhibit highly nonlinear kinematics and dynamics. In fact, in order to accommodate configurations in which gravity and inertia terms reach their minimum amplitude, the gain associated with the derivative feedback (D) must be set to a relatively large value, thereby leading to a generally over-damped behaviour that limits the performance. Nevertheless, in most current robotic applications, PD controllers are functional and sufficient due to the high reduction ratio of the transmissions used. However, this assumption is no longer valid for manipulators with low transmission ratios such as human-friendly manipulators or those intended to perform high accelerations like parallel robots.

Over the last few decades, a new control approach based on the so-called Model Predictive Control (MPC) algorithm was proposed. Arising from the work of Kalman (Kalman, 1960) in the 1960's, predictive control can be said to provide the possibility of controlling a system using a proactive rather than reactive scheme. Since this control method is mainly based on the recursive computing of the dynamic model of the process over a certain time horizon, it naturally made its first successful breakthrough in slow linear processes. Common current applications of this approach are typically found in the petroleum and chemical industries.

Several attempts were made to adapt this computationally intensive method to the control of robot manipulators. A little more than a decade ago, it was proposed to apply predictive control to nonlinear robotic systems (Berlin & Frank, 1991), (Compas et al., 1994). However, in the latter references, only a restricted form of predictive control was presented and the implementation issues – including the computational burden – were not addressed. Later, predictive control was applied to a broader variety of robotic systems such as a 2-DOF (degree-of-freedom) serial manipulator (Zhang & Wang, 2005), robots with flexible joints (Von Wissel et al., 1994), or electrical motor drives (Kennel et al., 1987). More recently, (Hedjar et al., 2005), (Hedjar & Boucher, 2005) presented simplified approaches using a limited Taylor expansion. Due to their relatively low computation time, the latter approaches open the avenue to real-time implementations. Finally, (Poignet & Gautier, 2000), (Vivas et al., 2003), (Lydoire & Poignet, 2005), experimentally demonstrated

predictive control on a 4-DOF parallel mechanism using a linear model in the optimization combined with a feedback linearization.
Several other control schemes based on the prediction of the torque to be applied at the actuators of a robot manipulator can be found in the literature. Probably the best-known and most commonly used technique is the so-called Computed Torque Method (Anderson, 1989), (Ubel et al., 1992). However, this control scheme has the disadvantage of not being robust to modelling errors. In addition to having the capability of making predictions over a certain time horizon, model predictive control contains a feedback mechanism compensating for prediction errors due to structural mismatch between the model and the process. These two characteristics make predictive control very efficient in terms of optimal control as well as very robust.
This chapter aims at providing an introduction to the application of model predictive control to robot manipulators despite their typical nonlinear dynamics and fast servo rate. First, an overview of the theory behind model predictive control is provided. Then, the application of this method to robot control is investigated. After making some assumptions on the robot dynamics, equations for the cost function to be minimized are derived. The solution of these equations leads to an analytic and computationally efficient expression for position and velocity control which are functions of a given prediction time horizon and of the dynamic model of the robot. Finally, several experiments using a 1-DOF pendulum and a 6-DOF cable-driven parallel mechanism are presented in order to illustrate the performance in terms of dynamics as well as the computational efficiency.

2. Overview

The very first predictive control schemes appeared around 1980 under the form of Dynamic Matrix Control (DMC) and Generalized Predictive Control (GPC). These two approaches led to a successful breakthrough in the chemical industry and opened the avenue to several new control algorithms later known as the family of model predictive control. A more detailed account of the history of model predictive control can be found in (Morari & Lee, 1999).
Model predictive control is based on three main key ideas. These ideas have been well summarized by Camacho and Bordons in their book on the subject (Camacho & Bordons, 2004). This method is in fact based on *the explicit use of a model to predict the process output at future time instants horizon.* The prediction is done via the *calculation of a control sequence minimizing an objective function.* It also has the particularity that it is based on a *receding strategy, so that at each instant the horizon is displaced toward the future, which involves the application of the first control signal of the sequence calculated at each step*. This last particularity partially explains why predictive control is sometime called receding horizon control.
As mentioned above, a predictive control scheme required the minimization of a quadratic cost function over a prediction horizon in order to predict the correct control input to be applied to the system. The cost function is composed of two parts, namely, a quadratic function of the deterministic and stochastic components of the process and a quadratic function of the constraints. The latter is one of the main advantages of this control method over many other schemes. It can deal at same time with model regulation and constraints. The constraints can be on the process as well as on the control output. The global function to be minimized can then be written in a general form as:

$$J = \sum_{n=1}^{H_p} (\mathbf{y}_{k+n} - \mathbf{r}_{k+n})^T \mathbf{Q} (\mathbf{y}_{k+n} - \mathbf{r}_{k+n}) + \sum_{m=1}^{H_c} \boldsymbol{\psi}^T \boldsymbol{\lambda} \boldsymbol{\psi} \tag{1}$$

where

k is the current time step,
H_p is the prediction horizon,
H_c is the control horizon,
$\mathbf{W}$ is a weighting matrix,
$\boldsymbol{\lambda}$ is a weighting matrix,
$\mathbf{r}_{k+n}$ is the reference input,
$\mathbf{y}_{k+n}$ is the output of the system,
$\boldsymbol{\psi}$ is a constraint function.

Although this function is the key of the effectiveness of the predictive control scheme in terms of optimal control, it is also its weakness in term of computational time. For linear processes, and depending on the constraint function, the optimal control sequence can be found relatively fast. However, for nonlinear model the problem is no longer convex and hence the computation of the function over the prediction horizon becomes computationally intensive and sometime very hard to solve explicitly.

3. Application to manipulators

This part aims at showing how Model Predictive Control can be efficiently applied to robot manipulators to suit their fast servo rate. Figure 1 provides a schematic representation of the proposed scheme, where $\mathbf{d}_k$ represents the error between the output of the system and the output of the model. In the next subsection the different parts of this scheme will be defined for velocity control as well as for position control.

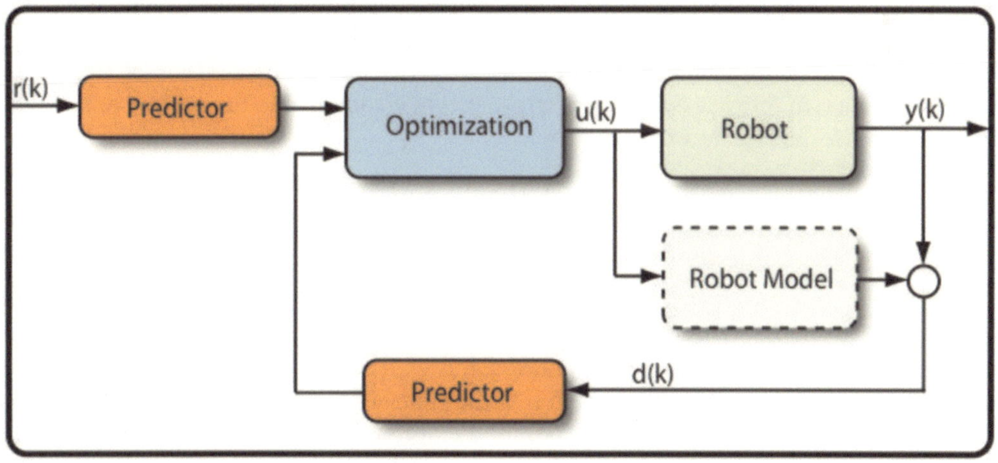

Figure 1. MPC applied to manipulator

3.1 Velocity control

Velocity control is rarely implemented in conventional industrial manipulators since the majority of the tasks to be performed by robots require precise position tracking. However, over the last few years, several researchers have developed a new generation of robots that are capable of working in collaboration with humans (Berbardt et al., 2004), (Peshkin & Colgate, 2001), (Al-Jarrah & Zheng, 1996). For this type of tasks, velocity control seems more appropriate (Duchaine & Gosselin, 2007) due to the fact that the robot is not constrained to given positions but rather has to follow the movement of the human collaborator. Also, velocity control has infinitely spatial equilibrium points which is a very safe intrinsic behaviour in a context where a human being is sharing the workspace of a robot. The predictive control approach presented in this chapter can be useful in this context.

3.1.1 Modelling

For velocity control, the reference input is usually relatively constant, especially considering the high servo rates used. Therefore, it is reasonable to assume that the reference velocity remains constant over the prediction horizon. With this assumption, the stochastic predictor of the reference velocity becomes:

$$\begin{aligned} \tilde{\mathbf{r}}(k+1) &= \mathbf{r}(k) \\ \tilde{\mathbf{r}}(k+2) &= \mathbf{r}(k) \\ &\vdots \\ \tilde{\mathbf{r}}(k+H_c) &= \mathbf{r}(k) \end{aligned} \tag{2}$$

where $\tilde{\mathbf{r}}(j)$ stands for the predicted value of $\mathbf{r}$ at time step j.

The error, $\mathbf{d}$, is obtained by computing the difference between the system's output and the model's output. Taking into account this difference in the cost function will help to increase the robustness of the control to model mismatch. The error can be decomposed in two parts. The first one is the error associated directly with model uncertainties. Often, this component will produce an offset proportional to the mismatch. The error may also include a zero-mean white noise given by the noise of the encoder or some random perturbation that cannot be included in the deterministic model. Since the error term is partially composed of zero-mean white noise, it is difficult to define a good stochastic predictor of the future values. However, in the case considered here, a future error equal to the present one will be simply assumed. This can be expressed as:

$$\begin{aligned} \tilde{\mathbf{d}}(k+1) &= \mathbf{d}(k) \\ \tilde{\mathbf{d}}(k+2) &= \mathbf{d}(k) \\ &\vdots \\ \tilde{\mathbf{d}}(k+H_c) &= \mathbf{d}(k) \end{aligned} \tag{3}$$

where $\tilde{\mathbf{d}}(j)$ is the predicted value of $\mathbf{d}$ at time step j.

In this chapter, a constraint on the variation of the control input signal ($\mathbf{u}$) over a prediction horizon will be used as a constraint function in the optimization. This is a typical constraint over the control input that helps smoothing the command and tends to maximize the effective life of the actuator, namely:

$$\psi = (\mathbf{u}_{k+m} - \mathbf{u}_{k+m-1}) \tag{4}$$

The model of the robot itself is directly linked with its dynamic behaviour. The dynamic equations of a robot manipulator can be expressed as:

$$\mathbf{\Gamma} = \mathbf{M}(q)\ddot{q} + \mathbf{h}(q, \dot{q}) + \mathbf{V}\dot{q} + \mathbf{g}(q) \tag{5}$$

Where

$\mathbf{\Gamma}$ is the vector of actuator forces/torques,
$\mathbf{M}(q)$ is the generalized inertia matrix,
$\ddot{q}$ is the vector of actuator accelerations,
$\mathbf{h}(q, \dot{q})$ is the vector of centrifugal and Corriolis forces,
$\mathbf{V}\dot{q}$ is the vector of viscous friction torques,
$\mathbf{g}(q)$ is the vector of gravity torques.

The acceleration resulting from a torque applied on the system can be found by inverting eq. (5), which leads to:

$$\ddot{q} = -\mathbf{M}(q)^{-1}[\mathbf{h}(q, \dot{q}) + \mathbf{V}\dot{q} + \mathbf{g}(q) - \mathbf{\Gamma}] \tag{6}$$

where q and $\dot{q}$ are the positions and velocities measured by the encoders. Assuming that the acceleration is constant over one time period, the above expression can be substituted into the equations associated with the motion of a body undergoing constant acceleration, which leads to:

$$\dot{q}_{k+1} = \dot{q}_k - \mathbf{M}(q)^{-1}[\mathbf{h}(q, \dot{q}) + \mathbf{V}\dot{q} + \mathbf{g}(q) - \mathbf{\Gamma}]T_s \tag{7}$$

where T_s is the sampling period. Since robots usually run on a discrete controller with a very small sampling period, assuming a constant acceleration over a sample period is a reasonable approximation that will not induce significant errors.

Eq. (7) represents the behaviour of the robot over a sampling period. However, in predictive control, this behaviour must be determined over a number of sampling periods given by the horizon of prediction. Since the dynamic model of the manipulator is nonlinear, it is not straightforward to compute the necessary recurrence over this horizon, especially considering the limited computational time available. This is one of the reasons why predictive control is still not commonly used for manipulator control.

Instead of computing exactly the nonlinear evolution of the manipulator dynamics, it can be more efficient to make some assumptions that will simplify the calculations. For the accelerations normally encountered in most manipulator applications, the gravitational term is the one that has the most impact on the dynamic model. The evolution of this term over time is a function of the position. The position is obtained by integrating the velocity over time. Even a large variation of velocity will not lead to a significant change of position since it is integrated over a very short period of time. From this point of view, the high sampling rate that is typically used in robot controllers allows us to assume that the nonlinear terms of the dynamic model are constant over a prediction horizon. Obviously, this assumption will induce some error, but this error can easily be managed by the error term included in the minimization.

It is known from the literature that for an infinite prediction horizon and for a stabilizable process, as long as the objective function weighting matrices are positive definite, predictive control will always stabilize the system (Qin & Badgwell, 1997). However, the simplifications that have been made above on the representation of the system prevent us from concluding on stability since the errors in the model will increase nonlinearly with an increasing prediction horizon. It is not trivial to determine the duration of the prediction horizon that will ensure the stability of the control method. The latter will depend on the dynamic model, the geometric parameters and also on the conditioning of the manipulator at a given pose.

3.1.2 The optimization cost function

From the above derivations, combining the deterministic and stochastic components and the constraint on the input variable leads to the general cost function to be optimized as a function of the prediction and control horizons. This function can be divided into two sums in order to manage distinctively the prediction horizon and control horizon. One has:

$$J = \sum_{n=1}^{H_c-1} [\mathbf{A}(n)^T \mathbf{W} \mathbf{A}(n) + \Delta\mathbf{\Gamma}_n^T \boldsymbol{\lambda} \Delta\mathbf{\Gamma}_n] + \sum_{n=H_c}^{H_p} [\mathbf{B}(n)^T \mathbf{W} \mathbf{B}(n) + \Delta\mathbf{\Gamma}_{H_c}^T \boldsymbol{\lambda} \Delta\mathbf{\Gamma}_{H_c}] \tag{8}$$

with

$$\mathbf{A}(n) = \dot{\boldsymbol{q}} + n\mathbf{M}^{-1}(\mathbf{\Gamma}_n - \mathbf{h}_N(\boldsymbol{q}, \dot{\boldsymbol{q}}))T_s - (\mathbf{r} - \mathbf{d}) \tag{9}$$

$$\mathbf{B}(n) = \dot{\boldsymbol{q}} + n\mathbf{M}^{-1}(\mathbf{\Gamma}_{H_c} - \mathbf{h}_N(\boldsymbol{q}, \dot{\boldsymbol{q}}))T_s - (\mathbf{r} - \mathbf{d}) \tag{10}$$

being the integration form for q_{k+n} of the linear equation (7) and where

$\mathbf{h}_N(\boldsymbol{\theta}, \dot{\boldsymbol{\theta}})$ The gravity, friction, centripetal and Coriolis term.
$\mathbf{W}$ diagonal matrix of weighting factors,
$\boldsymbol{\lambda}$ diagonal matrix of weighting factors,
$\mathbf{u}$ the vector of input variables (current) and
$\Delta\mathbf{\Gamma}_j$ $\mathbf{\Gamma}_j - \mathbf{\Gamma}_{j-1}$.

An explicit solution to the minimization of J can be found for given values of H_p and H_c. However, it is more difficult to find a general solution that would be a function of H_p and H_c. Nevertheless, a minimum of J can easily be found numerically. From eq. (8), it is clear that J is a quadratic function of Γ. Moreover, because of its physical meaning, the minimum of J is reached when the derivative of J with respect to Γ is equal to zero. The problem can thus be reduced to finding the root of the following equation:

$$\begin{aligned}\frac{\partial J}{\partial \mathbf{\Gamma}} &= \sum_{n=1}^{H_c-1} ([\mathbf{C}(n)\mathbf{D}(n) - (\mathbf{r} - \mathbf{d}))T_s] + 2\boldsymbol{\lambda}\Delta\mathbf{\Gamma}_n) \\ &+ \sum_{n=H_c}^{H_p} ([\mathbf{C}(n)\mathbf{E}(n) - (\mathbf{r} - \mathbf{d}))T_s] + 2\boldsymbol{\lambda}\Delta\mathbf{\Gamma}_{H_c}) = 0\end{aligned} \tag{11}$$

with

$$\mathbf{C}(n) = 2n\mathbf{W} \tag{12}$$

$$\mathbf{D}(n) = (\dot{\boldsymbol{q}} + n\mathbf{M}^{-1}(\mathbf{\Gamma}_n - \mathbf{h}_N(\boldsymbol{q}, \dot{\boldsymbol{q}}))T_s \tag{13}$$

$$\mathbf{E}(n) = (\dot{\boldsymbol{q}} + n\mathbf{M}^{-1}(\mathbf{\Gamma}_{H_c} - \mathbf{h}_N(\boldsymbol{q}, \dot{\boldsymbol{q}}))T_s \tag{14}$$

An exact and unique solution to this equation exists since it is linear. However, the computation of the solution involves the resolution of a system of linear equation whose size increases linearly with the control horizon. Another drawback of this approach is that the generalized inertia matrix must be inverted, which can be time consuming. The next section will present strategies to avoid these drawbacks.

3.1.3 Analytical Solution of the minimization problem

The previous section provided a general formulation of the MPC applied to robot manipulators with an arbitrary number of degrees of freedom and arbitrary chosen prediction and control horizons. However, in this section, only the prediction horizon will be considered, discarding the constraint function. This simplification of the general approach of the predictive control will make it possible to find an exact expression of the optimal control input signal for any prediction horizon thereby reducing drastically the computing time.

Many predictive schemes presented in the literature (Berlin & Frank, 1991), (Hedjar et al., 2005), (Hedjar & Boucher, 2005) consider only the prediction horizon and disregard the control horizon, which simplifies greatly the formulation. Also, the constraint imposed on the input variable can be eliminated. At high servo rates, neglecting this constraint does not have a major impact since the input signal does not usually vary much from one period to another. Thus, the aggressiveness of the control variable Γ that will result from the elimination of the constraint function can easily be compensated for by the use of a longer prediction horizon. The above simplifications lead to a new cost function given by:

$$J = \sum_{n=1}^{H_p} \mathbf{F}(n)^T \mathbf{F}(n) \tag{15}$$

where

$$\mathbf{F}(n) = \dot{\boldsymbol{q}} + n\mathbf{M}^{-1}(\boldsymbol{\Gamma} - \mathbf{h}_N(\boldsymbol{q}, \dot{\boldsymbol{q}}))T_s - (\mathbf{r} - \mathbf{d}). \tag{16}$$

Computing the derivative of eq. (15) with respect to Γ and setting it to zero, a general expression of the optimal control input signal as a function of the prediction horizon is obtained, namely:

$$\boldsymbol{\Gamma} = \mathbf{h}_N(\boldsymbol{q}, \dot{\boldsymbol{q}}) - \frac{3\mathbf{M}(\dot{\boldsymbol{q}} - \mathbf{r} + \mathbf{d})}{(1 + 2H_p)T_s} \tag{17}$$

The algebraic manipulations that lead to eq. (17) from eq. (15) are summarized in (Duchaine et al., 2007). Although this simplification leads to the loose of one of the main advantages of the MPC, the resulting control schemes will still exhibit good characteristics such as an easy tuning procedure, an optimal response and a better robustness to model mismatch compared to conventional computed torque control. It is noted also that this solution does not require the computation of the inverse of the generalized inertia matrix, thereby improving the computational efficiency. Moreover, since the solution is analytical, an online numerical optimization is no longer required.

3.2 Position control

The position-tracking scheme of control follows a formulation similar to the one that was presented above for velocity control. The main differences are the stochastic predictor of the future reference position and the deterministic model of the manipulator that must now predict the future positions instead of velocities.

3.2.1 Modelling

In the velocity control scheme, it was assumed that the reference input was constant over the prediction horizon. This assumption was justified by the high servo rate and by the fact that the velocity does not usually vary drastically over a sampling period even in fast trajectories. However, this assumption cannot be used for position tracking. In particular, in the context of human-robot cooperation, no trajectory is established *a priori* and the future reference input must be predicted from current positions and velocities. A simple approximation that can be made is to use the time derivative of the reference to linearly predict its future. This can be written as:

$$\begin{aligned} \tilde{\mathbf{r}}(k) &= \mathbf{r}(k) \\ \tilde{\mathbf{r}}(k+1) &= \mathbf{r}(k) + \Delta\mathbf{r} \\ &\vdots \\ \tilde{\mathbf{r}}(k+H_c) &= \mathbf{r}(k) + H_c\Delta\mathbf{r} \end{aligned} \tag{18}$$

where $\Delta\mathbf{r}$ is given by:

$$\Delta\mathbf{r} = \mathbf{r}(k) - \mathbf{r}(k-1) \tag{19}$$

Since the error term *d(k)* is again partially composed of zero-mean white noise, one will consider the future of this error equal to the present. Therefore, eq. (4) is also used here. As shown in the previous section, the joint space velocity can be predicted using eq. (7). Integrating the latter equation once more with respect to time --- and assuming constant acceleration ---, the prediction on the position is obtained as:

$$q_{k+1} = \boldsymbol{q}_k + \dot{\boldsymbol{q}}_k T_s - \frac{1}{2}\mathbf{M}(\boldsymbol{q})^{-1}[\mathbf{h}(\boldsymbol{q},\dot{\boldsymbol{q}}) + \mathbf{V}\dot{\boldsymbol{q}} + g(\boldsymbol{q}) - \boldsymbol{\Gamma}]T_s^2. \tag{20}$$

3.2.2 The optimization cost function

Including the deterministic model and the stochastic part inside the function to be minimized, the general function of predictive control for the manipulator is obtained:

$$J = \sum_{n=1}^{H_c-1} [\mathbf{G}(n)^T\mathbf{W}\mathbf{G}(n) + \Delta\boldsymbol{\Gamma}_n^T\boldsymbol{\lambda}\Delta\boldsymbol{\Gamma}_n] + \sum_{n=H_c}^{H_p} [\mathbf{L}(n)^T\mathbf{W}\mathbf{L}(n) + \Delta\boldsymbol{\Gamma}_{H_c}^T\boldsymbol{\lambda}\Delta\boldsymbol{\Gamma}_{H_c}] \tag{21}$$

with

$$\begin{aligned} \mathbf{G}(n) &= \mathbf{N}(n) + \tfrac{n^2}{2}\mathbf{M}^{-1}\left(\boldsymbol{\Gamma}_n - \mathbf{h}_N(\boldsymbol{q},\dot{\boldsymbol{q}})\right)T_s^2 \\ \mathbf{L}(n) &= \mathbf{N}(n) + \tfrac{n^2}{2}\mathbf{M}^{-1}\left(\boldsymbol{\Gamma}_{H_c} - \mathbf{h}_N(\boldsymbol{q},\dot{\boldsymbol{q}})\right)T_s^2 \\ \mathbf{N}(n) &= \boldsymbol{q} + n\dot{\boldsymbol{q}}T_s - (\tilde{\mathbf{r}} - \mathbf{d}). \end{aligned} \tag{22}$$

Taking the derivative of this function with respect to $\boldsymbol{\Gamma}$ and setting it to zero leads to a linear equation where the root is the minimum of the cost function:

$$\frac{\partial J}{\partial \boldsymbol{\Gamma}} = \sum_{n=1}^{H_c-1} (n^2\mathbf{W}\mathbf{M}^{-1}\mathbf{G}(n)T_s^2 + 2\boldsymbol{\lambda}\Delta\boldsymbol{\Gamma}_n) + \sum_{n=H_c}^{H_p} (n^2\mathbf{W}\mathbf{M}^{-1}\mathbf{L}(n)T_s^2 + 2\boldsymbol{\lambda}\Delta\boldsymbol{\Gamma}_{H_c}) = 0 \tag{23}$$

3.2.3 Exact Solution to the minimization

Since the above result requires the use of a numerical procedure and also the inversion of the inertia matrix, the same assumptions that were made for simplifying the cost function for velocity control will be used again here. These assumptions lead to a simplified predictive control law that allows to find a direct solution to the minimization without using a numerical procedure. This function can be written as:

$$J = \sum_{n=1}^{H_p} \mathbf{S}(n)^T \mathbf{S}(n) \tag{24}$$

where

$$\mathbf{S}(n) = \boldsymbol{q} + n\dot{\boldsymbol{q}}T_s + \frac{n^2}{2}\mathbf{M}^{-1}(\mathbf{\Gamma} - \mathbf{h}_N(\boldsymbol{q}, \dot{\boldsymbol{q}}))T_s^2 - (\tilde{\mathbf{r}} - \mathbf{d}) \tag{25}$$

Setting the derivative of this function with respect to $\mathbf{u}$ equal to zero and after some manipulations summarized in (Duchaine et al., 2007), the following optimal solution is obtained:

$$\mathbf{\Gamma} = \mathbf{h}_N(\boldsymbol{q}, \dot{\boldsymbol{q}}) - \frac{2\mathbf{M}(P_2\dot{\boldsymbol{q}}T_s + \boldsymbol{q} - \mathbf{r} + P_3\mathbf{\Delta r} + \mathbf{d})}{P_1 T_s^2} \tag{26}$$

with

$$\begin{aligned} P_1 &= \frac{(3H_p^2 + 3H_p - 1)}{5} \\ P_2 &= \frac{3H_p(H_p + 1)}{2(2H_p + 1)} \\ P_3 &= \frac{-3H_p^2 + H_p + 2}{4H_p + 2} \end{aligned} \tag{27}$$

where H_p is the horizon of prediction.

It is again pointed out that the direct solution of the minimization given by eq. (26) does not require the computation of the inverse of the inertia matrix.

4. Experimental demonstration

The predictive control algorithm presented in this chapter aims at providing a more accurate control of robots. The first goal of the experiment is thus to compare the performance of the predictive controller to the performance of a PID controller on an actual robot. The second objective is to verify that the simplifying assumptions that were made in this paper hold in practice. The argument in favour of the predictive controller is that it should lead to better performances than a PID control scheme since it takes into account the dynamics of the robot and its futur behaviour while requiring almost the same computation time. In order to illustrate this phenomenon, the control algorithms were first used to actuate a simple 1-dof pendulum. Then, the position and velocity control were implemented on a 6-DOF cable-driven parallel mechanism. The controllers were implemented on a real time QNX computer with a servo rate of 500 Hz --- a typical servo rate for robotics applications. The PID controllers were tuned experimentally by minimizing the square norm of the error of the motors summed over the entire trajectories.

4.1 Illustration with the 1-DOF pendulum

A simple pendulum attached to a direct drive motor was controlled using a PID scheme and the predictive controller. This system, which represents one of the worst candidates for PID controllers, has been used to demonstrate how our assumption on the dynamical model does not affect the capability of the proposed predictive controller to stabilize nonlinear systems. The use of a direct drive motor maximizes the impact of the nonlinear terms of the dynamic model, making the system difficult to control by a conventional regulator.
Also, the simplicity of the system helps to obtain accurate estimations of the parameters of the dynamic model that allow testing the ideal case. Despite the fact that its inertia remains constant over time, under constant angular velocity, the gravitational torque is the dominating term in the dynamic model. This setup also makes it possible to test the velocity control at high speed without having to consider angular limitations.
Figure 2 provides the response of the system (angular velocity) to a given sequence of input reference velocities for the different controllers. The predictive control was implemented according to eq. (17) and an experimentally determined prediction horizon of four was used for the tests. It can be easily seen that PID control is inappropriate for this nonlinear dynamic mechanism. The sinusoidal error corresponds to the gravitational torque that varies with the rotation of the pendulum. The predictive control follows the reference input more adequately as it anticipates the variation of this term.

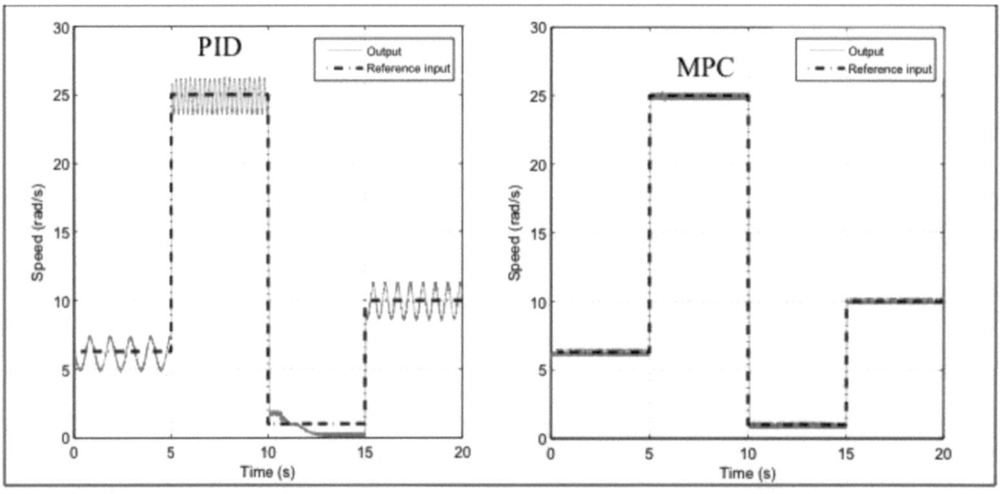

Figure 2. Speed response of the direct drive pendulum for PID and MPC control. (© 2007 IEEE)

4.2: 6-DOF cable-driven robot parallel

A 6-DOF cable-driven robot with an architecture similar to the one presented in (Bouchard & Gosselin, 2007) is used in the experiment. It is shown in Fig.3 where the frame is a cube with two-meters edges. The end-effector is suspended by six cables. The cables are wound on pulleys actuated by motors fixed at the top of the frame.

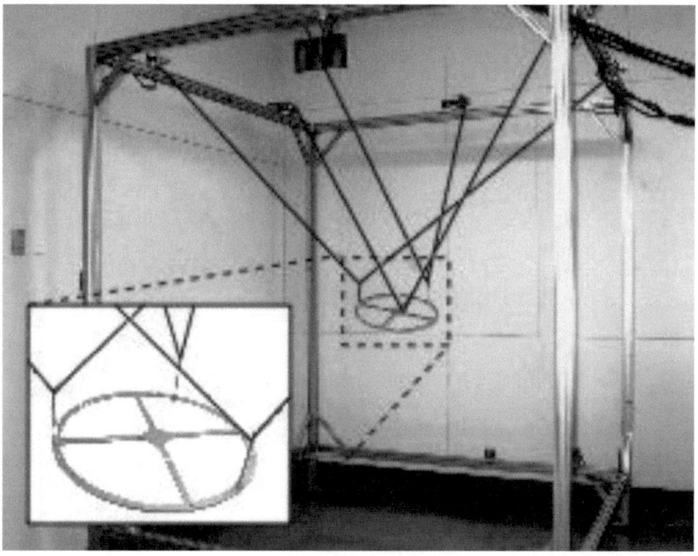

Figure 3. Cable robot used in the experiment. (© 2007 IEEE)

4.2.1 Kinematic modeling

For a given end-effector pose **x** the necessary cable lengths ρ can be calculated using the inverse kinematics. The length of cable *i* can be calculated by:

$$\rho_i^2 = \mathbf{v}_i^T \mathbf{v}_i \tag{28}$$

where

$$\mathbf{v}_i = \mathbf{a}_i - \mathbf{b}_i. \tag{29}$$

In eq. (29), $\mathbf{b}_i$ and $\mathbf{a}_i$ are respectively the position of the attachment point of cable *i* on the frame and on the end-effector, expressed in the global coordinate frame. Thus, vector $\mathbf{a}_i$ can be expressed as:

$$\mathbf{a}_i = \mathbf{c} + \mathbf{Q}\mathbf{a}'_i, \tag{30}$$

$\mathbf{a}'_i$ being the attachment point of cable *i* on the end-effector, expressed in the reference frame of the end-effector and **Q** being the rotation matrix expressing the orientation of the end-effector in the fixed reference frame. Vector **c** is defined as the position of the reference point on the end-effector in the fixed frame.

Considering a fixed pulley radius *r* the cable length can be related to the angular positions $\boldsymbol{\theta}$ of the actuators

$$\boldsymbol{\rho} = r\boldsymbol{\theta}. \tag{31}$$

Substituting ρ in eq. (28) and differentiating with respect to time, one obtains the velocity equation:

$$\mathbf{A}\dot{\mathbf{x}} = \mathbf{B}\boldsymbol{\theta} \tag{32}$$

where $\dot{\mathbf{x}}$ is the twist of the end-effector,

$$\dot{\mathbf{x}} = \begin{bmatrix} \dot{\mathbf{c}}^T & \boldsymbol{\omega}^T \end{bmatrix}^T \tag{33}$$

$$\boldsymbol{\theta} = \begin{bmatrix} \theta_1 & \cdots & \theta_6 \end{bmatrix}^T \tag{34}$$

$$\mathbf{B} = diag[r^2\theta_1, \cdots, r^2\theta_6] \tag{35}$$

$$\mathbf{A} = \begin{bmatrix} \mathbf{c}_1^T \\ \vdots \\ \mathbf{c}_6^T \end{bmatrix}, \tag{36}$$

vector $\mathbf{c}_i$ being

$$\mathbf{c}_i = \begin{bmatrix} (\mathbf{a}_i - \mathbf{b}_i) \\ (\mathbf{Q}\mathbf{a}_i') \times (\mathbf{a}_i - \mathbf{b}_i) \end{bmatrix} \tag{37}$$

and where $\boldsymbol{\omega}$ is the angular velocity of the end-effector.

4.2.2 Dynamic modeling

In this work, the cables are considered straight, massless and infinitely stiff. The assumption of straight cables is justified since the robot is small and the mass of the end-effector is much larger than the mass of the cables, which induces no sag. The measurements are made for chosen trajectories for which the cables are kept in tension at all times. The inertia of the wires is negligible compared to the combined inertia of the pulleys and end-effector. Although it is of research interest, the elasticity of the cables is not considered in the dynamic model for this work. The elastic behaviour is not exhibited strongly because of the high stiffness of the cables under relatively low accelerations (maximum 9.81 m/s^2) of a 600g end-effector.
Eq. (32) can be rearranged as

$$\dot{\mathbf{x}} = \mathbf{J}\dot{\boldsymbol{\theta}} \tag{38}$$

where

$$\mathbf{J} = \mathbf{A}^{-1}\mathbf{B}. \tag{39}$$

From eq. (38) and using the principle of virtual work, the following dynamic equation can be obtained:

$$\boldsymbol{\tau} = \mathbf{I}_p\ddot{\boldsymbol{\theta}} + \mathbf{K}_\nu\dot{\boldsymbol{\theta}} + \mathbf{J}^T\mathbf{M}(\ddot{\mathbf{x}} + \mathbf{w}_g) \tag{40}$$

where $\mathbf{I}_p$ is the inertia matrix of the pulleys and motors combined

$$\mathbf{I}_p = diag[I_{p1}, \cdots, I_{p6}], \tag{41}$$

$\mathbf{K}_v$ is the matrix of the viscous friction at the actuators

$$\mathbf{K}_\nu = diag[k_{\nu 1}, \cdots, k_{\nu 6}], \tag{42}$$

$\mathbf{M}_e$ is the inertia matrix of the end-effector

$$\mathbf{M}_e = \begin{bmatrix} diag[m_e \, m_e \, m_e] & [\mathbf{0}]_{(3\times3)} \\ [\mathbf{0}]_{(3\times3)} & \mathbf{I}_e \end{bmatrix}, \tag{43}$$

m_e being the mass of the end-effector and $\mathbf{I}_e$ its inertia matrix given by the CAD model. Vector $\mathbf{w}_g$ is the wrench applied by gravity on the end-effector.
By differentiating eq. (38) with respect to time, one obtains:

$$\ddot{\mathbf{x}} = \dot{\mathbf{J}}\dot{\boldsymbol{\theta}} + \mathbf{J}\ddot{\boldsymbol{\theta}}. \tag{44}$$

Substituting this expression for $\ddot{\mathbf{x}}$ in eq. (40), the dynamics can be expressed with respect to the joint variables $\boldsymbol{\theta}$:

$$\boldsymbol{\tau} = [\mathbf{I}_p + \mathbf{J}^T\mathbf{M}\mathbf{J}]\ddot{\boldsymbol{\theta}} + \mathbf{J}^T\mathbf{M}\dot{\mathbf{J}}\dot{\boldsymbol{\theta}} - \mathbf{K}_\nu\dot{\boldsymbol{\theta}} + \mathbf{J}^T\mathbf{M}\mathbf{w}_g. \tag{45}$$

Equation (45) has the same form as eq. (5) where

$$\begin{aligned} \mathbf{M}(\boldsymbol{\theta}) &= \mathbf{I}_p + \mathbf{J}^T\mathbf{M}_e\mathbf{J} \\ \mathbf{h}(\boldsymbol{\theta}, \dot{\boldsymbol{\theta}}) &= \mathbf{J}^T\mathbf{M}_e\dot{\mathbf{J}}\dot{\boldsymbol{\theta}} \\ \mathbf{V} &= \mathbf{K}_\nu \\ \mathbf{g}(\boldsymbol{\theta}) &= \mathbf{J}^T\mathbf{M}\mathbf{w}_g \end{aligned} \tag{46}$$

4.2.3 Trajectories

The trajectories are defined in the Cartesian space. For the experiment, the selected trajectory is a displacement of 0.95 meter along the vertical axis performed in one second. The displacement follows a fifth order polynomial with respect to time, with zero speed and acceleration at the beginning and at the end. This smooth displacement is chosen in order to avoid inducing vibrations in the robot since the elasticity of the cables is not taken into account in the dynamic model. As mentioned earlier, it was verified prior to the experiment that this trajectory does not require compression in any of the cables.
The cable robot is controlled using $\boldsymbol{\theta}$ the joint coordinates and velocity of the pulleys. The Cartesian trajectories are thus converted in the joint space using the inverse kinematics and the velocity equations. A numerical solution to the direct kinematic problem is also implemented to determine the pose from the information of the encoders. This estimated pose is used to calculate the terms that depend on $\mathbf{x}$.

4.3 Experimental results for the 6-DOF robot

4.3.1 Velocity control

Figure 4 provides the error between the response of the system (joint velocities) and the time derivative of the joint trajectory described above for the two different controllers. The corresponding control input signals are also shown on this figure. The predictive control algorithm was implemented according to eq. (17) and an experimentally determined horizon of prediction H_p=11 was used.
It can be observed that the magnitude of the error is smaller with the proposed predictive control than with the PID. The control input signals also appears to be smoother with the proposed approach than with the conventional linear controller. The PID suffers from the use of the second derivative of the encoder for the derivative gain (D) which reduces the

stability of the control. The predictive control, according to eq. (17) requires only the encoder signal and its first derivative.

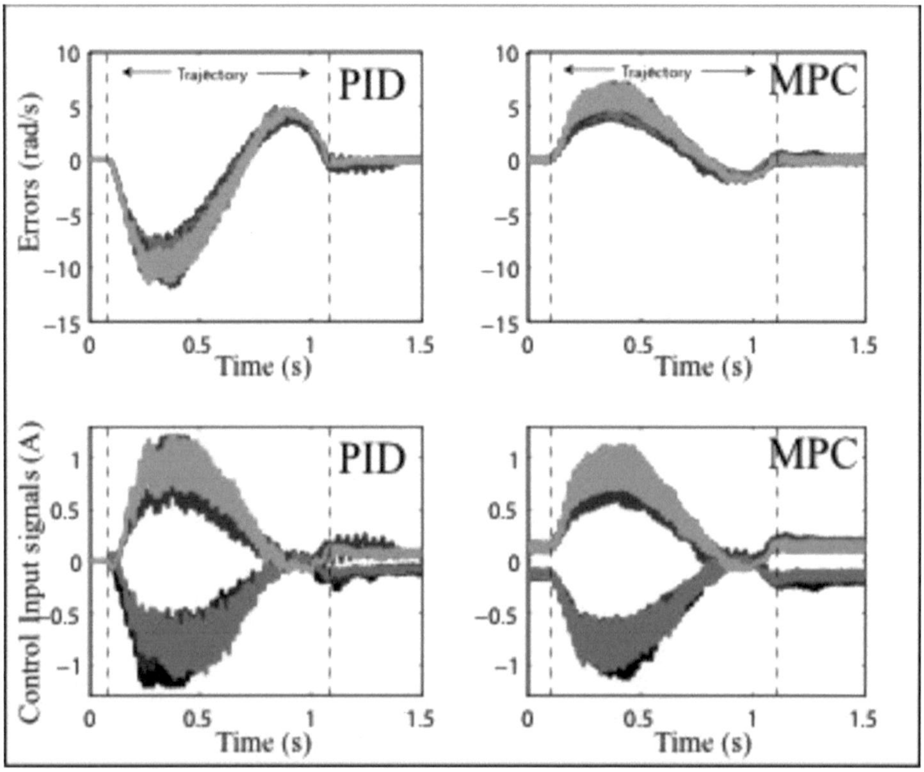

Figure 4. Velocity control - Error between the output and the reference input of the six motors for the PID (top left) and the predictive control (top right). The corresponding control input signals are shown at the bottom.

4.3.2 Position control

A predictive position controller was implemented according to equation (26). An experimentally determined horizon of prediction H_p=14 was used.

Figure 5 illustrates the capability of the proposed scheme to perform position tracking compared to a PID. The error over the trajectory for the six motors and the control input signals are presented on the same figure. It can be seen from these figures that the magnitude of the error is in the same range for the two control methods. The main difference occurs at the end of the trajectory where the PID leads to a small overshoot and takes some time to stabilize. Indeed, that is where the predictive control exhibits a clear advantage of performance over the PID. One can also note that during the trajectory, the PID error appears to have a more random distribution and variation than the errors obtained with the predictive control. In fact, for the latter, the errors follow exactly the velocity profile of the trajectory probably as a consequence of an inaccurate estimation of the friction parameter $\mathbf{K}_v$.

The PID is tuned to follow the trajectory as closely as possible. Even if the magnitude of the acceleration is the same at the end of the trajectory as at the beginning, the velocity is different. At the beginning, the velocity is small. The errors that feed the PID build up fast enough in a short amount of time to provide a good trajectory tracking. At the end of the trajectory, the velocity is higher, causing this time --- due to the integral term --- an overshoot at the end. This type of behaviour is common for a PID and is illustrated with experimental data on figure 6. If it is tuned to track closely a trajectory, there is an overshoot at the end. If it is tuned so there is no overshoot, the tracking is not as good.

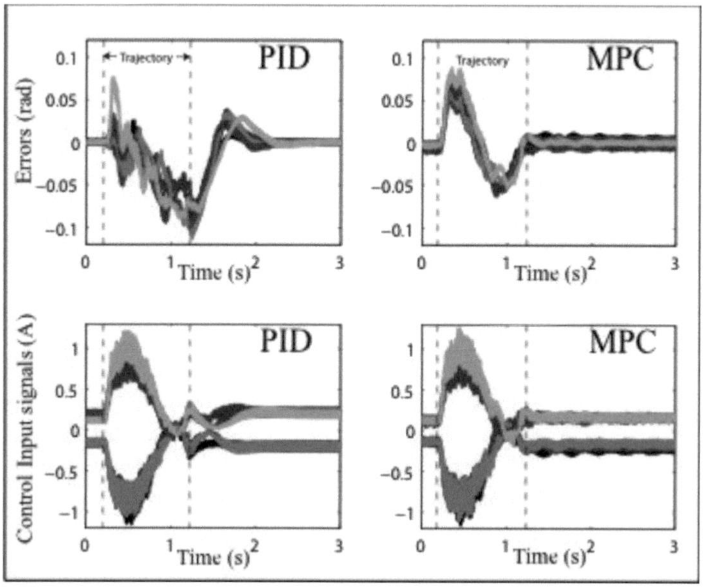

Figure 5. Position control - Error between the output and the reference input of the six motors for the PID (top left) and the predictive control (top right). The corresponding control input signals are shown at the bottom. (© 2007 IEEE)

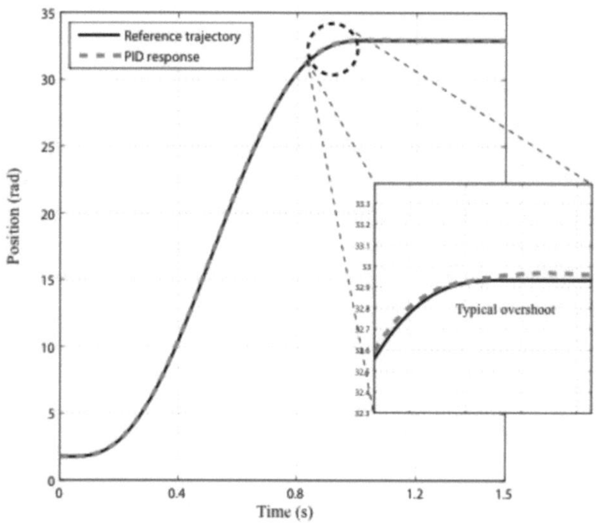

Figure 6. Position control - Angular position trajectory of a motor using the PID. (© 2007 IEEE)

The predictive controller does not suffer from this problem. It is possible to have a controller that tracks closely a fast trajectory, without overshooting at the end. The reason is that the controller anticipates the required torques, the future reference input and takes into account the dynamics and the actual conditions of the system. Experimentally, another advantage of the predictive controller is that it appears to be much easier to tune than the PID. With a good dynamic model, only the prediction horizon has to be adjusted in order to obtain a controller that is more or less aggressive.

4.3.3 Computation time

The computation time was also determined for each controller during position tracking. The results are given in table 1. The PID controller requires a longer calculation time at each step. The integrator in the RT-Lab/QNX computer consumes the most calculation time in the PID. Actually, if it is removed to obtain a PD controller, the computation time drops to 209 μs. Each controller requires computation times of the same order of magnitude. This means that it is fair to compare them with the same servo rate.

Controller	Mean Computation time (μs)
PID controller	506
Predictive control	281

Table 1. Computation time required for each controller

5. Conclusion

This chapter presented a simplified approach to predictive control adapted to robot manipulators. Control schemes were derived for velocity control as well as position tracking, leading to general predictive equations that do not require online optimization. Several justified simplifications were made on the deterministic part of the typical predictive control in order to obtain a compromise between the accuracy of the model and the computation time. These simplifications can be seen as a means of combining the advantages of predictive control with the simplicity of implementation of a computed torque method and the fast computing time of a PID.

Despite all these simplifications, experimental results on a 6-DOF cable-driven parallel manipulator demonstrated the effectiveness of the method in terms of performance. The method using the exact solution of the optimal control appears to alleviate two of the main drawbacks of predictive control for manipulators, namely: the complexity of the implementation and the computational burden.Further investigations should focus on the stability analysis using Lyapunov functions and also on the demonstration of the robustness of the proposed control law.

6. References

Al-Jarrah, O. M. & Zheng, Y. F. (1996). Arm-manipulator coordination for load sharing using compliant control. *IEEE 1996 International Conference on Robotics and automation*, pp. 1000–1005.

Anderson, R. J. (1989). Passive computed torque algorithms for robot. *Proceeding of the 28th Conference on Decision and Control* , pp. 1638–1644.

Berbardt, R.; Lu, D.; & Dong, Y. (2004). A novel cobot and control. *Proceeding of the 5th world congress on intelligent control*, pp. 4635– 4639.

Berlin, F. & Frank, P. M. (1991). Robust predictive robot control. *Proceeding of the 5th International Conference on Advanced Robotics* , pp. 1493–1496.

Bestaoui, Y. & Benmerzouk, D. (1995). A sensity analysis of the computed torque technique. *Proceeding of the American Control Conference* , pp. 4458–4459.

Bouchard, S. & Gosselin, C. M. (2007). Workspace optimization of a very large cable-driven parallel mechanism for a radiotelescope application. *Proceeding of the ASME International Design Engineering Technical Conferences, Mechanics and Robotics Conference* .

Camacho, EF & Bordons, C. (2004). *Model predictive control*, Springer.

Compas, J. M.; Decarreau, P.; Lanquetin, G.; Estival, J. & Richalet, J. (1994). Industrial application of predictive functionnal control to rolling mill, fast robot, river dam. *Proceedings of the 3th IEEE Conference on Control Applications* , pp. 1643–1655.

Duchaine, V. & Gosselin, C. (2007). General model of human-robot cooperation using a novel velocity based variable impedance control. *Proceeding of the IEEE World haptics 2007*, pp. 446–451.

Duchaine, V.; Bouchard, S. & Gosselin, C. (2007). Computationally Efficient Predictive Robot Control, *IEEE/ASME Transactions on Mechatronics*, Vol.12, pp. 570-578

Hedjar, R. & Boucher, P. (2005). Nonlinear receding-horizon control of rigid link robot manipulators. *International Journal of Advanced Robotic Systems*, pp. 15–24.

Hedjar, R.; Toumi, R.; Boucher, P. & Dumur, D. (2005). Finite horizon nonlinear predictive control by the taylor approximation: Application to robot tracking trajectory. *Int.J. Appl. Math.Sci*, pp. 527–540.

Kalman, R. (1960). Contributions to the theory of optimal control. *Bul l. Soc. Math. Mex.* (1960), pp. 102–119.

Kennel, R.; Linder, A. & Linke, M. (2001). Generalized predictive control (gpc)- ready for use in drive application? *32nd IEEE Power Electronics Specialists Conference PELS.*

Lydoire, F. & Poignet, P. (2005). Non linear model predictive control via interval analysis. *Proceeeding of the 44th IEEE Conference on Decision and Control*, pp. 3771–3776.

Morari, M. & Lee, J.H. (1999), Model predictive control: past, present and future, *Computer and Chemical Engineering*, Vol.23, pp. 667-682

Peshkin, M.; Colgate, J.; Wannasuphoprasit, W.; Moore, C. & Gillespie, R. (2001). Cobot architecture. *IEEE Transaction on Robotics and Automation* 17, pp. 377–390.

Poignet, P. & Gautier, M. (2000). Nonlinear model predictive control of a robot manipulator. *Proceedings of the 6th International Workshop on Advanced Motion Control*, 2000. , pp. 401–406.

Qin, S. & Badgwell, T. (1997). An overview of industrial model predictive control technology. *Chemical Process Control*-V, pp. 232–256.

Uebel, M., Minis, I., and Cleary, K. (1992). Improved computed torque control for industrial robots. *International Conference on Robotics and Automation*, pp. 528–533.

Vivas, A., Poignet, P., and Pierrot, F. (2003). Predictive functional control for a parallel robot. *Proceedings of the International Conference on Intelligent Robots and Systems*, pp. 2785–2790.

Von Wissel, D., Nikoukhah, R., and Campbell, S. L. (1994). On a new predictive control strategy: Application to a flexible-joint robot. *Proceedings of the IEEE Conference on Decision and Control*, pp. 3025–3026.

Zhang, Z., and Wang, W. (2005). Predictive function control of a two links robot manipulator. *Proceeding of the Int. Conf. on Mechatronics and Automation*, pp. 2004 - 2009.

8

Experimental Control of Flexible Robot Manipulators

A. Sanz and V. Etxebarria
Universidad del País Vasco
Spain

1. Introduction

Flexible-link robotic manipulators are mechanical devices whose control can be rather challenging, among other reasons because of their intrinsic under-actuated nature. This chapter presents various experimental studies of diverse robust control schemes for this kind of robotic arms. The proposed designs are based on several control strategies with well defined theoretical foundations whose effectiveness are demonstrated in laboratory experiments.

First it is experimented a simple control method for trajectory tracking which exploits the two-time scale nature of the flexible part and the rigid part of the dynamic equations of flexible-link robotic manipulators: a slow subsystem associated with the rigid motion dynamics and a fast subsystem associated with the flexible link dynamics. Two experimental approaches are considered. In a first test an LQR optimal design strategy is used, while a second design is based on a sliding-mode scheme. Experimental results on a laboratory two-dof flexible manipulator show that this composite approach achieves good closed-loop tracking properties for both design philosophies, which compare favorably with conventional rigid robot control schemes.

Next the chapter explores the application of an energy-based control design methodology (the so-called IDA-PBC, interconnection and damping assignment passivity-based control) to a single-link flexible robotic arm. It is shown that the method is well suited to handle this kind of under-actuated devices not only from a theoretical viewpoint but also in practice. A Lyapunov analysis of the closed-loop system stability is given and the design performance is illustrated by means of a set of simulations and laboratory control experiments, comparing the results with those obtained using conventional control schemes for mechanical manipulators.

The outline of the chapter is as follows. Section 2 covers a review on the modeling of flexible-link manipulators. In subsection 2.1 the dynamic modelling of a general flexible multilink manipulator is presented and in subsection 2.2 the methodology is applied to a laboratory flexible arm. Next, some control methodologies are outlined in section 3. Two control strategies are applied to a two-dof flexible robot manipulator. The first design, in subsection 3.1, is based on an optimal LQR approach and the second design, in subsection 3.2, is based on a sliding-mode controller for the slow subsystem. Finally, section 4 covers the IDA-PBC method. In subsection 4.1 an outline of the method is given and in subsection

4.2 the proposed control strategy based on IDA-PBC is applied to a laboratory arm. The chapter ends with some concluding remarks.

2. Modeling of flexible-link manipulator

2.1 The dynamic modeling of a general flexible multilink manipulator

In order to obtain a dynamic model for a multilink flexible robot manipulator, it becomes necessary the introduction of a convenient kinematic description of the manipulator, including the deformation of the links. In order to limit the complexity of the derivation, it is assumed that rigid motion and link deformation happen in the same plane, without torsional effects. A sketch of a two-link manipulator of this kind is shown in Fig. 1 with an appropriate coordinate frame assignment.

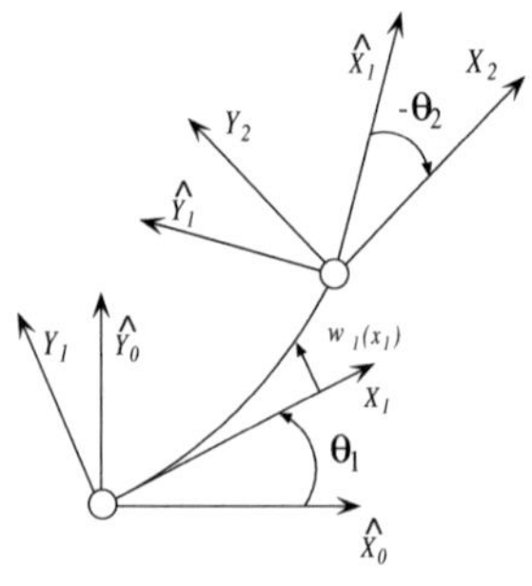

Figure 1. Planar two-link flexible manipulator

The rigid motion is described by the joint angles θ_1, θ_2, while $w_1(x_1)$ denotes the transversal deflection of link 1 at x_1 with $0\leq x_1 \leq l_1$, being l_1 the link length. This deflection is expressed as a superposition of a finite number of modes where the spatial and time variables are separated:

$$w_1(x,t) = \sum_{i=1}^{n_e} \varphi_i(x) q_i(t) \tag{1}$$

where $\varphi_i(x)$ is the mode shape and $q_i(t)$ is the mode amplitude.
To obtain a finite-dimensional dynamic model, the assumed modes link approximation can be used. By applying Lagrange's formulation the dynamics of any multilink flexible-link robot can be represented by

$$H(r)\ddot{r} + V(r,\dot{r}) + Kr + F(\dot{r}) + G(r) = B(r)\tau \tag{2}$$

where r(t) expresses the position vector as a function of the rigid variables θ and the deflection variables q. $H(r)$ represents the inertia matrix, $V(r,\dot{r})$ are the components of the Coriolis vector and centrifugal forces, K is the diagonal and positive definite link stiffness matrix, $F(\dot{r})$ is the friction matrix, and $G(r)$ is the gravity matrix. $B(r)$ is the input matrix, which depends on the particular boundary conditions corresponding to the assumed modes. Finally, τ is the vector of the input joint torques.
The explicit equations of motion can be derived computing the kinetic energy T and the potential energy U of the system and then forming the Lagrangian.

The kinetic energy of the entire system is

$$T = \sum_{i=1}^{n} T_{hi} + \sum_{i=1}^{n} T_{li} \tag{3}$$

where T_{hi} is the kinetic energy of the rigid body located at hub i of mass m_{hi} and moment of inertia I_{hi}; r_i indicates the absolute position in frame $(\hat{X}_0, \hat{Y}_0)$ of the origin of frame (X_i, Y_i) and $\dot{\alpha}_i$ is the absolute angular velocity of frame (X_i, Y_i).

$$T_{hi} = \frac{1}{2} m_{hi} \dot{r}_i^T \dot{r}_i + \frac{1}{2} I_{hi} \dot{\alpha}_i^2 \tag{4}$$

And T_{li} is the kinetic energy of the *ith* element where ρ_i is the mass density per unit length of the element and $\dot{p}_i$ is the absolute linear velocity of an arm point.

$$T_{li} = \frac{1}{2} \int_0^{li} \rho_i(x_i) \dot{p}_i(x_i)^T \dot{p}_i(x_i) dx_i \tag{5}$$

Since the robot moves in the horizontal plane, the potential energy is given by the elastic energy of the system, because the gravitational effects are supported by the structure itself, and therefore they disappear inside the formulation:

$$U = \sum_{i=1}^{n} U_{ei} \tag{6}$$

The elastic energy stored in link i is

$$U_{ei} = \frac{1}{2} \int (EI)_i(x_i) \left(\frac{d^2 \mathrm{w}_i(x_i)}{dx_i^2}\right)^2 dx_i \tag{7}$$

The total elastic energy of the system can be written as:

$$U = U_e = \frac{1}{2} q^T \cdot K \cdot q \tag{8}$$

where K is the diagonal stiffness matrix that only affects to the flexible modes.
As a result, the dynamical equation (2) can be partitioned in terms of the rigid, $\theta_i(t)$, and flexible, $q_{ij}(t)$, generalized coordinates.

$$\begin{pmatrix} H_{\theta\theta}(\theta,q) & H_{\theta q}(\theta,q) \\ H_{\theta q}^T(\theta,q) & H_{qq}(\theta,q) \end{pmatrix} \begin{pmatrix} \ddot{\theta} \\ \ddot{q} \end{pmatrix} + \begin{pmatrix} c_\theta(\theta,q,\dot{\theta},\dot{q}) \\ c_q(\theta,q,\dot{\theta},\dot{q}) \end{pmatrix} + \begin{pmatrix} 0 \\ D\dot{q} + Kq \end{pmatrix} = \begin{pmatrix} \tau \\ 0 \end{pmatrix} \tag{9}$$

Where $H_{\theta\theta}$, $H_{\theta q}$ y H_{qq} are the blocks of the inertia matrix H, which is symmetric and positive definite, c_θ, c_q are the components of the vector of Coriolis and centrifugal forces and D is the diagonal and positive semidefinite link damping matrix.

2.2 Application to a flexible laboratory arm

The above analysis will now be applied to the flexible manipulator shown in Fig. 2, whose geometric sketch is displayed in Fig. 3. As seen in the figures, it is a laboratory robot which moves in the horizontal plane with a workspace similar to that of common Scara industrial manipulators. Links marked 1 and 4 are flexible and the remaining ones are rigid. The flexible links' deflections are measured by strain gauges, and the two motor's positions are measured by standard optical encoders.

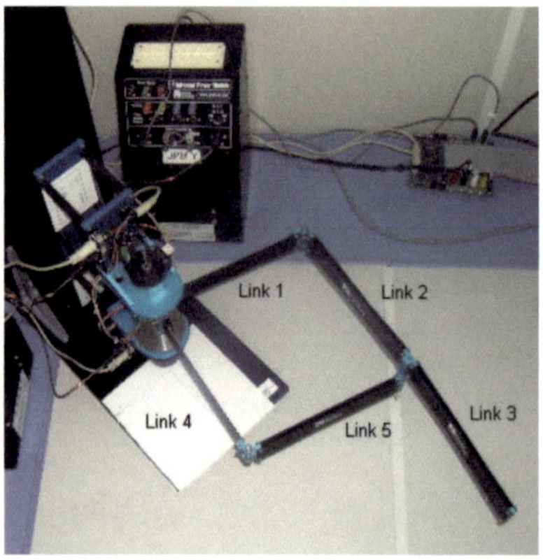

Figure 2. Photograph of the experimental flexible manipulator

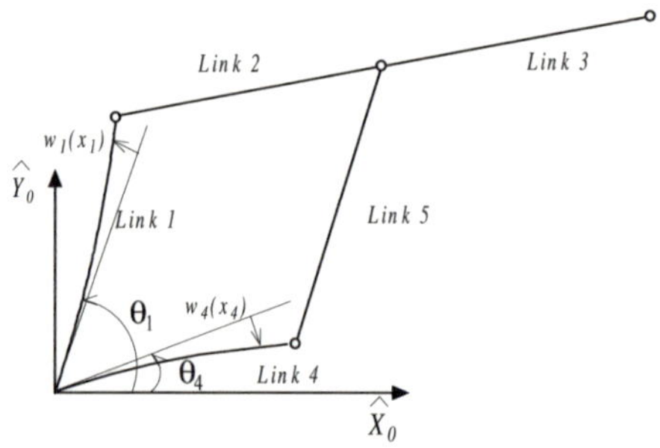

Figure 3. Planar two-dof flexible manipulator

The model is obtained by a standard Lagrangian procedure, where the inertia matrices have the structure:

$$H_{\theta\theta}(\theta,q)=\begin{pmatrix} h_{11} & h_{12} \\ h_{21} & h_{22} \end{pmatrix} \quad H_{\theta q}(\theta,q)=\begin{pmatrix} h_{13} & h_{14} \\ h_{23} & h_{24} \end{pmatrix} \quad H_{qq}(\theta,q)=\begin{pmatrix} h_{33} & h_{34} \\ h_{43} & h_{44} \end{pmatrix}$$

with:

$$\begin{aligned} h_{11} = \; & I_{h1} + m_{j1}(l^2 + \varphi_1^2 q_1^2) + m_{j2}(2l^2 + \varphi_1^2 q_1^2 + l^2 \varphi_4'^2 q_4^2) + I_{h5} + I_{h6} + \frac{1}{3} m_1 l^2 + \varsigma_{111} q_1^2 + m_2(2l^2 + 2\varphi_1^2 q_1^2) \\ & + m_5(\frac{1}{3} l^2 + \frac{1}{3} l^2 \varphi_4'^2 q_4^2) \\ h_{12} = \; & m_{j2}(2l^2 \cos(\theta_1 - \theta_4) + l^2 \varphi_1'^2 q_1 \sin(\theta_1 - \theta_4) - l\varphi_1 q_1 \sin(\theta_1 - \theta_4) + l\cos(\theta_1 - \theta_4)\varphi_1 \varphi_1' \, q_1^2 - l^2 \sin(\theta_1 - \theta_4)\varphi_4' \, q_4 \\ & + l\sin(\theta_1 - \theta_4)\varphi_4 q_4 + l\cos(\theta_1 - \theta_4)\varphi_4 \varphi_4' \, q_4^2) + 2m_2(l^2 \cos(\theta_1 - \theta_4) + l^2 \sin(\theta_1 - \theta_4)\varphi_1' \, q_1 - l\sin(\theta_1 - \theta_4)\varphi_1 q_1 \\ & + l\cos(\theta_1 - \theta_4)\varphi_1 \varphi_1' \, q_1^2) + m_5(\frac{1}{2} l^2 \cos(\theta_1 - \theta_4) - \frac{1}{2} l^2 \sin(\theta_1 - \theta_4)\varphi_4' \, q_4 + \frac{1}{2} l\sin(\theta_1 - \theta_4)\varphi_4 q_4 \\ & + \frac{1}{2} l\cos(\theta_1 - \theta_4)\varphi_4 \varphi_4' \, q_4^2) \end{aligned}$$

$$h_{21} = h_{12}$$

$$h_{22} = I_{h2} + m_{j2}(l^2 + l^2\varphi_1'^2 q_1^2) + I_{h3} + I_{h4} + m_{j1}(l^2 + \varphi_4^2 q_4^2) + m_{j2}(l^2 + \varphi_4^2 q_4^2) + 2m_2(\frac{4}{3}l^2 + \frac{4}{3}l^2\varphi_1'^2 q_1^2) + \frac{1}{3}m_4 l^2 + \varsigma_{411} q_4^2 + m_5(l^2 + \varphi_4^2 q_4^2)$$

$$h_{13} = m_{j1} l\varphi_1 + m_{j2}(l\varphi_1 + l^2\cos(\theta_1 - \theta_4)\varphi_1' - l\sin(\theta_1 - \theta_4)\varphi_1\varphi_1' q_1) + \upsilon_{11} + 2m_2(l\varphi_1 + l^2\cos(\theta_1 - \theta_4)\varphi_1' - l\sin(\theta_1 - \theta_4)\varphi_1\varphi_1' q_1)$$

$$h_{14} = I_{h5}\varphi_4' + m_{j2}(l^2\varphi_4' + l\cos(\theta_1 - \theta_4)\varphi_4 - l\sin(\theta_1 - \theta_4)\varphi_4\varphi_4' q_4) + I_{h6}\varphi_4' + m_5(\frac{1}{3}l^2\varphi_4' - \frac{1}{2}l\sin(\theta_1 - \theta_4)\varphi_4\varphi_4' q_4 + \frac{1}{2}l\cos(\theta_1 - \theta_4)\varphi_4)$$

$$h_{23} = I_{h2}\varphi_1' + m_{j2}(l^2\varphi_1' + l\cos(\theta_1 - \theta_4)\varphi_1 + l\sin(\theta_1 - \theta_4)\varphi_1\varphi_1' q_1) + I_{h3}\varphi_1' + 2m_2(\frac{4}{3}l^2\varphi_1' + l\sin(\theta_1 - \theta_4)\varphi_1\varphi_1' q_1 + l\cos(\theta_1 - \theta_4)\varphi_1)$$

$$h_{24} = m_{j1} l\varphi_4 + m_{j2}(l\varphi_4 + l^2\cos(\theta_1 - \theta_4)\varphi_4' + l\sin(\theta_1 - \theta_4)\varphi_4\varphi_4' q_4) + \upsilon_{41} + m_5(l\varphi_4 + \frac{1}{2}l^2\cos(\theta_1 - \theta_4)\varphi_4' + \frac{1}{2}l\sin(\theta_1 - \theta_4)\varphi_4\varphi_4' q_4)$$

$$h_{33} = m_{j1}\varphi_1^2 + I_{h2}\varphi_1'^2 + m_{j2}(\varphi_1^2 + l^2\varphi_1'^2 + 2l\cos(\theta_1 - \theta_4)\varphi_1\varphi_1') + I_{h3}\varphi_1'^2 + \varsigma_{111} + 2m_2(\varphi_1^2 + \frac{4}{3}l^2\varphi_1'^2 + 2l\cos(\theta_1 - \theta_4)\varphi_1\varphi_1')$$

$$h_{34} = h_{43} = 0$$

$$h_{44} = m_{j1}\varphi_4^2 + I_{h5}\varphi_4'^2 + m_{j2}(\varphi_4^2 + l^2\varphi_4'^2 + 2l\cos(\theta_1 - \theta_4)\varphi_4\varphi_4') + I_{h6}\varphi_4'^2 + \varsigma_{411} + m_5(\varphi_4^2 + \frac{1}{3}l^2\varphi_4'^2 + l\cos(\theta_1 - \theta_4)\varphi_4\varphi_4')$$

where φ_i is the amplitude of the first mode evaluated in the end of the link i, $\varphi_i' = (d\varphi_i / dx_i)\big|_{x_i = l_i}$; υ_{ij} is the deformation moment of order one of mode j of link i and ς_{ijk} is the cross moment of modes j and k of link i. Also l is the links length, m_{j1} is the elbow joint mass, m_{j2} is the tip joint mass, m_i (i=1,4) is the flexible link mass, m_k (k=2,3,5) is the rigid link mass and q_i is the deflection variable.

3. Control design

The control of a manipulator formed by flexible elements bears the study of the robot's structural flexibilities. The control objective is to move the manipulator within a specific trajectory but attenuating the vibrations due to the elasticity of some of its components.
Since time scales are usually present in the motion of a flexible manipulator, it can be considered that the dynamics of the system is divided in two parts: one associated to the manipulator's movement, considered slow, and another one associated to the deformation of the flexible links, much quicker. Taking advantage of this separation, diverse control strategies can be designed.

3.1 LQR control design

In a first attempt to design a control law, a standard LQR optimal technique is employed. The error feedback matrix gains are obtained such that the control law $\tau(t) = -K_c \Delta x$ minimizes the cost function:

$$J = \int_0^\infty (\Delta x^* Q \Delta x + \tau * R\tau) dt \tag{10}$$

where Q and R are the standard LQR weighting matrices; and Δx is the applicable state vector comprising either flexible, rigid or both kind of variables. This implies to minimize the tracking error in the state variables, but with an acceptable cost in the control effort. The resulting Riccati equations can conveniently be solved with state-of-the-art tools such as the Matlab Control System Toolbox .

The effectiveness of the proposed control schemes has been tested by means of real time experiments on a laboratory two-dof flexible robot. This manipulator, fabricated by Quanser Consulting Inc. (Ontario, Canada), has two flexible links joined to rigid links using high quality low friction joints. The laboratory system is shown in Fig. 2.

From the general equation (9) the dynamics corresponding to the rigid part can be extracted:

$$\begin{pmatrix} h_{11}(\theta,q) & h_{12}(\theta,q) \\ h_{21}(\theta,q) & h_{22}(\theta,q) \end{pmatrix} \begin{pmatrix} \ddot{\theta}_1 \\ \ddot{\theta}_4 \end{pmatrix} + \begin{pmatrix} c_{11}(\theta,q,\dot{\theta},\dot{q}) \\ c_{21}(\theta,q,\dot{\theta},\dot{q}) \end{pmatrix} = \begin{pmatrix} \tau_1 \\ \tau_4 \end{pmatrix} \tag{11}$$

The first step is to linearize the system (11) around the desired final configuration or reference configuration. Defining the incremental variable $\Delta\theta = \theta - \theta_d$ and with a state-space representation defined by $\Delta x = (\Delta\theta \quad \Delta\dot{\theta})^T$ the approximate model corresponding to the considered flexible manipulator's rigid part can be described as:

$$\Delta\dot{x} = \begin{pmatrix} 0 & 0 & 1 & 0 \\ 0 & 0 & 0 & 1 \\ 0 & 0 & 0 & 0 \\ 0 & 0 & 0 & 0 \end{pmatrix} \Delta x - \begin{pmatrix} 0 \\ H_{\theta\theta}^{-1} \end{pmatrix} \tau \tag{12}$$

where it has been taken into account the fact that the Coriolis terms are zero around the desired reference configuration.

Table 1 displays the physical parameters of the laboratory robot. Using these values, the following control laws have been designed. In all the experiments the control goal is to follow a prescribed cartesian trajectory while damping the excited vibrations as much as possible.

Property	Value
Motor 1 inertia , I_h	0.0081 Kgm^2
Link length, l	0.23 m
Flexible link mass, m_1,m_4	0.09 Kg
Rigid link mass, m_2,m_3,m_5	0.08 Kg
Elbow joint mass, m_{j1}	0.03 Kg
Tip joint mass, m_{j2}	0.04 Kg

Table 1. Flexible link parameters

The control goal is to track a cartesian trajectory which comprises 10 straight segments in about 12 seconds. Each stretch is covered in approximately 0.6 seconds, after which the arm remains still for the same time at the points marked 1 to 10 in Fig. 4. As a first test, Fig. 5 shows the system's response with a pure rigid LQR controller that is obtained solving the Riccati equations for R=*diag* (2 2) and Q=*diag*(20000 20000 10 10). The values of R and Q have been chosen experimentally, keeping in mind that if the values of Q are relatively large with respect to the values of R a quick evolution is specified toward the desired equilibrium $\Delta x = 0$ which causes a bigger control cost (see (10)). As seen in the figures, this kind of control, which is very commonly used for rigid manipulators, is able to follow the prescribed trajectory but with a certain error, due to the links' flexibilities.

It has been shown in the above test that, as expected, a flexible manipulator can not be very accurately controlled if a pure rigid control strategy is applied. From now on two composite control schemes (where both the rigid and the flexible dynamics are taken into account) will be described.

The non-linear dynamic system (9) is being linearized around the desired final configuration. The obtained model is:

$$\begin{pmatrix} H_{\theta\theta}(\theta,q) & H_{\theta q}(\theta,q) \\ H_{\theta q}^{T}(\theta,q) & H_{qq}(\theta,q) \end{pmatrix}\begin{pmatrix} \Delta\ddot{\theta} \\ \Delta\ddot{q} \end{pmatrix}+\begin{pmatrix} 0 & 0 \\ 0 & K \end{pmatrix}\begin{pmatrix} \Delta\theta \\ \Delta q \end{pmatrix}=\begin{pmatrix} \Delta\tau \\ 0 \end{pmatrix} \tag{13}$$

where again the Coriolis terms are zero around the given point.

Reducing the model to the maximum (taking a single elastic mode in each one of the flexible links), a state vector is defined as:

$$\Delta x = (\Delta\theta \quad \Delta q \quad \Delta\dot{\theta} \quad \Delta\dot{q})^{T} \tag{14}$$

and the approximate model of the considered flexible manipulator can be described as:

$$\Delta\dot{x} = \begin{pmatrix} 0 & 0 & I & 0 \\ 0 & 0 & 0 & I \\ 0 & L_2^{-1}H_{qq}^{-1}K & 0 & 0 \\ 0 & L_1^{-1}H_{\theta q}^{T-1}K & 0 & 0 \end{pmatrix}\Delta x + \begin{pmatrix} 0 \\ 0 \\ L_2^{-1}H_{\theta q}^{-1} \\ L_1^{-1}H_{\theta\theta}^{-1} \end{pmatrix}\Delta\tau \tag{15}$$

where

$$L_1^{-1} = (H_{\theta\theta}^{-1}H_{\theta q} - H_{\theta q}^{T-1}H_{qq})^{-1}$$
$$L_2^{-1} = (H_{\theta q}^{-1}H_{\theta\theta} - H_{qq}^{-1}H_{\theta q}^{T})^{-1}$$

Following the same LQR control strategy as in the previous lines, a scheme is developed making use of the values of Table 1, with R=*diag*(2 2) and Q=*diag*(19000 19000 1000000 1000 100 100 100 100). From measurements on the laboratory robot the following values for the elements' inertias $I_{h2} = I_{h3} = I_{h5} = I_{h6} = 1.10^{-5}$ Kg.m^2 have been estimated. Also the elastic constant K=113, and the space form in the considered vibration modes $\varphi_{11} = \varphi_{41} = 0.001$m and $\varphi'_{11} = \varphi'_{41} = 0.1$ have been measured.

In Fig. 6 the obtained results applying a combined LQR control are shown. The tip position exhibits better tracking of the desired trajectory, as it can be seen comparing Fig. 5(a) and Fig. 6(a). Even more important, it should be noted that the oscillations of elastic modes are now attenuated quickly (compare Fig. 5(b,c) and Fig. 6(b,c)).

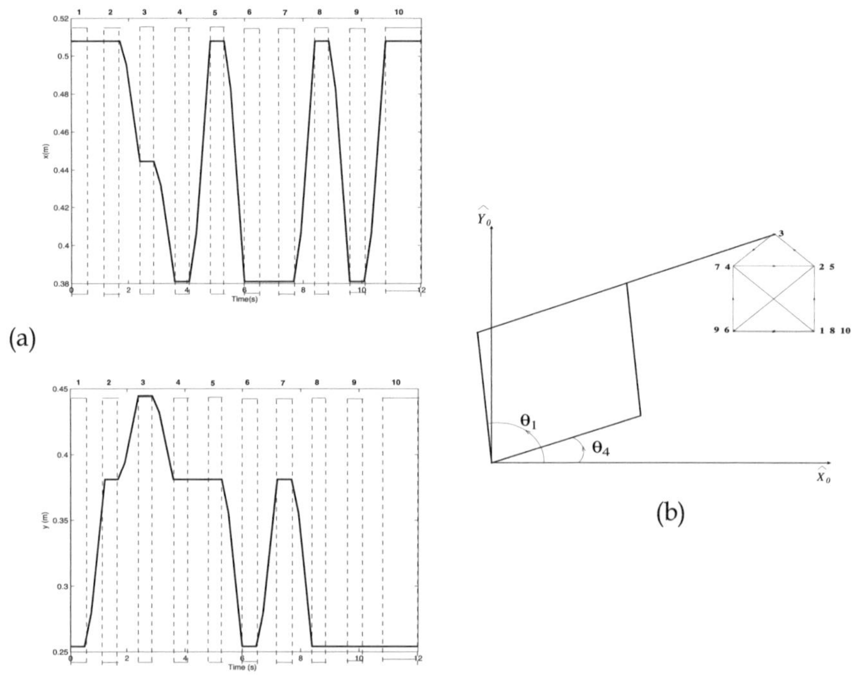

Figure 4. Cartesian trajectory of the manipulator. (a) Time evolution of the x and y coordinates. (b) Sketch of the manipulator and trajectory

3.2 Sliding-mode control design

The presence of unmodelled dynamics and disturbances can lead to poor or unstable control performance. For this reason the rigid part of the control can be designed using a robust sliding mode philosophy, instead of the previous LQR methods. This control design has been proved to hold solid stability and robustness theoretical properties in rigid manipulators (Barambones & Etxebarria, 2000), and is tested now in real experiments.
The rigid control law is thus changed to:

$$\tau_{ri} = \begin{cases} p_i \cdot \mathrm{sgn}(S_i) & if \ \ |S_i| > \beta_i \\ p_i \dfrac{S_i}{\beta_i} & otherwise \end{cases}$$

where $\beta = (\beta_1, \ldots, \beta_n)^T, \beta_i > 0$ is the width of the band for each sliding surface associated to each motor, p_i is a constant parameter and S is a surface vector defined by $S = \dot{E} + \lambda E$, being $E = \theta - \theta_d$ the tracking error (θ_d is the desired trajectory vector) and $\lambda = diag(\lambda_1, \ldots, \lambda_n), \lambda_i > 0$.

The philosophy of this control is to attract the system to S_i=0 by an energetic control $p \cdot \mathrm{sgn}(S_i)$ and once inside a band around S_i=0, to regulate the position using a smoother control, roughly coincident with the one designed previously. The sliding control can be tuned by means of the sliding gain p and by adjusting the width of band β which limits the area of entrance of the energetic part of the control. It should be kept in mind that the external action to the band should not be excessive, to avoid exciting the flexibilities of the fast subsystem as much as possible (which would invalidate the separation hypothesis between the slow dynamics and the fast one).

In this third experiment a second composite control scheme (sliding-mode for the slow part and LQR for the fast one) is tested.

Fig. 7 exhibits the experimental results obtained with the sliding control strategy described for the values β=(6 4 9 4) and λ=(10.64 6.89 32.13 12.19). β is the width of the band for each sliding surface associated to each motor, which limits the area of entrance of the energetic part of the control. If the values of β are too big, the action of the energetic control is almost negligible; on the contrary if this values are too small, the action of the external control is much bigger and the flexible modes can get excited. The width of the band is chosen by means of a trial-and-error experimental methodology. The values used of λ are such that the gain control values inside the band match the LQR gains designed in the previous section.

In the same way as with the combined LQR control, the rigid variable follows the desired trajectory, and again, the elastic modes are attenuated with respect to the rigid control, (compare Fig. 5(b,c) and Fig. 7(b,c)). Also, it can be seen that this attenuation is even better than in the previous combined LQR control case (compare Fig. 6(b,c) and Fig. 7(b,c)).

To gain some insight on the differences of the two proposed combined schemes, the contribution to the control torque is compared in the sliding-mode case (rigid) with the LQR combined control case (rigid), both to τ_1 (Fig. 8(a) and 8(b)). It is observed how in the intervals from 1 to 2s., 2 to 3s. (corresponding to the stretch between the points marked 2 and 3 in Fig.4) and 7 to 8s. (corresponding to the diagonal segment between the points marked 7 and 8 in Fig.4), the sliding control acts in a more energetic fashion, favoring this way the pursuit of the wanted trajectory. However, in the intervals from 5 to 6s. (point marked 5 in Fig.4), 6 to 7s. (marked 6), 9.5 to 10.5s. (marked 9) and 11 to 12s. (marked 10), in the case of the sliding control, a smoother control acts, and its control effort is smaller than in the case of the LQR combined control, causing this way smaller excitation of the oscillations in these intervals (as it can be seen comparing Fig. 7(b,c) with Fig. 6(b,c)). Note that the sliding control should be carefully tuned to achieve its objective (attract the system to S_i =0) but without introducing too much control activity which could excite the flexible modes to an unwanted extent.

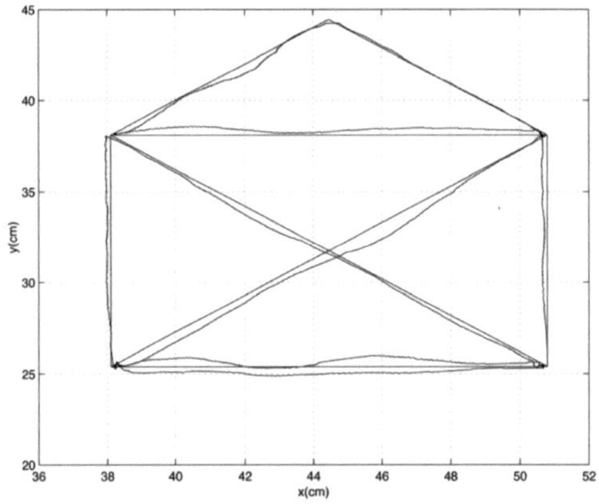

(a)

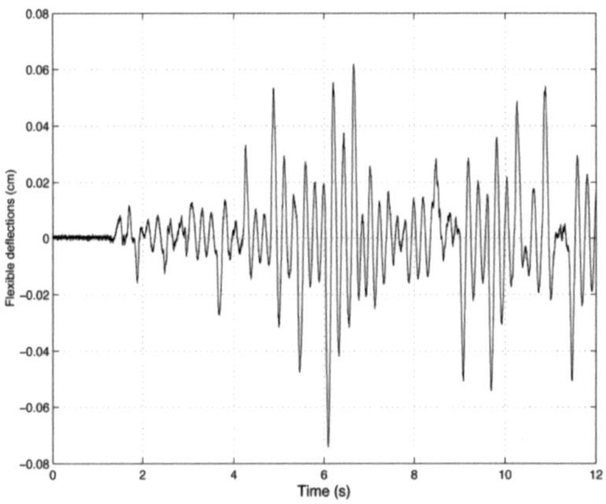

(b)

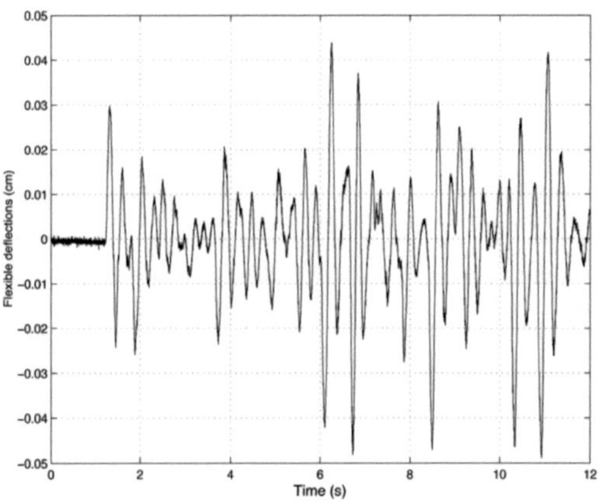

(c)

Figure 5. Experimental results for pure rigid LQR control. a) Tip trajectory and reference. b) Time evolution of the flexible deflections (link 1 deflections). c) Time evolution of the flexible deflections (link 4 deflections)

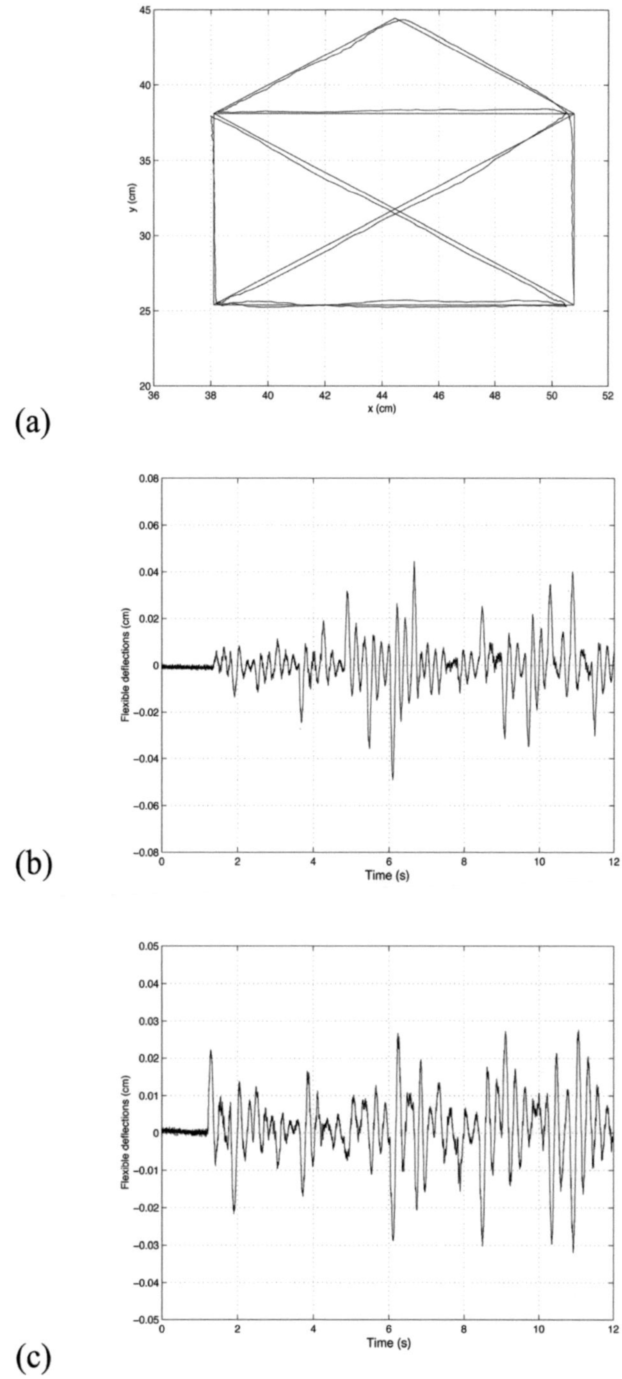

Figure 6. Experimental results for composite (slow-fast) LQR control. a) Tip trajectory and reference. b) Time evolution of the flexible deflections (link 1 deflections). c) Time evolution of the flexible deflections (link 4 deflections)

(a)

(b)

(c)

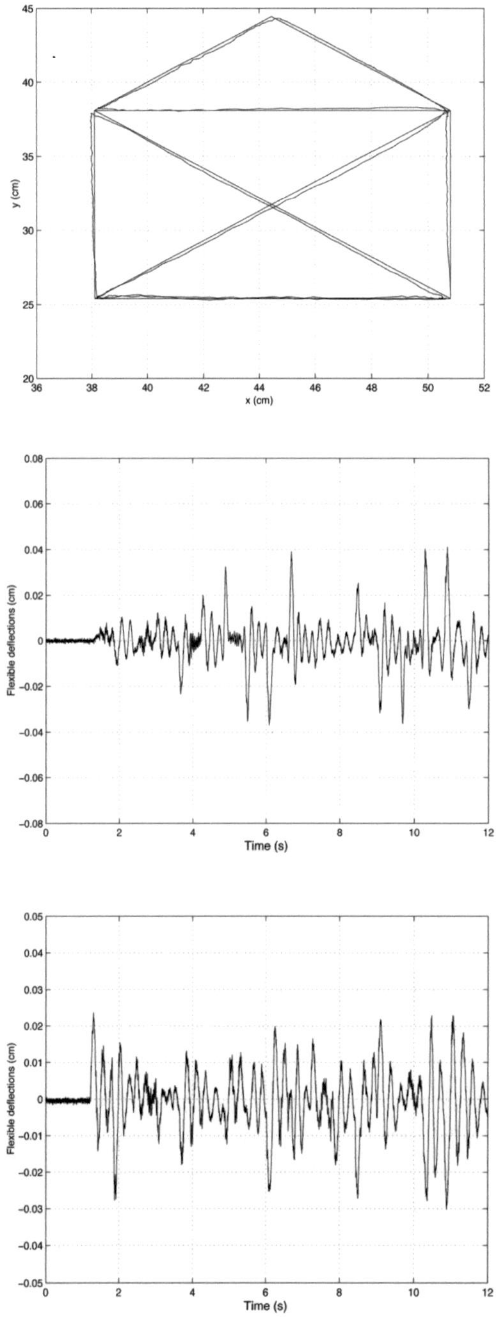

Figure 7. Experimental results for composite (slow-fast) sliding-LQR control. a) Tip trajectory and reference. b) Time evolution of the flexible deflections (link 1 deflections). c) Time evolution of the flexible deflections (link 4 deflections)

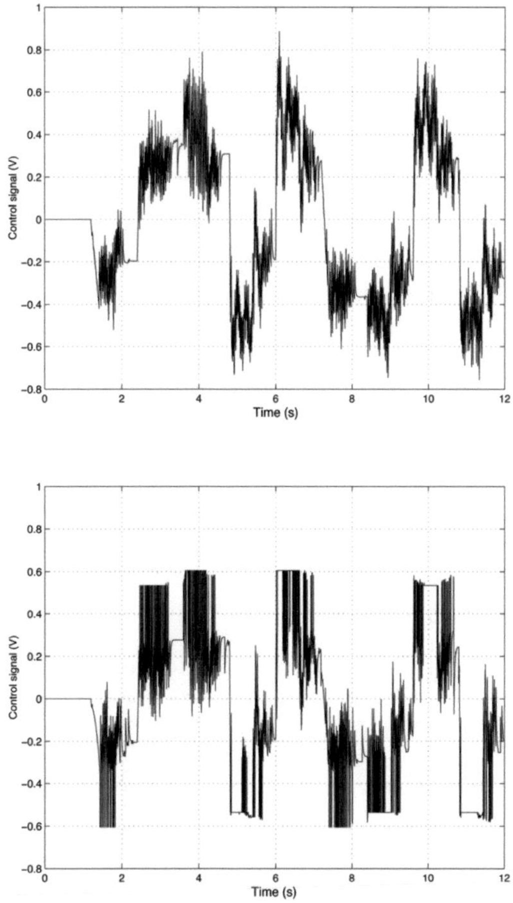

(a)

(b)

Figure 8. a) Contribution to the torque control τ_1 (rigid) with the LQR combined control squeme. b) Contribution to the torque control τ_1 (rigid) in the sliding-mode case

4. The IDA-PBC method

4.1 Outline of the method

The IDA-PBC (interconnection and damping assignment passivity-based control) method is an energy-based approach to control design (see (Ortega & Spong, 2000) and (Ortega et al., 2002) for complete details). The method is specially well suited for mechatronic applications, among others. In the case of a flexible manipulator the goal is to control the position of an under-actuated mechanical system with total energy:

$$H(q,p)=\frac{1}{2}p^T M^{-1}(q)p+U(q) \tag{16}$$

where $q\in\mathbb{R}^n$, $p\in\mathbb{R}^n$, are the generalized positions and momenta respectively, $M(q)=M^T(q)>0$ is the inertia matrix and U(q) is the potential energy.

If it is assumed that the system has no natural damping, then the equations of motion of such system can be written as:

$$\begin{pmatrix} q \\ p \end{pmatrix} = \begin{pmatrix} 0 & I_n \\ -I_n & 0 \end{pmatrix} \begin{pmatrix} \nabla_q H \\ \nabla_p H \end{pmatrix} + \begin{pmatrix} 0 \\ G(q) \end{pmatrix} u \tag{17}$$

The IDA-PBC method follows two basic steps:

1. Energy shaping, where the total energy function of the system is modified so that a predefined energy value is assigned to the desired equilibrium.
2. Damping injection, to achieve asymptotic stability.

The following form for the desired (closed-loop) energy function is proposed:

$$H_d(q,p) = \frac{1}{2} p^T M_d^{-1}(q) p + U_d(q) \tag{18}$$

where $M_d = M_d^T > 0$ is the closed-loop inertia matrix and U_d the potential energy function. It will be required that U_d have an isolated minimum at the equilibrium point q_*, that is:

$$q_* = \arg\min \quad U_d(q) \tag{19}$$

In PBC, a composite control law is defined:

$$u = u_{es}(q,p) + u_{di}(q,p) \tag{20}$$

where the first term is designed to achieve the energy shaping and the second one injects the damping.
The desired (closed-loop) dynamics can be expressed in the following form:

$$\begin{pmatrix} \dot{q} \\ \dot{p} \end{pmatrix} = (J_d(q,p) - R_d(q,p)) \begin{pmatrix} \nabla_q H_d \\ \nabla_p H_d \end{pmatrix} \tag{21}$$

where

$$J_d = -J_d^T = \begin{pmatrix} 0 & M^{-1} M_d \\ -M_d M^{-1} & J_2(q,p) \end{pmatrix} \tag{22}$$

$$R_d = R_d^T = \begin{pmatrix} 0 & 0 \\ 0 & G K_v G^T \end{pmatrix} \geq 0 \tag{23}$$

represent the desired interconnection and damping structures. J_2 is a skew-symmetric matrix, and can be used as free parameter in order to achieve the kinetic energy shaping (see Ortega & Spong, 2000)).
The second term in (20) , the damping injection, can be expressed as:

$$u_{di} = -K_v G^T \nabla_p H_d \tag{24}$$

where $K_v = K_v^T > 0$.

To obtain the energy shaping term u_{es} of the controller, (20) and (24), the composite control law, are replaced in the system dynamic equation (17) and this is equated to the desired closed-loop dynamics, (21):

$$\begin{pmatrix} 0 & I_n \\ -I_n & 0 \end{pmatrix}\begin{pmatrix} \nabla_q H \\ \nabla_p H \end{pmatrix} + \begin{pmatrix} 0 \\ G \end{pmatrix} u_{es} = \begin{pmatrix} 0 & M^{-1}M_d \\ -M_d M^{-1} & J_2(q,p) \end{pmatrix}\begin{pmatrix} \nabla_q H_d \\ \nabla_p H_d \end{pmatrix}$$

$$Gu_{es} = \nabla_q H - M_d M^{-1} \nabla_q H_d + J_2 M_d^{-1} p \tag{25}$$

In the under-actuated case, G is not invertible, but only full column rank. Thus, multiplying (25) by the left annihilator of G, $G^{\perp} G=0$, it is obtained:

$$G^{\perp}\left\{\nabla_q H - M_d M^{-1} \nabla_q H_d + J_2 M_d^{-1} p\right\} = 0 \tag{26}$$

If a solution (M_d, U_d) exists for this differential equation, then u_{es} will be expressed as:

$$u_{es} = (G^T G)^{-1} G^T (\nabla_q H - M_d M^{-1} \nabla_q H_d + J_2 M_d^{-1} p) \tag{27}$$

The PDE (26) can be separated in terms corresponding to the kinetic and the potential energies:

$$G^{\perp}\left\{\nabla_q (p^T M^{-1} p) - M_d M^{-1} \nabla_q (p^T M_d^{-1} p) + 2 J_2 M_d^{-1} p\right\} = 0 \tag{28}$$

$$G^{\perp}\left\{\nabla_q U - M_d M^{-1} \nabla_q U_d\right\} = 0 \tag{29}$$

So the main difficulty of the method is in solving the nonlinear PDE corresponding to the kinetic energy (28). Once the closed-loop inertia matrix, M_d, is known, then it is easier to obtain U_d of the linear PDE (29), corresponding to the potential energy.

4.2 Application to a laboratory arm

The object of the study is a flexible arm with one degree of freedom that accomplishes the conditions of Euler-Bernoulli (Fig.9). In this case, the elastic deformation of the arm $w(x,t)$ can be represented by means of an overlapping of the spatial and temporary parts, see equation (1).

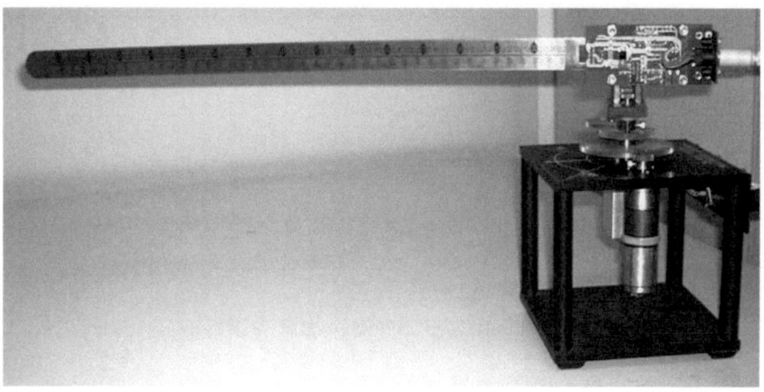

Figure 9. Photograph of the experimental flexible manipulator

If a finite number of modes m is considered, Lagrange equations lead to a dynamical system defined by m+1 second order differential equations:

$$\begin{pmatrix} I_t & 0 \\ 0 & I \end{pmatrix}\begin{pmatrix} \ddot{q}_0(t) \\ \ddot{q}_i(t) \end{pmatrix} + \begin{pmatrix} 0 & 0 \\ 0 & K_f \end{pmatrix}\begin{pmatrix} q_0(t) \\ q_i(t) \end{pmatrix} = \begin{pmatrix} 1 \\ \varphi'(0) \end{pmatrix} u \quad (30)$$

where q_0 (a 1x1 vector) and q_i (an mx1 vector) are the dynamic variables; $q_0(t)$ is the rigid generalized coordinate, $q_i(t)$ is the vector of flexible modes, I_t is the total inertia, K_f is the stiffness mxm matrix that depends on the elasticity of the arm, and it is defined as K_f=diag(ω_i^2), where ω_i is the resonance frequency of each mode; φ' are the first spatial derivatives of $\varphi_i(x)$ evaluated at the base of the robot. Finally u includes the applied control torques. Defining the incremental variables as $\tilde{q} = q - q_d$, where q_d is the desired trajectory for the robot such that $\dot{q}_{0d} \neq 0$, $\ddot{q}_{0d} = 0$ and q_{id}=0, then the dynamical model is given by:

$$\begin{pmatrix} I_t & 0 \\ 0 & I \end{pmatrix}\begin{pmatrix} \ddot{\tilde{q}}_0(t) \\ \ddot{\tilde{q}}_i(t) \end{pmatrix} + \begin{pmatrix} 0 & 0 \\ 0 & K_f \end{pmatrix}\begin{pmatrix} \tilde{q}_0(t) + q_{0d} \\ \tilde{q}_i(t) \end{pmatrix} = \begin{pmatrix} 1 \\ \varphi'(0) \end{pmatrix} u \quad (31)$$

The total energy of the mechanical system is obtained as the sum of the kinetic and potential energy:

$$H(\tilde{q}, p) = \frac{1}{2} p^T M^{-1} p + \frac{1}{2} \tilde{q}^T K \tilde{q} \quad (32)$$

where $p = (I_t \dot{\tilde{q}}_0 \quad I \dot{\tilde{q}}_i)^T$ and M and K are (m+1)x(m+1) matrices

$$M = \begin{pmatrix} I_t & 0 \\ 0 & I \end{pmatrix} \quad K = \begin{pmatrix} 0 & 0 \\ 0 & K_f \end{pmatrix}$$

The point of interest is $\tilde{q}_* = (0,0)$, which corresponds to zero tracking error in the rigid variable and null deflections.

The controller design will be made in two steps; first, we will obtain a feedback of the state that produces energy shaping in closed-loop to stabilize the position globally, then it will be injected the necessary damping to achieve the asymptotic stability by means of the negative feedback of the passive output.

The inertia matrix, M, that characterizes to the system, is independent of $\tilde{q}$, hence it follows that it can be chosen J_2=0, (see (Ortega & Spong, 2000)). Then, from (28) it is deduced that the matrix M_d should also be constant. The resulting representation capturing the first flexible mode is

$$M_d = \begin{pmatrix} a_1 & a_2 \\ a_2 & a_3 \end{pmatrix}$$

where the condition of positive definiteness of the inertia matrix, lead to the following inequations:

$$a_1 > 0 \quad a_1 a_3 > a_2^2 \quad (33)$$

Now, considering only the first flexible mode, equation (29) corresponding to the potential energy can be written as:

$$\left(\frac{a_2 - \varphi'(0)a_1}{I_t}\right)\frac{\partial U_d}{\partial \tilde{q}_0} + (a_3 - \varphi'(0)a_2)\frac{\partial U_d}{\partial \tilde{q}_1} = \tilde{q}_1\omega_1^2 \tag{34}$$

where $\tilde{q}_0 = q_0 - q_{0d}$ is the rigid incremental variable and $\tilde{q}_1 = q_1 - 0$ is the flexible incremental variable. The equation (34) is a trivial linear PDE whose general solution is

$$U_d = -\frac{I_t^2\omega_1^2(a_3 - \varphi'(0)a_2)}{2(-a_2 + \varphi'(0)a_1)^2}\tilde{q}_0^2 - \frac{I_t\,\omega_1^2}{(-a_2 + \varphi'(0)a_1)}\tilde{q}_0\tilde{q}_1 + F(z) \tag{35}$$

$$z = \gamma\tilde{q}_0 + \tilde{q}_1 \tag{36}$$

$$\gamma = \frac{I_t(a_3 - \varphi'(0)a_2)}{(-a_2 + \varphi'(0)a_1)} \tag{37}$$

where F is an arbitrary differentiable function that we should choose to satisfy condition (19) in the points $q*$.
Some simple calculations show that the necessary restriction $\nabla_q U_d(0) = 0$ is satisfied if $\nabla F(z(0)) = 0$, while condition $\nabla_q^2 U_d(0) > 0$ is satisfied if:

$$F'' > \frac{\omega_1^2}{(a_3 - \varphi'(0)a_2)} \tag{38}$$

Under these restrictions it can be proposed $F(z) = (1/2)K_1 z^2$, with $K_1 > 0$, which produces the condition:

$$a_3 > \varphi'(0)a_2 \tag{39}$$

So now, the term corresponding to the energy shaping in the control input (27) is given as

$$u_{es} = (G^T G)^{-1} G^T(\nabla_q U - M_d M^{-1}\nabla_q U_d) = K_{p1}\tilde{q}_0 + K_{p2}\tilde{q}_1 \tag{40}$$

where

$$K_{p1} = \frac{I_t(a_1 a_3 - a_2^2)}{(-a_2 + \varphi'(0)a_1)^2}(K_1(-a_3 + \varphi'(0)a_2) + \omega_1^2)$$

$$K_{p2} = \frac{a_1\omega_1^2 - K_1(a_1 a_3 - a_2^2)}{(-a_2 + \varphi'(0)a_1)}$$

The controller design is completed with a second term, corresponding to the damping injection. This is achieved via negative feedback of the passive output $G^T\nabla_p H_d$, (24). As

$H_d = (1/2)p^T M_d^{-1} p + U_d(\tilde{q})$, and U_d only depends on the $\tilde{q}$ variable, u_{di} only depends on the appropriate election of M_d :

$$u_{di} = K_{v1}\dot{\tilde{q}}_0 + K_{v2}\dot{\tilde{q}}_1 \tag{41}$$

with

$$K_{v1} = K_v \frac{I_t(-a_3 + \varphi'(0)a_2)}{(a_1 a_3 - a_2^2)}$$
$$K_{v2} = K_v \frac{(a_2 - \varphi'(0)a_1)}{(a_1 a_3 - a_2^2)}$$

A simple analysis on the constants K_{p1} and K_{v1} with the conditions previously imposed, implies that both should be negative to assure the stability of the system.
To analyze the stability of the closed-loop system we consider the energy-based Lyapunov function candidate (Kelly & Campa, 2005), (Sanz & Etxebarria, 2007)

$$\begin{aligned} V(\tilde{q},\dot{\tilde{q}}) = \frac{1}{2}p^T M_d^{-1} p + U_d &= \frac{1}{2}\frac{I_t^2\dot{\tilde{q}}_0^2 a_3 - 2I_t\dot{\tilde{q}}_0\dot{\tilde{q}}_1 a_2 + \dot{\tilde{q}}_1^2 a_1}{a_1 a_3 - a_2^2} - \frac{1}{2}\frac{I_t^2\omega_1^2(a_3 - \varphi'(0)a_2)\tilde{q}_0^2}{(-a_2 + \varphi'(0)a_1)^2} \\ &- \frac{I_t\omega_1^2\tilde{q}_0\tilde{q}_1}{-a_2 + \varphi'(0)a_1} + \frac{1}{2}K_1(\gamma\tilde{q}_0 + \tilde{q}_1)^2 \end{aligned} \tag{42}$$

which is globally positive definite, i.e: $V(0,0)=(0,0)$ and $V(\tilde{q},\dot{\tilde{q}}) > 0$ for every $(\tilde{q},\dot{\tilde{q}}) \neq (0,0)$.
The time derivative of (42) along the trajectories of the closed-loop system can be written as

$$\dot{V}(\tilde{q},\dot{\tilde{q}}) = -K_v \frac{(I_t\dot{\tilde{q}}_0(-a_3 + \varphi'(0)a_2) + (a_2 - \varphi'(0)a_1)\dot{\tilde{q}}_1)^2}{(a_1 a_3 - a_2^2)^2} \leq 0 \tag{43}$$

where $K_v>0$, so $\dot{V}$ is negative semidefinite, and $(\tilde{q},\dot{\tilde{q}}) = (0,0)$ is stable (not necessarily asymptotically stable).
By using LaSalle's invariant set theory, it can be defined the set R as the set of points for which $\dot{V} = 0$

$$R = \left\{ \begin{pmatrix} \tilde{q}_0 \\ \tilde{q}_1 \\ \dot{\tilde{q}}_0 \\ \dot{\tilde{q}}_1 \end{pmatrix} \in \mathbb{R}^4 : \dot{V}(\tilde{q},\dot{\tilde{q}}) = 0 \right\} = \left\{ \tilde{q}_0, \tilde{q}_1 \in \mathbb{R}^1 \quad \& \quad \gamma\dot{\tilde{q}}_0 + \dot{\tilde{q}}_1 = 0 \right\} \tag{44}$$

Following LaSalle's principle, given R and defined N as the largest invariant set of R, then all the solutions of the closed loop system asymptotically converge to N when $t \to \infty$.
Any trajectory in R should verify:

$$\gamma\dot{\tilde{q}}_0 + \dot{\tilde{q}}_1 = 0 \tag{45}$$

and therefore it also follows that:

$$\gamma\ddot{\tilde{q}}_0 + \ddot{\tilde{q}}_1 = 0 \tag{46}$$

$$\gamma\tilde{q}_0 + \tilde{q}_1 = k \tag{47}$$

Considering the closed-loop system and the conditions described by (46) and (47), the following expression is obtained:

$$(\frac{\gamma}{I_t} + \varphi'(0) - K_f)k + (\frac{\gamma\omega_1^2}{(-a_2 + \varphi'(0)a_1)} + \varphi'(0))\tilde{q}_0 = 0 \tag{48}$$

As a result of the previous equation, it can be concluded that $\tilde{q}_0(t)$ should be constant, in consequence $\dot{\tilde{q}}_0(t) = 0$, and replacing it in (45), it also follows that $\tilde{q}_1(t) = 0$. Therefore $\ddot{\tilde{q}}_0 = \ddot{\tilde{q}}_1 = 0$ and replacing these in the closed-loop system this leads to the following solution:

$$\begin{pmatrix} \tilde{q}_0(t) \\ \tilde{q}_1(t) \end{pmatrix} = \begin{pmatrix} 0 \\ 0 \end{pmatrix}$$

In other words, the largest invariant set N is just the origin $(\tilde{q}, \dot{\tilde{q}}) = (0,0)$, so we can conclude that any trajectory converge to the origin when $t \to \infty$, so the equilibrium is in fact asymptotically stable.

To illustrate the performance of the proposed IDA-PBC controller, in this section we present some simulations. We use the model of a flexible robotic arm presented in (Canudas et al., 1996) and the values of Table 2 which correspond to the real arm displayed in Fig. 9.

The results are shown in Figs. 10 to 12. In these examples the values a_1=1, a_2=0.01 and a_3=50 have been used to complete the conditions (33) and (39). In Fig. 10 the parameters are K_1=10 and K_v=1000. In Figs. 11 and 12 the effect of modifying the damping constant K_v is demonstrated. With a smaller value of K_v, K_v =10, the rigid variable follows the desired trajectory reasonably well. For K_v =1000, the tip position exhibits better tracking of the desired trajectory, as it can be seen comparing Fig. 10(a) and Fig. 11(a). Even more important, it should be noted that the oscillations of elastic modes are now attenuated quickly (compare Fig. 10(c) and Fig. 11(b)). But If we continue increasing the value of K_v, K_v =100000, the oscillations are attenuated even more quickly (compare Fig. 10(c) and Fig. 12(b)), but the tip position exhibits worse tracking of the desired trajectory.

Property	Value
Motor inertia, I_h	0.002 kgm^2
Link length, L	0.45 m
Link height, h	0.02 m
Link thickness, d	0.0008 m
Link mass, M_b	0.06 kg
Linear density, ρ	0.1333 kg/m
Flexural rigidity, EI	0.1621 Nm^2

Table 2. Flexible link parameters

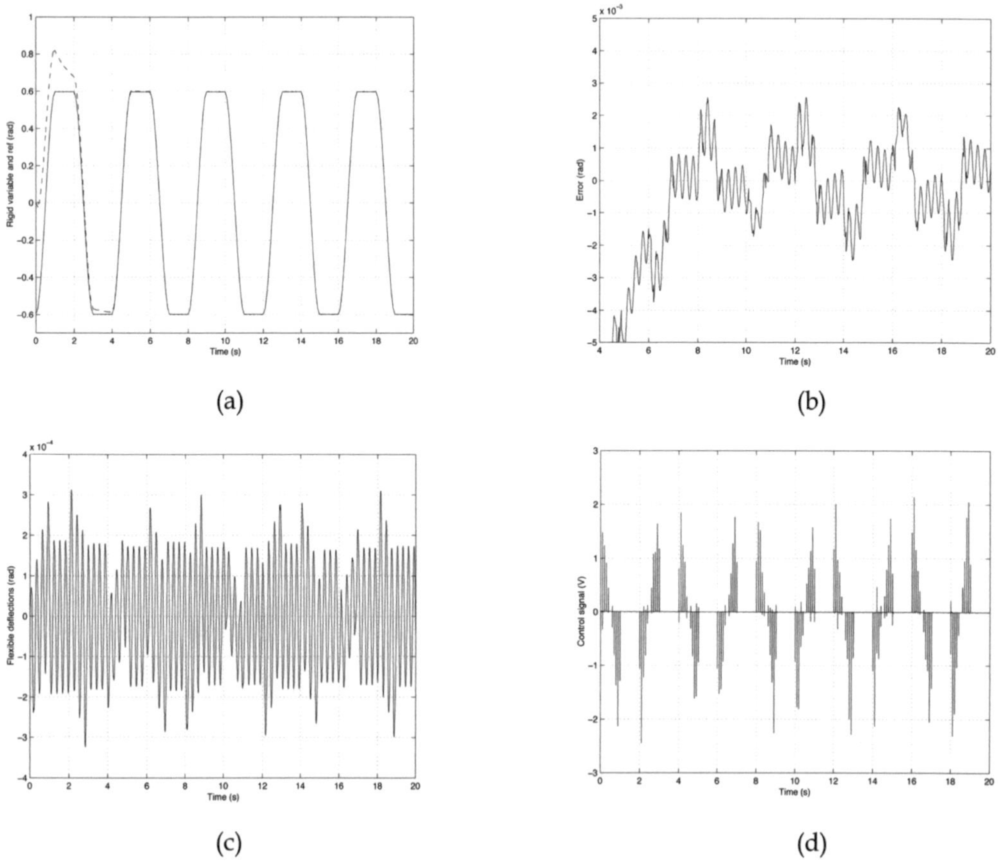

Figure 10. Simulation results for IDA-PBC control with K_v=1000: (a) Time evolution of the rigid variable q_0 and reference q_d__; (b) Rigid variable tracking error; (c) Time evolution of the flexible deflections; (d) Composite control signal

The effectiveness of the proposed control schemes has been tested by means of real time experiments on a laboratory single flexible link. This manipulator arm, fabricated by Quanser Consulting Inc. (Ontario, Canada), is a spring steel bar that moves in the horizontal plane due to the action of a DC motor. A potentiometer measures the angular position of the system, and the arm deflections are measured by means of a strain gauge mounted near its base (see Fig. 9). These sensors provide respectively the values of q_0 and q_1 (and thus $\tilde{q}_0$ and $\tilde{q}_1$ are also known, since q_{0d} and q_{1d} are predetermined).

The experimental results are shown on Figs. 13, 14 and 15. In Fig. 13 the control results using a conventional PD rigid control design are displayed:

$$u = K_P(q_0 - q_{0d}) + K_D(\dot{q}_0 - \dot{q}_{0d})$$

where K_P=-14 and K_D=-0.028. These gains have been carefully chosen, tuning the controller by the usual trial-an-error method. The rigid variable tracks the reference (with a certain error), but the naturally excited flexible deflections are not well damped (Fig. 13(b)).

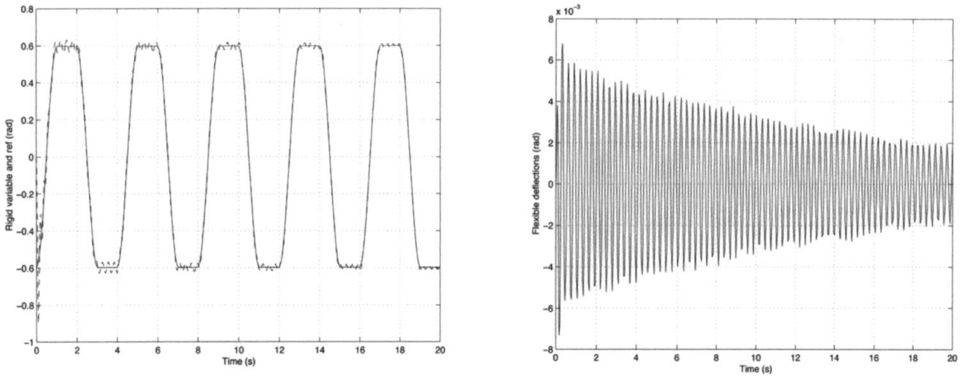

(a) (b)

Figure 11. Simulation results for IDA-PBC control with K_v=10: (a) Time evolution of the rigid variable q_0 and reference q_d__; (b) Time evolution of the flexible deflections

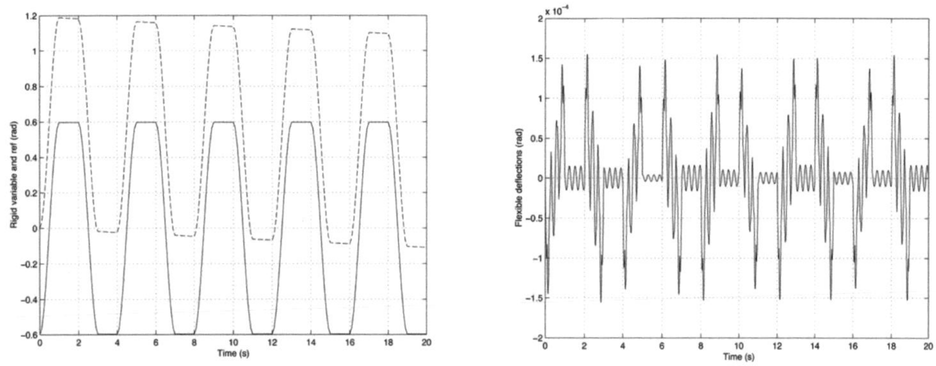

(a) (b)

Figure 12. Simulation results for IDA-PBC control with K_v=100000: (a) Time evolution of the rigid variable q_0 and reference q_d__; (b) Time evolution of the flexible deflections

In Fig. 14 the results using the IDA-PBC design philosophy are displayed. The values a_1=1, a_2=0.01, a_3=50, K_1=10, K_v=1000 and $\varphi'(0)=1.11$ have been used. As seen in the graphics, the rigid variable follows the desired trajectory, and moreover the flexible modes are now conveniently damped, (compare Fig. 13(b) and Fig. 14(c)). It is shown that vibrations are effectively attenuated in the intervals when q_d reaches its upper and lower values which go from 1 to 2 seconds, 3 to 4 s., 5 to 6 s., etc.

The PD controller might be augmented with a feedback term for link curvature:

$$u = K_P(q_0 - q_{0d}) + K_D(\dot{q}_0 - \dot{q}_{0d}) + K_C \mathrm{w}''$$

where w'' represents the link curvature at the base of the link. This is a much simpler version of the IDA-PBC and doesn't require time derivatives of the strain gauge signals.

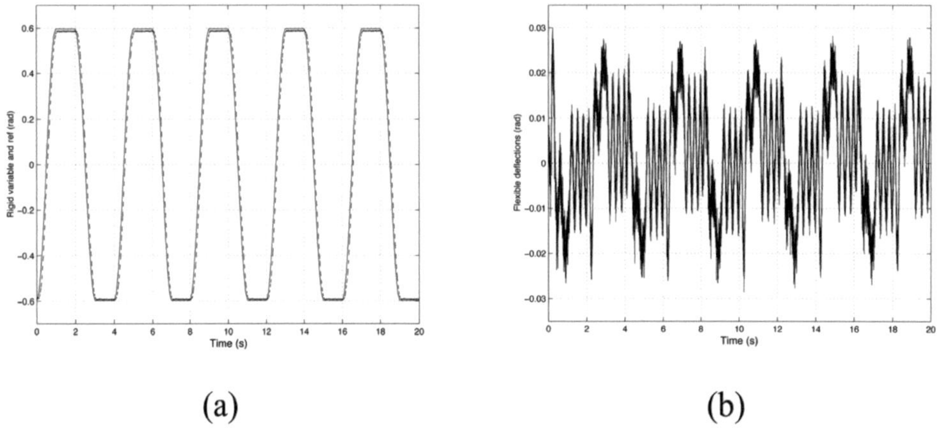

Figure 13. Experimental results for a conventional PD rigid controller:: (a) Time evolution of the rigid variable q_0 and reference q_d__; (b) Time evolution of the flexible deflections

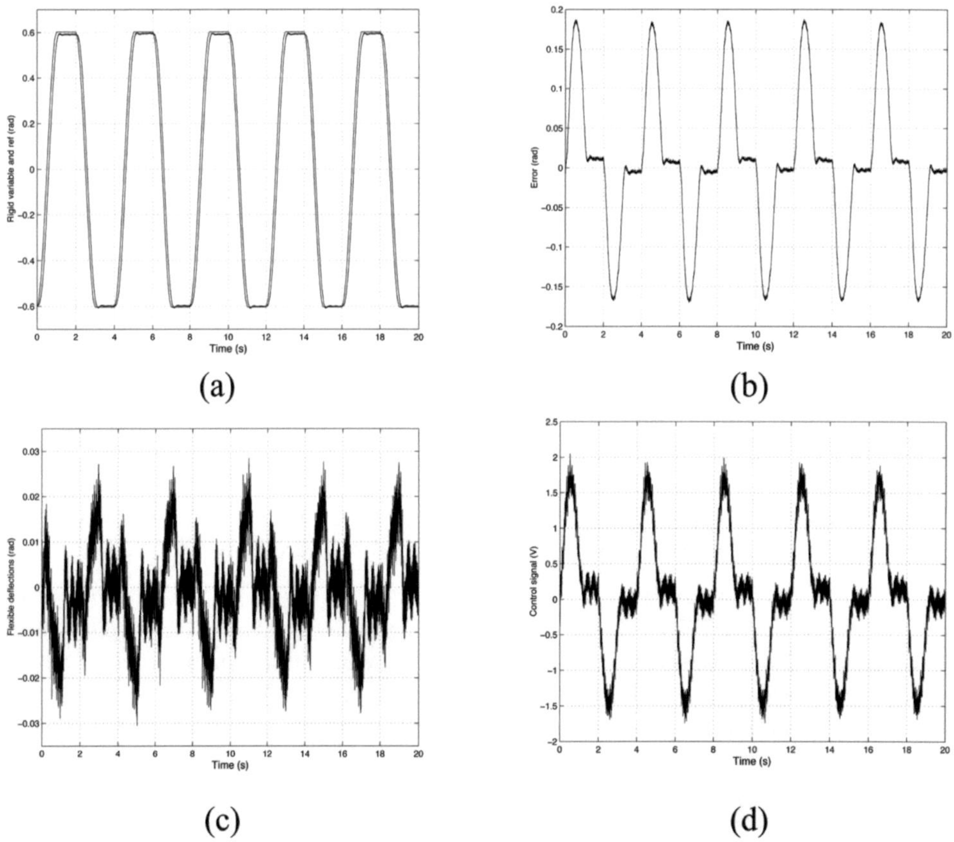

Figure 14. Experimental results for IDA-PBC control: (a) Time evolution of the rigid variable q_0 and reference q_d__; (b) Rigid variable tracking error; (c) Time evolution of the flexible deflections; (d) Composite control signal

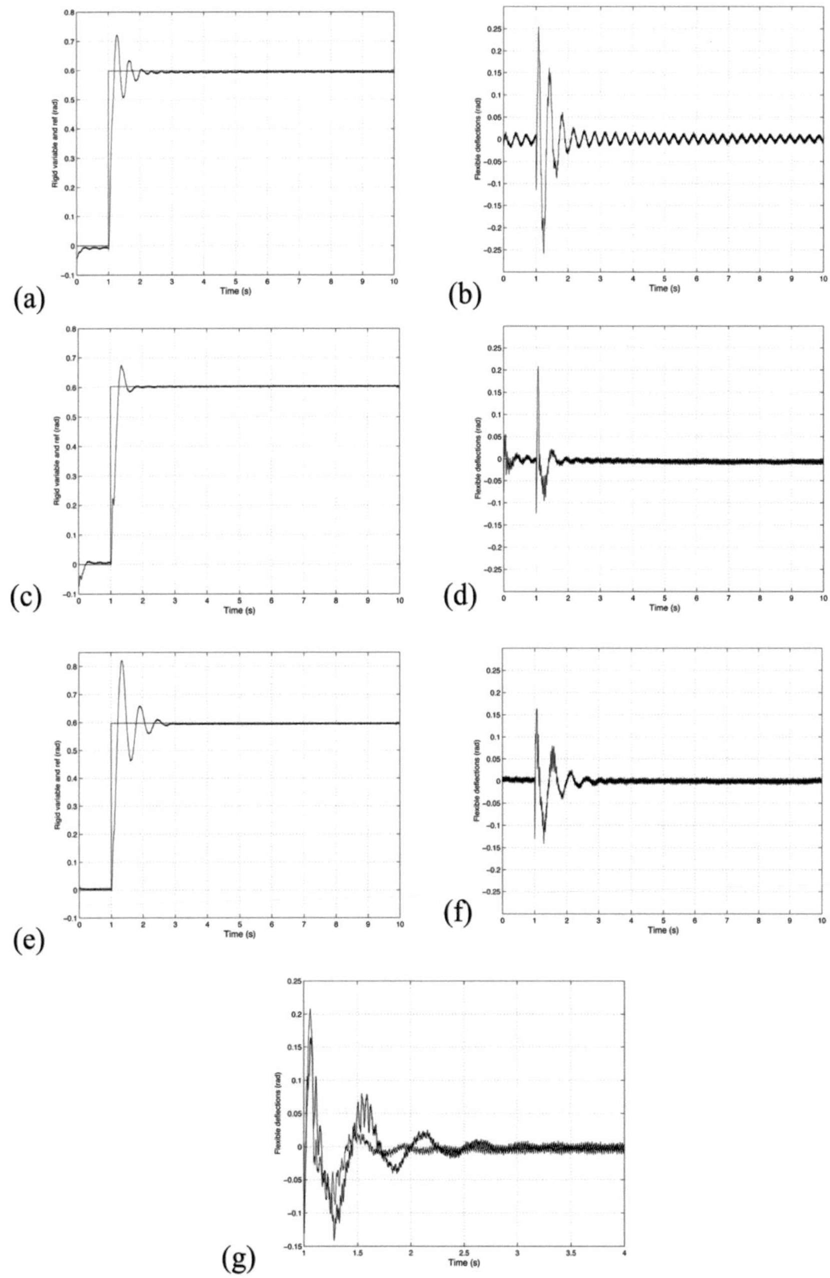

Figure 15. Performance comparison between a conventional rigid PD controller, augmented PD and IDA-PBC controller. (a) Time evolution of the rigid variable q_0 and reference q_d_(PD); (b) Time evolution of the flexible deflections (PD); (c) Time evolution of the rigid variable q_0 and reference q_d_(IDA-PBC); (d) Time evolution of the flexible deflections (IDA-PBC); (e) Time evolution of the rigid variable q_0 and reference q_d_(augmented PD); (f) Time evolution of the flexible deflections (augmented PD); (g) Comparison between the strain gauge signals with the IDA-PBC controller and the augmented PD controller

Fig. 15 shows the comparison between the conventional rigid PD controller, augmented PD and the IDA-PBC controller for a step input. The K_P and K_D gains are the same both for the PD and the augmented PD case. Gain K_C has been carefully tuned to effectively damp the flexible vibrations in the augmented PD control. Comparing Fig. 15(a,b) to Fig. 15(c,d) it is clearly seen how with the IDA-PBC controller the tip oscillation improves notably. Moreover, from Fig. 15(e,f) it can be observed that the augmented PD effectively attenuates the flexible vibrations, but at the price of worsening the rigid tracking. Also in Fig. 15(g), where the flexible responses of both the augmented PD and the IDA-PBC controllers have been amplified and plotted together for comparison, it can be observed that the effective attenuation time is longer in the augmented PD case (2 seconds) than in the IDA-PBC case (1 second).

Notice that the final structure of all the considered controls is similar but with very different gains and all of them imply basically the same implementation effort, although the IDA-PBC gains calculation may imply solving some involved equations. In this sense, the IDA-PBC method gives for this application an energy interpretation of the control gains values and it can be used, in this particular case, as a systematic method for tuning these gains (outperforming the trial-and-error method). Finally, it should be remarked that for a more complicated robot the structure of the IDA-PBC and the PD controls can be very different.

5. Conclusion

An experimental study of several control strategies for flexible manipulators has been presented in this chapter. As a first step, the dynamical model of the system has been derived from a general multi-link flexible formulation. If should be stressed the fact that some simplifications have been made in the modelling process, to keep the model reasonably simple, but, at the same time, complex enough to contain the main rigid and flexible dynamical effects. Three control strategies have been designed and tested on a laboratory two-dof flexible manipulator. The first scheme, based on an LQR optimal philosophy, can be interpreted as a conventional rigid controller. It has been shown that the rigid part of the control performs reasonably well, but the flexible deflections are not well damped. The strategy of a combined optimal rigid-flexible LQR control acting both on the rigid subsystem and on the flexible one has been tested next. The advantage that this type of combined control offers is that the oscillations of the flexible modes attenuate considerably, which demonstrates that a strategy of overlapping a rigid control with a flexible control is effective from an experimental point of view. The third experimented approach introduces a sliding-mode controller. This control includes two completely different parts: one sliding for the rigid subsystem and one LQR for the fast one. In this case the action of the energetic control turns out to be effective for attracting the rigid dynamics to the sliding band, but at the same time the elastic modes are attenuated, even better than in the LQR case. This method has been shown to give a reasonably robust performance if it is conveniently tuned. The IDA-PBC method is a promising control design tool based on energy concepts. In this chapter we have also presented a theoretical and experimental study of this method for controlling a laboratory single flexible link, valuing in this way the potential of this technique in its application to under-actuated mechanical systems, in particular to flexible manipulators. The study is completed with the Lyapunov stability analysis of the closed-loop system that is obtained with the proposed control law. Then, as an illustration, a set of simulations and laboratory control experiments have been presented. The experimented scheme, based on an IDA-PBC philosophy, has been shown to achieve good tracking properties on the rigid variable and definitely superior damping of the unwanted vibration of the flexible variables compared to conventional rigid robot control schemes. In view of the obtained experimental results, Fig. 13, 14 and 15, it has also been shown that the proposed IDA-PBC controller leads to a remarkable improvement in the tip oscillation of the robot arm with respect to a conventional PD or an augmented PD approach. It is also worth mentioning that for our application the proposed energy-based methodology, although includes some involved calculations, finally results in a simple controller as easily implementable as a PD. In summary, the experimental results shown in the chapter have illustrated the suitability of the proposed composite control schemes in practical flexible robot control tasks

6. References

O. Barambones and V. Etxebarria, Robust adaptive control for robot manipulators with unmodeled dynamics. *Cybernetics and Systems*, 31(1), 67-86, 2000.

C. Canudas de Wit, B. Siciliano and G. Bastin, Theory of Robot Control. *Europe: The Zodiac*, Springer, 1996.

R. Kelly and R. Campa, Control basado en IDA-PBC del péndulo con rueda inercial: Análisis en formulación lagrangiana. *Revista Iberoamericana de Automática e Informática Industrial*,1, pp. 36--42,January 2005.

P. Kokotovic, H. K. Khalil and J.O'Reilly, *Singular perturbation methods in control*. SIAM Press, 1999.

Y. Li, B. Tang, Z. Zhi and Y. Lu, Experimental study for trajectory tracking of a two-link flexible manipulator. *International Journal of Systems Science*, 31(1):3-9, 2000.

M. Moallem, K. Khorasani and R.V. Patel, Inversion-based sliding control of a flexible-link manipulator. *International Journal of Control*, 71(3), 477-490, 1998.

M. Moallem, R.V. Patel and K. Khorasani, Nonlinear tip-position tracking control of a flexible-link manipulator: theory and experiments. *Automatica*, 37, pp.1825--1834, 2001.

R. Ortega and M. Spong, Stabilization of underactuatedmechanical systems via interconnection and damping assignment, in *Proceedings of the 1st IFAC Workshop on Lagrangian and Hamiltonian methods in nonlinear systems*, Princeton, NJ,USA, 2000, pp. 74--79.

R. Ortega and M. Spong and F. Gómez and G. Blankenstein, Stabilization of underactuated mechanical systems via interconnection and damping assignment. *IEEE Transactions on Automatic Control*, AC-47, pp. 1218--1233, 2002.

R. Ortega and A. van der Schaft and B. Maschke and G. Escobar, Interconnection and damping assignment passivity-based control of port-controlled Hamiltonian systems. *Automatica*, 38, pp. 585--596, 2002.

A. Sanz and V. Etxebarria, Experimental Control of a Two-Dof Flexible Robot Manipulator by Optimal and Sliding Methods. *Journal of Intelligent and Robotic Systems*, 46: 95-110, 2006.

A. Sanz and V. Etxebarria, Experimental control of a single-link flexible robot arm using energy shaping. *International Journal of Systems Science*, 38(1): 61-71, 2007.

M. W. Vandegrift, F. L. Lewis and S. Q. Zhu, Flexible-link robot arm control by a feedback linearization singular perturbation approach. *Journal of Robotic Systems*, 11(7):591-603, 1994.

D. Wang and M. Vidyasagar, Transfer functions for a flexible link. *The International Journal of Robotic Research*, 10(5), pp. 540--549, 1991.

J.X. Xu and W.J. Cao, Direct tip regulation of a single-link flexible manipulator by adaptive variable structure control.*International Journal of Systems Science*, 32(1): 121-135, 2001.

J.H. Yang, F.L. Lian, L.C. Fu, Nonlinear adaptive control for flexible-link manipulators. *IEEE Transactions on Robotics and Automation*, 13(1): 140-148, 1997.

The Mathworks, Inc, Control System Toolbox.*Natick*, MA, 2002.

S. Yoshiki and K. Ogata and Y. Hayakawa, Control of manipulator with 2 flexible-links via energy modification method by using modal properties, in *Proceedings of the 2003 IEEE/ASME International Conference on Advanced Intelligent Mechatronics* (AIM 2003), pp. 1435--1441.

Improvement of Force Control in Robotic Manipulators Using Sensor Fusion Techniques

J. Gamez[1], A. Robertsson[2], J. Gomez Ortega[1] and R. Johansson[2]
[1]*System Engineering and Automation Department at Jaén University,*
[2]*Automatic Control Department at Lund University*
[1]*Spain,* [2]*Sweden*

1. Introduction

In the never-ending effort of the humanity to simplify their existence, the introduction of 'intelligent' resources was inevitable to come (IFR, 2001). One characteristic of these 'intelligent' systems is their ability to adapt themselves to the variations in the outside environment as the internal changes occurring within the system. Thus, the robustness of an 'intelligent' system can be measured in terms of the sensitivity and adaptability to such internal and external variations.
In this sense, robot manipulators could be considered as 'intelligent' systems; but for a robotic manipulator without sensors explicitly measuring positions or contact forces acting at the end-effector, the robot TCP has to follow a path in its workspace without regard to any feedback other than its joints shaft encoders or resolvers. This restrictive fact imposes severe limitations on certain tasks where an interaction between the robot and the environment is needed. However, with the help of sensors, a robot can exhibit an adaptive behaviour (Harashima and Dote, 1990), the robot being able to deal flexibly with changes in its environment and to execute complicated skilled tasks.
On the other hand, the manipulation can be controlled only after the interaction forces are managed properly. That is why force control is required in manipulation robotics. For the force control to be implemented, information regarding forces at the contact point has to be fed back to the controller and force/torque (F/T) sensors can deliver that information. But an important problem arises when we have only a force sensor. That is a dynamic problem: in the dynamic situation, not only the interaction forces and moments at the contact point but also the inertial forces of the tool mass are measured by the wrist force sensor (Gamez et al., 2008b). In addition, the magnitude of these dynamics forces cannot be ignored when large accelerations and fast motions are considered (Khatib, 1987). Since the inertial forces are perturbation forces to be measured or estimated in the robot manipulation, we need to process the force sensor signal in order to extract the contact force exerted by the robot.
Previous results, related to contact force estimation, can be found in (Uchiyama, 1979), (Uchiyama and Kitagaki, 1989) and (Lin, 1997). In all the cases, the dynamic information of the tool was considered but some of the involved variables, such as the acceleration of the tool, were simply estimated by means of the kinematic model of the manipulator. However, these estimations did not reflect the real acceleration of the tool and thus high accuracy

could not be expected, leading to poor experimental results. Kumar et al. (Kumar and Garg, 2004) applied the same optimal filter approach to multiple cooperative robots.
Later, in (Gamez et al., 2004), (Gamez et al., 2005b), (Gamez et al., 2006b) and (Gamez et al., 2007b), Gamez et al. presented a contact force estimator restricting initially the results to one and three dimensions. In these works, they proposed a new contact force estimator that fused the information of a wrist force sensor, an accelerometer placed at the robot tool and the joint position measurements, in order to differentiate the contact force measurements from the wrist force sensor signals. They proposed combining the F/T estimates from model based observers with F/T and acceleration sensor measurements in tasks with heavy tools interacting with the environment. Many experiments were carried out on different industrial platforms showing the noise-response properties of the model-based observer validating its behaviour. For these observers, self-calibrating algorithms were proposed to easily implement them in industrial robotic platforms (Gamez et al., 2008a); (Gamez et al., 2005a).
In addition, Kröger et al. (Kröger et al., 2006); (Kröger et al., 2007) also presented a contact F/T estimator based on this sensor fusion approach. In this work, they presented a contact F/T estimator based also on the integration of F/T and inertial sensors but they did not consider the stochastic properties of the system. Finally, a 6D contact force/torque estimator for robotic manipulators with filtering properties was recently proposed in (Gamez et al., 2008b).
This work describes how the force control performance in robotic manipulators can be increased using sensor fusion techniques. In particular, a new sensor fusion approach applied to the problem of the contact force estimation in robot manipulators is proposed to improve the manipulator-environment interaction. The presented strategy is based on the application of sensor fusion techniques that integrate information from three different sensors: a wrist force/torque (F/T) sensor, an inertial sensor attached to the end effector, and joint sensors. To experimentally evaluate the improvement obtained with this new estimator, the proposed methodology was applied to several industrial manipulators with fully open software architecture. Furthermore, two different force control laws were utilized: impedance control and hybrid control.

2. Problem Formulation

Whereas force sensors may be used to achieve force control, they may have drawbacks if used in harsh environments and their measurements are complex in the sense that they reflect forces other than contact forces. Furthermore, if the manipulator is working with heavy tools interacting with the environment, the magnitude of these dynamics forces cannot be ignored, forcing control engineers to consider some kind of compensator that eliminates the undesired measurements. In addition, these perturbations, particulary the inertial forces, are higher when large accelerations and fast motions are considered.
In this context, let us consider a robot manipulator where a force sensor has been placed at the robot wrist and that an inertial sensor has been attached to the robot tool to measure its linear acceleration and angular velocity (Fig. 1). Then, when the robot manipulator moves in either free or constrained space, a F/T sensor attached to the robot tip measures not only the contact forces and torques exerted to the environment but also the non-contact ones produced by the inertial and gravitational effects. That is,

$$\begin{pmatrix} \vec{F}_S \\ \vec{N}_S \end{pmatrix} = \begin{pmatrix} \vec{F}_E \\ \vec{N}_E \end{pmatrix} + \begin{pmatrix} \vec{F}_I \\ \vec{N}_I \end{pmatrix} + \begin{pmatrix} \vec{F}_G \\ \vec{N}_G \end{pmatrix} \tag{1}$$

where $\vec{F}_S$ and $\vec{N}_S$ are the forces and torques measured by the wrist sensor; $\vec{F}_E$ and $\vec{N}_E$ are the environmental forces and torques; $\vec{F}_I$ and $\vec{N}_I$ correspond to the inertial forces and torques produced by the tool dynamic and $\vec{F}_G$ and $\vec{N}_G$ are the forces and torques due to gravity. Normally, the task undertaken requires the control of the environmental force $\vec{F}_E$ and the torque $\vec{N}_E$.

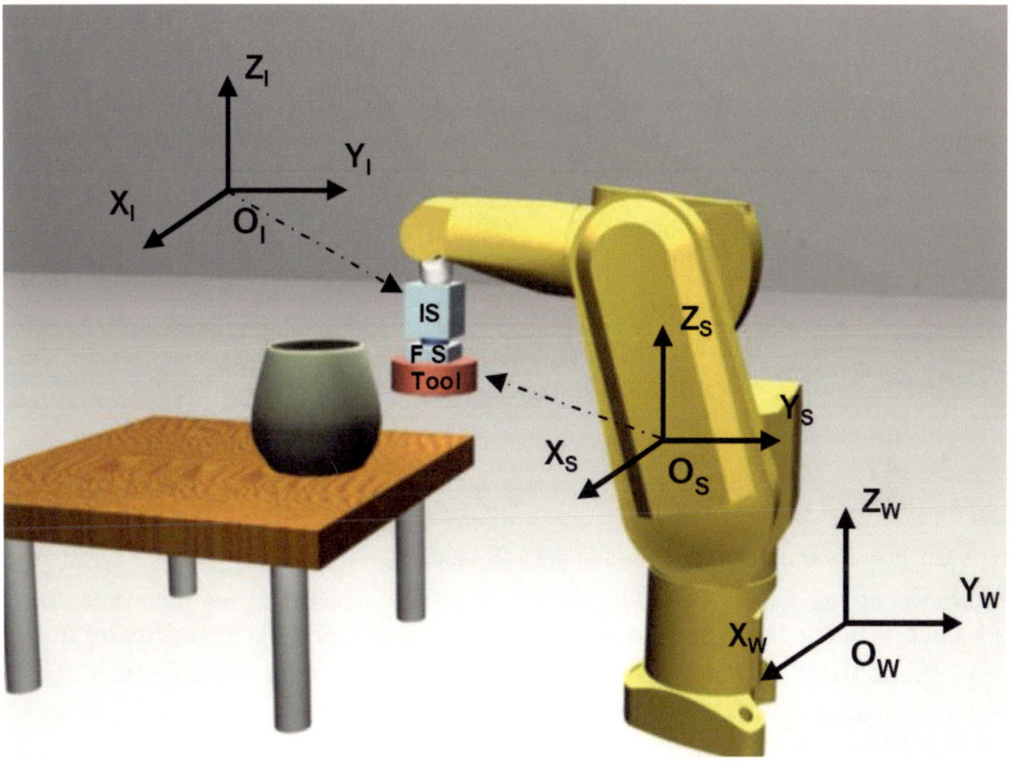

Figure 1. Coordinate frames of the robotic system. The sensor frames of the robot arm ($\{O_W X_W Y_W Z_W\}$), of the force-torque sensor (FS) ($\{O_S X_S Y_S Z_S\}$) and of the inertial sensor (IS) ($\{O_I X_I Y_I Z_I\}$) are shown

The main goal of this work is to evaluate how the implementation of a contact force/torque estimator, that distinguishes the contact F/T from the wrist force sensor measurement, can improve the force control loop in robotic manipulators. This improvement has to be based in the elimination of the non-desired effects of the non-contact forces and torques and, moreover, to the reduction of the high amount of noise introduced by some of the sensors used.

3. 6DOF Contact Force-Torque Estimator

3.1 Description of Coordinate Frames and Motion

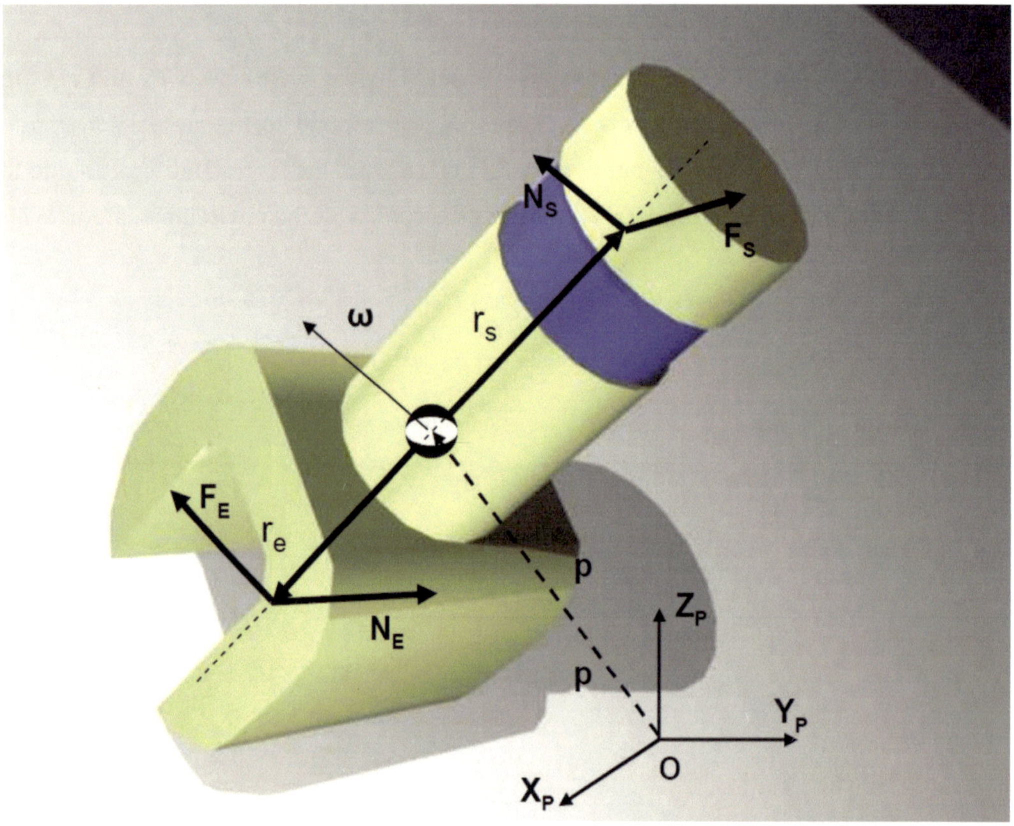

Figure 2. Force and moments applied to the robot tool

As shown in Fig. 1, let us consider an inertial sensor attached to the robot tool. Then, $\{O_S X_S Y_S Z_S\}$ and $\{O_I X_I Y_I Z_I\}$ correspond to the force sensor and the inertial sensor frames respectively. The world frame is an inertial coordinate system represented by $\{O_W X_W Y_W Z_W\}$ and without loss of generality we let it coincide with the robot frame for simplified notation. Let $\{{}^W R_S\}$ and $\{{}^S R_I\}$ denote the rotation matrices that relate the force sensor to the world frames and the inertial sensor to the force sensor frames, respectively. Assume that the force sensor is rigidly attached to the robot tip and that the inertial sensor can be placed either in the tool or at the robot wrist.

Let us also define the following frames to characterize the robot tool or end effector (Fig. 2): $\{O_E X_E Y_E Z_E\}$ is the end effector coordinate system by which the components of the external force and moment are represented and $\{O_P X_P Y_P Z_P\}$ represents the coordinate system fixed to the end effector so that $\{O_P\}$ coincides with the center of gravity and the $\{X_P\}$, $\{Y_P\}$, and $\{Z_P\}$ axes with the principal axes of the end effector.

3.2 Force-Torque Transformation

The set of forces F ($F \in \mathbb{R}^3$) and moments N ($N \in \mathbb{R}^3$) referred to frame A can be defined using *wrench* notation (Murray et al, 1994) as :

$$ {}^{A}U_{A} = \begin{pmatrix} {}^{A}F_{A} \\ {}^{A}N_{A} \end{pmatrix} \tag{2} $$

which can be transformed into a new frame B applying the following transformation

$$ \begin{pmatrix} {}^{B}F_{A} \\ {}^{B}N_{A} \end{pmatrix} = \begin{pmatrix} {}^{B}R_{A}^{T} & 0_{3\times3} \\ -{}^{B}R_{A}^{T}\,{}^{B}p^{x} & {}^{B}R_{A}^{T} \end{pmatrix} \begin{pmatrix} {}^{A}F_{A} \\ {}^{A}N_{A} \end{pmatrix} \tag{3} $$

where ${}^{B}R_{A}$ is the rotation matrix that relates frame A to frame B, and matrix ${}^{B}p^{x}$ is obtained as (Murray et al, 1994):

$$ {}^{B}p^{x} \times q = {}^{B}p^{x}q, \qquad {}^{B}p^{x} = \begin{pmatrix} 0 & -p_3 & p_2 \\ p_3 & 0 & -p_1 \\ -p_2 & p_1 & 0 \end{pmatrix} \tag{4} $$

being ${}^{B}p^{x}$ the position vector from $\{O_B\}$ to $\{O_A\}$ with ${}^{B}p^{x}$, $q \in \mathbb{R}^3$.

3.3 Robot Tool Dynamics Modeling

The F/T sensor measures not only the contact forces and torques, but also the disturbances produced by the non-contact forces and torques (inertial and gravitational effects) due to the robot tool dynamics. From Fig. 2, the tool dynamics is modeled using forces and torques that occur between the robot tool and the manipulator tip and that are measured by the wrist F/T sensor ${}^{S}F_{S}$ and ${}^{S}N_{S}$; and, on the other hand, the forces and moments exerted by the manipulator tool to its environment (${}^{E}F_{E}$ and ${}^{E}N_{E}$). Then, from a static equilibrium relation using wrench notation (Murray et al, 1994), the force and moment in $\{O_P\}$ are obtained as:

$$ {}^{P}U_{P} = \begin{pmatrix} {}^{P}F_{P} \\ {}^{P}N_{P} \end{pmatrix} = \begin{pmatrix} {}^{P}R_{S}^{T} & 0_{3\times3} \\ -{}^{P}R_{S}^{T}\,{}^{S}p^{x} & {}^{P}R_{S}^{T} \end{pmatrix} \begin{pmatrix} {}^{S}F_{S} \\ {}^{S}N_{S} \end{pmatrix} + \begin{pmatrix} {}^{P}R_{E}^{T} & 0_{3\times3} \\ -{}^{P}R_{E}^{T}\,{}^{E}p^{x} & {}^{P}R_{E}^{T} \end{pmatrix} \begin{pmatrix} {}^{E}F_{E} \\ {}^{E}N_{E} \end{pmatrix} $$

$$ = {}^{P}T_{S} \begin{pmatrix} {}^{S}F_{S} \\ {}^{S}N_{S} \end{pmatrix} + {}^{P}T_{E} \begin{pmatrix} {}^{E}F_{E} \\ {}^{E}N_{E} \end{pmatrix} \tag{5} $$

with ${}^{P}R_{S}$, ${}^{P}R_{E}$ being the rotation matrices that relate frames $\{S\}$ and $\{E\}$ to frame $\{P\}$; ${}^{S}p^{x}$ and ${}^{E}p^{x}$ are matrices of dimension 3×3 obtained from Eq. (4) where p is the position vector from $\{O_P\}$ to $\{O_S\}$ or from $\{O_P\}$ to $\{O_E\}$. Finally, applying the Newton-Euler equations, the dynamics of the robot tool are defined as

$$^{P}U_{P}=\begin{pmatrix} m^{W}R_{P}^{-1}{}^{W}R_{I}a-m^{W}R_{P}^{-1}g \\ \mathcal{I}^{P}R_{I}\dot{\omega}+^{P}R_{I}\omega\times\mathcal{I}^{P}R_{I}\omega \end{pmatrix}$$

$$=\begin{pmatrix} m^{W}R_{P}^{-1W}R_{I} & 0_{3x3} \\ 0_{3\times3} & \mathcal{I}^{P}R_{I} \end{pmatrix}\begin{pmatrix} a \\ \dot{\omega} \end{pmatrix}+\begin{pmatrix} -m^{W}R_{P}^{-1} & 0_{3\times3} \\ 0_{3\times3} & (^{P}R_{I}\omega)^{\times}\mathcal{I}^{P}R_{I} \end{pmatrix}\begin{pmatrix} g \\ \omega \end{pmatrix} \tag{6}$$

where m is the robot tool mass, $a=[\ddot{x},\ddot{y},\ddot{z}]^{T}$ is the robot tool acceleration measured in the inertial frame $\{I\}$; $g=[g_x,g_y,g_z]^{T}$ is the gravity acceleration vector; $\omega=[\omega_1,\omega_2,\omega_3]^{T}$ is the angular velocity of the robot tool with respect to frame $\{I\}$. Matrix $\mathcal{I}$ denotes the moment of inertia of the robot tool calculated with respect to the frame $\{P\}$ (Gamez et al., 2008b).

3.4 Contact Forces and Torques Estimator

The objective of the force observer is to estimate the environmental forces and torques by separating them from the end-effector inertial and gravitational forces and moments in the measurement given by the force sensor. As the sensor fusion approach considered integrates the information from different sensors, which may be quite noisy, it is necessary to consider an estimator with filtering properties.

To define the robot tool dynamics, Eqs. (5-6) are translated to state space formulation. Let then us consider the following state space vector

$$X=[x,y,z,\theta,\phi,\psi,\dot{x},\dot{y},\dot{z},\omega_1,\omega_2,\omega_3] \tag{7}$$

where x,y,z are the position coordinates of the robot tool center of mass ($p=(x,y,z)^{T}$) and θ,ϕ and ψ are the Euler angles that define the robot tool orientation ($o=(\theta,\phi,\psi)^{T}$). Both refer to the $O_W X_W Y_W Z_W$ frame. $\dot{x},\dot{y},\dot{z}$ correspond to the linear velocity of the robot tool center of mass.

Those variables that can be measured are regarded as outputs: the robot tool position and orientation (p,o), the F/T sensor signals (F_s,N_s), the angular velocities (ω) and the robot tool linear acceleration ($\ddot{p}$). (ω) and ($\ddot{p}$) are measured through the inertial sensor. In addition, the angular accelerations $\dot{\omega}$ of the robot tool are estimated using the angular velocities, also including them as outputs. Thus the output vector Y will be :

$$Y=\left(p^{T},o^{T},F_{s}^{T},N_{s}^{T},\omega^{T},\ddot{p}^{T},\dot{\omega}^{T}\right)^{T} \tag{8}$$

Then, we propose the following state space system

$$\begin{cases} \dot{X}=A(X)X+B_{S}U_{S}+B_{E}U_{E}+B_{G}g+\nu_{X} \\ Y=C(X)X+D_{S}U_{S}+D_{E}U_{E}+D_{G}g+\nu_{y} \end{cases} \tag{9}$$

where g is the gravity vector, $U_S = [{}^S F_S^T, {}^S N_S^T]^T$, and $U_E = [{}^E F_E^T, {}^E N_E^T]^T$. v_X and v_y are, respectively, the process and the output noises. Matrices $A(X)$, B_S, B_E, B_G, $C(X)$, D_S, D_E and D_G obtain their values from Eqs. (5) and (6) (Gamez et al., 2008b).
Note the non-linearity of the system through the entries A_1 and A_2 in the matrices $A(X)$ and $C(X)$.
In this context, an observer where input U_E has not been considered is proposed as (Gamez et al., 2004); (Gamez et al., 2005b):

$$\widehat{X} = A(X)\widehat{X} + B_S U_S + B_G g + L(t)(Y - \hat{Y}) \tag{10}$$

$$\hat{Y} = C(X)\widehat{X} + D_S U_S + D_G g \tag{11}$$

where $\widehat{X}$ corresponds to the state estimation of X and $L(t)$ is the observer gain for a given instant t: Then, the dynamics of the estimation error $\widetilde{X} = X - \widehat{X}$ are obtained as

$$\widetilde{X} = (A(X) - L(t)C(X))\widetilde{X} + (B_E - L(t)D_E)U_E$$
$$+ v_X - L(t)v_y \tag{12}$$

$$\widetilde{Y} = Y - \hat{Y} = C(X)\widetilde{X} + D_E U_E + v_y \tag{13}$$

Then, a dynamic force estimator is defined as:

$$\widehat{U}_E = D_E^{\dagger}(-C(X)\widetilde{X} + \widetilde{Y}) \tag{14}$$

where $D_E^{\dagger}$ is the Moore-Penrose pseudoinverse of D_E (Johansson, 1993).
Note that, from Eqs. (10) and (12), the core of the estimator is to combine F/T estimates from model based observers, where the input U_E has not been considered, with F/T, inertial and position measurements, in order to obtain a dynamic contact F/T estimator with low-pass properties.

3.5 Observer Gain Selection

While there are many application-specific approaches to estimating an unknown state from a set of process measurements, many of these methods do not inherently take into consideration the typically noisy nature of the measurements. In our case, while the contact F/T estimates vary with the robotic task, the fundamental sources of information are always the same: wrist F/T, inertial and position measurements. All of them are derived from noisy sensors. This noise is typically statistical in nature (or can be effectively modeled as such), which leads us to stochastic methods for addressing the problems. The solution adopted for this work is the Extended Kalman Filter (Jazwinski, 1970).
Considering that stochastic disturbances are present in our system (Eq. (9)) and supposing that the process noises v_X and v_y are white, Gaussian, zero mean, and independent with

constant covariance matrices Q and R respectively, then, there exists an optimal Extended Kalman Filter gain $L(t)$ for the state space system (12) that minimizes the estimation error variance due to the system noise (Grewal and Andrews, 1993) that, in our case, is directly related to minimizing the variance of the contact F/T estimation error. That is, as the variance of the F/T is not stationary under relevant conditions of application, there is no unique optimal observer gain.
Since matrices $A(X)$ and $C(X)$ from Eqs. (10) and (11) are non-linear, a linearization of the system for an online estimated trajectory $\overline{X}(t)$ is used to select $L(t)$ (see (Gamez et al., 2008b) for more details).

4. Improvement of Force Control in Robotic Manipulators

For the applied force control laws used to fully test the contact force observer, note that the dynamic interaction between a manipulator and an environment occurring during assembly tasks or manufacturing requires frequent exchanges between a free motion and a constrained motion making desirable to have a unified approach to position and force control. That is why all the results are mainly obtained using an impedance control as force control law. However, another different force control law has been used to validate the behavior of the contact force observer: Hybrid position-force control. To analyze the behavior of the force control loops where the force estimator is used, two industrial manipulators were used: an ABB system and a Stäubli manipulator.

4.1 Force control loop utilized

A brief explanation of the implemented control laws is developed as follows.

4.1.1 Impedance Control

Since the proposed observer must work in both free space and constrained space, it is interesting to use a force control law which allows a manipulator to handle both types of motion without switching control strategies during transitions. In this sense, the impedance control (Hogan, 1985) manages how stiff the end-point comes into contact with the environment and required no control mode switching. However, it can not command the contact forces directly. The motivation of the impedance control is that, instead of controlling position and force separately in task space, the desirable dynamic behavior of a manipulator is specified as a relation between force and motion, referred to as impedance, and accomplished based on an estimated or sensed contact force.
As pointed out in (Hogan, 1985), the environmental model is central to any force strategy. Basically, a second-order mass-spring-damper system is used to specify the target dynamics. However simpler models such as pure stiffness or combination of dampening and stiffness can also be used (Whitney, 1985). In this manner, the equation of the dynamic relationship between the end-effector position $\vec{p}=[x,y,z]$ and the contact force $\vec{F}_E$ used to control the force exerted to the environment is

$$\vec{F}_E = B(\dot{\vec{p}}_r - \dot{\vec{p}}) + K(\vec{p}_r - \vec{p}) \tag{15}$$

where the diagonal matrices B and K contain the impedance parameters along Cartesian axes representing the damping and stiffness of the robot respectively, $\vec{p}$ is the steady state nominal equilibrium position of the end effector in the absence of any external forces and $\vec{p}_r$ is the reference position. As $\vec{p}_r$ is software specified, it may reach positions beyond the reachable workspace or inside the environment during contact with the environment. Note that for a desired static behavior of the interaction, the impedance control is reduced to a stiffness control (Chiaverini et al., 1999).
To carry out the impedance control, a LQR controller was used to make the impedance relation variable go to zero for the three axes (x, y, z) (Johansson and Robertsson, 2003). The control law applied was

$$u = -L \cdot \vec{p} + f \cdot \widehat{\vec{F}}_E + l_r \vec{p}_r \tag{16}$$

with f as the force gains in the impedance control, $\widehat{\vec{F}}_E$ the estimated environmental force, which, in our case, was estimated using the contact force observer, $\vec{p}_r$ the position reference and l_r the position gain constants, L being calculated considering the linear model of the robot system.

4.1.2 Hybrid Control

Consider the contact force $\vec{F}_E$ and the end-effector position $\vec{p}$, expressed in the world frame $O_W X_W Y_W Z_W$, which are estimated through the force observer and measured through the joint sensors and the kinematic model respectively. Denoting the desired values for $\vec{F}_E$ and $\vec{p}$ as $\vec{F}_E^d$ and $\vec{p}^d$, respectively, we obtain expressions for the position and force errors after considering their directions to be controlled:

$$\vec{p}^e = (I - S)[\vec{p}^d - \vec{p}] \tag{17}$$

$$\vec{F}_E^e = (S)[\vec{F}_E^d - \vec{F}_E] \tag{18}$$

where I is the unity matrix and S is the selection matrix for the force controlled directions. Using Eqs. (17) and (18), a hybrid control law can be given by:

$$\tau(t) = \tau_P(t) + \tau_F(t) \tag{19}$$

$$\tau_P(t) = K_{P_p} \vec{p}^e + K_{P_d} \dot{\vec{p}}^e \tag{20}$$

$$\tau_F(t) = K_{F_i} \int_0^t \vec{F}_E^e dt \tag{21}$$

where $\tau_P(t)$ compensates for the position error and $\tau_F(t)$ compensates for the force error. Applying this control law one can expect that the component of the desired position $\vec{p}^d$ and the component of the desired force $\vec{F}_c^d$ are closely tracked.

4.2 Experimental Platforms

Two different platforms were used to validate the behavior of the resulting contact force observers. The first one, placed in the Robotics Lab at the Department of Automatic Control, Lund University, is composed of the following devices and sensors (Fig. 3): an ABB IRB 2400 robot, a JR3 wrist force sensor, a compliant grinding tool that links the robot tip and the tool, and a capacitive 3D accelerometer.

This platform was based on an ABB robot. A totally open architecture is its main characteristic, permitting the implementation and evaluation of advanced control strategies. The controller was implemented in Matlab/Simulink using the Real Time Workshop of Matlab, and later compiled and linked to the Open Robot Control System (Nilsson and Johansson, 1999); (Blomdell et al., 2005). The wrist sensor used was a DSP-based force/torque sensor of six degrees of freedom from JR3. The tool used for our experiments was a grinding tool with a weight of 12 kg. The mechanical device Optidrive-in itself a linear force sensor-the purpose of which was to provide to the tool additional compliance in the contact with the environment, was considered as a spring-damping system and provided a measure of the force exerted between its extremes. The accelerometers were placed on the tip of the tool to measure its acceleration. The accelerometer and Optidrive signals were read by the robot controller in real time via analog inputs.

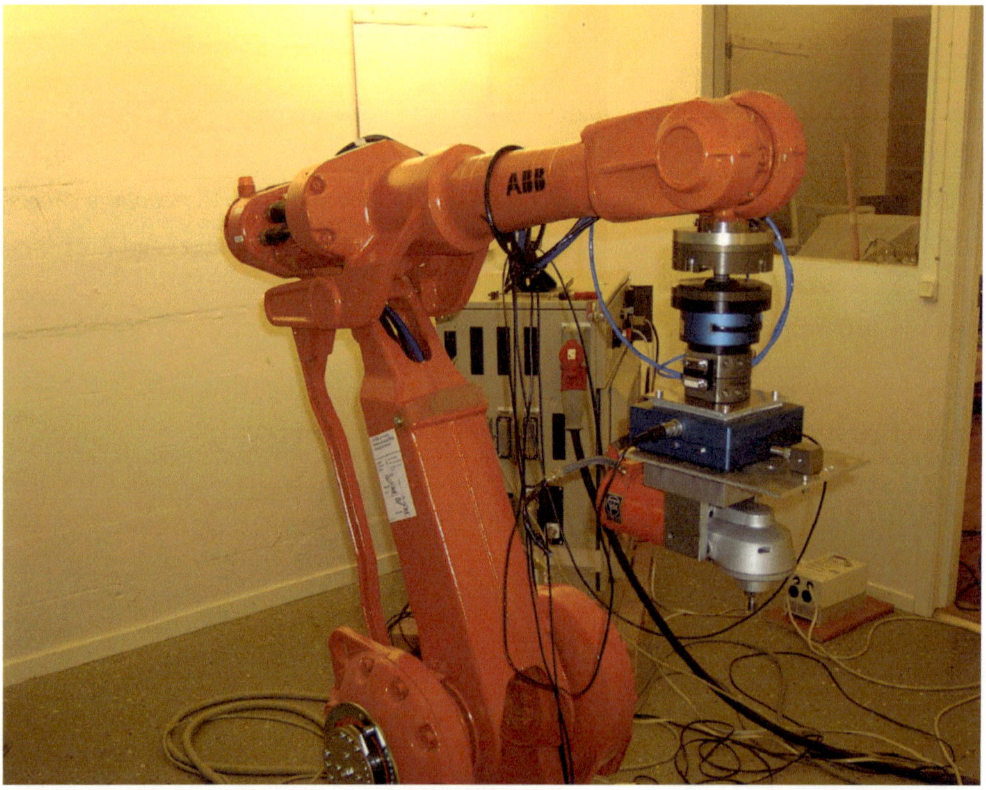

Figure 3. The ABB experimental setup. An ABB industrial robot IRB 2400 with an open control architecture system is used. The wrist sensor is placed between the robot TCP and a compliance tool where the grinding tool is attached. The accelerometer is placed on the tip of the grinding tool

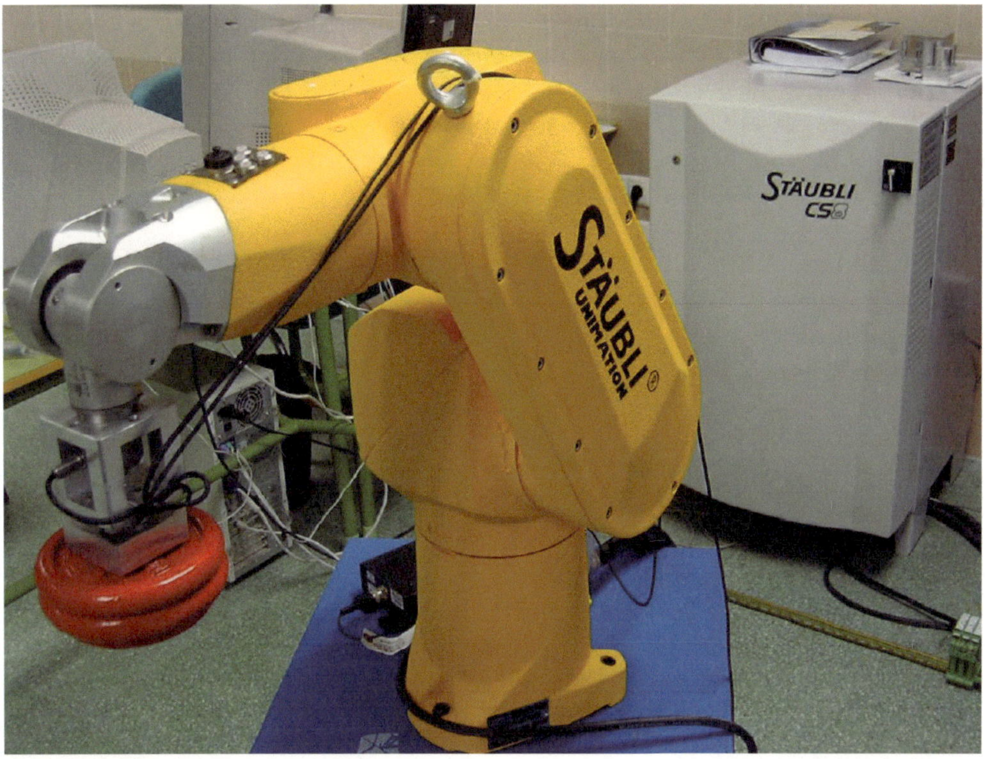

Figure 4. Stäubli Platform. An RX60 robot with an open control architecture system is used. An ATI wrist force sensor is attached to the robot tip. The accelerometer is placed between the robot TCP and the force sensor

Regarding the second robotic platform (Fig. 4), it consisted of a Stäubli manipulator situated in the Robotics and Automation Lab at Jaen University, Spain. It is based on a RX60 which has been adapted in order to obtain a completely open platform (Gamez et al., 2007a). This platform allows the implementation of complex algorithms in Matlab/Simulink where, once they have been designed, they were compiled using the Real Time Toolbox and downloaded to the robot controller. The task created by Simulink communicates with other tasks through shared memory. Note that the operative system that managed the robot controller PC is VxWorks (Wind River, 2005).

For the environment, in both situations a vertical screen made of cardboard, whose stiffness was determined experimentally, was used to represent the physical constraint.

With regard to the sensors, an ATI wrist force sensor together with an accelerometer was attached to the robot flange. Both sensors were read via analog inputs (Gamez et al., 2003). The acceleration sensor is a capacitive PCB sensor with a real bandwidth of 0-250Hz.

4.3 Experimental Results

In this subsection, the results obtained from the contact force observers to both robotic platforms are described. Firstly, the results obtained applying the estimator to the Stäubli platform are presented in Fig. 5. The first experiment consisted of a linear movement in the axis z of three phases: an initial movement in free space (from t =5s to t = 6.2s), a contact transition (from t =6.2s to t = 6.4s) and a movement in constrained space (from t =6.4s to t = 9s). The force controller was an impedance one. In this case, it can be compared how the observer eliminates the inertial effects and the noise introduced by the sensors. Regarding the noise spectra, in Fig. 6, the power spectrum density for the composed signal $u - m\ddot{\xi}_z$ (left) and the observer output power spectrum density (right) are shown. Note how the observer cuts off the noise introduced by the sensors.

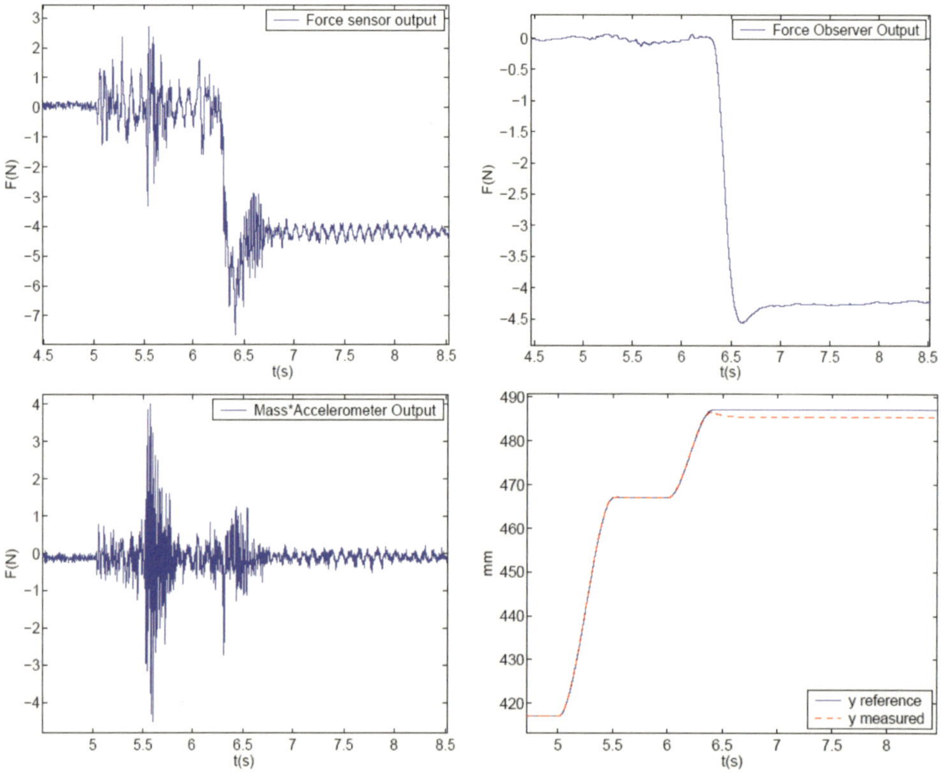

Figure 5. Force measurement from the wrist sensor ATI (upper-left), force observer output (upper-right), accelerometer output (lower-left) and real and measured position of the robot tip for y-axis (lower-right)

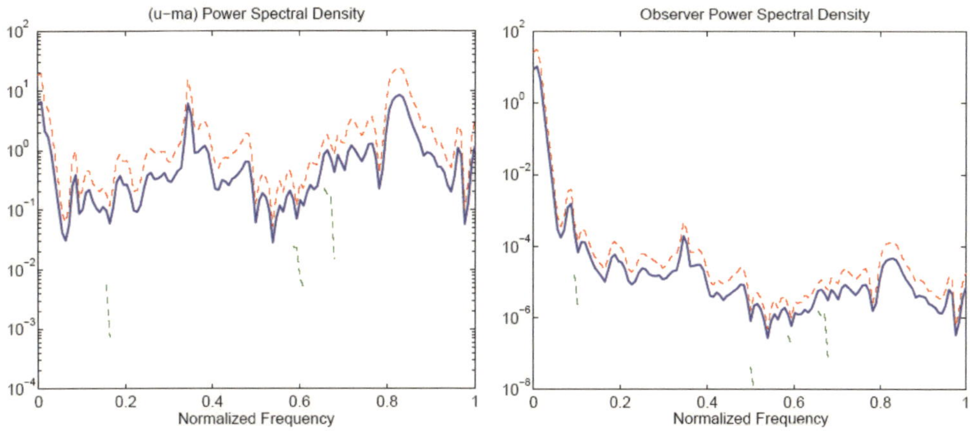

Figure 6. Power spectrum density for the composed signal $u - m\ddot{z}$ (left) and observer output power spectrum density (right). The sample frequency is 250 Hz

On the other hand, another experiment was executed consisting of an oscillation movement for one of the Cartesian axis. Results are depicted in Fig. 7. From this figure it can be verified how the observer eliminates the force oscillations due to the inertial forces that are measured by the force sensor. Note that the movement follows a cubic spline, which means that the acceleration is almost a straight line.

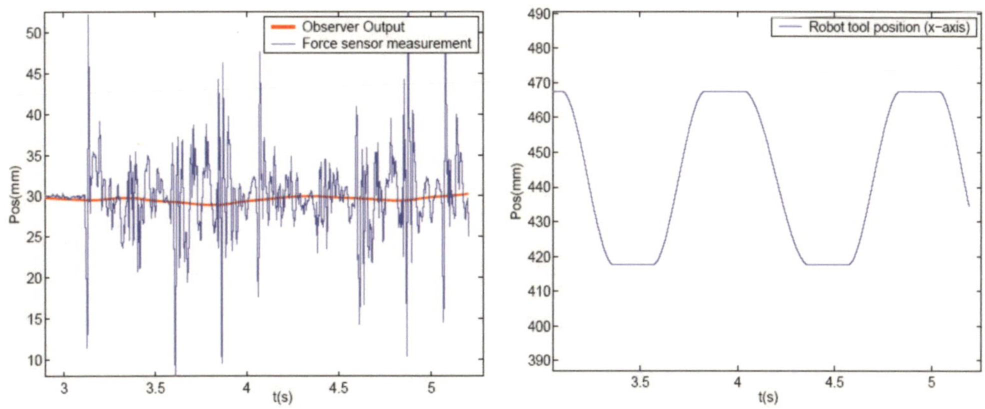

Figure 7. Force sensor measurement and Observer output (left) and robot tool position (right) measured during an oscillation movement in free space

Another experiment carried out consisted of the same former phases but the force control loop is a hybrid control one. In Fig. 8, the ATI force sensor measurement and the observer output are shown. From this figure it is possible to verify how the observer eliminates the force oscillations due to the inertial forces at the beginning of the experiment (where the robot is moving in free space).

Regarding the experiments carried out on the ABB robot, they consisted of three phases for all axes (x, y, z) as well: an initial movement in free space, a contact transition and later, a movement in constrained space. The velocity during the free movement was 300 mm/s and the angle of impact was 30 degrees with respect to the normal of the environment.

The experiments for axis x are shown in Fig. 9, which depicts, at the top, the force measurement from the ATI sensor (left) and the force observer output (right) while at the bottom, the acceleration of the tool getting into contact with the environment (left), and the observer compensation (right). Note that the observer eliminates the inertial effects.

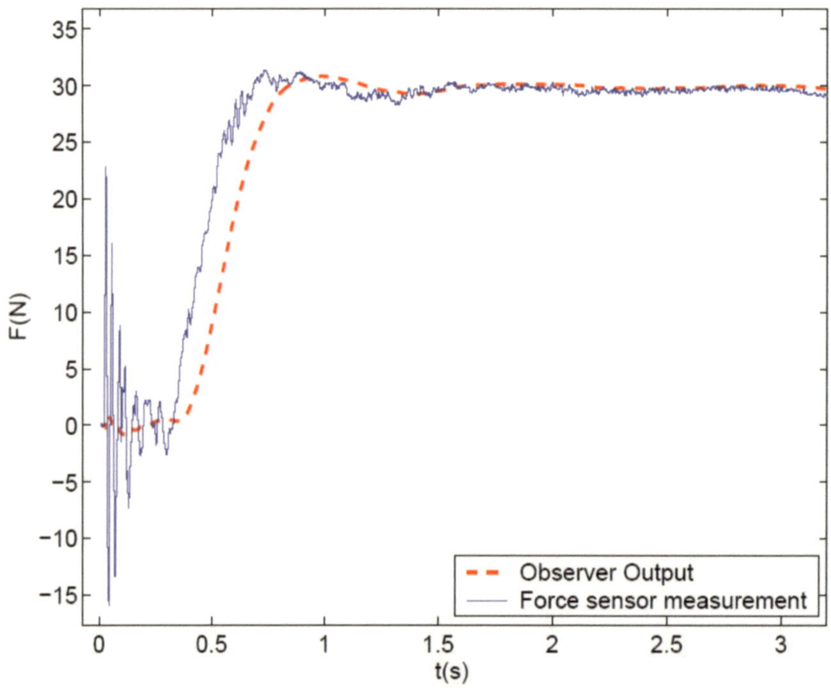

Figure 8. Force sensor and observer output in a hybrid force/position control loop

To test the torque estimation effects, Fig. 10 shows an experiment where, first the robot tool followed an angular trajectory in free space varying the Euler angle ϕ and, later, the manipulator gets into contact with the environment. Fig. 10 (a) and (b) show how the inertial disturbances can be indirectly measured through the angular velocity sensor. Fig. 10 (c) depicts a comparison between the torque sensor measurement N_{S_y} and the contact torque estimator N_{E_y}. It can be observed how the estimator eliminates the disturbances produced by the inertial torques due to the robot tool angular velocity. Fig. 10 (d), (e) and (f) show the power spectrum density of the torque measurement, the angular velocity measurement and the contact torque estimator, respectively. In these spectra, the dynamic filtering properties of the proposed estimator can be noted since the proposed observer filters the noise introduce mainly by the torque sensor at 0.15 of the normalized frequency.

Another strategy could be that, starting from the premise we have attached an accelerometer to the robot tool to determine the tool acceleration, why not to to subtract the tool acceleration multiplied by the tool mass from the force sensor measurement. However, as both the accelerometer and the force sensors are quite noisy elements, it must be pointed out that we would have a final signal with too much noise with simple addition of accelerometer sensors. Then, a next step to be applied is, after subtracting from the force sensor measurement the tool acceleration pondered by the tool mass, why not to apply a standard low-pass filter to eliminate the noise introduced by the sensors. To this purpose, we compare the performance of the proposed force observer against different standard low-pass filters in order to validate the proposed contact force estimator. The most common types of low-pass filters-i.e. Butterworth filter, Chebyshev filters (type I and II) and, finally, an elliptic filter-will be compared using the ABB platform as benchmark.

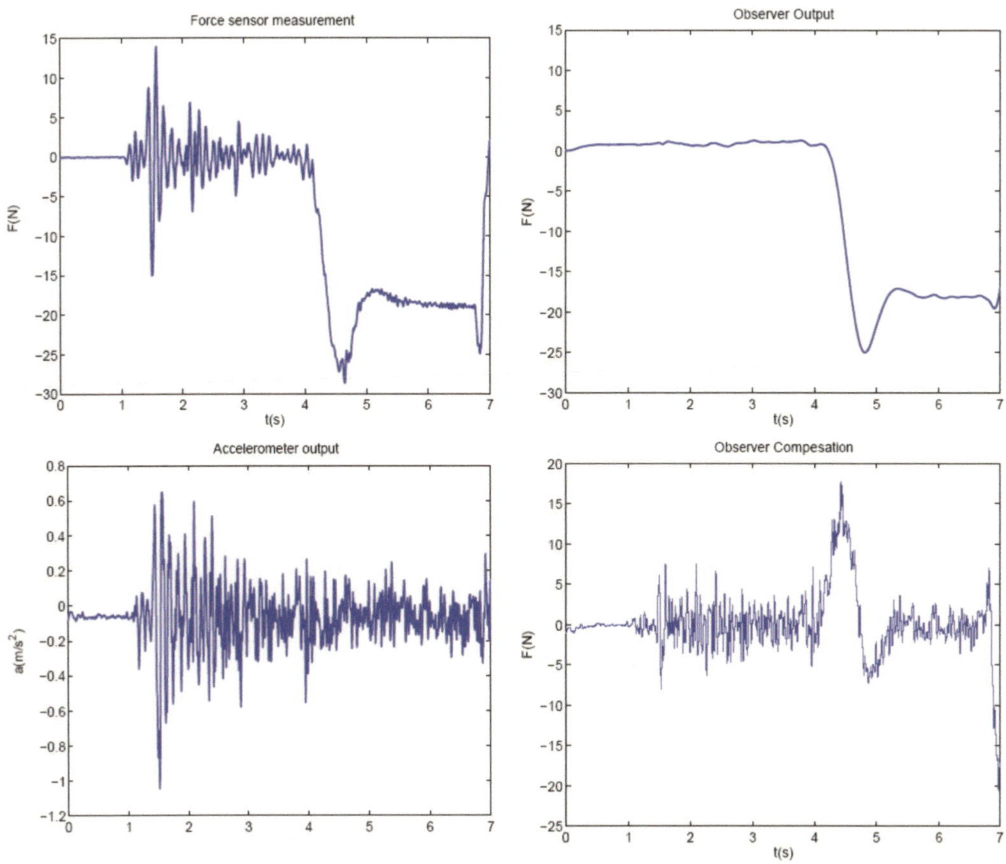

Figure 9. Force measurement from the wrist sensor (upper-left), force observer output (upper-right), acceleration of the robot TCP (lower-left) and observer compensation (lower-right). These results were obtained for x-axis

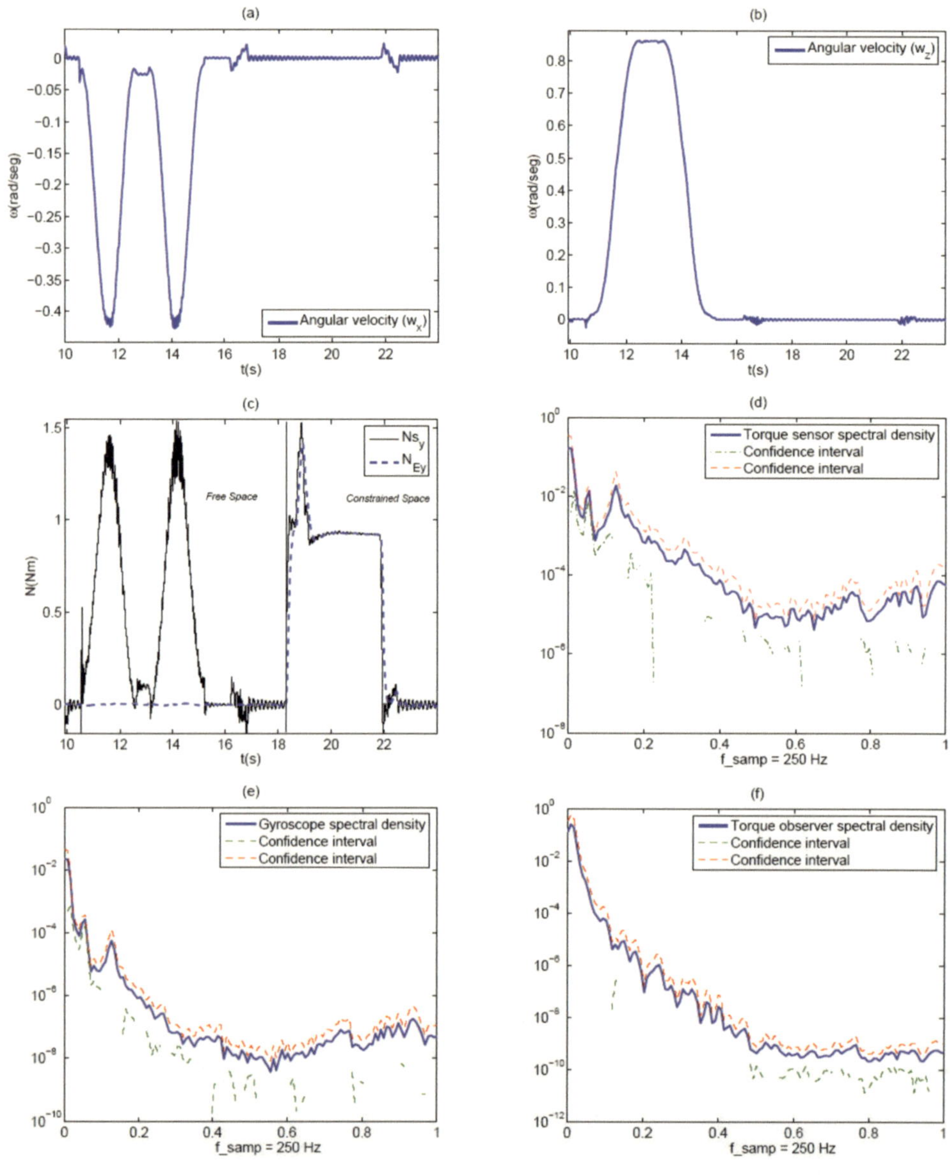

Figure 10. (a) Angular velocity w_x measured with the gyroscope. (b) Angular velocity w_z measured with the gyroscope. (c) Comparison between the torque sensor measurement (N_{S_y}) and the contact torque observer (N_{E_y}) for axis y . (d) Power spectrum density of the torque sensor for axis y . (e) Power spectrum density of the gyroscope sensor for axis x . (f) Power spectrum density of the contact torque observer for axis y . The sampling frequency is 250 Hz

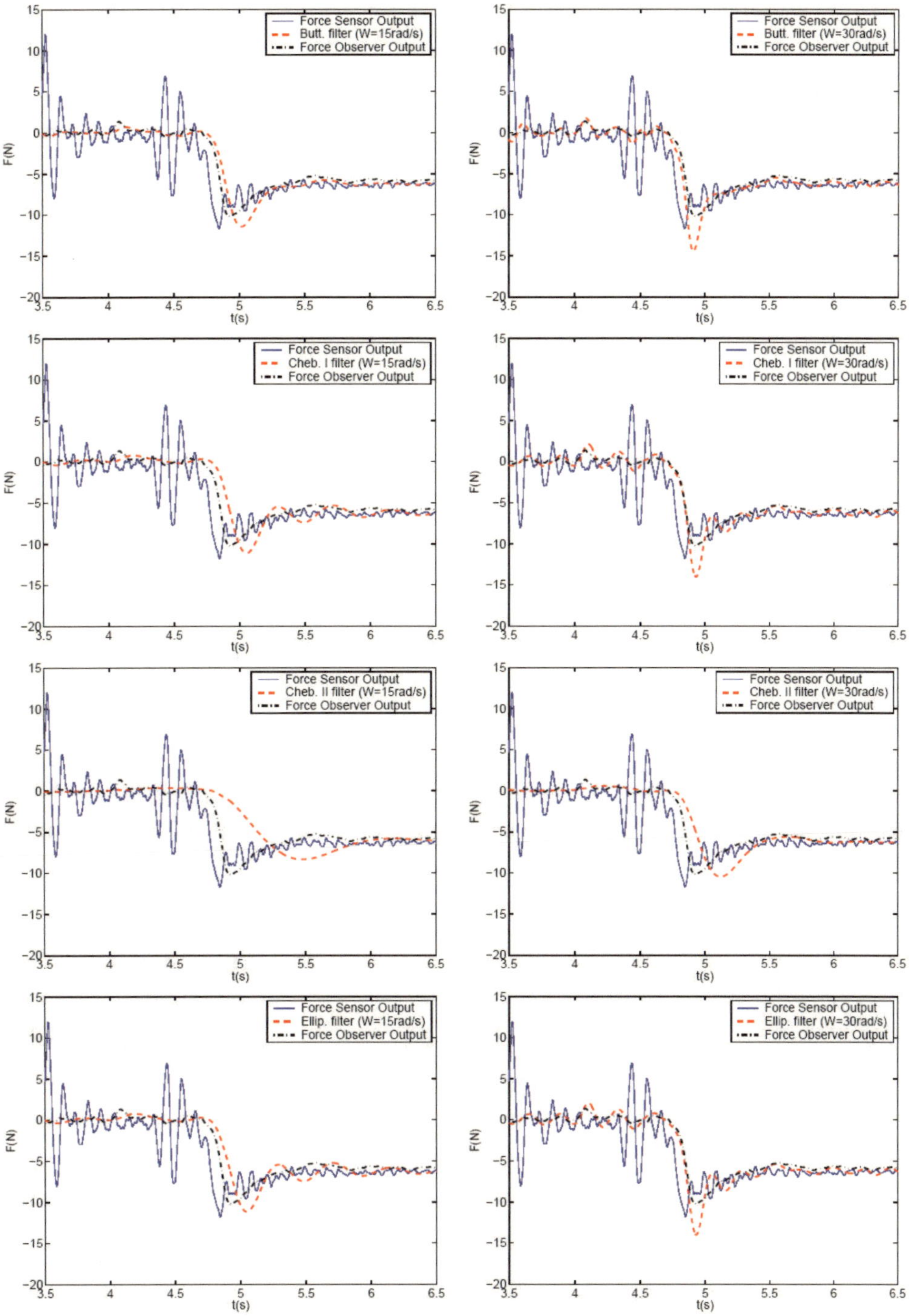

Figure 11. Comparison between the force sensor measurements, the observer outputs and different low-pass filters with different cut-off frequencies (ABB system)

Fig. 11 shows a comparison between the force sensor measurements, the observer output and different low-pass filters with different cut-off frequencies. These figures show results from different configurations of the filters (degree and cut-off frequency) applied to the ABB platform. Analyzing the data, this solution presents two main drawbacks against the proposed contact force observer: depending on the cut-off frequency of the low-filter, they introduce a quite considerable delay; another problem is that these filters also suffers from an overshoot problem.
Finally, another strategy is that before considering the idea of placing an accelerometer at the robot tool to measure its acceleration and, using this measurement to compensate for the inertial forces, the tool acceleration was estimated through the joint measurements and the kinematics robot model (Gamez, 2006a).

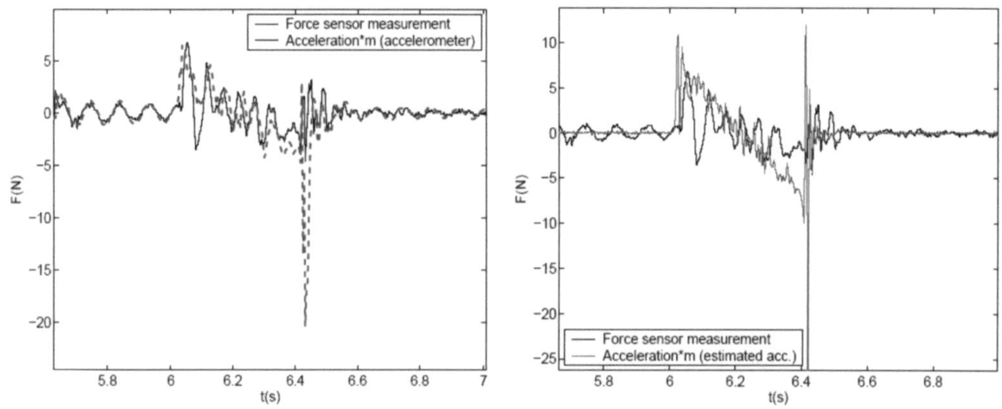

Figure 12. Comparison between the force sensor measurement with the accelerometer output multiplied by the tool mass (upper) and the acceleration estimate using the kinematic model multiply by the tool mass (lower). These experiments were carried out in the Stäubli platform

Figs. 12 show a comparison between the force sensor measurement and the accelerometer output (left), and a comparison between the force sensor signal and the accelerometer estimate (right) for the Stäubli platform. The picture on the left depict how the acceleration measurement follows accurately the force sensor measurement divided by the tool mass. However, it can be appreciated that high accuracy can not be expected because the acceleration estimate, independently from the acceleration estimation algorithm used, does not represent accurately the true acceleration (right).

5. Conclusions

A new contact force observer approach, that fusions the data from three different kind of sensors-that is, resolvers/encoders mounted at each joint of the robot with six degrees of freedom, a wrist force sensor and accelerometers-with the goal of obtaining a suitable contact force estimator has been developed.
We have shown all the advantages and drawbacks of the new sensor fusion approach, comparing the performance of the proposed observers to other alternatives such as the use of the kinematic robot model to estimate the tool acceleration or the use of standard low pass filters.

From the experimental results, it can be concluded that the final observer approach helps to improve the performance as well as stability and robustness for the impact transition phase since it eliminates the perturbations introduced by the inertial forces.
The observer obtained has been validated on several industrial robots applying two well known and widely used control approaches: an impedance control and a hybrid position-force control.
Several successful experiments were performed to validate the performance of the proposed architecture, with a particular interest in compliant motion control. These experiments showed the actual possibility to easily test advanced control algorithms and to integrate new sensors in the developed benchmark as it is the case of the accelerometer.

6. References

Blomdell, A., Bolmsjö, G., Brogårdh, T., Cederberg, P., Isaksson, M. Johansson, R., Haage, M., Nilsson, K., Olsson, M., Olsson, T,. Robertsson, A. and Wang, J.J. (2005). Extending an Industrial Robot Controller---Implementation and Applications of a Fast Open Sensor Interface. *IEEE Robotics and Automation Magazine*, 12(3):85-94. (Blomdell et al., 2005)

Chiaverini, S. , Siciliano, B. and Villani, L. (1999). A Survey of Robot Interaction Control Schemes with Experimental Comparison. *IEEE Trans. Mechatronics*, 4(3):273-285. (Chiaverini et al., 1999)

Gamez, J., Gomez Ortega, J., Gil, M. and Rubio, F. R. (2003). Arquitectura abierta basada en un robot Stäubli RX60 (In Spanish). *Jornadas de Automatica*, pages CD-ROM. (Gamez et al., 2003)

Gamez, J., Robertsson, A., Gomez, J. and Johansson, R. (2004). Sensor Fusion of Force and Acceleration for Robot Force Control. *Int. Conf. Intelligent Robots and Systems (IROS 2004)*, pages 3009-3014, Sendai, Japan. (Gamez et al., 2004)

Gamez, J., Robertsson, A., Gomez, J. and Johansson, R. (2005). Automatic Calibration Procedure for a Robotic Manipulator Force Observer. *IEEE Int. Conf. on Robotics and Automation (ICRA2005)*, pages 2714-2719, Barcelona, Spain. (Gamez et al., 2005a)

Gamez, J., Robertsson, A., Gomez, J. and Johansson, R. (2005). Force and Acceleration Sensor Fusion for Compliant Robot Motion Control. *IEEE Int. Conf. on Robotics and Automation (ICRA2005)*, pages 2720-2725, Barcelona, Spain. (Gamez et al., 2005b)

Gamez, J. (2006). *Sensor Fusion of Force, Acceleration and Position for Compliant Robot Motion Control*. Phd Thesis, Jaen University, Spain. (Gamez, 2006a)

Gamez, J., Robertsson, A., Gomez, J. and Johansson, R. (2006). Generalized Contact Force Estimator for a Robot Manipulator. *IEEE Int. Conf. on Robotics and Automation (ICRA2006)*, pages 4019-4025, Orlando, Florida. (Gamez et al., 2006b)

Gamez, J., Robertsson, A., Gomez, J. and Johansson, R. (2007) Design and Validation of an Open Architecture for an Industrial Robot Control. *IEEE International Symposium on Industrial Electronics (IEEE ISIE 2007)*, pages 2004-2009, Vigo, Spain. (Gamez et al., 2007a)

Gamez, J., Robertsson, A., Gomez, J. and Johansson, R. (2007). Estimación de la Fuerza de Contacto para el control de robots manipuladores con movimientos restringidos (In Spanish). *Revista Iberoamericana de Automática e Informática Industrial*, 4(1):70-82. (Gamez et al., 2007b)

Gamez, J., Robertsson, A., Gomez, J. and Johansson, R. (2008). Self-Calibrated Robotic Manipulator Force Observer (In press). *Robotics and Computer Integrated Manufacturing*. (Gamez et al., 2008a)

Gamez, J., Robertsson, A., Gomez, J. and Johansson, R. (2008). Sensor Fusion for Compliant Robot Motion Control. *IEEE Trans. on Robotics*, 24(2):430-441. (Gamez et al., 2008b)

Grewal, M. S. and Andrews, A. P. (1993). *Kalman Filtering. Theory and Practice*. Prentice Hall, New Jersey, USA. (Grewal and Andrews, 1993)

Harashima, F. and Dote, Y. (1990). Sensor Based Robot Systems. *Proc. IEEE Int. Workshop Intelligent Motion Control*, pages 10-19, Istanbul, Turkey. (Harashima and Dote, 1990)

Hogan, N. (1985). Impedance control: An approach to manipulation, Parts 1-3. *J. of Dynamic Systems, Measurement and Control. ASME*, 107:1-24,. (Hogan, 1985)

IFR. (2001). World Robotics 2001, statistics, analysis and forecasts. *United Nations and Int. Federation of Robotics.* (IFR, 2001)

Jazwinski, A.H. (1970). *Stochastic Processes and Filtering Theory*. Academic Press, New York. (Jazwinski, 1970)

Johansson, R. (1993). *System Modeling and Identification*. Prentice Hall, Englewood Cliffs, NJ. (Johansson, 1993)

Johansson, R. and Robertsson (2003), A. Robotic Force Control using Observer-based Strict Positive Real Impedance Control. *IEEE Proc. Int. Conf. Robotics and Automation*, pages 3686-3691, Taipei, Taiwan. (Johansson and Robertsson, 2003)

Khatib, O. (1987). A unified approach for motion and force control of robot manipulators: The operational space formulation. *IEEE J. of Robotics and Automation*, 3(1):43-53. (Khatib, 1987)

Kröger, T., Kubus, D. and Wahl, F.M. (2006). 6D Force and Acceleration Sensor Fusion for Compliant Manipulation Control. *Int. Conf. Intelligent Robots and Systems (IROS 2006)*, pages 2626-2631, Beijing, China. (Kröguer et al., 2006)

Kröger, T. , Kubus, D. and Wahl, F.M. (2007). Force and acceleration sensor fusion for compliant manipulation control in 6 degrees of freedom. *Advanced Robotics*, 21:1603-1616(14), October. (Kröguer et al., 2007)

Kumar, M., Garg, D. (2004). Sensor Based Estimation and Control of Forces and Moments in Multiple Cooperative Robots. *Trans. ASME J. of Dynamic Systems, Measurement and Control*, 126(2):276-283. (Kumar and Garg, 2004)

Lin, S.T. (1997). Force Sensing Using Kalman Filtering Techniques for Robot Compliant Motion Control. *J. of Intelligent and Robotics Systems*, 135:1-16. (Lin, 1997)

Murray, R. M., Li, Z. and Sastry S.S., A Mathematical Introduction to Robotic Manipulation, *CRC Press*, Boca Raton, FL, 1994. (Murray et al, 1994)

Nilsson, K. and Johansson, R. (1999). Integrated architecture for industrial robot programming and control. *J. Robotics and Autonomous Systems*, 29:205-226. (Nilsson and Johansson, 1999)

Wind River (2005). *VxWorks: Reference Manual*. Wind River Systems. (Wind River, 2005)

Uchiyama, M. (1979). A Study of Computer Control Motion of a Mechanical Arm (2nd report, Control of Coordinate Motion utilizing a Mathematical Model of the Arm). *Bulletin of JSME*, pages 1648-1656. (Uchiyama, 1979)

Uchiyama, M. and Kitagaki, K. (1989). Dynamic Force Sensing for High Speed Robot Manipulation Using Kalman Filtering Techniques. *Proc. of Conf. Decision and Control*, pages 2147-2152, Tampa, Florida. (Uchiyama and Kitagaki, 1989)

Whitney, D. (1985). A Historical Perspective of Force Control. *Proc. IEEE Int. Conf. Robotics and Automation*, pages 262 - 268. (Whitney, 1985)

10

Adaptive Neural Network Based Fuzzy Sliding Mode Control of Robot Manipulator

Ayca Gokhan AK and Galip Cansever
Marmara University, Yildiz Technical University
Turkey

1. Introduction

Robotic manipulators are highly nonlinear, highly time-varying and highly coupled. Moreover, there always exists uncertainty in the system model such as external disturbances, parameter uncertainty, and sensor errors. All kinds of control schemes, including the classical Sliding Mode Control (SMC) have been proposed in the field of robotic control during the past decades (Guo & Wuo, 2003).

SMC has been developed and applied to nonlinear system for the last three decades (Utkin, 1977, Edwards & Spurgeon, 1998). The main feature of the SMC is that it uses a high speed switching control law to drive the system states from any initial state onto a switching surface in the state space, and to maintain the states on the surface for all subsequent time (Hussain & Ho, 2004). So, there are two phase in the sliding mode control system: Reaching phase and sliding phase. In the reaching phase, corrective control law will be applied to drive the representation point everywhere of state space onto the sliding surface. As soon as the representation point hit the surface the controller turns the sliding phase on, and applies an equivalent control law to keep the state on the sliding surface (Lin & Chen, 1994).

The advantage of SMC is its invariance against parametric uncertainties and external disturbances. One of the SMC disadvantages is the difficulty in the calculation of the equivalent control. Neural network is used to compute of equivalent control. Especially multilayer feed forward neural network has been used. A few examples can be given as (Ertugrul & Kaynak, 1998, Tsai et al., 2004, Morioka et al., 1995). A Radial Basis Function Neural Networks (RBFNN) with a two-layer data processing structure had been adopted to approximate an unknown mapping function. Hence, similar to multilayered feed forward network trained with back propagation algorithm, RBFNN also known to be good universal approximator. The input data go through a non-linear transformation, Gaussian basis function, in the hidden layer, and then the functions responses are linearly combined to form the output (Huang et al., 2001). Also, back propagation NN has the disadvantages of slower learning speed and local minima converge.

By introducing the fuzzy concept to the sliding mode and fuzzifying the sliding surface, the chattering in SMC system can be alleviated, and fuzzy control rules can be determined systematically by the reaching condition of the SMC. There has been much research involving designs for fuzzy logic based on SMC, which is referred to as fuzzy sliding mode control (Lin & Mon, 2003).

In this study, a fuzzy sliding mode controller based on RBFNN is proposed for robot manipulator. Fuzzy logic is used to adjust the gain of the corrective control of the sliding mode controller. The weights of the RBFNN are adjusted according to some adaptive algorithm for the purpose of controlling the system states to hit the sliding surface and then slide along it.
The paper is organized as follows: In section 2 model of robot manipulator is defined. Adaptive neural network based fuzzy sliding mode controller is presented in section 3. Robot parameters and simulation results obtained for the control of three link scara robot are presented in section 4. Section 5 concludes the paper.

2. Model of Robot Manipulator

The dynamic model of an n-link robot manipulator may be expressed in the following Lagrange form:

$$M(q)\ddot{q}+C(q,\dot{q})\dot{q}+G(q)+F(q,\dot{q})=u(t) \tag{1}$$

where $q,\dot{q},\ddot{q}\in R^{n}$ are the joint position, velocity, and acceleration vectors, respectively; $M(q)\in R^{nxn}$ denotes the inertia matrix; $C(q,\dot{q})\in R^{nxn}$ expresses the coriolis and centrifugal torques, $G(q)\in R^{n}$ is the gravity vector; $F(q,\dot{q})\in R^{nxn}$ is the unstructured uncertainties of the dynamics including friction and other disturbances; $u(t)\in R^{nx1}$ is the actuator torque vector acting on the joints.

3. Adaptive Neural Network Based Fuzzy Sliding Mode Control

Sliding Mode Control

If the desired system states are available, the desired angle of the manipulator joint are denoted by q_d. The control objective is to drive the joint position q to the desired position q_d. The tracking error equation can be written as follows:

$$e=q-q_d \tag{2}$$

Define the sliding surface of the sliding mode control design should satisfy two requirements, i.e., the closed-loop stability and performance specifications (Chen & Lin, 2002). A conventional sliding surface corresponding to the error state equation can be represented as:

$$s=\dot{e}+\lambda e \tag{3}$$

where λ is a positive constant.
In general, sliding mode control law can be represented as:

$$u=u_{eq}+u_c \tag{4}$$

where u_{eq} is the equivalent control law for sliding phase motion and u_c is the corrective control for the reaching phase motion. The control objective is to guarantee that the state trajectory can converge to the sliding surface. So, corrective control u_c is chosen as follows:

$$u_c = Ksign(s) \tag{5}$$

where K is a positive constant. The *sign* function is a discontinuous function as follows:

$$sign(s) = \begin{cases} 1 & s > 0 \\ 0 & s = 0 \\ -1 & s < 0 \end{cases} \tag{6}$$

Notice that (5) exhibits high frequency oscillations, which is defined as chattering. Chattering is undesired because it may excite the high frequency response of the system. Common methods to eliminate the chattering are usually adopting the following. i) Using the saturation function. ii) Inserting a boundary layer (Tsai et al., 2004). In this paper, shifted sigmoid function is used instead of *sign* function:

$$h(s_i) = \frac{2}{1+e^{-s_i}} - 1 \tag{7}$$

3.2 Fuzzy Sliding Mode Controller

Control gain K is fuzzified with the fuzzy system that shown in Fig. 1 (Guo & Wuo, 2003).

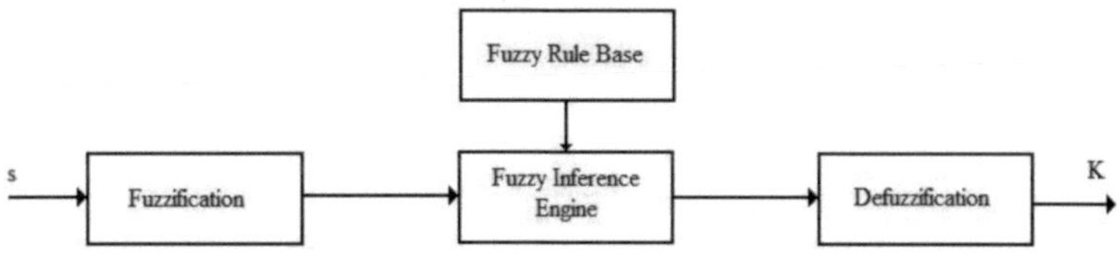

Figure 1. Diagram for a fuzzy system

The rules in the rule base are in the following form:

$$\text{IF } s_i \text{ is } A_i^m, \text{THEN } K_i \text{ is } B_i^m$$

where A_i^m and B_i^m are fuzzy sets. In this paper it is chosen that s_i has membership functions: NB, NM, NS, Z, PS, PM, PB and K_i has membership functions: Z, PS, PM, PB; where N stands for negative, P positive, B big, M medium, S small, Z zero. They are all Gaussian membership functions defined as follows:

$$\mu_A(x_i) = \exp\left(-\left(\frac{x_i - \alpha}{\sigma}\right)^2\right) \tag{8}$$

From the knowledge of the fuzzy systems, K_i can be written as

$$K_i = \frac{\sum_{m=1}^{M} \theta_{\lambda_i}^{m} \mu_{A^m}(s_i)}{\sum_{m=1}^{M} \mu_{A^m}(s_i)} = \theta_{ki}^{T} \psi_{ki}(s_i) \tag{9}$$

where $\theta_{ki} = \left[\theta_{ki}^{1},...,\theta_{ki}^{m},...,\theta_{ki}^{M}\right]^T$ is the vector of the center of the membership functions of K_i, $\psi_{ki}(s_i) = \left[\psi_{ki}^{1}(s_i),...,\psi_{ki}^{m}(s_i),...\psi_{ki}^{M}(s_i)\right]^T$ is the vector of the height of the membership functions of K_i in which $\psi_{ki}^{m}(s_i) = \mu_{A^m}(s_i) / \sum_{m=1}^{M} \mu_{A^m}(s_i)$, and M is the amount of the rules.

3.3 Radial Basis Function Network

A whole knowledge of the system dynamics and the system parameters is required to be able to compute the equivalent control (Ertugrul & Kaynak, 1998). This is practically very difficult. To be able to solve this problem, neural network can used to compute the equivalent control.

A RBFNN is employed to model the relationship between the sliding surface variable, s, and equivalent control, u_{eq}. In other words, sliding variable, s, will be used as the input signal for establishing a RBFNN model to calculate the equivalent control.

The Gaussian function is used as the activation function of each neuron in the hidden layer (10). The excitation values of these Gaussian functions are distances between the input values of the sliding surface value, s, and the central positions of the Gaussian functions (Huang et al., 2001).

$$g(x) = \exp\left(-\frac{\left\| s - c_j \right\|^2}{\sigma_j^2} \right) \tag{10}$$

where j is the j. neuron of the hidden layer, c_j is the central position of neuron j. σ_j is the spread factor of the Gaussian function.

Weightings between input and hidden layer neurons are specified as constant 1. Weightings between hidden and output layer neurons (w_j) are adjusted based on adaptive rule.

The output of the network is:

$$u_{eq} = \sum_{j=1}^{n} w_j g_j(s) \tag{11}$$

Based on the Lyapunov theorem, the sliding surface reaching condition is $s\dot{s} < 0$. If control input chooses to satisfy this reaching condition, the control system will converge to the origin of the phase plane. Since a RBFNN is used to approximate the non-linear mapping between the sliding variable and the control law, the weightings of the RBFNN should be regulated based on the reaching condition, $s\dot{s} < 0$. An adaptive rule is used to adjust the weightings for searching the optimal weighting values and obtaining the stable converge property. The adaptive rule is derived from the steep descent rule to minimize the value of $s\dot{s}$ with respect to w_j. Then the updated equation of the weighting parameters is (Huang et al., 2001):

$$\dot{w}_j = -\tau \frac{\partial s(t)\dot{s}(t)}{\partial w_j(t)} \tag{12}$$

where τ is adaptive rate parameter. Using the chain rule, (12) can be rewritten as follows:

$$\dot{w}_j = \eta s(t) g(s) \tag{13}$$

where η is the learning rate parameter. The structure of RBFNN is shown in Fig. 2.

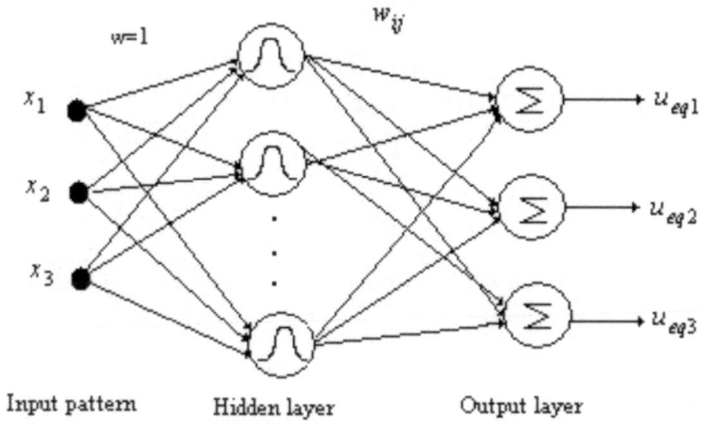

Figure 2. Radial Basis Function Neural Network (RBFNN)

3.3 Proposed Controller

The configuration of proposed controller is shown in Fig. 3. The control law for proposed controller is as (4) form. K gain of the corrective control u_c is adjusted with fuzzy logic and is the equivalent control u_{eq} is computed by RBFNN. The structure of RBFNN using this study has three inputs and three outputs. The number of hidden nodes is equal to 5. The weights of the RBFNN are changed with adaptive algorithm in adaptive law block. Outputs of the corrective control and RBFNN are summed and then applied to robotic manipulator.

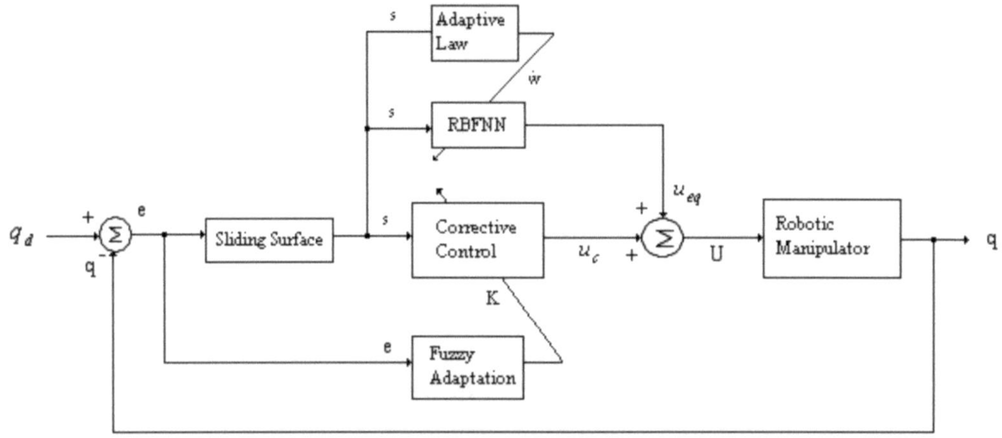

Figure 3. Block diagram of the proposed control system

4. Application

4.1 Robot Parameters

A three link scara robot parameters are utilized in this study to verify the effectiveness of the proposed control scheme. The dynamic equations which derived via the Euler-Lagrange method are presented as follows (Wai & Hsich, 2002):

$$\begin{bmatrix} M_{11} & M_{12} & M_{13} \\ M_{21} & M_{22} & M_{23} \\ M_{31} & M_{32} & M_{33} \end{bmatrix} \begin{bmatrix} \ddot{q}_1 \\ \ddot{q}_2 \\ \ddot{q}_3 \end{bmatrix} + l_1 l_2 \sin(q_2) \begin{bmatrix} C_{11} & C_{12} & C_{13} \\ C_{21} & C_{22} & C_{23} \\ C_{31} & C_{32} & C_{33} \end{bmatrix} \begin{bmatrix} \dot{q}_1 \\ \dot{q}_2 \\ \dot{q}_3 \end{bmatrix} + \begin{bmatrix} 0 \\ 0 \\ -m_3 g \end{bmatrix} + f(\dot{q}) + t_1(t) = u(t) \tag{14}$$

where,

$$M_{11} = l_1^2\left(\frac{m_1}{3} + m_2 + m_3\right) + l_1 l_2 (m_2 + 2m_3)\cos(q_2) + l_2^2\left(\frac{m_2}{3} + m_3\right)$$

$$M_{13} = M_{23} = M_{31} = M_{32} = 0$$

$$M_{12} = -l_1 l_2\left(\frac{m_2}{2} + m_3\right)\cos(q_2) - l_2^2\left(\frac{m_2}{3} + m_3\right) = M_{21}$$

$$M_{22} = l_2^2\left(\frac{m_2}{3} + m_3\right)$$

$$M_{33} = m_3$$

$$C_{11} = -\dot{q}_2 (m_2 + 2m_3)$$

$$C_{12} = -\dot{q}_2\left(\frac{m_2}{2} + m_3\right)$$

$$C_{22} = -\dot{q}_2\left(\frac{m_2}{2} + m_3\right)$$

$$C_{13} = C_{22} = C_{23} = C_{31} = C_{32} = C_{33} = 0$$

in which q_1, q_2, q_3, are the angle of joints 1,2 and 3; m_1, m_2, m_3 are the mass of the links 1,2 and 3; l_1, l_2, l_3 are the length of links 1,2 and 3; g is the gravity acceleration. Important parameters that affect the control performance of the robotic system are the external disturbance $t_1(t)$, and friction term $f(\dot{q})$.

The system parameters of the scara robot are selected as following:

$$l_1 = 1.0m \;\; l_2 = 0.8m \;\; l_3 = 0.6m \;\; m_1 = 1.0kg \;\; m_2 = 0.8kg \;\; m_3 = 0.5kg \;\; g = 9.8$$

External disturbances are selected as:

$$t_1(t) = \begin{bmatrix} 5\sin(2t) \\ 5\sin(2t) \\ 5\sin(2t) \end{bmatrix} \tag{15}$$

In addition, friction forces are also considered in this simulation and are given as following:

$$f(\dot{q}) = \begin{bmatrix} 12\dot{q}_1 + 0.2sign(\dot{q}_1) + 3\sin(3t) \\ 12\dot{q}_2 + 0.2sign(\dot{q}_2) + 3\sin(3t) \\ 12\dot{q}_3 + 0.2sign(\dot{q}_3) + 3\sin(3t) \end{bmatrix} \tag{16}$$

4.2 Simulation Results

Central positions of the Gaussian function c_j are selected from -2 to 2. Spread factors σ_j are specified from 0.1 to 0.7. Initial weights of network are adjusted to zero.

The proposed fuzzy SMC based on RBFNN in Fig. 3 is applied to control the Scara robot manipulator.

The desired trajectories for three joint to be tracked are,

$$q_d(t) = 1 + 0.1\sin(pi * t) \tag{17}$$

Tracking position, tracking error, control torque and sliding surface of joint 1 is shown in Fig. 4 a, b, c and d respectively. Tracking position, tracking error, control torque and sliding surface of joint 2 is shown in Fig. 5 a, b, c and d respectively. Fig 6 a, b, c and d shows the tracking position, tracking error control torque and sliding surface of joint 3. Position of joint 1, 2 and 3 is reached the desired position at 2s, 1.5s and 0.5s, respectively.

In Fig 4c, 5c and 6c it can be seen that there is no chattering in the control torques of joints. Furthermore, sliding surfaces in Fig 4d, 5d, 6d converge to zero. It is obvious that the chattering in the sliding surfaces is eliminated.

5. Conclusion

The dynamic characteristics of a robot manipulator are highly nonlinear and coupling, it is difficult to obtain the complete model precisely. A novel fuzzy sliding mode controller based on RBFNN is proposed in this study. To verify the effectiveness of the proposed control method, it is simulate on three link scara robot manipulator.

RBFNN is used to compute the equivalent control. An adaptive rule is employed to on-line adjust the weights of RBFNN. On-line weighting adjustment reduces data base requirement. Adaptive training algorithm were derived in the sense of Lyapunov stability analysis, so that system-tracking stability of the closed-loop system can be guaranteed whether the uncertainties or not. Using the RBFNN instead of multilayer feed forward network trained with back propagation provides shorter reaching time.

In the classical SMC, the corrective control gain may choose larger number, which causes the chattering on the sliding surface. Or, corrective control gain may choose smaller number, which cause increasing of reaching time and tracking error. Using fuzzy controller to adjust the corrective control gain in sliding mode control, system performance is improved. Chattering problem in the classical SMC is minimized.

It can be seen from the simulation results, the joint position tracking response is closely follow the desired trajectory occurrence of disturbances and the friction forces.

Simulation results demonstrate that the adaptive neural network based fuzzy sliding mode controller proposed in this paper is a stable control scheme for robotic manipulators.

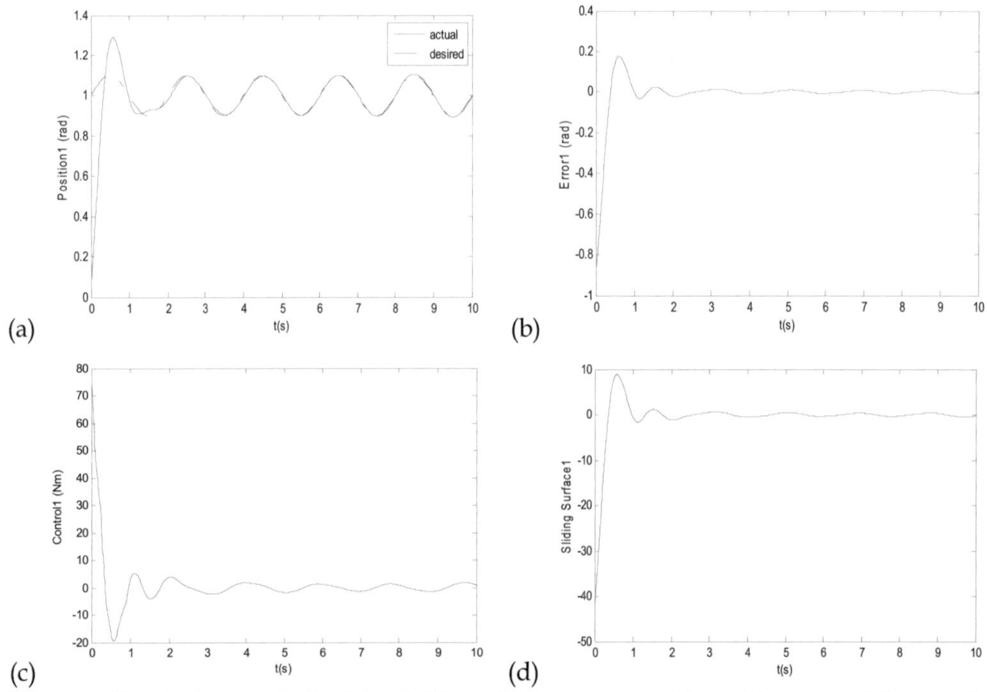

Figure 4. Simulation results for joint 1: (a) tracking response; (b) tracking error; (c) control torque; (d) sliding surface

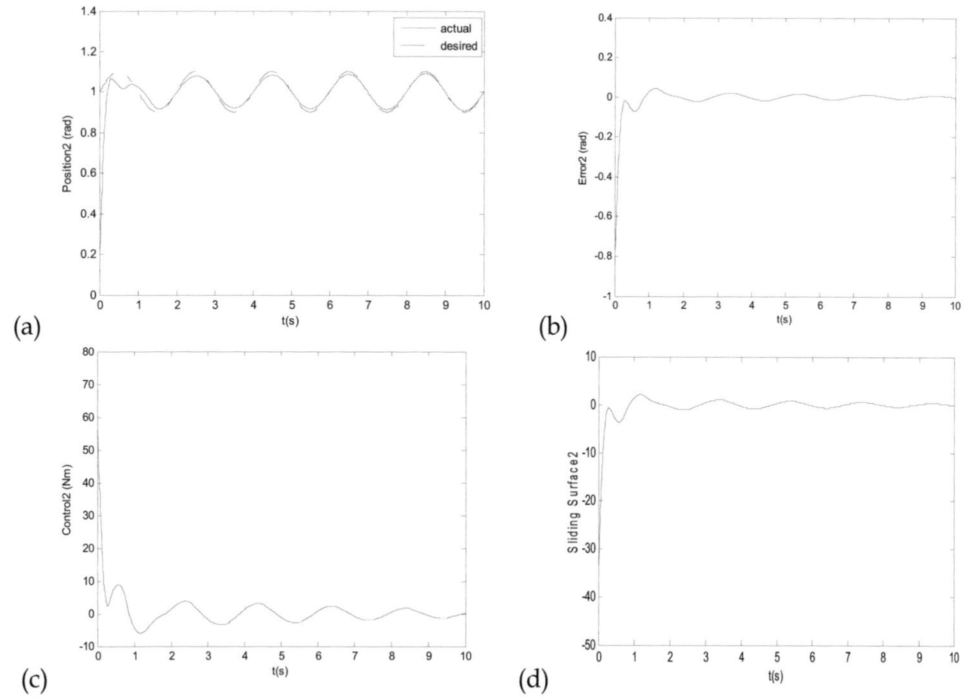

Figure 5. Simulation results for joint 2: (a) tracking response; (b) tracking error; (c) control torque; (d) sliding surface

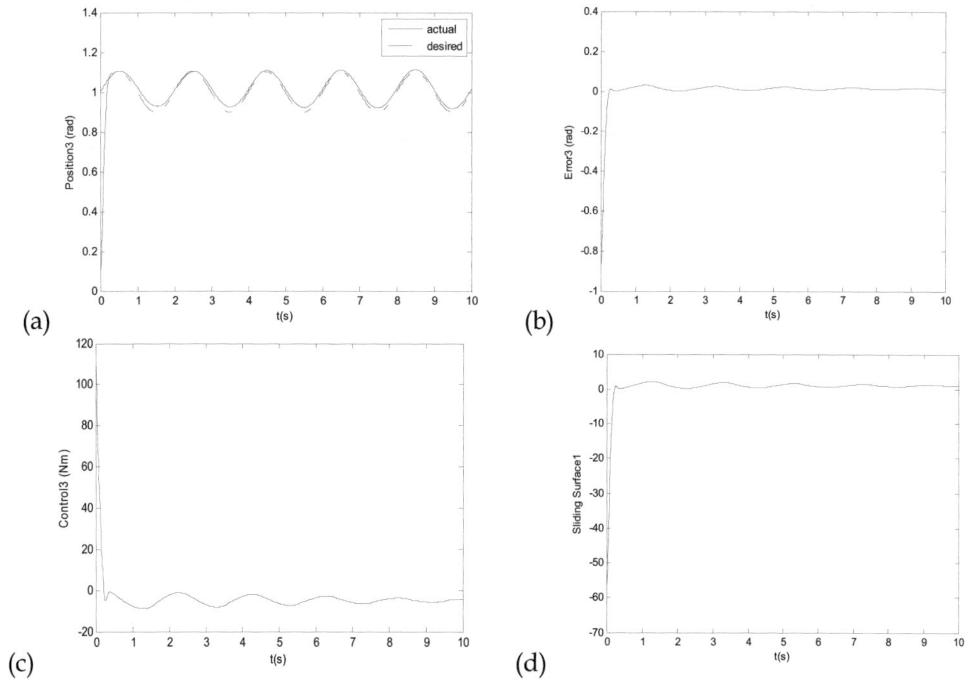

Figure 6. Simulation results for joint 3: (a) tracking response; (b) tracking error; (c) control torque; (d) sliding surface

6. References

Guo, Y., Woo, P. (2003). An Adaptive Fuzzy Sliding Mode Controller for Robotic Manipulator. *IEEE Trans. on System, Man and Cybernetics-Part A: Systems and Humans*, Vol.33, NO.2.

Utkin, V.I., (1997). Variable Structure Systems with Sliding Modes. *IEEE Trans. on Automatic Control AC*-22 212-222.

Edwards, C. and Spurgeon, K. (1998). *Sliding Mode Control*. Taylor&Fransis Ltd.

Hussain, M.A., Ho, P.Y. (2004). Adaptive Sliding Mode Control with Neural Network Based Hybrid Models. *Journal of Process Control* 14 157-176.

Lin, S. and Chen, Y. (1994). RBF-Network-Based Sliding Mode Control. *System, Man and Cybernetics, 'Humans, Information and Technology' IEEE Conf.* Vol.2.

Ertugrul, M., Kaynak, O. (1998). Neural Computation of the Equivalent Control in Sliding Mode for Robot Trajectory Control Applications. *IEEE International Conference on Robotics&Automation*.

Tsai, C., Chung, H.and Yu, F. (2004). Neuro-Sliding Mode Control with Its Applications to Seesaw Systems. *IEEE Trans. on Neural Networks*, Vol.15, NO1.

Morioka, H., Wada, K., Sabanociv, A., Jezernik, K., (1995). Neural Network Based Chattering Free Sliding Mode Control. SICE'95. *Proceeding of the 34th Annual Conf.*

Huang, S., Huang, K., Chiou, K. (2001). Development and Application of a Novel Radial Basis Function Sliding Mode Controller. *Mechatronics*. pp 313- 329.

Lin, C., Mon, Y. (2003). Hybrid adaptive fuzzy controllers with application to robotic systems. *Fuzzy Set and Systems*. 139. pp 151-165.

Chen, C. and Lin, C. (2002). A Sliding Mode Approach to Robotic Tracking Problem. *IFAC*.

Wai, R.J. and Hsich, K.Y. (2002). Tracking Control Design for Robot Manipulator via Fuzzy Neural Network. *Proceedings of the IEEE International Conference on Fuzzy Systems*. Volume 2, pp 1422–1427.

11

Impedance Control of Flexible Robot Manipulators

Zhao-Hui Jiang
Hiroshima Institute of Technology
Japan

1. Introduction

Force control is very important for many practical manipulation tasks of robot manipulators, such as, assembly, grinding, deburring that associate with interaction between the end-effector of the robot and the environment. Force control of the robot manipulator can be break down into two categories: Position/Force Hybrid Control and Impedance Control. The former is focused on regulation of both position and manipulation force of the end-effector on the tangent and perpendicular directions of the contact surface. The later, however, is to realize desired dynamic characteristics of the robot in response to the interaction with the environment. The desired dynamic characteristics are usually prescribed with dynamic models of systems consisting of mass, spring, and dashpot. Literature demonstrates that various control methods in this area have been proposed during recent years, and the intensive research has mainly focused on rigid robot manipulators. On the research area of force control for flexible robots, however, only a few reports can be found.

Regarding force control of the flexible robot, a main body of the research has concentrated on the position/force hybrid control. A force control law was presented based on a linearized motion equation[l]. A quasi-static force/position control method was proposed for a two-link planar flexible robot [2]. For serial connected flexible-macro and rigid-micro robot arm system, a position/force hybrid control method was proposed [3]. An adaptive hybrid force/position controller was resented for automated deburring[4]. Two-time scale position/force controllers were proposed using perturbation techniques[5],[6]. Furthermore, a hybrid position-force control method for two cooperated flexible robots was also presented [7]. The issue on impedance control of the flexible robot attracts many attentions as well. But very few research results have been reported. An impedance control scheme was presented for micro-macro flexible robot manipulators [8]. In this method, the controllers of the micro arm (rigid arm) and macro arm (flexible arm) are designed separately, and the micro arm is controlled such that the desired local impedance characteristics of end-effector are achieved. An adaptive impedance control method was proposed for n-link flexible robot manipulators in the presence of parametric uncertainties in the dynamics, and the effectiveness was confirmed by simulation results using a 2-link flexible robot [9].

In this chapter, we deal with the issue of impedance control of flexible robot manipulators. For a rigid robot manipulator, the desired impedance characteristics can be achieved by changing the dynamics of the manipulator to the desired impedance dynamics through nonlinear feedback control[13]. This method, however, cannot be used to the flexible robot because the number of its control inputs is less than that of the dynamic equations so that the whole dynamics including the flexible links cannot be changed simultaneously as desired by using any possible control methods. To overcome this problem, we establish an impedance control system using a trajectory tracking method rather than the abovementioned direct impedance control method. We aimed at desired behavior of the end-effector of the flexible robot. If the end-effector motion precisely tracks the impedance trajectory generated by the impedance dynamics under an external force, we can say that the desired impedance property of the robot system is achieved. Prom this point of view, the control design process can be divided into two steps. First, the impedance control objective is converted into the tracking of the impedance trajectory in the presence of the external force acting at the end-effector. Then, impedance control system is designed based on any possible end-effector trajectory control methods to guarantee that end-effector motion converses and remains on the trajectory. In this chapter, detail design of impedance control is discussed. Stability of the system is verified using Lyapunov stability theory. As an application example, impedance control simulations of a two-link flexible robot are carried out to demonstrate the usage of the control method.

2. Kinematics and dynamics

We consider a flexible robot manipulator consisted of n flexible links driven by n rigid joints. Let $\theta \in R^n$ be the joint variables, $\mathbf{e} \in R^k$ be the vector of the link flexible displacements. We assume that there is no redundant degree of freedom in the joints, and define vector $\mathbf{P} \in R^n$ to describe the end-effector position and orientation in the n-dimension workspace. The kinematical relationship among $\mathbf{P}$, θ and $\mathbf{e}$ is non-linear, and can be given as

$$\mathbf{P} = \mathbf{f}(\theta, \mathbf{e}) \tag{1}$$

The relationship among the velocities $\dot{\theta}$, $\dot{\mathbf{e}}$, and $\dot{\mathbf{P}}$ is a linear function given as

$$\dot{\mathbf{P}} = \mathbf{J}_\theta \dot{\theta} + \mathbf{J}_e \dot{\mathbf{e}} \tag{2}$$

where $\mathbf{J}_\theta \in R^{n\times n}$ and $\mathbf{J}_e \in R^{n\times k}$ are the Jacobian matrices of f with respect to θ and $\mathbf{e}$. They are defined as $\mathbf{J}_\theta = \partial \mathbf{f}/\partial\theta$, and $\mathbf{J}_e = \partial \mathbf{f}/\partial \mathbf{e}$. By taking derivation of (2) with respect to time, we have

$$\ddot{\mathbf{P}} = \mathbf{J}_\theta \ddot{\theta} + \mathbf{J}_e \ddot{\mathbf{e}} + \dot{\mathbf{J}}_\theta \dot{\theta} + \dot{\mathbf{J}}_e \dot{\mathbf{e}} \tag{3}$$

The dynamics of the flexible robot can be derived using Lagrange approach and FEM(Finite Element Method). In the case that there is an external force $\mathbf{F}_E \in R^n$ acting at the end-effector, the approximated model with FEM can be given as

$$\mathbf{M}_{11}\ddot{\theta} + \mathbf{M}_{12}\ddot{\mathbf{e}} + \mathbf{D}_{11}\dot{\theta} + \mathbf{D}_{12}\dot{\mathbf{e}} + \mathbf{L}_1 = \tau - \mathbf{J}_\theta^T \mathbf{F}_E \tag{4}$$

$$\mathbf{M}_{21}\ddot{\theta} + \mathbf{M}_{22}\ddot{\mathbf{e}} + \mathbf{D}_{21}\dot{\theta} + \mathbf{D}_{22}\dot{\mathbf{e}} + \mathbf{L}_2 = -\mathbf{J}_e^T\mathbf{F}_E \tag{5}$$

where, (4) and (5) are the motion equations with respect to the joints and links of the robot. $\mathbf{M}_{11} \in R^{n\times n}$, $\mathbf{M}_{12} \in R^{n\times k}$, $\mathbf{M}_{21} \in R^{k\times n}$, and $\mathbf{M}_{22} \in R^{k\times k}$ are inertia matrices; $\mathbf{D}_{11}\dot{\theta} \in R^n$, $\mathbf{D}_{12}\dot{\mathbf{e}} \in R^n$, $\mathbf{D}_{21}\dot{\theta} \in R^k$, and $\mathbf{D}_{22}\dot{\mathbf{e}} \in R^n$ contain the Coriolis, centrifugal, forces; $\mathbf{L}_1 \in R^n$ and $\mathbf{L}_2 \in R^k$ denote elastic forces and gravity; τ presents the input torque produced by the joint actuators.
System representation (4) and (5) can be expressed in a compact form as follows:

$$\mathbf{M}\ddot{\mathbf{x}} + \mathbf{D}\dot{\mathbf{x}} + \mathbf{L} = \mathbf{u} - \mathbf{J}^T\mathbf{F}_E \tag{6}$$

where $\dot{\mathbf{x}}^T = [\dot{\theta}^T, \dot{e}^T]^T \in R^m \quad (m = n + k)$, $\ddot{\mathbf{x}} = d\dot{\mathbf{x}}/dt$, $\mathbf{u} = [\tau^T, \mathbf{0}]^T \in R^m$, and $\mathbf{M} \in R^{m\times m}$, $\mathbf{D} \in R^{m\times m}$ and $\mathbf{L} \in R^k$ are consisted of the elements $\mathbf{M}_{ij}$, $\mathbf{D}_{ij}$,, and $\mathbf{L}_i$ $(i,j = 1,2)$ respectively. $\mathbf{J}$ is defined as $\mathbf{J} = [\mathbf{J}_\theta, \mathbf{J}_e]$.
Remark 1: In motion equation (6), the inertia matrix $\mathbf{M}$ is a symmetric and positive definite matrix, and $\dot{\mathbf{M}} - 2\mathbf{H}$ is skew symmetric. These properties will be used in stability analysis of the control system in Section 4.

3. Objective of impedance control and problem formulation

3.1 Desirable impedance dynamics

When an external force acts on the end-effector of the flexible robot, the objective of impedance control is to cause the end-effector to respond to the force according to some desirable impedance dynamics. The desirable impedance dynamics can be designed as following second-order equation

$$\mathbf{M}_d(\ddot{\mathbf{P}}_d - \ddot{\mathbf{P}}_c) + \mathbf{D}_d(\dot{\mathbf{P}}_d - \dot{\mathbf{P}}_c) + \mathbf{K}_d(\mathbf{P}_d - \mathbf{P}_c) = -\mathbf{F}_E \tag{7}$$

where, $\mathbf{P}_c$, $\dot{\mathbf{P}}_c$, $\ddot{\mathbf{P}}_c$ are the commanded end-effector position, velocity and acceleration in the workspace; $\mathbf{P}_d$, $\dot{\mathbf{P}}_d$, $\ddot{\mathbf{P}}_d$ are the responses to the external force and commanded position, velocity and acceleration; $\mathbf{M}_d$, $\mathbf{D}_d$, and $\mathbf{K}_d$ are moment of inertia, damping, and stiffness matrices defined by user so that (7) possesses the desired dynamic characteristics. These matrices are usually defined as constant and diagonal ones.

3.2 The Main idea of impedance control using trajectory tracking approach

For a rigid robot manipulator, the desired impedance described above can be achieved by using impedance control to change dynamics of the robot manipulator [13]. This method, however, is difficult to be used to a flexible robot because one cannot change the whole dynamics of the robot including the flexible links at same time using any possible control method. To overcome this problem, the impedance control system is implemented using a trajectory tracking method rather than using the direct method-the abovementioned impedance control method. Let's consider behavior of the end-effector of the flexible robot. If the end-effector motion correctly tracks the desired impedance trajectories, we can say that the desired end-effector impedance is achieved. Prom this point of view, impedance control can be achieved by tracking the desired impedance trajectory generated by the designed impedance dynamics (7) in the presence of the external force acting at the end-effector. Therefore, the impedance control design can be changed to the trajectory tracking control design. Accordingly, impedance control is implemented using end-effector trajectory control algorithms. Based on this methodology, the design of the control system is divided into three stages. In the first stage, impedance control of the robot is converted into tracking an on-line impedance trajectory generated by the impedance dynamics designed to possess with terms of the desired moment of inertia, damping, and stiffness as (7). On this stage, on-line impedance trajectory generating algorithms is established, and $\mathbf{P}_d$, $\dot{\mathbf{P}}_d$, $\ddot{\mathbf{P}}_d$ are calculated numerically at every sampling time in response to the external force which may vary. The second stage is concerned on the designing of a manifold to prescribe the desirable performance of impedance trajectory tracking and link vibration damping. The third stage is to derive a control scheme such that the end-effector motion of the robot precisely tracks the impedance trajectory. Figure 1 illustrates the concept of impedance control via trajectory tracking.

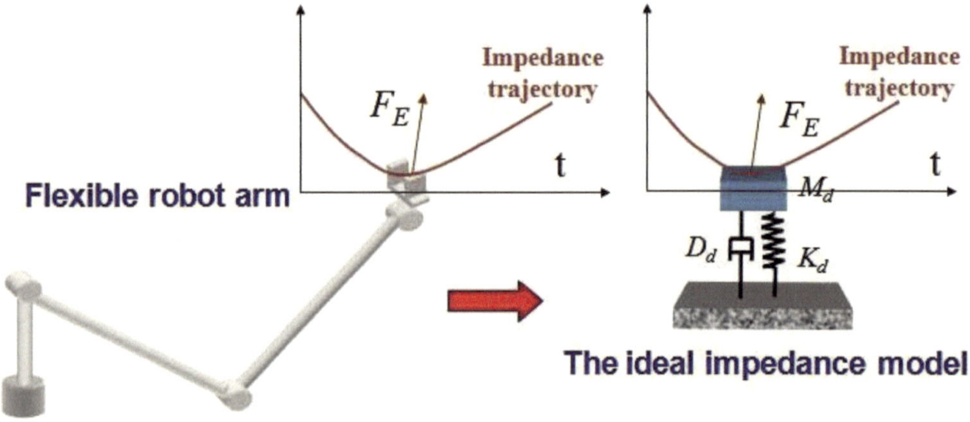

Figure 1. Impedance trajectory generated by the impedance model and tracked by the flexible robot

3.3 The manifold design

In order to achieve good performance in tracking the desired impedance trajectory while keeping the whole system stable, the control system has to include two basic functions. One is link vibration suppressing function and the other one is end-effector trajectory tracking function. We design an ideal manifold
to formulate the desired performance of the system with respect to these two functions. In so doing, let II be the ideal manifold, and divide it into two sub-manifolds HI and n_2 according to the functions abovementioned, and design each of them separately. Now, we define error vectors as follows:

$$\delta\mathbf{P} = \mathbf{P} - \mathbf{P}_d \tag{8}$$

$$\delta\mathbf{e} = \mathbf{e} - \mathbf{e}_d \tag{9}$$

We design the first submanifold to prescribe the desirable performance of end-effector trajectory tracking as follows

$$\mathbf{\Pi}_1 : \{(\delta\mathbf{P}, \delta\dot{\mathbf{P}}) : \delta\dot{\mathbf{P}} + \Lambda_1\delta\mathbf{P} = \mathbf{0}\} \tag{10}$$

where $\Lambda_1 \in R^{n\times n}$ is a constant positive definite matrix.
To describe a desired link flexural behavior, the second submanifold is specified as

$$\mathbf{\Pi}_2 : \{(\delta\mathbf{e}, \delta\dot{\mathbf{e}}) : \delta\dot{\mathbf{e}} + \Lambda_2\delta\mathbf{e}) = \mathbf{0}\} \tag{11}$$

where $\Lambda_2 \in R^{k\times k}$ is a constant positive definite matrix.
By combining $\mathbf{\Pi}_1$ and $\mathbf{\Pi}_2$ together, we have the complete manifold $\mathbf{\Pi}$ as follows

$$\mathbf{\Pi} = \mathbf{\Pi}_1 \cup \mathbf{\Pi}_2 \tag{12}$$

In order to formulate $\mathbf{\Pi}$ in a compact form, we define a new vector $\mathbf{S} \in R^m$ as

$$\mathbf{S} = \begin{bmatrix} \delta\mathbf{P} + \Lambda_1\delta\mathbf{P} \\ \delta\mathbf{e} + \Lambda_2\delta\mathbf{P} \end{bmatrix} \tag{13}$$

Taking (2), (8), and (9) into consideration, we have

$$\mathbf{S} = \begin{bmatrix} \mathbf{J}_\theta & \mathbf{J}_e \\ \mathbf{0} & \mathbf{I} \end{bmatrix} \begin{bmatrix} \dot{\theta} \\ \dot{\mathbf{e}} \end{bmatrix} + \begin{bmatrix} -\dot{\mathbf{P}}_d + \Lambda_1\delta\mathbf{P} \\ -\dot{\mathbf{e}}_d + \Lambda_2\delta\mathbf{e} \end{bmatrix} \tag{14}$$

Obviously, the ideal manifold $\mathbf{\Pi}$ is equivalent to

$$\mathbf{S} = \mathbf{0} \tag{15}$$

We define a transformation matrix as

$$\mathbf{T} = \begin{bmatrix} \mathbf{J}_\theta & \mathbf{J}_e \\ \mathbf{0} & \mathbf{I} \end{bmatrix} \tag{16}$$

and rewrite (14) as

$$\mathbf{S} = \mathbf{T}\dot{\mathbf{x}} + \mathbf{b} \tag{17}$$

where

$$\mathbf{b} = \begin{bmatrix} -\dot{\mathbf{P}}_d + \Lambda_1\delta\mathbf{P} \\ -\dot{\mathbf{e}}_d + \Lambda_2\delta\mathbf{e} \end{bmatrix} \tag{18}$$

Taking time differentiation of $\mathbf{S}$, we have

$$\dot{\mathbf{S}} = \mathbf{T}\ddot{\mathbf{x}} + \Lambda\mathbf{T}\dot{\mathbf{x}} + \mathbf{h} \tag{19}$$

where

$$\Lambda = \begin{bmatrix} \mathbf{\Lambda}_1 & \mathbf{0} \\ \mathbf{0} & \mathbf{\Lambda}_2 \end{bmatrix} \tag{20}$$

$$\mathbf{h} = \begin{bmatrix} -\ddot{\mathbf{P}}_d - \Lambda_1 \mathbf{P}_d + \dot{\mathbf{J}}_\theta \dot{\theta} + \dot{\mathbf{J}}_e \dot{\mathbf{e}} \\ -\ddot{\mathbf{e}}_d + \Lambda_2 \dot{\mathbf{e}} \end{bmatrix} \tag{21}$$

Remark 2: It should be pointed out that the condition for existence of submanifold (10) is the Condition for Compensability of Flexible Robots.

The concept and theory of compensability of flexible robots are introduced and systematically analyzed by [10], [11], and [12]. The sufficient and necessary condition of compensability of a flexible robot is given as

$$\mathbf{J}_e \subseteq \mathbf{J}_\theta \tag{22}$$

In fact, submanifold (10) is equivalent to

$$\mathbf{J}_\theta \delta\dot{\theta} + \mathbf{J}_e \delta\dot{\mathbf{e}} + \dot{\mathbf{J}}_\theta \delta\theta + \dot{\mathbf{J}}_e \delta\mathbf{e} + \Lambda_1 \mathbf{J}_\theta \delta\theta + \Lambda_1 \mathbf{J}_e \delta\mathbf{e} = \mathbf{0} \tag{23}$$

If (23) exists, it should also exist at the equilibrium state $\dot{\theta} = \mathbf{0}$, $\dot{\mathbf{e}} = \mathbf{0}$, $\delta\dot{\theta} = \mathbf{0}$, $\delta\dot{\mathbf{e}} = \mathbf{0}$, that is

$$\mathbf{J}_\theta \delta\theta + \mathbf{J}_e \delta\mathbf{e} = \mathbf{0} \tag{24}$$

The existence of surface (23) means its equilibrium exists, that is, for any possible flexural displacements $\delta\mathbf{e}$ a joint variable $\delta\theta$ exists such that (24) is satisfied. Therefore, the condition of existence of (24) is the condition of existence of such $\delta\theta$. From [10], [11], and [12] we know the condition of existence of $\delta\theta$ is the compensability condition given by (22).

4. Impedance Control Scheme

In the following, we discuss the derivation of impedance control scheme and the verification of stability of the system. From (17) and (19), we have

$$\dot{\mathbf{x}} = \mathbf{T}^{-1}\mathbf{S} - \mathbf{T}^{-1}\mathbf{b} \tag{25}$$

$$\ddot{\mathbf{x}} = \mathbf{T}^{-1}\dot{\mathbf{S}} - \mathbf{T}^{-1}\Lambda\mathbf{S} - \mathbf{T}^{-1}(\mathbf{h} - \Lambda\mathbf{b}) \tag{26}$$

Using the transformation defined above, the system representation (6) can be rewritten as

$$\bar{\mathbf{M}}\dot{\mathbf{S}} + \bar{\mathbf{H}}\mathbf{S} + \mathbf{Q}^T\mathbf{w} + \mathbf{Q}^T\mathbf{J}^T\mathbf{F}_E = \mathbf{Q}^T\mathbf{u} \tag{27}$$

where $\bar{\mathbf{M}} = \mathbf{Q}^T\mathbf{M}\mathbf{Q}$, $\bar{\mathbf{H}} = \mathbf{Q}^T\mathbf{H}\mathbf{Q}$,

$$\mathbf{w} = \mathbf{M}\mathbf{Q}(\Lambda\mathbf{S} + \mathbf{h} - \Lambda\mathbf{b}) - \bar{\mathbf{H}}\mathbf{b} + \mathbf{Q}^T\mathbf{L} \tag{28}$$

and

$$\mathbf{Q} = \mathbf{T}^{-1} = \begin{bmatrix} \mathbf{J}_\theta^{-1} & -\mathbf{J}_\theta^{-1}\mathbf{J}_e \\ \mathbf{0} & \mathbf{I} \end{bmatrix} \tag{29}$$

In (27), $\mathbf{w}$ contains Coriolis, centrifugal and stiffness forces and the nonlinear forces resulting from the coordinate transformation from $\mathbf{X}$ to $\mathbf{S}$. w needed to be compensated in the end-effector trajectory tracking control. However, some elements of $\mathbf{w}$ with respect to the dynamics of the flexible links cannot be compensated by the control torque τ directly. To deal with them, we design an indirect compensating scheme such that stability of the system would be maintained. To do so, we partition $\mathbf{w}$ into the directly compensable term and not directly compensable term as follows

$$\mathbf{w} = [\mathbf{w}_1^T, \mathbf{w}_2^T]^T \tag{30}$$

where $\mathbf{w}_1 \in R^n$, and $\mathbf{w}_2 \in R^k$.

Note that in (27) the ideal manifold is expressed as the origin of the system. Eventually, the purpose of the control design is to derive such a control scheme to insure the motion of the system converging to the origin $\mathbf{S} = \mathbf{0}$. We design the control scheme with two controllers as

$$\tau = \tau_1 + \tau_2 \tag{31}$$

where τ_1 is designed for tracking the desired end-effector impedance trajectories, and τ_2 is for link vibration suppressing. In detail, τ_1 and τ_2 are given as follows

$$\tau_1 = \mathbf{w}_1 - \mathbf{\Gamma}_1 - K_1\bar{\mathbf{S}}_1 + \mathbf{J}_\theta{}^T\mathbf{F}_E \tag{32}$$

$$\tau_2 = \frac{v_2}{\bar{\mathbf{S}}_1^T v_2}\bar{\mathbf{S}}_2^T(\mathbf{w}_2 - \mathbf{\Gamma}_2 - K_2\bar{\mathbf{S}}_2 + \mathbf{J}_e{}^T\mathbf{F}_E) \tag{33}$$

where $K_1 \in R^{n \times n}$ and $K_2 \in R^{k \times k}$ are the feedback gain matrices to be chosen as constant positive definite ones; $v_2 \in R^n$ is an arbitrary non-zero vector, $\mathbf{\Gamma}_1$, $\mathbf{\Gamma}_2$ and $\bar{\mathbf{S}}_1$, $\bar{\mathbf{S}}_2$ are the elements of two new vectors defined as bellow

$$\mathbf{\Gamma} = [\mathbf{\Gamma}_1^T, \mathbf{\Gamma}_2^T]^T = \mathbf{M}\dot{\mathbf{Q}}\mathbf{S} \tag{34}$$

where $\mathbf{\Gamma}_1 \in R^n$, $\mathbf{\Gamma}_2 \in R^k$, and

$$\bar{\mathbf{S}} = [\bar{\mathbf{S}}_1^T, \bar{\mathbf{S}}_2^T]^T = \mathbf{Q}\mathbf{S} \tag{35}$$

where $\bar{\mathbf{S}}_1 \in R^n$, $\bar{\mathbf{S}}_2 \in R^k$.

From the control design expressed above, the following result can be obtained.

Theorem For the flexible robot manipulator that the dynamics is described by (6), the motion of the robot uniformly and asymptotically converges to the ideal manifold (15) under control schemes given by (31) ~ (33).

Proof. From Remark 1, we know $\mathbf{M}$ is positive definite matrix, therefore $\bar{\mathbf{M}} = \mathbf{Q}^T\mathbf{M}\mathbf{Q}$ is also a positive definite matrix. Thus, we choose the following positive scalar function as a Lyapunov function candidate

$$\mathbf{V} = \frac{1}{2}\mathbf{S}^T\bar{\mathbf{M}}\mathbf{S} \tag{36}$$

Taking differentiation of $\mathbf{V}$ with respect to time yields

$$\dot{\mathbf{V}} = \mathbf{S}^T\bar{\mathbf{M}}\dot{\mathbf{S}} + \frac{1}{2}\mathbf{S}^T\dot{\bar{\mathbf{M}}}\mathbf{S} \tag{37}$$

where

$$\dot{\bar{\mathbf{M}}} = \dot{\mathbf{Q}}^T\mathbf{M}\mathbf{Q} + \mathbf{Q}^T\dot{\mathbf{M}}\mathbf{Q} + \mathbf{Q}^T\mathbf{M}\dot{\mathbf{Q}} \tag{38}$$

By substituting (27) into (37) we have

$$\begin{aligned}\dot{\mathbf{V}} &= \mathbf{S}^T(-\mathbf{Q}^T\mathbf{w} - \bar{\mathbf{H}}\mathbf{S} - \mathbf{Q}^T\mathbf{J}^T\mathbf{F}_E + \mathbf{Q}^T\mathbf{u}) \\ &+\frac{1}{2}\mathbf{S}^T(\dot{\mathbf{Q}}^T\mathbf{M}\mathbf{Q} + \mathbf{Q}^T\dot{\mathbf{M}}\mathbf{Q} + \mathbf{Q}^T\mathbf{M}\dot{\mathbf{Q}})\mathbf{S}\end{aligned} \tag{39}$$

Since $\dot{\mathbf{M}} - 2\mathbf{H}$ is a skew symmetric matrix, it is easy to show $\mathbf{Q}^T\dot{\mathbf{M}}\mathbf{Q} - 2\bar{\mathbf{H}}$ is also a skew symmetric matrix. Thus, (39) can be rewritten as

$$\begin{aligned}\dot{\mathbf{V}} &= \bar{\mathbf{S}}^T(-\mathbf{w} - \mathbf{J}^T\mathbf{F}_E + \mathbf{u}) + \bar{\mathbf{S}}^T\mathbf{M}\dot{\mathbf{Q}}\mathbf{S} \\ &= \bar{\mathbf{S}}_1^T(-\mathbf{w}_1 - \mathbf{J}_\theta{}^T\mathbf{F}_E + \mathbf{\Gamma}_1) \\ &\quad +\bar{\mathbf{S}}_2^T(-\mathbf{w}_2 - \mathbf{J}_e{}^T\mathbf{F}_E + \mathbf{\Gamma}_2) + \bar{\mathbf{S}}_1\tau\end{aligned} \tag{40}$$

Combining (31) and (40) yields

$$\dot{\mathbf{V}} = -\bar{\mathbf{S}}^T K\bar{\mathbf{S}} \tag{41}$$

where $\mathbf{K} = \begin{bmatrix} \mathbf{K}_1 & \mathbf{0} \\ \mathbf{0} & \mathbf{K}_2 \end{bmatrix}$.

Notice both $\mathbf{K}_1$ and $\mathbf{K}_2$ are chosen as positive definite matrices, then $\mathbf{K}$ is a positive definite matrix too. We have

$$\dot{\mathbf{V}} < 0 \tag{42}$$

Thus system (27) is asymptotically stable.

Since M is always bounded, and system (27) is stable, the Lyapunov function (36) is bounded and satisfies

$$\frac{1}{2}\lambda_{min}\left\|\mathbf{S}\right\|^2 \leq V \leq \frac{1}{2}\lambda_{max}\left\|\mathbf{S}\right\|^2 \tag{43}$$

where, $\lambda_{min}, \lambda_{max} > 0$ are the minimum and maximum singular values of $\mathbf{M}$, $\|*\|$ denotes the Euclidean norm of a vector. Similarly, from (41) we obtain

$$-\zeta_{max}\left\|\mathbf{S}\right\|^2 \leq \dot{V} \leq -\zeta_{min}\left\|\mathbf{S}\right\|^2 \tag{44}$$

where, $\zeta_{min}, \zeta_{max} > 0$ are the minimum and maximum singular values of matrix $\mathbf{Q}^T\mathbf{K}\mathbf{Q}$.

By combining (43) and (44), we have

$$\dot{V} \le -\frac{2\zeta_{min}}{\lambda_{max}} V \tag{45}$$

Taking time integration of (45) and noting left side inequality of (43) yields

$$\|\mathbf{S}\| \le \sqrt{\frac{2V_0}{\lambda_{min}}} e^{-\frac{2\zeta_{min}}{\lambda_{max}} t} \tag{46}$$

where V_0 is the initial value of the Lyapunov function. It is shown that system (27) uniformly asymptotically converges to $\mathbf{S} = \mathbf{0}$, i. e. the motion of system (6) uniformly asymptotically converge to the designed ideal manifold.

It should be emphasized that when all the parameters are unknown, joint space tracking control can still be implemented using suitable adaptive and/or robust approaches. For end-effector trajectory control, however, at less the kinematical parameters, such as link lengths and twist angles which determine connections between links, are necessary when the direct measurement of end-effector motion is not available. Unlike the joint variables which are independent from kinematical parameters, end-effector positions and velocities are synthesized with the joint variables, link flexural variables and such kinematical parameters, so that unless an end-effector position sensor is employed, the kinematical parameters cannot be separated from the end-effector motion in a task space control design.

5. An application example: impedance control of a 2-link flexible robot

As an illustration, we apply the impedance control method to a 2-link flexible robot manipulator shown in Fig. 2. The dynamic model used for the control design and simulations was derived using Hamilton's Principle and Finite Element Method. Two elements for each flexible link were specified to approximate elastic deflections and rotation angle caused by the deflections. The approximated system has 10 degree of freedom.

The coordinate systems are set as shown in Figure 3. $O_0 - X_0Y_0$ is the base coordinate system with its origin O_0 locating on joint 1, and axes X_0 and Y_0 are along the vertical direction and horizontal direction respectively. $O_1 - X_1Y_1$, and $O_2 - X_2Y_2$ are joint 2 coordinate frame and end-effector coordinate frame. Their origins O_1 and O_2 lay on joint 2 i.e. the distal end of link, and end-effector, respectively.

Using 3x3 homogenous transformation matrices, the end-effector position vector in the base coordinate system can be given as

$$\mathbf{P} = \mathbf{T}_1\mathbf{T}_{1e}\mathbf{T}_2\mathbf{q}_2 \tag{47}$$

In the above equation, $\mathbf{T}_1$ and $\mathbf{T}_1$ are rotation matrices of joint 1 and 2 give as

$$\mathbf{T}_1 = \begin{bmatrix} C_1 & -S_1 & 0 \\ S_1 & C_1 & 0 \\ 0 & 0 & 1 \end{bmatrix} \tag{48}$$

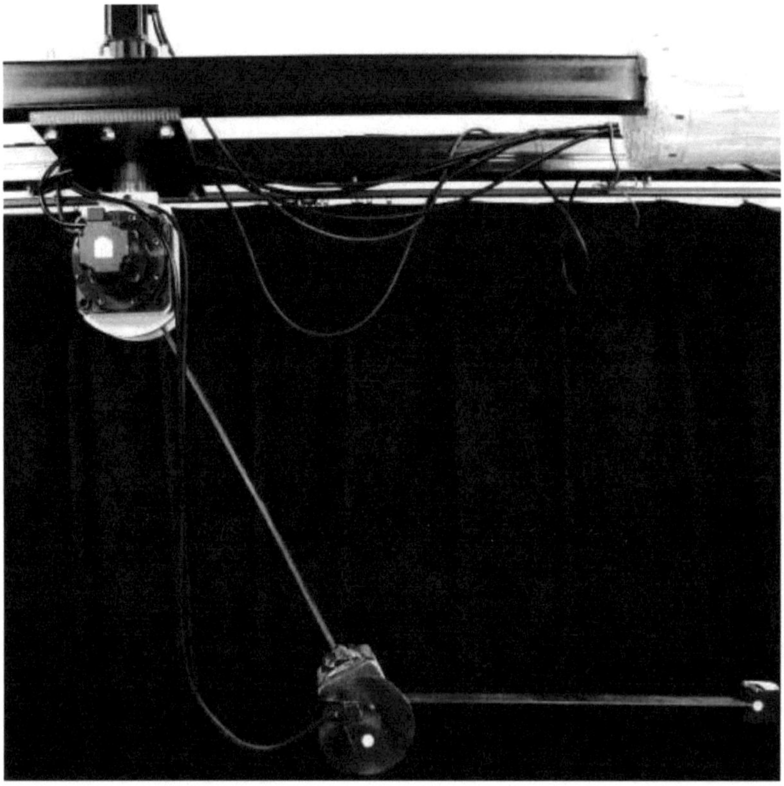

Figure 2. The flexible robot manipulator

$$\mathbf{T}_2 = \begin{bmatrix} C_2 & -S_2 & 0 \\ S_2 & C_2 & 0 \\ 0 & 0 & 1 \end{bmatrix} \tag{49}$$

where $S_1 = \sin\theta_1$, $C_1 = \cos\theta_1$, $S_2 = \sin\theta_2$, and $C_2 = \cos\theta_2$ with θ_1 and θ_2 being the joint angles. $\mathbf{T}_{1e}$ denotes link 1 transformation matrix as

$$\mathbf{T}_{1e} = \begin{bmatrix} C_\phi & -S_\phi & L_1 \\ S\phi & C\phi & \delta_1 \\ 0 & 0 & 1 \end{bmatrix} \tag{50}$$

where $S_\phi = \sin\phi_1$, $C_\phi = \cos\phi_1$, L_1 denotes the length of linkl and δ_1 denotes the deflection at the distal end of link 1. $\mathbf{q}_2$ is a vector defined as

$$\mathbf{q}_2 = [L_2, \delta_2, 1]^T \tag{51}$$

with L_2 and δ_2 denoting the length of Iink2 and deflection at the distal end of Iink2.

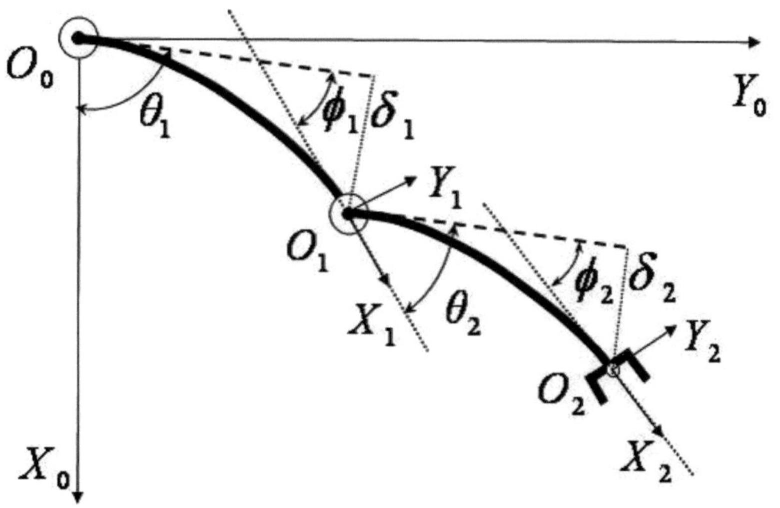

Figure 3. Coordinate systems of two-link robot manipulator

Using the above kinematical analysis results, one can obtain Jacobian matrices with respect to the joint velocity $[\dot{\theta}_1, \dot{\theta}_2]^T$ and link flexural velocity $[\dot{\delta}_1, \dot{\phi}_1, \dot{\delta}_2]^T$ as follows.

$$\mathbf{J}_\theta = \begin{bmatrix} -L_1S_1 - \delta_1C_1 - L_2S_{12\phi} - \delta_2C_{12\phi} & -L_2S_{12\phi} - \delta_2C_{12\phi} \\ L_1C_1 - \delta_1S_1 + L_2C_{12\phi} - \delta_2S_{12\phi} & L_2C_{12\phi} - \delta_2S_{12\phi} \end{bmatrix} \quad (52)$$

and

$$\mathbf{J}_e = \begin{bmatrix} -S_1 & -L_2S_{12\phi} - \delta_2C_{12\phi} & -S_{12\phi} \\ C_1 & L_2C_{12\phi} - \delta_2S_{12\phi} & C_{12\phi} \end{bmatrix} \quad (53)$$

where $S_{12\phi} = \sin(\theta_1 + \theta_2 + \phi_1)$, $C_{12\phi} = \cos(\theta_1 + \theta_2 + \phi_1)$. The parameters of the flexible robot are shown in Table 1.

	link1	link2	joint1	joint2
L	2.0	2.0		
EJ	200	100		
ρ	2	1		
i			0.01	0.005

L–length(m); EJ–bending rigidity$(N-m^2)$;
ρ–density(kg/m); i–moment of inertia$(kg-m^2)$

Table 1. The parameters of the robot

The simulations are carried out with no natural damping of the flexural behavior of the links. The external force acted at the end-effector is $\mathbf{F}_E$= $[1.0,1.0]^T$. The inertia, damping, stiffness matrices of the desired impedance dynamics are designed as

$$\mathbf{M}_d = \begin{bmatrix} 1.0 & 0 \\ 0 & 1.0 \end{bmatrix}, \quad \mathbf{D}_d = \begin{bmatrix} 2.8 & 0 \\ 0 & 2.8 \end{bmatrix}, \quad \mathbf{K}_d = \begin{bmatrix} 4.0 & 0 \\ 0 & 4.0 \end{bmatrix} \tag{54}$$

The parameters of the controller are designed as follows Λ_1 = *diag* (10.0 ,10.0), Λ_2 = *diag*(10.0,10.0,10.0,10.0,10.0,10.0,10.0,10.0), $\mathbf{K}_1$ = *diag*(100.0,100.0), $\mathbf{K}_2$ = *diag*(100.0,100.0,100.0,100.0,100.0,100.0,100.0,100.0).

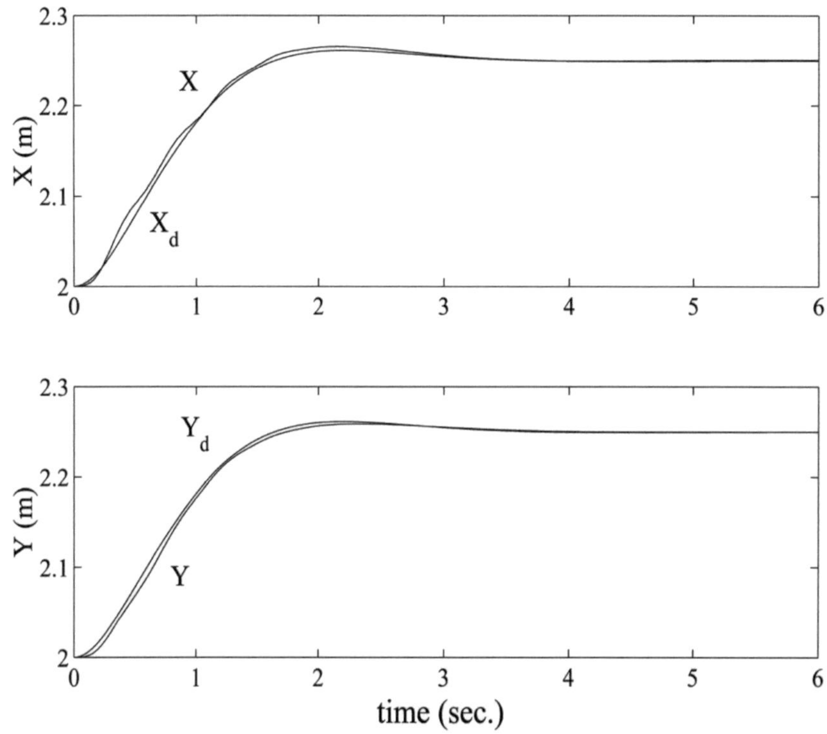

Figure 4. Targeted impedance trajectories and control results

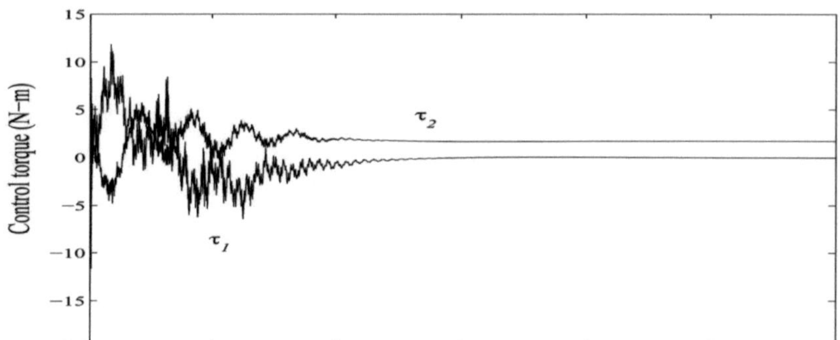

Figure 5. Control torque of jointl and joint2

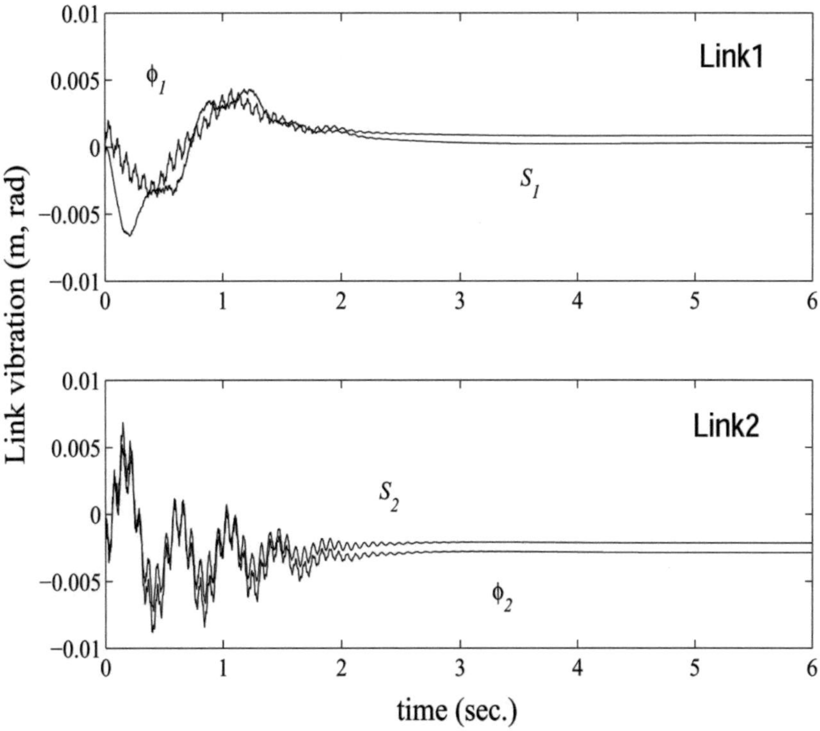

Figure 6. Link deflections and rotations caused by the deflections

Fig 4. shows the desired impedance trajectories and tracking results. The control torque is shown in Fig 5. Fig 6. shows the flexural deflections and the rotation of the cross section caused by the deflections at distal ends of link 1 and link 2.

The simulations validate that the impedance control method using a trajectory tracking approach can achieve good performance of impedance control. By using the proposed control method vibrations of the flexible links are damped effectively stability of the system is guaranteed.

6. Conclusions

An impedance control method for flexible robot manipulators was discussed. The impedance control objective was converted into tracking the end-effector impedance trajectories generated by the designed impedance dynamics. An ideal manifold related to the desired impedance trajectory tracking was designed to prescribe the desirable characteristics of the system. The impeadnce control scheme was derived to govern the motion of the robot system converging and remaining to the ideal manifold in the presence of parametric uncertainties. Using Lyapunov theory it was shown that under the impedance control scheme the motion of the robot is exponentially stable to the designed ideal manifold. Impedance control simulations were carried out using a two-link flexible robot manipulator. The result confirmed the effectiveness and good performance of the proposed adaptive impedance control method.

7. References

B. C. Chiou and M. Shahinpoor, Dynamic Stability Analysis of a Two-Link Force-Controlled Flexible Manipulator, *ASME J. of Dynamic Systems, Measurement, and Control*. 112 (1990) 661-666 [I]

F. Matsuno, T. Asano and Y. Sakawa, Modeling and Quasi-Static Hybrid Position/Force Control of Constrained Planar Two-Link Flexible Manipulators, *IEEE, Trans, on Robotics and Automation*, 10 (3) (1994) 287-297 [2]

T. Yoshikawa, et. al, Hybrid Position/Force Control of Flexible-Macro/Rigid-Micro Manipulator Systems, *IEEE, Trans, on Robotics and Automation*, 12 (4) (1996) 633-640[3]

I. C. Lin and L. C. Fu, Adaptive Hybrid Force/Position Control of Flexible Manipulator for Automated Deburring with on-Line Cutting Trajectory Modification, *Int. Proc. of the IEEE Conf. On Robotics and Automation*, (1998) 818-825[4]

P.Rocco and W.J.Book, Modeling for Two-Time Scale Force/Position Control of Flexible Robots, *Int. Proc. of the IEEE Conf. On Robotics and Automation*, (1996) 1941-1946 [5]

B.Siciliano and L.Villani, Two-Time Scale Force/Position Control of Flexible Manipulators, *Int. Proc. of the IEEE Conf. On Robotics and Automation*, (2001) 2729-2734 [6]

M.Yamano, J.Kim and M.Uchiyama, Hybrid Position/Force Control of Two Cooperative Flexible Manipulators Working in 3D Space, *Int. Proc. of the IEEE Conf. On Robotics and Automation*, (1998) 1110-1115 [7]

Z. H. Jiang, Impedance Control of Micro-Macro Robots with Structural Flexibility, *Transactions of Japan Society of Mechanical Engineers*, 65 (631) (1999) Series C, 142-149 [8]

Z.H.Jiang, Impedance Control of Flexible Robot Arms with Parametric Uncertainties, *Journal of Intelligent and Robot Systems*, Vol.42 (2005) 113-133 [9]

Z.H.Jiang, M.Uchiyama and K.Hakomori: Active Compensating Control of Flexural Error of Elastic Robot Manipulators, *Proc. IMACS/IFAC Int. Symp. on Modeling and Simulation of Distributed Parameter Systems*, 413-418, 1987, [10]

Z.H.Jiang and M.Uchiyama: Compensability of Flexible Robot Arms, *Journal of Robotics Society of Japan*, (6) 5, 1988, pp.416-423 [11]

M.Uchiyama, Z.H.Jiang: Compensability of End-Effector Position Errors for Flexible Robot Manipulators, *Proc. of the 1991 American Control Conference*, Vol.2, 1991, 1873-1878 [12]

N. Hogan, Impedance Control: An Approach to Manipulation, Part I, II, III *ASME J. of Dynamic Systems Measurement and Control*, 107 (1) (1985) 1-24 [13]

12

Simple Effective Control for Robot Manipulators with Friction

Maolin Jin, Sang Hoon Kang and Pyung Hun Chang
Korea Advanced Institute of Science and Technology (KAIST)
South Korea

1. Introduction

Friction accounts for more than 60% of the motor torque, and hard nonlinearities due to Coulomb friction and stiction severely degrade control performances as they account for nearly 30% of the industrial robot motor torque. Although many friction compensation methods are available (Armstrong-Helouvry et al., 1994; Lischinsky et al., 1999; Bona & Indri, 2005;), here we only introduce relatively recent research works. Model-based friction compensation techniques (Mei et al., 2006; Bona et al., 2006; Liu et al., 2006) require *prior* experimental identification, and the drawback of these off-line friction estimation methods is that they can't adapt when the friction effects vary during the robot operations (Visioli et al., 2001). The adaptive compensation methods (Marton & Lantos, 2007; Xie, 2006) take the modelling error of Lugre friction model into account, but they still require the complex *prior* experimental identification. Consequently, implementation of these schemes is highly complicated and computationally demanding due to the identification of many parameters and the calculation of the nonlinear friction model. Joint-torque sensory feedback (JTF) (Aghili & Namvar, 2006) compensates friction without the dynamic models, but high price torque sensors are needed for the implementation of JTF.

To give satisfactory solution to robot control, in general, we should continue our research in modelling the robot dynamics and nonlinear friction so that they become applicable to wider problem domain. In the mean time, when the parameters of robot dynamics are poorly understood, and for the ease of practical implementation, simple and efficient techniques may represent a viable alternative.

Accordingly, the authors developed a non-model based control technique for robot manipulators with nonlinear friction (Jin et al., 2006; Jin et al., 2008). The nonlinear terms in robot dynamics are classified into two categories from the time-delay estimation (TDE) viewpoint: soft nonlinearities (gravity, viscous friction, and Coriolis and centrifugal torques, disturbances, and interaction torques) and hard nonlinearities (due to Coulomb friction, stiction, and inertia force uncertainties). TDE is used to cancel soft nonlinearities, and ideal velocity feedback (IVF) is used to suppress the effect of hard nonlinearities. We refer the control technique as time-delay control with ideal velocity feedback (TDCIVF). Calculation of complex robot model and nonlinear friction model is not required in the TDCIVF. The TDCIVF control structure is transparent to designers; it consists of three elements that have clear meaning: a soft nonlinearity cancelling element, a hard nonlinearity suppressing element, and a target dynamics injecting element. The TDCIVF turns out simple in form and easy to tune; specifying desired dynamics for robots is all that is necessary for a user to do. The same control structure can be used for both free space motion and constrained motion of robot manipulators.

This chapter is organized as follows. After briefly review TDCIVF (Jin et al., 2006; Jin et al., 2008), the implementation procedure and discussion are given in section 2. In section 3, simulations are carried out to examine the cancellation effect of soft nonlinearities using TDE and the suppression effect of hard nonlinearities using IVF. In section 4, the TDCIVF is compared with adaptive friction compensation (AFC) (Visioli et al., 2001; Visioli et al., 2006; Jatta et al., 2006), a non-model based friction compensation method that is similar to the TDCIVF in that it requires neither friction identification nor expensive torque sensor. Cartesian space formulation is presented in section 5. An application of human-robot cooperation is presented using a two-degree-of-freedom (2-DOF) SCARA-type industrial robot in section 6. Finally, section 7 concludes the paper.

2. Time-Delay Control with Ideal Velocity Feedback

2.1 Classification of Robot Dynamics from TDE Viewpoint

The dynamics equation of n-DOF robot manipulator in joint space coordinates is given by:

$$\boldsymbol{\tau} = \mathbf{M}(\boldsymbol{\theta})\ddot{\boldsymbol{\theta}} + \mathbf{V}(\boldsymbol{\theta},\dot{\boldsymbol{\theta}}) + \mathbf{G}(\boldsymbol{\theta}) + \mathbf{F} + \mathbf{D} - \boldsymbol{\tau}_s , \tag{1}$$

where $\boldsymbol{\tau} \in \Re^n$ denotes actuator torque and $\boldsymbol{\tau}_s \in \Re^n$ interaction torque; $\boldsymbol{\theta},\dot{\boldsymbol{\theta}},\ddot{\boldsymbol{\theta}} \in \Re^n$ denote the joint angle, the joint velocity, and the joint acceleration, respectively; $\mathbf{M}(\boldsymbol{\theta}) \in \Re^{n\times n}$ a positive definite inertia matrix; and $\mathbf{V}(\boldsymbol{\theta},\dot{\boldsymbol{\theta}}) \in \Re^n$ Coriolis and centrifugal torque; $\mathbf{G}(\boldsymbol{\theta}) \in \Re^n$ a gravitational force; $\mathbf{F} \in \Re^n$ stands for the friction term including Coulomb friction, viscous friction and stiction; and $\mathbf{D} \in \Re^n$ continuous disturbance.

Introducing a constant matrix $\overline{\mathbf{M}} \in \Re^{n\times n}$, one can obtain another expression of (1) as follows:

$$\boldsymbol{\tau} = \overline{\mathbf{M}}\ddot{\boldsymbol{\theta}} + \mathbf{N}(\boldsymbol{\theta},\dot{\boldsymbol{\theta}},\ddot{\boldsymbol{\theta}}) , \tag{2}$$

where $\mathbf{N}(\boldsymbol{\theta},\dot{\boldsymbol{\theta}},\ddot{\boldsymbol{\theta}})$ includes all the nonlinear terms of robot dynamics as

$$\mathbf{N}(\boldsymbol{\theta},\dot{\boldsymbol{\theta}},\ddot{\boldsymbol{\theta}}) = [\mathbf{M}(\boldsymbol{\theta}) - \overline{\mathbf{M}}]\ddot{\boldsymbol{\theta}} + \mathbf{V}(\boldsymbol{\theta},\dot{\boldsymbol{\theta}}) + \mathbf{G}(\boldsymbol{\theta}) + \mathbf{F} + \mathbf{D} - \boldsymbol{\tau}_s . \tag{3}$$

$\mathbf{N}(\boldsymbol{\theta},\dot{\boldsymbol{\theta}},\ddot{\boldsymbol{\theta}})$ can be classified into two categories as follows (Jin et al., 2008):

$$\mathbf{N}(\boldsymbol{\theta},\dot{\boldsymbol{\theta}},\ddot{\boldsymbol{\theta}}) = \mathbf{S}(\boldsymbol{\theta},\dot{\boldsymbol{\theta}}) + \mathbf{H}(\boldsymbol{\theta},\dot{\boldsymbol{\theta}},\ddot{\boldsymbol{\theta}}), \tag{4}$$

$$\mathbf{S}(\boldsymbol{\theta},\dot{\boldsymbol{\theta}}) = \mathbf{V}(\boldsymbol{\theta},\dot{\boldsymbol{\theta}}) + \mathbf{G}(\boldsymbol{\theta}) + \mathbf{F}_v(\dot{\boldsymbol{\theta}}) + \mathbf{D} - \boldsymbol{\tau}_s , \tag{5}$$

$$\mathbf{H}(\boldsymbol{\theta},\dot{\boldsymbol{\theta}},\ddot{\boldsymbol{\theta}}) = \mathbf{F}_c(\dot{\boldsymbol{\theta}}) + \mathbf{F}_{st}(\boldsymbol{\theta},\dot{\boldsymbol{\theta}}) + [\mathbf{M}(\boldsymbol{\theta}) - \overline{\mathbf{M}}]\ddot{\boldsymbol{\theta}}, \tag{6}$$

where $\mathbf{F}_v, \mathbf{F}_c, \mathbf{F}_{st} \in \Re^n$ denote viscous friction, Coulomb friction, stiction, respectively. When the time-delay L is sufficiently small, it is very reasonable to assume that $\mathbf{S}(\boldsymbol{\theta},\dot{\boldsymbol{\theta}})$ is closely approximated by $\mathbf{S}(\boldsymbol{\theta},\dot{\boldsymbol{\theta}})_{t-L}$ shwon in Fig. 1. It is expressed by

$$\hat{\mathbf{S}}(\boldsymbol{\theta},\dot{\boldsymbol{\theta}}) = \mathbf{S}(\boldsymbol{\theta},\dot{\boldsymbol{\theta}})_{t-L} \cong \mathbf{S}(\boldsymbol{\theta},\dot{\boldsymbol{\theta}}) . \tag{7}$$

But $\mathbf{H}(\boldsymbol{\theta},\dot{\boldsymbol{\theta}},\ddot{\boldsymbol{\theta}})$ is not compatible with the TDE technique, as

$$\hat{\mathbf{H}}(\boldsymbol{\theta},\dot{\boldsymbol{\theta}},\ddot{\boldsymbol{\theta}}) = \mathbf{H}(\boldsymbol{\theta},\dot{\boldsymbol{\theta}},\ddot{\boldsymbol{\theta}})_{t-L} \neq \mathbf{H}(\boldsymbol{\theta},\dot{\boldsymbol{\theta}},\ddot{\boldsymbol{\theta}}) . \tag{8}$$

In this paper, $\hat{\bullet}$ denotes estimated value of $\bullet$, and $\bullet_{t-L}$ denotes time delayed value of $\bullet$. Incidentally, time-delay estimation (TDE) is originated from pioneering works called time-delay control (TDC) (Youcef-Toumi & Ito, 1990; Hsia et al., 1991). However, little attention has been paid to hard nonlinearities such as Coulomb friction and stiction forces in previous researches on TDC.

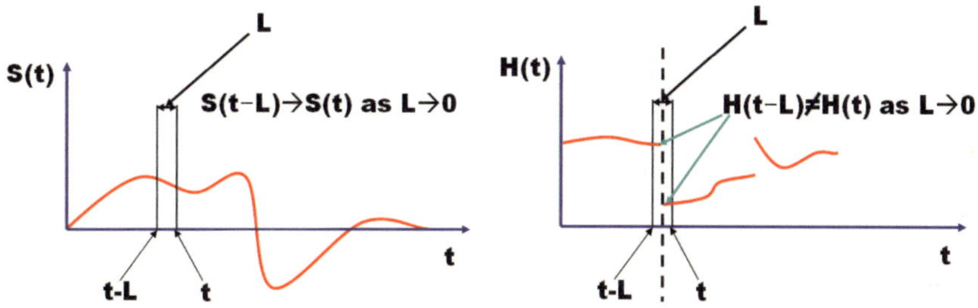

(a) TDE of soft nonlinearities (b) TDE of hard nonlinearities

Figure 1. Description of TDE of soft and hard nonlinearities

2.2 Target Dynamics

The control objective is to make the robot to achieve the following *target impedance dynamics*:

$$\mathbf{M}_d(\ddot{\boldsymbol{\theta}}_d - \ddot{\boldsymbol{\theta}}) + \mathbf{B}_d(\dot{\boldsymbol{\theta}}_d - \dot{\boldsymbol{\theta}}) + \mathbf{K}_d(\boldsymbol{\theta}_d - \boldsymbol{\theta}) + \boldsymbol{\tau}_s = \mathbf{0} , \tag{9}$$

where $\mathbf{M}_d \in \Re^{n\times n}, \mathbf{K}_d \in \Re^{n\times n}$, $\mathbf{B}_d \in \Re^{n\times n}$, denote desired mass, spring, and damper, respectively; $\boldsymbol{\theta}_d, \dot{\boldsymbol{\theta}}_d, \ddot{\boldsymbol{\theta}}_d \in \Re^n$ the desired position, desired velocity, desired acceleration, respectively. When the robot is performing free space motion, the sensed torque $\boldsymbol{\tau}_s$ is 0, and (9) is reduced to the well-known *error dynamics* of motion control; it is expressed by

$$\ddot{\boldsymbol{\theta}}_d - \ddot{\boldsymbol{\theta}} + \mathbf{K}_V(\dot{\boldsymbol{\theta}}_d - \dot{\boldsymbol{\theta}}) + \mathbf{K}_P(\boldsymbol{\theta}_d - \boldsymbol{\theta}) = \mathbf{0} \tag{10}$$

with

$$\mathbf{K}_V \triangleq \mathbf{M}_d^{-1}\mathbf{B}_d, \ \mathbf{K}_P \triangleq \mathbf{M}_d^{-1}\mathbf{K}_d . \tag{11}$$

2.3 Actual Dynamics When Only TDE is Used

The robot dynamics equation can be rewritten (2) as

$$\tau = \bar{\mathbf{M}}\ddot{\theta} + \mathbf{S}(\theta,\dot{\theta}) + \mathbf{H}(\theta,\dot{\theta},\ddot{\theta}) . \tag{12}$$

The control input using TDE is

$$\tau = \bar{\mathbf{M}}\mathbf{u} + \hat{\mathbf{S}}(\theta,\dot{\theta}) + \hat{\mathbf{H}}(\theta,\dot{\theta},\ddot{\theta}) . \tag{13}$$

where

$$\mathbf{u} = \ddot{\theta}_d + \mathbf{M}_d^{-1}[\mathbf{K}_d(\theta_d - \theta) + \mathbf{B}_d(\dot{\theta}_d - \dot{\theta}) + \tau_s] . \tag{14}$$

In (13), $\hat{\mathbf{S}}(\theta,\dot{\theta}) + \hat{\mathbf{H}}(\theta,\dot{\theta},\ddot{\theta})$ is obtained from the TDE of $\mathbf{S}(\theta,\dot{\theta}) + \mathbf{H}(\theta,\dot{\theta},\ddot{\theta})$, expressed by

$$\hat{\mathbf{S}}(\theta,\dot{\theta}) + \hat{\mathbf{H}}(\theta,\dot{\theta},\ddot{\theta}) = \mathbf{S}(\theta,\dot{\theta})_{t-L} + \mathbf{H}(\theta,\dot{\theta},\ddot{\theta})_{t-L} = \tau_{t-L} - \bar{\mathbf{M}}\ddot{\theta}_{t-L} . \tag{15}$$

With the combination of (7),(12)-(15), and (17), one can obtain actual impedance error dynamics as

$$\mathbf{M}_d(\ddot{\theta}_d - \ddot{\theta}) + \mathbf{B}_d(\dot{\theta}_d - \dot{\theta}) + \mathbf{K}_d(\theta_d - \theta) + \tau_s = \mathbf{M}_d\varepsilon. \tag{16}$$

where the TDE error ε is defined as

$$\varepsilon \triangleq \overline{\mathbf{M}}^{-1}[\mathbf{H}(\theta,\dot{\theta},\ddot{\theta}) - \mathbf{H}(\theta,\dot{\theta},\ddot{\theta})_{t-L}] . \tag{17}$$

2.4 Hard Nonlinearity Compensation

TDCIVF uses *ideal velocity feedback (IVF)* term in order to suppress ε that causes the deviation of resulting dynamics from the target impedance dynamics, as

$$\tau = \tau_{t-L} - \bar{\mathbf{M}}\ddot{\theta}_{t-L} + \bar{\mathbf{M}}\{\ddot{\theta}_d + \mathbf{M}_d^{-1}[\tau_s + \mathbf{K}_d(\theta_d - \theta) + \mathbf{B}_d(\dot{\theta}_d - \dot{\theta})] + \Gamma(\dot{\theta}_{ideal} - \dot{\theta})\} \tag{18}$$

where $\Gamma \in \Re^{n \times n}$ denotes a positive definite diagonal matrix, and

$$\dot{\theta}_{ideal} = \int\{\ddot{\theta}_d + \mathbf{M}_d^{-1}[\mathbf{K}_d(\theta_d - \theta) + \mathbf{B}_d(\dot{\theta}_d - \dot{\theta}) + \tau_s]\}dt . \tag{19}$$

If we considere the integral sliding surface as

$$\mathbf{s} = \int\{\ddot{\theta}_d - \ddot{\theta} + \mathbf{M}_d^{-1}[\tau_s + \mathbf{K}_d(\theta_d - \theta) + \mathbf{B}_d(\dot{\theta}_d - \dot{\theta})]\}dt. \tag{20}$$

Then, (19) reduces to $\mathbf{s} = \dot{\theta}_{ideal} - \dot{\theta}$, and (18) can be expressed by

$$\tau = \tau_{t-L} - \bar{\mathbf{M}}\ddot{\theta}_{t-L} + \bar{\mathbf{M}}\{\ddot{\theta}_d + \mathbf{M}_d^{-1}[\tau_s + \mathbf{K}_d(\theta_d - \theta) + \mathbf{B}_d(\dot{\theta}_d - \dot{\theta})] + \Gamma\mathbf{s}\} . \tag{21}$$

The *s*-trajectory, which is integral error of target dynamics, represents a time-varying measure of impedance error (Slotine, 1985). According to Barbalat's Lemma, as the sliding surface s(t)→0, we can expect that $\dot{\mathbf{s}}$ (t)→0, which implies achieving desired impedance (9) as time→ ∞ . When the robot is performing free space motion, the sensed torque/force, τ_s is 0; $\dot{\theta}_{ideal} - \dot{\theta}$ →0 implies $\theta \rightarrow \theta_d$.

Stability analysis can be found in (Jin et al., 2008).

2.5 The Simple, Transparent Control Structure

The TDCIVF has transparent structure as shown in (22). It consists of three elements that have *clear meaning*: a soft nonlinearity cancelling element, a hard nonlinearity suppressing element, and a target dynamics injecting element.

$$\boldsymbol{\tau} = \underbrace{\boldsymbol{\tau}_{t-L} - \bar{\mathbf{M}}\ddot{\boldsymbol{\theta}}_{t-L}}_{\text{cancelling soft nonlinearities}} + \underbrace{\bar{\mathbf{M}}\{\ddot{\boldsymbol{\theta}}_d + \mathbf{M}_d^{-1}[\boldsymbol{\tau}_s + \mathbf{K}_d(\boldsymbol{\theta}_d - \boldsymbol{\theta}) + \mathbf{B}_d(\dot{\boldsymbol{\theta}}_d - \dot{\boldsymbol{\theta}})]\}}_{\text{injecting target impedance dynamics}} + \underbrace{\bar{\mathbf{M}}\boldsymbol{\Gamma}(\dot{\boldsymbol{\theta}}_{ideal} - \dot{\boldsymbol{\theta}})}_{\text{suppressing the effect of hard nonlinearities}} . \quad (22)$$

Among the nonlinear terms in the robot dynamics, soft nonlinearities (gravity, viscous friction, and Coriolis and centrifugal torques, disturbances, and interaction torques) are cancelled by TDE, $\boldsymbol{\tau}_{t-L} - \bar{\mathbf{M}}\ddot{\boldsymbol{\theta}}_{t-L}$; and hard nonlinearities (due to Coulomb friction, stiction, and inertia force uncertainties) are suppressed by IVF $\boldsymbol{\Gamma} \cdot (\dot{\boldsymbol{\theta}}_{ideal} - \dot{\boldsymbol{\theta}})$; thus, calculations of complex robot dynamics as well as that of nonlinear friction are unnecessary, and the controller is easy to implement. Only two gain matrices, $\bar{\mathbf{M}}$ and $\boldsymbol{\Gamma}$, must be tuned for the control law. The gains have clear meaning: $\bar{\mathbf{M}}$ is for soft nonlinearities cancellation and noise attenuation, and $\boldsymbol{\Gamma}$ is for hard nonlinearities suppression. Consequently, the TDCIVF turns out simple in form and easy to tune; specifying the target dynamics for robot is all that is necessary for a user to do.

2.6 Implementation Procedure

The implementation procedure is straightforward as follows:

1. Describe desired trajectory $\boldsymbol{\theta}_d$ according to task requirements.
2. Select $\mathbf{K}_d$ and $\mathbf{M}_d$ according to task requirements and the approximate value of robot inertia.
3. Select $\mathbf{B}_d$ by applying critically damped condition .
4. Set $\boldsymbol{\Gamma}$ =0 and tune $\bar{\mathbf{M}}$. Start with a small positive value, and increase it. Stop increasing $\bar{\mathbf{M}}$ when robot joints become noisy. The elements of $\bar{\mathbf{M}}$ can be tuned separately.
5. Tune $\boldsymbol{\Gamma}$. Increase it from zero to an appropriate value. The elements of $\boldsymbol{\Gamma}$ can be tuned separately.

Incidentally, the digital implementation of TDCIVF is as follows:

1. Following numerical integration is used to calculate ideal velocity:

$$\dot{\boldsymbol{\theta}}_{ideal}(t) = \dot{\boldsymbol{\theta}}_{ideal}(t-L) + L\{\ddot{\boldsymbol{\theta}}_d + \mathbf{M}_d^{-1}\,[\mathbf{K}_d(\boldsymbol{\theta}_d - \boldsymbol{\theta}) + \mathbf{B}_d(\dot{\boldsymbol{\theta}}_d - \dot{\boldsymbol{\theta}}) + \boldsymbol{\tau}_s\,]\}. \quad (23)$$

2. $\ddot{\boldsymbol{\theta}}_{t-L}$ and $\dot{\boldsymbol{\theta}}$ are calculated by numerical differentiation (24) to implement the TDCIVF.

$$\ddot{\boldsymbol{\theta}}_{t-L} = (\dot{\boldsymbol{\theta}}_t - \dot{\boldsymbol{\theta}}_{t-L})/L \text{ and } \dot{\boldsymbol{\theta}} = (\boldsymbol{\theta}_t - \boldsymbol{\theta}_{t-L})/L. \quad (24)$$

2.7 Drawbacks

The drawback of TDCIVF inherits from TDE. TDE term $\boldsymbol{\tau}_{t-L} - \bar{\mathbf{M}}\ddot{\boldsymbol{\theta}}_{t-L}$ in (22) needs numerical differentiations (24) which may amplify the effect of noise and deteriorate the control performance. Thus, the elements of $\bar{\mathbf{M}}$ are lowered for practical use to filter the noise (Jin et al., 2008).

3. Simulation Studies

Simulations are carried out to examine the cancellation effect of soft nonlinearities using TDE and the suppression effect of hard nonlinearities using IVF. Viscous friction parameters (soft nonlinearities), and Coulomb friction parameters (hard nonlinearities) are varied while other parameters remain constant in simulations to see how TDE affects the compensation of nonlinear terms.

If $\boldsymbol{\Gamma}$ =0 in the TDCIVF (22), we can obtain the formulation as follows:

$$\tau = \tau_{t-L} - \bar{M}\ddot{\theta}_{t-L} + \bar{M}\{\ddot{\theta}_d + M_d^{-1}[\tau_s + K_d(\theta_d - \theta) + B_d(\dot{\theta}_d - \dot{\theta})]\}. \tag{25}$$

This formulation is referred to as internal force based impedance control (IFBIC) in (Bonitz & Hsia, 1996; Lasky & Hsia, 1991), where the compensation of the hard nonlinearities was not considered.

For simplicity and clarity, a single arm with soft and hard nonlinearities is considered as shown in Fig. 2. The simulation parameters are as follows: The mass of the link is m=8.163Kg, the link length is l=0.35m, and the inertia is I=1.0Kgm^2; the stiffness of the external spring as disturbance is $K_{disturbance}$=10 Nm/rad; the acceleration due to gravity is g=9.8Kgm/s^2. The parameters of environment are K_e=12000Nm/rad and B_e=0.40Nms/rad; its location is at θ_e =0.2rad. The sampling frequency of the simulation is 1 KHz.

The dynamics of a single arm is

$$\tau = I\ddot{\theta} + G(\theta) + F_v(\dot{\theta}) + F_c(\dot{\theta}) + d \tag{26}$$

where

$$G(\theta) = mgl\sin(\theta) \tag{27}$$

$$d = -K_{disturbance}\theta \tag{28}$$

$$F_v(\dot{\theta}) = -C_v\dot{\theta} \tag{29}$$

$$F_c(\dot{\theta}) = -C_c\,\mathrm{sgn}(\dot{\theta}). \tag{30}$$

Soft nonlinearities and hard nonlinearities can be expressed by

$$S(\theta,\dot{\theta}) = G(\theta) + F_v(\dot{\theta}) + d\,, \tag{31}$$

$$H(\theta,\dot{\theta},\ddot{\theta}) = F_c(\dot{\theta}). \tag{32}$$

For comparisons, identical parameters are used for the target dynamics in free space motion, as follows: M_d =1.0Kgm^2, K_d =100Nm/rad, and B_d =20Nms/rad.

The desired trajectory is given by (33)

$$\theta_d = A[1 - \exp(\text{-}\omega t)]\sin(\omega t) \tag{33}$$

where $\omega = 2\pi / p$, p =5s, A=0.15rad.

The simulation results are arranged in Figs. 3-7. The tracking errors of IFBIC with fixed Coulomb friction coefficients and various viscous friction coefficients are shown in Fig. 4 (a). The viscous friction has little effect on the tracking error because it can be cancelled by TDE. The tracking errors of IFBIC with various Coulomb friction coefficients and fixed viscous friction coefficients are shown in Fig. 4 (b). The larger Coulomb friction results in the larger tracking error. These results imply that TDE can not cancel hard nonlinearities. Maximum absolute errors and Mean absolute errors show that Coulomb friction severely deteriorates tracking errors whereas viscous friction has little effect on tracking errors shown in Figs. 4 (c) and (d).

Fig. 5 (a) shows soft nonlinearities $S(\theta,\dot{\theta})$ and the TDE error of soft nonlinearities $S(\theta,\dot{\theta}) - S(\theta,\dot{\theta})_{t-L}$. Because soft nonlinearities are continuous functions, the TDE error of soft nonlinearities is almost zero, and TDE works well on the cancellation of soft nonlinearities. Fig. 5 (b) shows hard nonlinearities $H(\theta,\dot{\theta},\ddot{\theta})$ and the TDE error of hard nonlinearities $H(\theta,\dot{\theta},\ddot{\theta}) - H(\theta,\dot{\theta},\ddot{\theta})_{t-L}$. Because hard nonlinearities are discontinuous functions, the TDE error of hard nonlinearities cannot be ignored. The TDE error of Coulomb friction can be regarded as a pulse type disturbance when the velocity changes its sign. The TDCIVF can reduce the tracking error due to Coulomb friction by increasing $\mathbf{\Gamma}$, as shown in Fig. 6.

The element of control input of the TDCIVF due to IVF term, $\bar{\mathbf{M}}\mathbf{\Gamma}(\dot{\boldsymbol{\theta}}_{ideal} - \dot{\boldsymbol{\theta}})$, is plotted in Fig. 7. The IVF term is almost zero in the presence of only soft nonlinearities in Figs. 7 (a) and (b); however, it activates in the presence of hard nonlinearities Figs. 7 (c) and (d).

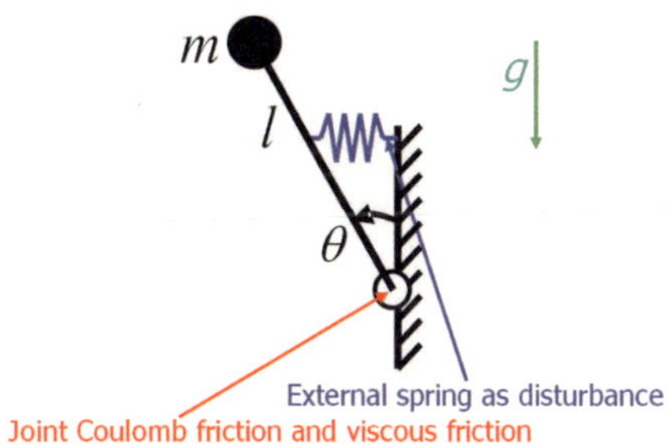

Figure 2. A single arm with friction

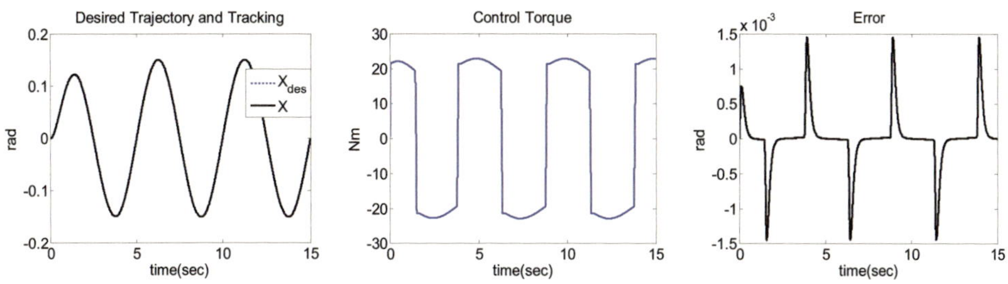

Figure 3. Simulation results with Coulomb friction coefficient C_c =20 and viscous friction coefficient C_v =15

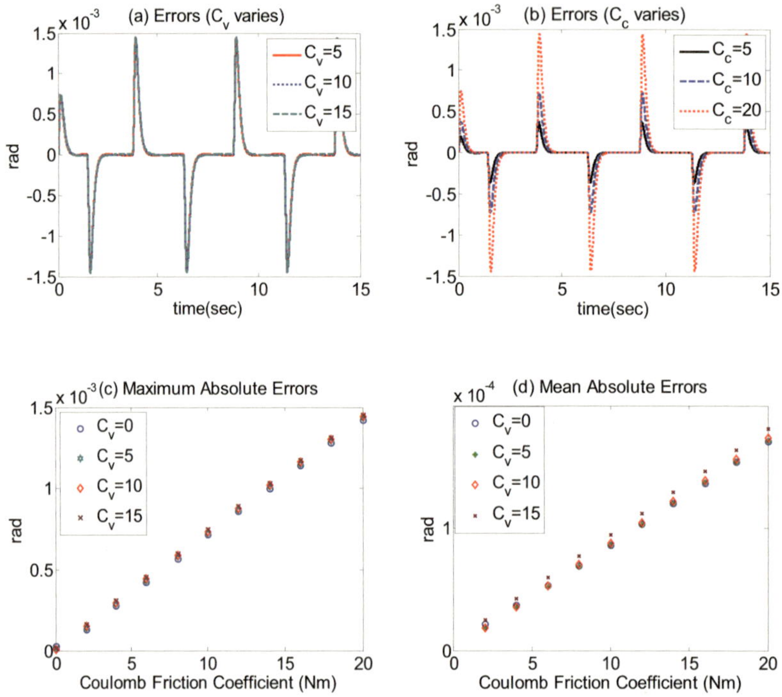

Figure 4. Tracking errors of IFBIC: (a) with fixed Coulomb friction (C_c =20) and various C_v . (b) with fixed viscous friction (C_v =10) and various C_c . (c) and (d) Maximum (Mean) absolute errors with various C_c and C_v . It shows that Coulomb friction severely deteriorates the tracking performance, and viscous friction has little effect on tracking errors

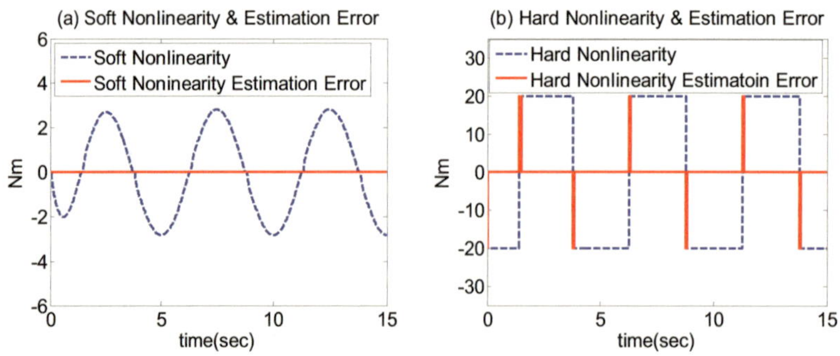

Figure 5. Nonlinear terms and their estimation errors. (a) Soft nonlinearity and its estimation error. (b) Hard nonlinearity and its estimation error. (C_c =20, C_v =15.)

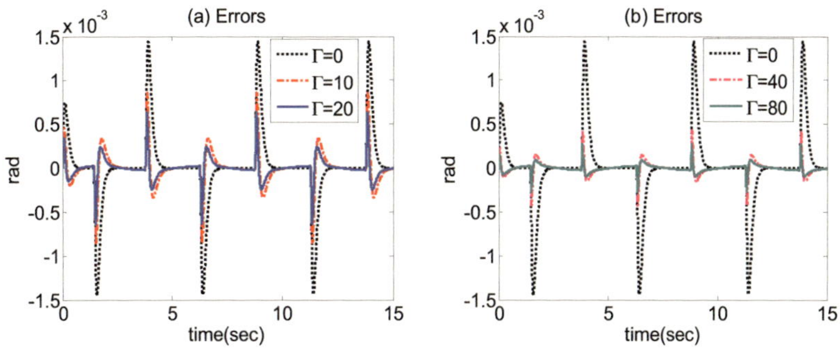

Figure 6. The effect of gain $\mathbf{\Gamma}$ in TDCIVF (C_c =20, C_v =10). (a) $\mathbf{\Gamma}$ =0, 10, 20; (b) $\mathbf{\Gamma}$ =0, 40, 80

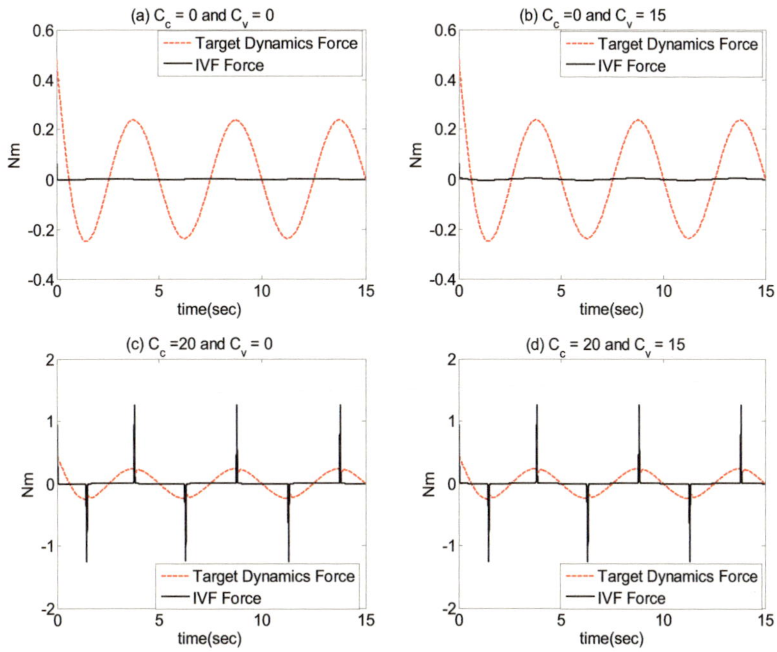

Figure 7. TDCIVF control input: Target dynamics injection torque and ideal velocity feedback torque. The IVF does not activate when hard nonlinearity is 0 (C_c =0) in (a) and (b); The IVF activates when hard nonlinearities do exist (C_c =20) in (c) and (d)

4. Comparison with Adaptive Friction Compensation Method

Adaptive friction compensation (Visioli et al., 2001; Visioli et al., 2006; Jatta et al., 2006) is regarded as a promising non-model based technique and it provides simple, effective online friction compensation. Hence, the TDCIVF is compared with AFC, a recently developed friction compensation method.

4.1 Adaptive Friction Compensation

First we briefly review the AFC (Visioli et al., 2001; Visioli et al., 2006; Jatta et al., 2006). As shown in Fig. 11, the AFC algorithm assumes that the difference between torque estimation error u_i^t and its estimation u_i depends only on the friction for each joint *i*. Thus, the estimation error between the actual friction torque $f_i^a(\dot{q}_i)$ and estimated one $f_i(\dot{q}_i)$ can be approximately considered equal to the output of the PID controller, i.e.:

$$f_i^a(\dot{q}_i) - f_i(\dot{q}_i) \cong u_i^r. \tag{34}$$

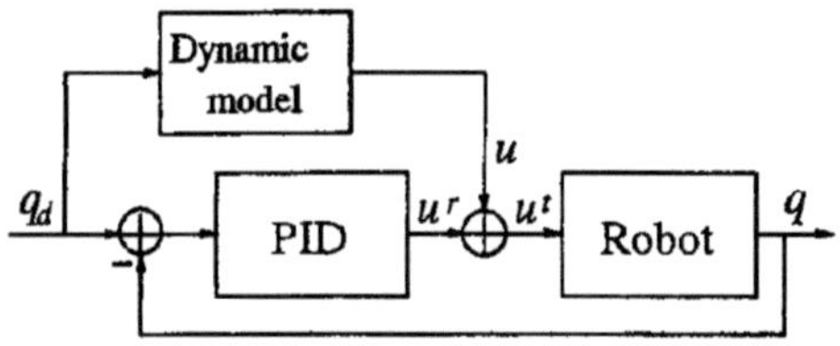

Figure 11. The general model for adaptive friction compensation

The friction terms are approximated by polynomial functions of degree *h*. Positive and negative velocities might be considered separately to obtain better results in case the actual friction functions is not symmetrical; hence,

$$f_i(\dot{q}_i) = \begin{cases} p_{i0}^- + p_{i1}^-\dot{q}_i + \cdots + p_{ih}^-\dot{q}_i^h & \text{if } \dot{q}_i < 0 \\ p_{i0}^+ + p_{i1}^+\dot{q}_i + \cdots + p_{ih}^+\dot{q}_i^h & \text{if } \dot{q}_i > 0 \end{cases} \quad (i = 1, \cdots, n). \tag{35}$$

Formally, the method can be described as follows. Define

$$\mathbf{p}_i \triangleq [p_{i0}^- p_{i1}^- \cdots p_{ih}^-] = \begin{cases} \left[p_{i0}^- p_{i1}^- \cdots p_{ih}^- \right] & \text{if } \dot{q}_i < 0 \\ \left[p_{i0}^+ p_{i1}^+ \cdots p_{ih}^+ \right] & \text{if } \dot{q}_i > 0 \end{cases} \tag{36}$$

and

$$\mathbf{v}_i \triangleq \begin{bmatrix} 1 & \dot{q}_i & \cdots & \dot{q}_i^h \end{bmatrix}. \tag{37}$$

The algorithm can be described as follows:

a. For $(i = 1, \cdots, n)$
   1. Measure $q_i(k)$ and $\dot{q}_i(k)$.
   2. Calculate $\ddot{q}_i(k)$ by differentiation and filtering.
   3. Set $e_i(k) = u_i^r(k)$.
   4. Calculate $\Delta\mathbf{p}_i(k) = \eta e_i(k)\mathbf{v}_i(k) + \alpha\Delta\mathbf{p}_i(k-1)$.
   5. Set $\mathbf{p}_i(k) = \mathbf{p}_i(k-1) + \Delta\mathbf{p}_i(k)$.
   6. Calculate $u_i^t = u_i + u_i^r$.

b. Apply the reference command torque signal u_i^t. The parameter η determines the velocity of the descent to the minimum and therefore the adaptation velocity.

Parameter α is the momentum coefficient, which helps to prevent large oscillations of the weight values and to accelerate the descent when the gradient is small.

4.2 Comparisons

For simplicity and clarity, we present experimental results using a single arm robot with friction. As shown in Fig. 12 (a), a single arm robot is commanded to move very fast in free space repeatedly (t=0-6s). A 5th polynomial desired trajectory is used for 8 path segments listed in Table 1, where both the initial position and the final position of each segment are listed along with the initial time and the final time. The velocity and the acceleration at the beginning and end of each path segment are set to zero.

TIME(S)	0.0	0.5	1.5	2.5	3.5	4.5	5.5	6.0	7.0
X(RAD)	0.0	0.1	-0.1	0.1	-0.1	0.1	-0.1	0.0	0.0

Table 1. The desired trajectory

To implement AFC, the method described in previous section was followed exactly.
The PID control is expressed as

$$u(t) = K\left(e(t) + T_D\dot{e}(t) + T_I^{-1}\int_0^t e(\sigma)d\sigma\right) \tag{38}$$

where K, T_D, and T_I are gains. The gains are selected as K =1000, T_D =0.05, and T_I =0.2. AFC is implemented with the PID. Parameters of AFC are $\eta = 0.001$, and $\alpha = 0.7$.

The TDCIVF, thanks to the impedance control formulation property, becomes a motion control formulation when the interaction torque $\boldsymbol{\tau}_s$ is omitted in the target dynamics (9). The motion control formulation for 1 DOF is

$$\tau = \tau_{t-L} - \bar{M}\ddot{\theta}_{t-L} + \bar{M}[\ddot{\theta}_d + K_V(\dot{\theta}_d - \dot{\theta}) + K_P(\theta_d - \theta) + \Gamma(\dot{\theta}_{ideal} - \dot{\theta})] \tag{39}$$

where

$$\dot{\theta}_{ideal} = \int[\ddot{\theta}_d + K_P(\theta_d - \theta) + K_V(\dot{\theta}_d - \dot{\theta})]\,dt. \tag{40}$$

The desired error dynamics of TDCIVF is

$$\ddot{\theta}_d - \ddot{\theta} + K_V(\dot{\theta}_d - \dot{\theta}) + K_P(\theta_d - \theta) = 0 \tag{41}$$

or

$$\ddot{\theta}_d - \ddot{\theta} + 2\zeta\omega_d(\dot{\theta}_d - \dot{\theta}) + \omega_d^2(\theta_d - \theta) = 0 \tag{42}$$

with

$$K_V = 2\zeta\omega_d, K_P = \omega_d^2. \tag{43}$$

The gains of TDCIVF and the speed of the target error dynamics are selected as follows: $\bar{M}$ =0.1, Γ =50.0, and ω_d =10; and ζ =1 for the fastest no overshoot performance. Shown in Fig. 12 (b), the tracking error of TDCIVF is the smallest, which confirms that the TDCIVF is superior to AFC.

AFC attempts to adapt the polynomial coefficients of friction. The adaptation takes quite some sampling time to update the friction parameters due to its property of neural network (Visioli et al., 2001) that is normally slower than TDE as discussed in (Lee & Oh, 1997). In contrast, the TDCIVF directly cancels most of nonlinearities using TDE and immediately activates the IVF when TDE is not sufficient. AFC must tune three gains of the PID by trial and error (it takes a quite some time to tune gains heuristically), and two additional parameters of AFC (η and α). In the TDCIVF, after specifying the convergence speed of the error dynamics by selecting ω_d and ζ, the tuning of the gains $\bar{M}$ and Γ is systematic and straightforward as discussed in section 2.6. Consequently, the TDCIVF is an easier, more systematic, and more robust friction compensation method than AFC.

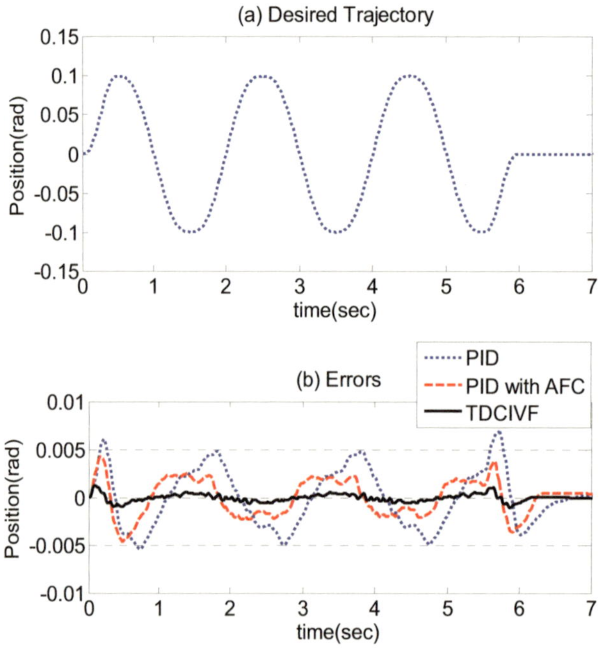

Figure 12. Experiment results on single arm with friction. (a) Desired trajectory. (b) Comparison of tracking performance. (dotted: PID, dashed: PID with AFC, and solid: TDCIVF.)

5. Cartesian Space Formulation

The Cartesian state space equation of an n-DOF robot manipulator is given by

$$\mathbf{F}_u = \mathbf{M}_x(\boldsymbol{\theta})\ddot{\mathbf{x}} + \mathbf{V}_x(\boldsymbol{\theta},\dot{\boldsymbol{\theta}}) + \mathbf{G}_x(\boldsymbol{\theta}) + \mathbf{F}_x(\boldsymbol{\theta},\dot{\boldsymbol{\theta}}) - \mathbf{F}_s \tag{44}$$

where $\mathbf{F}_u$ denotes a fore-torque vector acting on the end-effector of the robot, $\mathbf{F}_s$ the interaction force, and $\mathbf{x}$, $\dot{\mathbf{x}}$, $\ddot{\mathbf{x}}$ is an appropriate Cartesian vector representing position and orientation of the end-effector; $\boldsymbol{\theta},\dot{\boldsymbol{\theta}},\ddot{\boldsymbol{\theta}} \in \Re^n$ denote the joint angle, the joint velocity, and the joint acceleration, respectively; $\mathbf{M}_x(\boldsymbol{\theta})$ is the Cartesian mass matrix; $\mathbf{V}_x(\boldsymbol{\theta},\dot{\boldsymbol{\theta}})$ is a vector of velocity terms in Cartesian space, and $\mathbf{G}_x(\boldsymbol{\theta})$ is a vector of gravity terms in Cartesian space. $\mathbf{F}_x(\boldsymbol{\theta},\dot{\boldsymbol{\theta}})$ is a vector of friction terms in Cartesian space (Craig, 1989). Note that the fictitious forces acting on the end-effector, $\mathbf{F}_u$, could in fact be applied by the actuators at the joints using the relationship

$$\boldsymbol{\tau} = \mathbf{J}^T(\boldsymbol{\theta})\mathbf{F}_u \tag{45}$$

where $\mathbf{J}(\boldsymbol{\theta})$ is the Jacobian. The derivation of $\mathbf{M}_x(\boldsymbol{\theta})$, $\mathbf{V}_x(\boldsymbol{\theta},\dot{\boldsymbol{\theta}})$, $\mathbf{G}_x(\boldsymbol{\theta})$, $\mathbf{F}_x(\boldsymbol{\theta},\dot{\boldsymbol{\theta}})$ are given in (Craig, 1989), and expressed by

$$\mathbf{M}_x(\boldsymbol{\theta}) = \mathbf{J}^{-T}(\boldsymbol{\theta})\mathbf{M}(\boldsymbol{\theta})\mathbf{J}^{-1}(\boldsymbol{\theta}) \tag{46}$$

$$\mathbf{V}_x(\boldsymbol{\theta},\dot{\boldsymbol{\theta}}) = \mathbf{J}^{-T}(\boldsymbol{\theta})\left(\mathbf{V}(\boldsymbol{\theta},\dot{\boldsymbol{\theta}}) - \mathbf{M}(\boldsymbol{\theta})\mathbf{J}^{-1}(\boldsymbol{\theta})\dot{\mathbf{J}}(\boldsymbol{\theta})\dot{\boldsymbol{\theta}}\right) \tag{47}$$

$$\mathbf{G}_x(\boldsymbol{\theta}) = \mathbf{J}^{-T}(\boldsymbol{\theta})\mathbf{G}(\boldsymbol{\theta}) \tag{48}$$

$$\mathbf{F}_x(\boldsymbol{\theta},\dot{\boldsymbol{\theta}}) = \mathbf{J}^{-T}(\boldsymbol{\theta})\mathbf{F}(\boldsymbol{\theta},\dot{\boldsymbol{\theta}}) \tag{49}$$

$$\mathbf{F}_x(\boldsymbol{\theta},\dot{\boldsymbol{\theta}}) = (\mathbf{F}_v)_x + (\mathbf{F}_c)_x + (\mathbf{F}_{st})_x \tag{50}$$

where $(\mathbf{F}_v)_x$, $(\mathbf{F}_c)_x$, and $(\mathbf{F}_{st})_x$ denote viscous friction, Coulomb friction, and stiction terms in Cartesian space, respectively.

Introducing a matrix $\bar{\mathbf{M}}_x(\boldsymbol{\theta})$, another expression of (44) is obtained as follows:

$$\mathbf{F}_u = \bar{\mathbf{M}}_x(\boldsymbol{\theta})\ddot{\mathbf{x}} + \mathbf{N}_x(\boldsymbol{\theta},\dot{\boldsymbol{\theta}},\ddot{\boldsymbol{\theta}}) \tag{51}$$

where $\mathbf{N}_x(\boldsymbol{\theta},\dot{\boldsymbol{\theta}},\ddot{\boldsymbol{\theta}})$ and $\bar{\mathbf{M}}_x(\boldsymbol{\theta})$ are expressed by

$$\mathbf{N}_x(\boldsymbol{\theta},\dot{\boldsymbol{\theta}},\ddot{\boldsymbol{\theta}}) = [\mathbf{M}_x(\boldsymbol{\theta}) - \bar{\mathbf{M}}_x(\boldsymbol{\theta})]\ddot{\mathbf{x}} + \mathbf{V}_x(\boldsymbol{\theta},\dot{\boldsymbol{\theta}}) + \mathbf{G}_x(\boldsymbol{\theta}) + \mathbf{F}_x(\boldsymbol{\theta},\dot{\boldsymbol{\theta}}) - \mathbf{F}_s \tag{52}$$

$$\bar{\mathbf{M}}_x(\boldsymbol{\theta}) \triangleq \mathbf{J}^{-T}(\boldsymbol{\theta})\bar{\mathbf{M}}\mathbf{J}^{-1}(\boldsymbol{\theta}) \tag{53}$$

where $\bar{\mathbf{M}}$ is constant diagonal matrix.

The nonlinear term $\mathbf{N}_x(\boldsymbol{\theta},\dot{\boldsymbol{\theta}},\ddot{\boldsymbol{\theta}})$ can be classified into two categories: soft nonlinearities and hard nonlinearities, as follows:

$$\mathbf{N}_x(\boldsymbol{\theta},\dot{\boldsymbol{\theta}},\ddot{\boldsymbol{\theta}}) = \mathbf{S}_x(\boldsymbol{\theta},\dot{\boldsymbol{\theta}}) + \mathbf{H}_x(\boldsymbol{\theta},\dot{\boldsymbol{\theta}},\ddot{\boldsymbol{\theta}}) \tag{54}$$

$$\mathbf{S}_x(\boldsymbol{\theta},\dot{\boldsymbol{\theta}}) = \mathbf{V}_x(\boldsymbol{\theta},\dot{\boldsymbol{\theta}}) + \mathbf{G}_x(\boldsymbol{\theta}) + (\mathbf{F}_v)_x(\boldsymbol{\theta},\dot{\boldsymbol{\theta}}) - \mathbf{F}_s \tag{55}$$

$$\mathbf{H}_x(\boldsymbol{\theta},\dot{\boldsymbol{\theta}},\ddot{\boldsymbol{\theta}}) = (\mathbf{F}_c)_x(\boldsymbol{\theta},\dot{\boldsymbol{\theta}}) + (\mathbf{F}_{st})_x(\boldsymbol{\theta},\dot{\boldsymbol{\theta}}) + [\mathbf{M}_x(\boldsymbol{\theta}) - \bar{\mathbf{M}}_x(\boldsymbol{\theta})]\ddot{\mathbf{x}}. \tag{56}$$

The control objective in Cartesian space is to achieve following target impedance dynamics

$$\mathbf{M}_{xd}(\ddot{\mathbf{x}}_d - \ddot{\mathbf{x}}) + \mathbf{B}_{xd}(\dot{\mathbf{x}}_d - \dot{\mathbf{x}}) + \mathbf{K}_{xd}(\mathbf{x}_d - \mathbf{x}) + \mathbf{F}_s = 0 \tag{57}$$

where $\mathbf{F}_s$ denotes the interaction force; $\mathbf{M}_{xd}$, $\mathbf{B}_{xd}$, and $\mathbf{K}_{xd}$ denote the desired mass, the desired damping, and the desired stiffness in Cartesian space, respectively; and $\mathbf{x}_d, \dot{\mathbf{x}}_d, \ddot{\mathbf{x}}_d$ the desired position, the desired velocity, the desired acceleration in Cartesian space, respectively; $\mathbf{x}, \dot{\mathbf{x}}, \ddot{\mathbf{x}}$ denote the position, the velocity, and the acceleration in Cartesian space, respectively.
The Cartesian formulation of TDCIVF can be derived as follows:

$$\mathbf{F}_u = \bar{\mathbf{M}}_x(\boldsymbol{\theta})\mathbf{u}_x + \hat{\mathbf{S}}_x(\boldsymbol{\theta}, \dot{\boldsymbol{\theta}}) + \hat{\mathbf{H}}_x(\boldsymbol{\theta}, \dot{\boldsymbol{\theta}}, \ddot{\boldsymbol{\theta}}) \tag{58}$$

where

$$\mathbf{u}_x = \ddot{\mathbf{x}}_d + \mathbf{M}_{xd}^{-1}\left[\mathbf{K}_{xd}(\mathbf{x}_d - \mathbf{x}) + \mathbf{B}_{xd}(\dot{\mathbf{x}}_d - \dot{\mathbf{x}}) + \mathbf{F}_s\right]. \tag{59}$$

The estimation of $\hat{\mathbf{S}}_x(\boldsymbol{\theta}, \dot{\boldsymbol{\theta}}) + \hat{\mathbf{H}}_x(\boldsymbol{\theta}, \dot{\boldsymbol{\theta}}, \ddot{\boldsymbol{\theta}})$ is given by

$$\hat{\mathbf{S}}_x(\boldsymbol{\theta}, \dot{\boldsymbol{\theta}}) + \hat{\mathbf{H}}_x(\boldsymbol{\theta}, \dot{\boldsymbol{\theta}}, \ddot{\boldsymbol{\theta}}) = \mathbf{S}_x(\boldsymbol{\theta}, \dot{\boldsymbol{\theta}})_{t-L} + \mathbf{H}_x(\boldsymbol{\theta}, \dot{\boldsymbol{\theta}}, \ddot{\boldsymbol{\theta}})_{t-L} \tag{60}$$

and

$$\mathbf{S}_x(\boldsymbol{\theta}, \dot{\boldsymbol{\theta}})_{t-L} + \mathbf{H}_x(\boldsymbol{\theta}, \dot{\boldsymbol{\theta}}, \ddot{\boldsymbol{\theta}})_{t-L} = (\mathbf{F}_u)_{t-L} - \bar{\mathbf{M}}_x(\boldsymbol{\theta})\mathbf{x}_{t-L}. \tag{61}$$

Here, TDE error $\boldsymbol{\varepsilon}$ is defined as

$$\boldsymbol{\varepsilon} \triangleq \bar{\mathbf{M}}_x^{-1}(\boldsymbol{\theta})[\mathbf{H}_x(\boldsymbol{\theta}, \dot{\boldsymbol{\theta}}, \ddot{\boldsymbol{\theta}}) - \mathbf{H}_x(\boldsymbol{\theta}, \dot{\boldsymbol{\theta}}, \ddot{\boldsymbol{\theta}})_{t-L}]. \tag{62}$$

Then, with the combination of (44), (54), (58), (59), (60), (62), impedance error dynamics is

$$\mathbf{M}_{xd}(\ddot{\mathbf{x}}_d - \ddot{\mathbf{x}}) + \mathbf{B}_{xd}(\dot{\mathbf{x}}_d - \dot{\mathbf{x}}) + \mathbf{K}_{xd}(\mathbf{x}_d - \mathbf{x}) + \mathbf{F}_s = \mathbf{M}_{xd}\boldsymbol{\varepsilon}. \tag{63}$$

$\boldsymbol{\varepsilon}$ causes the resulting dynamics to deviate from the target impedance dynamics. To suppress $\boldsymbol{\varepsilon}$, ideal velocity feedback term is introduced, with $\dot{\mathbf{x}}_{ideal}$ defined here as

$$\dot{\mathbf{x}}_{ideal} \triangleq \int\left\{\ddot{\mathbf{x}}_d + \mathbf{M}_{xd}^{-1}\left[\mathbf{K}_{xd}(\mathbf{x}_d - \mathbf{x}) + \mathbf{B}_{xd}(\dot{\mathbf{x}}_d - \dot{\mathbf{x}}) + \mathbf{F}_s\right]\right\}dt. \tag{64}$$

Combining previous formulations, the control law is

$$\begin{aligned} \mathbf{F}_u = {} & (\mathbf{F}_u)_{t-L} - \bar{\mathbf{M}}_x(\boldsymbol{\theta})\mathbf{x}_{t-L} \\ & + \bar{\mathbf{M}}_x(\boldsymbol{\theta})\left\{\ddot{\mathbf{x}}_d + \mathbf{M}_{xd}^{-1}[\mathbf{K}_{xd}(\mathbf{x}_d - \mathbf{x}) + \mathbf{B}_{xd}(\dot{\mathbf{x}}_d - \dot{\mathbf{x}}) + \mathbf{F}_s]\right\} \\ & + \bar{\mathbf{M}}_x(\boldsymbol{\theta})\boldsymbol{\Gamma}(\dot{\mathbf{x}}_{ideal} - \dot{\mathbf{x}}) \end{aligned} \tag{65}$$

where $\boldsymbol{\Gamma}$ is a gain matrix.

6. Human-Robot Cooperation

The experiment scenario for a human-robot cooperation task is as follows: The robot end-effector draws a circle in 4s in free space and afterwards stands still before it is pushed by a human. A human pushes the robot end-effector forth and back in Y direction while the robot tries to keep in contact with it. This is to simulate the situation, for example, when a worker and a robot work together to move or install an object. In this cooperation task, it is often desired for the robot end-effector to behave like a low-stiffness spring, and the target dynamics is described as follow:

$$\mathbf{M}_{xd} = \begin{bmatrix} 20 & 0 \\ 0 & 20 \end{bmatrix} \text{Kg and } \mathbf{K}_{xd} = \begin{bmatrix} 2000 & 0 \\ 0 & 500 \end{bmatrix} \text{N/m.}$$

ζ =1.0 for free space motion and ζ =6.0 for constrained space. The Y direction stiffness of the target impedance is low (500N/m) for the compliant interaction with human.
Experimental results are displayed in Fig. 8 and Fig. 9. The end-effector of the robot acts like a soft spring-mass-damper system to a human being. It neither destabilizes the robot nor exerts an excessive force to the human being. In free space task, shown in Fig. 10 (b), the TDCIVF is better than IFBIC at tracking the circle. In the robot-human cooperation task, shown in Fig. 10 (d), the TDCIVF shows the smallest impedance error.
This experiment illustrates that human-robot cooperative tasks can be accomplished successfully under TDCIVF. The TDCIVF shows robust compliant motion control while possessing soft compliance.

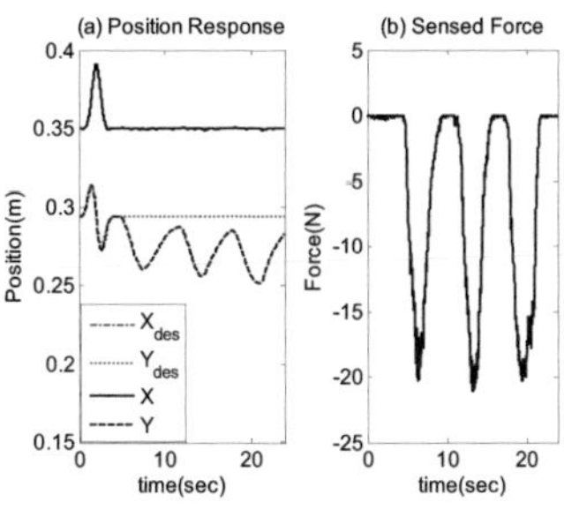

Figure 8. Human-Robot Cooperation: Position and force responses of IFBIC

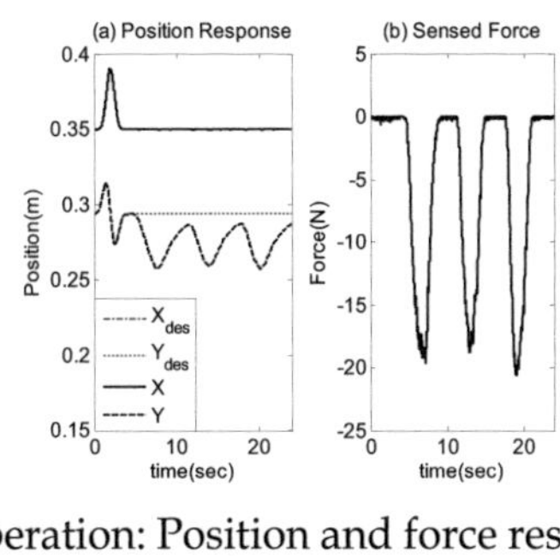

Figure 9. Human-Robot Cooperation: Position and force responses of TDCIVF

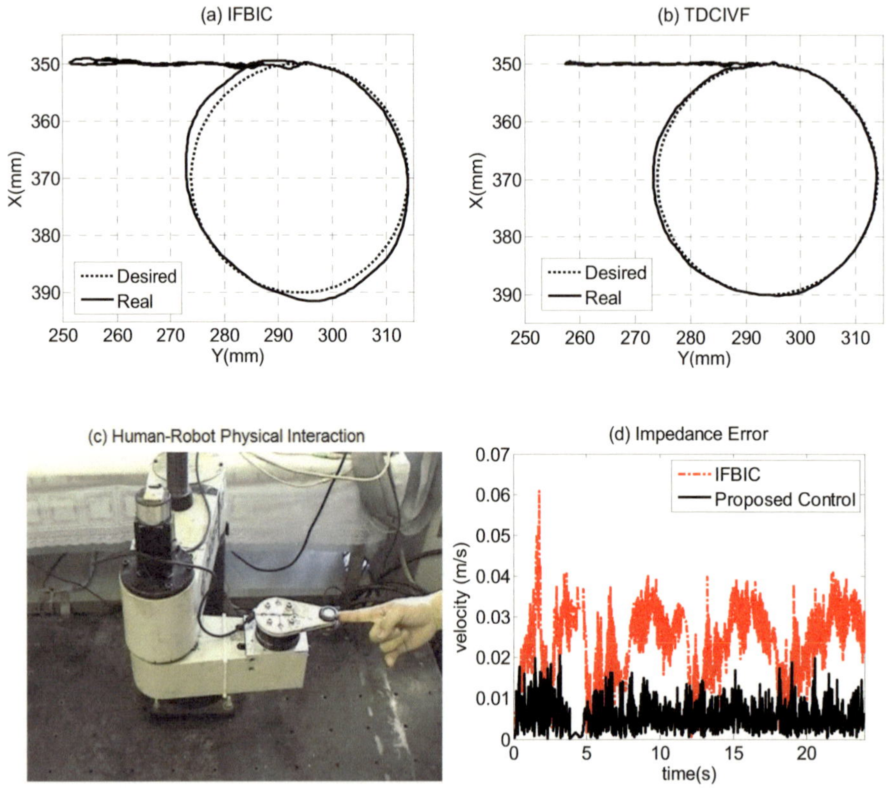

Figure 10. Human-Robot Cooperation. (a) IFBIC: only TDE is used. (b) TDCIVF: both TDE and IVF is used. (c) Experiment scienario. (d) Comparison of impedance error

7. Conclusion

The TDCIVF has following properties:

1. Transparent control structure.
2. Simplicity of gain tuning.
3. Non-model based online friction compensation.
4. Robustness against both soft nonlinearities and hard nonlinearities.

The overall implementation procedure of TDCIVF is straightforward and practical. The cancellation effect of soft nonlinearities using TDE and the suppression effect of hard nonlinearities using ideal velocity feedback are confirmed through simulation studies. The TDCIVF can reduce the effect of hard nonlinearities by raising the gain $\mathbf{\Gamma}$.

Compared with AFC, a recently developed non-model based method, the TDCIVF provides a more systematic, easier, and more robust friction compensation method. The human-robot cooperation experiment shows that the TDCIVF is practical for realizing compliant manipulation of robots.

The following directions can be considered for future research:

1. The term caused by hard nonlinearity, $\boldsymbol{\varepsilon}$ in (16) may be regarded as perturbation to the target impedance dynamics. It is noteworthy for further research that the TDE error $\boldsymbol{\varepsilon}$ can be suppressed by other compensation methods.

2. The terms in robot dynamics considered in this TDCIVF are: the gravity term, viscous friction term, the Coriolis and centrifugal term, the disturbance term, and the environmental force term (soft nonlinearities); and the Coulomb friction term, the stiction term, and the inertia uncertainty term (hard nonlinearities). *From the TDE viewpoint*, consideration and compensation of other terms such as *saturation*, *backlash* and *hysteresis terms* can be investigated in depth.
3. *Without modelling robot dynamics and nonlinear friction*, the TDCIVF provides robust compensation and fast accurate trajectory tracking. Recently, the Lugre friction mode based control has attracted research activities because the Lugre model is known as accurate friction model in friction compensation literature. It will be interesting to experimentally compare the TDCIVF (which uses no friction model) with the Lugre friction model based control.

8. References

Aghili F. & Namvar M. (2006). Adaptive control of manipulators using uncalibrated joint-torque sensing, *IEEE Transactions on Robotics*, Vol. 22, No. 4, pp. 854-860.

Armstrong-Helouvry B.; Dupont P. & de Wit C. C. (1994). A survey of models, analysis tools and compensation methods for the control, *Automatica*, Vol. 30, No. 1, pp. 1083-1138.

Bona B. & Indri M. (2005). Friction compensation in robotics: An overview, *Proceedings of 44th IEEE International Conference on Decision and Control and 2005 European Control Conference (CDC2005)*, pp. 4360- 4367.

Bona B.; Indri M. &, Smaldone N. (2006). Rapid prototyping of a model-based control with friction compensation for a direct-drive robot, *IEEE-ASME Transactions on Mechatronics*, Vol. 11, No. 5, pp. 576-584.

Bonitz R. C. & Hsia T. C. (1996). Internal force-based impedance control for cooperating manipulators, *IEEE Transactions on Robotics and Automation*, Vol. 12, No. 1, pp. 78-89.

Craig J.J. (1989). *Introduction to Robotics: Mechnaics and Control*, Addison-Wesley, ISBN-13: 9780201095289.

Hsia T. C.; Lasky T. A. & Guo Z. (1991). Robust independent joint controller design for industrial robot manipulators, *IEEE Transactions on Industrial Electronics*, Vol. 38, No. 1, pp. 21-25.

Jatta F.; Legnani G. & Visioli A. (2006). Friction compensation in hybrid force/velocity control of industrial manipulators, *IEEE Transactions on Industrial Electronics*, Vol. 53, No. 2, pp. 604-613.

Jin M.; Kang S. H. & Chang P. H. (2006). A robust compliant motion control of robot with certain hard nonlinearities using time delay estimation. *Proceedings - IEEE International Symposium on Industrial Electronics (ISIE2006)*, Montreal, Quebec, Canada, pp. 311-316.

Jin M.; Kang S. H. & Chang P. H. (2008). Robust compliant motion control of robot with nonlinear friction using time delay estimation. *IEEE Transactions on Industrial Electronics*, Vol. 55, No. 1, pp. 258-269.

Lasky T. A. & Hsia T. C. (1991). On force-tracking impedance control of robot manipulators, *Proceedings - IEEE International Conference on Robotics and Automation (ICRA'91)*, pp. 274-280.

Lee J.-W. & J.-H. Oh (1997). Time delay control of nonlinear systems with neural network modelling, *Mechatronics*, Vol. 7, No. 7, pp. 613-640.

Lischinsky P.; de Wit C. C. & Morel G. (1999). Friction compensation for an industrial hydraulic robot, *IEEE Control System Magazine*, Vol 19, No. 1, pp. 25-32.

Liu G.; Goldenberg A. A. & Zhang Y. (2004), Precise slow motion control of a direct-drive robot arm with velocity estimation and friction compensation, *Mechatronics*, Vol. 14, No. 7, pp. 821-834.

Marton L. & Lantos B. (2007). Modeling, identification, and compensation of stick-slip friction, *IEEE Transactions on Industrial Electronics*, Vol. 54, No. 1, pp. 511-521.

Mei Z.-Q.; Xue Y.-C. & Yang R.-Q. (2006). Nonlinear friction compensation in mechatronic servo systems, *International Journal of Advanced Manufacturing Technology*, Vol. 30, No. 7-8, pp. 693-699.

Slotine J.-J. E. (1985). Robustness issues in robot control, *Proceedings - IEEE International Conference on Robotics and Automation (ICRA'1985)*, Vol. 3, pp. 656-661.

Xie W.-F. (2007). Sliding-mode-observer-based adaptive control for servo actuator with friction, *IEEE Transactions on Industrial Electronics*, Vol. 54, No. 3, pp. 1517-1527.

Visioli A.; Adamini R. & Legnani G. (2001). Adaptive friction compensation for industrial robot control, *Proceedings-IEEE/ASME International Conference on Advanced Intelligent Mechatronics (AIM2001)*, Vol. 1, Jul. 2001, pp. 577-582.

Visioli A.; Ziliani G. & Legnani G. (2006). Friction compensation in hybrid force/velocity control for contour tracking tasks, In: *Industrial Robotics: Theory, Modeling and Control*, Sam Cubero (Ed.), pp. 875-894. Pro Literatur Verlag, Germany / ARS, Austria, ISBN 3-86611-285-8.

Youcef-Toumi K. & Ito O. (1990). A time delay controller design for systems with unknown dynamics, *Journal of Dynamic Systems Measurement and Control-Transactions of the ASME*, Vol. 112, No. 1, pp. 133-142.

13

On transpose Jacobian control for monocular fixed-camera 3D Direct Visual Servoing

Rafael Kelly[1], Ilse Cervantes[2], Jose Alvarez-Ramirez[3], Eusebio Bugarin[1] and Carmen Monroy[4]
[1]Centro de Investigación Científica y de Educacion Superior de Ensenada, [2]Instituto Potosino de Investigación Científica y Tecnológica, [3]Universidad Autónoma Metropolitana, [4]Instituto Tecnológico de la Laguna Mexico

1. Introduction

This chapter deals with direct visual servoing of robot manipulators under monocular fixed-camera configuration. Considering the image-based approach, the complete nonlinear system dynamics is utilized into the control system design. Multiple coplanar object feature points attached to the end-effector are assumed to introduce a family of transpose Jacobian based control schemes. Conditions are provided in order to guarantee asymptotic stability and achievement of the control objective. This chapter extends previous results in two direction: first, instead of only planar robots, it treats robots evolving in the 3D Cartesian space, and second, a broad class of controllers arises depending on the free choice of proper artificial potential energy functions. Experiments on a nonlinear direct-drive spherical wrist are presented to illustrate the effectiveness of the proposed method.
Robots can be made flexible and adaptable by incorporating sensory information from multiple sources in the feedback loop (Hutchinson et al., 1996), (Corke, 1996). In particular, the use of vision provides of non-contact measures that are shown to be very useful, if not necessary, when operating under uncertain environments. In this context, visual servoing can be classified in two configurations: Fixed-camera and camera-in-hand. This chapter addresses the fixed-camera configuration to visual servoing of robot manipulators.
There exists in the literature a variety of visual control strategies that study the robot positioning problem. Some of these strategies deal with the control with a single camera of planar robots or robots constrained to move on a given plane (Espiau, 1992), (Kelly, 1996), (Maruyama & Fujita, 1998), (Shell et al., 2002), (Reyes & Chiang, 2003). Although a number of practical applications can be addressed under these conditions, there are also applications demanding motions in the 3D Cartesian space. The main limitation to
Work partially supported by CONACYT grant 45826
extend these control strategies to the non-planar case is the extraction of the 3D pose of the robot end-effector or target from a planar image. To overcome this problem several monocular and binocular strategies have been reported mainly for the camera-in-hand configuration, (Lamiroy et al., 2000), (Hashimoto & Noritsugu, 1998), (Fang & Lin, 2001), (Nakavo et al., 2002). When measuring or estimating 3D positioning from visual information

there are several performance criteria that must be considered. An important performance criterion that must be considered is the number and configuration of the image features. In particular, in (Hashimoto & Noritsugu, 1998), (Yuan, 1989) a minimum point condition to guarantee a nonsingular image Jacobian is given.

Among the existing approaches to solve the 3D regulation problem, it has been recognized that image-based schemes posses some degree of robustness against camera miscalibrations. It is worth noticing, that most reported 3D control strategies satisfy one of the following characteristics: 1) the control law is of the so-called type "kinematic control", in which the control input are the joint velocity instead the joint torques), (Lamiroy et al., 2000, (Fang & S-K. Lin, 2001), (Nakavo et al., 2002); 2) the control law requires the computation of the inverse kinematics of the robot (Sim et al., 2002). These approaches have the disadvantage of neglecting the dynamics of the manipulator for control design purposes or may have the drawback of the lack of an inverse Jacobian matrix of the robot.

In this chapter we focus on the fixed-camera, monocular configuration for visual servoing of robot manipulators. Specifically, we address the regulation problem using direct vision feedback into the control loop. Following the classification in (Hager, 1997), the proposed control strategy belongs to the class of controllers having the following features: **a)** direct visual servo where the visual feedback is converted to joint torques instead of joint or Cartesian velocities; **b)** endpoint closed-loop system where vision provides the end-effector and target positions, and **c)** image-based control where the definition of the servo errors is taken directly from the camera image.

Specifically, a family of transpose Jacobian control laws is introduced which is able to provide asymptotic stability of the controlled robot and the accomplishment of the regulation objective. In this way, a contribution of this note is to extend the results in (Kelly, 1996) to the non-planar case and to establish the conditions for such an extension. The performances of two controllers arose from the proposed strategy are illustrated via experimental work on a 3 DOF direct-drive robot wrist.

The chapter is organized in the following way. Section II provides a brief description of the robot manipulator dynamics and kinematics. Section III describes the camera-vision system and presents the control problem formulation. Section IV introduces a family of control systems based in the transpose Jacobian control philosophy. Section V is devoted to illustrate the performance of the proposed controllers via experimental work. Finally, Section VI presents brief concluding remarks.

Notations. Throughout this chapter, $\|.\|$ denotes the Euclidean norm, x^T denotes the transpose of vector x, and $\nabla_x s(x)$ denotes the gradient of the scalar differentiable function $s(x)$.

2. Robot dynamics and kinematics

The dynamics of a n-DOF robot manipulator moving in the 3D Cartesian space is given by (Sciavicco & Siciliano, 1996)

$$M(q)\ddot{q} + C(q,\dot{q})\dot{q} + g(q) = \tau \tag{1}$$

where $q \in \Re^n$ is the vector of joint positions, $\tau \in \Re^n$ is the vector of applied joint torques, $M(q) \in \Re^{n \times n}$ is the symmetric and positive definite inertia matrix, $C(q,\dot{q}) \in \Re^{n \times n}$ is the matrix associated to the Coriolis/centrifugal torques, and $g(q) \in \Re^n$ is the vector of gravitational torques originated by the gravitational potential function $\mathsf{U}_g(q)$ (*i.e.* $g(q) = \nabla_q \mathsf{U}_g(q)$). The configuration space where the admissible joint coordinates lie is denoted by $\mathsf{C} \subset \Re^n$.

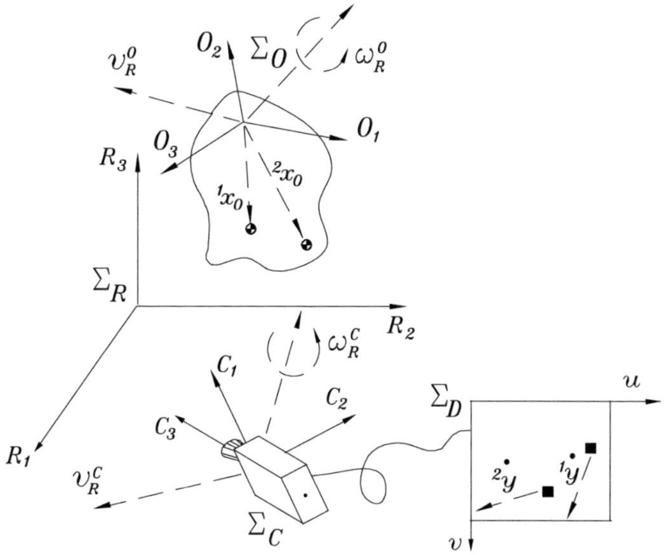

Figure 1. General differential imaging model

On the other hand, consider a fixed *robot* frame Σ_R and a moving frame Σ_O attached at the robot end-effector grasping an object which is assumed to be a rigid body (see Figure 1). The position of the end-effector (object) frame Σ_O relative to Σ_R is denoted by $O_R^O(q)$ and the corresponding rotation matrix by $R_R^O(q)$. A local description for the pose (position and orientation) of the end-effector frame is considered by using a minimal representation of the orientation. In this way, it is possible to describe the end-effector frame Σ_O pose by means of a vector $s \in \Re^m$ with $m \leq 6$:

$$s = s(q)$$

where $s(q)$ represents the direct kinematics. We regard to the robot, this chapter assumes that it is nonredundant in the sense $m = n$.

The *differential kinematics* describes the Cartesian end-effector velocity as a function of the joint velocity $\dot{q}$. Let ϑ_R^O and ω_R^O be the linear and angular velocity, respectively, of the frame Σ_O with respect to Σ_R. They are related to the joint velocity via the robot differential kinematics

$$\begin{bmatrix} \vartheta_R^O \\ \omega_R^O \end{bmatrix} = J_G(q)\dot{q}$$

where $J_G(q) \in IR^{6 \times n}$ is the robot geometric Jacobian.

The nonsingular robot configurations are the subset of the configuration space C where the robot Jacobian is full rank, that is

$$\mathsf{C}_{NS} = \{q \in \mathsf{C} : rank\{J_G(q)\} = n\}.$$

It is convenient to define the set of admissible nonsingular end-effector poses as

$$\mathsf{P} = \{s \in \Re^n : \exists\, q \in \mathsf{C}_{NS} : s = s(q)\}$$

At regular configurations $q \in \mathsf{C}_{NS}$ of nonredundant robots, there exists no self-motion around these configurations. Thus, no regular configurations can lie on a continuous path in the configuration space that keeps the kinematics function at the same value (Seng et al., 1997). This yields the following

Lemma 1. For nonredundant robots and an admissible end-effector pose $p \in \mathsf{P}$, the corresponding solutions $q \in \mathsf{C}_{NS}$ of the inverse kinematics are isolated.

3. Fixed-camera imaging model

Consider a video camera standing in a fixed place with frame Σ_C which captures images of a subspace of the robot workspace.

Moreover, Frame Σ_C is located relative to Σ_R as described by a position vector, O_R^C, and a rotation matrix, R_R^C. In order to get a confident measure of the position and orientation of the robot end-effector, the grasping object is provided with p feature points. Such points are referred as *object feature points*, denoted –with respect to Σ_O– by ${}^1x_O, {}^2x_O, \cdots, {}^px_O$. The description of the object feature points ${}^ix_O$ for $i = 1, \cdots, p$ relative to frames Σ_R and Σ_C are denoted by ${}^ix_R$ and ${}^ix_C$ respectively. For each object feature point ${}^ix_O$ there is a corresponding *image* feature point ${}^iy = [{}^iu, {}^iv]^T$.

A. Imaging model

The *imaging model* gives the image feature vector ${}^iy$ of the corresponding object features ${}^ix_O$ (referred to Σ_O). This model involves transformations and a perspective projection. The whole imaging model is described by (Hutchison, 1996)

$${}^ix_R = R_R^O(q)\,{}^ix_O + O_R^O(q) \tag{2}$$

$${}^ix_C = R_R^{CT}\left[{}^ix_R - O_R^C\right] \tag{3}$$

$${}^iy = \frac{\lambda\,\alpha}{{}^ix_{C_3} - \lambda}\begin{bmatrix} {}^ix_{C_1} \\ {}^ix_{C_2} \end{bmatrix} - \alpha\begin{bmatrix} O_{I_1} \\ O_{I_2} \end{bmatrix} + \begin{bmatrix} u_0 \\ v_0 \end{bmatrix} \tag{4}$$

where α is a scale factor, λ is the lens focal length, $[O_{I_1} \;\; O_{I_2}]^T$ is the image plane offset and $[u_0 \;\; v_0]^T$ is the image center.

It is worth noticing that the model requires the robot kinematics $s(q)$ in terms of $R_R^O(q)$ and $O_R^O(q)$. Hence, the image feature point ${}^iy$ is a function of the end-effector pose s and finally of the joint position vector q. With abuse of notation, when required ${}^{i}y(s)$ or ${}^{i}y(q)$ may be written.

According to the exterior orientation calibration problem addressed in (Yuan, 1989), in general the computation of the pose of a rigid object in the 3D Cartesian space relative to a camera based in the 2D image of known coplanar feature points located on the object is solved by using 4 noncollinear feature points. A consequence of this result is stated in the following

Lemma 2. Consider the *extended* imaging model corresponding to p points

$$\begin{bmatrix} {}^{1}y \\ \vdots \\ {}^{p}y \end{bmatrix} = \begin{bmatrix} {}^{1}y(s) \\ \vdots \\ {}^{p}y(s) \end{bmatrix}.$$

At least $p=4$ coplanar but noncollinear points are required generically to have isolated end-effector pose s solutions of the inverse extended imaging model.

Due to this result, the following assumption is in order A1 Attached to the object are $p=4$ coplanar but noncollinear feature points.

B. Differential imaging model

The *differential imaging model* describes how the image feature ${}^{i}y$ changes as a function of the joint velocity $\dot{q}$. This is obtained by differentiating the imaging model which after tedious by straightforward manipulations can be written as

$${}^{i}\dot{y} = A_i({}^{i}y, {}^{i}x_{C_3}, q, {}^{i}x_O)\dot{q} \tag{5}$$

where matrix $A_i(\cdot) \in \Re^{2\times n}$ is given in (6) at the top of next page. $J_{image}(\cdot) \in \Re^{2\times 6}$ in (7) denotes the image Jacobian (Hutchinson et al., 1996) where the focal length λ was neglected (i.e. ${}^{i}x_{C_3} - \lambda = {}^{i}x_{C_3}$) and the misalignment vector O_I was absorbed by the computer image center vector $[u_0 \ \ v_0]^T$. The matrix $S(\cdot) \in \Re^{3\times 3}$ is a skew-symmetric matrix defined as

$$S(x) = \begin{bmatrix} 0 & -x_3 & x_2 \\ x_3 & 0 & -x_1 \\ -x_2 & x_1 & 0 \end{bmatrix}.$$

C. Control objective

For the sake of notation, define the *extended* image feature vector $y \in \Re^8$ and the *desired* image feature vector $y_d \in \Re^8$ as

$$y = \begin{bmatrix} {}^{1}y \\ {}^{2}y \\ {}^{3}y \\ {}^{4}y \end{bmatrix} \quad \textit{and} \quad y_d = \begin{bmatrix} {}^{1}y \\ {}^{2}y_d \\ {}^{3}y_d \\ {}^{4}y_d \end{bmatrix}$$

where ${}^{i}y_d$ for $i=1,2,3,4$ are the *desired* image feature points. For convenience we define the set of admissible desired image features as

$$\mathrm{Y}_d = \{ y_d \in \Re^8 : \exists s \in \mathrm{P} : y_d = y(s) \}.$$

In words, the admissible desired image features are those obtained from the projection of the object feature points for any admissible nonsingular end-effector pose.
One way to get such a y_d is the "teach-by-showing" strategy (Weis et al., 1987). This approach utilizes manual positioning of the end-effector at a desired pose −neither joint position nor end-effector pose measurement is needed− and the vision system extracts the corresponding image feature points.
Let $\tilde{y} = y_d - y$ be the image feature error. The image-based control objective considered in this chapter is to achieve asymptotic matching between the actual and desired image feature points, that is

$$A_i({}^i y, {}^i x_{C_3}, q, {}^i x_O) = \left[-J_{image}({}^i y, {}^i x_{C_3})[I_3 0_{3x3}] R_R^{C^T} \quad J_{image}({}^i y, {}^i x_{C_3})[I_3 0_{3x3}] R_R^{C^T} S(R_R^O(q)\, {}^i x_O) \right] J_G(q) \tag{6}$$

$$J_{image}({}^i y, {}^i x_{C_3}) = \begin{bmatrix} -\frac{\alpha\lambda}{{}^i x_{C_3}} & 0 & \frac{{}^i u - u_0}{{}^i x_{C_3}} & \frac{[{}^i u - u_0][{}^i v - v_0]}{\alpha\lambda} & -\alpha\lambda - \frac{[{}^i u - u_0]^2}{\alpha\lambda} & [{}^i v - v_0] \\ 0 & -\frac{\alpha\lambda}{{}^i x_{C_3}} & \frac{{}^i v - v_0}{{}^i x_{C_3}} & \alpha\lambda + \frac{[{}^i v - v_0]^2}{\alpha\lambda} & -\frac{[{}^i u - u_0][{}^i v - v_0]}{\alpha\lambda} & -[{}^i u - u_0] \end{bmatrix} \tag{7}$$

$$\lim_{t \to \infty} \tilde{y}(t) = 0. \tag{8}$$

It is worth noticing that in virtue of assumption A1 about the selection of 4 coplanar but noncollinear object feature points and Lemma 2, the end-effector poses s that yield $\tilde{y} = 0$ are isolated and they are nonsingular. On the other hand, this conclusion and Lemma 1 allow to affirm that the corresponding joint configurations q achieving such poses is also isolated and nonsingular. This is summarized in the following
Lemma 3 The joint configuration solutions $q \in \mathrm{C}_{NS}$ of

$$\tilde{y}(s(q)) = 0 \tag{9}$$

are isolated.
For convenience, define the nonempty set Q as the collection of such points −isolated solutions−, that is,

$$\mathrm{Q} = \{ q \in \mathrm{C}_{NS} : \tilde{y}(s(q)) = 0 \}.$$

Therefore, the following equivalence arises straightforwardly

$$\tilde{y}(s(q)) = 0 \Leftrightarrow q \in \mathrm{Q}. \tag{10}$$

4. Transpose Jacobian Control

In order to solve the control problem stated above, we have focus our selves on the so-called Transpose Jacobian (TJ) control schemes originally introduced by (Takegaki & Arimoto, 1981). Such controllers have the advantage of non-requiring the robot's inverse kinematics

nor the inverse Jacobian matrix. Motivated by (Kelly, 1996), the objective of this section is to propose a family of TJ controllers which be able to attain the control objective (8).
For the sake of notation, define the *extended* Jacobian $J(q)$ as

$$J(q) = \begin{bmatrix} A_1(q) \\ A_2(q) \\ A_3(q) \\ A_4(q) \end{bmatrix} \in \Re^{8\times n}$$

where with abuse of notation $A_i(q) = A_i({}^i y, {}^i x_{C_3}, q, {}^i x_O)$.

On the other hand, notice that the *extended* differential imaging model can be written as

$$\dot{y} = J(q)\dot{q}. \tag{11}$$

Inspired from (Kelly, 1996), (Takegaki & Arimoto, 1981), (Miyazaki & Arimoto, 1985), (Kelly, 1999), we propose the following class of transpose Jacobian-based control laws

$$\tau = J(q)^T \nabla_{\tilde{y}} \mathsf{U}(\tilde{y}) - K_v \dot{q} + g(q), \tag{12}$$

where $\mathsf{U}(\tilde{y})$ is a continuously differentiable positive definite function called *artificial potential energy*, and $K_v \in \Re^{n\times n}$ is a symmetric positive definite matrix.

We digress momentarily to establish the following result whose proof is in the appendix.
Lemma 4. Consider the equation

$$J(q)^T \nabla_{\tilde{y}} \mathsf{U}(\tilde{y}(q)) = 0.$$

Then, each $q^* \in \mathsf{Q}$ is an isolated solution, that is there exists $\delta > 0$ such that in $\|q^* - q\| < \delta$

$$J(q)^T \nabla_{\tilde{y}} \mathsf{U}(\tilde{y}(q)) = 0 \Leftrightarrow q = q^*.$$

The main result of this chapter is stated in the following
Proposition 1. Consider the robot camera-system (1)-(5) and the family of transpose Jacobian control schemes (12). Then, the closed-loop control system satisfies the control objective (8) locally.
Proof: The closed-loop system analysis is carried out by invoking the Krasovskii–LaSalle's theorem. The theorem statement is presented in (Vidyasagar, 1993) ; this allows to study asymptotic stability of autonomous systems.
The closed-loop system dynamics is obtained by substituting the control law (12) into the robot dynamics (1). This leads to the autonomous system

$$M(q)\ddot{q} + C(q,\dot{q})\dot{q} = J(q)^T \nabla_{\tilde{y}} \mathsf{U}(\tilde{y}) - K_v \dot{q} \tag{13}$$

which can be written as

$$\frac{d}{dt}\begin{bmatrix} q \\ \dot{q} \end{bmatrix} = \begin{bmatrix} \dot{q} \\ M(q)^{-1}\left[J(q)^T \nabla_{\tilde{y}} \mathsf{U}(\tilde{y}(q)) - K_v \dot{q} - C(q,\dot{q})\dot{q}\right] \end{bmatrix}.$$

Since the artificial potential energy $\mathsf{U}(\tilde{y})$ was assumed to be positive definite, then it has an isolated minimum point at $\tilde{y}=0$, so its gradient $\nabla_{\tilde{y}}\mathsf{U}(\tilde{y})$ has an isolated critical point at $\tilde{y}=0$. This implies that it vanishes at the isolated point $\tilde{y}=0$.

Therefore, in virtue of (10) we have that $[q^T \ \dot{q}^T]^T = \left[q^T \ 0^T\right]^T$ with $q \in \mathsf{Q}$ are equilibria of the closed-loop system. Furthermore, because the elements of Q are isolated, then the corresponding equilibria are isolated too. However, other equilibria may be possible for $q \notin \mathsf{Q}$ whether $J(q)^T \nabla_{\tilde{y}} \mathsf{U}(\tilde{y}(q)) = 0$.

Let q^* be any point in Q and define a region D around the isolated equilibrium $[q^T \ \dot{q}^T]^T = [q^{*T} \ 0^T]^T$ as

$$D = \left\{ \begin{bmatrix} q \\ \dot{q} \end{bmatrix} \in \mathsf{C} \times \Re^n : \left\| \begin{matrix} q^* - q \\ \dot{q} \end{matrix} \right\| < \varepsilon \right\}$$

where $\varepsilon > 0$ is sufficiently small constant such that a unique equilibrium lie in D, and in the region D the artifical potential energy $\mathsf{U}(\tilde{y}(q))$ be positive definite with respect to $q^* - q$.

As a consequence, in the set $\left\{q : \left\| q^* - q \right\| < \varepsilon \right\}$, the unique solution of $J(q)^T \nabla_{\tilde{y}} \mathsf{U}(\tilde{y}(q)) = 0$ is $q = q^*$.

To carry out the stability analysis of the equilibria $[q^T \ \dot{q}^T]^T = [q^{*T} \ 0^T]^T$, consider the following Lyapunov function candidate

$$V(q^* - q, \dot{q}) = \frac{1}{2} \dot{q}^T M(q) \dot{q} + \mathsf{U}(\tilde{y}(q))$$

which is a positive definite function in D.

The time-derivative of $V(q, \dot{q})$ along the trajectories of the system (13), is

$$\dot{V}(q, \dot{q}) = -\dot{q}^T K_v \dot{q}$$

where $\dot{\tilde{y}} = -J\dot{q}$ and the property $\dot{q}^T \left[\frac{1}{2} \dot{M}(q) - C(q, \dot{q}) \right] \dot{q} = 0$ have been used.

Since by design matrix K_v is negative definite, hence $\dot{V}(q, \dot{q})$ is negative semidefinite function in D. In agreement with the Lyapunov's direct method, this is sufficient to guarantee that the equilibria $[q^T \ \dot{q}^T]^T = [q^{*T} \ 0^T]^T$ with $q^* \in \mathsf{Q}$ are stable.

Furthermore, the autonomous nature of the closed-loop system (13) allows the use of the Krasovskii–LaSalle's theorem (Vidyasagar, 1993) to study asymptotic stability. Accordingly, define the set

$$\begin{aligned} \Omega &= \left\{ \begin{bmatrix} q \\ \dot{q} \end{bmatrix} \in D : \dot{V}(q, \dot{q}) = 0 \right\}, \\ &= \left\{ q \in \mathsf{C} : \left\| q^* - q \right\| < \varepsilon ; \dot{q} = 0 \right\} \end{aligned}$$

The next step is to find the largest invariant set in Ω. Since in Ω we have $\dot{q} = 0$, therefore from the closed-loop system (13) it remains to determine the constant solutions q satisfying

$\|q^* - q\| < \varepsilon$ and $J(q)^T \nabla_{\tilde{y}} \mathsf{U}(\tilde{y}(q)) = 0$. As already shown in Lemma 4, if the region D is enough small, then the unique solution is $q = q^*$.
Since ε was selected such that no equilibria exists in D other than
Following arguments in (Kelly, 1999), we utilize the fact that $\mathsf{U}(\tilde{y})$ is a nonnegative function with respect to $\tilde{y}$. As a consequence of (9) we have the important conclusion that $\mathsf{U}(\tilde{y}(q))$ has minimum points if and only if $q \in \mathsf{Q}$.
This proves that the equilibria $[q^T \ \dot{q}^T]^T = [q^{*T} \ 0^T]^T$ with $q^* \in \mathsf{Q}$ are locally asymptotically stable, so they are attractive. As a consequence, $\lim_{t\to\infty} q(t) = q^*$, thus invoking (10) the control objective (8) is achieved. This concludes the proof.
Remark 1. So far, one objective of this chapter has been to make an extension of the existing results on transpose Jacobian controllers to the case of a 3D position and orientation. It is well known that a drawback of such control strategies is that they induce a steady-state error under uncertain gravitational torques. We propose the following control law

$$\tau = J^T \left[K_P \tilde{y} + K_I \int_0^t \tilde{y}(\sigma)\, d\sigma \right] - K_v \dot{q} + \hat{g}(q) \tag{14}$$

where K_I is a diagonal positive definite matrix of suitable dimensions and $\hat{g}(q)$ is the available estimate of the gravitational torques. It can be shown that the use of low integral gains in control law (14) guarantees the asymptotic stability of closed-loop system and the accomplishment of the control objective (8) in spite of uncertain gravitational torques. Furthermore, notice that if in the control law (14), $\hat{g}(q) = 0$ is chosen, the controller becomes a class of PID controller with a PI visual action and joint-coordinate damping. It can be shown also that the region of attraction of the controller (12) increases as the integral gain $K_I = \varepsilon \bar{K}_I$ decreases, being the main limitation to enlarge this region, the singularities of the Jacobian matrix J.
Remark 2. In practice, the Jacobian matrix J can be uncertain due to uncertain manipulator kinematics, camera position and orientation, and camera intrinsic parameters. In such a case, only an estimate $\hat{J}$ is available. It can be shown that by following the analysis methodology due to (Deng et al., 2002), (Cheah et al., 2003), (Cheah & Zhao, 2004) one can conclude that the feedback control law (12) (alternately (14)) can tolerate small deviation of the Jacobian matrix in the sense that

$$\|J - \hat{J}\| \le \rho$$

with ρ sufficiently small and $\|\cdot\|$ the induced matrix norm, in a neighborhood of the operating point. In the following section, we will show experimental runs to illustrate the performance of the controller for this case.
Remark 3. In the case of redundant robots $n > 6$, it may not exist a unique equilibrium point in joint space even if the control objective (8) is attained. In such case a space of dimension $n-6$ may be arbitrarily assigned. In that case, asymptotic stability to a invariant set (instead to a point) can only be concluded. This means that phenomena such as self-motion may be present.

Figure 2. Direct--drive spherical wrist

5. Experiments

This section is devoted to show the experimental evaluation of the proposed control scheme. The experimental setup is composed by a vision system and a direct-drive nonlinear mechanism.

Image acquisition and processing was performed using a Pentium based computer working at 450 MHz under Linux. The image acquisition was performed using a Panasonic GP-MF502 camera model coupled to a TV acquisition board. Image processing consisted of image thresholding and detection of an object position through the determination of its centroid. As shown in Figure 3, the four object feature points are on a plane attached at the end-effector. Their position with respect to the end-effector frame is

$$\begin{aligned} {}^1x_O &= [-0.031\ 0.032\ 0]^T[\text{m}], \\ {}^2x_O &= [0.036\ 0.037\ 0]^T[\text{m}], \\ {}^3x_O &= [-0.034\ -0.033\ 0]^T[\text{m}], \\ {}^4x_O &= [0.032\ -0.035\ 0]^T[\text{m}]. \end{aligned}$$

Figure 3 also shows four marks `+' corresponding to the desired image features obtained following the "teach-by-showing" strategy; they are given by

$$\begin{aligned} {}^1y_d &= [364\ 197]^T[pixels], \\ {}^2y_d &= [353\ 157]^T[pixels], \\ {}^3y_d &= [403\ 208]^T[pixels], \\ {}^4y_d &= [394\ 171]^T[pixels]. \end{aligned}$$

The capture of the image and the extracted information are updated at the standard rate of 30 frames per second. The centroid data is transmitted to the mechanism control computer via the standard serial port.

The camera intrinsic parameters are the following: $\lambda = 0.008$ [m], $\alpha = 72000$ [pixels/m], $u_0 = 320$ [pixels], $v_0 = 240$ [pixels] y $O_I = 0$. The camera pose during the experimental tests was

$$O_R^C = [0\ \ 0\ \ 1.32]^T[\text{m}],$$

for its position and the following rotation matrix for the orientation

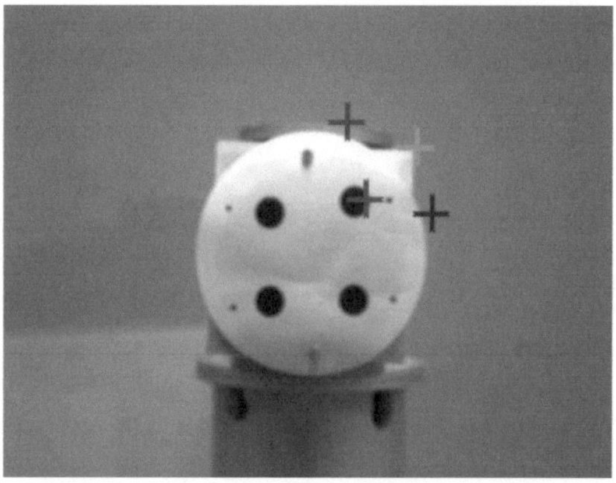

Figure 3. Initial configuration `●' and desired image feature points `+'

$$R_R^C = [1000 - 1000 - 1].$$

The robot manipulators is a direct-drive 3 degrees of freedom spherical wrist shown in Figure 2. This direct-drive wrist was built at CICESE Research Center and it is equipped with joint position sensors, motor drivers, a host computer and software environment which generates a user-friendly interface. Further information regarding the robot wrist can be found in **Error! Reference source not found.**. The geometric Jacobian is given by

$$J_G = \begin{bmatrix} -ds_1s_2 & dc_1c_2 & 0 \\ dc_1s_2 & ds_1c_2 & 0 \\ 0 & -ds_2 & 0 \\ 0 & -s_1 & c_1s_2 \\ 0 & c_1 & s_1s_2 \\ 1 & 0 & c_2 \end{bmatrix}$$

which has singular configurations at $q_2 = n\pi$.

A. PD transpose Jacobian control

Consider the following global positive definite artificial potential energy function

$$\mathrm{U}(\tilde{y}) = \frac{1}{2}\tilde{y}^T K_p \tilde{y}$$

where $K_p \in \Re^8 \times 8$ is a symmetric positive definite matrix. Since its gradient is $\nabla_y \mathrm{U}(\tilde{y}) = K_p \tilde{y}$, so the control law (12) becomes the PD transpose Jacobian control

$$\tau = J(q)^T K_p \tilde{q} - K_v \dot{q} + g(q) \tag{15}$$

where $K_v \in \Re^{n \times n}$ is a symmetric positive definite matrix.

The controller parameters were chosen by a tedious trail and error procedure in such a way to obtain the smaller steady state imaging error by keeping two practical constraints: maintaining the control actions within the actuator torque capabilities and avoiding large velocities to allow the vision system compute the centroids at the specified rate. The experiments were conducted with $K_p = diag\{5,5,5,1,1,3,5,5\} \times 10^{-4}$ and $K_v = diag\{0.9,0.7,0.17\}$.

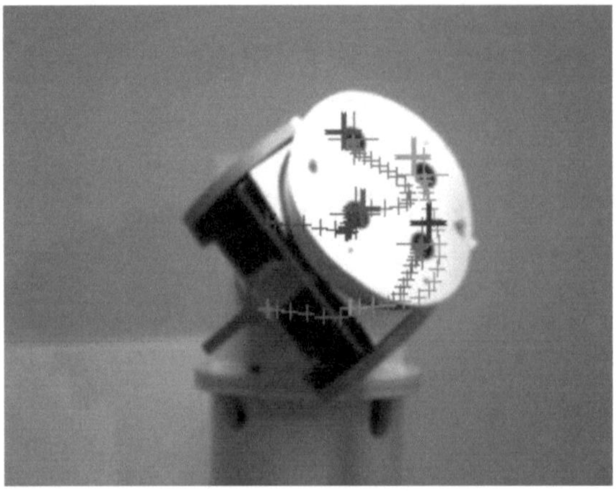

Figure 4. Trace of the image feature points: PD transpose Jacobian control

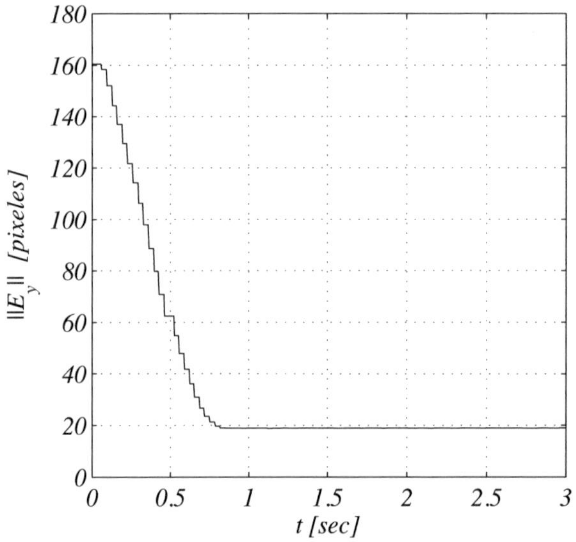

Figure 5. Norm of the image feature error: PD transpose Jacobian control

B. Tanh-D transpose Jacobian control

To deal with friction at the robot joint, let us consider the following artificial potential energy function (Kelly, 1999), (Kelly et al., 1996)

$$\mathrm{U}(\tilde{y}) = \sum_{i=1}^{8} \frac{k_{pi}}{\lambda_i} \ln\{cosh(\lambda_i \tilde{y}_i)\}$$

where k_{pi} and λ_i are positive constants. It can be shown that this is a globally positive definite function whose gradient is $\nabla_{\tilde{y}} \mathrm{U}(\tilde{y}) = K_p tanh(\Lambda \tilde{y})$ where $K_p = diag\{k_{p1},\cdots,k_{p8}\}$, $\Lambda = diag\{\lambda_1,\cdots,\lambda_8\} \in \Re^{8\times 8}$ are diagonal positive definite matrices. For a given vector $x \in \Re^8$, function $tanh(x)$ is defined as

$$tanh(x) = \begin{bmatrix} tanh(x_1) \\ \vdots \\ tanh(x_8) \end{bmatrix}$$

where $tanh(\cdot)$ is the hyperbolic tangent function.

With the above choice of the artificial potential energy, the control law (12) produces the Tanh-D transpose Jacobian control

$$\tau = J^T K_p tanh(\Lambda \tilde{y}) - K_v \dot{q} + g(q) \tag{16}$$

where $K_v \in \Re^{n \times n}$ is a symmetric positive definite matrix.

The first term on the right hand side of control law (16) offers more flexibility than the simpler $J^T K_p \tilde{y}$ in (15) to deal with practical aspects such as friction at the robot joints and torque limit in the robot actuators (Kelly, 1999).

The experiments were carried out with the following controller parameters $K_p = 0.015I$, $K_v = diag\{1.0, 0.7, 0.3\}$ and $\Lambda = 0.2I$.

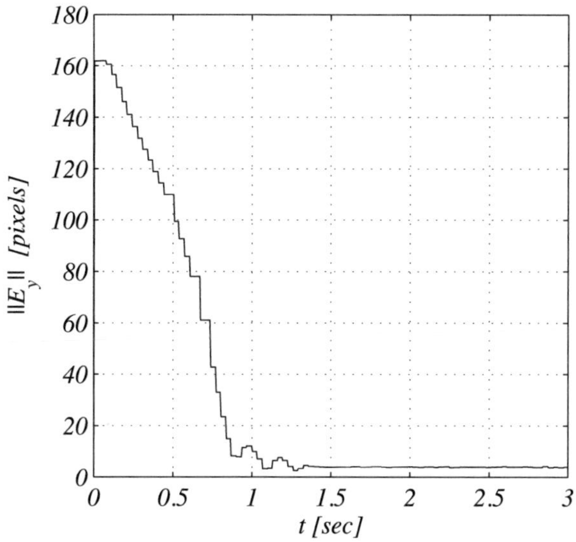

Figure 6. Norm of image feature error: Tanh-D transpose Jacobian control

Figure 6 depicts the norm of the image feature error $\tilde{y}$. Notice that the norm of the error has a smooth evolution with no overshoot reaching a steady state value of 4.0 pixels in about 1.5 seconds. Although it is expected to have asymptotically a zero error, in practice a small remaining error is present. This behavior is mainly due to uncertainties in the camera pose measurement.

C. Tanh-D approximate transpose Jacobian control

The experiment conducted with control laws (15) and (16) used complete knowledge of the experimental setup, then all the information was available to compute the extended Jacobian J. More specifically, the evolution of the deep associated at each object feature point was computed on-line from (3) thanks to the use of the robot kinematics and measurement of joint positions q for ${}^i x_R$.

In the more realistic case where uncertainties are present in the camera calibration, then it is no anymore possible to invoke (3) for computing the deep ${}^i x_{C3}$ of the object feature points. This implies that the extended Jacobian J cannot be computed exactly. Notwithstanding, as pointed out previously, it is still possible to preserve asymptotic stability in case of an approximate Jacobian $\hat{J}$ be utilized in lieu of the exact one.

The experiment was carried out by utilizing a constant deep ${}^i x_{C3} = 0.828$ [m] for the four object feature points $i = 1,2,3,4$. This yields an approximated Jacobian $\hat{J}$ to be plugged into the control law (16)

$$\tau = \hat{J}^T K_p tanh(\Lambda \tilde{y}) - K_v \dot{q} + g(q).$$

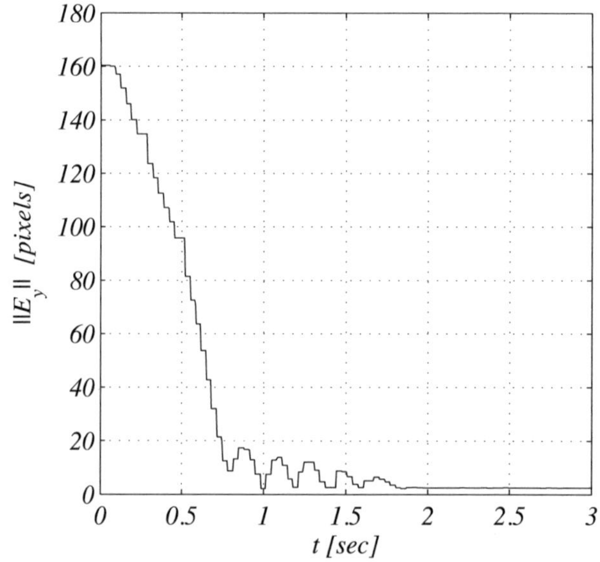

Figure 7. Norm of image feature error: Approximate Jacobian

The time evolution of the norm of the image error vector is depicted in Figure 7. It can be observed that although the response has a small oscillatory transient, it reaches an steady state error of 2.5 pixels around 2 seconds. This error is smaller than the obtained during experiments utilizing the exact Jacobian.

6. Conclusions

In this chapter we have provided some insights on the stability of robotic systems with visual information for the case of a monocular fixed-eye configuration on n DOF robot manipulators. General equations describing the fixed-camera imaging model are given and a family of transpose Jacobian control schemes is introduced to solve the regulation problem. Conditions are provided to show that such controllers guarantee asymptotic stability of the closed-loop system.

In order to illustrate the effectiveness of the proposed control schemes, experiments were conducted on a nonlinear direct-drive mechanical wrist. Both, the case of exact Jacobian as well as approximate Jacobian were evaluated. Basically they present similar performance, although the steady state image error was smaller when the approximate Jacobian was tested.

7. Appendix

Proof of Lemma 4. According to (10) we have that $q \in \mathsf{Q}$ are isolated solutions of $\tilde{y}(q) = 0$. Denote by q^* any element of Q. Since by design the artificial potential energy $\mathsf{U}(\tilde{y})$ is definite positive, then $\mathsf{U}(\tilde{y}(q))$ is definite positive with respect to $q^* - q$. This implies that $\mathsf{U}(\tilde{y}(q))$ has an isolated minimum point at $q = q^*$. Hence, its gradient with respect to q has an isolated critical point at $q = q^*$, i.e. there is $\delta > 0$ such that in $\|q^* - q\| < \delta$

$$\nabla_q \mathsf{U}(\tilde{y}(q)) = 0 \Leftrightarrow q = q^*.$$

Finally, the desired result follows from

$$\nabla_q \mathsf{U}(\tilde{y}(q)) = -J(q)^T \nabla_{\tilde{y}} \mathsf{U}(\tilde{y}(q)).$$

8. References

Hutchinson, S., Hager, G. and Corke, P. (1996), A tutorial on visual servoing. *IEEE Transactions on Robotics and Automation*, pp. 651-670, Vol. 12, No.5, 1996.

Corke, P. (1996), *Visual control of robots*, Research Studies Press Ltd.

Espiau, B., Chaumette, F. and Rives, P. (1992),A new approach to visual servoing in robotics, *IEEE Trans. on Robotics and Automation*, Vol. 8, No. 3, pp. 313-326, june 1996.

Kelly, R. (1996), Robust asymptotically stable visual servoing of planar robots. *IEEE Trans. on Robotics and Automation*, Vol. 12, pp. 759-766, October 1996 .

Maruyama, A. & Fujita, M. (1998), Robust control for planar manipulators with image feature parameter potential, *Advanced Robotics*, Vol. 12, No. 1, pp. 67-80.

Shen, Y., Xiang, G., Liu, Y. & Li, K. (2002), Uncalibrated visual servoing of planar robots, *Proc. IEEE Int. Conf. on Robotics and Automation*, Washington, DC, pp. 580-585, May 2002.

Reyes, F. & Chiang, L. (2003), Image--to--space path planning for a SCARA manipulator with single color camera, *Robotica*, Vol. 21, pp. 245-254, 2003.

***A. Behal, P. Setlur, W. Dixon and D. E. Dawson (2005), Adaptive position and orientatio regulation for the camera-in-hand problem, *Journal of Robotic Systems*, vol. 22, No. 9, pp. 457--473, September 2005.

Lamiroy, B., Puget, C. & Horaud, R. (2000), What metric stereo can do for visual servoing, *Proc. Int. Conf. Intell. Robot. Syst.*, pp. 251-256, 2000.

Hashimoto, K. & Noritsugu, T. (1998), Performance and sensitivity in visual servoing, *Proc. of the Int. Conf. on Robotics and Automation*, Leuven, Belgium, pp. 2321-2326, 1998.

Fang, C-J & Lin, S-K, (2001) A performance criterion for the depth estimation of a robot visual control system, *Proc. Int Conf. Robot. Autom*, Seoul, Korea, pp.1201-1206, 2001.

Nakavo, Y., Ishii, I. & Ishikawa, M. (2002), 3D tracking using high-speed vision systems, *Proc. Int Conf. Intell.Robot. Syst.*, Lausanne, Switzerland, pp.360-365, 2002.

Chen, J., Dawson , D. M., Dixon, W. E. & Behal, A. (2005), Adaptive Homography-Based Visual Servo Tracking for Fixed and Camera-in-Hand Configurations, *IEEE Transactions on Control Systems Technology*, Vol. 13, No. 5, pp. 814-825, 2005.

Yuan ,J., (1989) A general photogrammetric method for determining object position and orientation, *IEEE Transaction on Robotics and Automation*, Vol.5, No. 2, pp. 129-142, 1989.

Sim, T. P., Hong, G. S. & Lim, K. B. (2002), A pragmatic 3D visual servoing system *Proc. Int Conf. Robot. Autom*, Washington, D.C., USA, pp.4185-4190, 2002.

Hager, G. D. (1997) , A modular system for robust positioning using feedback from stereo vision, *IEEE Trans. on Robotics and Automation*, Vol. 13, No. 4, pp. 582-595, August 1997.

Sciavicco, L. and Siciliano, B. (1996), *Modeling and Control of Robot Manipulators*, McGraw-Hill, 1996.

Seng, J., Chen, Y. C., & O'Neil, K.A. (1997), On the existence and the manipulability recovery rate of self--motion at manipulator singularities, *International Journal of Robotics Research*, Vol. 16, No. 2, pp. 171-184, April 1997.

Weiss, L. E. , Sanderson, A. C. & Neuman, C. P. (1987), Dynamic sensor--based control of robots with visual feedback, *IEEE J. of Robotics and Automation*, Vol. 3, pp. 404-417, September 1987.

Takegaki, M., & Arimoto, S. (1981), A new feedback method for dynamic control of manipulators, *Trans. of ASME, Journal of Dynamic Systems, Measurement and Control*, vol. 102, No. 6, pp. 119-125, June 1981.

Miyazaki, F., & Arimoto, S. (1985), Sensory feedback for robot manipulators, *Journal of Robotic Systems*, Vol. 2, No. 1, pp. 53-71, 1985.

Kelly, R. (1999), Regulation of manipulators in generic task space: An energy shaping plus damping injection approach, *IEEE Trans. on Robotics and Automation*, Vol. 15, No. 2, pp. 381-386, April 1999.

Vidyasagar, M. (1993), *Nonlinear systems analysis*, Prentice Hall, 1993.

Deng, L. , Janabi-Sharifi F. & Wilson, W. (2002), Stability and robustness of visual servoing methods, *Proc. of the 2002 IEEE International Conference on Robotics and Automation*, Washington, DC, pp. 1604--1609, 11-15 May 2002.

Cheah, C.C., Hirano, M., Kawamura, S. & Arimoto, S. (2003) Approximate Jacobian Control for Robots with Uncertain Kinematics and Dynamics, *IEEE Trans. on Robotics and Automation*, Vol. 19, No. 4, pp 692- 702, 2003.

Cheah, C. & Zhao, Y. (2004), Inverse Jacobian regulator for robot manipulators: Theory and experiments, *43rd IEEE Conference on Decision and Control*, Paradise Island, Bahamas, pp. 1252-1257, December 2004.

Campa R., Kelly, R., & Santibañez, V. (2004), Windows--based real--time control of direct--drive mechanisms: Platform description and experiments, *Mechatronics*, Vol. 14, No. 9, pp. 1021-1036, November 2004 .

Kelly, R., Shirkey, P., Spong, M. W. (1996), Fixed--camera visual servo control for planar robots, *Proc. IEEE Int. Conf. on Robotics and Automation*, pp. 2643-2649, Minneapolis, Minnesota, April 1996.

14

Novel Framework of Robot Force Control Using Reinforcement Learning

Byungchan Kim[1] and Shinsuk Park[2]
[1]*Center for Cognitive Robotics Research, Korea Institute of Science and Technology*
[2]*Department of Mechanical Engineering, Korea University*
Korea

1. Introduction

Over the past decades, robotic technologies have advanced remarkably and have been proven to be successful, especially in the field of manufacturing. In manufacturing, conventional position-controlled robots perform simple repeated tasks in static environments. In recent years, there are increasing needs for robot systems in many areas that involve physical contacts with human-populated environments. Conventional robotic systems, however, have been ineffective in contact tasks. Contrary to robots, humans cope with the problems with dynamic environments by the aid of excellent adaptation and learning ability. In this sense, robot force control strategy inspired by human motor control would be a promising approach.

There have been several studies on biologically-inspired motor learning. Cohen et al. suggested impedance learning strategy in a contact task by using associative search network (Cohen et al., 1991). They applied this approach to wall-following task. Another study on motor learning investigated a motor learning method for a musculoskeletal arm model in free space motion using reinforcement learning (Izawa et al., 2002). These studies, however, are limited to rather simple problems. In other studies, artificial neural network models were used for impedance learning in contact tasks (Jung et al., 2001; Tsuji et al., 2004). One of the noticeable works by Tsuji et al. suggested on-line virtual impedance learning method by exploiting visual information. Despite of its usefulness, however, neural network-based learning involves heavy computational load and may lead to local optimum solutions easily.

The purpose of this study is to present a novel framework of force control for robotic contact tasks. To develop appropriate motor skills for various contact tasks, this study employs the following methodologies. First, our robot control strategy employs impedance control based on a human motor control theory - the equilibrium point control model. The equilibrium point control model suggests that the central nervous system utilizes the *spring*-like property of the neuromuscular system in coordinating multi-DOF human limb movements (Flash, 1987). Under the equilibrium point control scheme, force can be controlled separately by a series of equilibrium points and modulated stiffness (or more generally impedance) at the joints, so the control scheme can become simplified considerably. Second, as the learning framework, reinforcement learning (RL) is employed to optimize the performance of contact task. RL can handle an optimization problem in an

unknown environment by making sequential decision policies that maximize external reward (Sutton et al., 1998). While RL is widely used in machine learning, it is not computationally efficient since it is basically a Monte-Carlo-based estimation method with heavy calculation burden and large variance of samples. For enhancing the learning performance, two approaches are usually employed to determine policy gradient. One approach provides the baseline for gradient estimator for reducing variance (Peters et al., 2006), and the other suggests Bayesian update rule for estimating gradient (Engel et al., 2003). This study employs the former approach for constructing the RL algorithm.
In this work, episodic natural actor-critic algorithm based on the RLS filter was implemented for RL algorithm. Episodic Natural Actor-Critic method proposed by Peters et al. is known effective in high-dimensional continuous state/action system problems and can provide optimum closest solution (Peters et al., 2005). A RLS filter is used with the Natural Actor-Critic algorithm to further reduce computational burden as in the work of Park *et al.* (Park et al., 2005). Finally, different task goals or performance indices are selected depending on the characteristics of each task. In this work, the performance indices for two contact tasks were chosen to be optimized: point-to-point movement in an unknown force field, and catching a flying ball. The performance of the tasks was tested through dynamic simulations.
This paper is organized as follows. Section 2 introduces the equilibrium point control model based impedance control methods. In section 3, we describe the details of motor skill learning based on reinforcement learning. Finally, simulation results and discussion of the results are presented.

2. Impedance control based on equilibrium point control model

Mechanical impedance of a robot arm plays an important role in the dynamic interaction between the robot arm and its environment in contact. Impedance control is a widely-adopted control method to execute robotic contact tasks by regulating its mechanical impedance which characterizes the dynamic behavior of the robot at the port of interaction with its environment. The impedance control law may be described as follows (Asada et al., 1986):

$$\boldsymbol{\tau}_{actuator} = -\mathbf{J}^T(\mathbf{q})\left[\mathbf{K}_C(\mathbf{x}-\mathbf{x}_d)+\mathbf{B}_C\dot{\mathbf{x}}\right] \tag{1}$$

Where $\boldsymbol{\tau}_{actuator}$ represents the joint torque exerted by the actuators, and the current and desired positions of the end-effector are denoted by vectors $\mathbf{x}$ and $\mathbf{x}_d$, respectively. Matrices $\mathbf{K}_C$ and $\mathbf{B}_C$ are stiffness and damping matrices in Cartesian space. This form of impedance control is analogous to the equilibrium point control, which suggests that the resulting torque by the muscles is given by the deviations of the instantaneous hand position from its corresponding equilibrium position. The equilibrium point control model proposes that the muscles and neural control circuits have "spring-like" properties, and the central nervous system may generate a series of equilibrium points for a limb, and the "spring-like" properties of the neuromuscular system will tend to drive the motion along a trajectory that follows these intermediate equilibrium postures (Park et al., 2004; Hogan, 1985). Fig. 1 illustrates the concept of the equilibrium point control. Impedance control is an extension of the equilibrium point control in the context of robotics, where robotic control is achieved by imposing the end-effector dynamic behavior described by mechanical impedance.

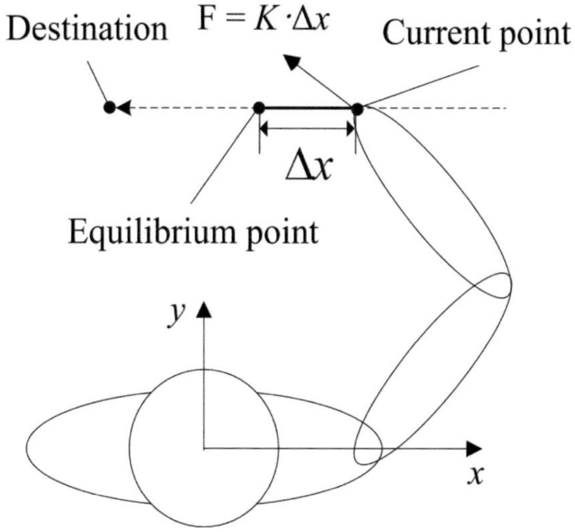

Figure 1. Conceptual model of equilibrium point control hypothesis

For impedance control of a two-link manipulator, stiffness matrix $\mathbf{K}_C$ is formed as follows:

$$\mathbf{K}_C = \begin{bmatrix} K_{11} & K_{12} \\ K_{21} & K_{22} \end{bmatrix} \tag{2}$$

Stiffness matrix $\mathbf{K}_C$ can be decomposed using singular value decomposition:

$$\mathbf{K}_C = \mathbf{V}\mathbf{\Sigma}\mathbf{V}^T \tag{3}$$

, where $\mathbf{\Sigma} = \begin{bmatrix} \lambda_1 & 0 \\ 0 & \lambda_2 \end{bmatrix}$ and $\mathbf{V} = \begin{bmatrix} \cos(\theta) & -\sin(\theta) \\ \sin(\theta) & \cos(\theta) \end{bmatrix}$

In equation (3), orthogonal matrix **V** is composed of the eigenvectors of stiffness matrix $\mathbf{K}_C$, and the diagonal elements of diagonal matrix **Σ** consists of the eigenvalues of stiffness matrix $\mathbf{K}_C$. Stiffness matrix $\mathbf{K}_C$ can be graphically represented by the *stiffness ellipse* (Lipkin et al., 1992). As shown in Fig. 2, the eigenvectors and eigenvalues of stiffness matrix correspond to the directions and lengths of principal axes of the stiffness ellipse, respectively. The characteristics of stiffness matrix $\mathbf{K}_C$ are determined by three parameters of its corresponding stiffness ellipse: the magnitude (the area of ellipse: $2\lambda_1\lambda_2$), shape (the length ratio of major and minor axes: λ_1/λ_2), and orientation (the directions of principal axes: θ). By regulating the three parameters, all the elements of stiffness matrix $\mathbf{K}_C$ can be determined.

In this study, the stiffness matrix in Cartesian space is assumed to be symmetric and positive definite. This provides a sufficient condition for static stability of the manipulator when it interacts with a passive environment (Kazerooni et al., 1986). It is also assumed that damping matrix $\mathbf{B}_C$ is approximately proportional to stiffness matrix $\mathbf{K}_C$. The ratio $\mathbf{B}_C/\mathbf{K}_C$ is chosen to be a constant of 0.05 as in the work of Won for human arm movement (Won, 1993).

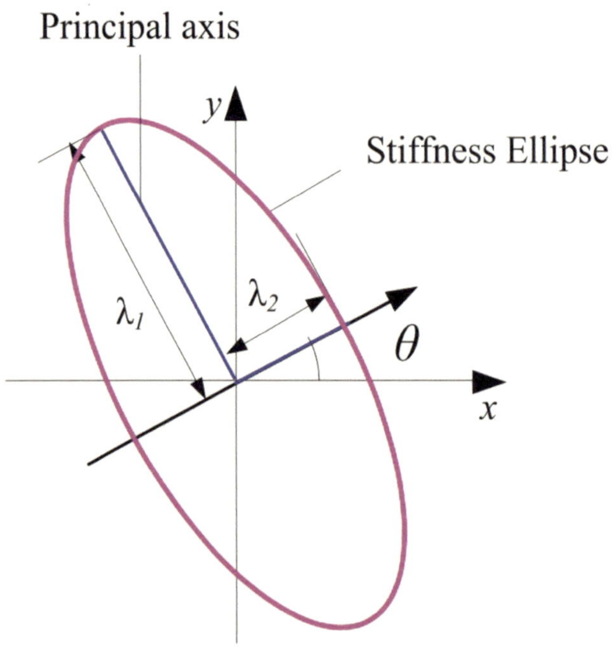

Figure 2. A graphical representation of the end-effector's stiffness in Cartesian space. The lengths λ_1 and λ_2 of principal axes and relative angle θ represent the magnitude and the orientation of the end-effectors stiffness, respectively

For trajectory planning, it is assumed that the trajectory of equilibrium point for the end-effector, which is also called the *virtual trajectory*, has a minimum- jerk velocity profile for smooth movement of the robot arm (Flash et al., 1985). The virtual trajectory is calculated from the start point x_i to the final point x_f as follows:

$$x(t) = x_i + (x_f - x_i)(10(\frac{t}{t_f})^3 - 15(\frac{t}{t_f})^4 + 6(\frac{t}{t_f})^5) \tag{4}$$

, where t is a current time and t_f is the duration of movement.

3. Motor Skill Learning Strategy

A two-link robotic manipulator for two-dimensional contact tasks was modeled as shown in Fig. 3. The robotic manipulator is controlled using the impedance control method based on the equilibrium point control hypothesis as described in Section 2. The stiffness and damping of the manipulator are modulated during a contact task, while the trajectory of equilibrium point is given for the task. The manipulator learns the impedance modulation strategy for a specific task through reinforcement learning. The state vector is composed of the joint angles and velocities at the shoulder and elbow joints. The action vector changes the three parameters of stiffness ellipse: the magnitude (the area of ellipse), shape (the length ratio of major and minor axes), and orientation (the direction of principal axes). This section describes the learning method based on reinforcement learning for controlling task impedance of the two-link manipulator in performing two-dimensional contact tasks.

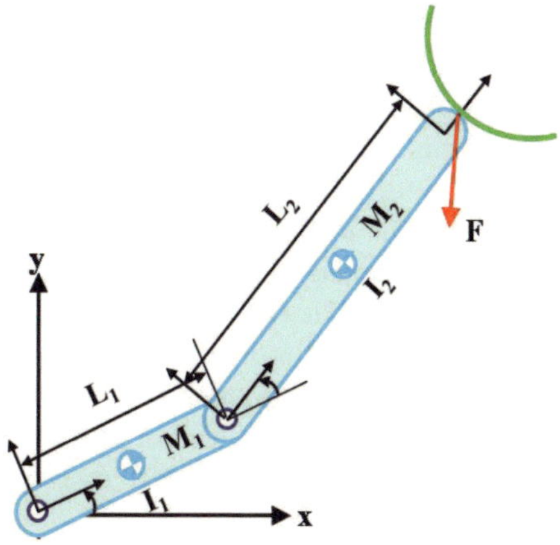

Figure 3. Two-link robotic manipulator (L_1 and L_2: Length of the link, M_1 and M_2: Mass of the link, I_1 and I_2: Inertia of the link)

3.1 Reinforcement Learning

The main components of RL are the decision maker, the agent, and the interaction with the external environment. In the interaction process, the agent selects action $\mathbf{a}_t$ and receives environmental state $\mathbf{s}_t$ and scalar-valued reward r_t as a result of the action at discrete time t. The reward r_t is a function that indicates the action performance. The agent strives to maximize reward r_t by modulating policy $\pi(\mathbf{s}_t,\mathbf{a}_t)$ that chooses the action for a given state $\mathbf{s}_t$. The RL algorithm aims to maximize the total sum of future rewards or the expected return, rather than the present reward. A discounted sum of rewards during one episode (the sequence of steps to achieve the goal from the start state) is widely used as the expected return:

$$
\begin{aligned}
R_t &= \sum_{i=0}^{T} \gamma^k \cdot r_{t+i+1} \\
V^{\pi}(\mathbf{s}) &= E_{\pi}\left\{ \sum_{i=0}^{T} \gamma^k \cdot r_{t+i+1} \middle| \mathbf{s}_t = \mathbf{s} \right\}
\end{aligned}
\tag{5}
$$

Here, γ is the discounting factor ($0 \le \gamma \le 1$), and $V^{\pi}(\mathbf{s})$ is the value function that represents an expected sum of rewards. The update rule of value function $V^{\pi}(\mathbf{s})$ is given as follows:

$$
V^{\pi}(\mathbf{s}_t) \leftarrow V^{\pi}(\mathbf{s}_t) + \alpha\left(r_t + \gamma \cdot V^{\pi}(\mathbf{s}_{t+1}) - V^{\pi}(\mathbf{s}_t)\right) \tag{6}
$$

In equation (6), the term $r_t + \gamma \cdot V^{\pi}(\mathbf{s}_{t+1}) - V^{\pi}(\mathbf{s}_t)$ is called the Temporal Difference (TD) error. The TD error indicates whether action $\mathbf{a}_t$ at state $\mathbf{s}_t$ is good or not. This updated rule is repeated to decrease the TD error so that value function $V^{\pi}(\mathbf{s}_t)$ converges to the maximum point.

3.2 RLS-based episodic Natural Actor-Critic algorithm

For many robotic problems, RL schemes are required to deal with continuous state/action space since the learning methods based on discrete space are not applicable to high dimensional systems. High-dimensional continuous state/action system problems, however, are more complicated to solve than discrete state/action space problems. While Natural Actor-Critic (NAC) algorithm is known to be an effective approach for solving continuous state/action system problems, this algorithm requires high computational burden in calculating inverse matrices. To relieve the computational burden, Park et al. suggested modified NAC algorithm combined with RLS filter. The RLS filter is used for adaptive signal filtering due to its fast convergence speed and low computational burden while the possibility of divergence is known to be rather low (Xu et al., 2002). Since this filter is designed for infinitely repeated task with no final state (*non-episodic* task), this approach is unable to deal with *episodic* tasks.

This work develops a novel NAC algorithm combined with RLS filter for *episodic* tasks. We named this algorithm the "RLS-based eNAC (*episodic* Natural Actor-Critic) algorithm." The RLS-based eNAC algorithm has two separate memory structures: the actor structure and the critic structures. The actor structure determines policies that select actions at each state, and the critic structure criticizes the selected action of the actor structure whether the action is good or not.

In the actor structure, the policy at state $\mathbf{s}_t$ in episode e is parameterized as $\pi(\mathbf{a}_t \mid \mathbf{s}_t)=p(\mathbf{a}_t \mid \mathbf{s}_t, \boldsymbol{\psi}_e)$, and policy parameter vector $\boldsymbol{\psi}_e$ is iteratively updated after finishing one episode by the following update rule:

$$\boldsymbol{\psi}_{e+1} \leftarrow \boldsymbol{\psi}_e + \alpha \nabla_{\boldsymbol{\psi}_e} J(\boldsymbol{\psi}_e) \tag{7}$$

, where $J(\boldsymbol{\psi}_e)$ is the objective function to be optimized (*value function* $V^{\pi}(s)$), and $\nabla_{\boldsymbol{\psi}_e} J(\boldsymbol{\psi}_e)$ represents the gradient of objective function $J(\boldsymbol{\psi}_e)$. Peters et al. derived the gradient of the objective function based on the natural gradient method originally proposed by Amari (Amari, 1998). They suggested a simpler update rule by introducing the natural gradient vector w_e as follows:

$$\boldsymbol{\psi}_{e+1} \leftarrow \boldsymbol{\psi}_e + \alpha \nabla_{\boldsymbol{\psi}_e} J(\boldsymbol{\psi}_e) \approx \boldsymbol{\psi}_e + \alpha \mathbf{w}_e \tag{8}$$

, where a denotes the learning rate ($0 \le a \le 1$).

In the critic structure, the least-squares (LS) TD-Q(1) algorithm is used for minimizing the TD error which represents the deviation between the expected return and the current prediction value (Boyan, 1999). By considering the Bellman equation in deriving LSTD-Q(1) algorithm, the action-value function $Q^{\pi}(\mathbf{s}, \mathbf{a})$ can be formulated as follows (Sutton et al., 1998):

$$\begin{aligned} Q^{\pi}(\mathbf{s},\mathbf{a}) &= \sum_{\mathbf{s}'} P_{\mathbf{ss}'}^{\mathbf{a}} \left[R_{\mathbf{ss}'}^{\mathbf{a}} + \gamma V^{\pi}(\mathbf{s}') \right] \\ &= E_{\pi} \left\{ r(\mathbf{s}_t) + \gamma V^{\pi}(\mathbf{s}_{t+1}) \middle| \mathbf{s}_t = \mathbf{s}, \mathbf{a}_t = \mathbf{a} \right\} \end{aligned} \tag{9}$$

Equation (9) can be approximated as

$$Q^{\pi}(\mathbf{s}, \mathbf{a}) \approx r(\mathbf{s}_t) + \gamma V(\mathbf{s}_{t+1}) \tag{10}$$

where $V(\mathbf{s}_{t+1})$ approximates value function $V^{\pi}(\mathbf{s}_{t+1})$ as iterative learning is repeated. Peters *et al.* introduced advantage value function $A^{\pi}(\mathbf{s}, \mathbf{a}) = Q^{\pi}(\mathbf{s}, \mathbf{a}) - V^{\pi}(\mathbf{s})$ and assumed that the function can be approximated using the compatible function approximation $A^{\pi}(\mathbf{s}_t, \mathbf{a}_t) = \nabla_{\psi_e} \log \pi(\mathbf{a}_t | \mathbf{s}_t)^T \mathbf{w}_e$ (Peters et al., 2003). With this assumption, we can rearrange the discounted summation of (10) for one episode trajectory with N states like below:

$$\sum_{t=0}^{N} \gamma^t A^{\pi}(\mathbf{s}_t, \mathbf{a}_t) = \sum_{t=0}^{N} \gamma^t (Q^{\pi}(\mathbf{s}_t, \mathbf{a}_t) - V^{\pi}(\mathbf{s}_t))$$

$$\sum_{t=0}^{N} \gamma^t \nabla_{\psi_e} \log \pi(\mathbf{a}_t | \mathbf{s}_t)^T \mathbf{w}_e = \sum_{t=0}^{N} \gamma^t (r(\mathbf{s}_t, \mathbf{a}_t) + \gamma V(\mathbf{s}_{t+1}) - V(\mathbf{s}_t)) \tag{11}$$

$$= \sum_{t=0}^{N} \gamma^t r(\mathbf{s}_t, \mathbf{a}_t) + \cancel{\gamma^{N+1} V(\mathbf{s}_{N+1})} - V(\mathbf{s}_0)$$

$$\sum_{t=0}^{N} \gamma^t \nabla_{\psi_e} \log \pi(\mathbf{a}_t | \mathbf{s}_t)^T \mathbf{w}_e + V(\mathbf{s}_0) = \sum_{t=0}^{N} \gamma^t r(\mathbf{s}_t, \mathbf{a}_t)$$

where the term $\gamma^{N+1} V(\mathbf{s}_{N+1})$ is negligible for its small effect. By letting $V(s_0)$ the product of 1-by-1 critic parameter vector $\mathbf{v}$ and 1-by-1 feature vector [1], we can formulate the following regression equation:

$$\sum_{t=0}^{N} \gamma^t [\nabla_{\psi_e} \log \pi(\mathbf{a}_t | \mathbf{s}_t)^T, 1] \begin{bmatrix} \mathbf{w}_e \\ v_e \end{bmatrix} = \sum_{t=0}^{N} \gamma^t r(\mathbf{s}_t, \mathbf{a}_t) \tag{12}$$

Here, natural gradient vector $\mathbf{w}_e$ is used for updating policy parameter vector $\boldsymbol{\theta}_e$ (equation (8)), and value function parameter v_e is used for approximating value function $V^{\pi}(\mathbf{s}_t)$. By letting $\tilde{\boldsymbol{\varphi}}_e = \sum_{t=0}^{N} \gamma^t [\nabla_{\psi_e} \log \pi(\mathbf{a}_t | \mathbf{s}_t)^T, 1]^T$, $\chi_e = [\mathbf{w}_e^T, v_e]^T$, and $\tilde{r} = \sum_{t=0}^{N} \gamma^t r(\mathbf{s}_t, \mathbf{a}_t)$, we can rearrange equation (12) as a least-square problem as follow:

$$\mathbf{G}_e \chi_e = \mathbf{b}_e \tag{13}$$

where $\mathbf{G}_e = \tilde{\boldsymbol{\varphi}}_e \tilde{\boldsymbol{\varphi}}_e^T$ and $\mathbf{b}_e = \tilde{\boldsymbol{\varphi}}_e \tilde{r}$. Matrix $\mathbf{G}_e$ for episode e is updated to yield the solution vector χ_e using the following update rule:

$$\begin{aligned} \mathbf{G}_e &\leftarrow \mathbf{G}_e + \delta \mathbf{I} \\ \mathbf{P}_e &= \mathbf{G}_e^{-1} \\ \chi_e &= \mathbf{P}_e \mathbf{b}_e \end{aligned} \tag{14}$$

, where δ is a positive scalar constant, and $\mathbf{I}$ is an identity matrix. By adding the term $\delta\mathbf{I}$ in (14), matrix $\mathbf{G}_e$ becomes diagonally-dominant and non-singular, and thus invertible (Strang, 1988). As the update progresses, the effect of perturbation is diminished. Equation (14) is the conventional form of critic update rule of eNAC algorithm.

The main difference between the conventional eNAC algorithm and the RLS-based eNAC algorithm is the way matrix $\mathbf{G}_e$ and solution vector χ_e are updated. The RLS-based eNAC algorithm employs the update rule of RLS filter. The key feature of RLS algorithm is to exploit $\mathbf{G}_{e-1}^{-1}$, which already exists, for the estimation of $\mathbf{G}_e^{-1}$. Rather than calculating $\mathbf{P}_e$

(inverse matrix of $\mathbf{G}_e$) using the conventional method for matrix inversion, we used the RLS filter-based update rule as follows: (Moon et al., 2000):

$$\mathbf{P}_e = \frac{1}{\beta}\left(\mathbf{P}_{e-1} - \frac{\mathbf{P}_{e-1}\tilde{\boldsymbol{\varphi}}_e\tilde{\boldsymbol{\varphi}}_e^T\mathbf{P}_{e-1}}{\beta+\tilde{\boldsymbol{\varphi}}_e^T\mathbf{P}_{e-1}\tilde{\boldsymbol{\varphi}}_e}\right)$$

$$\mathbf{k}_e = \frac{\mathbf{P}_{e-1}\tilde{\boldsymbol{\varphi}}_e}{\beta+\tilde{\boldsymbol{\varphi}}_e^T\mathbf{P}_{e-1}\tilde{\boldsymbol{\varphi}}_e} \tag{15}$$

$$\boldsymbol{\chi}_e = \boldsymbol{\chi}_{e-1} + \mathbf{k}_e(\tilde{r}_e - \boldsymbol{\varphi}_e^T\boldsymbol{\chi}_{e-1})$$

Where forgetting factor β ($0\leq\beta\leq1$) is used to accumulate the past information in a discount manner. By using the RLS-filter based update rule, the inverse matrix can be calculated without too much computational burden. This is repeated until the solution vector $\boldsymbol{\chi}_e$ converges. It should be noted, however, that matrix $\mathbf{G}_e$ should be obtained using (14) for the first episode (episode 1).

The entire procedure of the RLS-based episodic Natural Actor-Critic algorithm is summarized in Table 1.

```
Initialize each parameter vector:
  ψ = ψ_0, G = 0, P = 0, b = 0, s_t = 0
for each episode,
  Run simulator:
    for each step,
      Take action a_t, from stochastic policy π,
      then, observe next state s_{t+1}, reward r_t.
    end
  Update Critic structure:
    if first update,
      Update critic information matrices,
      following the initial update rule in (13) (14).
    else
      Update critic information matrices,
      following the recursive least-squares update rule in (15).
    end
  Update Actor structure:
    Update policy parameter vector following the rule in (8).
repeat until converge
```

Table 1. RLS-based episodic Natural Actor-Critic algorithm

3.3 Stochastic Action Selection

As discussed in section 2, the characteristics of stiffness ellipse can be changed by modulating three parameters: the magnitude, shape, and orientation. For performing contact tasks, policy π is designed to plan the change rate of the magnitude, shape, and orientation of stiffness ellipse at each state by taking actions ($\mathbf{a}_t = [a_{mag}, a_{shape}, a_{orient}]^T$). Through the sequence of an episode, policy π determines those actions. The goal of the learning algorithm is to find the trajectory of the optimal stiffness ellipse during the episode. Fig. 4 illustrates the change of stiffness ellipse corresponding to each action.

Policy π is in the form of Gaussian density function. In the critic structure, compatible function approximation $\nabla_{\psi} \log \pi(\mathbf{a}_t|\mathbf{s}_t)^T \mathbf{w}$ is derived from the stochastic policy π based on the algorithm suggested by Williams (Williams, 1992). Policy π for each component of action vector $\mathbf{a}_t$ can be described as follows:

$$\pi(a_t|\mathbf{s}_t) = N(a_t|\mu,\sigma) = \frac{1}{\sigma\sqrt{2\pi}}\exp(\frac{-(a_t-\mu)^2}{2\sigma^2}) \tag{16}$$

Since the stochastic action selection is dependent on the conditions of Gaussian density function, the action policy can be designed by controlling the mean μ and the variance σ of equation (16). These variables are defined as:

$$\mu = \xi \boldsymbol{\omega}^T \mathbf{s}_t$$
$$\sigma = \tilde{\xi}\left(0.001 + \frac{1}{1+\exp(-\eta)}\right)$$

Since the stochastic action selection is dependent on the conditions of Gaussian density function, the action policy can be designed by controlling the mean μ and the variance σ of equation (16). These variables are defined as:

$$\mu = \xi \boldsymbol{\omega}^T \mathbf{s}_t$$
$$\sigma = \tilde{\xi}\left(0.001 + \frac{1}{1+\exp(-\eta)}\right) \tag{17}$$

As can be seen in the equation, mean μ is a linear combination of the vector $\boldsymbol{\omega}$ and state vector $\mathbf{s}_t$ with mean scaling factor ξ. Variance σ is in the form of the sigmoid function with the positive scalar parameter η and variance scaling factor $\tilde{\xi}$. As a two-link manipulator model is used in this study, the components of state vector $\mathbf{s}_t$ include the joint angles and velocities of shoulder and elbow joints ($\mathbf{s}_t = [x_1, x_2, \dot{x}_1, \dot{x}_2, 1]^T$, where the fifth component of 1 is a bias factor). Therefore, natural gradient vector $\mathbf{w}$ (and thus policy parameter vector $\boldsymbol{\psi}$) is composed of 18 components as follows:

$$\mathbf{w} = [\boldsymbol{\omega}_{mag}^T, \boldsymbol{\omega}_{shape}^T, \boldsymbol{\omega}_{orient}^T, \eta_{mag}, \eta_{shape}, \eta_{orient}]^T$$

, where, $\boldsymbol{\omega}_{mag}$, $\boldsymbol{\omega}_{shape}$, and $\boldsymbol{\omega}_{orient}$ are 5-by-1 vectors and η_{mag}, η_{shape}, and η_{orient} are parameters corresponding to the three components of action vector ($\mathbf{a}_t = [a_{mag}, a_{shape}, a_{orient}]^T$).

4. Contact Task Applications

In this section, the method developed in the previous section is applied to two contact tasks: point-to-point movement in an unknown force field, and catching a flying ball. The two-link manipulator developed in Section 3 was used to perform the tasks in two-dimensional space. The dynamic simulator is constructed by MSC.ADAMS2005, and the control algorithm is implemented using Matlab/Simulink (Mathworks, Inc.).

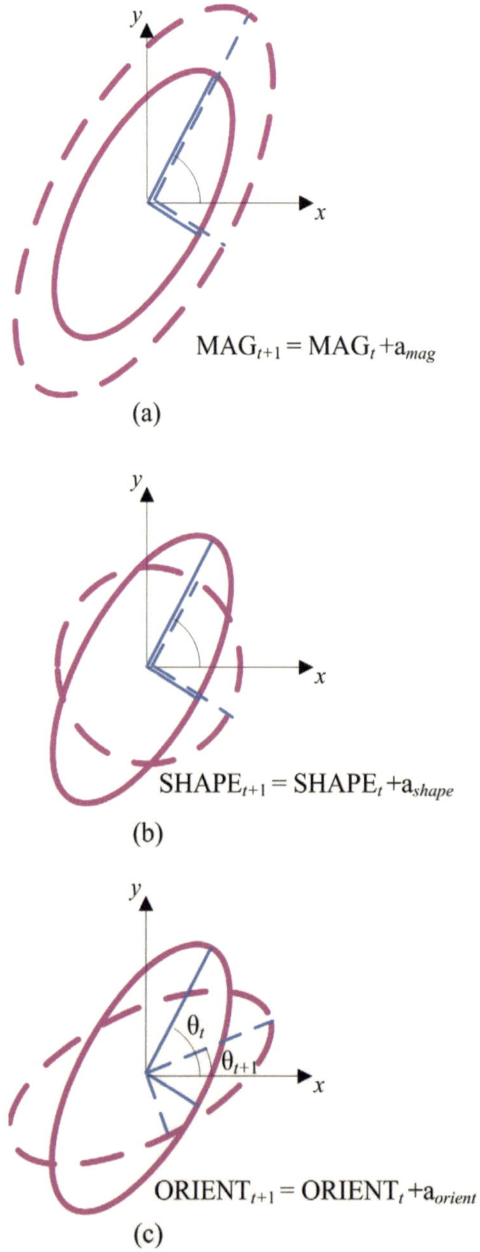

Figure 4. Some variation examples of stiffness ellipse. (a) The magnitude (area of ellipse) is changed (b) The shape (length ratio of major and minor axes) is changed (c) The orientation (direction of major axis) is changed (solid line: stiffness ellipse at time *t*, dashed line: stiffness ellipse at time *t*+1)

4.1 Point-to-Point Movement in an Unknown Force Field

In this task, the two-link robotic manipulator makes a point-to-point movement in an unknown force field. The velocity-dependent force is given as:

$$\mathbf{F}_{viscous} = \begin{bmatrix} -10 & 0 \\ 0 & -10 \end{bmatrix} \begin{Bmatrix} \dot{x} \\ \dot{y} \end{Bmatrix} \tag{18}$$

The movement of the endpoint of the manipulator is impeded by the force proportional to its velocity, as it moves to the goal position in the force field. This condition is identical to that of biomechanics study of Kang et al. (Kang et al., 2007). In their study, the subject moves ones hand from a point to another point in a velocity-dependent force field, which is same as equation (18), where the force field was applied by an special apparatus (KINARM, BKIN Technology).

For this task, the performance indices are chosen as follows:

1. The root mean square of the difference between the desired position (virtual trajectory) and the actual position of the endpoint (Δ_{rms})
2. The magnitude of the time rate of torque vector of two arm joints ($\|\dot{\boldsymbol{\tau}}\| = \sqrt{\dot{\tau}_1^{\,2} + \dot{\tau}_2^{\,2}}$)

The torque rate represents power consumption, which can also be interpreted as metabolic costs for human arm movement (Franklin et al., 2004; Uno et al., 1989). By combining the two performance indices, two different rewards are formulated for one episode as follows:

$$reward1 = \kappa_1 - \sum_{t=1}^{N} (\Delta_{rms})_t$$

$$reward2 = w_1\left(\kappa_1 - \sum_{t=1}^{N} (\Delta_{rms})_t\right) + w_2\left(\kappa_2 - \sum_{t=1}^{N} \|\dot{\boldsymbol{\tau}}\|_t\right)$$

, where w_1 and w_2 are the weighting factors, and κ_1 and κ_2 are constants. The reward is a weighted linear combination of time integrals of two performance indices.

The learning parameters were chosen as follows: α = 0.05, β = 0.99, γ = 0.99. The change limits for action are set as [-10, 10] degrees for the orientation, [-2, 2] for the major/minor ratio, and [-200π, 200π] for the area of stiffness ellipse. The initial ellipse before learning was set to be circular with the area of 2500π.

The same physical properties as in (Kang et al., 2007) were chosen for dynamic simulations (Table 2).

	Length(m)	Mass(Kg)	Inertia(Kg·m^2)
Link 1	0.11	0.20	0.0002297
Link 2	0.20	0.18	0.0006887

Table 2. Physical properties of two link arm model

Fig. 5 shows the change of stiffness ellipse trajectory before and after learning. Before learning, the endpoint of the manipulator was not even able to reach the goal position (Fig. 5 (a)). Figs. 5 (b) and (c) compare the stiffness ellipse trajectories after learning using two different rewards (*reward1* and *reward2*). As can be seen in the figures, for both rewards the major axis of stiffness ellipse was directed to the goal position to overcome resistance of viscous force field.

Fig. 6 compares the effects of two rewards on the changes of two performance indices (Δ_{rms} and $\|\dot{\tau}\|$) as learning iterates. While the choice of reward does not affect the time integral of Δ_{rms}, the time integral of $\|\dot{\tau}\|$ was suppressed considerably by using *reward2* in learning. The results of dynamic simulations are comparable with the biomechanics study of Kang et al.. The results of their study suggest that the human actively modulates the major axis toward the direction of the external force against the motion, which is in accordance with our results.

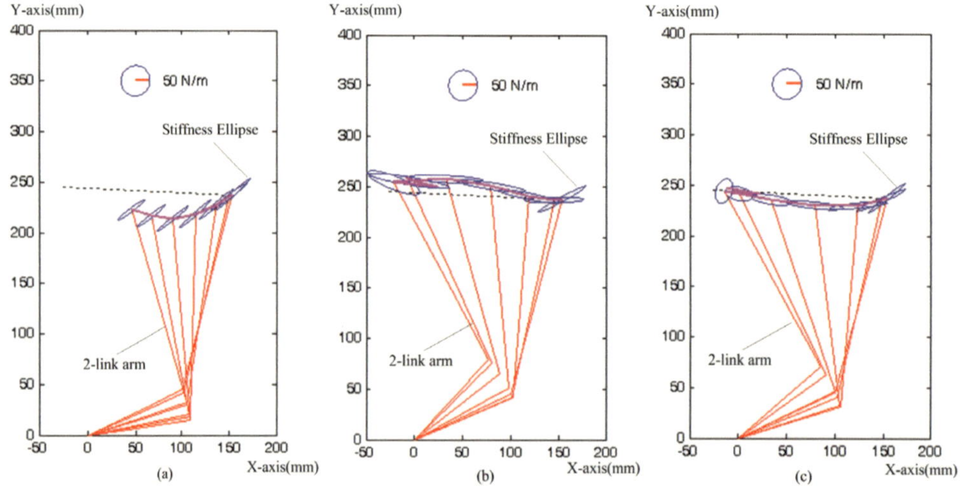

Figure 5. Stiffness ellipse trajectories. (dotted line: virtual trajectory, solid line: actual trajectory). (a) Before learning. (b) After learning (*reward1*). (c) After learning (*reward2*)

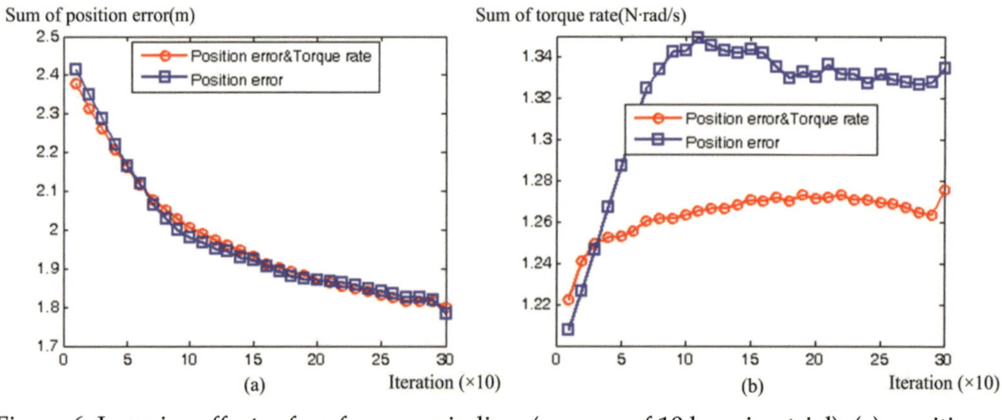

Figure 6. Learning effects of performance indices (average of 10 learning trial). (a) position error. (b) torque rate

4.2 Catching a Flying Ball

In this task, the two-link robotic arm catches a flying ball illustrated in Fig. 7. The simulation was performed using the physical properties of the arm as listed in Table 2. The main issues in ball-catching task would be how to detect the ball trajectory and how to reduce the impulsive force between the ball and the end-effector. This work focuses on the latter and assumes that the ball trajectory is known in advance.

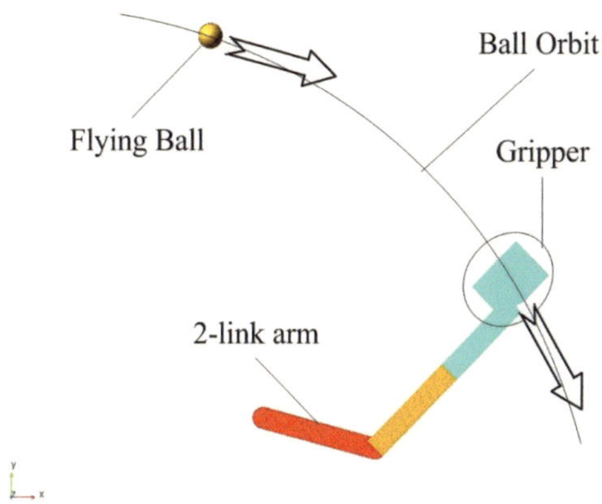

Figure 7. Catching a flying ball

When a human catches a ball, one moves ones arm backward to reduce the impulsive contact force. By considering the human ball-catching, the task is modeled as follow: A ball is thrown to the end-effector of the robot arm. The time for the ball to reach the end-effector is approximately 0.8sec. After the ball is thrown, the arm starts to move following the parabolic orbit of the flying ball. While the end-effector is moving, the ball is caught and then moves to the goal position together. The robot is set to catch the ball when the end-effector's moving at its highest speed to reduce the impulsive contact force between the ball and the end-effector. The impulsive force can also be reduced by modulating the stiffness ellipse during the contact.

The learning parameters were chosen as follows: a = 0.05, β = 0.99, γ = 0.99. The change limits for action are set as [-10, 10] degrees for the orientation, [-2, 2] for the major/minor ratio, and [-200π, 200π] for the area of stiffness ellipse. The initial ellipse before learning was set to be circular with the area of 10000 π.

For this task, the contact force is chosen as the performance index:

$$\left\|\mathbf{F}_{contact}\right\| = \sqrt{F_x^{\,2} + F_y^{\,2}}$$

The reward to be maximized is the impulse (time integral of contact force) during contact:

$$reward = \kappa - \sum_{t=1}^{N} \left\|\mathbf{F}_{contact}\right\|_t \Delta t_{\,t}$$

where κ is a constant. Fig. 8 illustrates the change of stiffness during contact after learning. As can be seen in the figure, the stiffness is tuned soft in the direction of ball trajectory, while the stiffness normal to the trajectory is much higher. Fig. 9 shows the change of the impulse as learning continues. As can be seen in the figure, the impulse was reduced considerably after learning.

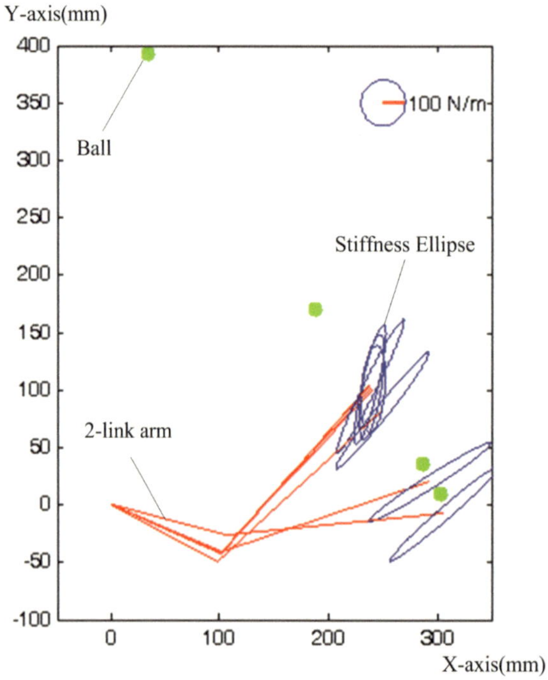

Figure 8. Catching a flying ball

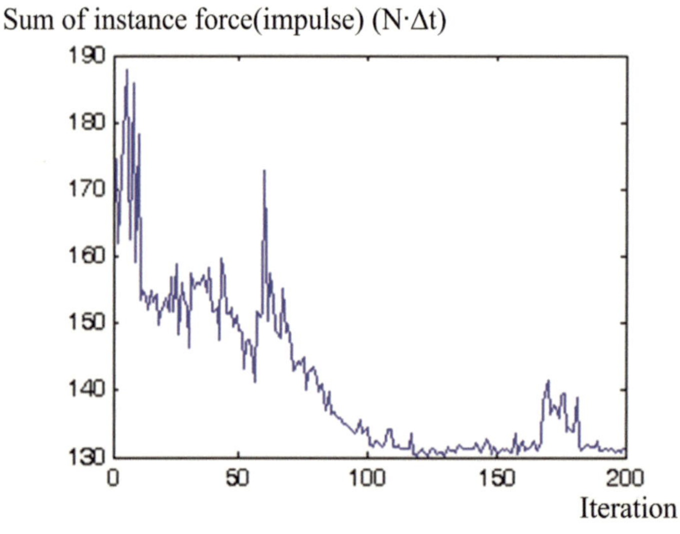

Figure 9. Catching a flying ball

5. Conclusions

Safety in robotic contact tasks has become one of the important issues as robots spread their applications to dynamic, human-populated environments. The determination of impedance control parameters for a specific contact task would be the key feature in enhancing the robot performance. This study proposes a novel motor learning framework for determining impedance parameters required for various contact tasks. As a learning framework, we employed reinforcement learning to optimize the performance of contact task. We have demonstrated that the proposed framework enhances contact tasks, such as door-opening, point-to-point movement, and ball-catching.

In our future works we will extend our method to apply it to teach a service robot that is required to perform more realistic tasks in three-dimensional space. Also, we are currently investigating a learning method to develop motor schemata that combine the internal models of contact tasks with the actor-critic algorithm developed in this study.

6. References

Amari, S. (1998). Natural gradient works efficiently in learning, *Neural Computation*, Vol. 10, No. 2, pp. 251-276, ISSN 0899-7667

Asada, H. & Slotine, J-J. E. (1986). *Robot Analysis and Control*, John Wiley & Sons, Inc., ISBN 978-0471830290

Boyan. J. (1999). Least-squares temporal difference learning, *Proceeding of the 16th International Conference on Machine Learning*, pp. 49-56

Cohen, M. & Flash, T. (1991). Learning impedance parameters for robot control using associative search network, *IEEE Transactions on Robotics and Automation*, Vol. 7, Issue. 3, pp. 382-390, ISSN 1042-296X

Engel, Y.; Mannor, S. & Meir, R. (2003). Bayes meets bellman: the gaussian process approach to temporal difference learning, *Proceeding of the 20th International Conference on Machine Learning*, pp. 154-161

Flash, T. & Hogan, N. (1985). The coordination of arm movements: an experimentally confirmed mathematical model, *Journal of Neuroscience*, Vol. 5, No. 7, pp. 1688-1703, ISSN 1529-2401

Flash, T. (1987). The control of hand equilibrium trajectories in multi-joint arm movement, *Biological Cybernetics*, Vol. 57, No. 4-5, pp. 257-274, ISSN 1432-0770

Franklin, D. W.; So, U.; Kawato, M. & Milner, T. E. (2004). Impedance control balances stability with metabolically costly muscle activation, *Journal of Neurophysiology*, Vol. 92, pp. 3097-3105, ISSN 0022-3077

Hogan, N. (1985). Impedance control: An approach to manipulation: part I. theory, part II. implementation, part III. application, *ASME Journal of Dynamic System, Measurement, and Control*, Vol. 107, No. 1, pp. 1-24, ISSN 0022-0434

Izawa, J.; Kondo, T. & Ito, K. (2002). Biological robot arm motion through reinforcement learning, *Proceedings of the IEEE International Conference on Robotics and Automation*, Vol. 4, pp. 3398-3403, ISBN 0-7803-7272-7

Jung, S.; Yim, S. B. & Hsia, T. C. (2001). Experimental studies of neural network impedance force control of robot manipulator, *Proceedings of the IEEE International Conference on Robotics and Automation*, Vol. 4, pp. 3453-3458, ISBN 0-7803-6576-3

Kang, B.; Kim, B.; Park, S. & Kim, H. (2007). Modeling of artificial neural network for the prediction of the multi-joint stiffness in dynamic condition, *Proceeding of IEEE/RSJ International Conference on Intelligent Robots and Systems*, pp. 1840-1845, ISBN 978-1-4244-0912-9, San Diego, CA, USA

Kazerooni, H.; Houpt, P. K. & Sheridan, T. B. (1986). The fundamental concepts of robust compliant motion for robot manipulators, *Proceedings of the IEEE International Conference on Robotics and Automation*, Vol. 3, pp. 418-427, MN, USA

Lipkin, H. & Patterson, T. (1992). Generalized center of compliance and stiffness, *Proceedings of the IEEE International Conference on Robotics and Automation*, Vol. 2, pp. 1251-1256, ISBN 0-8186-2720-4, Nice, France

Moon, T. K. & Stirling, W. C. (2000). *Mathematical Methods and Algorithm for Signal Processing*, Prentice Hall, Upper Saddle River, NJ, ISBN 0-201-36186-8

Park, J.; Kim, J. & Kang, D. (2005). An rls-based natural actor-critic algorithm for locomotion of a two-linked robot arm, *Proceeding of International Conference on Computational Intelligence and Security, Part I, LNAI*, Vol. 3801, pp. 65-72, ISSN 1611-3349

Park, S. & Sheridan, T. B. (2004). Enhanced human-machine interface in braking, *IEEE Transactions on Systems, Man, and Cybernetics, - Part A: Systems and Humans*, Vol. 34, No. 5, pp. 615-629, ISSN 1083-4427

Peters, J.; Vijayakumar, S. & Schaal, S. (2005). Natural actor-critic, *Proceeding of the 16th European Conference on Machine Learning, LNCS*, Vol. 3720, pp.280-291, ISSN 1611-3349

Peters J. & Schaal, S. (2006). Policy gradient methods for robotics, *Proceeding of IEEE/RSJ International Conference on Intelligent Robots and Systems*, pp. 2219-2225, ISBN 1-4244-0259-X, Beijing, China

Strang, G. (1988). *Linear Algebra and Its Applications*, Harcourt Brace & Company

Sutton, R. S. & Barto, A. G. (1998). *Reinforcement Learning: An Introduction*, MIT Press, ISBN 0-262-19398-1

Tsuji, T.; Terauchi, M. & Tanaka, Y. (2004). Online learning of virtual impedance parameters in non-contact impedance control using neural networks, *IEEE Transactions on Systems, Man, and Cybernetics, - Part B: Cybernetics*, Vol. 34, Issue 5, pp. 2112-2118, ISSN 1083-4419

Uno, Y.; Suzuki, R. & Kawato, M. (1989). Formation and control of optimal trajectory in human multi-joint arm movement: minimum torque change model, *Biological Cybernetics*, Vol. 61, No. 2, pp. 89-101, ISSN 1432-0770

Williams, R. J. (1992). Simple statistical gradient-following algorithms for connectionist reinforcement learning, *Machine Learning*, Vol. 8, No. 3-4, pp. 229-256. ISSN 1573-0565

Won, J. (1993). The Control of Constrained and Partially Constrained Arm Movement, *S. M. Thesis*, Department of Mechanical Engineering, MIT, Cambridge, MA, 1993.

Xu, X.; He, H. & Hu, D. (2002). Efficient reinforcement learning using recursive least-squares methods, *Journal of Artificial Intelligent Research*, Vol. 16, pp. 259-292

15

Link Mass Optimization Using Genetic Algorithms for Industrial Robot Manipulators

Serdar Kucuk and Zafer Bingul
Kocaeli University
Turkey

1. Introduction

Obtaining optimum energy performance is the primary design concern of any mechanical system, such as ground vehicles, gear trains, high speed electromechanical devices and especially industrial robot manipulators. The optimum energy performance of an industrial robot manipulator based on the minimum energy consumption in its joints is required for developing of optimum control algorithms (Delingette et al., 1992; Garg & Ruengcharungpong, 1992; Hirakawa & Kawamura, 1996; Lui & Wang, 2004). The minimization of individual joint torques produces the optimum energy performance of the robot manipulators. Optimum energy performance can be obtained to optimize link masses of the industrial robot manipulator. Having optimum mass and minimum joint torques are the ways of improving the energy efficiency in robot manipulators. The inverse of inertia matrix can be used locally minimizing the joint torques (Nedungadi & Kazerouinian, 1989). This approach is similar to the global kinetic energy minimization.

Several optimization techniques such as genetic algorithms (Painton & Campbell, 1995; Chen & Zalzasa, 1997; Choi et al., 1999; Pires & Machado, 1999; Garg & Kumar, 2002; Kucuk & Bingul, 2006; Qudeiri et al., 2007), neural network (Sexton & Gupta, 2000; Tang & Wang, 2002) and minimax algorithms (Pin & Culioli, 1992; Stocco et al., 1998) have been studied in robotics literature. Genetic algorithms (GAs) are superior to other optimization techniques such that genetic algorithms search over the entire population instead of a single point, use objective function instead of derivatives, deals with parameter coding instead of parameters themselves. GA has recently found increasing use in several engineering applications such as machine learning, pattern recognition and robot motion planning. It is an adaptive heuristic search algorithm based on the evolutionary ideas of natural selection and genetic. This provides a robust search procedure for solving difficult problems. In this work, GA is applied to optimize the link masses of a three link robot manipulator to obtain minimum energy. Rest of the Chapter is composed of the following sections. In Section II, genetic algorithms are explained in a detailed manner. Dynamic equations and the trajectory generation of robot manipulators are presented in Section III and Section IV, respectively. Problem definition and formulation is described in Section V. In the following Section, the rigid body dynamics of a cylindrical robot manipulator is given as example. Finally, the contribution of this study is presented in Section VII.

2. Genetic algorithms

GAs were introduced by John Holland at University of Michigan in the 1970s. The improvements in computational technology have made them attractive for optimization problems. A genetic algorithm is a non-traditional search method used in computing to find exact or approximate solutions to optimization and search problems based on the evolutionary ideas of natural selection and genetic. The basic concept of GA is designed to simulate processes in natural system necessary for evolution, specifically those that follow the principles first laid down by Charles Darwin of survival of the fittest. As such they represent an intelligent exploitation of a random search within a defined search space to solve a problem. The obtained optima are an end product containing the best elements of previous generations where the attributes of a stronger individual tend to be carried forward into following generation. The rule is survival of the fittest will. The three basic features of GAs are as follows:

i. Description of the objective function
ii. Description and implementation of the genetic representation
iii. Description and implementation of the genetic operators such as reproduction, crossover and mutation.

If these basic features are chosen properly for optimization applications; the genetic algorithm will work quite well. GA optimization possesses some unique features that separate from the other optimization techniques given as follows:

i. It requires no knowledge or gradient information about search space.
ii. It is capable of scanning a vast solution set in a short time.
iii. It searches over the entire population instead of a single point.
iv. It allows a number of solutions to be examined in a single design cycle.
v. It deals with parameter coding instead of parameters themselves.
vi. Discontinuities in search space have little effect on overall optimization performance.
vii. It is resistant to becoming trapped in local optima.

These features provide GA to be a robust and useful optimization technique over the other search techniques (Garg & Kumar, 2002). However, there are some disadvantages to use genetic algorithms.

i. Finding the exact global optimum in search space is not certain.
ii. Large numbers of fitness function evaluations are required.
iii. Configuration is not straightforward and problem dependent.

The representation or coding of the variables being optimized has a large impact on search performance, as the optimization is performed on this representation of the variables. The two most common representations, binary and real number codings, differ mainly in how the recombination and mutation operators are performed. The most suitable choice of representation is based upon the type of application. In GAs, a set of solutions represented by chromosomes is created randomly. Chromosomes used here are in binary codings. Each zero and one in chromosome corresponds to a gene. A typical chromosome can be given as

10000010

An initial population of random chromosomes is generated at the beginning. The size of the initial population may vary according to the problem difficulties under consideration. A different solution to the problem is obtained by decoding each chromosome. A small initial with composed of eight chromosomes can be denoted as the form

10110011
11100110
10110010
10001111
11110111
11101111
10110110
10111111

Note that, in practice both the size of the population and the strings are larger then those of mentioned above. Basically, the new population is generated by using the following fundamental genetic evolution processes: reproduction, crossover and mutation.

At the reproduction process, chromosomes are chosen based on natural selections. The chromosomes in the new population are selected according to their fitness values defined with respect to some specific criteria such as roulette wheel selection, rank selection or steady state selection. The fittest chromosomes have a higher probability of reproducing one or more offspring in the next generation in proportion to the value of their fitness.

At the crossover stage, two members of population exchange their genes. Crossover can be implemented in many ways such as having a single crossover point or many crossover points which are chosen randomly. A simple crossover can be implemented as follows. In the first step, the new reproduced members in the mating pool are mated randomly. In the second step, two new members are generated by swapping all characteristics from a randomly selected crossover point. A good value for crossover can be taken as 0.7. A simple crossover structure is shown below. Two chromosomes are selected according to their fitness values. The crossover point in chromosomes is selected randomly. Two chromosomes are given below as an example

10110011
11100110

After crossover process is applied, all of the bits after the crossover point are swapped. Hence, the new chromosomes take the form

101*00110
111*10011

The symbol "*" corresponds the crossover point. At the mutation process, value of a particular gene in a selected chromosome is changed from 1 to 0 or vice versa., The probability of mutation is generally kept very small so these changes are done rarely. In general, the scheme of a genetic algorithm can be summarized as follows.

i. Create an initial population.
ii. Check each chromosome to observe how well is at solving the problem and evaluate the fitness of the each chromosome based on the objective function.
iii. Choose two chromosomes from the current population using a selection method like roulette wheel selection. The chance of being selected is in proportion to the value of their fitness.
iv. If a probability of crossover is attained, crossover the bits from each chosen chromosome at a randomly chosen point according to the crossover rate.
v. If a probability of mutation is attained, implement a mutation operation according to the mutation rate.
vi. Continue until a maximum number of generations have been produced.

3. Dynamic Equations

A variety of approaches have been developed to derive the manipulator dynamics equations (Hollerbach, 1980; Luh et al., 1980; Paul, 1981; Kane and Levinson, 1983; Lee et al., 1983). The most popular among them are Lagrange-Euler (Paul, 1981) and Newton-Euler methods (Luh et al., 1980). Energy based method (LE) is used to derive the manipulator dynamics in this chapter. To obtain the dynamic equations by using the Lagrange-Euler method, one should define the homogeneous transformation matrix for each joint. Using D-H (Denavit & Hartenberg, 1955) parameters, the homogeneous transformation matrix for a single joint is expressed as,

$$ {}^{i-1}_{\ \ i}T = \begin{bmatrix} c\theta_i & -s\theta_i & 0 & a_{i-1} \\ s\theta_i c\alpha_{i-1} & c\theta_i c\alpha_{i-1} & -s\alpha_{i-1} & -s\alpha_{i-1}d_i \\ s\theta_i s\alpha_{i-1} & c\theta_i s\alpha_{i-1} & c\alpha_{i-1} & c\alpha_{i-1}d_i \\ 0 & 0 & 0 & 1 \end{bmatrix} \tag{1} $$

where a_{i-1}, α_{i-1}, d_i, θ_i, c_i and s_i are the link length, link twist, link offset, joint angle, $\cos\theta_i$ and $\sin\theta_i$, respectively. In this way, the successive transformations from base toward the end-effector are obtained by multiplying all of the matrices defined for each joint.
The difference between kinetic and potential energy produces Lagrangian function given by

$$ L(q,\dot{q}) = K(q,\dot{q}) - P(q) \tag{2} $$

where q and $\dot{q}$ represent joint position and velocities, respectively. Note that, q_i is the joint angle θ_i for revolute joints, or the distance d_i for prismatic joints. While the potential energy (P) is dependent on position only, the kinetic energy (K) is based on both position and velocity of the manipulator. The total kinetic energy of the manipulator is defined as

$$ K(q,\dot{q}) = \frac{1}{2}\sum_{i=1}^{n}(v_i)^T m_i v_i + (\omega_i)^T I_i \omega_i \tag{3} $$

where m_i is the mass of link i and I_i denotes the 3x3 inertia tensor for the center of the link i with respect to the base coordinate frame. I_i can be expressed as

$$I_i = {}^{0}_{i}R \, I_m \, {}^{0}_{i}R^T \tag{4}$$

where ${}^{0}_{i}R$ represents the rotation matrix and I_m stands for the inertia tensor of a rigid object about its center of mass . The terms v_i and ω_i refer to the linear and angular velocities of the center of mass of link i with respect to base coordinate frame, respectively.

$$v_i = A_i(q)\dot{q} \text{ and } \omega_i = B_i(q)\dot{q} \tag{5}$$

where $A_i(q)$ and $B_i(q)$ are obtained from the Jacobian matrix, $J_i(q)$. If v_i and ω_i in equations 5 are substituted in equation 3, the total kinetic energy is obtained as follows.

$$K(q,\dot{q}) = \frac{1}{2}\dot{q}^T \sum_{i=1}^{n} \left[A_i(q)^T m_i A_i(q) + B_i(q)^T I_i B_i(q)\right]\dot{q} \tag{6}$$

The equation 6 can be written in terms of manipulator mass matrix and joint velocities as

$$K(q,\dot{q}) = \frac{1}{2}\dot{q}^T M(q)\dot{q} \tag{7}$$

where M(q) denotes mass matrix given by

$$M(q) = \sum_{i=1}^{n} \left[A_i(q)^T m_i A_i(q) + B_i(q)^T I_i B_i(q)\right] \tag{8}$$

The total potential energy is determined as

$$P(q) = -\sum_{i=1}^{n} m_i g^T h_i(q) \tag{9}$$

where g and $h_i(q)$ denotes gravitational acceleration vector and the center of mass of link i relative to the base coordinate frame, respectively. As a result, the Lagrangian function can be obtained by combining the equations 7 and 9 as follows.

$$L(q,\dot{q}) = \frac{1}{2}\dot{q}^T M(q)\dot{q} + \sum_{i=1}^{n} m_i g^T h_i(q) \tag{10}$$

The equations of motion can be derived by substituting the Lagrangian function in equation 10 into the Euler-Langrange equations

$$\frac{d}{dt}\frac{\partial L(q,\dot{q})}{\partial \dot{q}} - \frac{\partial L(q,\dot{q})}{\partial q} = \tau \tag{11}$$

to create the dynamic equations with the form

$$\sum_{j=1}^{n} M_{ij}(q)\ddot{q}_j + \sum_{k=1}^{n}\sum_{j=1}^{n} C^i_{kj}(q)\dot{q}_k\dot{q}_j + G_i(q) = \tau_i \qquad 1 \le i,j,k \le n \tag{12}$$

where, τ is the nx1 generalized torque vector applied at joints, and q, $\dot{q}$ and $\ddot{q}$ are the nx1 joint position, velocity and acceleration vectors, respectively. $M(q)$ is the nxn mass matrix, $C(q,\dot{q})$ is an nx1 vector of centrifugal and Coriolis terms given by

$$C^i_{kj}(q) = \frac{\partial}{\partial q_k} M_{ij}(q) - \frac{1}{2}\frac{\partial}{\partial q_i} M_{kj}(q) \tag{13}$$

$G(q)$ is an nx1 vector of gravity terms of actual mechanism expressed as

$$G_i(q) = -\sum_{k=1}^{3}\sum_{j=1}^{n} g_k m_j A^j_{ki}(q) \tag{14}$$

The friction term is omitted in equation 12. The detailed information about Lagrangian dynamic formulation can be found in text (Schilling, 1990).

4. Trajectory Generation

In general, smooth motion between initial and final positions is desired for the end-effector of a robot manipulator since jerky motions can cause vibration in the manipulator. Joint and Cartesian trajectories in robot manipulators are two common ways to generate smooth motion. In joint trajectory, initial and final positions of the end-effector are converted into joint angles by using inverse kinematics equations. A time (t) dependent smooth function is computed for each joint. All of the robot joints pass through initial and final points at the same time. Several smooth functions can be obtained from interpolating the joint values. A 5th order polynomial is defined here under boundary conditions of joints (position, velocity, and acceleration) as follows.

$$x(0) = \theta_i \qquad x(t_f) = \theta_f$$
$$\dot{x}(0) = \dot{\theta}_i \qquad \dot{x}(t_f) = \dot{\theta}_f$$
$$\ddot{x}(0) = \ddot{\theta}_i \qquad \ddot{x}(t_f) = \ddot{\theta}_f$$

These boundary conditions uniquely specify a particular 5th order polynomial as follows.

$$x(t) = s_0 + s_1 t + s_2 t^2 + s_3 t^3 + s_4 t^4 + s_5 t^5 \tag{15}$$

The desired velocity and acceleration calculated, respectively

$$\dot{x}(t) = s_1 + 2s_2 t + 3s_3 t^2 + 4s_4 t^3 + 5s_5 t^4 \tag{16}$$

and

$$\ddot{x}(t) = 2s_2 + 6s_3 t + 12s_4 t^2 + 20s_5 t^3 \tag{17}$$

where $s_0, s_1, ..., s_5$ are the coefficients of the polynomial and given by

$$s_0 = \theta_i \tag{18}$$

$$s_1 = \dot{\theta}_i \tag{19}$$

$$s_2 = \frac{\ddot{\theta}_i}{2} \tag{20}$$

$$s_3 = \frac{20(\theta_f - \theta_i) - (8\dot{\theta}_f + 12\dot{\theta}_i)t_f + (\ddot{\theta}_f - 3\ddot{\theta}_i)t_f^2}{2t_f^3} \tag{21}$$

$$s_4 = \frac{30(\theta_i - \theta_f) + (14\dot{\theta}_f + 16\dot{\theta}_i)t_f + (3\ddot{\theta}_i - 2\ddot{\theta}_f)t_f^2}{2t_f^4} \tag{22}$$

and

$$s_5 = \frac{12(\theta_f - \theta_i) - 6(\dot{\theta}_f + \dot{\theta}_i)t_f - (\ddot{\theta}_i - \ddot{\theta}_f)t_f^2}{2t_f^5} \tag{23}$$

where θ_i, θ_f, $\dot{\theta}_i$, $\dot{\theta}_f$, $\ddot{\theta}_i$, $\ddot{\theta}_f$ denote initial and final position, velocity and acceleration, respectively. t_f is the time at final position. Fig. 1 illustrates the position velocity and acceleration profiles for a single link robot manipulator. Note that the manipulator is assumed to be motionless at initial and final positions.

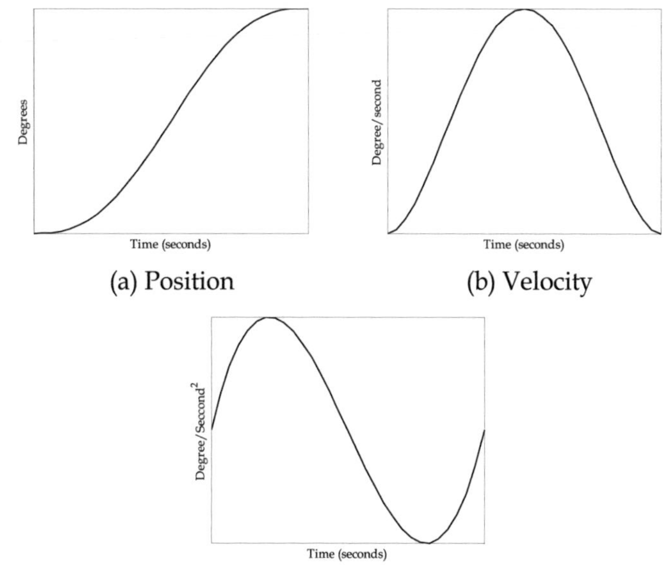

(a) Position (b) Velocity

(c) Acceleration

Figure 1. Position, velocity and acceleration profiles for a single link robot manipulator

5. Problem Formulation

If the manipulator moves freely in the workspace, the dynamic equation of motion for n DOF robot manipulator is given by the equation 12. This equation can be written as a simple matrix form as

$$M(q)\ddot{q}+C(q,\dot{q})+G(q)=\tau \tag{24}$$

Note that, the inertia matrix M(q) is a symmetric positive definite matrix. The local optimization of the joint torgue weighted by inertia results in resolutions with global characteristics. That is the solution to the local minimization problem (Lui & Wang, 2004). The formulation of the problem is

$$\text{Minimize} \quad \tau^T M^{-1}\tau \;\; (m_1, m_2, m_3)$$

Subjected to

$$J(q)\ddot{q}+\dot{J}(q)\dot{q}-\ddot{r}=0 \tag{25}$$

where f(q) is the position vector of the end-effector. J(q) is the Jacobian matrix and defined as

$$J(q)=\frac{\partial f(q)}{\partial q} \tag{26}$$

The total kinetic energy is given by

$$\frac{1}{2}\int_{t_0}^{t_f}\dot{q}^T M\dot{q} \tag{27}$$

The objective function in equation 25 also minimizes this kinetic energy.

6. Example

A three-link robot manipulator including two revolute joints and a prismatic joint is chosen as an example to examine the optimum energy performance. The transformation matrices are obtained using well known D-H method (Denavit & Hartenberg, 1955). For simplification, each link of the robot manipulator is modelled as a homogeneous cylindrical or prismatic beam of mass, *m* which is located at the centre of each link. The kinematics and dynamic representations for the three-link robot manipulator are shown in Fig. 2.

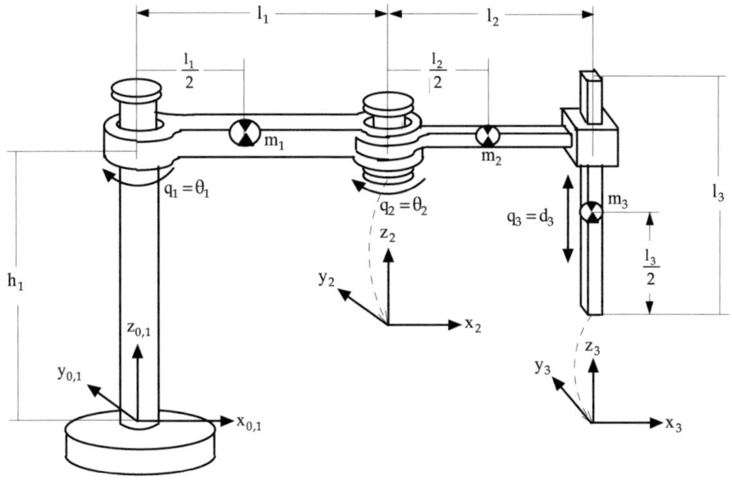

Figure 2. Kinematics and dynamic representations for the three-link robot manipulator

The kinematics and dynamic link parameters for the three-link robot are given in Table 1.

i	θ_i	α_{i-1}	a_{i-1}	d_i	q_i	m_i	L_{C_i}	I_{m_i}
1	θ_1	0	0	h_1	θ_1	m_1	$\frac{l_1}{2}$	$I_{m_1} = \text{diag}(I_{xx_1}, I_{yy_1}\ I_{zz_1})$
2	θ_2	0	l_1	0	θ_2	m_2	$\frac{l_2}{2}$	$I_{m_2} = \text{diag}(I_{xx_2}, I_{yy_2}, I_{zz_2})$
3	0	0	l_2	d_3	d_3	m_3	$\frac{l_3}{2}$	$I_{m_3} = \text{diag}(I_{xx_3}, I_{yy_3}, I_{zz_3})$

Table 1. The kinematics and dynamic link parameters for the three-link robot manipulator

where m_i and L_{C_i} stand for link masses and the centers of mass of links, respectively. The matrix I_m includes only the diagonal elements I_{xx}, I_{yy} and I_{zz}. They are called the principal moment of inertia about x, y and z axes. Since the mass distribution of the rigid object is symmetric relative to the body attached coordinate frame, the cross products of inertia are taken as zero. The transformation matrices for the robot manipulator illustrated in Fig. 2 are obtained using the parameters in Table 1 as follows.

$$ {}^0_1T = \begin{bmatrix} \cos\theta_1 & -\sin\theta_1 & 0 & 0 \\ \sin\theta_1 & \cos\theta_1 & 0 & 0 \\ 0 & 0 & 1 & h_1 \\ 0 & 0 & 0 & 1 \end{bmatrix} \tag{28} $$

$$ {}^1_2T = \begin{bmatrix} \cos\theta_2 & -\sin\theta_2 & 0 & l_1 \\ \sin\theta_2 & \cos\theta_2 & 0 & 0 \\ 0 & 0 & 1 & 0 \\ 0 & 0 & 0 & 1 \end{bmatrix} \tag{29} $$

and

$$ {}^{2}_{3}T = \begin{bmatrix} 1 & 0 & 0 & l_2 \\ 0 & 1 & 0 & 0 \\ 0 & 0 & 1 & d_3 \\ 0 & 0 & 0 & 1 \end{bmatrix} \tag{30} $$

The inertia matrix for the robot manipulator can be derived in the form

$$ M = \begin{bmatrix} M_{11} & M_{12} & 0 \\ M_{12} & M_{22} & 0 \\ 0 & 0 & m_3 \end{bmatrix} \tag{31} $$

where

$$ M_{11} = \frac{1}{4} m_1 l_1^2 + I_{zz_1} + m_2 (\frac{1}{4} l_2^2 + l_1 l_2 \cos\theta_2 + l_1^2) + I_{zz_2} + m_3 (l_2^2 + 2 l_1 l_2 \cos\theta_2 + l_1^2) + I_{zz_3}, \tag{32} $$

$$ M_{12} = m_2 (\frac{1}{4} l_2^2 + \frac{1}{2} l_1 l_2 \cos\theta_2) + I_{zz_2} + m_3 (l_2^2 + l_1 l_2 \cos\theta_2) + I_{zz_3} \tag{33} $$

and

$$ M_{22} = \frac{1}{4} m_2 l_2^2 + I_{zz_2} + l_2^2 m_3 + I_{zz_3} \tag{34} $$

Generalized torque vector applied at joints are

$$ \tau = [\tau_1 \ \tau_2 \ \tau_3]^T \tag{35} $$

where

$$ \begin{aligned} \tau_1 = & \left[\frac{1}{4} m_1 l_1^2 + I_{zz_1} + m_2 (\frac{1}{4} l_2^2 + l_1 l_2 \cos\theta_2 + l_1^2) + I_{zz_2} + m_3 (l_2^2 + 2 l_1 l_2 \cos\theta_2 + l_1^2) + I_{zz_3} \right] \ddot{\theta}_1 \\ & + \left[m_2 (\frac{1}{4} l_2^2 + \frac{1}{2} l_1 l_2 \cos\theta_2) + I_{zz_2} + m_3 (l_2^2 + l_1 l_2 \cos\theta_2) + I_{zz_3} \right] \ddot{\theta}_2 \\ & - l_1 l_2 \sin\theta_2 (m_2 + 2 m_3) \dot{\theta}_1 \dot{\theta}_2 - l_1 l_2 \sin\theta_2 (\frac{1}{2} m_2 + m_3) \dot{\theta}_2^2, \end{aligned} \tag{36} $$

$$ \tau_2 = \left[m_2 (\frac{1}{4} l_2^2 + \frac{1}{2} l_1 l_2 \cos\theta_2) + I_{zz_2} + m_3 (l_2^2 + l_1 l_2 \cos\theta_2) + I_{zz_3} \right] \ddot{\theta}_1 $$

$$+\left[\frac{1}{4}m_2l_2^2+I_{zz_2}+l_2^2m_3+I_{zz_3}\right]\ddot{\theta}_2+\left[\frac{1}{2}l_1l_2m_2\sin\theta_2+l_1l_2m_3\sin\theta_2\right]\dot{\theta}_1^2 \tag{37}$$

$$\tau_3=m_3\ddot{d}_3+gm_3 \tag{38}$$

As a result, the objective function is obtained as follows.

$$\tau^TM^{-1}\tau=\left(-\tau_1\frac{m_3l_2^2+I_{zz_3}}{m_3l_1^2(m_3l_2^2\cos^2\theta_2-l_2^2m_3-I_{zz_3})}+\tau_2\frac{m_3l_2^2+m_3l_1l_2\cos\theta_2+I_{zz_3}}{m_3l_1^2(m_3l_2^2\cos^2\theta_2-l_2^2m_3-I_{zz_3})}\right)\tau_1$$

$$+\left(\tau_1\frac{m_3l_2^2+m_3l_1l_2\cos\theta_2+I_{zz_3}}{m_3l_1^2(m_3l_2^2\cos^2\theta_2-l_2^2m_3-I_{zz_3})}-\tau_2\frac{m_3l_2^2+2m_3l_1l_2\cos\theta_2+m_3l_1^2+I_{zz_3}}{m_3l_1^2(m_3l_2^2\cos^2\theta_2-l_2^2m_3-I_{zz_3})}\right)\tau_2$$

$$+\left(\frac{1}{m_3}\tau_3\right)\tau_3 \tag{39}$$

6.1 Optimization with Genetic Algorithms

The displacement of the end-effector with respect to the time is illustrated in Fig. 3. The end-effector starts from the initial position p_{xi}=44, p_{yi}=0, p_{zi}=18 at t=0 second, and reach the final position p_{xf}=13.2721, p_{yf}=12.7279, p_{zf}=2 at t=3 seconds in Cartesian space.

When the end-effector perform a motion from the initial position p_{xi}=44, p_{yi}=0, p_{zi}=18 to the final position p_{xf}=13.2721, p_{yf}=12.7279, p_{zf}=2 in 3 seconds in Cartesian space, each joint has position velocity and acceleration profiles obtained from the 5th order polynomial as shown in Fig. 4.

Ten samples obtained from Fig. 4 at time intervals between 0 and 3 seconds for the optimization problem are given in Table 2. Pos.-i, Vel.-i and Acc.-i represent position, velocity and acceleration of ith joint of the robot manipulator, respectively (1≤ i ≤3). The letter "S" represents sample.

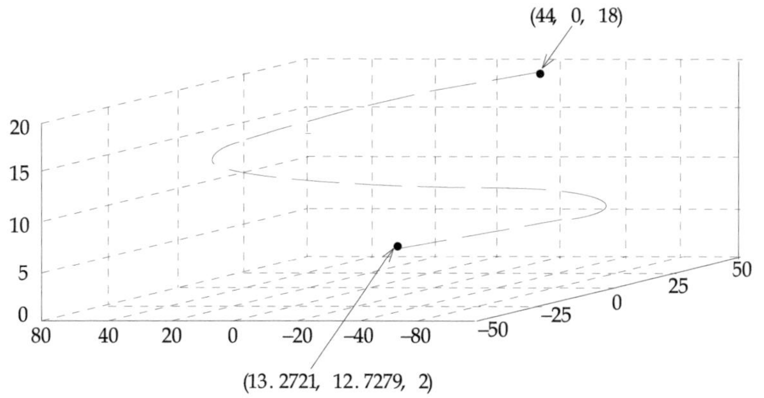

Figure 3. The displacement of the end-effector with respect to the time

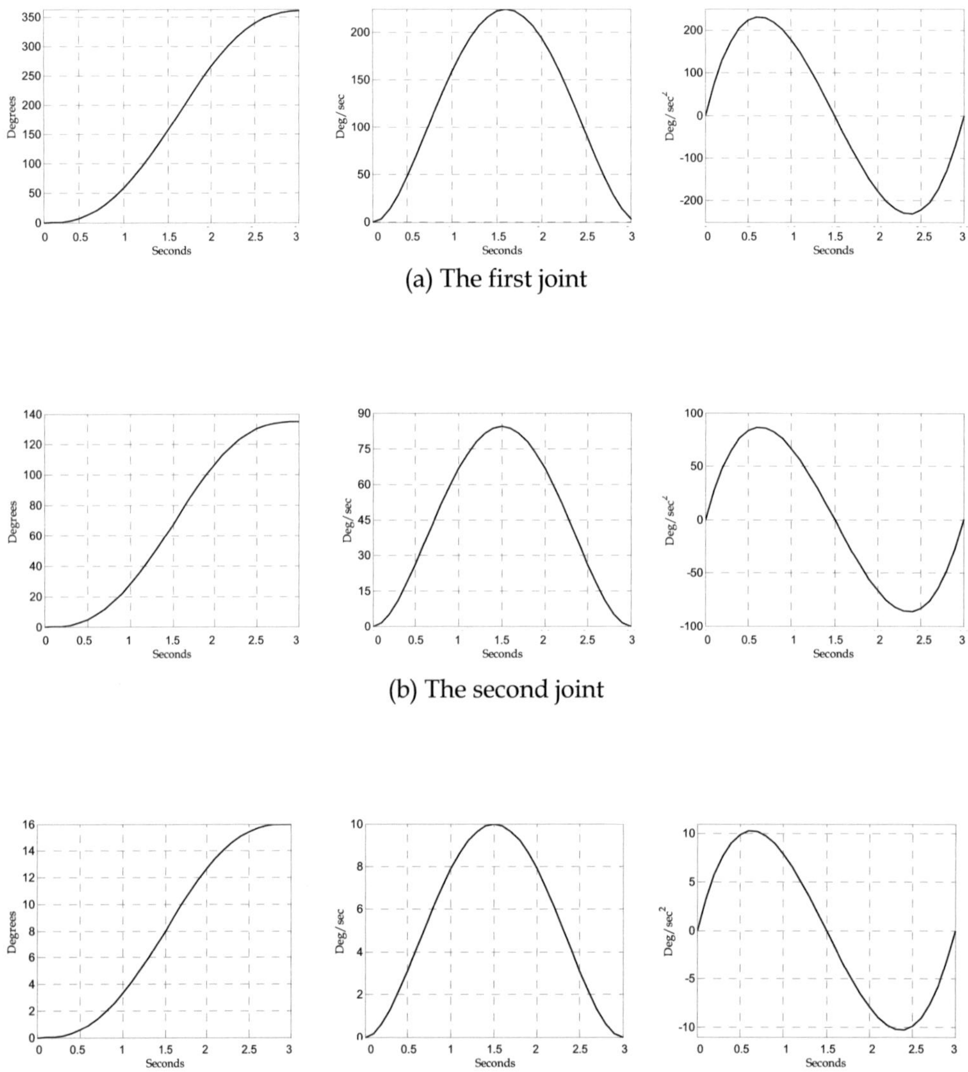

(a) The first joint

(b) The second joint

(c) The third joint

Figure 4. The position velocity and acceleration profiles for first, second and third joints

S	Pos.-1	Vel.-1	Acc.-1	Pos.-2	Vel.-2	Acc.-2	Pos.-3	Vel.-2	Acc.-3
1	0.126	3.737	72.177	0.047	1.401	27.066	0.005	0.166	3.207
2	6.917	48.071	203.37	2.594	18.02	76.266	0.307	2.136	9.039
3	31.220	115.20	228.97	11.707	43.20	85.866	1.387	5.120	10.176
4	75.555	177.77	177.77	28.333	66.66	66.666	3.358	7.901	7.901
5	135.53	217.07	78.577	50.823	81.40	29.466	6.023	9.647	3.492
6	202.43	223.00	-39.822	75.912	83.62	-14.933	8.997	9.911	-1.769
7	265.82	194.13	-48.622	99.684	72.80	-55.733	11.814	8.628	-6.605
8	316.12	137.67	-19.022	118.54	51.62	-82.133	14.050	6.118	-9.734
9	347.22	69.444	-22.222	130.20	26.04	-83.333	15.432	3.086	-9.876
10	359.03	13.937	-29.422	134.63	5.226	-48.533	15.957	0.619	-5.752

Table 2. Position, velocity and acceleration samples of first, second and third joints

In robot design optimization problem, the link masses m_1, m_2 and m_3 are limited to an upper bound of 10 and to a lower bound of 0. The objective function does not only depend on the design variables but also the joint variables (q_1, q_2, q_3) which have lower and upper bounds ($0<q_1<360$, $0<q_2<135$ and $0<q_3<16$), on the joint velocities ($\dot{q}_1, \dot{q}_2, \dot{q}_3$) and joint accelerations ($\ddot{q}_1, \ddot{q}_2, \ddot{q}_3$). The initial and final velocities of each joint are defined as zero. In order to optimize link masses, the objective function should be as small as possible at all working positions, velocities and accelerations. The following relationship was adapted to specify the corresponding fitness function.

$$\tau^T M^{-1} \tau = \tau^T M^{-1} \tau(1) + \tau^T M^{-1} \tau(2) +, \cdots, + \tau^T M^{-1} \tau(9) + \tau^T M^{-1} \tau(10) \tag{40}$$

In the GA solution approach, the influences of different population sizes and mutation rates were examined to find the best GA parameters for the mass optimization problem where the minimizing total kinetics energy. The GA parameters, population sizes and mutation rates were changed between 20-60 and 0.005-0.1 where the number of iterations was taken 50 and 100, respectively. The GA parameters used in this study were summarized as follows.

Population size :20, 40, 60

Mutation rate :0.005, 0.01, 0.1

Number of iteration :50, 100

All results related these GA parameters are given in Table 3. As can be seen from Table 3, population sizes of 20, mutation rate of 0.01 and iteration number of 50 produce minimum kinetic energy of 0. 892 10^{-14} where the optimum link masses are m_1=9.257, m_2=1.642 and m_3=4.222.

Population size	20	20	20	20	20	20	40	40	40
Mutation rate	0.005	0.01	0.1	0.005	0.01	0.1	0.005	0.01	0.1
Number of iteration	50	50	50	100	100	100	50	50	50
Fitness Value (10^{-14})	1.255	0.892	5.27	1.619	3.389	554.1	1.850	8.517	6.520
m_1	0.977	9.257	6.559	3.421	3.167	0.146	7.507	4.271	9.462
m_2	1.915	1.642	0.508	1.661	0.508	0.039	0.185	0.254	0.156
m_3	1.877	4.222	0.205	0.889	1.446	0.009	2.717	0.332	0.664

Population size	40	40	40	60	60	60	60	60	60
Mutation rate	0.005	0.01	0.1	0.005	0.01	0.1	0.005	0.01	0.1
Number of iteration	100	100	100	50	50	50	100	100	100
Fitness Value (10^{-14})	1.339	2.994	13.67	4.191	1.876	12.35	2.456	4.362	19.42
m_1	3.040	3.773	9.677	5.347	3.949	0.078	6.432	6.041	1.349
m_2	0.048	0.889	0.117	0.449	1.397	0.078	0.713	0.381	0.058
m_3	3.958	0.351	0.263	0.772	0.498	0.351	1.388	0.811	0.215

Table 3. The GA parameters, fitness function and optimized link masses

7. Conclusion

In this chapter, the link masses of the robot manipulators are optimized using GAs to obtain the best energy performance. First of all, fundamental knowledge about genetic algorithms, dynamic equations of robot manipulators and trajectory generation were presented in detail. Second of all, GA was applied to find the optimum link masses for the three-link serial robot manipulator. Finally, the influences of different population sizes and mutation rates were searched to achieve the best GA parameters for the mass optimization. The optimum link masses obtained at minimum torque requirements show that the GA optimization technique is very consistent in robotic design applications. Mathematically simple and easy coding features of GA also provide convenient solution approach in robotic optimization problems.

8. References

Denavit, J.; Hartenberg, R. S. (1955). A kinematic notation for lower-pair mechanisms based on matrices, *Journal of Applied Mechanics*, pp. 215-221, vol. 1, 1955

Hollerbach, J. M. (1980). A Recursive Lagrangian Formulation of Manipulator Dynamics and A Comparative Study of Dynamic Formulation Complexity, *IEEE Trans. System., Man, Cybernetics.*, pp. 730-736, vol. SMG-10, no. 11, 1980

Luh, J. Y. S.; Walker, M. W. & Paul, R. P. C. (1980). On-Line Computational Scheme for Mechanical Manipulators, *Trans. ASME, J. Dynamic Systems, Measurement and Control,* vol. 102, no. 2, pp. 69-76, 1980

Paul, R. P. (1981). Robot Manipulators: Mathematics, Programming, and Control, MIT Press, Cambridge Mass., 1981

Kane, T. R.; Levinson, D. A. (1983). The Use of Kane's Dynamic Equations in Robotics, *Int. J. of Robotics Research,* vol. 2, no. 3, pp. 3-20, 1983

Lee, C. S. G.; Lee, B. H. & Nigam, N. (1983). Development of the Generalized D'Alembert Equations of Motion for Robot Manipulators, *Proc. Of 22nd Conf. on Decision and Control,* pp. 1205-1210, 1983, San Antonio, TX

Nedungadi, A.; Kazerouinian, K. (1989). A local solution with global characteristics for joint torque optimization of redundant manipulator, *Journal of Robotic Systems,* vol. 6, 1989

Schilling, R. (1990). Fundamentals of Robotics Analysis and Control, Prentice Hall, 1990, New Jersey

Delingette, H.; Herbert, M. & Ikeuchi, K. (1992). Trajectory generation with curvature constraint based on energy minimization, *Proceedings of the IEEE/RSJ International Workshop on Intelligent Robots and Systems*, pp. 206–211, Japan, 1992, Osaka

Garg, D.; Ruengcharungpong, C. (1992). Force balance and energy optimization in cooperating manipulation, *Proceedings of the 23rd Annual Pittsburgh Modeling and Simulation Conference,* pp. 2017–2024, vol. 23, Pt. 4, USA, 1992, Pittsburgh

Pin, F.; Culioli, J. (1992). Optimal positioning of combined mobile platform-manipulator systems for material handling tasks, *Journal of Intelligent Robotic System, Theory and Application 6,* pp. 165–182, 1992

Painton, L.; Campbell, J. (1995). Genetic algorithms in optimization of system reliability, *IEEE Transactions on Reliability,* pp. 172– 178, 44 (2), 1995

Hirakawa, A.;Kawamura, A. (1996). Proposal of trajectory generation for redundant manipulators using variational approach applied to minimization of consumed electrical energy, *Proceedings of the Fourth International Workshop on Advanced Motion Control,* Part 2, pp. 687–692, , Japan, 1996, Tsukuba

Chen, M.; Zalzasa, A. (1997). A genetic approach to motion planning of redundant mobile manipulator systems considering safety and configuration, *Journal of Robotic Systems*, pp. 529-544, vol. 14, 1997

Stocco, L.; Salcudean, S. E. & Sassani, F. (1998). Fast constraint global minimax optimization of robot parameters, *Robotica,* pp. 595-605, vol. 16, 1998

Choi, S.; Choi, Y. & Kim, J. (1999). Optimal working trajectory generation for a biped robot using genetic algorithm, *Proceedings of IEEE/RSJ International Conference on Intelligent Robots and Systems,* pp. 1456-1461, vol. 3, 1999

Pires, E.; Machado, J. (1999). Trajectory planner for manipulators using genetic algorithms, *Proceeding of the IEEE Int. Symposium on Assembly and Task Planning,* pp. 163-168, 1999

Sexton, R.; Gupta, J. (2000). Comparative evaluation of genetic algorithm and back propagation for training neural networks, *Information Sciences*, pp. 45–59, 2000

Garg, D.; Kumar, M. (2002). Optimization techniques applied to multiple manipulators for path planning and torque minimization, *Journal for Engineering Applications of Artificial Intelligence*, pp. 241-252, vol. 15, 2002

Tang, W.; Wang, J. (2002). Two recurrent neural networks for local joint torque optimization of kinematically redundant manipulators, *IEEE Trans. Syst. Man Cyber,* Vol. 32(5), 2002

Lui, S.; Wang, J. (2004). A dual neural network for bi-criteria torque optimization of redundant robot manipulators, *Lecture Notes in Computer Science,* vol. 3316, pp. 1142-1147, 2004

Kucuk S.; Bingul Z. (2006). Link mass optimization of serial robot manipulators using genetic algorithms, *Lecture Notes in Computer Science, Knowledge-Based Intelligent Information and Engineering Systems*, vol. 4251, pp. 138-144, 2006

Qudeiri, J. A.; Yamamoto, H. & Ramli, R. (2007). Optimization of operation sequence in CNC machine tools using genetic algorithm, *Journal of Advance Mechanical Design, Systems, and Manufacturing,* pp. 272-282, vol. 1, no. 2, 2007

16

FPGA-Realization of a Motion Control IC for Robot Manipulator

Ying-Shieh Kung and Chia-Sheng Chen
Southern Taiwan University
Taiwan

1. Introduction

Robotic control is currently an exciting and highly challenging research focus. Several solutions for implementing the control architecture for robots have been proposed (Kabuka et al., 1988; Yasuda, 2000; Li et al., 2003; Oh et al., 2003). Kabuka et al. (1998) apply two high-performance floating-point signal processors and a set of dedicated motion controllers to build a control system for a six-joint robots arm. Yasuda (2000) adopts a PC-based microcomputer and several PIC microcomputers to construct a distributed motion controller for mobile robots. Li et al. (2003) utilize an FPGA (Field Programmable Gate Array) to implement autonomous fuzzy behavior control on mobile robot. Oh et al. (2003) present a DSP (Digital Signal Processor) and a FPGA to design the overall hardware system in controlling the motion of biped robots. However, these methods can only adopt PC-based microcomputer or the DSP chip to realize the software part or adopt the FPGA chip to implement the hardware part of the robotic control system. They do not provide an overall hardware/software solution by a single chip in implementing the motion control architecture of robot system.

For the progress of VLSI technology, the Field programmable gate arrays (FPGAs) have been widely investigated due to their programmable hard-wired feature, fast time-to-market, shorter design cycle, embedding processor, low power consumption and higher density for implementing digital system. FPGA provides a compromise between the special-purpose ASIC (application specified integrated circuit) hardware and general-purpose processors (Wei et al., 2005). Hence, many practical applications in motor control (Zhou et al., 2004; Yang et al., 2006; Monmasson & Cirstea, 2007) and multi-axis motion control (Shao & Sun, 2005) have been studied, using the FPGA to realize the hardware component of the overall system.

The novel FPGA (Field Programmable Gate Array) technology is able to combine an embedded processor IP (Intellectual Property) and an application IP to be an SoPC (System-on-a-Programmable-Chip) developing environment (Xu et al., 2003; Altera, 2004; Hall & Hamblen, 2004), allowing a user to design an SoPC module by mixing hardware and software. The circuits required with fast processing but simple computation are suitable to be implemented by hardware in FPGA, and the highly complicated control algorithm with heavy computation can be realized by software in FPGA. Results that the software/hardware co-design function increase the programmable, flexibility of the designed digital system and reduce the development time. Additionally, software/hardware parallel processing enhances the controller performance. Our previous works (Kung & Shu, 2005; Kung et al., 2006; Kung &

Tsai, 2007) have successfully applied the novel FPGA technology to the servo system of PMSM drive, robot manipulator and X-Y table.
To exploit the advantages, this study presents a fully digital motion control IC for a five-axis robot manipulator based on the novel FPGA technology, as in Fig. 1(a). Firstly, the mathematical model and the servo controller of the robot manipulator are described. Secondly, the inverse kinematics and motion trajectory is formulated. Thirdly, the circuit being designed to implement the function of motion control IC is introduced. The proposed motion control IC has two IPs. One IP performs the functions of the motion trajectory for robot manipulator. The other IP performs the functions of inverse kinematics and five axes position controllers for robot manipulator. The former is implemented by software using Nios II embedded processor due to the complicated processing but low sampling frequency control (motion trajectory: 100Hz). The latter is realized by hardware in FPGA owing to the requirements of high sampling frequency control (position loop:762Hz, speed loop:1525Hz, PWM circuit: 4~8MHz) but simple processing. To reduce the usage of the logic gates in FPGA, an FSM (Finite State Machine) method is presented to model the overall computation of the inverse kinematics, five axes position controllers and five axes speed controllers. Then the VHDL (VHSIC Hardware Description Language) is adopted to describe the circuit of FSM. Hence, all functionality required to build a fully digital motion controller for a five-axis robot manipulator can be integrated in one FPGA chip. As the result, hardware/software co-design technology can make the motion controller of robot manipulator more compact, flexible, better performance and less cost. The FPGA chip employed herein is an Altera Stratix II EP2S60F672C5ES (Altera, 2008), which has 48,352 ALUTs (Adaptive Look-Up Tables), maximum 718 user I/O pins, total 2,544,192 RAM bits. And a Nios II embedded processor which has a 32-bit configurable CPU core, 16 M byte Flash memory, 1 M byte SRAM and 16 M byte SDRAM is used to construct the SoPC development environment. Finally, an experimental system included by an FPGA experimental board, five sets of inverter and a Movemaster RV-M1 micro robot produced by Mitsubishi is set up to verify the correctness and effectiveness of the proposed FPGA-based motion control IC.

2. Important

The typical architecture of the conventional motion control system for robot manipulator is shown in Fig. 1(b), which consists of a central controller, five sets of servo drivers and one robot manipulator. The central controller, which usually adopts a float-pointed processor, performs the function of motion trajectory, inverse kinematics, data communication with servo drivers and the external device. Each servo driver uses a fixed-pointed processor, some specific ICs and an inverter to perform the functions of position control at each axis of robot manipulator and the data communication with the central controller. Data communication media between them may be an analog signal, a bus signal or a serial asynchronous signal. Therefore, once the central controller receives a motion command from external device. It will execute the computation of motion trajectory and inverse kinematics, generate five position commands and send those signals to five servo drivers. Each servo driver received the position command from central controller will execute the position controller to control the servo motor of manipulator. As the result, the motion trajectory of robot manipulator will follow the prescribed motion command. However, the motion control system in Fig .1 (b) has some drawbacks, such as large volume, easy effected by the noise, expensive cost, inflexible, etc. In addition, data communication and handshake protocol between the central controller and servo drivers slow down the system executing speed. To improve aforementioned drawbacks, based on the novel FPGA technology, the central controller and the controller part of five servo drivers in Fig. 1 (b) are integrated into a motion control IC in this study, which is shown in Fig. 1(a). The proposed motion control IC has two IPs. One IP performs the functions of the motion trajectory by software. The other IP performs the functions of inverse kinematics and five axes position controllers by hardware. Hence, hardware/software co-design technology can make the motion controller of robot manipulator more compact, flexible, better performance and less cost. The detailed design methodology is described in the following sections, and the contributions of this study will be illustrated in the conclusion section.

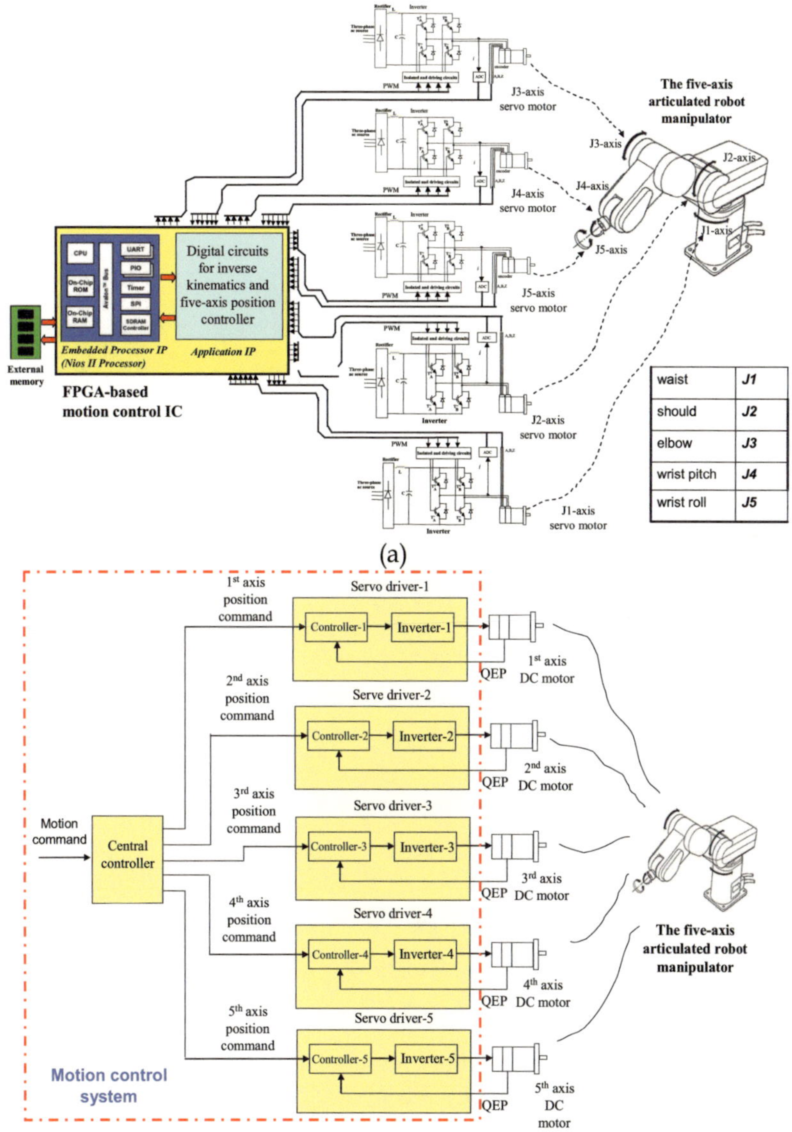

(b)

Figure 1. (a) The proposed architecture of an FPGA-based motion control system for robot manipulator (b) the conventional motion control system for robot manipulator

3. System description and design method of robot manipulator

3.1 Mathematical model of a robot manipulator with actuator

The dynamic equation of the n-link manipulator is given by (Lewis et al., 1993):

$$M(\theta)\ddot{\theta} + V_m(\theta,\dot{\theta})\dot{\theta} + F(\dot{\theta}) + G(\theta) = \tau \tag{1}$$

where $M(\theta)$ denotes the inertial matrix; $V_m(\theta,\dot{\theta})$ denotes the Coriolis/centripetal vector; $G(\theta)$ denotes the gravity vector; and $F(\dot{\theta})$ denotes the friction vector. The terms θ, $\dot{\theta}$, $\ddot{\theta}$ and τ represent the n-vector of the position, the velocity, the acceleration and the generalized forces, respectively; and are set to R^n. The dynamic model of the ith axis DC motor drive system for the robot manipulator is expressed as

$$L_{a_i}\frac{di_i}{dt} + R_{a_i} i_i = v_i - K_{b_i}\dot{\theta}_{M_i} \tag{2}$$

$$\tau_{M_i} = K_{M_i} i_i \tag{3}$$

$$J_{M_i}\ddot{\theta}_{M_i} + B_{M_i}\dot{\theta}_{M_i} = \tau_{M_i} - \tau_{L_i} \tag{4}$$

where L_{a_i}, R_{a_i}, i_i, v_i, K_{b_i}, K_{M_i} denote inductance, resistance, current, voltage, voltage constant and current constant, respectively. Terms J_{M_i}, B_{M_i}, τ_{M_i} and τ_{L_i} denote the inertial, viscous and generated torque of the motor, and the load torque to the ith axis of the robot manipulator, respectively. Consider a friction F_{M_i} added to each arm link and replace the load torque τ_{L_i} by an external load torque through a gear ratio r_i, giving $\tau_{L_i} = r_i\tau_i$ and assume that the inductance term in (2) is small and can be ignored. Then, (2) to (4) can be arranged and simplified by the following equation.

$$J_{M_i}\ddot{\theta}_{M_i} + (B_{M_i} + \frac{K_{M_i}K_{b_i}}{R_{a_i}})\dot{\theta}_{M_i} + F_{M_i} + r_i\tau_i = \frac{K_{M_i}}{R_{a_i}} v_i \tag{5}$$

Therefore, the dynamics of the dc motors that drive the arm links are given by the *n* decoupled equations

$$J_M\ddot{\theta}_M + B\dot{\theta}_M + F_M + R\tau = K_M v \tag{6}$$

with

$$\begin{aligned} &\theta_M = vec\{\theta_{M_i}\},\ J_M = diag\{J_{M_i}\} \\ &B = diag\{B_{M_i} + K_{b_i}K_{M_i}/R_{a_i}\} \triangleq diag\{B_i\} \\ &R = diag\{r_i\},\ K_M = diag\{K_{M_i}/R_{a_i}\} \\ &v = vce(v_i),\ \tau = vec(\tau_i),\ F_M = vec(F_{M_i}) \end{aligned} \tag{7}$$

The dynamic equations including the robot manipulator and DC motor can be obtained by combining (1) and (6). The gear ratio of the coupling from the motor *i* to arm link *i* is given by r_i, which is defined as

$$\theta_i = r_i\theta_{M_i} \text{ or } \theta = R\theta_M \tag{8}$$

Hence, substituting (8) into (6), then into (1), it can be obtained by

$$(J_M + R^2 M(\theta))\ddot{\theta} + (B + R^2 V_m(\theta,\dot{\theta}))\dot{\theta} + RF_M(\dot{\theta}) + R^2 F(\dot{\theta}) + R^2 G(\theta) = RK_M v \tag{9}$$

The gear ratio r_i of a commercial robot manipulator is usually small (< 1/100) to increase the torque value. Therefore, the formula can be simplified. From (9), the dynamic equation combining the motor and arm link is expressed as

$$(J_{M_i} + r_i^2 m_{ii})\ddot{\theta}_i + B_i\dot{\theta}_i + r_i F_{M_i} = \frac{r_i K_{M_i}}{R_{a_i}} v_i - r_i^2 d_i \tag{10}$$

where

$$d_i = \sum_{j \neq i} m_{ij}\ddot{\theta}_j + \sum_{j,k} V_{jki}\dot{\theta}_j\dot{\theta}_k + F_i + G_i \tag{11}$$

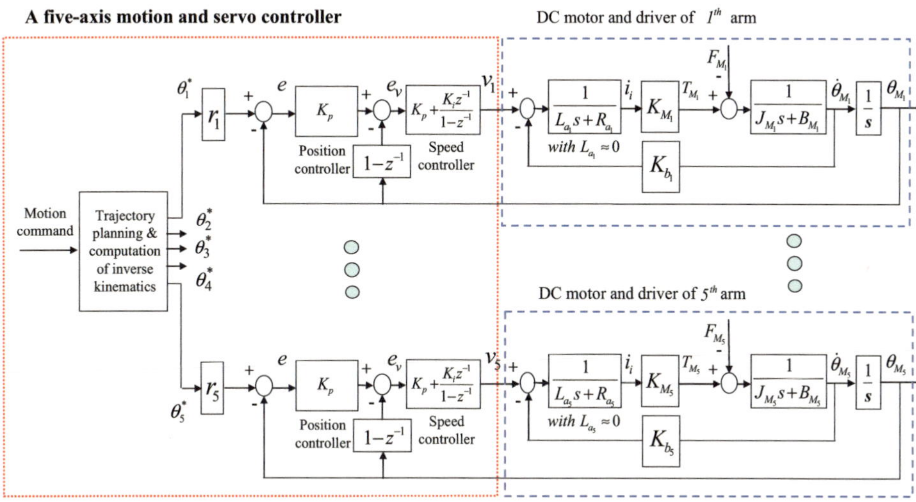

Figure 2 The block diagram of motion and servo controller with DC motor for a five-axis robot manipulator

Since the gear ratio r_i is small in a commercial robot manipulator, the r_i^2 term in (10) can be ignored, and (10) can be simplified as

$$J_{M_i}\ddot{\theta}_i + B_i\dot{\theta}_i + r_i F_{M_i} = \frac{r_i K_{M_i}}{R_{a_i}} v_i \tag{12}$$

Substituting (7) and (8) into (12), yields

$$J_{M_i}\ddot{\theta}_{M_i} + (B_{M_i} + \frac{K_{M_i}K_{b_i}}{R_{a_i}})\dot{\theta}_{M_i} + F_{M_i} = \frac{K_{M_i}}{R_{a_i}} v_i \tag{13}$$

If the gear ratio is small, then the control block diagram combining arm link, motor actuator, position loop P controller, speed loop PI controller, inverse kinematics and trajectory planning for the servo and motion controller in a five-axis robot manipulator can be represented in Fig. 2.
In Fig.2, the digital PI controllers in the speed loop are formulated as follows.

$$v_p(n) = k_p e(n) \tag{14}$$

$$v_i(n) = v_i(n-1) + k_i e(n-1) \tag{15}$$

$$v(n) = v_p(n) + v_i(n) \tag{16}$$

Where the k_p, k_i are P-controller gain and I-controller gain, the $e(n)$ is the error between command and sensor value and the $v(n)$ is the PI controller output.

3.2 Computational of the inverse kinematics for robot manipulator

Figure 3 shows the link coordinate system of the five-axis articulated robot manipulator using the Denavit-Hartenberg convention. Table 1 illustrates the values of the kinematics parameters. The inverse kinematics of the articulated robot manipulator will transform the coordinates of robot manipulator from Cartesian space R^3 (x,y,z) to the joint space R^5 $(\theta_1,\theta_2,\theta_3,\theta_4,\theta_5)$ and it is described in detail by author of (Schilling, 1998). The computational procedure is as follows.

Step1:
$$\theta_1 = \theta_5 = a\tan 2(y,x) \tag{17}$$

Step2:
$$\mathrm{b} = \sqrt{\mathrm{x}^2 + \mathrm{y}^2} \tag{18}$$

Step3:
$$\theta_3 = \cos^{-1}\left(\frac{b^2 + (d_1 - d_5 - z)^2 - a_2^2 - a_3^2}{2a_2 a_3}\right) \tag{19}$$

Step4:
$$S_2 = (a_2 + a_3 \cos\theta_3)(d_1 - d_5 - z) - a_3 b \sin\theta_3 \tag{20}$$

Step5:
$$C_2 = (a_2 + a_3 \cos\theta_3)b + a_3 \sin\theta_3 (d_1 - d_5 - z) \tag{21}$$

Step6:
$$\theta_2 = a\tan 2(S_2, C_2) \tag{22}$$

Step7:
$$\theta_4 = -\theta_2 - \theta_3 \tag{23}$$

where a_1, a_5, d_1, d_5 are Denavit-Hartenberg parameters and *atan2* function refers to Table 2.

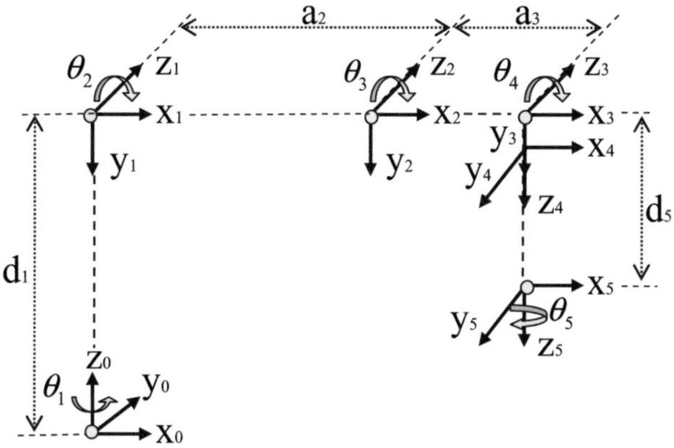

Figure 3. Link coordinates of a five-axis robot manipulator

	θ	d	a	α
1	θ_1	d_1 (300 mm)	(0 mm)	$\alpha_1(-\pi/2)$
2	θ_2	(0 mm)	a_2 (250 mm)	0
3	θ_3	(0 mm)	a_3 (160 mm)	0
4	θ_4	(0 mm)	(0 mm)	$\alpha_4(-\pi/2)$
5	θ_5	d_5 (72 mm)	(0 mm)	0

Table 1. Denavit-Hartenberg parameters of the robot manipulator

Case	Quadrants	*atan2* (*y*, *x*)
$x > 0$	1, 4	$\arctan(y/x)$
$x = 0$	1, 4	$[sgn(y)]\pi/2$
$x < 0$	2, 3	$\arctan(y/x) + [sgn(y)]\pi$

Table 2. Four-quadrant *arctan* function (Schilling, 1998)

3.3 Planning of the motion trajectory

3.3.1The computation of the point-to-point motion trajectory

To consider the smooth motion when manipulating the robot, the point-to-point motion scheme with constant acceleration/deceleration trapezoid velocity profile is applied in Fig.2. In this scheme, the designed parameters at each axis of robot are the overall displacement $\Delta\theta_i^*$ (degree), the maximum velocity W_i (degree/s), the acceleration or deceleration period T_{acc} and the position loop sampling interval t_d. Therefore, the instantaneous position command q at each sampling interval, based on the trapezoid velocity profile, can be computed as follows:

Step 1: Compute the overall running time. First, compute the maximum running time without considering the acceleration and deceleration:

$$T_1 = Max(\Delta\theta_1^* / W_1, \Delta\theta_2^* / W_2, \Delta\theta_3^* / W_3, \Delta\theta_4^* / W_4, \Delta\theta_5^* / W_5) \tag{24}$$

This T_1 must exceed acceleration time T_{acc}. The acceleration and deceleration design is then considered, and the overall running time is

$$T = Max\ (T_1, T_{acc}) + T_{acc} \tag{25}$$

Step 2: Adjust the overall running time to meet the condition of a multiple of the sampling interval.

$$N' = [T_{acc} / t_d] \text{ and } N = [T / t_d] \tag{26}$$

Where N represents the interpolation number and [.] refers to the Gauss function with integer output value. Hence,

$$T'_{acc} = N' * t_d \text{ and } T' = N * t_d \tag{27}$$

Step 3: Modify the maximum velocity

$$L \triangleq [\Delta\theta_1^*, \Delta\theta_2^*, \Delta\theta_3^*, \Delta\theta_4^*, \Delta\theta_5^*]^T \tag{28}$$

$$W' \triangleq [W_1', W_2', W_3', W_4', W_5']^T = L / (T' - T'_{acc}) \tag{29}$$

Step 4 : Calculate the acceleration/deceleration value

$$A = W' / T'_{acc} \tag{30}$$

Step 5 : Calculate the position command, $q = [q_1, q_2, q_3, q_4, q_5]$, at the mid-position

a. Acceleration region:

$$q = q_{r0} + \frac{1}{2} * A * t^2 \tag{31}$$

where $t = n * t_d$, $0 < n \le N'$, and q_{r0} denotes the initial position.

b. Constant speed region:

$$q = q_{r1} + W' * t \tag{32}$$

where $t = n * t_d$, $0 < n \le N1$, q_{r1} denotes the final position in the acceleration region and $N1 = N - 2 * N'$.

c. Deceleration region:

$$q = q_{r2} + (W' * t - \frac{1}{2} * A * t^2) \tag{33}$$

where $t = n * t_d$, $0 < n \le N'$ and q_{r2} denotes the final position in the constant speed region.

Therefore, applying this point-to-point motion control scheme, all joins of robot will rotate with simultaneous starting and stopping.

3.3.2 The computation of the linear and circular motion trajectory

Typical motion trajectories have linear and circular trajectory.

a. Linear motion trajectory

The formulation of linear motion trajectory is as follows:

$$x_i = x_{i-1} + \Delta s \, sin\gamma \, cos\,\alpha \tag{34}$$

$$y_i = y_{i-1} + \Delta s \, sin\gamma \, cos\beta \tag{35}$$

$$z_i = z_{i-1} + \Delta s \, cos\gamma \tag{36}$$

where Δs, γ, α, β, x_i, y_i, z_i are path increment, angle between z-axis and path vector, angle between x-axis and the vector that path vector projects to x-y plane, angle between y-axis and the vector that path vector projects to x-y plane, x-axis, y-axis and z-axis trajectory command, respectively. The running speed of the linear motion is determined by Δs.

b. Circular motion trajectory with constant value in Z-axis

The formulation of circular motion trajectory with constant value in Z-axis is as follows:

$$\varphi_i = \varphi_{i-1} + \Delta\varphi \tag{37}$$

$$x_i = R\sin(\varphi_i) \tag{38}$$

$$y_i = R\cos(\varphi_i) \tag{39}$$

$$z_i = z_{i-1} \tag{40}$$

where φ_i, $\Delta\varphi$, R, x_i, y_i and z_i are angle, angle increment, radius, x-axis, y-axis and z-axis trajectory command, respectively. The running speed of the circle motion is determined by $\Delta\varphi$.

4. The proposed FPGA-based motion IC for robot manipulator

Figure 4 illustrates the internal architecture of the proposed FPGA-based motion control IC for a five-axis robot manipulator. The motion control IC, which comprises a Nios II embedded processor IP and a motion control IP, is designed based on the SoPC technology, which is developed by Altera Cooperation. The FPGA chip employed herein is an Altera Stratix II EP2S60F672C5ES, which has 48,352 ALUTs, maximum 718 user I/O pins, total 2,544,192 RAM bits. And a Nios II embedded processor has a 32-bit configurable CPU core,

16 M byte Flash memory, 1 M byte SRAM and 16 M byte SDRAM. A custom software development kit (SDK) consists of a compiled library of software routines for the SoPC design, a Make-file for rebuilding the library, and C header files containing structures for each peripheral.

The Nios II embedded processor depicted in Fig. 4 performs the function of motion command, computation of point-to-point motion trajectory as well as linear and circular motion trajectory in software. The clock frequency of the Nios II processor of is 50MHz. Figure 5 illustrates the flow charts of the main program, subroutine of point-to-point trajectory planning and the interrupt service routine (ISR), where the interrupt interval is designed with 10ms. However, point-to-point motion and trajectory tracking motion are run by different branch program. Because the inverse kinematics module is realized by hardware in FPGA, the computation in Fig. 5 needed to compute the inverse kinematics has to firstly send the x, y, z value from I/O pin of Nios II processor to the inverse kinematics module, wait 2μs then read back the $\theta_1^* \sim \theta_5^*$. All of the controller programs in Fig. 5 are coded in the C programming language; then, through the complier and linker operation in the Nios II IDE (Integrated Development Environment), the execution code is produced and can be downloaded to the external Flash or SDRAM via JTAG interface. Finally, this execution code can be read by Nios II processor IP via bus interface in Fig. 4. Using the C language to develop the control algorithm not only has the portable merit but also is easier to transfer the mature code, which has been well-developed in other devices such as DSP device or PC-based system, to the Nios II processor.

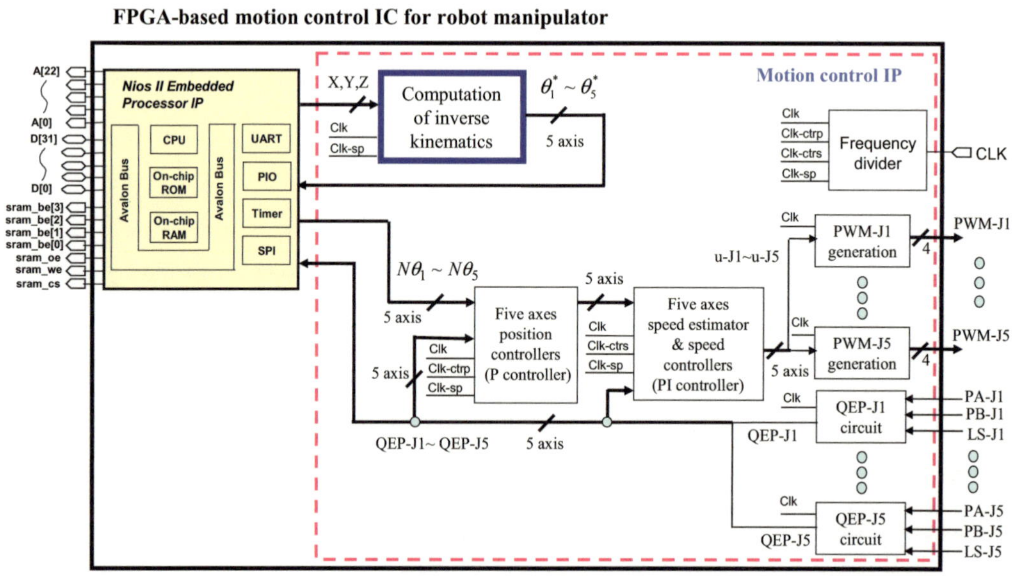

Figure 4. Internal architecture of the proposed FPGA-based motion control IC for robot manipulator

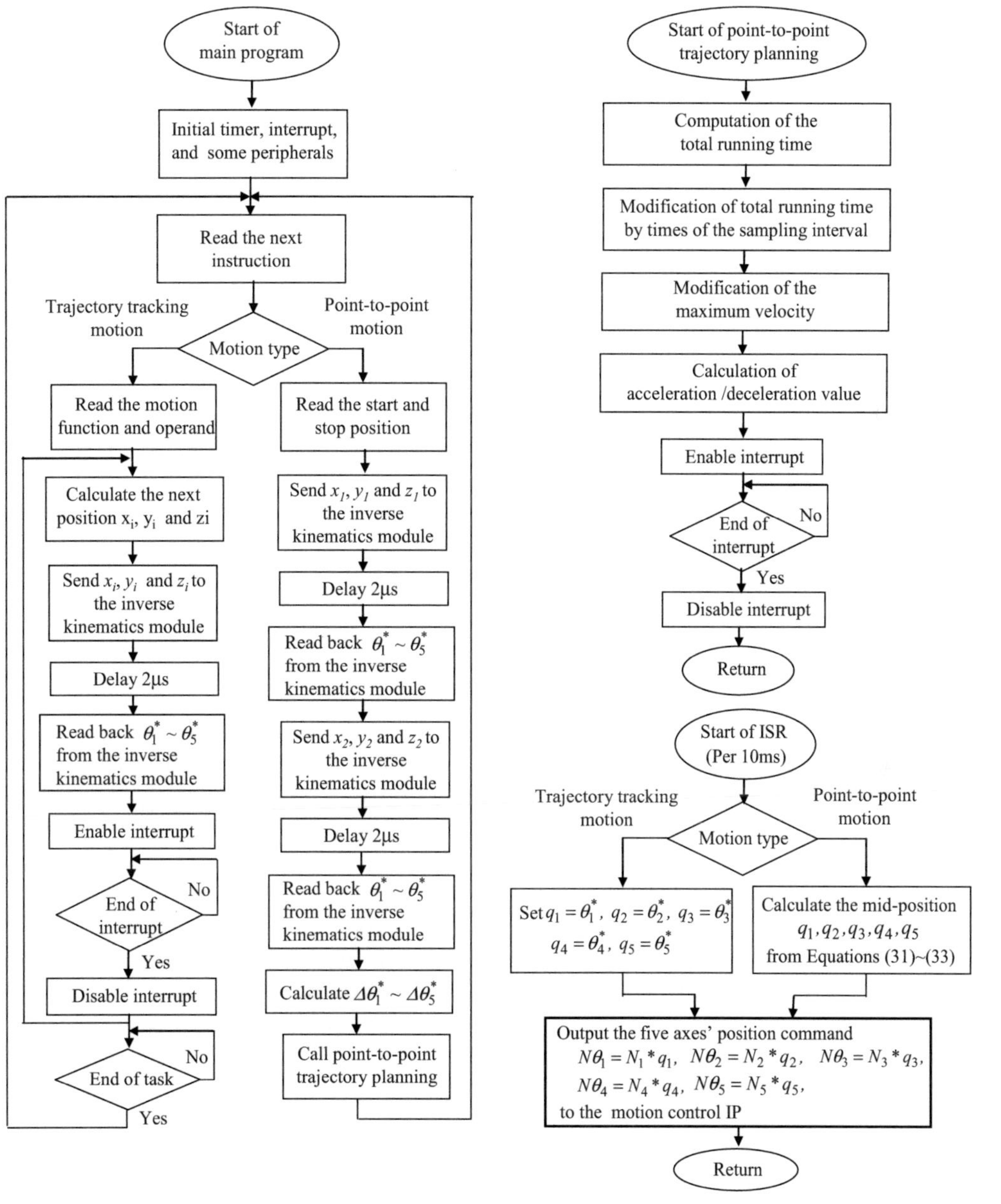

Figure 5. Flow charts of main and ISR program in Nios II embedded processor

The motion control IP implemented by hardware, as depicted in Fig. 4, is composed of a frequency divider, an inverse kinematics module, a five-axis position controller module, a five-axis speed estimated and speed controller module, five sets of QEP (Quadrature

Encoder Pulse) module and five sets of PWM (Pulse Width Modulation) module. The frequency divider generates 50MHz (*clk*), 762Hz (*clk_ctrp*), 1525Hz (*clk_ctrs*) and 50MHz (*clk_sp*) to supply all module circuits of the motion control IP in Fig.4. To reduce the usage in FPGA, the FSM method is proposed to design the circuit in the position/speed controller module and the inverse kinematics module, as well as the VHDL is used to describe the behaviors. Figure 6(a) and 6(b) respectively describe the behavior of the single axis *P* controller in the position controller module and the single axis *PI* controller in the speed controller module. There are 4 steps and 12 steps to complete the computation of the single axis *P* controller and the single axis *PI* controller, respectively. And only one multiplier, one adder and one comparator circuits are used in Fig.6. The data type is 16-bit length with Q15 format and 2's complement operation. To prevent numerical overflow and alleviate windup phenomenon, the output values of I controller and *PI* controller are both limited within a specific range. The step-by-step computation of single axis controller is extended to five axes controller and illustrated in Fig.7. We can find that, it needs 20 steps to complete the computation of overall five axes P controllers in position control in Fig. 7(a), and 60 steps to complete the computation of overall five axes PI controllers in speed control in Fig. 7(b). However, in Fig.7(a) and Fig.7(b), it still need only one multiplier, one adder and one comparator circuits to complete the overall circuit, respectively. Furthermore, because the execution time of per step machine is designed with 20ns (*clk_sp* : 50MHz clock), the total execution time of five axes controller needs 0.4μs and 1.2 μs in position and speed loop control, respectively. And, the sampling frequency in position and speed loop control are respectively designed with 762 Hz (1.31ms) and 1525 Hz (0.655ms) in our proposed servo system of robot manipulator. It apparently reveals that it is enough time to complete the five axes controller during the control interval. Designed with FSM method, although five axes controllers need much computation time than the single axis controller, but the resource usage in FPGA is the same. The circuit of the *QEP* module is shown in Fig.8(a), which consists of two digital filters, a decoder and an up-down counter. The filter is used for reducing the noise effect of the input signals *PA* and *PB*. The pulse count signal *PLS* and the rotating direction signal *DIR* are obtained using the filtered signals through the decoder circuit. The *PLS* signal is a four times frequency pulses of the input signals *PA* or *PB*. The *Qep* value can be obtained using *PLS* and *DIR* signals through a directional up-down counter. Figure 8(b) and (c) are the simulation results for *QEP* module. Figure 8(b) shows the count-up mode where *PA* signal leads to *PB* signal, and Fig. 8(c) shows the count-down mode while *PA* signal lags to *PB* phase signal. The circuit of the *PWM* module is shown in Fig.9(a), which consists of two up-down counters, two comparators, a frequency divider, a dead-band generating circuit, four *AND* logic gates and a *NOR* logic gate. The first up-down counter generates an 18 kHz frequency symmetrical triangle wave. The counter value will continue comparing with the input signal *u* and generating a *PWM* signal through the comparator circuit. To prevent the short circuit occurred while both gates of the inverter are triggered-on at the same time; a circuit with 1.28μs dead-band value is designed. Finally, after the output signals of *PWM* comparator and dead-band comparator through *AND* and *NOR* logic gates circuit, four *PWM* signals with 18 *kHz* frequency and 1.28μs dead-band values are generated and sent out to trigger the IGBT device. Figure 9(b) shows the *PWM* simulated waveforms, and this result demonstrates the correctness of the *PWM* module.

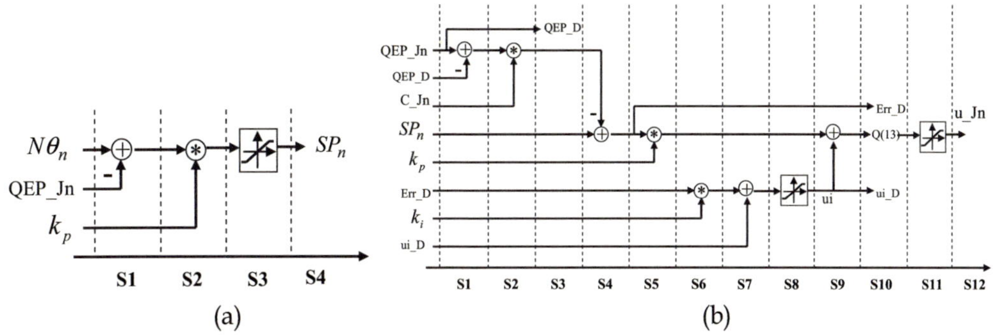

Figure 6 (a) The single axis P controller circuit in the position controller module (b) Speed estimation and single axis PI controller circuit in the speed controller module

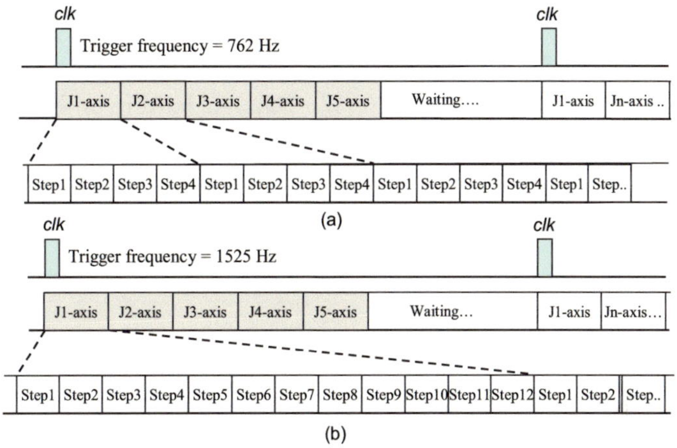

Figure 7. Sequence control signal for five axes (a) position controller (b) speed controller

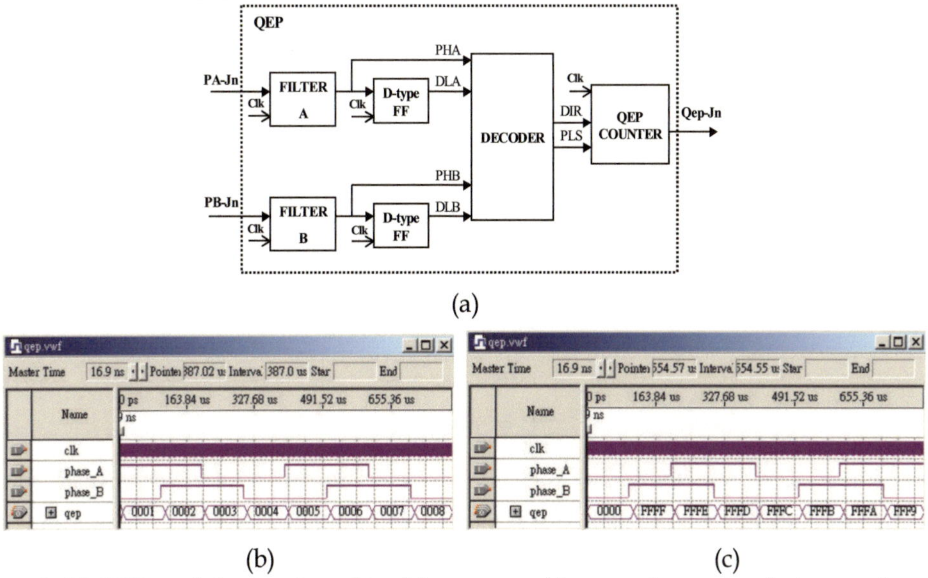

Figure 8. (a) QEP module circuit and its (b) count-up (c) count-down simulation result

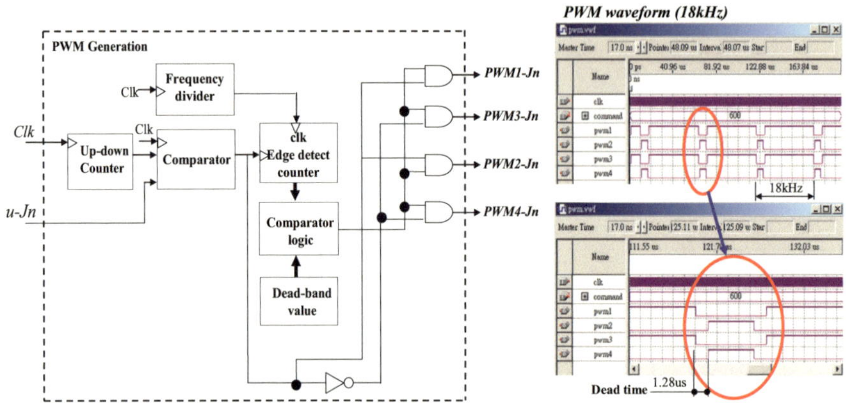

(a) (b)

Figure 9. (a) PWM module circuit and its (b) simulation result

From the equations of inverse kinematics in (17)~(23), it is hard to understand the designed code of VHDL. Therefore, we reformulate it as follows.

$$v_1 = d_1 - d_5 - z \tag{41}$$

$$\theta_1 = a\tan 2(y, x) \tag{42}$$

$$v_2 = b = \sqrt{x^2 + y^2} \tag{43}$$

$$\theta_5 = \theta_1 \tag{44}$$

$$v_3 = (v_2^2 + v_1^2 - a_2^2 - a_3^2) / 2a_2a_3 \tag{45}$$

$$\theta_3 = \cos^{-1}(v_3) \tag{46}$$

$$v_4 = sin\theta_3 \tag{47}$$

$$v_5 = a_3 \, cos\theta_3 + a_2 \tag{48}$$

$$v_6 = a_3 \, cos\theta_3 \tag{49}$$

$$v_7 = a_3 \, sin\theta_3 \tag{50}$$

$$\begin{aligned} S_2 &= (a_2 + a_3\cos\theta_3)(d_1 - d_5 - z) - a_3 b \sin\theta_3 \\ &= v_5 \cdot v_1 - v_7 \cdot v_2 \end{aligned} \tag{51}$$

$$\begin{aligned} C_2 &= (a_2 + a_3 \, cos\theta_3) b + a_3 \, sin\theta_3 (d_1 - d_5 - z) \\ &= v_5 \cdot v_2 + v_7 \cdot v_1 \end{aligned} \tag{52}$$

$$\theta_2 = a\tan 2(S_2, C_2) \tag{53}$$

$$\theta_4 = -\theta_2 - \theta_3 \tag{54}$$

where $v_1 \sim v_7, S_2, C_2$ are temporary variables. The computation of inverse kinematics in (41)~(54) by parallel operation method is resource consumption in FPGA; therefore, an FSM is adopted to model the inverse kinematics and illustrated in Fig. 10, then the VHDL is adopted to describe the circuit of FSM. The data type is 16-bit length with Q15 format and 2's complement operation. There are total 42 steps with 20ns/step to perform the overall computation of inverse kinematics. The circuits in Fig.10 need two multipliers, one divider, two adders, one component for square root function, one component for arctan function, one component for arcos function, one look-up-table for sin function and some comparators for atan2 function. The multiplier, adder, divider and square root circuit are Altera LPM (Library Parameterized Modules) standard component but the arcos and arctan function are our developed component. The divider and square root circuits respectively need 5 steps executing time in Fig. 10; the arcos and arctan component need 9 steps executing time; others circuit, such as adder, multiplier, LUT etc., needs only one step executing time. Furthermore, to realize the overall computation of the inverse kinematics needs 840ns (20ns/step* 42 steps) computation time. In our previous experience, the computation for inverse kinematics in Nios II processor using C language by software will take 5.6ms. Therefore, the computation of the inverse kinematics realized by hardware in FPGA is 6,666 times faster than those in software by Nios II processor. Finally, to test the correctness of the computation of inverse kinematics in Fig.10, three cases at Cartesian space (0,0,0), (200,100,300) and (300,300,300) are simulated and transformed to the joint space. The simulation results are shown in Fig. 11 that *com_x, com_y, com_z* denote input commands at Cartesian space, and those *com_1* to *com_5* denote output position commands at joint space. The simulation result in Figure 11 demonstrates the correctness of the proposed design method of the inverse kinematics.

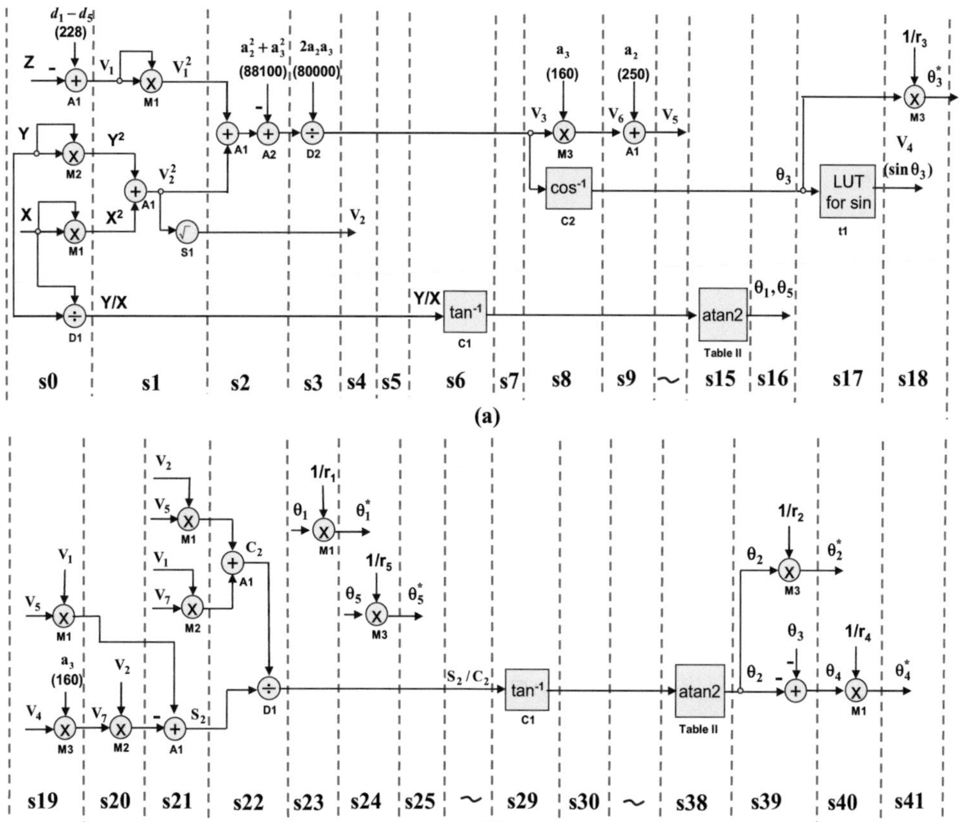

Figure 10. Circuit of computing the inverse kinematics using finite state machine

Finally, the FPGA utility of the motion control IC for robot manipulator in Fig. 4 is evaluated and the result is listed in Table 3. The overall circuits included a Nios II embedded processor (5.1%, 2,468 ALUTs) and a motion control IP (14.4%, 6,948 ALUTs) in Fig. 4, use 19.5% (9,416 ALUTs) utility of the Stratix II EP2S60. Nevertheless, for the cost consideration, the less expensive device - Stratix II EP2S15 (12,480 ALUTs and 419,329 RAM bits) is a better choice. In Fig. 4, the software/hardware program in parallel processing enhances the controller performance of the motion system for the robot manipulator.

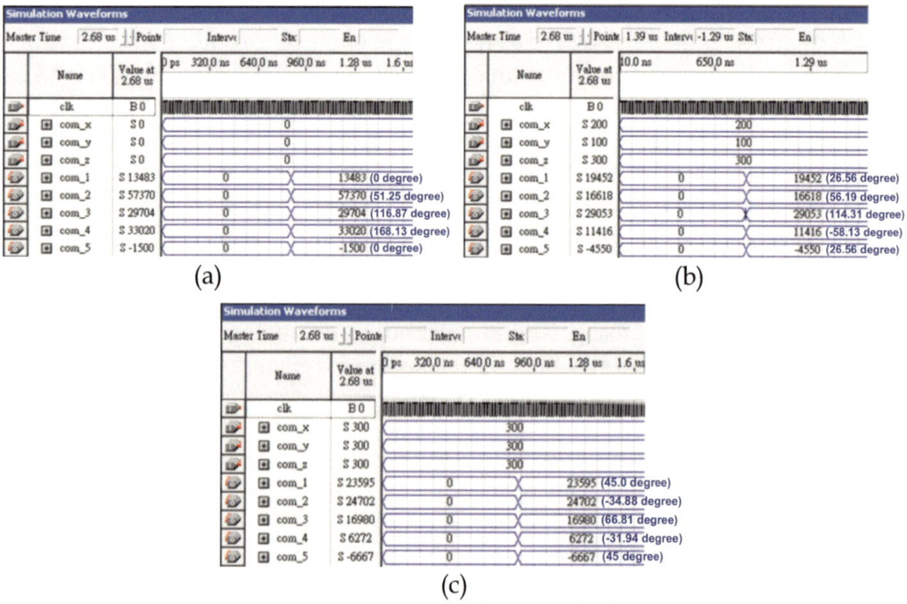

Figure 11. Simulation result of Cartesian space (a) (0,0,0) (b) (200,100,300) (c) (300,300,300) to join space

Module Circuit	Resource usage in FPGA (ALUTs)
Nios II embedded processor IP	2,468
Five sets of PMW module	316
Five sets of QEP module	285
Position controller module	206
Speed controller module	748
Inverse kinematics module	5,322
Others circuit	71
Total	**9,416**

Table 3. Utility evaluation of the motion control IC for robot manipulator in FPGA

5. Experimental system and results

Figure 12 presents the overall experimental system which includes an FPGA experimental board, five sets of inverter, five sets rectifier, a power supplier system and a Mitsubishi Movemaster RV-M1 micro articulated robot. The micro articulated robot has five servo axes (excluding the hand) and its specification is shown in Fig.13. Each axis is driven by a 24V DC servo motor with a reduction gear. The operation ranges of the articulated robot are wrist roll ±180 degrees (J5-axis), wrist pitch ±90 degrees (J4-axis), elbow rotation 110 degrees (J3-axis), shoulder rotation 130 degrees (J2-axis) and waist rotation 300 degrees (J1-axis). The gear ratios for J1 to J5 axis of the robot are 1:100, 1:170, 1:110, 1:180 and 1:110, respectively. Each DC motor is attached an optical encoder. Through four times frequency circuit, the encoder pulses generate 800pulses/cycle at J1 to J3 axis and 380pulses/cycle at J4 and J5 axis. The maximum path velocity is 1000mm/s and the lifting capacity is 1.2kg including the hand. The total weight of this robot is 19 kg. The inverter has 4 sets of IGBT type power transistors. The collector-emitter voltage of the IGBT is rating 600V, the gate-emitter voltage is rating ±12V, and the collector current in DC is rating 25A and in short time (1ms) is 50A. The photo-IC, Toshiba TLP250, is used for gate driving circuit of IGBT. Input signals of the inverter are PWM signals from FPGA chip. The FPGA-Altera Stratix II EP2S60F672C5ES in Fig. 1(a) is used to develop a full digital motion controller for robot manipulator. A Nios II embedded processor can be download to this FPGA chip.

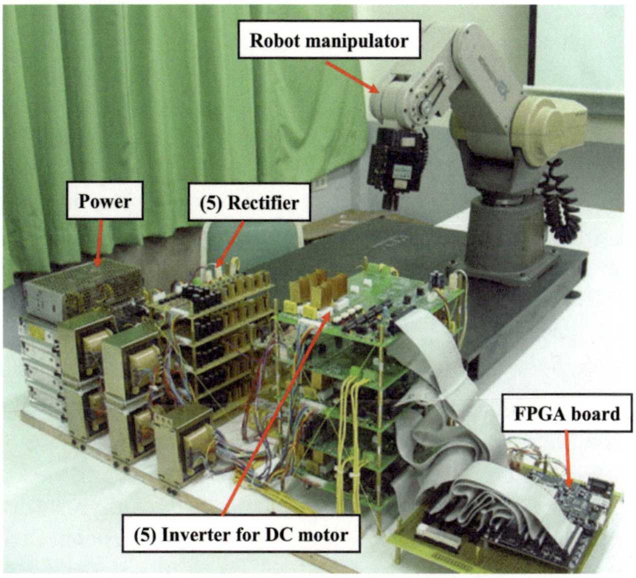

Figure 12. Experimental system

In Fig.12 or Fig.4, the realization in PWM switching frequency, dead-band of inverter, position and speed control sampling frequency are set at 18k Hz, 1.28μs, 762 Hz and 1525 Hz, respectively. Moreover, in the position loop P controller design, the controller parameters at each axis of robot manipulator are selected with identical values by P-gain with 2.4. However, in the speed loop PI controller design, the controller parameters at each J1~J5 axis are selected with different values by [3.17, 0.05], [3.05, 0.12], [2.68, 0.07], [2.68, 0.12] and [2.44, 0.06], respectively.

To confirm the effectiveness of the proposed motion control IC, the square-wave position command with ±3 degrees amplitude and 2 seconds period is firstly adopted to test the dynamic response performance. At the beginning of the step response testing, the robot manipulator is moved to a specified attitude for joints J1-J5 rotating at the [9°, 40°, 60°, 45°, 10°] position. Figure 14 shows the experimental results of the step response under these design conditions, where the rise time of the step responses are with 124ms, 81ms, 80ms, 151ms and 127 ms for axis J1-J5, respectively. The results also indicate that these step responses have almost zero steady-state error and no oscillation. Next, to test the performance of a point-to-point motion control for the robot manipulator, a specified path is run where the robot moves from the point 1 position, (94.3, 303.5, 403.9) mm to the point 2 position, (299.8, 0, 199.6) mm, then back to point 1. After through inverse kinematics computation in (17) ~ (23), moving each joint rotation angle of the robot from point 1 to point 2 need rotation of -72.74° (-16,346 Pulses), 23.5° (8,916 Pulses), 32.16° (8,173 Pulses), -56.72° (-11,145 Pulses) and -71.18° (-8,173 Pulses), respectively. Additionally, a point-to-point control scheme with constant acceleration/deceleration trapezoid velocity profile adopts to smooth the robot manipulator movement. Applying this motion control scheme, all joins of robot rotate with simultaneous starting and simultaneous stopping time. The acceleration/deceleration time and overall running time are set to 256ms and 1s, and the computation procedure in paragraph 3.3.1 is applied. Figure 15 shows the tracking results of using this design condition at each link. The results indicate that the motion of each robot link produces perfect tracking with the target command in the position or the velocity response. Furthermore, the path trajectory among the position command, actual position trajectory and point-to-point straight line in Cartesian space R^3 (x,y,z) are compared, with the results shown in Fig. 16. Analytical results indicate that the actual position trajectory can precisely track the position command, but that the middle path between two points can not be specified in advanced. Next, the performance of the linear trajectory tracking of the proposed motion control IC is tested, revealing that the robot can be precisely controlled at the specified path trajectory or not. The linear trajectory path is specified where the robot manipulator moves from the starting position, (200, 100, 400) mm to the ending position, (300, 0, 300) mm, then back to starting position. The linear trajectory command is generated 100 equidistant segmented points from the starting to the ending position. Each middle position at Cartesian space R^3 (x,y,z) will be transformed to joint space $(\theta_1^*,\theta_2^*,\theta_3^*,\theta_4^*,\theta_5^*)$, through the inverse kinematics computation, then sent to the servo controller of the robot manipulator. The tracking result of linear trajectory tracking is displayed in Fig. 17. The overall running time is 1 second and the tracking errors are less than 4mm. Similarly, the circular trajectory command generated by 300 equidistant segmented points with center (220,200,300) mm and radius 50 mm is tested again and its tracking result is shown in Fig. 18. The overall running time is 3 second and the trajectory tracking errors at each axis are less than 2.5mm. Figures 17~18 show the good motion tracking results under a prescribed planning trajectory. Therefore, experimental results from Figs. 14~18, demonstrate that the proposed FPGA-based motion control IC for robot manipulator ensures effectiveness and correctness

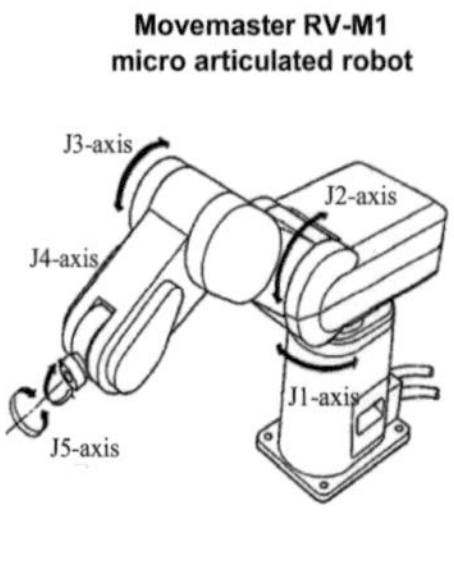

Name of link	No.	Max. working range (degree)	Encoder pulse x 4 (ppr)	Gear ratio
waist	J1	300° (67416 pulse)	800	1:100
should	J2	130° (49311 pulse)	800	1:170
elbow	J3	110° (27958 pulse)	800	1:110
wrist pitch	J4	±90° (±17676 pulse)	380	1:180
wrist roll	J5	±180° (±20667 pulse)	380	1:110

Figure 13. Mitsubishi Movemaster RV-M1 micro articulated robot

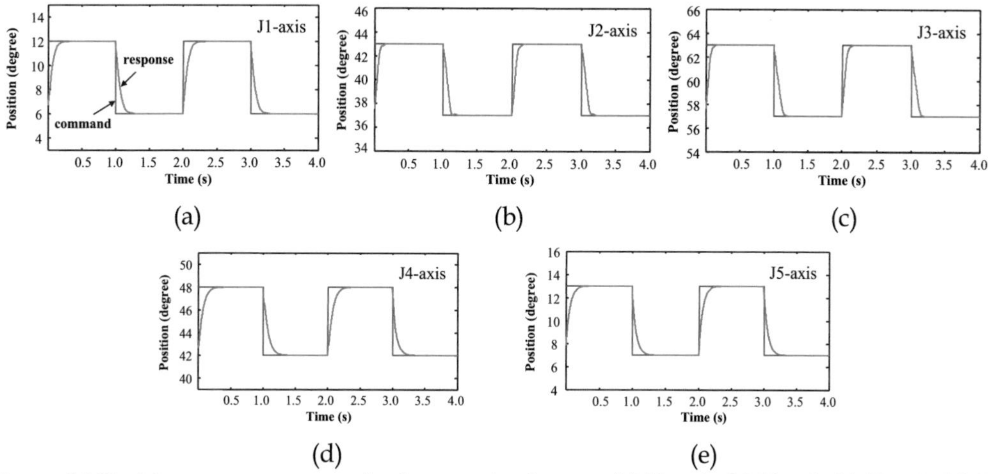

Figure 14 Position step response of robot manipulator at (a) J1 axis (b) J2 axis (c) J3 axis (d) J4 axis (e) J5 axis

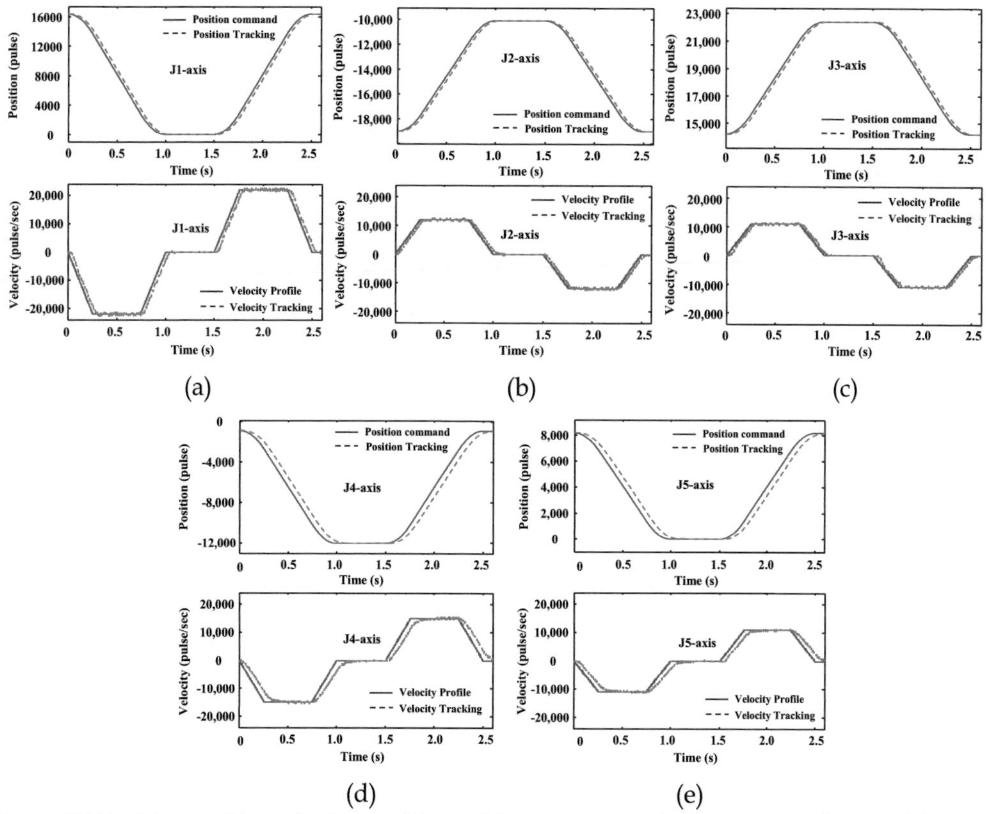

Figure 15 Position and its velocity profile tracking response of robot manipulator at (a) J1 axis (b) J2 axis (c) J3 axis (d) J4 axis (e) J5 axis

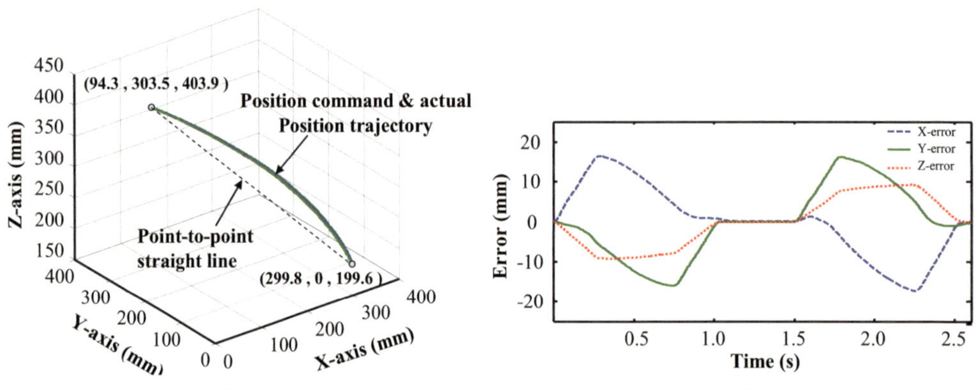

(a) (b)

Figure 16 (a) Point-to-point path tracking result of robot manipulator (b) tracking error

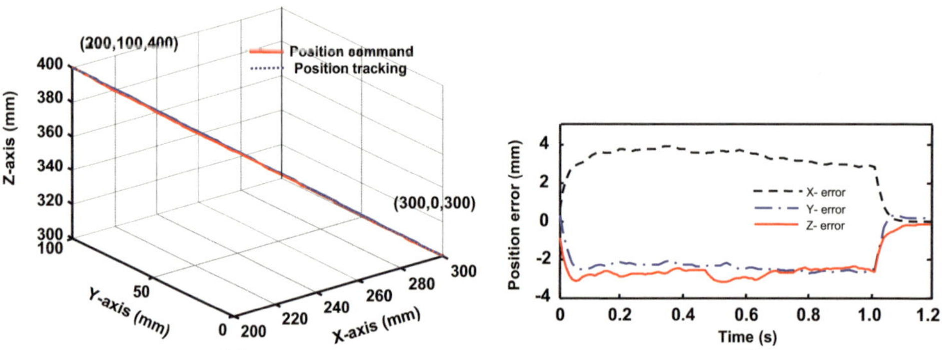

(a) (b)

Fig. 17 (a) Linear trajectory tracking result of robot manipulator (b) tracking error

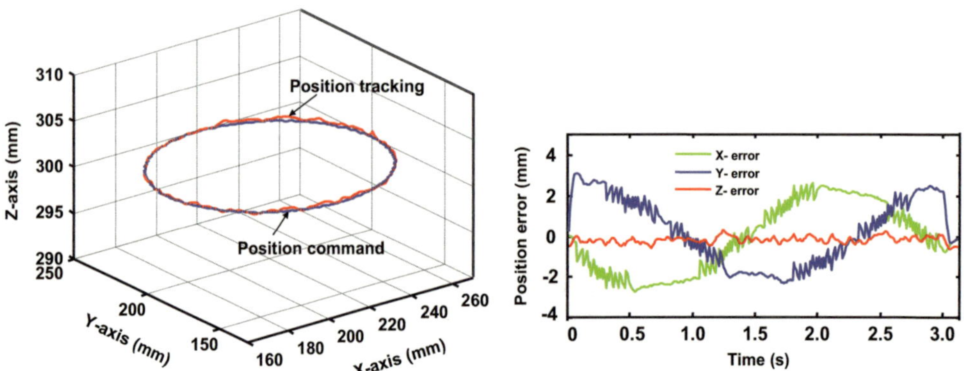

(a) (b)

Figure 18 (a) Circular trajectory tracking result of robot manipulator (b) tracking error

6. Conclusion

This study presents a motion control IC for robot manipulator based on novel FPGA technology. The main contributions herein are summarized as follows.

1. The functionalities required to build a fully digital motion controller of a five-axis robot manipulator, such as the function of a motion trajectory planning, an inverse kinematics, five axes position and speed controller, five sets of PWM and five sets of QEP circuits have been integrated and realized in one FPGA chip.
2. The function of inverse kinematics is successfully implemented by hardware in FPGA; as the result, it diminishes the computation time from 5.6ms using Nios II processor to 840ns using FPGA hardware, and increases the system performance.
3. The software/hardware co-design technology under SoPC environment has been successfully applied to the motion controller of robot manipulator.

Finally, the experimental results by the step response, the point-to-point motion trajectory response and the linear and circular motion trajectory response, have been revealed that based on the novel FPGA technology, the software/hardware co-design method with parallel operation ensures a good performance in the motion control system of robot manipulator.

Compared with DSP, using FPGA in the proposed control architecture has the following benefits.

1. Inverse kinematics and servo position controllers are implemented by hardware and the trajectory planning is implemented by software, which can all be programmable design. Therefore, the flexibility of designing a specified function of robot motion controller is greatly increased.
2. Parallel processing in each block function of the motion controller makes the dynamic performance of the robot's servo drive increasable.
3. In the commercial DSP product, it is difficult to integrate all the functions of implementing a five-axis motion controller for robot manipulator into only one chip.

7. References

Altera Corporation, (2004). *SOPC World.*

Altera (2008): www.altera.com

Hall, T.S. & Hamblen, J.O. (2004). System-on-a-programmable-chip development platforms in the classroom, *IEEE Trans. on Education*, Vol. 47, No. 4, pp.502-507.

Monmasson, E. & Cirstea, M.N. (2007) FPGA design methodology for industrial control systems - a review, *IEEE Trans. Industrial Electronics*, Vol. 54, No. 4, pp.1824-1842.

Kabuka, M.; Glaskowsky, P. & Miranda, J. (1988). Microcontroller-based Architecture for Control of a Six Joints Robot Arm, *IEEE Trans. on Industrial Electronics*, Vol. 35, No. 2, pp. 217-221.

Kung, Y.S. & Shu, G.S. (2005). Design and Implementation of a Control IC for Vertical Articulated Robot Arm using SOPC Technology, *Proceeding of IEEE International Conference on Mechatronics*, 2005, pp. 532~536.

Kung, Y.S.; Tseng, K.H. & Tai, F.Y. (2006) FPGA-based servo control IC for X-Y table, *Proceedings of the IEEE International Conference on Industrial Technology*, pp. 2913-2918.

Kung, Y.S. & Tsai, M.H. (2007). FPGA-based speed control IC for PMSM drive with adaptive fuzzy control, *IEEE Trans. on Power Electronics*, Vol. 22, No. 6, pp. 2476-2486.

Lewis, F.L.; Abdallah C.T. & Dawson, D.M. (1993). *Control of Robot Manipulators,* Macmillan Publishing Company.

Li, T.S.; Chang S.J. & Chen, Y.X. (2003) Implementation of Human-like Driving Skills by Autonomous Fuzzy Behavior Control on an FPGA-based Car-like Mobile Robot, *IEEE Trans. on Industrial Electronics*, Vol. 50, No.5, pp. 867-880.

Oh, S.N.; Kim, K.I. & Lim, S. (2003). Motion Control of Biped Robots using a Single-Chip Drive, *Proceeding of IEEE International Conference on Robotics & Automation*, pp. 2461~2469.

Schilling, R. (1998). *Fundamentals of Robotics – Analysis and control,* Prentice-Hall International.

Shao, X. & Sun, D. (2005). A FPGA-based Motion Control IC Design, *Proceeding of IEEE International Conference on Industrial Technology*, pp. 131-136.

Wei, R.; Gao, X.H.; Jin, M.H.; Liu, Y.W.; Liu, H.; Seitz, N.; Gruber, R. & Hirzinger, G. (2005). FPGA based Hardware Architecture for HIT/DLR Hand, *Proceeding of IEEE/RSJ International Conference on intelligent Robots and System*, pp. 523~528.

Xu, N.; Liu, H.; Chen, X. & Zhou, Z. (2003). Implementation of DVB Demultiplexer System with System-on-a-programmable-chip FPGA, *Proceeding of 5th International Conference on ASIC*, Vol. 2, pp. 954-957.

Yang, G.; Liu, Y.; Cui, N. & Zhao, P. (2006). Design and Implementation of a FPGA-based AC Servo System, *Proceedings of the Sixth World Congress on the Intelligent Control and Automation,* Vol. 2, pp. 8145-8149.

Yasuda, G. (2000). Microcontroller Implementation for Distributed Motion Control of Mobile Robots, *Proceeding of International workshop on Advanced Motion Control,* pp. 114-119.

Zhou, Z.; Li, T.; Takahahi, T. & Ho, E. (2004). FPGA realization of a high-performance servo controller for PMSM, *Proceeding of the 9th IEEE Application Power Electronics conference and Exposition,* Vol.3, pp. 1604-1609.

17

Experimental Identification of the Inverse Dynamic Model: Minimal Encoder Resolution Needed Application to an Industrial Robot Arm and a Haptic Interface

Marcassus Nicolas[1], Alexandre Janot[2], Pierre-Olivier Vandanjon[3] and Maxime Gautier[1]
[1]IRCCyN & Nantes University, [2]Haption S.A. & CEA, LIST, Service de Robotique Interactive, [3]Laboratoire Central des Ponts et Chaussées France

1. Introduction

To develop appropriate control laws and use fully the capacities of robots, a precise modelization is needed. Classic models such as ARX or ARMAX can be used but in the robotic field the Inverse Dynamic Model (IDM) gives far better results. In this model, the motor torques depend on the acceleration, speed and position of each joint, and of the physical parameters of the link of the robots (inertia, mass gravity, stiffness and friction).
The parametric identification estimates the values of these last parameters. These estimations can also help to improve the mechanical conception during retro-engineering steps... It comes that the identification process must be as accurate and reliable as possible.
The most popular identification methods are based on the least-squares (LS) regression "because of its simplicity" (Atkeson et al., 1986), (Swevers et al., 1997), (Ha et al., 1989), (Kawasaki & Nishimura 1988), (Khosla & Kanade 1985), (Kozlowski 1998), (Prüfer et al., 1994) and (Raucent et al., 1992). In the last two decades, the IRCCyN robotic team has designed an identification process using IDM of robots and LS regressions which will be developed in the second part of this chapter. This technique was applied and improved on several robots and prototypes (see Gautier et al., 1995 - Gautier & Poignet 2002 for example). More recently, this method was also successfully applied on haptic devices (Janot et al., 2007).
However, it is very difficult to know how much these methods are dependent on the measurement accuracy, especially when the identification process takes place when the system is controlled by feedback. So, we ignore the necessary resolution they require to produce good quality and reliable results.
Some identification techniques seem robust with respect to measurement noises. They are called "robust identification methods". But even if they give reliable results, they are only applied on linear systems and, overall, they are very time consuming as can be seen in (Hampel, 1971) and (Hubert, 1981). Finally, it seems difficult to apply them on robots and we do not know how much they are robust with respect to these noises.

Another simple and adequate way consists in derivating the CESTAC method (Contrôle et Estimation Stochastique des Arrondis de Calculs developed in Vignes & La Porte, 1974) which is based on a probabilistic approach of round-off errors using a random rounding mode. The third part of this chapter introduces the design and the application of a derivate of the CESTAC method enabling us to estimate the minimal resolution needed for an accurate parametric identification.
This theoretical technique was successfully applied on a 6 degrees of freedom (DOF) industrial arm (Marcassus et al., 2007) and a 3 DOF haptic device (Janot et al., 2007), the major results obtained will be used to illustrate the use of this new tool of reliability.

2. Inverse dynamic model and Least Squares estimation

2.1 General Inverse Dynamic Model

The IDM calculates the joint torques as a function of joint positions, velocities and accelerations. It is usually represented by the following equation:

$$\Gamma=A(q)\ddot{q}+H(q,\dot{q})+F_v\dot{q}+F_s\,\mathrm{sign}(\dot{q}) \tag{1}$$

where Γ is the torques vector of the actuators, q, and are respectively the joint positions, velocities and accelerations vector of each links, $A(q)$ is the inertia matrix of the robot, $H(q,)$ is the vector regrouping Coriolis, centrifugal and gravity torques applied on the links, F_v and F_s are respectively the viscous and Coulomb friction matrices.
The parameters used in this model are $XX_j, XY_j, XZ_j, YY_j, YZ_j, ZZ_j$ the components of the inertia tensor of link j, noted ${}^j\mathbf{J}_j$, the mass of the link j called M_j, the inertia of the actuator noted Ia_j, the first moments vector of link j around the origin of frame j noted ${}^j\mathbf{MS}_j=\left[MX_j MY_j\ MZ_j\right]^T$, FV_j and FS_j respectively the viscous and Coulomb friction coefficients and an offset of current measurement noted $OFFSET_j$.
The kinetic and potential energies being linear with respect to the inertial parameters, so is the dynamic model (Gautier & Khalil, 1990). Equation (1) can thus be rewritten as:

$$\Gamma=D_s(q,\dot{q},\ddot{q})\chi \tag{2}$$

where $D_s(q,\dot{q},\ddot{q})$ is a linear regressor and χ is a vector composed of the inertial parameters, it is written:

$$\chi=\left[\chi^{1T}\quad \chi^{2T}\quad \cdots\quad \chi^{nT}\right]^T \tag{3}$$

with χ^j the dynamic parameters of link j and its actuator written:

$$\chi^j = [\,XX_j, XY_j, XZ_j, YY_j, YZ_j, ZZ_j, MX_j, MY_j, MZ_j, M_j, Ia_j, FV_j, FS_j, OFFSET_j]^T \tag{4}$$

To calculate the dynamic model we do not need all these parameters but only a set of base parameters which are the ones necessary for this computation. They can be deduced from

the classical parameters by eliminating those which have no effect on the dynamic model and by regrouping some others. Actually, they represent the only identifiable parameters.
Two main methods have been designed for calculating them: a direct and recursive method based on calculation of the energy (Gautier & Khalil, 1990) and a method based on QR numerical decomposition (Gautier, 1991). The numerical method is particularly useful for robots consisting of closed loops.
By considering only the b base parameters, equation (2) has to be rewritten as follows:

$$\Gamma = D(q,\dot{q},\ddot{q})\,\chi_b \tag{5}$$

where $D(q,\dot{q},\ddot{q})$ is the linear regressor and χ_b is the vector composed of the base parameters.

2.1 Least Squares Method

2.1.1 General theory

Generally, ordinary LS technique is used to estimate the base parameters by solving an over-determined linear system obtained from the sampling of the dynamic model, along a specifically dedicated trajectory (q, $\dot{q}$, $\ddot{q}$), (Gautier et al., 1995) or (Khalil et al. 2007).

X being the b minimum parameters vector to be identified, Y the torques measurements vector, W the observation matrix and ρ the vector of errors, the system is described as follows:

$$Y(\Gamma) = W(q,\dot{q},\ddot{q})X + \rho \tag{6}$$

$\hat{X}$ being the solution of the LS regression, it minimizes the 2-norm of the errors vector ρ. W is a r×b full rank and well conditioned matrix, obtained by tracking exciting trajectories and by considering the base parameters, r being the number of samplings along a given trajectory, r>>b.
Hence, there is only one solution $\hat{X}$, (Gautier, 1997) :

$$\hat{X} = \left(W^T W\right)^{-1} W^T Y = W^{+} Y \tag{7}$$

with W^+ the pseudo-invert matrix of W.
Standard deviations of the identified parameters, $\sigma_{\hat{X}_i}$, are estimated using classical and simple results from statistics considering that the matrix W is deterministic and ρ is a zero-mean additive independent noise with a standard deviation such as:

$$C_\rho = E(\rho\rho^T) = \sigma_\rho^2 I_r \tag{8}$$

where E is the expectation operator and I_r the r×r identity matrix. An unbiased estimation of σ_ρ is:

$$\sigma_{\rho}^{2}=\frac{\left\|Y-W\hat{X}\right\|^{2}}{r-b} \tag{9}$$

The variance-covariance matrix of the standard deviation is calculated as follows:

$$C_{\hat{X}\hat{X}}=\sigma_{\rho}^{2}\left(W^{T}W\right)^{-1} \tag{10}$$

Then $\sigma_{\hat{X}_j}^{2}=C_{\hat{X}\hat{X}_{jj}}$ is the jth diagonal coefficient of $C_{\hat{X}\hat{X}}$, and $\%\sigma_{\hat{X}_{jr}}$ the relative standard deviation of the jth parameter is given by:

$$\%\sigma_{\hat{X}_{jr}}=100\frac{\sigma_{\hat{X}_j}}{\left|\hat{X}_j\right|} \tag{11}$$

However, in practice, W is not deterministic. This problem can be solved by filtering the measurement vector Y and the columns of the observation matrix W.

2.2.2 Data Filtering

Vectors Y and q are measures sampled during an experimental test. We know that the LS method is sensitive to outliers and leverage points so a median filter is applied to eliminate them.

The derivatives $\dot{q}$ and $\ddot{q}$ are obtained without phase shift using a centered finite difference of the vector q. Knowing that q is perturbed by high frequency noises, which will be amplified by the numeric derivations, a lowpass filter, whose order is greater than 2, is applied on q and $\dot{q}$. The choice of the cut-off frequency ω_f is very sensitive because the filtered data $(q_f,\dot{q}_f,\ddot{q}_f)$ must be equal to the vector $(q,\dot{q},\ddot{q})$ in the range $[0, \omega_f]$ in order to avoid distortion of the dynamic regressor. So the filter must have a flat amplitude characteristic without phase shift in the range $[0, \omega_f]$, with the rule of thumb $\omega_f>10*\omega_{dyn}$, where ω_{dyn} represents the dynamic frequency of the system. Considering an off-line identification, it is easily achieved with a non-causal zero-phase digital filter by processing the input data through an IIR lowpass Butterworth filter in both the forward and reverse directions.

Since the measurement vector Y and matrix W have been filtered, a new filtered linear system is defined:

$$Y_f=W_fX+\rho_f \tag{12}$$

Because there is no more signal in the range $[\omega_f, \omega_s/2]$, where ω_s is the sampling frequency, the vector Y_f and the columns of W_f are resampled at a lower rate after a lowpass filtering, keeping one sample over nd samples, in order to obtain the final system to be identified. This process called parallel filtering is done thanks to the "decimate" function available in the Signal Processing Toolbox of Matlab. To have the same cut-off frequency ωf for the lowpass filter, we choose:

$$n_d=0.8\omega_s/2\omega_f \tag{13}$$

So the final linear system is:

$$Y_{fd} = W_{fd} X + \rho_{fd} \tag{14}$$

2.2.3 Exciting Trajectories

Knowing the base parameters, exciting trajectories must be designed. In fact, they represent the trajectories which excite well these parameters. Design and calculations of these trajectories can be found in (Gautier & Khalil, 1991).
If the trajectories are enough exciting, the conditioning number of W, (denoted cond(W)) is close to 1. However, in practice, this conditioning number can reach few hundreds depending of the high number of base parameters.
If the trajectories are not enough exciting, the system is ill conditioned, undesirable gatherings occur between inertial parameters and finally the results have no sense whatever the encoder resolutions.

3. Theory of the CESTAC method

From a theoretical point of view, the LS assumptions are violated in practical applications. In equation (6), the observation matrix **W** is built from the joint positions **q** which are measured and from $\dot{\mathbf{q}}$, $\ddot{\mathbf{q}}$ which are often computed numerically from **q**. Therefore the observation matrix is noisy. Moreover identification process takes place when the robot is controlled by feedback. It is well known that these violations of assumption imply that the LS estimator is biased.
An adequate and simple way to evaluate the robustness of the LS estimator with respect to the quantization errors (which contribute significantly to the bias of the estimator) consists in deriving the CESTAC method which is based on a probabilistic approach of round-off errors using a random rounding mode defined below:
Definition: each real number *x*, which is not a floating-point number, is bounded by two consecutive floating-point numbers denoted respectively X^- (rounded down) and X^+ (rounded up). The random rounding mode defines the floating-point number X representing *x* as being one of two values X^- or X^+ with probability 1/2.
Thus, with this random rounding mode, the same program run several times provides different results, due to different round-off errors. Under some assumptions, X can be considered as a quasi-Gaussian distribution (Jezequel, 2004).
In our case, the position is perturbed by the encoder resolution. This measurement can be counted up (q^+) as it can be counted down (q^-) at each sample. So, we can derive the CESTAC method by building a new position:

$$q_{CESTAC} = \begin{cases} q^- = q - \Delta \text{ with } P(q^-) = 50\% \\ q^+ = q + \Delta \text{ with } P(q^+) = 50\% \end{cases} \tag{15}$$

Equation (15) defines our rounding mode. Then, $\mathbf{q}_{CESTAC}$ is filtered thanks to the data filtering previously described. Finally, we build a new linear regression called $\mathbf{W}_{CESTAC}$:

$$\mathbf{W}_{CESTAC} = \mathbf{W}(\mathbf{q}_{CESTAC}, \dot{\mathbf{q}}_{CESTAC}, \ddot{\mathbf{q}}_{CESTAC}) \tag{16}$$

Each column of $\mathbf{W}_{CESTAC}$ is filtered by using the decimation filter as explain in section xx. Hence, we obtain a new estimation of base parameters denoted $\mathbf{X}_{CESTAC}$:

$$\mathbf{X}_{CESTAC} = \mathbf{W}_{CESTAC}^{+}\mathbf{Y} \tag{17}$$

We run this rounding mode N times and because of equation (15), we obtain different results. Hence, one obtains N estimation vectors denoted $\mathbf{X}_{CESTAC/k}$. The mean value is computed thanks to (18):

$$\hat{\mathbf{X}}_{CESTAC} = \frac{1}{N}\sum_{k=1}^{N}\mathbf{X}_{CESTAC/k} \tag{18}$$

The standard deviations are given by (19):

$$\boldsymbol{\sigma}^2 = \frac{1}{N-1}\sum_{k=1}^{N}\left(\mathbf{X}_{CESTAC/k} - \hat{\mathbf{X}}_{CESTAC}\right)^2 \tag{19}$$

Then, the relative standard deviations are calculated:

$$\%\sigma_{\hat{X}^j_{CESTAC}} = 100\left(\sigma_{\hat{X}^j_{CESTAC}} \Big/ \left|\hat{X}^j_{CESTAC}\right|\right) \tag{20}$$

We calculate the relative estimation error of the main parameters given by the following equation:

$$\%e^j = 100\left|1 - \hat{X}^j_{CESTAC} \Big/ \hat{X}^j_{LS}\right| \tag{21}$$

where:

- X^j_{CESTAC} is the jth main parameter identified by means of the CESTAC method,
- X^j_{LS} is the initial value of the jth main parameter identified through LS technique.

Finally, we calculate the relative variation of the norm of the residual torque:

$$\%e(\|\boldsymbol{\rho}\|) = 100\left|1 - \|\boldsymbol{\rho}_{CESTAC}\| / \|\boldsymbol{\rho}_{LS}\|\right| \tag{22}$$

where:

- $\|\boldsymbol{\rho}_{LS}\| = \left\|\mathbf{Y} - \mathbf{W}\hat{\mathbf{X}}_{LS}\right\|$,
- $\|\boldsymbol{\rho}_{CESTAC}\| = \left\|\mathbf{Y} - \mathbf{W}\hat{\mathbf{X}}_{CESTAC}\right\|$.

If the contribution of the noise from the observation matrix **W** is negligible compared to modeling errors and current measurements noises, then $\%e^j$ will be practically negligible (less than 1%), as $\%e(|\boldsymbol{\rho}|)$ (less than 1%). In this case, the bias of the LS estimator proves to be negligible.

Otherwise, the relative error $\%e^j$ can not be neglected (greater than 5%), as $\%e(|\boldsymbol{\rho}|)$ (greater than 5%). This means that the LS estimator is biased and the results are controversial.

The previous reasoning can be justified by considering the following equation:

$$\mathbf{Y} = (\mathbf{W} + \partial\mathbf{W})\mathbf{X} + \partial\mathbf{Y} \tag{23}$$

where:

- $\partial\mathbf{W}$ represents the variation of $\mathbf{W}$ due to joint positions, velocities and accelerations errors,
- $\partial\mathbf{Y}$ represents modeling errors and current measurements noises,
- $\mathbf{X}$ is the "true" solution.
- $\mathbf{Y}$ is the "perfect" measurements vector

Hence, one obtains:

$$\boldsymbol{\rho} = \partial\mathbf{W}\mathbf{X} + \partial\mathbf{Y} \tag{24}$$

This gives:

$$\|\boldsymbol{\rho}\| = \|\partial\mathbf{W}\mathbf{X} + \partial\mathbf{Y}\| \leq \|\partial\mathbf{W}\mathbf{X}\| + \|\partial\mathbf{Y}\| \tag{25}$$

While a rounding mode is defined for q and if no variations are observed, then, it comes that the solution $\mathbf{X}$ is weakly sensitive to quantizing noises. It comes also that the norm of the residual vector the is poorly correlated with $\partial\mathbf{W}$. We can write that:

$$\|\boldsymbol{\rho}\| \approx \|\partial\mathbf{Y}\| \tag{26}$$

And, this implies that $\mathbf{W}$ is practically uncorrelated with $\boldsymbol{\rho}$. Finally, the LS estimator is practically unbiased.

4. Experimental results

4.1 6 DOF industrial robot arm

The robot to be identified is presented Figure 1, it is a Stäubli TX40. Its structure is a classical anthropomorphic arm with a 6 DOF serial architecture and its characteristics can be found at the Stäubli web site. The initial encoder resolution is less than 2.10-4 degree per count so the original resolution is more than enough to provide valuable measures.

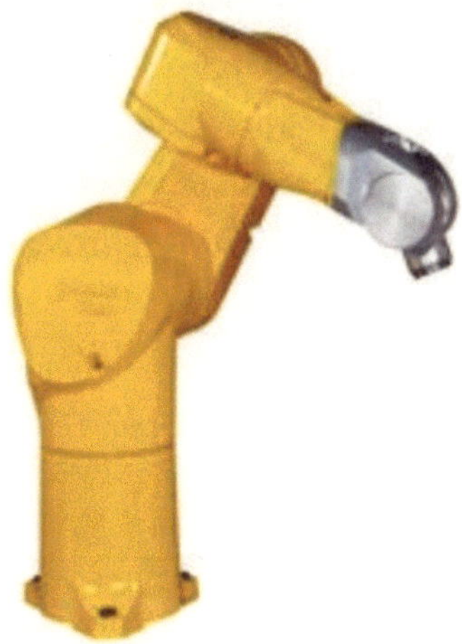

Figure 1. Stäubli TX40 to be identified

j	a(j)	μ(j)	σ(j)	α(j)	d(j)	θ(j)	r(j)
1	0	1	0	0	0	t1	0
2	1	1	0	π/2	0	t2	0
3	2	1	0	0	D3	t3	0
4	3	1	0	- π/2	0	t4	RL4
5	4	1	0	π/2	0	t5	0
6	5	1	0	- π/2	0	t6	0

Table 1. Modified Dennavit Hartenberg Geometric Parameters for the Stäubli TX40

In order to establish the IDM, firstly we define the Modified Dennavit and Hartenberg (DHM) geometric parameters (Khalil & Kleinfinger, 1986), Table 1, based on the schematic of Figure 2. Next, the linear regressor W and the base parameters are computed thanks to the software SYMORO+ (Khalil & Creusot 1997).

We notice that this robot has 60 base parameters, some inertial parameters are gathered with others, the letter "R" is added at the end of the regrouped parameters.

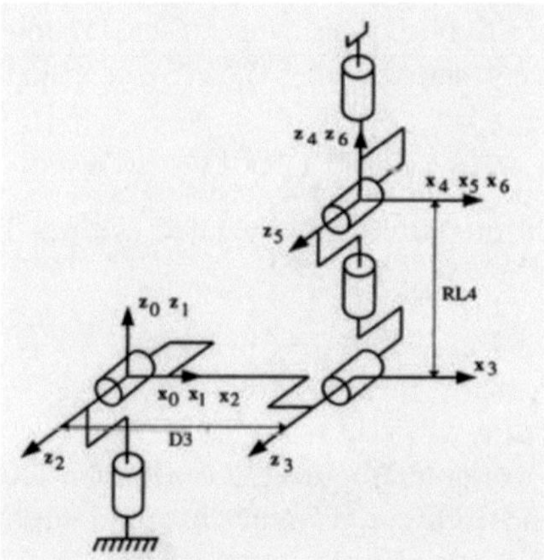

Figure 2. DHM frames of the Stäubli TX40

The gathering rules give us the following analytic formulas:

$$
\begin{aligned}
&ZZ1R=Ia1 + d3^2*(M3+M4+M5+M6) + YY2 + YY3 + ZZ1\\
&XX2R=-d3^2*(M3+M4+M5+M6) + XX2 - YY2\\
&XZ2R=-d3*MZ3 + XZ2\\
&ZZ2R=Ia2 + d3^2*(M3+M4+M5+M6) + ZZ2\\
&MX2R=d3*(M3+M4+M5+M6) + MX2\\
&XX3R=2*MZ4*RL4 + (M4+M5+M6)*RL4^2 + XX3 - YY3 + YY4\\
&ZZ3R=2*MZ4*RL4 +(M4+M5+M6)*RL4^2 + YY4 + ZZ3\\
&MY3R=MY3 - MZ4 -(M4+M5+M6)*RL4\\
&XX4R=XX4 - YY4 + YY5\\
&ZZ4R=YY5 + ZZ4\\
&MY4R=MY4 + MZ5\\
&XX5R=XX5 - YY5 + YY6
\end{aligned}
\qquad (27)
$$

$$ZZ5R=YY6 + ZZ5$$

$$MY5R=MY5 - MZ6$$

$$XX6R=XX6 - YY6$$

4.1.1 Identification Results using the LS Technique

The exciting trajectories, illustrated on Figure 3, consist of polynomial positions which are designed to have constant velocities over a period (the gravity and friction parameters are thus well excited) and to reach maximum admissible accelerations (the inertia parameters are also well excited).

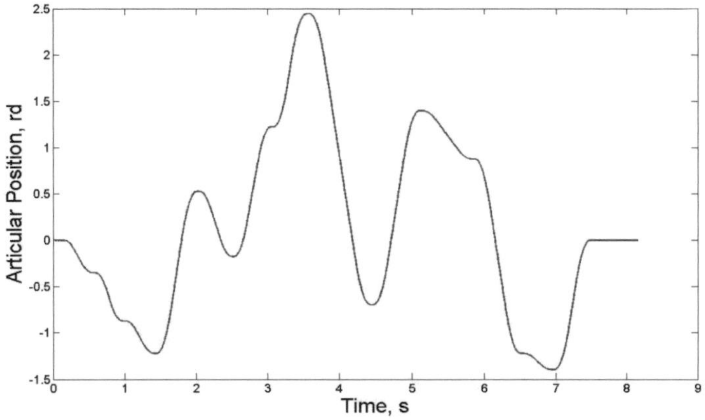

Figure 3. Typical Exciting Trajectory

In our case, W is a 42620x60 matrix and its conditioning number is close to 200, so the trajectories are enough exciting.

The cut-off frequencies of the Butterworth filter and the decimate filter are close to 50Hz. This value was found thanks to a spectral analysis.

Finally, only 28 essential parameters are enough to characterize the dynamic model of the TX40.

Our computations give us the whole 60 parameters, but at the end of our algorithm the parameter with the higher relative standard deviation is removed. Then the LS Method is applied on the new dynamic model until the relative standard deviation of the error vector is above a threshold: $\sigma_{\rho e} \geq 1.04\sigma_\rho$, σ_ρ being the initial relative standard deviation.

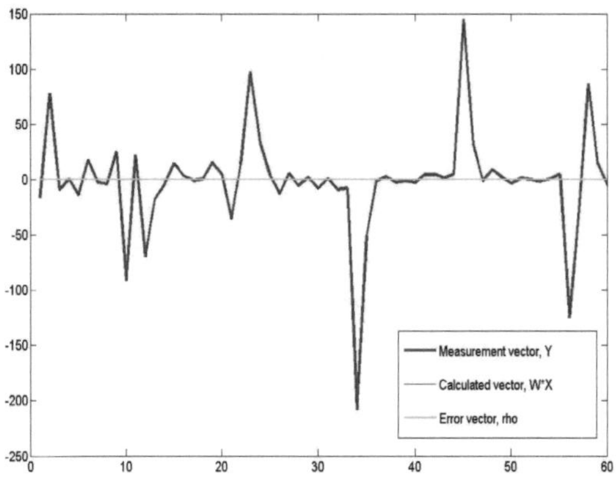

Figure 4. Comparison between the measurement vector and the computed results

Figure 4 shows the measurement vector Y_{fd}, the results of W_{fd} $*X_{fd}$ and the error vector ρ_{fd}. It is obvious that the measured torques vector is matched by the estimated torques vector which validates the identification.
The results of the identification are summed up in Table 2, only the most significant parameters for the purpose of this paper are written.

Mechanical Parameters	Value	Relative deviation $\%\sigma_j$
ZZ1R (Kgm^2)	1.230	0.483
FV1 ($Nm/rd.s^{-1}$)	8.070	0.468
FS1 (Nm)	5.780	1.610
MX2R (Kgm)	2.090	0.681
FV2 ($Nm/rd.s^{-1}$)	5.550	0.632
FS2 (Nm)	7.540	1.024
XX3R (Kgm^2)	0.127	5.306
MX3 (Kgm)	0.045	11.865
Ia3 (Kgm^2)	0.084	4.013
FV3 ($Nm/rd.s^{-1}$)	2.240	0.963
FS3 (Nm)	5.860	0.992
Ia4 (Kgm^2)	0.027	4.825
FV4 ($Nm/rd.s^{-1}$)	1.130	0.604
FS4 (Nm)	2.310	0.967
OFFSET4 (Nm)	0.096	15.960
Ia5 (Kgm^2)	0.053	5.442
FV5 ($Nm/rd.s^{-1}$)	2.010	1.114
FS5 (Nm)	3.780	1.411
Ia6 (Kgm^2)	0.014	8.324
FV6 ($Nm/rd.s^{-1}$)	0.721	1.809
OFFSET6 (Nm)	0.166	16.30

Table 2. Reference Values of the Mechanical Parameters through the LS Technique

4.1.2 Identification with various lower resolution encoders

The identification protocol of the derivate CESTAC method is designed as explained in section 3. Considering the initial encoder resolution, for the first identification with a reduced resolution, we decide to use $\Delta 1=2\pi/10000$. Then the resolution is decreased by 1000 down to $2\pi/1000$, $2\pi/500$ is also used. Only $\Delta 2=2\pi/5000$, $\Delta 3=2\pi/2000$ and $\Delta 4=2\pi/1000$, the more relevant, are presented in Table 3.

The exciting trajectories are the identical to those used for the LS technique. The cut-off frequencies of the Butterworth filter and of the decimate filter are closed to 50Hz and the parameters previously identified as essential are the same from those exposed in Table 2.
All tests are carried out 5 times, and each result of these tests is used to compute a set of dynamic parameters. The results summed up in Table 3 are the mean values of the results of each identification.

Mechanical Parameters	Value with Δ1	Value with Δ2	Value with Δ3	Value with Δ4
ZZ1R (Kgm^2)	1.200	1.141	0.850	0.406
FV1 ($Nm/rd.s^{-1}$)	8.120	8.202	8.175	8.168
FS1 (Nm)	5.620	5.421	5.399	5.319
MX2R (Kgm)	2.110	2.132	2.269	2.454
FV2 ($Nm/rd.s^{-1}$)	5.570	5.613	5.658	5.589
FS2 (Nm)	7.570	7.389	7.196	7.092
XX3R (Kgm^2)	0.122	0.116	0.084	0.065
MX3 (Kgm)	0.055	0.049	0.046	0.060
Ia3 (Kgm^2)	0.086	0.089	0.100	0.092
FV3 ($Nm/rd.s^{-1}$)	2.440	2.471	2.525	2.480
FS3 (Nm)	5.770	5.632	5.475	5.441
Ia4 (Kgm^2)	0.027	0.028	0.024	0.019
FV4 ($Nm/rd.s^{-1}$)	1.150	1.180	1.182	1.195
FS4 (Nm)	2.220	2.100	2.091	2.014
OFFSET4 (Nm)	0.071	0.032	0.040	0.029
Ia5 (Kgm^2)	0.051	0.054	0.043	0.023
FV5 ($Nm/rd.s^{-1}$)	2.080	2.178	2.232	2.243
FS5 (Nm)	3.590	3.387	3.184	3.086
Ia6 (Kgm^2)	0.014	0.014	0.013	0.009
FV6 ($Nm/rd.s^{-1}$)	0.744	0.770	0.793	0.782
OFFSET6 (Nm)	0.120	0.128	0.133	0.147
$\%e(\lvert\rho\rvert)$	<1%	<2%	<10%	~10%

Table 3. Identified Values of the Mechanical Parameters with Various Resolution Encoders

Primary analysis of Table 3 give that Δ4 is the threshold beyond which the identification becomes irrelevant.

4.1.3 Identification with lower resolution torque sensors

The protocol is the same as in the previous sections. The ratio between the torque measurement resolution Φ_i is of 1/5.

Mechanical Parameters	Value with Φ 1	Value with Φ 2	Value with Φ 3	Value with Φ 4
ZZ1R (Kgm²)	1.200	1.192	1.184	1.170
FV1 (Nm/rd.s⁻¹)	8.192	8.180	8.204	8.188
FS1 (Nm)	5.388	5.457	5.386	5.417
MX2R (Kgm)	2.112	2.114	2.114	2.116
FV2 (Nm/rd.s⁻¹)	5.582	5.581	5.594	5.605
FS2 (Nm)	7.400	7.371	7.388	7.375
XX3R (Kgm²)	0.117	0.120	0.115	0.115
MX3 (Kgm)	0.049	0.046	0.048	0.056
Ia3 (Kgm²)	0.086	0.087	0.089	0.092
FV3 (Nm/rd.s⁻¹)	2.461	2.463	2.486	2.480
FS3 (Nm)	5.678	5.674	5.587	5.643
Ia4 (Kgm²)	0.028	0.028	0.029	0.028
FV4 (Nm/rd.s⁻¹)	1.173	1.174	1.175	1.173
FS4 (Nm)	2.164	2.133	2.127	2.127
OFFSET4 (Nm)	0.055	0.064	0.061	0.070
Ia5 (Kgm²)	0.056	0.055	0.056	0.055
FV5 (Nm/rd.s⁻¹)	2.174	2.160	2.197	2.206
FS5 (Nm)	3.391	3.443	3.311	3.291
Ia6 (Kgm²)	0.014	0.014	0.014	0.015
FV6 (Nm/rd.s⁻¹)	0.764	0.762	0.785	0.761
OFFSET6 (Nm)	0.134	0.139	0.144	0.123
%e(|ρ|)	<1%	<2%	<3%	<4%

Table 4. Identified values of the mechanical parameters with various torque sensors resolution

Table 4 shows that there is no "a priori" lower boundary for the torque measurement resolution.

4.1.4 Development

It appears that the results exposed in Table 2 and those exposed in the columns 1,2 and 3 of Table 3 are close to each others, values in Table 4 are also generally close to the ones in Table 2. All the relative estimation errors, given by (21), have been calculated.

Of the various statistic indicators which can be calculated from these results, the relative error of ρ, the error vector, is the best one because it indicates the general accuracy of the identification from a global point of view.

The last line of Table 3 indicates that while the resolution is higher than 1000 counts per revolution the LS identification is consistent and excepted for few parameters, the identified values can be considered as correct.

Overall, it comes that the variations of parameters for the resolutions Δ1, Δ2 and Δ3 are not very important, excepted for MX3, OFFSET4 and OFFSET6 whom respective %ej exceed 20%. The observed major variations for MX3, OFFSET4 and OFFSET6 can be explained by the fact that these parameters do not contribute significantly to the dynamic of the system, and also that they are very sensitive to noises.

On the other hand the last line of Table 4 indicates that despite the variations of resolution of torque measurement, the identification still gives accurate results. Indeed, the variations of main inertia parameters (i.e. ZZ_j and XX_j included in the essential parameters) are less than 5%, while those of the main gravity parameters are less than 3% (excepted the variation of MX3) and the variations of friction parameters are less than 5%.

4.2 3 DOF haptic device : a medical interface

4.2.1 Presentation and modeling

The CEA LIST has recently developed a 6 DOF high fidelity haptic device for telesurgery (Gosselin et al., 2005). As serial robots are quite complex to actuate while fully parallel robots exhibit a limited workspace, this device makes use of a redundant hybrid architecture composed of two 3 DOF branches connected via a platform supporting a motorized handle, having thus a total of 7 motors. Each branch is composed of a shoulder (link 1), an arm (link 2) and a forearm (link 3) actuated by a parallelogram loop (link 5 and 6). To provide a constant orientation between the support of the handle (link 4) and the shoulder, a double parallelogram loop is used (see Figure).

Figure 6 presents the DHM frames and Table **5** the DHM parameters. Our purpose is to apply the CESTAC method to the serial upper branch of the interface (the handle is disconnected). We note that q_1, q_2 and q_5 are the active joint positions. The complete modeling of the branches can be found in (Janot et al., 2007a).

Figure 5. CEA LIST high fidelity haptic interface. Description of the upper branch

"Exciting" trajectories consist of triangular at constant low velocities and sinusoidal positions with various frequencies and amplitudes. The cut off frequency of the Butterworth filter and of decimation filter equals 10Hz. **W** is a (16000x30) matrix and its conditioning number is close to 150. The trajectories are thus enough exciting for identifying the base parameters of the upper branch.

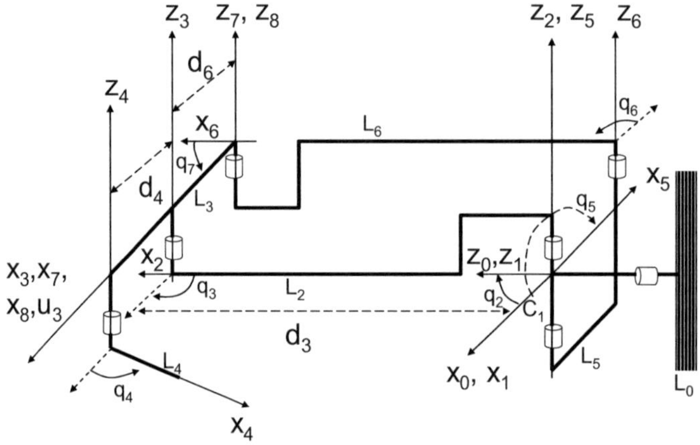

Figure 6. DHM frames for modeling the upper branch of the medical interface

j	a_j	μ_j	σ_j	γ_j	b_j	α_j	d_j	θ_j	r_j
1	0	1	0	0	0	0	0	q_1	0
2	1	1	0	0	0	-90°	0	q_2	0
3	2	0	0	0	0	0	d_3	q_3	0
4	3	0	0	0	0	0	d_4	q_4	0
5	1	1	0	0	0	-90°	0	q_5	0
6	5	0	0	0	0	0	d_6	q_6	0
7	6	0	0	0	0	0	d_3	q_7	0
8	3	0	2	0	0	0	$-d_6$	0	0

Table 5. DHM parameters of the medical interface

The joint torque is calculated through current measurement. We have $\Gamma_a = NK_TI$, where Γ_a is the joint torque, N the transmission gear ratio, K_T the motor torque constant and I the current of the motor.

The results are summed up in Table 6. The relative standard deviations are not given, although they have also been calculated, because they don't give important information. The physical meaning of these parameters is explained in section 2.1. The subscript R means that this parameter is regrouped with others. The symbol * means that these values have been identified through a specific tests.

The parameters having small influence have been removed. We checked that when identified they have a large relative deviation, and that when removed from the identification model, the estimation of the other parameters is not perturbed. Moreover, the norm of the residual torque does not vary (its variation is less than 1%). Finally, only 14 parameters are enough for characterizing the dynamic model of the 3 DOF branch. They are the main parameters of the interface. Direct comparisons have been performed. These tests consist in comparing the measured and the estimated torques just after the identification procedure. An example is given Fig. 4.

Now, we identify the base parameters thanks to the derivate CESTAC method as explained in section 3. For each motor, the encoder resolution equals 4000 counts per revolution. Thus, we have $\Delta = 2\pi/4000$. The identification process is running 5 times. The results are also summed up in Table . It appears that the results are close to each others. The calculation of the relative estimation error given by (21) shows that the relative errors of the main parameters are less than 1%. Moreover, the variation of the residual torque proves to be negligible. In this case, the LS estimator is practically unbiased and its use is justified. So, the values given in Table can be considered as the "real" values.

Now, the limit of the encoder resolution beyond which the bias of the LS estimator can not be neglected is estimated. To do so, we carry out several identification tests using the derivate CESTAC method by increasing Δ in equation (15).

Finally, $\%e_j$ is calculated with the LS estimations given in Table 7. In other words, one supposes that position measurement is increasingly corrupted and we analyze the LS estimator behavior.

In our case, small variations appear when Δ equals $2\pi/100$: $\%e_j$ reaches 1% for main inertia parameters while it reaches 3% for the main gravity parameters. Strong and unacceptable variations occur if Δ is less than $2\pi/80$. Indeed, $\%e_j$ reaches 5% for main inertia parameters while it reaches 6% for the main gravity parameters. In addition, the variation of the residual torque is greater than 10%. In this case, the bias of the LS can not be neglected and the results are controversial. The results are summed Table .

These results prove that the estimation of the base parameters is reliable and practically unbiased. Thus, by writing the dynamic model in the operational space, the apparent mass, stiffness and friction felt by the user can be calculated with a good accuracy. One can assess the quality of the interface as made in (Janot et al., 2007b) for example.

Mechanical Parameters	CAD Value	LS Identified Value	CESTAC Identified Value
ZZ_{1R} (Kgm^2)	0.050	0.051	0.051
MY_{1R} (Kgm)	0.025	0.024	0.024
f_{C1} (Nm)	0.12*	0.12	0.12
XX_{2R} (Kgm^2)	-0.023	-0.023	-0.023
ZZ_{2R} (Kgm^2)	0.03	0.029	0.029
MX_{2R} (Kgm)	-0.02	-0.019	-0.019
f_{C2R} (Nm)	0.11*	0.11	0.11
$offset_2$ (Nm)	0.03	0.020	0.020
XX_{3R} (Kgm^2)	-0.012	-0.011	-0.011
ZZ_{3R} (Kgm^2)	0.014	0.014	0.014
MX_{3R} (Kgm)	0.04	0.039	0.039
MX_{5R} (Kgm)	0.07	0.068	0.068
f_{C5R} (Nm)	0.11*	0.11	0.11
$offset_5$ (Nm)	0.03	0.030	0.030

Table 6. Identified values for the upper branch

Parameters	$\Delta=2\pi/4000$	$\Delta=2\pi/100$	$\Delta=2\pi/80$
$\%e(ZZ_{1R})$	<1%	1%	4%
$\%e(MY_{1R})$	<1%	2%	6%
$\%e(f_{C1})$	<1%	<1%	1%
$\%e(XX_{2R})$	<1%	2%	5%
$\%e(ZZ_{2R})$	<1%	1%	4%
$\%e(MX_{2R})$	<1%	3%	6%
$\%e(f_{C2R})$	<1%	<1%	<1%
$\%e(offset_2)$	<1%	<1%	5%
$\%e(XX_{3R})$	<1%	2%	5%
$\%e(ZZ_{3R})$	<1%	<1%	<1%
$\%e(MX_{3R})$	<1%	<1%	2%
$\%e(MX_{5R})$	<1%	<1%	2%
$\%e(f_{C5R})$	<1%	<1%	<1%
$\%e(offset_5)$	<1%	<1%	<1%
$\%e(\lvert\rho\rvert)$	<1%	5%	>10%

Table 7. Relative errors with respect to the encoder resolution

4.2.3 Discussion

The derivate CESTAC method enables us to evaluate the bias of the LS estimator. When the relative errors are very small, its use is fully justified. Otherwise, the LS estimator is biased and the results are controversial.

In the case of the haptic device, small variations appear when Δ equals $2\pi/100$. This value is 40 times as great as the initial value. From (Diolaiti et al., 2005), we know that the stability of the haptic rendering depends explicitly on the encoder resolution. Hence, one can not have a poor encoder resolution. It is one of particularities of haptic devices compared to classical industrial robots. So, due to the high resolution of the encoders, there is a high probability that the LS estimator is practically unbiased. However, an external verification using the derivate CESTAC method is useful.

In addition, we have also tried different probability distributions in equation (15): a uniform distribution and a Gaussian distribution. For the haptic device, one obtains the following results: small variations appear when Δ equals $2\pi/100$ for a uniform distribution while they appear when Δ equals $2\pi/200$ for a Gaussian distribution.

The use of a Gaussian distribution means that the probability such that the position error varies between $-\Delta$ and $+\Delta$ is close to 67%. Compared with a uniform distribution and the rounding mode defined in this paper, this distribution is pessimistic. It can be considered as the worst of cases. Hence, a classical Gaussian can be used to define the rounding mode. Equation (15) becomes:

$$q_{CESTAC} = q + N\left(0, \Delta^2\right) \tag{28}$$

where:

- q is the measured motor position,
- Δ the encoder resolution,
- $N(0,\Delta^2)$ a Gaussian distribution.

5. Conclusion

Considering the importance for the robotic applications of a precise parametric modelization, especially to optimize the performances, it is clearly important that the method used for this modelization have too be accurate and unbiased. The CESTAC method let us decrease virtually the resolution of the robot sensors in order to evaluate the bias of the LS estimation.

The different results presented throughout the last pages show that the identification process give reliable results and that this technique does not require unusually accurate encoder resolution according to the values found for the lower resolution limit.

It is also important to recall that this technique makes it possible to assess the qualities and drawbacks of haptic interfaces during reengineering process.

In the future the CESTAC method will be mixed up with the Direct and Inverse Dynamic Model method (DIDIM) to build up an identification technique requiring less measurements and hence being less dependent of experimental conditions.

6. References

Atkeson C.G., An C.H. & Hollerbach J.M. (1986). Estimation of Inertial Parameters of Manipulator Loads and Links, *Int. J. of Robotics Research*, vol. 5(3), 1986, pp. 101-119

Diolaiti, N.; Niemeyer, G.; Barbagli F. & J. K. Jr. Salisbury (2006). Stability of haptic rendering: discretization, quantization, time delay and Coulomb effect, *IEEE Transactions on Robotics and Automation*, vol. 22(2), April 2006, pp. 256-268

Gautier, M. & Khalil, W. (1990). Direct calculation of minimum set of inertial parameters of serial robots, *IEEE Transactions on Robotics and Automation*, Vol. 6(3), June 1990

Gautier, M. (1991). Numerical calculation of the base inertial parameters, *Journal of Robotics Systems*, Vol. 8, August 1991, pp. 485-506

Gautier M. & Khalil W. (1991). Exciting trajectories for the identification of base inertial parameters of robots, *Proc. of the 30th Conf. on Decision and Control*, pp. 494-499, Brighton, England, December 1991

Gautier, M.; Khalil, W. & Restrepo, P. P. (1995). Identification of the dynamic parameters of a closed loop robot, *Proc. IEEE on Int. Conf. on Robotics and Automation*, pp. 3045-3050, Nagoya, May 1995

Gautier, M. (1997). Dynamic identification of robots with power model, *Proc. IEEE Int. Conf. on Robotics and Automation*, pp. 1922-1927, Albuquerque, 1997

Gautier, M. & Poignet, P. (2002). Identification en boucle fermée par modèle inverse des paramètres physiques de systèmes mécatroniques, *JESA, Journal Européen Des Systèmes Automatisés*, vol. 36(3), 2002, pp. 465–480

Gosselin, F.; Bidard, C. & Brisset, J. (2005). Design of a high fidelity haptic device for telesurgery, *IEEE Int. Conf. on Robotics and Automation*, pp. 206-211, Barcelone 2005

Ha I.J., Ko M.S. & Kwon S.K. (1989). An Efficient Estimation Algorithm for the Model Parameters of Robotic Manipulators, *IEEE Trans. On Robotics and Automation*, vol. 5(6), 1989, pp. 386-394

Hampel, F.R. (1971). A general qualitative definition of robustness, *Annals of Mathematical Statistics*, 42, 1971, pp. 1887-1896

Huber, P.J. (1981). *Robust statistics*, Wiley, New-York, 1981

Janot, A.; Bidard, C.; Gosselin, F.; Gautier, M. & Perrot Y. (2007a). Modeling and Identification of a 3 DOF Haptic Interface, *IEEE Int. Conf. on Robotics and Automation*, pp. 4949-4955, Roma, 2007

Janot, A.; Anastassova, M.; Vandanjon, P.O. & Gautier, M. (2007b). Identification process dedicated to haptic devices, *Proc. IEEE on Int. Conf. on Intelligent Robots and Systems*, San Diego, October 2007

Jezequel, F. (2004). Dynamical control of converging sequences computation, *Applied Numerical Mathematics*, vol. 50, 2004, pp. 147-164

Kawasaki H. & Nishimura K. (1988). Terminal-Link Parameter Estimation and Trajectory Control, *IEEE Trans. On Robotics and Automation*, vol. 4(5), 1988, pp. 485-490

Khalil, W. & Kleinfinger, J.F. (1986). A new geometric notation for open and closed loop robots, *IEEE Int. Conf. On Robotics and Automation*, pp. 1147-1180, San Francisco USA, April 1986

Khalil, W. & Creusot, D. (1997). SYMORO+: A system for the symbolic modeling of robots, *Robotica*, vol. 15, pp. 153-161, 1997

Khosla P.K. & Kanade T. (1985). Parameter Identification of Robot Dynamics, *Proc. 24th IEEE Conf. on Decision Control*, Fort-Lauderdale, December 1985

Kozlowski K. (1998). Modelling and Identification in Robotics, *Springer Verlag London Limited*, Great Britain, 1998.

Marcassus, N.; Janot, A.; Vandanjon, P.O. & Gautier, M. (2007). Minimal resolution for an accurate parametric identification - Application to an industrial robot arm, *Proc. IEEE on Int. Conf. on Intelligent Robots and Systems*, San Diego, October 2007

Prüfer M., Schmidt C. & Wahl F. (1994). Identification of Robot Dynamics with Differential and Integral Models: A Comparison, *Proc. 1994 IEEE Int. Conf. on Robotics and Automation*, San Diego, California, USA, May 1994, pp. 340-345

Raucent B., Bastin G., Campion G., Samin J.C. & Willems P.Y. (1992). Identification of Barycentric Parameters of Robotic Manipulators from External Measurements, *Automatica*, vol. 28(5), 1992, pp. 1011-1016

Swevers J., Ganseman C., Tückel D.B., de Schutter J.D. & Van Brussel H. (1997). Optimal Robot excitation and Identification, *IEEE Trans. On Robotics and Automation*, vol. 13(5), 1997, pp. 730-740

Vignes, J. & La Porte, M. (1974). Error Analysis in Computing, *Information Processing'74*, north-Holland, Amsterdam, 1974

18

Towards Simulation of Custom Industrial Robots

Cosmin Marcu and Radu Robotin
Technical University of Cluj-Napoca, Department of Automation
Romania

1. Introduction

The simulation tools used in industry require good knowledge of specific programming languages and modeling tools. Most of the industrial robots manufacturers have their own simulation applications and offline programming tools. Even if these applications are very complex and provide many features, they are manufacturer-specific and cannot be used for modeling or simulating custom industrial robots. In some cases the custom industrial robots can be designed using modeling software applications and then simulated using a programming language linked with the virtual model. This requires the use of specific programming languages, a good knowledge of the modeling software and experience in designing the mechanical part of the robot.

Researches within the robots simulation field have been made by various research groups, especially in the field of mobile robots simulation. The results of their researches were complex simulation tools like: *SimRobot* (Laue *et al.*, 2005) capable to simulate arbitrary defined robots in the 3D space together with the sensorial and the actuation systems; *USARSim* (Wang *et al.*, 2003) or the *Urban Search and Rescue Simulator* using as main kernel the *Unreal* game which is very efficient and capable of solving rendering, animation and modelling problems; *UCHILSIM* (Zagal & del Solar, 2004) is a simulator which reproduces the dynamics *AIBO* robots and also simulates the interaction of these robots with objects in the working space; Rohrmeier's industrial robot simulator (Rohrmeier, 2000) simulates serial robots using VRML in a web graphical interface.

Our primary objective was to design and build a simulation system containing a custom industrial robot and an open architecture robot controller in the first part and a simulation software package using well known and open source programming languages in the last part.

The custom industrial robot is a RPPR robot having cylindrical coordinates. In the first part of the project we designed and modeled the robotic structure and we obtained the forward kinematics, inverse kinematics and dynamic equations. From the mechanical point of view, the first joint contains a harmonic drive unit actuated by a DC motor. The two prismatic joints (vertical and horizontal) contain ball-screw mechanisms, both actuated by DC motors. The fourth joint is represented by a DC motor directly linked with the robot gripper.

Another objective was to build a reprogrammable open architecture controller in order to be able to test different types of sensors and communication routines. The robot controller contains a miniature modular computer, a PIC microcontroller interface board, 4 DC motor driver boards, 4 rotary encoders and 9 digital sensors. For reading the encoders and sensors data we built a multiplexer board capable of reading up to 16 digital or analog inputs in the same time.

From the software point of view the simulation system contains the software applications running on the miniature computer, the program running on the PIC microcontroller board and a remote visual application running on a desktop PC which contains the 3D graphical simulation module.

This chapter presents the technical information and the techniques we used in creating the simulation system together with the experimental results we obtained.

2. The Robot Modelling

2.1 Forward kinematics

In order to build the simulation system we considered a 4 degrees-of-freedom robot structure having two rotation joints and two translation joints. The kinematic scheme of the robot is presented in Fig. 1.

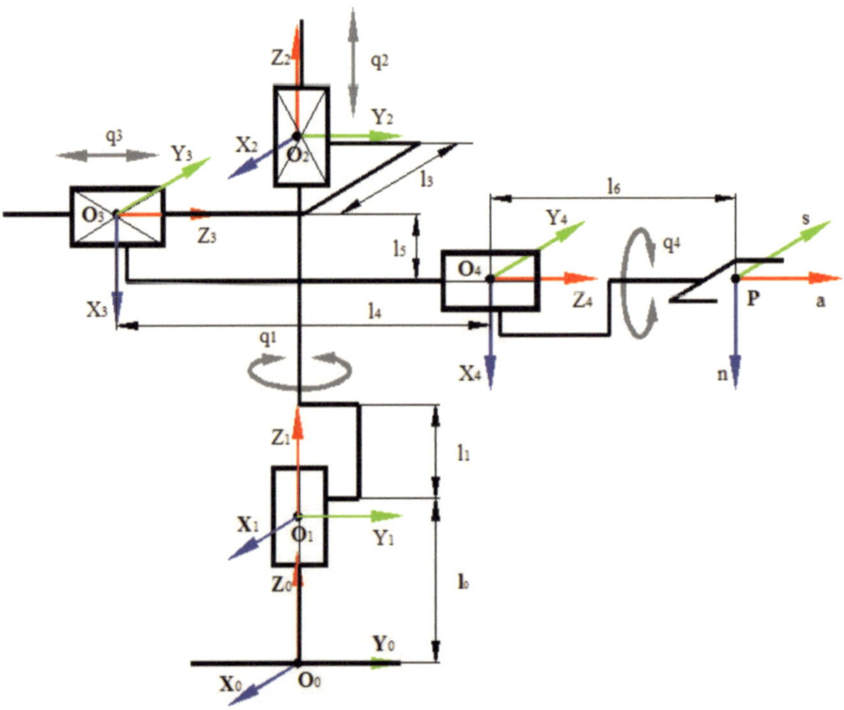

Figure 1. The kinematic scheme of the robot

To obtain the equations which determine the position and the orientation of the gripper relative to the robot base we applied the modified Denavit-Hartenberg method or Craig's method. Therefore, the motion axis for each joint will be Z_n, where n is the index of the corresponding joint. The displacement of each joint is denoted with q and it's measured in millimetres for the translation joints and radians for the rotation joints. The table of D-H parameters will be as follows:

Joint	a_{i-1}	α_{i-1}	d_i	θ_i
1	0	0	l_0	q_1
2	0	0	l_1+q_2	0
3	l_3	$-\pi/2$	$-q_3$	$\pi/2$
4	l_5	0	l_4	q_4
P	0	0	l_6	0

Table 1. Denavit-Hartenberg parameters of the robot

To obtain the position and orientation matrix for each joint relative to the previous joint we applied equation (1):

$$ {}^{i-1}_{i}[T]=\begin{pmatrix} \cos(\theta_i) & \sin(\theta_i)\cdot\cos(\alpha_{i-1}) & \sin(\theta_i)\cdot\sin(\alpha_{i-1}) & -a_{i-1}\cdot\cos(\theta_i) \\ -\sin(\theta_i) & \cos(\theta_i)\cdot\cos(\alpha_{i-1}) & \cos(\theta_i)\cdot\sin(\alpha_{i-1}) & a_{i-1}\cdot\sin(\theta_i) \\ 0 & -\sin(\alpha_{i-1}) & \cos(\alpha_{i-1}) & -d_i \\ 0 & 0 & 0 & 1 \end{pmatrix} \tag{1}$$

The position and orientation matrix of each joint relative to the robot base or the fixed coordinate system *{0}* is calculated using (2):

$$ {}^{0}_{n}[T]=\prod_{i=1}^{n} {}^{i-1}_{i}[T] \tag{2}$$

Using the D-H parameters from Table 1 and applying equations (1) and (2) we obtained the position and orientation matrix of the gripper relative to the fixed coordinate system *{0}* or robot base:

$$ {}^{0}_{P}[T]=\begin{pmatrix} -cq_1\cdot sq_4 & -cq_1\cdot cq_4 & -sq_1 & (q_3-l_4-l_6)\cdot sq_1+l_3\cdot cq_1 \\ -sq_1\cdot sq_4 & -sq_1\cdot cq_4 & cq_1 & -(q_3-l_4-l_6)\cdot cq_1+l_3\cdot sq_1 \\ -cq_4 & sq_4 & 0 & l_0+l_1-l_5+q_2 \\ 0 & 0 & 0 & 1 \end{pmatrix} \tag{3}$$

In equation (3) $\sin(q_n)$ and $\cos(q_n)$ are simplified with sq_n and cq_n.

Next step in modelling the robot structure is the determination of velocities and accelerations equations for each joint relative to the robot base.

In order to determine the velocity and acceleration of the gripper we set the angular and linear velocity and acceleration of the fixed coordinate system as being zero.

$$ \begin{aligned} {}^{0}\overline{\omega}_0 &= [0\ \ 0\ \ 0]^T; \quad {}^{0}\dot{\overline{\omega}}_0 = [0\ \ 0\ \ 0]^T; \\ {}^{0}\overline{v}_0 &= [0\ \ 0\ \ 0]^T; \quad {}^{0}\dot{\overline{v}}_0 = [0\ \ 0\ \ g]^T; \end{aligned} \tag{4}$$

To determine the velocity and acceleration of each joint we applied Negrean's iterative method (Negrean *et al.*, 1997), where ${}^{i}\overline{\omega}_i$, ${}^{i}\dot{\overline{\omega}}_i$, ${}^{i}\overline{v}_i$ and ${}^{i}\dot{\overline{v}}_i$ are calculated as follows:

$$ {}^{i}\overline{\omega}_i = {}^{i}_{i-1}[R]\,{}^{i-1}\overline{\omega}_{i-1} + \begin{cases} \dot{q}_i\cdot{}^{i}\overline{k}_i & ,\text{if } i=\text{Rot.} \\ 0 & ,\text{if } i=\text{Trans.} \end{cases} \tag{4}$$

$$ {}^{i}\overline{v}_i = {}^{i}_{0}[R]\cdot{}^{0}\overline{v}_i = {}^{i}_{i-1}[R]\{{}^{i-1}\overline{v}_{i-1} + {}^{i-1}\overline{\omega}_{i-1}\times{}^{i-1}\overline{r}_i\} + \begin{cases} 0 & ,\text{if } i = R \\ \dot{q}_i\cdot{}^{i}\overline{k}_i & ,\text{if } i = T \end{cases} \tag{5} $$

$$ {}^{i}\dot{\overline{\omega}}_i = {}^{i}_{i-1}[R]\cdot{}^{i-1}\dot{\overline{\omega}}_{i-1} + \begin{cases} {}^{i}_{i-1}[R]\cdot{}^{i-1}\overline{\omega}_{i-1}\times\dot{q}_i\cdot{}^{i}\overline{k}_i + \ddot{q}_i\,{}^{i}\overline{k}_i & ,\text{if } i = R \\ 0 & ,\text{if } i = T \end{cases} \tag{6} $$

$$ \begin{aligned} {}^{i}\dot{\overline{v}}_i = {}&{}^{i}_{i-1}[R]\{{}^{i-1}\dot{\overline{v}}_{i-1} + {}^{i-1}\dot{\overline{\omega}}_{i-1}\times{}^{i-1}\overline{r}_i + {}^{i-1}\overline{\omega}_{i-1}\times{}^{i-1}\overline{\omega}_{i-1}\times{}^{i-1}\overline{r}_i\} + \\ &+ \begin{cases} 0 & ,\text{if } i = R \\ 2\cdot{}^{i}\overline{\omega}_i\times\dot{q}_i\cdot{}^{i}\overline{k}_i + \ddot{q}_i\,{}^{i}\overline{k}_i & ,\text{if } i = T \end{cases} \end{aligned} \tag{7} $$

Where:

$$ {}^{i}\overline{k}_i = \begin{cases} [1\ \ 0\ \ 0]^T & \text{if motion axis} = X; \\ [0\ \ 1\ \ 0]^T & \text{if motion axis} = Y; \\ [0\ \ 0\ \ 1]^T & \text{if motion axis} = Z. \end{cases} \tag{8} $$

In equations (4)-(7) the parameters $\dot{q}_i$ and $\ddot{q}_i$ represent the 1st and 2nd time derivative of the *i* joint displacement, ${}^{i}_{1-1}[R]$ the transposed rotation matrix of joint *i* relative to joint *i-1* and ${}^{i-1}\overline{r}_i$ the position column matrix of joint *i*.

Using the D-H equations and applying equations (4)-(8) we obtained the gripper velocities and accelerations relative to the last joint coordinate system:

$$ {}^{P}\omega_P = \begin{pmatrix} -\dot{q}_1\cdot cq_4 \\ \dot{q}_1\cdot sq_4 \\ \dot{q}_4 \end{pmatrix};\ {}^{P}\overline{v}_P = \begin{pmatrix} -\dot{q}_2\cdot cq_4 + (l_4 + l_6 - q_3)\cdot\dot{q}_1\cdot sq_4 \\ \dot{q}_2\cdot sq_4 + (l_4 + l_6 - q_3)\cdot\dot{q}_1\cdot cq_4 \\ l_3\cdot\dot{q}_1 + \dot{q}_3 \end{pmatrix} \tag{9,10} $$

$$ {}^{P}\dot{\overline{\omega}}_P = \begin{pmatrix} \dot{q}_1\cdot\dot{q}_4\cdot sq_4 - \ddot{q}_1\cdot cq_4 \\ \dot{q}_1\cdot\dot{q}_4\cdot cq_4 + \ddot{q}_1\cdot sq_4 \\ \ddot{q}_4 \end{pmatrix} \tag{11} $$

$$ {}^{P}\dot{\overline{v}}_P = \begin{pmatrix} (l_4 + l_6 - q_3)\cdot\ddot{q}_1\cdot sq_4 - \ddot{q}_2\cdot cq_4 + l_3\cdot\dot{q}_1^{\,2}\cdot sq_4 + 2\cdot\dot{q}_1\cdot\dot{q}_3\cdot sq_4 - g\cdot cq_4 \\ (l_4 + l_6 - q_3)\cdot\ddot{q}_1\cdot cq_4 + \ddot{q}_2\cdot sq_4 + l_3\cdot\dot{q}_1^{\,2}\cdot cq_4 + 2\cdot\dot{q}_1\cdot\dot{q}_3\cdot cq_4 + g\cdot sq_4 \\ l_3\cdot\ddot{q}_1 + \ddot{q}_3 - (l_4 + l_6 - q_3)\cdot q_1^{\,2} \end{pmatrix} \tag{12} $$

The linear velocity and acceleration of the gripper relative to the fixed coordinate system is obtained by multiplying the rotation matrix of the gripper with the ${}^{P}\overline{v}_P$ and ${}^{P}\dot{\overline{v}}_P$ matrices. Therefore, ${}^{0}\overline{v}_P$ and ${}^{0}\dot{\overline{v}}_P$ matrices have the following form:

$$ {}^{0}\overline{v}_P = \begin{pmatrix} -\dot{q}_1\cdot(A\cdot cq_1 + l_3\cdot sq_1) - \dot{q}_3\cdot sq_1 \\ -\dot{q}_1\cdot(A\cdot sq_1 + l_3\cdot cq_1) + \dot{q}_3\cdot cq_1 \\ \dot{q}_2 \end{pmatrix} \tag{13} $$

$$ {}^{0}\dot{\bar{v}}_P = \begin{pmatrix} -\ddot{q}_1 \cdot (A \cdot cq_1 + l_3 \cdot sq_1) - \ddot{q}_3 \cdot sq_1 + \dot{q}_1^{\,2} \cdot (A \cdot sq_1 - l_3 \cdot cq_1) - 2 \cdot \dot{q}_1 \cdot \dot{q}_3 \cdot cq_1 \\ -\ddot{q}_1 \cdot (A \cdot sq_1 - l_3 \cdot cq_1) + \ddot{q}_3 \cdot cq_1 - \dot{q}_1^{\,2} \cdot (A \cdot cq_1 + l_3 \cdot sq_1) - 2 \cdot \dot{q}_1 \cdot \dot{q}_3 \cdot sq_1 \\ \ddot{q}_2 + g \end{pmatrix} \quad (14) $$

Because the last joint is rotating around its Z axis, parameter q_4 is not influencing the velocity and the acceleration of the gripper relative to the fixed coordinate system.

2.2 Inverse kinematics

For a desired position, velocity and acceleration in order to find the joints angular or linear displacements and their 1st and 2nd time derivative we determined the inverse kinematics equations of the robot. The equations look as follow:

$$ sq_1 = \frac{x}{(q_3 - l_6 - l_4)} - \frac{x \cdot l_3^{\,2} - y \cdot l_3 \cdot (q_3 - l_6 - l_4)}{(q_3 - l_6 - l_4) \cdot \left((q_3 - l_6 - l_4)^2 + l_3^{\,2}\right)} $$

$$ cq_1 = \frac{x \cdot l_3 - y \cdot (q_3 - l_6 - l_4)}{(q_3 - l_6 - l_4)^2 + l_3^{\,2}} $$

$$ q_1 = a\tan 2(sq_1, cq_1) \quad (15) $$

$$ q_2 = z + l_5 - l_0 - l_1 \quad (16) $$

$$ q_3 = -\sqrt{x^2 + y^2 - l_3^{\,2}} + l_6 + l_4 \quad (17) $$

Fig. 2 presents the variation of q_1, q_2 and q_3 displacements function of a desired trajectory for a fixed *y* coordinate. The *x* and *z* coordinates are increased simultaneously from 50 [mm] to 250 [mm].

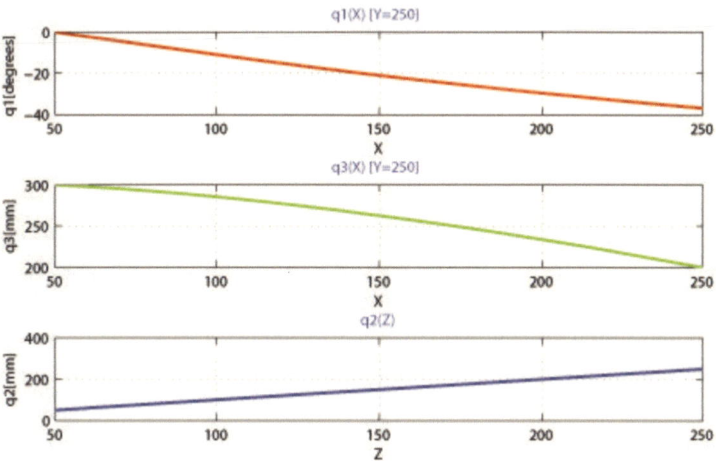

Figure 2. Inverse kinematics - displacements variation function of *{x,y,z}*

Equation (13) was used to obtain the 1st derivative of the joints displacements considering a desired linear velocity for the gripper. The inverse kinematics equations for velocities will be as follow:

$$\dot{q}_1 = \frac{v_x \cdot cq_1 + v_y \cdot sq_1}{l_4 + l_6 - q_3} \tag{18}$$

$$\dot{q}_2 = v_z \tag{19}$$

$$\dot{q}_3 = \left(v_y \cdot cq_1 - v_x \cdot sq_1\right) + \frac{l_3}{l_4 + l_6 - q_3} \cdot \left(v_y \cdot sq_1 + v_x \cdot cq_1\right) \tag{20}$$

To obtain the 2nd derivative of the joins displacements or the joints accelerations for a desired gripper acceleration we used equations (14). The resulting equations are:

$$\ddot{q}_1 = -\frac{l_3 \cdot \dot{q}_1^2 + 2 \cdot \dot{q}_1 \cdot \dot{q}_3 + a_x \cdot cq_1 + a_y \cdot sq_1}{l_4 + l_6 - q_3} \tag{21}$$

$$\ddot{q}_2 = a_z - g \tag{22}$$

$$\ddot{q}_3 = \left(a_y \cdot cq_1 - a_x \cdot sq_1\right) + \left(l_4 + l_6 - q_3\right) \cdot \dot{q}_1^2 + \frac{l_3 \cdot \left(a_y \cdot sq_1 + a_x \cdot cq_1 + \dot{q}_1^2 + 2 \cdot \dot{q}_1 \cdot \dot{q}_3\right)}{l_4 + l_6 - q_3} \tag{23}$$

2.3 Robot mechanics

In order to build the mechanical structure of the robot we initially designed and 3D modelled the robot components. The mechanical structure chosen for this project is a 4 degrees-of-freedom (DOF) industrial robot having a rotation joint in the robot base represented by a gear unit, two translation joints using ball screw-nut mechanisms and a rotation joint represented by a DC geared motor, directly linked with a parallel gripper.

The mechanical structure was designed taking into consideration one of the most important factors which are influencing the robot dynamics: the friction. In order to reduce the friction and increase the robot performance we used very accurate mechanisms like: a harmonic drive for the robot base, linear ball bearings and ball screw-nut mechanisms for the translation joints. The most important physical properties of these mechanisms are:

- zero or very low backlash;
- high accuracy;
- high toque capacities;
- high efficiency.

The harmonic drive has a reduction ratio of 100:1 and the ball screw mechanisms have the lead of 5 [mm]. The 4th joint uses a DC motor and a gripper from a refurbished robot.

Table 2 presents some mechanical properties that we established for the robot joints, based on the virtual model initially designed and the mechanical properties of the actuators, gear units and translation mechanisms.

Joint	Type	Stroke [degrees], [mm]	Max. Speed [^{0}/s],[mm/s]
1	Rotation	180^0	~18 [^{0}/s]
2	Translation	310 [mm]	20 [mm/s]
3	Translation	350 [mm]	20 [mm/s]
4	Rotation	360^0	180 [^{0}/s]

Table 2. The robot joints mechanical properties

Fig. 3 presents the 3D model of the robot arm and the real robot arm (first three joints).

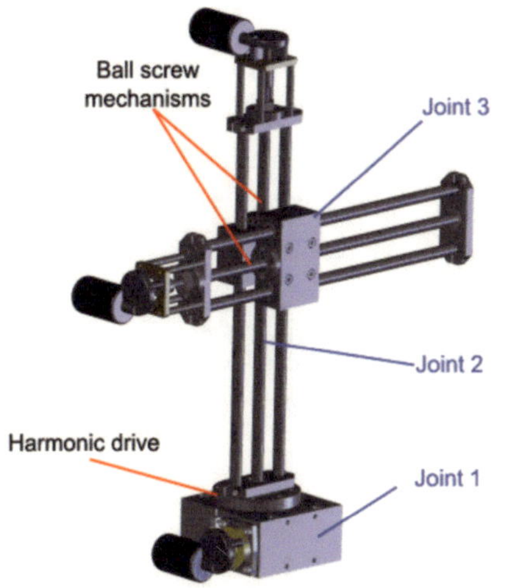

Figure 3. The virtual and the real robot arm

3. The Robot Controller

By definition, an industrial robot must be a re-programmable multi-functional manipulator designed to perform a variety of tasks (Dai Gil, 2005). The component which makes a robot reprogrammable and multi-functional is the robot controller. The robot controller, also known as the "robot brain", represents the component which gives functionality and autonomy to a certain robotic system. Basically, the robot controller is formed of two main parts: controller hardware and controller software. The complexity and the configuration of a controller differ from robot to robot. For simple tasks and robot structures the controllers may have simple configurations.

Nowadays, the production of industrial robot controllers is taken over by the robots manufacturers mostly because the industrial robots are used in mass production applications where high-level controllers are needed. For simple robotic systems, where simple mechanical structures are used, the robot controller can be made using cheap and small electronic devices. When designing controllers for research purposes it is recommended to focus on open architecture and/or modular design. Using open architecture and/or modular design lead to multi-purpose controllers.

In order to develop robot control algorithms, the robot controller must ensure a high degree of efficiency, modularity and scalability (Rusu et al., 2006; Lazea et al., 2006).

In our project the robot controller has an open architecture design both from the software and hardware point of view. The controller hardware consists of: a re-programmable processing unit board, a re-programmable microcontroller board, a 16-to-1 multiplexer board and four programmable motor controller boards. The controller software consists of:

- TCP/IP client-server application for processing unit;
- Serial communication application for processing unit;
- CCSC (*Custom Computer Services C*) application for PIC microcontroller.

3.1 The controller architecture

The robot controller was designed taking into consideration our previous researches made within the open architecture robot design area, especially the results that we obtained in the *"ZeeRO"* mobile robot project (Rusu et al., 2006). Therefore, the robot controller should be:

- open and modular in order to provide sufficient programmable Input/Output ports;
- customizable in order to permit both software and hardware upgrades;
- cheap and small in order to be affordable for any research applications.

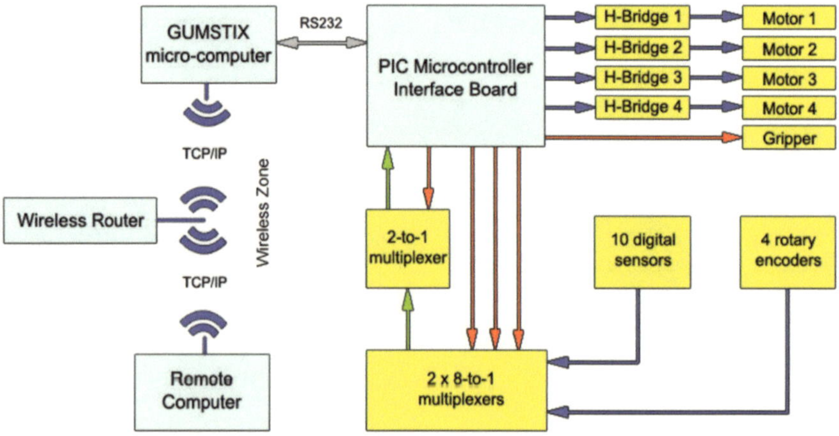

Figure 4. The robot controller architecture

The processing unit of the controller has two processing layers:

- high level processing layer, represented by the Gumstix micro-computer;
- low level processing layer, represented by the PIC microcontroller and the multiplexer board.

The communication between the two processing layers is handled by a RS232 connection. In order to be able to read as many input ports as possible we used a multiplexer board. The multiplexer board contains two 8-to-1 interconnected digital multiplexers and one 2-to-1 master multiplexer. The board is able to handle 16 input ports and is controlled by the PIC microcontroller. In our application, the multiplexer board handles 10 digital sensors (stroke end limit switches) and 4 rotary encoders.

The power circuit (the motors) is controlled by the PIC microcontroller via four H-Bridges. The H-Bridges are *Devantech MD03* type and they are able to control up to 100W (current - 20, voltage 50V) DC motors. The connection between the H-Bridges and the PIC microcontroller is made using two I/O lines, communication being handled by a I2C protocol (built in PIC microcontroller). The PIC microcontroller is able to control up to 12 MD03 H-Bridges.

The high level processing layer is represented by a Gumstix miniature computer (Fig. 5). The Gumstix micro-computer is a small device powered by a PXA255 Intel XScale processor, running at 400 MHz. The most important characteristics of this micro-computer are its modularity and size.

The Gumstix micro-computer is powered by Linux operating system which can be considered as an advantage, especially in developing open source applications.

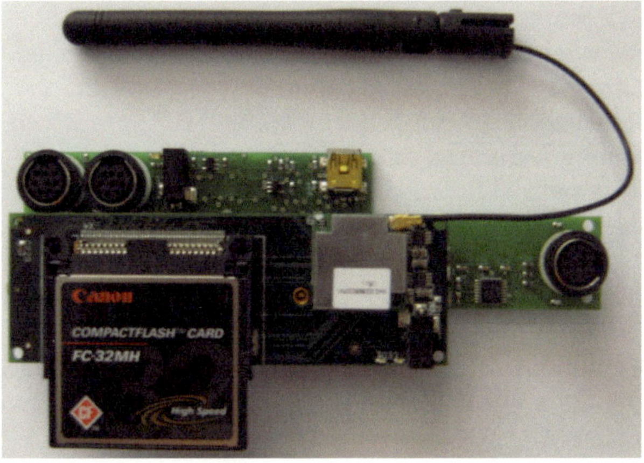

Figure 4. The GUMSTIX micro-computer

One of the serial ports is used to make the connection with the PIC microcontroller. The other two serial ports can be used to connect other devices. In our case, we reserved the two serial ports to connect two serial cameras (*CmuCam3*) for future researches.

The wireless expansion board is used to communicate with remote computers using the TCP-IP communication protocol.

The low level processing layer is represented by the PIC microcontroller board together with the multiplexer board and the H-bridges. From the hardware point of view, the low level processing layer is the interface between the controller "brain" and the robot sensors and motors.

The PIC microcontroller board (Fig. 5) contains a PIC16F876 microcontroller running at 20MHz, 15 I/O configurable pins and one serial port connected to the Gumstix micro-computer. One of the output pins of the microcontroller is reserved to control the gripper state (on/off state).

Figure 5. The low-level processing layer hardware

3.2 The controller software

The robot controller contains two main software packages and one additional application. The main packages are the Gumstix micro-computer programs and the PIC microcontroller program. The additional application is the remote client software which can be used by any remote computer from the same local area network with the Gumstix micro-computer.
The Gumstix micro-computer software contains two main routines:

- the TCP-IP routine, for communication with remote clients via wireless TCP-IP protocol;
- the RS232 routine, for communication with the PIC microcontroller via a serial port.

The application is written in ANSI C under Linux using standard *POSIX pthread* functions to provide support for multithreading, as in [5]. Because the Gumstix microprocessor has 32-bit *RISC* architecture, the programs need to be compiled using Advanced RISC Machine (*ARM*) GNU C Compiler (*ARM-GCC*).
Because the Gumstix memory is limited, the Linux libraries are not installed in the Gumstix flash memory. In order to compile the programs, the ARM Linux kernel was compiled on a stand-alone computer then used for C programs compilations. Also, the programs include all the libraries needed by the functions used. The disadvantage of this method is the executable size which is much bigger that an executable which uses precompiled libraries.
The RS232 routine receives the data sent by the PIC microcontroller through the serial port. The data received is in string format (array of characters) and contains information about the current sensors and motors state. For example, if the PIC microcontroller turns off the robot base motor, the RS232 routine receives "M1{0} - off". The number between braces represents the rotation sense of the motor, "0" for clockwise and "1" for counter clockwise. The RS232 routine processes the message received and sends a command back to the PIC microcontroller if needed. In some cases the PIC microcontroller acts autonomous, even if the message is sent to the Gumstix micro-computer. For example, if one of the stoke limit switches from the same axis with a running motor turns *"On"*, the PIC microcontroller turns off the motor in order to avoid collisions between the mechanical parts of that axis. After that, the microcontroller sends the proper message to the Gumstix RS232 routine.
The wireless TCP-IP routine handles the connections from remote computers. If a connection is open, the routine will automatically send data to the remote computer.
The TCP-IP connection is made through a wireless access point which allows connections from multiple remote computers, both from local area network and external network.
We used the TCP-IP connection for monitoring purposes and for reading the data used in the simulator. In some cases, the program code can be easily modified and recompiled in order to give full control of the robot to the remote client.
The PIC microcontroller program is written in CCS C programming language which provides the same syntax as ANSI C. The program contains two main routines:

- the sensors reading routine;
- the motors control routine.

The sensors reading routine controls the multiplexer board through 4 output pins and 1 input pin. One of the output pins controls the 2-to-1 multiplexer (the master multiplexer). Function of the pin value (logic 1 or 0), the master multiplexer reads the output pin of the proper 8-to-1 multiplexer (slave multiplexer). The other 3 output pins of the PIC microcontroller are used to control the inputs of the slave multiplexers. The state of the

input pins is read continuously. Function of the input pins state, the routine either sends a message to the Gumstix micro-computer or stops the motors if needed.
The motors control routine controls the four Devantech H-Bridges using the I2C protocol. Each H-Bridge can open a serial communication line using a data register. The data register can be configured using the 4-mode switches placed on each H-Bridge. The PIC microcontroller selects which H-Bridge to control with the help of the built-in *i2c_start*, *i2c_write* and *i2c_read* functions.

4. The Simulation Software Application

According to Zaratti, a simulation tool for industrial robots must accomplish the following requirements (Zaratti et al., 2006):

- Flexibility: the tool should allow the simulation of various types of robots, sensors and actuators. The working space of the robot(s) should allow easy modeling;
- Physical realism: to obtain good results, the interaction between the robot and the simulated working space should be modeled by using physical law and rigid body dynamics laws;
- Visual realism: the entire system should be as realistic as possible in order to obtain a correct representation of the data collected from the sensors;
- Efficiency: the simulation must be made in the most efficient way, preferable at a maximum refresh rate;
- Modularity: the simulator should allow the modification of the working space and the robot components;
- Effective control: the simulator should be able to interact with the robot controller software packages.

The requirements above can be accomplished when talking about well-known robot types from the market. Even in this case the effective control is very hard to be made and the programmer(s) should be familiar with a large number of robots, robot controllers, sensors and actuators. More than that, in case of a custom robot the simulator must be flexible enough in order to simulate custom components. Therefore, we proposed that in case of a custom industrial robot the simulator should be designed and built from scratch. The idea of building a simulator from scratch may be discouraging but analyzing the variety of methods for creating a 3D simulator it becomes an encouraging idea. Since our simulation system contains a single industrial robot, the first requirement would be only partially accomplished.
In order to be able to accomplish all the requirements we decided to create the simulation software by modeling with open-source programming languages. Therefore, we used Linux operating system, Qt4 programming language to design the interface and OpenGL to create the 3D simulation scene.

4.1 The simulator main window

The main window of the simulator (Fig. 6) was designed and built in Qt4 programming language (Blanchette & Summerfield, 2006). Qt4 is a cross-platform application framework used desktop or embedded development. We choose this platform because of the advantages provided:

- free and open source for Linux operating system;

- rich set of application building blocks;
- easy to use and learn;
- implemented in C++ and provides support for C++ development;
- has implemented support for OpenGL.

Figure 6. The main window

The main window program consists of four primary classes, five secondary classes and two experimental classes (Fig. 6).

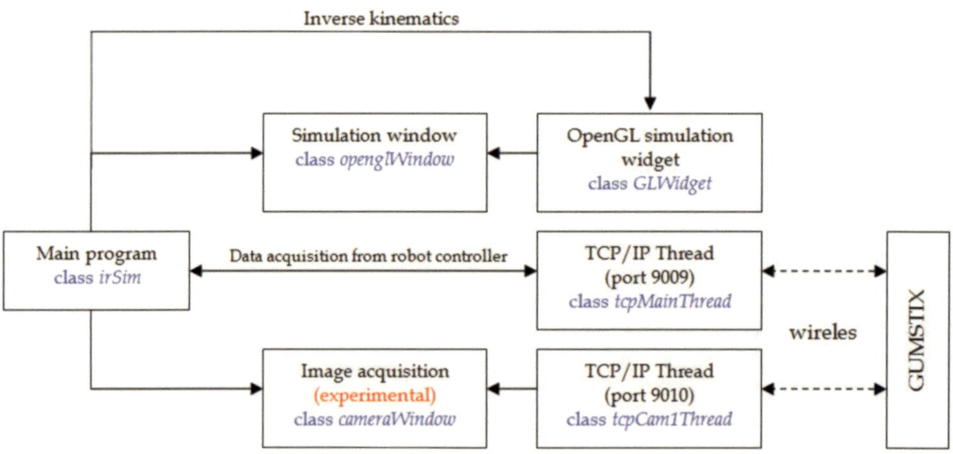

Figure 7. Main program classes and data flow

The data exchange between the program classes is made with the help of Qt *signals* and *slots*. The *signals* and the *slots* are predefined or user defined functions interconnected by the Qt *connect* function. When a *signal* is emitted by a function or a class, the corresponding *slot*

function loaded. From the programming point of view this represents an advantage and leads to an easy exchange of data between the classes.
The right side of the main window consists of information read by the *tcpMainThread* class from the Gumstix micro-computer. The *tcpMainThread* acts like a TCP/IP client class. The TCP/IP server resides on the robot controller and sends the sensors and motors status on connection start and every time a sensor or a motor changes its state (if a client is connected).
Based on the joints position read by the multiplexer and PIC microcontroller, the "Current Position" group indicates the current gripper position relative to the robot base. The gripper position is calculated with the forward kinematics equations implemented as variables in the Gumstix program and sent to the client main window through TCP/IP.
The "Sensor Status" group contains five pairs of edit boxes indicating the stroke limit sensors status for each of the four joints and for the robot gripper (*open* and *closed* state).
Function of the TCP/IP connection status, the "WiFi Connection" group can indicate three types of status messages: "Connected", "Not connected" or a connection error message (Fig. 8).

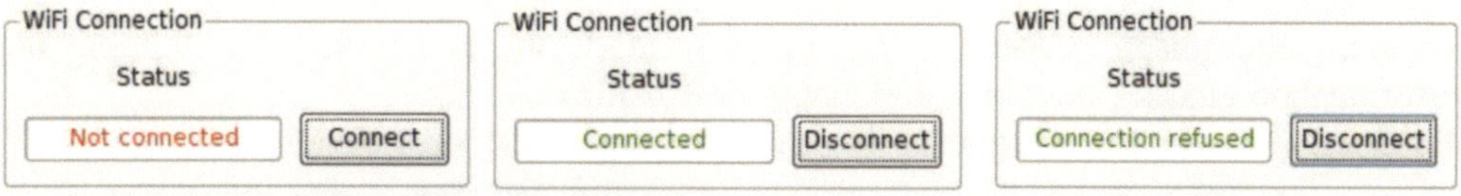

Figure 8. WiFi TCP/IP connection status

The simulation program is able to run in two modes: real mode and simulation mode. In real mode, the program read the information from the robot controller via TCP/IP and any change of data from the main window is changing the position of the virtual model from the simulation window. In simulation mode, the "*Set Position*" and "*Motion Axis*" groups are disabled (Fig. 9), the simulation window is automatically opened and the virtual model of the robot moves from initial position to the position given by the "*Set Position*" parameters.

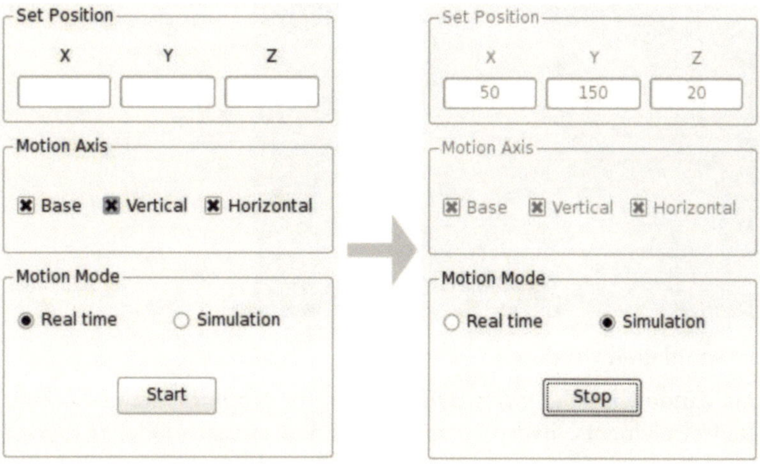

Figure 9. The simulation modes

4.2 The simulation window

The simulation window is *OpenGL* based and contains a primary class named *openglWindow* which is a child class of the main program class *irSim*. Therefore, the elements from this window can be easily controlled by the main window. The *openglWindow* is linked with a secondary class (child class) named *glWidget* which handles the graphical simulation process with the help of the *OpenGL* functions.

OpenGL is the premier environment for developing portable, interactive 2D and 3D graphics applications. Since its introduction in 1992, OpenGL has become the industry's most widely used and supported 2D and 3D graphics application programming interface (API), bringing thousands of applications to a wide variety of computer platforms. OpenGL fosters innovation and speeds application development by incorporating a broad set of rendering, texture mapping, special effects, and other powerful visualization functions. Developers can leverage the power of OpenGL across all popular desktop and workstation platforms, ensuring wide application deployment.

OpenGL routines simplify the development of graphics software—from rendering a simple geometric point, line, or filled polygon to the creation of the most complex lighted and texture-mapped *NURBS* curved surface. OpenGL gives software developers access to geometric and image primitives, display lists, modeling transformations, lighting and texturing, anti-aliasing, blending, and many other features.

Our simulation window reproduces the robot structure at a small scale (Fig. 10).

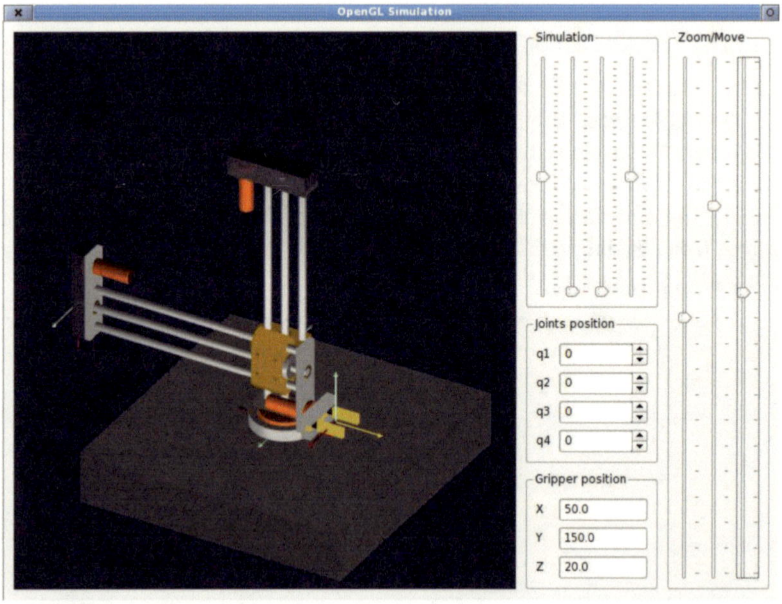

Figure 10. The simulation window

The simulation window is split into two main parts: the graphical simulation widget and the simulation control elements. Every component of the virtual model is represented by an OpenGL list and the position and orientation of the components is based on then values set by the simulation control elements. By using OpenGL lists the objects are precompiled and the simulation process is performed faster. Each list contains sets of functions which generate the objects from OpenGL primitives: points, lines, quads, disks, etc. The process of

building the lists is difficult and requires the exact knowledge of the object dimensions. Anyway, once created, the objects can be easily modified by changing their dimensions and visual properties.
Our robot simulation process can be performed automatically or manually. The simulation is performed automatically when the main program is set on real motion mode and is connected to the robot controller. Also, the simulation performs automatically when the OpenGL window is opened by the main window simulation mode. To simulate the robot manually the main program should not be connected to the robot controller. In this case, the virtual model of the robot can be moved in any possible position by changing the values of the joints positions. The joints positions can be changed both from the "*Simulation*" group slide bars and from the "*Joints position*" group spin boxes. At any change of the joints position the gripper coordinates are displayed in the "*Gripper position*" group edit boxes. The gripper position is calculated automatically using the forward kinematics equations, no matter if the simulation process is performed manually or automatically.
When the gripper coordinates are received from the main window the joints positions are calculated using the inverse kinematics equations and the virtual model of the robot is positioned accordingly (Fig. 11).

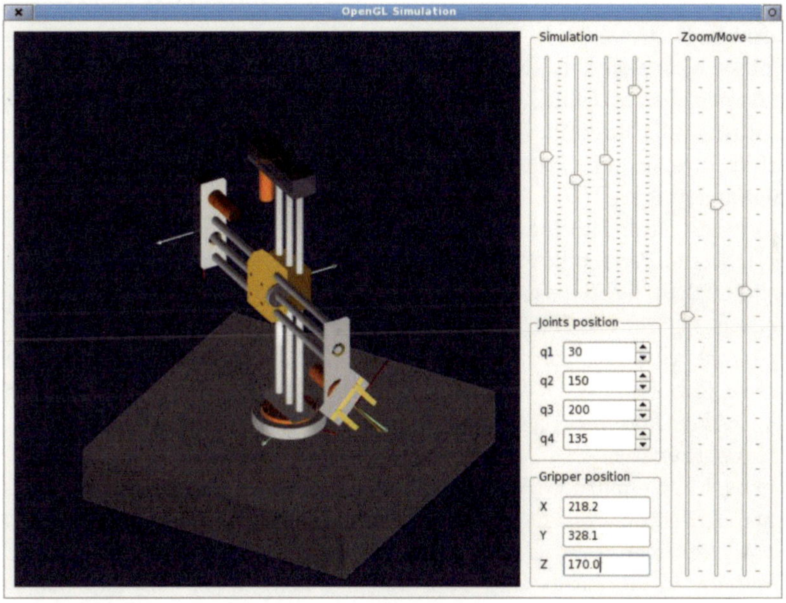

Figure 11. Robot virtual model in a particular position

5. Conclusions

In order to create a simulator for custom industrial robots, it is very important to know the forward and inverse kinematics equations of the robot structure, the controller output data and the limitations of the robot mechanical components. In this paper we presented the steps for building a simulation program for a custom industrial robot. The first step was the robot modeling where we obtained the forward and inverse kinematics equations used as motion laws both for the simulated and for the real robot. The second step was the design of an open architecture robot controller able to communicate with TCP/IP clients. The last step was the design of an open source flexible simulation program, able to reproduce the real robot as realistic as possible. By designing and building this simulation system we implemented a free and easy to use simulation method which can be easily implemented in other research areas. The use of OpenGL features lead to a flexible simulation program which can be easily modified function of the system needs. With regards to the information presented in this chapter we conclude that the simulation system of the robot is a great step forward for us in our research within the field of industrial robot simulation.

6. References

Blanchette, J. & Summerfield, M. (2006). *C++ GUI Programming with QT4*, Prentice Hall Publishing, ISBN 0-13-187249-4.

Dai Gil, L., (2005). *Axiomatic Design and Fabrication of Composite Structures*, Oxford University Press, ISBN 0195278777, pp. 513.

Gerkey, B.P., Vaughan, R.T. & Howard, A. (2003). The Player/Stage Project: Tools for Multi-robot and Distributed Sensors Systems, *Proceedings of the International Conference on Advanced Robotics (ICAR)*, pp. 317-323, ISBN 972-96889-9-0, June 30 - July 3, 2003, Coimbra, Portugal

Laue, T., Spiess, K. & Rofer, T. (2005). SimRobot - A General Physical Robot Simulator and its Application in RoboCup, *Proceedings of RoboCup Symposium*, Universitat Bremen

Lazea, Gh., Rusu, R.B., Robotin, R., Sime, R. (2006). The ZeeRO mobile robot - A modular architecture, *RAAD 2006 International Workshop on Robotics*, ISBN 963-7154-48-5, Balaton, Hungary

Negran, I., Vuscan, I. & Haiduc, N. (1997). *Robotics – Kinematics and Dynamic Modelling*, Didactic and Pedagogic Publishing, ISBN 973-30-5309-8, Bucharest, Romania

Rohrmeier, M. (2000). Web Based Robot Simulation using VRML, *Simulation Conference Proceedings*, ISBN: 0-7803-6579-8, vol. 2, pp. 1525-1528, Orlando, U.S.A

Rusu, R.B., Lazea, Gh., Robotin R. & Marcu C. (2006). Towards Open Architectures for Mobile Robots: ZeeRO, *IEEE-TTTC International Conference on Automation, Quality and Testing, Robotics AQTR 2006 (THETA 15)*, pp. 260-265, 25-28 May, 2006, Cluj-Napoca, Romania

Wang, J., Lewis, M., Gennari, J. (2003). A game engine based simulation of the NIST Urban Search & Rescue arenas, *Proceedings of the 2003 Winter Simulation Conference*

Zagal, J. C. & del Solar, J. R. (2004). UCHILSIM: A Dynamically and Visually Realistic Simulator for the RoboCup Four Legged League. *In RoboCup Symposium 2004*, Robot Soccer World Cup VIII.

Zaratti, M., Fratarcangeli, M. & Iocchi, L. (2006). A 3D Simulator of Multiple Legged Robots based on USARSim, *Proceedings of RoboCup Syposium*, Bremen, Germany

19

Design and Simulation of Robot Manipulators using a Modular Hardware-in-the-loop Platform

Adrian Martin and M. Reza Emami
University of Toronto Institute for Aerospace Studies
Canada

1. Introduction

The need for developing high quality systems with short and cost-effective design schedules has created an ongoing demand for efficient prototyping and testing tools (Wheelright & Clark, 1992). In many engineering applications failure of a system can have severe consequences, from loss of hardware and capital to complete mission failure, and can even result in the loss of human life (Ledin, 1999). The earliest form of prototyping, *physical prototyping,* began with the development of the first system, and it refers to fabricating a physical system to evaluate performance and test design alterations. There have been many advances in this field, such as the use of scaled models (Faithfull et al., 2001), but in most cases the time and cost involved in building complete physical prototypes are prohibitive. With the advent of computers a new form of prototyping, termed *analytical prototyping,* has become a second viable option (Ulrich & Eppinger, 2000). Computer models are generally inexpensive to develop and can be quickly modified to experiment with various aspects of the system. However, this flexibility often comes at the cost of approximations used to model complex physical phenomena, which in turn lead to inaccuracies in the model and system behaviour. A prototyping tool that has been gaining significant popularity in recent years is *hardware-in-the-loop simulation,* which can effectively combine the advantages of the two traditional prototyping methods. The underlying concept of hardware-in-the-loop (HIL) simulation is to use physical hardware for system components that are difficult or impossible to model and link them to a computer model that simulates the other aspects of the system. This technique has been successfully applied to development and testing in a wide range of engineering fields, including aerospace (Leitner, 1996), automotive (Hanselman, 1996), controls (Linjama et al., 2000), manufacturing (Stoeppler et al., 2005), and naval and defence (Ballard et al., 2002).

This research investigates the application of HIL simulation as a tool for the design and testing of serial-link industrial manipulators, and proposes a generic and modular robotic hardware-in-the-loop simulation (RHILS) architecture. The RHILS architecture was implemented in the simulation of a standard industrial manipulator and evaluated on its ability to simulate the robot and its usefulness as a design tool.

The remainder of this section briefly reviews the state-of-the-art in HIL simulation across a broad range of fields, highlighting some of the key benefits and considerations, and then summarizes the current work of other researchers in the specific field of robotic manipulators. Section 2 presents the details of the RHILS architecture and an analysis of the load emulation mechanism. The hardware setup designed to evaluate the effectiveness and viability of the RHILS architecture is outlined in section 3. This setup was used to simulate a 5-d.o.f. industrial manipulator during a real-world design scenario and the comparison between the RHILS setup and a complete physical prototype is presented in section 4. The conclusions from this research are presented in section 5, discussing the strengths and weaknesses of the RHILS platform implementation and the direction of current research.

1.1 Hardware-in-the-Loop Simulation

Hardware-in-the-loop simulations have been used successfully in a number of engineering fields, but found their first application in aerospace flight control systems (Maclay, 1997). The increasing importance of several factors has led to an increase in the use of HIL simulation as a tool for system design, testing, and training. These factors are listed in (Maclay, 1997) as: reducing development time, exhaustive testing requirements for safety critical applications, unacceptably high cost of failure, and reduced costs of the hardware necessary to run the simulation. These factors are mentioned repeatedly throughout the literature along with a number of other benefits of HIL simulation, which are demonstrated in the following paragraphs.

By using physical hardware as part of a computer simulation it is possible to reduce the complexity of the simulation and incorporate factors that would otherwise be difficult or impossible to model. The effectiveness of this technique was shown by the contact dynamics HIL simulator developed by the Canadian Space Agency. This simulation was of the manipulator motion for the Special Purpose Dexterous Manipulator (SPDM) to be installed on the International Space Station. The setup linking a computer simulation of the space manipulator to a physical hydraulic robot described in (Aghili & Piedboeuf, 2002) was successful in simulating the free motion and contact dynamics of the manipulator, a task that would be impossible with a pure computer simulation. Another benefit of HIL simulation exemplified in (Aghili & Piedboeuf, 2002) is the ability to simulate a 0-g environment, desirable since many space robots are incapable of operating in 1-g conditions and so physical prototyping is impractical.

The use of HIL simulation in machine tools and manufacturing systems is discussed in (Stoeppler et al., 2005). In addition to mentioning the cost, safety, and development time benefits, (Stoeppler et al., 2005) talks about how it can be used to safely and economically test new ideas and allow the concurrent design of hardware and software components.

An innovative application is detailed in (Faithfull et al., 2001), where they present the success of a HIL simulation used in conjunction with scaled physical prototyping during the design of a 4x4 electric vehicle. The HIL simulation proved to be an effective design tool, and had the added benefit of improving the credibility of the results when presented to both technical and non-technical persons (Faithfull et al., 2001).

Other applications of HIL simulation include embedded computing and field robotics. In embedded computing HIL simulation is used because many of the systems are safety critical and require thorough and accurate testing (Ledin, 1999). In the field of robotics the applications range from underwater vehicle testing (Lane et al., 2001), to aerospace robotic manipulators (Aghili & Piedboeuf, 2002), to multi-agent mobile robot systems (Hu, 2005).

When implementing software for HIL simulations such as those described above it is often beneficial to use object-oriented or graphical modelling techniques (Kasper et al., 1997).

Programming tools such as Dymola, based on Modelica®, make it easier to model complex mathematical systems by facilitating the decomposition of the system into components and allowing them to be connected using a graphical interface (Aronsson & Fritzson, 2001). This object-oriented approach has the additional benefit of translating easily to multiprocessor systems which can avoid computational limits that may otherwise prevent real-time simulations of complex systems (Kasper et al., 1997).

It is obvious that HIL simulation has been successfully applied in many areas and proven a useful design tool which reduced development time and costs (Stoeppler et al., 2005; Hu, 2005), and with the ever improving performance of today's computers it is possible to build HIL simulations without specialized and costly hardware (Stoeppler et al., 2005). However, (Ma et al., 2004) offers the caution that, as with any type of simulation, it is necessary to extensively validate the results before making use of the simulation. Based on the validation of the SPDM Task Verification Facility, (Ma et al., 2004) proposes a two-step methodology: the first step is verification at a general, higher level, while the second step is verification at a more detailed engineering level. With these considerations taken into account HIL simulation can be an extremely powerful design and testing tool.

1.2 Robotic Hardware-in-the-Loop Simulation

HIL simulation is receiving growing interest from researchers in the field of robotics, and has been applied from a number of different perspectives. These approaches include: *robot-in-the-loop* simulations, such as the platform used for the task verification of the SPDM at the Canadian Space Agency (Piedboeuf et al., 1999) or the use of both real and simulated mobile robots interacting with a virtual environment (Hu, 2005); *controller-in-the-loop* simulations, where a real control system interacts with a computer model of the robot (Cyril et al., 2000); and *joint-in-the-loop* simulations, which use a computer model to compute the dynamic loads seen at each joint and then emulate those loads on the real actuators (Temeltas et al., 2002). Each of these approaches applies the HIL concept slightly differently, but all have produced positive results.

In the recent work (Aghili, 2006) a hardware setup similar to that which was developed for this research is described. It focuses on the simulation of a simple 2-d.o.f. planar manipulator and includes environmental controls which allow testing in space-like thermal/vacuum conditions.

2. Platform Architecture

2.1 RHILS Architecture

The RHILS platform architecture developed for this research allows for simultaneous design and testing of both the joint hardware and control system of a robot manipulator. The architecture is designed to be adequately generic so that it can be applied to any serial-link robot manipulator system, and focuses on modularity and extensibility in order to facilitate concurrent engineering of a wide range of manipulators. This section presents a detailed breakdown of the main blocks of the architecture.

The architecture is separated into four subsystems: (a) the *User Interface*, (b) the *Computer Simulation*, (c) *Hardware Emulation*, and (d) the *Control System*, which are described below with reference to Fig. 1. These subsystems are further partitioned into two major categories: RHILS Platform components (indicated with a white background), and Test System components (indicated with a grey background). The RHILS Platform components are generic and should remain largely consistent over multiple applications, while the Test System components are part of the system being designed and/or tested on the platform. Depending on how much of the system is implemented in hardware versus how much is simulated it is possible to tailor the setup to all phases of the design cycle, and the architecture is designed to make adjusting this ratio as easy as possible.

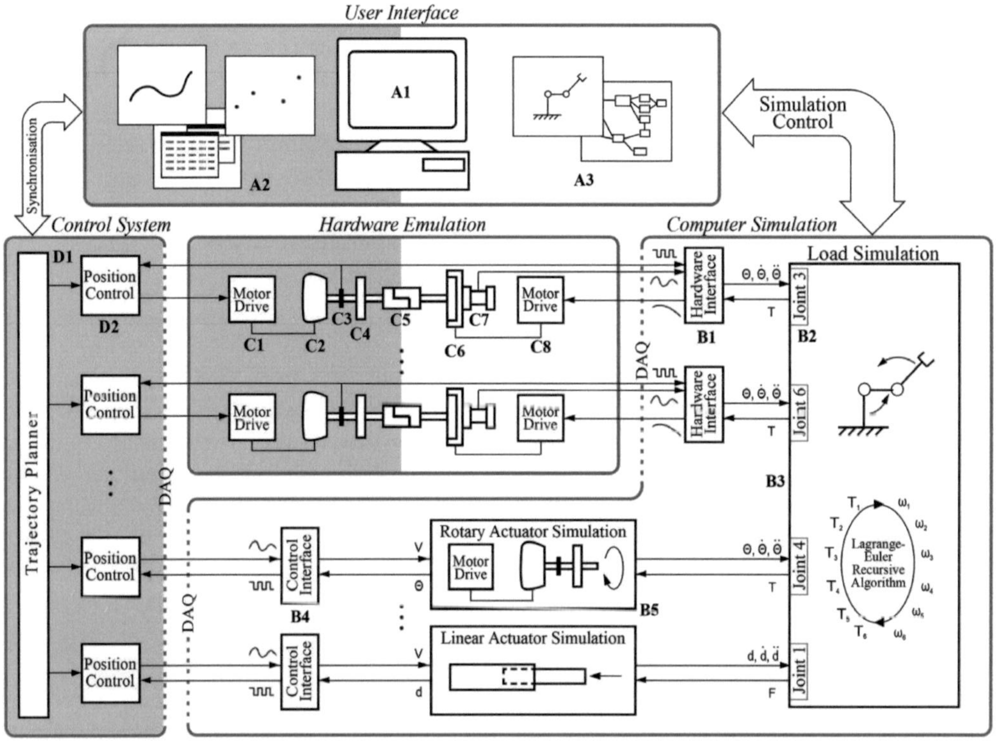

A1 User interface host computer

A2 Control system user interface and trajectory setup

A3 Simulation user interface and scheduler

B1 Motor interface block, converts between actual hardware signals and the standardized form used in the simulation

B2 Joint assignment for the module

B3 Inverse dynamics simulation

B4 Control interface block, converts between actual control signals and the standardized form used with simulated actuators

B5 Simulated model of an actuator, for cases where the hardware module is unavailable, impractical, or unnecessary

C1 Drive electronics for Test Motor

C2 Test Motor

C3 Differential rotary encoder

C4 Harmonic drive transmission

C5 Detachable coupling to allow test hardware to be swapped in and out

C6 Load Motor

C7 Reaction torque transducer, for closed loop control and data acquisition

C8 Drive electronics for Load Motor

D1 Trajectory planner

D2 Position controller

A grey background indicates that section is part of the system being designed and tested using the RHIL platform

Figure 1. RHILS Platform Architecture

A. User Interface Block

This block contains the most overlap between the RHILS Platform and the Test System. Because it is necessary to synchronize initial conditions before starting a simulation, this block acts as an intermediary between the custom control system and the generic simulation. On the RHILS Platform side robot configurations and parameters are chosen, as well as specifying any external conditions, for example zero-gravity or end-effector payloads, that will be used during a simulation. For the Test System side any configurable control parameters are set in the control system, such as the planned trajectories and feedback loop gains. Finally, the duration of the simulation and the type of data logging to be performed are selected.

B. Computer Simulation Block

The *Computer Simulation* performs three primary roles. Its first and most obvious task, represented by the *Load Simulation* block, is to run the inverse dynamics computations based on the instantaneous position, velocity, and acceleration of each joint, and solve for the dynamic load applied to each joint actuator. Due to the recursive algorithm used for computing the inverse dynamics (Li & Sankar, 1992) on the dedicated kernel, it is possible to specify any reasonable number of joints in any configuration and still attain the computational efficiency necessary to run the simulation in real-time. The second task is to convert the hardware signals read in and sent out through a data acquisition board into the standardized format used by the load simulation, which is shown by the *Hardware Interface* blocks. These hardware interface blocks play a key role in the modularity of the architecture since they allow different hardware to be used without significant changes to the simulation. The third task of the *Computer Simulation* is to simulate any joints that do not have a corresponding hardware module. In some situations it may be desirable to have one or more joint actuators without a hardware component, for example when the hardware is unavailable, too costly, or simply unnecessary. Then the computer simulation must model the joint and interface directly with the control system, shown in the *Actuator Simulation* and *Control Interface* blocks. This third task makes it possible to utilize the RHILS platform at early stages of the design as well as making it more cost effective to set up tests if only one section of the manipulator is under study.

C. Hardware Emulation Block

The *Hardware Emulation* system consists of separate modules for each joint, and each module interfaces with both the *Control System* and the *Computer Simulation*. These modules are further separated into two parts: a *Test Module*, the joint actuator that is being designed/tested, and a *Load Module*, the load-emulating device that mimics the dynamic loads that would be seen in a real system. The *Test Module* includes not only the real actuator, but also the transmission system, position/speed sensors, and motor drive that would be used in the real manipulator, all of which can lead to significant inaccuracies in a pure computer-based simulation. The *Test Module* interfaces directly with the *Control System*, which controls the motor as if it were part of a physical robot. The *Load Module* is coupled to the output of the transmission system, ideally without the use of a secondary transmission that may introduce unwanted uncertainty in the load emulation mechanism. For the range required by most applications, it was found that *torque motors* can supply the necessary torque directly and have other desirable features including consistent torque at low speeds, low inertia, and proper heat dissipation characteristics. The *Load Module* is controlled through a feedback loop that follows the torque calculated by the *Computer Simulation* block.

This torque represents the arm dynamics that must be reflected on each joint actuator to have a genuine simulation of the real system. To emulate the dynamic torque accurately closed-loop control is needed, which requires that the torque generated by the *Load Module* be identified. This is done through a unique torque sensor setup described in section 3.1.

D. Control System Block

This block can range from running in software on a standard PC to running on dedicated custom hardware depending on the nature and requirements of the application. It is possible to use the real control system for the robot, since as far as the control system is concerned it is connected to the real actuators in a physical robot. This has significant benefits over running a simulated or modified version of the control system: in many applications intense testing of the final control system is required, which can now begin before the final hardware is complete without building expensive prototypes. On the other hand, when the control system is not the focus of the design the flexibility of this architecture allows any simple controller to be quickly implemented and used.

2.2 Load Emulation

Using a recursive inverse dynamics algorithm it is possible to efficiently calculate the torque seen by the n joints of the manipulator given the current position, velocity, and acceleration for each joint (Li & Sankar, 1992). This estimated torque, τ^*, must then be applied to the joint by the *Load Module*. The physical setup of a *Load* and *Test Module* pair is shown in Fig. 2. This load emulation setup is similar to that used in (Aghili, 2006), yet has some differences that reduce the cost and complexity of the hardware. Specifically, a reaction torque sensor is placed behind the load motor rather than a rotary torque sensor mounted on the shaft, and no velocity sensor is used. The following analysis and experimental results in later sections show that the load emulator can achieve similar performance despite the simpler hardware. In the free body diagram of the load motor shown in Fig. 2 q, $\dot{q}$, and $\ddot{q}$ are the position, velocity, and acceleration, respectively, τ_l is the generated torque, $f_l(\dot{q})$ is the friction, J_l is the rotating moment of inertia, and τ is the torque transferred to the *Test Module*. With the assumption that there is no flexibility in the shaft or coupling the dynamic equation for the load motor is given by

$$J_l\ddot{q} = \tau_l - f_l(\dot{q}) + \tau \quad (1)$$

The reaction torque sensor mounted behind the load motor measures τ_m: the torque applied between the stator and rotor. Using (1) this can be written in two ways,

$$\tau_m = \tau_l - f_l(\dot{q}) = -\tau + J_l\ddot{q} \quad (2)$$

If $\ddot{q}^*$ is taken as an estimate of the current acceleration, either directly measured or calculated by other methods such as those discussed in section 3.3, then a torque error term can be specified as

$$e = \tau_m^* - \tau_m = (-\tau^* + J_l\ddot{q}^*) - (-\tau + J_l\ddot{q}) \quad (3)$$

or

$$e = \tau - \tau^* + e_{acc} \quad (4)$$

where e_{acc} is an error term related to the accuracy of the acceleration estimate. When an accelerometer is used to accurately measure the acceleration this term disappears. For the acceleration estimation techniques used in this research, given that the derivative of a random variable with Gaussian distribution is also Gaussian (Solak et al., 2003), the acceleration error can be considered to be a zero-mean Gaussian noise whose variance was determined to be negligible from experimental results, as discussed in section 3.3. Thus, as $e \rightarrow 0$, $\tau \approx \tau^*$.

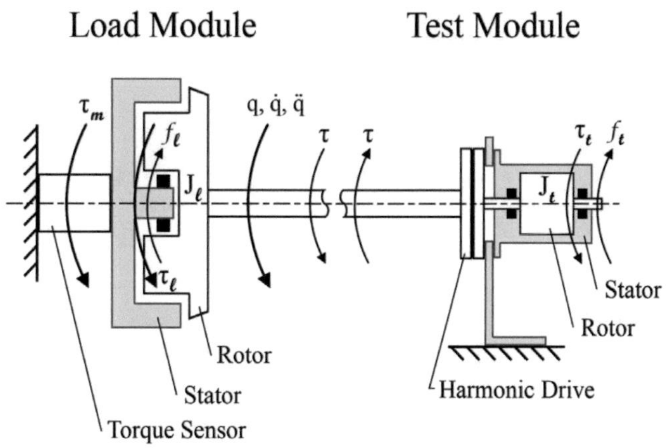

Figure 2. Load and Test Modules

2.2.1 Control

The following control law is proposed for the load motor to ensure that the torque error approaches zero over time:

$$\tau_l = -\tau^* + J_l \ddot{q}^* + f_l^*(\dot{q}) + K_P e + K_I \int e dt + K_D \tfrac{de}{dt} \tag{5}$$

where *K's* are the PID control gains and $f_l^*(\dot{q})$ is estimated using an appropriate friction model. The analysis of the control system is simplified if $f_l^*(\dot{q})$ is assumed to be exactly $f_l(\dot{q})$, and so two cases will be discussed.

Case 1: When a perfect friction model is used, the torque error for each joint exponentially decays to zero with the appropriate selection of K_P, K_I, and K_D.

Proof: Substituting (5) into (1) and applying (3) yields

$$0 = e + K_I \int e dt + K_P e + K_D \tfrac{de}{dt} \tag{6}$$

By taking the time derivative of (6) a second order differential equation is obtained:

$$0 = K_I e + (1 + K_P)\dot{e} + K_D \ddot{e} \tag{7}$$

with the solution

$$e = c_1 \exp(r_1 t) + c_2 \exp(r_2 t) \tag{8}$$

where c's are constants determined by the initial conditions and r's are the roots of the equation

$$K_I r^2 + (1 + K_P) r + K_D = 0 \,. \tag{9}$$

By selecting K_P, K_I, and K_D such that $\mathrm{Re}\{r_1, r_2\} < 0$, e is guaranteed to exponentially decay to zero.

Case 2: When friction error is considered, (6) becomes

$$e_{fric} = e + K_I \int e dt + K_P e + K_D \tfrac{de}{dt} \tag{10}$$

where e_{fric} is the difference between the estimated and true friction. Similarly, (7) becomes

$$\dot{e}_{fric} = K_I e + (1 + K_P)\dot{e} + K_D \ddot{e} \,. \tag{11}$$

Since $\dot{e}_{fric}$ cannot be expressed as a simple function of time it is non-trivial to solve the differential equation (11) and it is left to experimental results to show that e remains within acceptable limits.

2.2.2 Noise Rejection

In the presence of a disturbance torque, d, (6) becomes

$$d = e + K_I \int e dt + K_P e + K_D \tfrac{de}{dt} \,. \tag{12}$$

As before, taking the derivative with respect to time yields

$$\dot{d} = K_I e + (1 + K_P)\dot{e} + K_D \ddot{e} \,. \tag{13}$$

From here (13) can be transformed to the Laplace domain

$$e = \frac{s}{K_I + (1 + K_P)s + K_D s^2} d \,. \tag{14}$$

Considering the norm of the transfer function (14), one can obtain

$$\|e\| \le \frac{\omega}{K_I + K_D \omega^2} \|d\| \,. \tag{15}$$

From the above equation it is possible to see that K_I is key in rejecting low frequency noise, while K_D determines how quickly the second order term of the denominator becomes dominant and rejects high frequency noise. The effects of the remaining high band disturbance, i.e. jerks and vibrations, may be further reduced if the *Test Module* uses a flexible transmission system such as a harmonic drive.

2.3 Computer Simulation

The computer simulation portion of the hardware-in-the-loop setup is first constructed on the *User Interface PC* and then compiled and downloaded to a second dedicated PC that executes the simulation in real-time. The *User Interface PC* runs Microsoft Windows®, and uses MATLAB® to construct a Simulink® model for the simulation. The graphical interface of Simulink® simplifies the configuration of the model, and allows the simulation to be implemented in an easy to read manner by reusing a small set of standard blocks. Further configuration of the simulation is provided by a text file specifying the kinematic model of the robot as well as other simulation parameters such as gravity conditions or end-effector payloads. After configuration, the model is compiled through Real-Time Workshop® into a real-time executable and sent via TCP/IP to the *Simulation PC*. The *Simulation PC* is a barebones computer, little more than a processor and several interface cards, running the xPC Target® real-time kernel. The specialized hard real-time kernel is required because it is necessary to have repeatable and guaranteed latencies, which is not possible in soft real-time or non-real-time environments such as Microsoft Windows®. The *Simulation PC* uses a standard Ethernet port to communicate with the *User Interface PC*, and contains a data acquisition board for communicating with the hardware. After an experiment is run, data can be transmitted back to the *User Interface PC* for post processing and analysis.

The base model constructed in Simulink® is simple, consisting of several blocks for reading in data from the joint hardware modules, an inverse dynamics block that calculates the dynamic torque on each joint, and finally several output blocks to control the load emulation hardware. An example of a model for a 6-d.o.f. manipulator is shown in Fig. 3. The number of joints and physical parameters are specified in a configuration file, which then dynamically updates the inverse dynamics block in Simulink®. Changing the number of joints or specifying a new joint configuration involves editing the configuration file and adding, removing, or reordering the relevant hardware input and output blocks. Hardware input and output blocks depend on the specific hardware used for each joint module, but are generally the same format and require only minor customization. Adding new joints is often simply a matter of copying and pasting one of the current blocks and updating a few parameters such as encoder resolution and output torque range.

The role of a hardware input block is to read in the necessary data from the hardware to determine the instantaneous position, velocity, and acceleration of that joint. This can be done in a number of ways depending on the type of sensors available for each joint. The hardware input block shown in Fig. 4 uses an encoder to determine position and a differentiation and filtering technique to obtain velocity and acceleration estimates. The advantages and disadvantages of this technique are discussed in section 3.3.

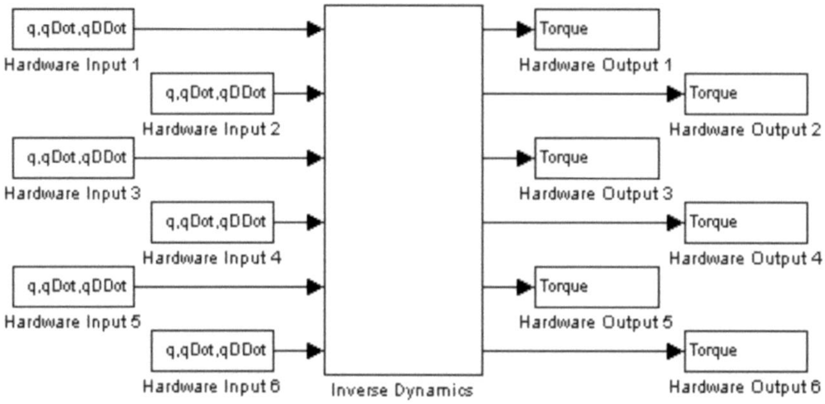

Figure 3. Simulink® Model: 6-d.o.f. Manipulator

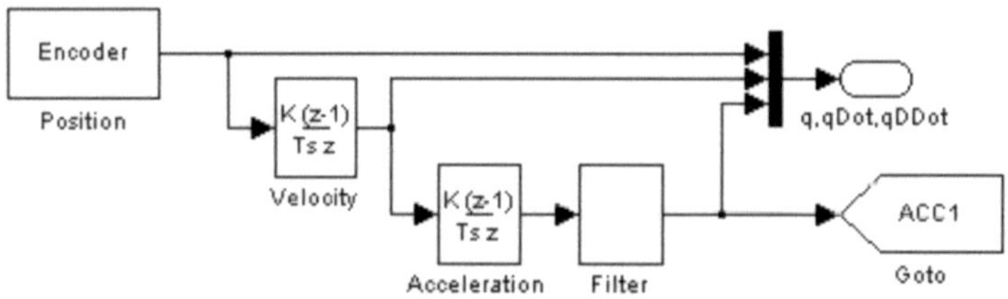

Figure 4. Simulink® Model: Hardware Input Block

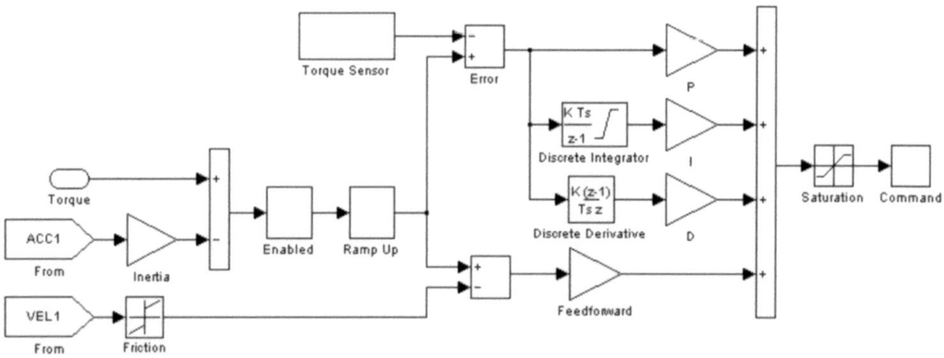

Figure 5. Simulink® Model: Hardware Output Block

The hardware output blocks are required to control each of the *Load Modules* and ensure they apply the torque computed by the inverse dynamics block. Fig. 5 shows how the computed torque is used in conjunction with the signal from the torque sensors to execute closed-loop control of the motor, significantly improving the accuracy of the system. The gains of this controller largely depend on the hardware of the *Load Module* and independent of the *Test Module* hardware, meaning that the simulation does not have to be modified if several different hardware possibilities are being tested. Three additional blocks have been added to the Simulink® model: (i) an "enable" switch has been placed on the incoming torque to allow the torque to be turned on and off without having to restart the simulation, (ii) a "ramp up" block prevents a rapid jump in torque when the simulation initially starts up, and (iii) a saturation block is used to limit the output within the acceptable range of the hardware.

The computational requirements of the RHILS platform are relatively modest due to the simplicity of the simulation and the non-iterative nature of the algorithms used. The *User Interface PC* can be any standard PC capable of running the MATLAB® environment and communicating with the *Control System* and *Computer Simulation PC*. Because the *User Interface PC* communicates with the RHILS platform through the network it does not have to be in the same physical location. It is possible to operate the platform remotely and perform any of the tests without being physically present, although the addition of a visual link such as a webcam might add to the user experience. The *Computer Simulation* can be run on a low-end and, with the exception of the data acquisition board, inexpensive PC. Using the xPC Target® real-time kernel leads to very efficient processing without any of the overhead that is present when running a standard operating system. Additionally, it is possible to guarantee that the code is executed with consistent latencies and is not interrupted by other processes. After performing benchmarking tests with the complete Simulink® model, a Pentium® III 600 MHz processor was chosen. Because of the efficiency of the recursive algorithm (Li & Sankar, 1992) even this low-end processor is able to run the simulation for a 5-d.o.f. system at rates of almost 10 kHz, far faster than the desired 1 kHz. The *Computer Simulation PC* uses a standard Ethernet port to communicate with the *User Interface PC*, and contains a *MultiQ-3* data acquisition board (Quanser, Inc.) for communicating with the hardware. The data acquisition board is by far the most expensive piece of computer hardware used in the simulation and for each joint requires: an encoder input, an analog-to-digital converter to read the torque sensor, and a digital-to-analog converter to send the command signal to the motor drives.

3. Hardware Setup

3.1 Design

The fundamental hardware components of the RHILS platform are a set of joint modules, represented in the *Hardware Emulation* section of Fig. 1. Each joint module is conceptually broken into two parts: a *Test Module* and a *Load Module*. This division is mirrored physically by separating the mounting structures of the two modules and joining them with a single shaft coupling. By separating the modules in this way it is possible to reuse *Load Modules* between multiple simulations or evaluate a number of different hardware configurations for the *Test Modules* with minimum impact on the platform. Each piece of the mounting hardware is designed to allow alignment flexibility so that hardware components can be rapidly moved into place by hand and fixed in position without relying on tight machining tolerances.

The *Test Module* consists of as much of the physical joint actuator as can reasonably be incorporated into the RHILS platform, typically including the motor drive, motor, transmission system, and sensors. The closer the module is to the hardware used in the real robot the less work needs to be done in the computer simulation and the more accurate the performance of the platform will be. The role of the *Load Module* is to emulate the torque that is generated through the kinematics and dynamics of the manipulator. Ideally it should do this with as little impact on the system as possible and without introducing any further

complexities to the computer simulation. To accomplish this, a second transmission system should be avoided and the motor must be able to supply the required torque directly. Additionally, the load emulation device should have low inertia and be able to generate sufficient torque even at low speeds. Due to these requirements most standard motors cannot be used, however, *torque motors* (or ring motors) are well suited for this application and have the additional benefits of low power consumption and good heat dissipation. It was noted above that a low inertia is desirable, and this is the case when simulating small or lightweight joints. However, in certain cases an inertia which is too low may in fact be a disadvantage: for a very high torque joint designed to drive heavy robots a low inertia may allow the joint to accelerate artificially quickly, before the RHILS platform has a chance to calculate and apply the correct torque in response. This has the potential to significantly reduce the effectiveness of the simulation, but can easily be corrected by adding an appropriate inertial load to the joint module to bring it within the range of the real robot.
One of the key challenges for the joint emulating modules is the control of the torque applied by the *Load Motor*. This torque represents the arm dynamics that would be seen by each joint actuator, and requires the control of only the portion of the torque that is generated by the *Load Motor* rather than the coupled torque between the *Load* and *Test Motors*. Various methods of controlling this torque have been suggested in the literature with some drawbacks. For instance, in (Sandholdt et al., 1996) no feedback is used for the load emulator under the assumption that the motor torque follows the input command based on predetermined calibrations performed off-line. Since the load motor is directly coupled to the test motor with potentially significant dynamics, there is no guarantee that the load motor follows the input torque commands satisfactorily. Another approach is to use a complex control scheme that implicitly estimates the load torque based on the current measurements from both motors (Akpolat et al, 1999). The third approach is to mount a rotary torque sensor on the shaft to measure the coupled torque (Aghili, 2006), but in addition to the significant cost of a rotary torque sensor a somewhat more complex control strategy is required to effectively control each joint.
The RHILS platform uses an innovative torque sensing technique that requires no complex modelling or indirect calculations. Instead of mounting a rotary torque sensor on the shaft, a reaction torque sensor is installed between the load motor case (stator) and its mounting fixture. Consequently, the measured signal is directly proportional to the load motor torque, not the coupled torque between the two motors. By decoupling the torque at the measurement stage, there is no need for approximation or estimation techniques involving actuator parameters to track the load motor torque. This also helps to make the architecture independent of the joint actuators being studied since no information about the *Test Module* is required to implement closed loop control of the applied torque.

3.2 Implementation

The hardware of the RHILS platform can be thought of in two parts: the load emulation modules and the test system components. The load emulation portion of the platform is largely generic and independent from the system being designed and tested, while the test system components depend on the hardware of the robot manipulator being simulated. This section details the hardware used for the *Load Modules* in the implementation of the platform.

Two different models of torque motor were selected for use in the platform to provide a range of torque generation capability. A midsize torque motor, the *PSR200* from IntelLiDrives Inc., was selected as the basic motor for the platform and used in two of the joints. The *PSR200* has continuous and peak torques of 17 and 45 Nm, respectively. Since often one of the joints in a robot manipulator will undergo significantly higher torque than the other joints due to its physical configuration a single larger motor was selected. The heavier *SRT-67 HC* (IntelLiDrives Inc.) is capable of a continuous torque of 67 Nm and a peak torque of 185 Nm. Both types of motor are mounted onto flanges at their base and connected to the torque sensors; a second flange and shaft is mounted to the front and supported by a *UCPE205-16* pillow block bearing (KML Bearing & Equipment Ltd.). The motor drives were chosen from the *DR100EE* series by Advanced Motion Controls Inc., and are capable of operating in current, velocity, and position mode.

To measure the torque output of the load motors the RHILS platform uses *TFF350-1300* reaction torque sensors with *JM-2A/AD* amplifier modules (Futek Inc.). Because of their extremely close proximity to the torque motors the sensors are subject to a large amount of electromagnetic interference. To counteract this problem two additional hardware components were used: AC power filters and ferrite suppression cores. The *FN2070-16/07* single phase AC power filters (Schaffner Holding AG) were connected between the wall socket and the motor drive, greatly reducing the line noise and preventing the motor drives from interfering with each other. The suppression cores (Fair-Rite Products Corp.) were used in two places: first around the power lines between the torque motors and their drives, and then around the shielded cable of the torque sensors. After taking these steps the electromagnetic noise on the torque signals was greatly reduced.

3.3 Verification

The two most basic functions of the joint hardware are for it to accurately apply the commanded torque signal and to obtain the position, velocity, and acceleration information necessary for the inverse dynamics calculations. A series of tests were carried out to verify the ability of the platform to follow a torque command and several methods for determining the position, velocity, and acceleration were evaluated.

The purpose of the first tests was to demonstrate the ability of the *Load Modules* to follow a command signal, as well as verify the necessity of closed-loop control to accurately apply the torque. The command torque signals were: (a) constant, (b) sinusoidal, and (c) random. Fig. 6(a) shows the command signal, open-loop torque, and closed-loop torque for the constant test. The benefit of closed-loop control is immediately obvious as it eliminates the steady state error that is present in the open-loop torque. The same set of curves is shown for the sinusoidal case in Fig. 6(b). Again, the closed-loop torque is much closer to command than the open-loop. The last torque following test, shown in Fig. 6(c), demonstrates the platform's ability to follow a torque signal generated by the inverse dynamics simulation during a pseudo-random move. Table 1 shows the mean error and root-mean-square (RMS) error for each torque signal, both as a percentage of the maximum torque range.

In order to compute the dynamic torque on the manipulator it is necessary to determine the position, velocity, and acceleration of each joint. Using an angular accelerometer is ideal from a measurement standpoint, since it directly reads the acceleration and can be integrated to determine the velocity with good accuracy. However, mounting the accelerometer presents some challenges and complicates the mechanical design. Also, purchasing accelerometers and the additional electronics for each joint would add significantly to the cost of the platform. Because of these drawbacks a differentiation technique using the position sensor was explored, although for an application in industry accelerometers may in fact be a viable option.

Command Signal	Mean Error [%]	RMS Error [%]
Constant (open-loop)	4.8352	4.8826
Sinusoidal (open-loop)	2.8856	3.3982
Constant	0.4769	0.5883
Sinusoidal	1.4633	1.8412
Random	1.6597	2.0128

Table 1. Torque Following Error

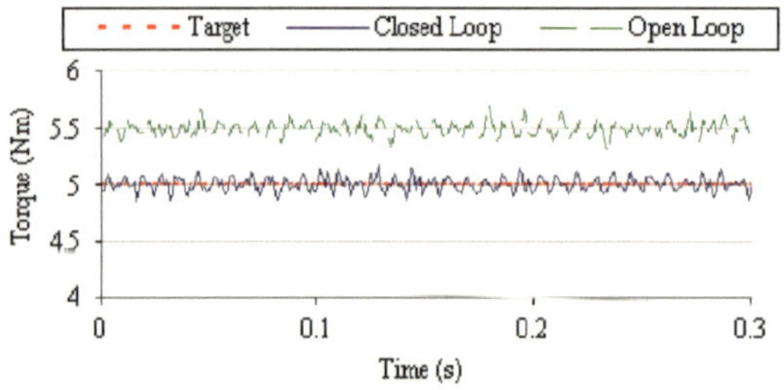

(a) Constant

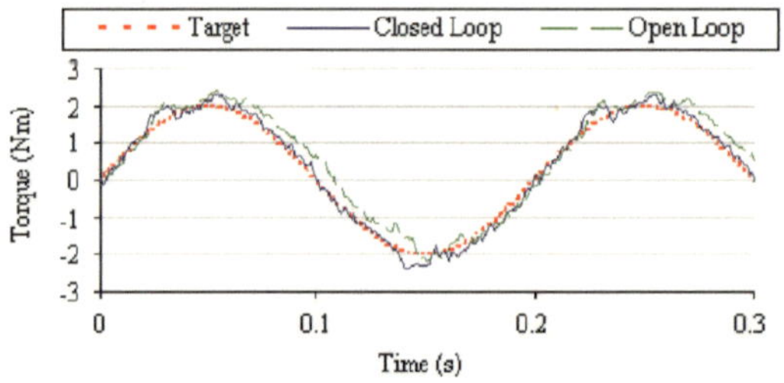

(b) Sinusoidal

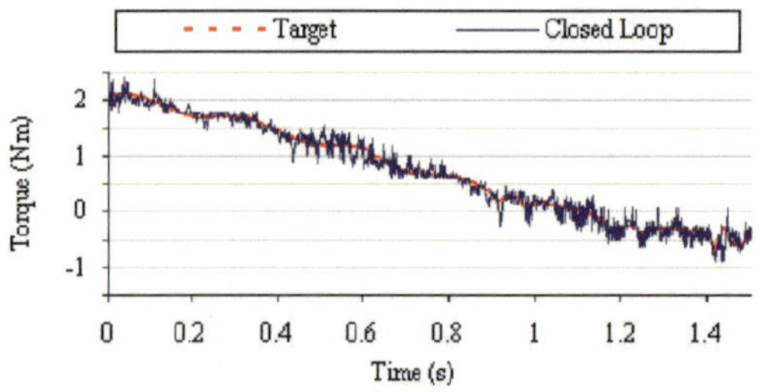

(c) Random

Figure 6. Torque Following

The issue of noise amplification during the derivative process is well known, and in order to produce a useable signal some form of filtering is required. However, the filter itself introduces a lag in the system, and so a compromise must be reached between the amount of unfiltered noise and the signal delay. A number of filtering techniques were considered based on the selection criteria, and finally a Butterworth Infinite Impulse Response (IIR) filter (Rabiner & Gold, 1975) with a 47ms delay was chosen. Various filters were applied after taking the position of from one of joints during a simple step trajectory from 0° to 180° at maximum speed and acceleration and again during a random trajectory. The filtered acceleration for the initial portion of the step trajectory is shown in Fig. 7 using a 10ms-delay Equiripple Finite Impulse Response (FIR) filter and the 47ms-delay Butterworth IIR filter. The superior filtering of the Butterworth can be observed, and measuring the length of the acceleration shows that the 47ms delay corresponds to roughly 7% of the duration. The power spectral density curves in Fig. 8 show that the position contains no significant frequencies passed 5 Hz, and how the Butterworth filter helps reduce higher frequencies that still dominate after the Equiripple filter.

To verify that the acceleration estimated through differentiation was accurate a second technique was used. Using the following equation it is possible to estimate the acceleration at each time step based on the current from each motor

$$\ddot{q}^*[n] = \frac{\tau_t/\eta - \tau_l - \dot{q}^*[n-1]\left(f_t(\dot{q})/\eta^2 + f_l(\dot{q})\right)}{J_t/\eta^2 + J_l}. \tag{16}$$

Velocity can then be found by integration. This method is not useful in the final platform since the control law for τ_l, (5), requires $\ddot{q}^*$ creating an algebraic loop. However, performing a number of tests using this method was useful in verifying the results of the differentiation method. Fig. 9 shows experimental data from the joint for a step trajectory, where the solid and dashed lines show the acceleration obtained from the differentiation and current methods, respectively. The initial acceleration is shown in Fig. 9(a), while the deceleration when the joint reaches its set point is shown in Fig. 9(b). As further validation Fig. 10 shows the similarity in power spectral density between the computed and current-based acceleration.

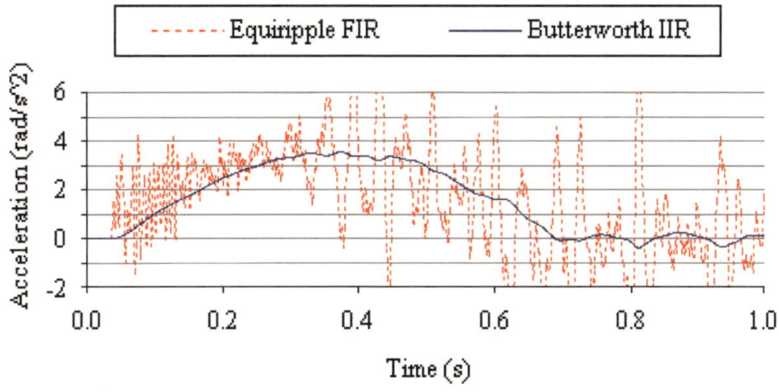

Figure 7. Acceleration Approximation: Differentiation Method

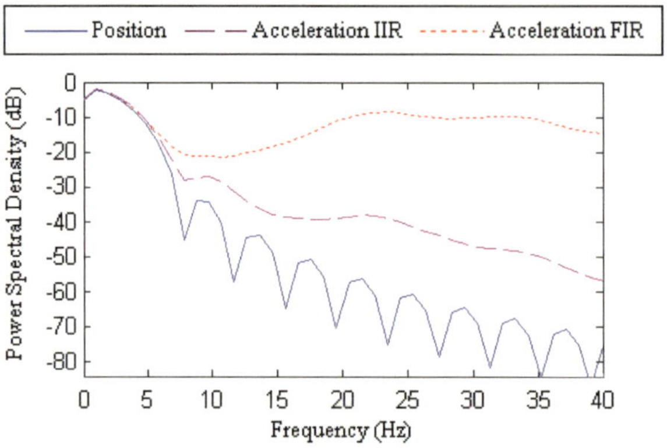

Figure 8. Acceleration Approximation: Differentiation Method PSD

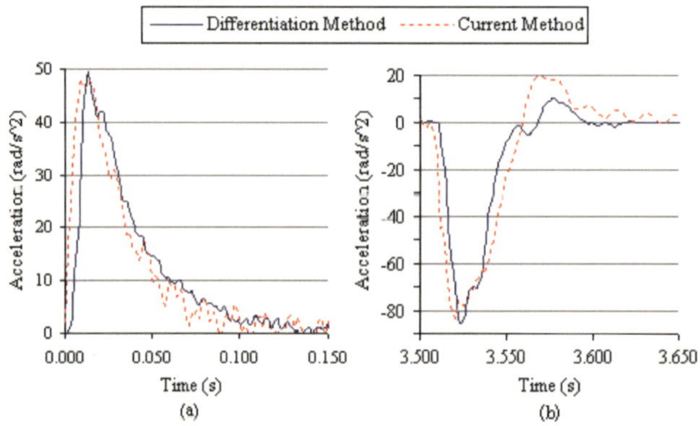

Figure 9. Acceleration Approximation: Current Method

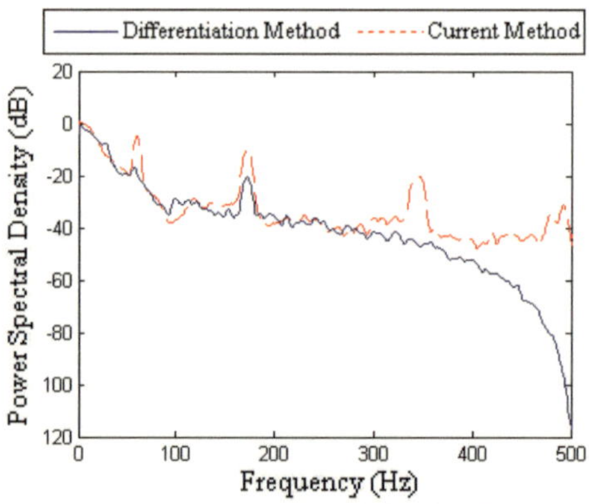

Figure 10. Acceleration Approximation: Current Method PSD

Returning to the discussion of e_{acc} from (4), using the acceleration error from Fig. 9 during an extremely aggressive step trajectory and the heaviest motor inertia, that of joint 2, the maximum and mean values of e_{acc} are 0.322 Nm and 0.093 Nm, respectively. Fig. 6(a) shows the noise picked up by the torque sensors to has a magnitude of 0.4 Nm, and so the impact of e_{acc} on the system is minimal.

4. Case Study

In order to evaluate the true potential of the RHILS platform it was applied to the simulation of a generic industrial robot manipulator, namely the CRS CataLyst-5 from Thermo Fisher Scientific Inc. The CataLyst-5, pictured in Fig. 11, is a midsized 5-d.o.f. manipulator with a maximum payload of 1 kg and a reach of 660 mm (Thermo, 2007). The goal of this implementation was to explore possible improvements to the CataLyst-5 and evaluate the true potential of the RHILS platform in real world applications. The selection of the CataLyst-5 was made for several reasons: (a) it is a general-purpose yet fully featured 5-d.o.f. manipulator such as those found in many labs and assembly lines, (b) it has an established and proven design, meaning that much of the necessary information is already available, and (c) the manufacturer was in the process of modifying the design and replacing the joint motors, providing a unique opportunity to test the capability of the platform as a design tool in a short time frame without the complications of trying to simulate a completely new robot. The exact hardware and control system from the robot were installed on the RHILS platform and the results were compared with the physical prototype in a series of tests. Fig. 12 shows the first three joints of the RHILS setup, where each *Test Module* includes the motor, belt transmission, and harmonic drive from the real manipulator.

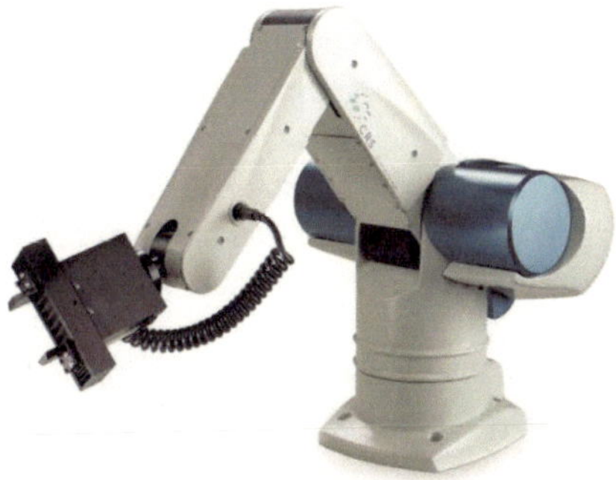

Figure 11. CRS CataLyst-5 Manipulator

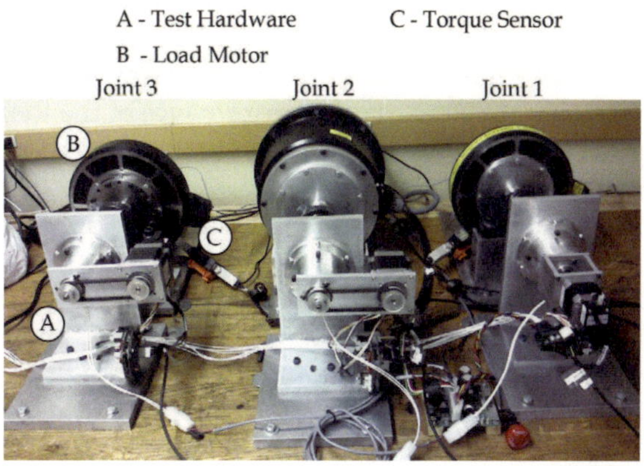

Figure 12. RHILS Setup

A number of tests were carried out to verify the functionality of the RHILS platform and then to evaluate its merit as a simulation and design tool. These experiments were divided into three phases: (a) platform functionality, a set of tests to demonstrate the ability of the platform to function as a complete system, (b) platform validation, a performance comparison between the platform and the physical prototype, and (c) design capabilities, an exploration of potential areas of manipulator and control system design that can benefit from RHILS. The following sections present the methodology behind these tests and a cross section of the results. A complete analysis is available in (Martin, 2007).

4.1 Platform Functionality

A series of experiments were conducted to demonstrate the ability of the platform to function as a complete system under various conditions. These tests involved the real-time cycle of monitoring the position of each joint, differentiating to approximate the velocity and acceleration signals, using this data in the inverse dynamics simulation to compute the torque at each joint, and finally, applying the computed torque to the joint using the *Load Modules*.

Initially, a single joint was subjected to several trajectories, including: (a) a "step," constrained by the maximum acceleration and velocity of the joint, (b) sinusoids at various amplitudes and frequencies, and (c) a pseudo-random trajectory. This "random" trajectory was generated by summing several sinusoids with various frequencies, the highest frequency being limited to half the resonant frequency of the robot. Given that the approximate range of resonant frequencies for standard manipulators is 5-25 Hz (Craig, 1989), 5 Hz was selected as the highest frequency in the random signal. Further limitations were placed on the signal based on maximum allowable accelerations and velocities for the joints. The output torque was monitored during each of the trajectories and found to follow the command torque sufficiently well.

After successfully simulating these basic conditions, all the joints were tested simultaneously as the robot moved between several positions as it would during a standard operation. The RHILS platform proved capable of estimating the joint velocities and accelerations, calculating the resultant torque at each joint, and using the *Load Modules* to accurately apply the torque throughout the experiment.

4.2 Platform Validation

In order to validate the RHILS platform a number of experiments were conducted on both the platform and the current physical prototype of the CataLyst-5. By comparing the two sets of data it is possible to see how closely the performance of the RHILS platform matches that of the physical prototype. Because of the limited sensors installed on the prototype it was impossible to directly compare joint torque. However, the position error data that was compared is a key factor in determining the tuning and performance of the controller and similar trends are a strong indicator of matching performance. An additional set of data was gathered from the RHILS platform without any dynamic load emulation, i.e. no load was supplied by the *Load Modules* and the joints were allowed to spin freely. This third set of data serves two purposes: (i) it provides a comparison for how the joints perform when operating freely, and (ii) it demonstrates that a full RHILS platform is necessary to accurately simulate the robot. Three sets of tests were carried out: (a) sinusoidal trajectories, first on each joint individually and then on all joints simultaneously, (b) random trajectories, again on each joint individually and then on all joints simultaneously, and (c) a standard series of motions such as the robot would perform in a real world application. Each test was run at maximum speed and acceleration carrying a 1.043 kg payload, the most extreme operating conditions of the robot, and executed multiple times to verify the repeatability of the results. Some of the results include values in encoder pulses, for the waist joint 1 pulse = 0.0025° while for the shoulder and elbow joints 1 pulse = 0.00125°.

The sinusoidal trajectories were carried out at a frequency determined by the maximum velocity and acceleration of each joint, with an amplitude of 40° for the waist and 20° for the shoulder and elbow joints. First, each joint was tested individually and then the sinusoidal trajectories were repeated with all joints moving simultaneously. Fig. 13 shows (a) the position of each joint, (b) the position error, and (c) the joint torque. An additional overlay is provided to give a visual indication of the joint configuration at various points in time.

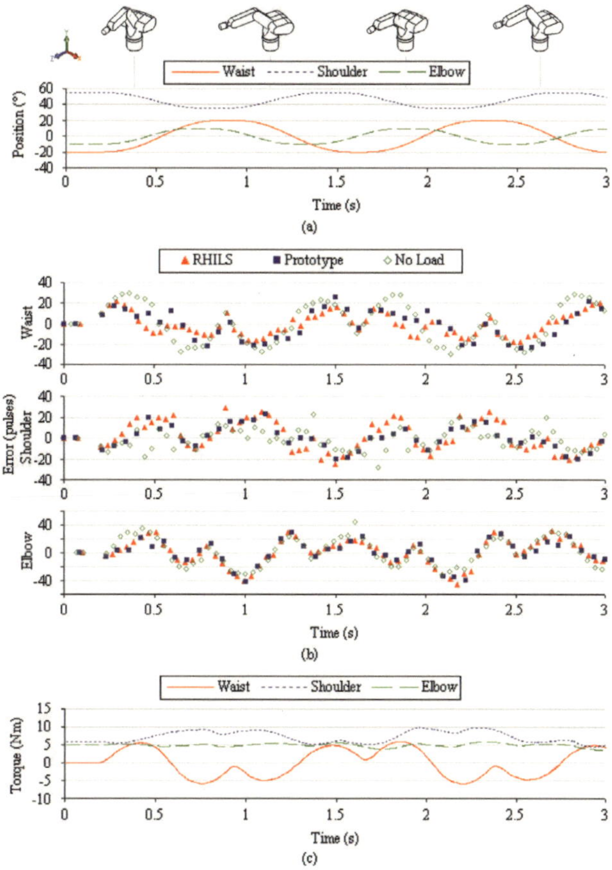

Figure 13. Platform Validation: Sinusoidal Trajectory

The position error in (b) is a useful metric for the performance of the control system, but for the current purpose it is sufficient to observe the similarities between the performance of the physical prototype and the RHILS platform and the distinctly different performance from the free spinning joint, particularly for the shoulder joint. It is also interesting to note the non-symmetry of the torque curves in (c), which is a result of the coupling between the joints. The one anomaly in the data, which persists throughout the subsequent tests, is the overall performance of the waist joint. By studying the shape of the error curve it is possible to analyze the motion and determine the reason for these discrepancies. In Fig. 13(b) the curves for the waist match well for the initial portion of the upswing as the joint leads its commanded position, and then in the middle of the move the prototype curve dips slightly and maintains a smaller lead while the RHILS curve dips significantly and in fact begins to lag behind the commanded position. This trend repeats again as the joint moves in the opposite direction and with each cycle. This additional lag suggests a number of possible causes for the discrepancies: (i) the rotating inertia in the simulation is too high, (ii) the waist motor of the RHILS platform is underperforming that of the prototype, or (iii) there is some mechanical defect in the joint module that is putting additional load on the joint at high speeds. For the remaining joints the difference in performance between the RHILS platform and the prototype is comparable to the variation found when running multiple repetitions of a test on the prototype.

Similar tests were done using pseudo-random trajectories, first with each joint tested individually and then with all the joints simultaneously. Due to the random nature of the applied torque the error curves followed a much less predictable pattern, yet were still very similar between the RHILS platform and the prototype.

The last set of experiments used to evaluate the validity of the RHILS platform was to run the manipulator through a standard set of movements such as it would perform during everyday operation. This test consisted of five moves starting from a "ready" position with the shoulder vertical and the elbow bent at 90°. The robot then moved to simulate a "pick and place" operation: reaching out and picking up an object then depositing the object at a different location. Fig. 14 shows (a) position, (b) position error, and (c) joint torque data, along with a visual representation of the robot at various points in time. The position error graphs have been dissected in order to highlight the time periods relevant to each joint. Observing the waist for the initial portion of this operation as the arm reaches out to pick up the object, both the RHILS platform and the prototype exhibit a squared-off pattern, consisting of three plateaus each lasting roughly a third of the move. Conversely, the free spinning joint exhibits a much smoother behaviour as would be expected from a simple rotating inertial load. Similarly, for the shoulder the performance of the free spinning joint is much smoother and flatter than the peaks and troughs of both the prototype and RHILS platform.

4.3 Design Capabilities

One of the main goals of the RHILS platform is to provide a design tool that can be used to improve various aspects of the robot. Two of the major areas of the design that can benefit from the RHILS platform are: (i) the control system of the manipulator and (ii) the physical parameters of the robot. In the case of the control system the RHILS platform allows testing to begin early in the design stage, before the final hardware design is complete and without building expensive physical prototypes. For designing physical parameters the RHILS platform has the ability to adjust many of the important parameters simply by editing a configuration file, and has the additional advantage of being able to test the extreme limits of the design without risking damage to a prototype robot. Three sets of tests were carried out to evaluate the design capabilities of the platform. The first was done to compare the performance of the RHILS platform to the prototype when the same control system modifications were made on each. The second set involved comparing the performance of the robot carrying different payloads, demonstrating that different conditions on the prototype could be replicated on the RHILS platform. The third set of tests is meant to show how the RHILS platform can go beyond the capabilities of a physical prototype by simulating the robot with a lighter weight but longer arm, something that would require significant time and effort to duplicate on a physical prototype.

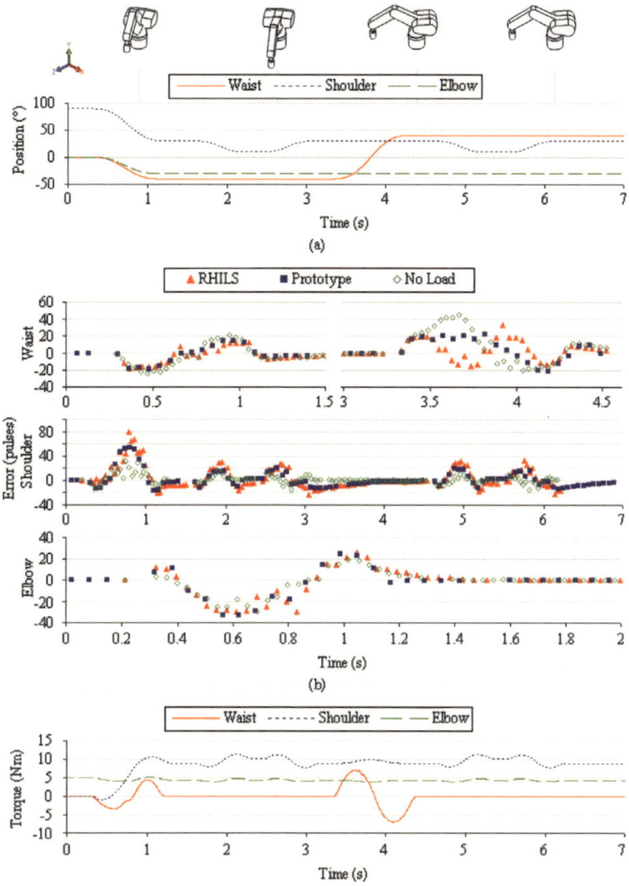

Figure 14. Platform Validation: "Pick-and-Place" Operation

Control system design and tuning is an important and often lengthy part of the development of a robot, and the earlier in the design this process can start the better the final product will be. In order to verify that modifications to the control system on the RHILS platform result in similar performance changes when applied to the real robot a number of tests were done; first with an un-tuned control system and then repeated with a tuned control system. Simple moves were carried out on each joint, and then a more complex move was done involving all the joints. The manipulator started with the waist rotated 90° and the shoulder vertical with the elbow bent at a right angle, then the was waist rotated back to 0° while the elbow and wrist rotated into a vertical position, and finally, the arm moved to the fully outstretched configuration. Fig. 15 shows: (a) position, (b) position error for both the tuned and un-tuned controllers, and (c) joint torque. The drastically different performance between the two controller tunings is obvious, and consistent between the RHILS platform and the prototype. These position error curves are a useful metric for fine-tuning the controller and its performance can be improved by observing the position lead and lag over the course of a move. For instance large initial leads may indicate that the feed-forward gains are set too high, while significant lags might indicate that the proportional or integral gains are too low. The fact that the RHILS platform is able to replicate these curves demonstrates that it could be used to tune the controller and the final results applied to the real robot with a reasonable degree of confidence.

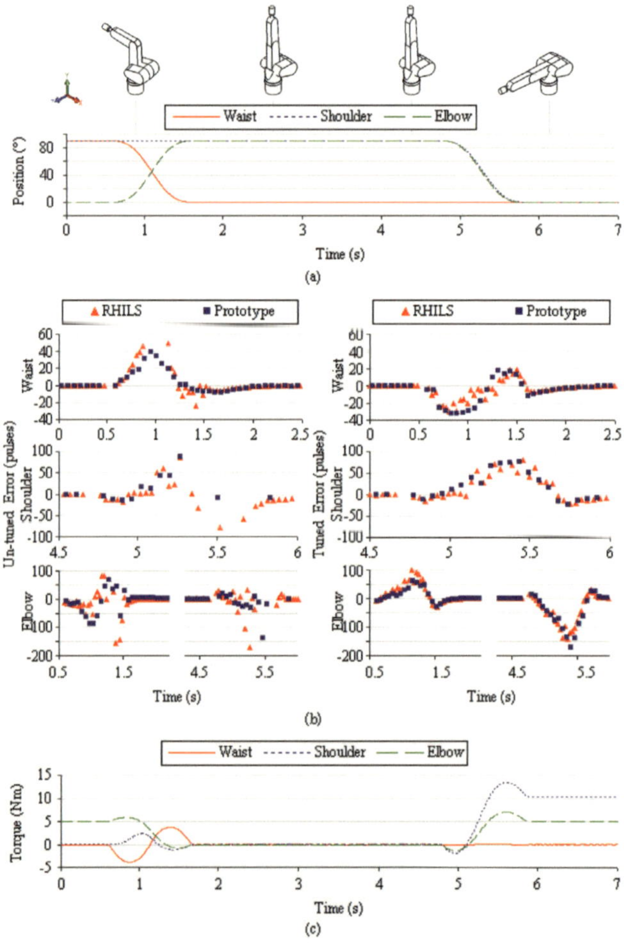

Figure 15. Design Capabilities: Controller Tuning

A second set of experiments was conducted to show how physical changes to the prototype could be easily replicated on the RHILS platform. This was done by attaching several different payloads to the prototype and running it through the same "pick and place" operation described in section 4.2, then updating the physical model and running the same motions on the RHILS platform. Three payloads, 0.0 kg (no payload), 0.420 kg, and 1.043 kg, were used during this test. Looking at the shoulder joint, Fig. 16 shows the position, position error, and torque for all three payloads. Due to the combination of high gear ratio and the mass of the arm relative to the payloads the differences in performance in each case are quite subtle, with the most marked changes occurring during the 5-6 second period. In the torque curves, the difference between each of the three cases is relatively small and it is easy to see why the position error graphs are similar.

A final set of tests was conducted to demonstrate the capability of the RHILS platform to simulate significant modifications to the physical structure of the robot. The weight of the arm was reduced by 30% while its length was increased by 20%. This type of test would be difficult and expensive to perform with a physical prototype, essentially requiring that most of the prototype be rebuilt, but can be done on the RHILS platform with minor modifications to the configuration file and no additional cost. Using a modified configuration file the platform was run through a series of simple motions and the differences in resultant joint torque were easily observed. Although it may not be realistic to assume the length of the arm can be increased while reducing the weight and still maintain the necessary stiffness, this type of test demonstrates how the RHILS platform can be used to rapidly evaluate different physical parameters or even joint configurations early in the design phase.

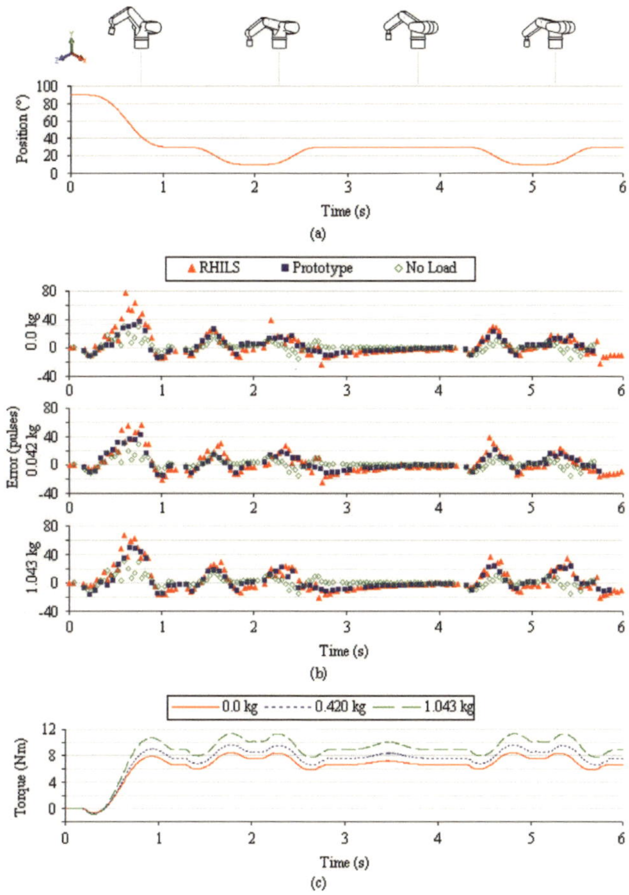

Figure 16. Design Capabilities: Payload Simulation, Shoulder Joint

5. Conclusion

The RHILS architecture and suitable hardware implementation were proposed as a coherent strategy for applying HILS techniques to the simulation and design of robot manipulators. First, it was shown that the architecture was able to run on a standard computer with modest processing power, and that the hardware joint modules were able to accurately apply torque commands of various types and approximate the joint acceleration using a differentiation and filtering technique. The architecture was then applied to the simulation of a standard industrial manipulator. A series of tests were carried out to evaluate: (i) the basic functionality of the platform - its ability to simulate a 5-d.o.f. manipulator in real-time, (ii) the accuracy of the simulation - the performance of the RHILS platform relative to a physical prototype of the robot, and (iii) the design capabilities of the platform - its potential when applied to the design of both the controller and the physical structure of the manipulator. The validation of the RHILS platform was successful, and the platform exhibited similar performance characteristics to the prototype under both normal and aggressive operating conditions. The RHILS platform also demonstrated significant promise as a control system design and tuning tool. Further, it was shown that modifications to the physical design can be rapidly evaluated on the RHILS platform without the need for expensive modifications to a physical prototype.

Future development of the RHILS architecture will take several directions. Improvements are being made to the current hardware in order to facilitate rapid switching of test modules in keeping with the ideal of modularity. The platform is also being used as a testbed for concurrent design strategies for reconfigurable robots using automated optimization techniques (Chhabra & Emami, 2008).

6. References

Aghili, F. & Piedboeuf, J.-C. (2002). Contact dynamics emulation for hardware-in-loop simulation of robots interacting with environment, *Proceedings. ICRA '02. IEEE International Conference on Robotics and Automation*, Vol. 1, pp. 523-529, 2002

Aghili, F. (2006). A Mechatronic Testbed for Revolute-Joint Prototypes of a Manipulator. *IEEE Transactions on Robotics*, Vol. 22, No. 6, pp. 1265-1273

Akpolat, Z. H.; Asher, G. M. & Clare, J. C. (1999). Dynamic Emulation of Mechanical Loads Using a Vector-Controlled Induction Motor-Generator Set. *IEEE Transactions on Industrial Electronics*, Vol. 46, No. 2, pp. 370-379

Aronsson, P. & Fritzson, P. (2001). Parallel Code Generation in MathModelica / An Object Oriented Component Based Simulation Environment, *Proceedings, Parallel / High Performance Object-Oriented Scientific Computing, Workshop, POOSC01 at OOPSLA01*, 2001

Ballard, B. L.; Elwell Jr., R. E.; Gettier, R. C.; Horan, F. P.; Krummenoehl, A. F. & Schepleng, D. B. (2002). Simulation Approaches for Supporting Tactical System Development. *John Hopkins APL Technical Digest (Applied Physics Laboratory)*, Vol. 23, No. 2-3, pp. 311-324

Chhabra, R. & Emami, M. R. (2008). Concurrent Design of Robot Manipulators Using Hardware-in-the-loop Simulation, *2008 IEEE International Conference on Technologies for Practical Robot Applications (TePRA)*, Massachusetts, USA, November 2008

Craig, J. (1989). *Introduction to Robotics Mechanics and Control*, 2nd Ed., Addison-Wesley Publishing Company Inc., USA, p. 323

Cyril, X.; Jaar, G. & St-Pierre, J. (2000). Advanced Space Robotics Simulation for Training and Operations, *AIAA Modeling and Simulation Technologies Conference*, pp. 1-6, Denver, USA, August 2000

Faithfull, P. T.; Ball, R. J. & Jones, R. P. (2001). An Investigation into the Use of Hardware-in-the-loop Simulation with a Scaled Physical Prototype as An Aid to Design. *Journal of Engineering Design*, Vol. 12, No. 3, pp. 231-243

Hanselman, H. (1996). Hardware-in-the-loop Simulation Testing and its Integration into a CACSD Toolset, *IEEE International Symposium on Computer-Aided Control System Design*, pp. 15-18, September 1996

Hu, X. (2005). Applying Robot-in-the-loop Simulation to Mobile Robot Systems, *12th International Conference on Advanced Robotics (ICAR 2005)*

Kasper, R.; Koch, W. & Sczyrba, C. (1997). Object-Oriented Software Techniques in Simulation of Mechatronic Systems: Towards Distributed Hardware-in-the-Loop-Simulation using Standard Hard- and Software, *Proceedings of the 15th IMACS World Congress on Scientific Computation, Modelling and Applied Mathematics*, Berlin, 1997

Lane, D.; Falconer, G. J. & Randall, G. (2001). Interoperablitiy and Synchronisation of Distributed Hardware-in-the-Loop Simulation for Underwater Robot Development, *Proceedings of the 2001 IEEE International Conference on Robotics & Automation*, 2001

Ledin, J. (1999). Hardware-in-the-loop simulation. *Embedded Systems Programming*, Vol. 12, No. 2, pp. 42-53

Leitner, J. (1996). Space Technology Transition Using Hardware in the Loop Simulation, *Proceedings of the 1996 Aerospace Applications Conference*, Vol. 2, pp. 303-311, February 1996

Li, C.-J. & Sankar, T. S. (1992). Fast inverse dynamics computation in real-time robot control. *Mechanism & Machine Theory*, Vol. 27, No. 6, pp. 741-750

Linjama, M.; Virvalo, T.; Gustafsson, J., Lintula, J.; Aaltonen, V. & Kivikoski, M. (2000). Hardware-in-the-loop Environment for Servo System Controller Design, Tuning, and Testing. *Microprocessors and Microsystems*, Vol. 24, No. 1, pp. 13-21

Ma, O.; Wang, J.; Misra, S. & Liu, M. (2004). On the validation of SPDM task verification facility. *Journal of Robotic Systems*, Vol. 21, No. 5, pp. 219-235

Maclay, D. (1997). Simulation gets into the loop. *IEE Review*, Vol. 43, No. 3, pp. 109-112

Martin, A. (2007). *Development of a Modular Hardware-in-the-Loop Simulation Platform for Synthesis and Analysis of Robot Manipulators*, M.A.Sc. Thesis Report, University of Toronto, April 2007

Piedboeuf, J.-C.; Carufel, J. de; Aghili, F. & Dupuis, E. (1999). Task Verification Facility for the Canadian Special Purpose Dexterous Manipulator, *Proceedings of the 1999 IEEE International Conference on Robotics and Automation*, pp. 1077–1083, USA, 1999

Rabiner, L.R. & Gold, B. (1975). *Theory and Application of Digital Signal Processing*, Prentice-Hall , Englewood Cliffs, NJ, p. 227

Sandholdt, P.; Ritchie, E. & Pedersen, J. K. (1996). A Dynamometer Performing Dynamical Emulation of Loads with Non-Linear Friction, *IEEE International Symposium on Industrial Electronics*, Vol. 2, pp. 873-878, 1996

Solak, E.; Murray-Smith, R.; Leithead, W. E.; Leith, D. J. & Rasmussen, C. E. (2003). Derivative observations in Gaussian Process Models of Dynamic Systems. *Advances in Neural Information Processing Systems 15*, Eds. Becker, S.; Thrun, S. & Obermayer, K., MIT Press

Stoeppler, G.; Menzel, T. & Douglas, S. (2005). Hardware-in-the-loop simulation of machine tools and manufacturing systems. *Computing & Control Engineering Journal*, Vol. 16, No. 1, pp. 10-15

Temeltas, H.; Gokasan, M.; Bogosyan, S. & Kilic, A. (2002). Hardware in the Loop Simulation of Robot Manipulators through Internet in Mechatronics Education, *The 28th Annual Conference of the IEEE Industrial Electronics Society*, Vol. 4, pp. 2617-2622, Sevilla, Spain, November 2002

Thermo Fisher Scientific Inc. (2007). CRS CataLyst-5 Robot System, Website: http://www.thermo.com/com/cda/product/detail/0,1055,21388,00.html, February 2007

Ulrich, K. & Eppinger, S. (2000). *Design and Development*, 2nd Edition, McGraw-Hill

Wheelright, S. & Clark, K. (1992). Revolutionizing Product Development- Quantum Leaps in Speed, Efficiency and Quality, The Free Press, New York

20

Soft-computing Techniques for the Trajectory Planning of Multi-robot Manipulator Systems

Emmanuel A. Merchán-Cruz[1], Alan S. Morris[2] and Javier Ramírez-Gordillo[1]
[1]Escuela Superior de Ingeniería Mécanica y Eléctrica, Unidad Azcapotzalco, IPN
[2]Automatic Control & Systems Engineering Dep., The University of Sheffield
[1]México, [2]United Kingdom

1. Introduction

The planning of the motion of robots has become increasingly more complex as robots are used in a wide spectrum of applications, from extremely repetitive tasks in assembly lines to assistance in delicate surgical procedures (Tombropoulos et al., 1999; Coste-Maniere et al., 2003). Whichever scenario, the planning has to consider the fact that the robot will be interacting with other elements in its environment avoiding collision with other objects, whether these remain static or in motion, while executing a given task.

In planning the motion for robots, it is a common misassumption that path planning and trajectory planning are synonymous. Motion planning, as defined by Sugihara and Hwang in (Hwang & Ahuja, 1992; Sugihara & Smith, 1997), is subdivided into path planning and trajectory planning. (Fraichard & Laugier, 1992) state the following distinction: path planning is the search of a continuous sequence of collision-free configurations between a start and a goal configuration, whereas trajectory planning is also concerned with the time history or scheduling of this sequence of configurations as well.

Considering that the motion of the elements that form the kinematic chain of robot manipulators is described by a system of non-linear equations that relate the motion in the Cartesian space of the end-effector as a consequence of the individual variations of the links of the manipulator due to the angular displacements at each joint. When solving for a particular position of the end-effector in the Cartesian space, the inverse kinematics problem, a set of configurations has to be calculated to position the tip of the manipulator at that desired point. Due to the natural dexterity of robot manipulators, the space of solution is non-linear and multidimensional, where more than a single solution exists to solve a particular point in the Cartesian space and choosing the appropriate solution requires an optimisation approach.

Taking this into consideration, the solution of the motion planning problem of robot manipulators is an ideal candidate for the use of soft-computing techniques such as genetic algorithms and fuzzy logic, as both approaches are known to perform well under multidimensional non-linear spaces without the need for complex mathematic manipulation to find a suitable solution (Zadeh, 1965; Mamdani, 1974; Goldberg, 1983; Bessiere et al., 1993; Doyle, 1995; Doyle & Jones, 1996).

Following this line of thought, the application of soft-computing techniques is considered to improve the performance of trajectory planning. The approach followed to solve this problem considers that the manipulators have to reach a specified goal or target defined in coordinates of theirs workspace, rather than a goal or target configuration as widely presented in numerous papers. This is because when the manipulator is forced to reach a certain configuration, the trajectory planning problem is greatly simplified by constraining the system to satisfy those specific values. By specifying the target in coordinates in the workspace instead, the system can solve for this condition assuming a number of different solutions, thus, dealing now with an optimisation problem.

1.1 Motion planning

Motion planning for robot manipulators has been extensively studied during the last two decades. A comprehensive survey on the common techniques used to solve this problem can be found in (Latombe, 1991; Hwang & Ahuja, 1992; Ata, 2007). Depending on the nature of the problem, some authors classify the motion planning problem in two categories: global and local motion planning (Lozano-Peréz, 1983; Latombe, 1991; Althoefer, 1996).

Global motion planning requires a complete description of the workspace of the robot where all obstacles are clearly identified before searching a path. This takes place in systems where the manipulators operate in a highly structured and controlled environment in which all possible obstacles are static and known in advance, the planning here is performed off-line allowing for an optimal trajectory to be found.

On the other hand, local motion planning calculates iteratively the next best position for the robot that best reduces the error to the goal while avoiding collision with any obstacles in the workspace. Conversely to global approaches, a local approach does not require prior knowledge of the system in terms of possible obstacles in the workspace. Local motion planning approaches have to process information which describes the vicinity of the manipulator and modifies its trajectory in order to avoid collision with any obstacle nearby while minimising the error to the target. Local planners are commonly used in dynamic environments where limited information on moving obstacles is available (Brooks, 1986; Lee, 1996; Liu, 2003).

The algorithms for motion planning are often categorised as either complete or heuristic (Chen & Hwang, 1998; Masehian & Sedighizadeh, 2007). Complete algorithms can take a considerable amount of computation time before finding a solution if there is one, or fail to prove that there is no solution. Heuristic algorithms are fast but are not guaranteed to find a solution even if there is one (Eldershaw, 2001; Isto, 2003). Their attractiveness resides in the fact that they will either find a solution or fail in a space of seconds.

The development of the configuration space method or C-space by (Lozano-Peréz & Wesley, 1979) marked a standard which has been used in various path-planning algorithms, as can be seen in (Wise & Bowyer, 2000). In the C-space, each node represents a possible configuration of the robot, and a path is specified by following a line that connects feasible configurations from a start to a goal configuration. Prior to this, all path planning methods were dependent on the use of the Cartesian space to represent a manipulator and its environment. The configuration space approach reduces the problem to that of moving a point through an n-dimensional space, where n corresponds to the number of degrees of freedom (dof) of the manipulator. Obstacles in the manipulator's environment are mapped into the C-space and are represented as a forbidden region known as configuration space obstacles or CS-obstacles where a collision with the manipulator occurs. A typical technique to navigate in the C-space consists of exploring a uniform grid in C-space using a best-first search algorithm guided by a goal-oriented potential field (Barraquand & Latombe, 1991).

1.2 The Multi-mover's problem

When there is more than one robot working in a common workspace, the planning of motion of such a system is referred to as the multi-movers problem. The multi-movers problem is that of finding paths for the robots between their initial and final configurations whilst avoiding one another as well as any obstacles.

Regarding the motion planning into systems where multiple robots share the same environment, three approaches can be identified: the centralized approach, the hierarchical or prioritized and decoupled planning.

The centralized approach consists of treating the various robots as if they were one single robot, considers the Cartesian product of their individual C-spaces called composite C-space. The forbidden region is the space in which one robot intersects an obstacle or two robots intersect each other. Algorithms based on this approach require a great amount of computational resources to store the resulting information of the representation of the composite C-space and for solving the path over this space (Schwartz & Sharir, 1983; Fortune et al., 1987).

The hierarchical or prioritized approach presented by Freund and Hoyer in (Freund & Hoyer, 1985), plans the motions of the robots accordingly to a certain priority assigned to each robot depending on their specified tasks. A common configuration under this approach in multi-robot arm systems is the Master-Slave configuration, where the motion of the slave is dependent on the motion of the master, either to avoid collision with the master in pick-and-place like tasks, or to assist in the manipulation of a common object.

In the decoupled planning, the paths for each robot are planned independently and are later synchronized by scheduling the robots' motion to ensure that no collision takes place while executing the desired task. Lee and Lee (Lee & Lee, 1987), use this technique for a system consisting of two manipulators, where the speed of one of them is fixed while the speed of the other is modified in order to synchronize the previously planned paths, obtaining a collision-free interaction. The collisions here are studied using a bidimensional graph called "Collision map" where the path of the second robot is represented against time and collision regions are identified. The manipulators are modelled by spheres and the motion of the robots is restricted to straight-line paths. An extension of this method is presented in (Chang et al., 1994), where the robots are represented as polyhedral and the minimal delay-time value, necessary for collision free coordination, is determined.

The use of evolutionary techniques in the motion coordination of two manipulators is explored by (Ridao et al., 2001). As with other decoupled planning based approaches, the problem is broken into path planning, where collision-free paths for the robots are found independently and considering only fixed obstacles in the workspace; and trajectory planning where the paths are synchronized to ensure a collision-free interaction. Here an evolutionary algorithm gives an initial approximate solution in the C-space and from this solution a heuristic local search algorithm consisting of a monotonous random walk is used to find an optimum solution.

1.3 Soft Computing Techniques

Soft computing is a keyword in information technologies, and refers to a synthesis of methodologies from fuzzy logic, neural networks and evolutionary algorithms used to solve non-linear systems where conventional methods fail to provide a feasible solution.
As defined by Zadeh (Zadeh, 1997), "Soft computing differs from conventional (hard) computing in that, unlike hard computing, it is tolerant of imprecision, uncertainty and partial truth. … At this juncture, the principal constituents of soft computing (SC) are fuzzy logic (FL), neural network theory (NN) and probabilistic reasoning (PR), with the latter subsuming belief networks, genetic algorithms, chaos theory and parts of learning theory. What is important to note is that SC is not a melange of FL, NN and PR. Rather; it is a partnership in which each of the partners contributes a distinct methodology for addressing problems in its domain. In this perspective, the principal contributions of FL, NN and PR are complementary rather than competitive".

1.3.1. Genetic Algorithms

Developed by Holland (Holland, 1975), and further implemented by Goldberg (Goldberg, 1983), a genetic algorithm (GA) is a search strategy that mimics the theory of evolution of life in nature, belonging to the class of stochastic search methods. The main difference between GAs and other search methods is that, while most stochastic search methods operate on a single solution to the problem at hand, genetic algorithms operate on a population of solutions.
For any given problem to be solved by a GA, an initial population of possible solutions has to be created. These solutions are coded into Chromosomes of a fixed or variable length. The coding can be done in any representation although binary representation is commonly used. Each of the chromosomes are evaluated and assigned a fitness value accordingly to a fitness function. The basic operations in a GA are: Reproduction, Crossover and Mutation.

1.3.2. Fuzzy Logic

Proposed in (Zadeh, 1965) and successfully implemented for the first time in (Mamdani, 1974), fuzzy logic is an extension of Boolean logic which allows the processing of "vague" or uncertain information. Classic logical systems based on Boolean logic classify elements as members or not of a particular set, while Fuzzy-based systems can establish a degree of pretence of an element to any given set within a membership range between [0, 1], providing a way to identify and rank an intermediate value. A membership value of zero indicates that the element is entirely outside the set, whereas a one indicates that the element lies entirely inside a given set. Any value between the two extremes indicates a degree of partial membership to the set. Fuzzy logic employs the classical set operations of union (OR), intersection (AND), and complementation (NOT), but their meaning varies from those in classic logic.

1.4 The Potential Field Approach

In order to avoid collisions between manipulators and other objects present within their workspace it is necessary to characterize these possible obstacles in a way that the information from this representation is of use to prevent any possible collision. The potential field approach, introduced in (Khatib, 1986), combines "attractive" and "repulsive" potential field to model the workspace and the obstacles therein. Figure 1 illustrates a line obstacle

and a goal at (10,10). As it can be appreciated, the global minimum of this combined field corresponds to the specified goal.
The attractive potential, *Pa*, is given by reducing the Euclidean error to the desired goal, while the repulsive potential, *Po*, is given by:

$$Po = \sqrt{Po_x^2 + Po_y^2} \tag{1}$$

subject to the following conditions:

$$\begin{aligned} &\text{if } DO > (s+r) && \begin{cases} Po_x = 0 \\ Po_y = 0 \end{cases} \\ &\text{if } r \le DO \le (s+r) && \begin{cases} Po_x = \beta(s+r-OD)\cos\phi \\ Po_y = \beta(s+r-OD)\sin\phi \end{cases} \\ &\text{if } DO < r && \begin{cases} Po_x = m\cos\phi \\ Po_y = m\sin\phi \end{cases} \end{aligned} \tag{2}$$

where:
DO = distance to obstacle
s = distance of influence of the potential field of the obstacle
r = distance considered as inminent contact
$Po_x, Po_y = x, y$ components of the Potential Field vector
β = scaling factor of the Potential Field for the influence zone
m = scaling factor of the Potential Field for the contact zone
ϕ = potential field direction

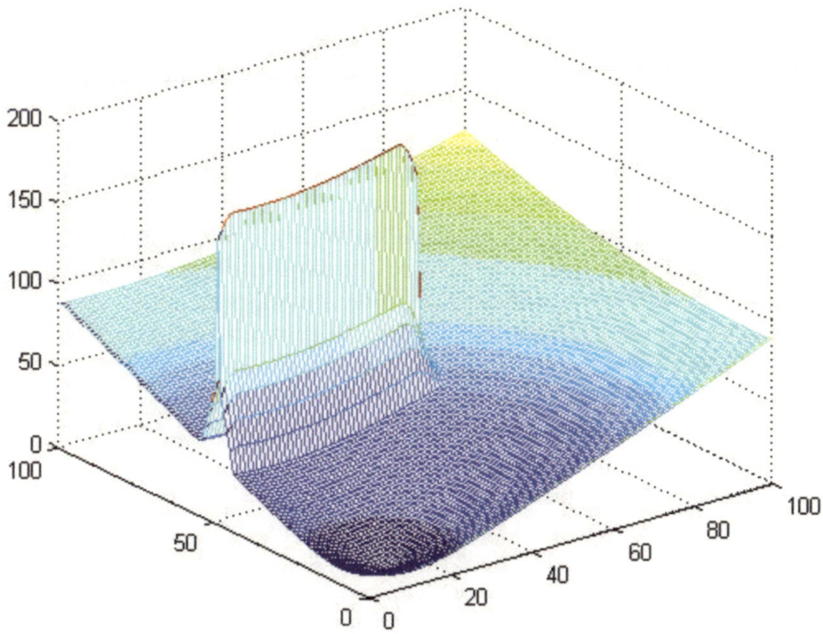

Figure 1. Combined potential field example

The resulting potential field P is obtained by the combination of these two potentials by:

$$P = Pa + Po \tag{2}$$

This representation provides enough information to guide the manipulator, through a search algorithm, from a starting point to a desired final position. However, path planners based on this representation have to deal with the problem of local minima. To overcome the local minima problem different approaches have been suggested for both, the construction of the potential field and the search algorithm itself. An alternative in the construction of the potential field is the use of harmonic functions to build local minima free potential fields as proposed by Kim and Connolly (Connolly et al., 1990; Kim & Kolsha, 1992; Connolly & Grupen, 1993). The drawback of this approach for its practical implementation of path planners is that it requires the implementation of an iterative numerical algorithm for the solution of Laplace's equation, which results on a high computational cost.
The search algorithms used to navigate a potential field follow the gradient descent of the field until reaching the goal. A comprehensive summary of these methods can be found in (Latombe, 1991). A common characteristic of these methods is the addition of a procedure to escape local minima.

2. A GA based approach to path planning

Path planning of robots can either be carried out in the workspace or in the corresponding configuration space (C-space) of the manipulator. The C-space approach simplifies the problem to that of moving a point in a n-dimensional space where the point represents a particular configuration of the robot, n being equal to the number of degrees of freedom (dof's) of the robot. In order to avoid collisions when solving a path in the C-space, it is necessary to map the obstacles present in the workspace into C-space obstacles, and the actual mapping grows in complexity as the dimensionality of the C-space considered is higher. Before any mapping begins, it is necessary to determine the resolution of the C-space as its construction involves a sampling process. The amount of computational time and space required to process and store the C-space acts as a constraint on the implementation of algorithms based on this approach. Furthermore, if the obstacles change their position and/or orientation in the workspace, recalculation of the C-space is necessary.
On the other hand, path planning in the workspace requires appropriate obstacle representation and collision avoidance/detection strategies to allow the solution for a suitable path. Once a path has been determined, the computation of the inverse kinematics is required in the case of robot manipulators to find the suitable sets of configurations that are required to follow the specified path. At this point, it is necessary the implementation of a collision avoidance algorithm to ensure that the links of the manipulator will not collide with an obstacle while following the desired path.
In the path planning of robot manipulators, the C-space method requires the start and goal configuration of the manipulator to be defined. This reduces the complexity of the space of solution of the problem to that of a single possible solution, since the desired positions of the links of the manipulator are already known. Conversely, path planning in the workspace requires the initial configuration of the robot and of the goal, expressed in workspace coordinates, for the end-effector of the manipulator. The characteristic dexterity of robot

manipulators produces a solution space with multiple solutions depending of the number of dof's of the manipulator.

It follows that the motion planning of a robot manipulator can be classified as an optimisation problem, for which the path to be obtained is subject to different criteria of optimisation (such as path length, collision avoidance, etc.) while being subject to kinematic and/or dynamic constraints. This makes the path planning problem in robotics an ideal candidate for the implementation of genetic and/or evolutionary algorithms (Parker et al., 1989; Rana, 1996).

2.1 Problem specification

The C-space method previously introduced consists in solving the path planning problem in the configuration space of the manipulator. Being a configuration based approach, it is necessary to specify the goal as a desired configuration of the links of the manipulator thereby "reducing" the problem of that of moving a point through the C-space until the goal is reached. The major drawbacks of this method are the mapping process of the obstacles into the C-space which becomes a very time consuming process as the dimensionality of the C-space increases, the search for feasible paths in a high dimensional space and the large amount of computational resources to store the resulting C-spaces. Another disadvantage of C-space based algorithms is that they are only suitable if the obstacles are static or, in the case of moving obstacles, their trajectories are known in advance, since every time an obstacle changes its position or configuration it has to be re-mapped into the C-space.

Figure 2(a) illustrates a 2 dof manipulator and a static line obstacle. The corresponding C-space of this system is presented in Figure 2(b). The obstacle can either be modelled using the potential field approach in the workspace (Figure 3(a)) or it can be mapped into the C-space (Figure 3(b)).

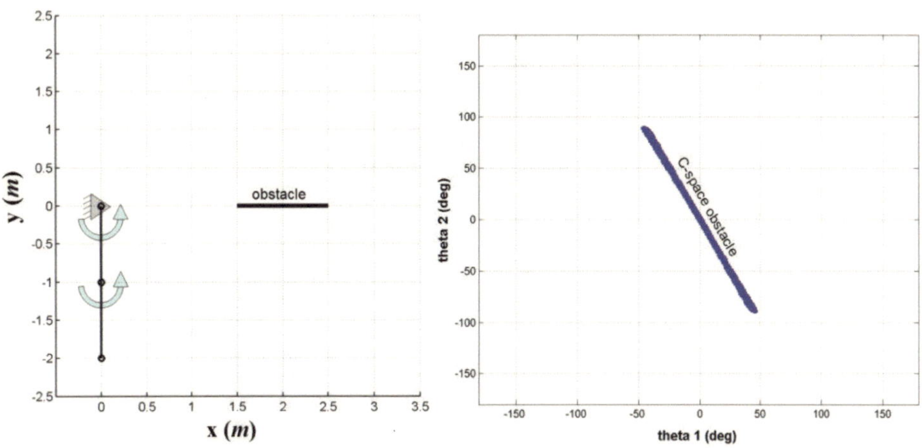

(a) (b)

Figure 2. (a) 2 dof manipulator and static obstacle; (b) Mapped obstacle in the C-space

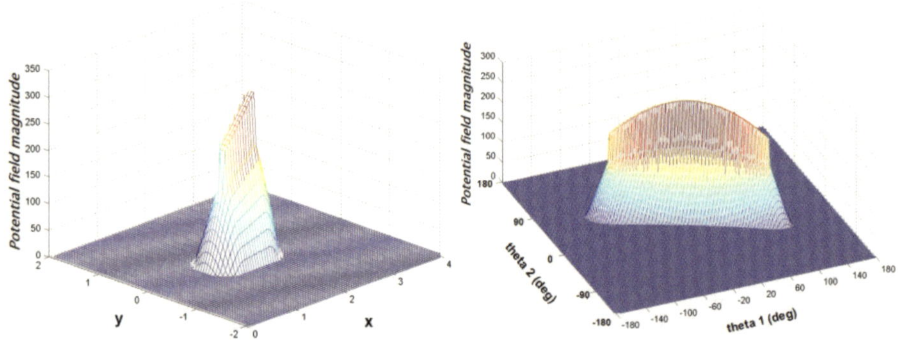

(a) (b)

Figure 3. (a) Potential Field of the line obstacle; (b) Potential Field in the 2 dof C-Space

For example, if the manipulator in Figure 2(a) is desired to reach an arbitrarily chosen goal at (1.7, 0.5) in the workspace, the problem could be solved using a GA with a fitness function that reduces the error to the goal and is also affected by the magnitude of the potential field due to the presence of an obstacle. Equation 3 describes a possible fitness function where both requirements are met.

$$f_i = \frac{1}{Po + e^{error}} \tag{3}$$

where:

$Po =$ Potential Field due to the Obstacle

$error =$ Error to Goal

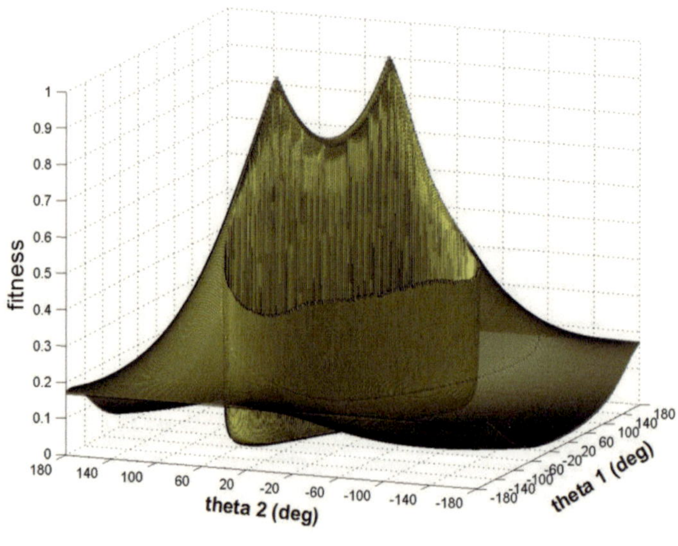

Figure 4. Fitness space for the 2 dof C-Space

The fitness space for the GA in the C-space is illustrated in Figure 4. It can be appreciated that, being a 2 dof system, two possible solutions to the problem exist which can be identified as the maxima of the fitness space. The fitness around the obstacle drops to zero.
Figure 5(a) shows a 3dof planar manipulator and three static obstacles in its workspace. The C-space corresponding to this manipulator, Figure 5(b), is a three-dimensional space in which the obstacles are forbidden regions represented as 3d obstacles corresponding to those configurations where a collision with the obstacle occurs.

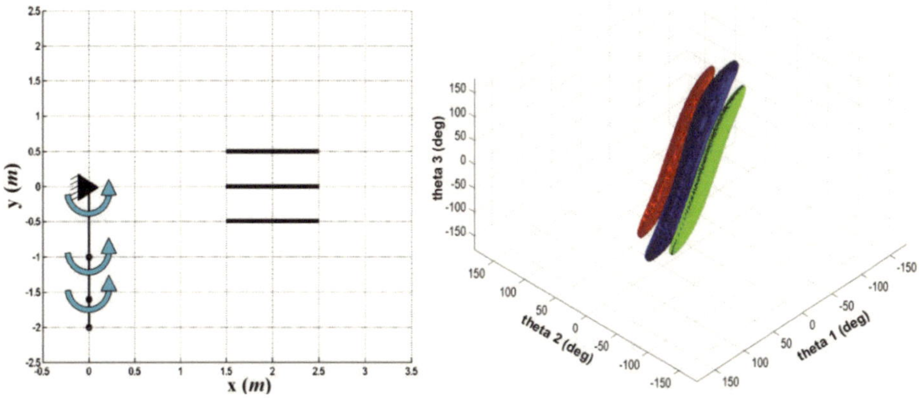

(a) (b)

Figure 5. (a) 3dof Manipulator and static obstacles; (b) Obstacles in the 3dof C-space

The path planning using the potential field method in the C-space illustrated in Figure 5(b), requires tensor notation to represent the magnitude and direction of the potential field for each set of configurations that describe the obstacles in this three-dimensional space.
The problem of path planning of robot manipulators is an ideal candidate for the implementation of a GA based approach due the multidimensionality and inherited non-linearity of the system. A GA based approach for the solution of the problem directly in the workspace of the manipulator eliminates the mapping process of obstacles and the solution is open to any configuration that, free of collision, takes the manipulator to its desired goal rather than restricting the solution to a single configuration.

2.2 Genetic Algorithms and Fuzzy Logic in path planning

2.2.1. Genetic Algorithms based approaches

The application of GA's in path planning has previously been studied by some authors, from where two approaches can be clearly identified, those based on the definition and optimisation of paths throughout via points in the C-space, and those based on the "fitting" of a desired path in the workspace through GA optimisation of the manipulator's variables until the resulting path is closely fitted to the desired one. Methods based on either approach are considered global and the motion planning takes place off-line as a result of the number of computations involved.
A GA based path planner for a bi-dimensional space considering static obstacles is presented in (Doyle, 1995; Doyle & Jones, 1996). Real coded chromosomes are used to represent the coordinates of via points that define the segments of the possible paths with

the following structure: $[(x_1y_1)(x_2y_2)\cdots(x_ny_n)]$ where n is the total number of defined via points.

The population is evaluated by considering if the path segments are free from collision and the total length of the path. This algorithm is later adapted to the path planning on the C-space of a 2 dof manipulator with static obstacles by changing the chromosome structure to $[(q_{11}q_{21})(q_{12}q_{22})\cdots(q_{1n}q_{2n})]$ being $(q)$the manipulator articulations and n the maximum number of configurations. This approach uses a single point crossover operator at the gene boundaries.

A similar approach for the path planning in the C-space of 2 and 3 dof manipulators is presented in (Rana, 1996; Rana & Zalzala, 1996). The problem is solved by generating an initial population of trajectories through randomly distributed via points. The GA then optimises the path taking into consideration the total length of the path, the uniform distribution of via points to ensure a "smooth" path and a penalty function when a via point or path segment lies within a C-obstacle.

A variation of this approach is presented in (Eldershaw, 2001), where the planning in the C-space is obtained by optimising the location of via points for a fixed number of path segments. As with the previous approaches, the solutions are evaluated taking into consideration whether a path is free from collisions in the C-space and the total length of the evaluated path. These algorithms use the same chromosome representation.

In a different approach for the path planning in the C-space, (Bessiere, Ahuactzin et al., 1993) propose a path planner that uses a local GA to define Manhattan motions to a series of sub-goals or landmarks. A second GA planner then attempts to connect the landmarks with the goal. In a system with n dof's a Manhattan motion, M, consists of moving each degree of freedom $q_i$successively once by a defined Δq. This method finds feasible paths, but due to the nature of the stepped Manhattan motions, the paths obtained tend not to be very fine in their execution.

On the other hand, the implementation of GA's in trajectory planning of robots has been previously investigated by several authors (Davidor, 1991; Doyle & Jones, 1996; Rana & Zalzala, 1996; Wang & Zalzala, 1996; Chen & Zalzala, 1997; Kubota et al., 1997; Nearchou & Aspragathos, 1997; Gill & Zomaya, 1998).

In (Parker, Khoogar *et al.*, 1989; Davidor, 1991; Chen & Zalzala, 1997; Gill & Zomaya, 1998) GA's are used to optimise the search for a solution of the inverse kinematics using the Jacobian matrix to relate the differential variations in the joint space to the workspace.

Chen and Zalzala focused their study on the position and configuration of a mobile robot. Like other conventional approaches based on the solution of the inverse kinematics, a singularity in the solution takes place when the Jacobian equals zero and therefore cannot be inverted.

In Davidor's approach, a desired path for the end-effector of the manipulator is pre-defined in the workspace. A number of trajectories in the C-space are randomly generated and are evaluated with respect to the trajectory they describe in the workspace against the required path. New paths are evolved by selecting analogous crossover points that will fit the evaluated trajectories closer to the desired one. The process of generating a new population starts by copying the joint displacements of the selected trajectory segments into a new string, producing chromosomes of different length in the same population. This process is repeated until an acceptable fit to the desired path is obtained. In general, the GA is applied to solve the inverse kinematics to pre-defined end-effector paths of the manipulator.

In (Gill & Zomaya, 1998), either the potential field or the cell decomposition approach are used to obtain the next best position from a starting configuration to a desired goal. This algorithm considers an alternative temporary goal if it gets stuck in a local minimum. Once the next position has been calculated, a GA is used to search for a solution to the inverse kinematics that best reduces the error between the current and desired end-effector position calculated in the previous step. To avoid the problem of singularities, the search scope of the GA for the solution of the inverse kinematics is restricted to a closed range of possible values for the increments of the joint positions of the manipulator.

The methods presented by Davidor and Gill & Zomaya can only be applied in non-redundant manipulators. In (Nearchou & Aspragathos, 1997), a similar path fitting method using GAs for redundant manipulators is presented. This approach was the first to solve the problem of path planning in the workspace of the manipulator rather than the C-space. A set of via points is established in the workspace and the GA is used to solve the inverse kinematics of the manipulator for the required positions. From the reported results, it can be appreciated that, when applying a path fitting algorithm to a redundant manipulator, the set of configurations for the joints of the manipulator to follow the desired path are not optimal with respect to the displacement to which the links are subjected. This is a result of the inherited dexterity of a redundant manipulator, where a single point in the workspace of the manipulator can be reached by a number of different configurations of the robot.

To avoid the problem of singularities, some authors like (Doyle & Jones, 1996; Rana & Zalzala, 1996; Wang & Zalzala, 1996; Kubota, Arakawa *et al.*, 1997; Gill & Zomaya, 1998) base their approaches on the direct kinematics. (Doyle & Jones, 1996) use a GA to search an optimum path for the robot in the C-space. (Rana & Zalzala, 1996) also use GA's to find an optimum path in the C-space of non-redundant manipulators and also take into consideration the required time to perform a task to find a near-optimal solution; the path is represented as a string of via points connected through cubic splines.

In (Merchán-Cruz & Morris, 2004) the application of GA's is extended to obtain collision free trajectories of a system of two manipulators, either non-redundant or redundant, using potential field approach. In this approach each manipulators is considered as a moving obstacle by the other and collision is avoided. The GA carries parallel optimization to find the best configuration for collision free as well as minimizing the error to their respective goals. Later, in (Merchán-Cruz et al., 2007) it is explained how the inherited monotony of the trajectories described by the robot manipulators, either when moving along a collision free path, or when avoiding an obstacle, can be exploited to enhance the performance of a GA based trajectory planner.

2.2.2 Fuzzy Logic based approaches

The use of an artificial potential field (APF) to describe the workspace of a manipulator is naturally appealing for the implementation of a fuzzy collision avoidance strategy since the value of the potential field increases or decreases accordingly to the nearness of obstacles or to the goal. In contrast to many methods, robot motion planning based on an artificial potential field (APF) is able to consider simultaneously the problems of collision avoidance and trajectory planning as this approach can drive the robot to its desired goal while keeping it from colliding with static or dynamic obstacles present in the workspace.

However, the use of this method is often associated with local minima problems which cause the planning algorithms to get stuck in a sub-optimal solution. While some previous research has been done in constructing potential fields free from local minima, other research has focused on the development of strategies to identify and move away from local minima within the planning algorithm.

Following on from the original approach proposed in (Khatib, 1986), (Volpe & Khosla, 1990) developed elliptical potential functions called superquadric artificial potential field functions which, from the results, do not generate local minima when used to represent the workspace in the cartesian space, but the problem of local minima is still present when used with the C-space representation. Later on, (Koditschek, 1991; Koditschek, 1991; Rimon & Koditschek, 1992) introduced analytical potential fields in the C-space for ball or star shaped obstacles which limited its application to other obstacle arrangements. The use of harmonic functions to construct local-minima free potential fields is explored in (Kim & Kolsha, 1992; Connolly, 1994), where the potential field is constructed for the C-space of the manipulator. In a more recent approach, another alternative in the construction of the potential field presented in (Vadakkepat et al., 2000) is the use of an evolutionary algorithm for the optimisation in the construction of the potential field. This approach starts by building a potential field using simple expressions, from where a local minima-free potential field is evolved. The size and resolution of the potential field reflects on the necessary number of generations needed to obtain an optimal field, and its application is restricted to static workspaces due to the computations involved if a dynamic obstacle is considered.

With regards to the algorithms for motion planning of robot manipulators using fuzzy logic, very little work can be referred to in comparison to the research done for mobile robots. The first reported approach of a fuzzy navigator for robot manipulators is that reported in (Althoefer & Fraser, 1996) and further discussed in (Althoefer, 1997; Althoefer et al., 1998; Althoefer et al., 2001). In this approach, the goal is specified as a particular arm configuration to which the manipulator is required to reach, that is, the goal is specified in the joint space of the robot. A fuzzy navigator is presented consisting of individual fuzzy units that govern the motions of each link individually and determine the distance to an obstacle by constantly scanning a predetermined rectangular surrounding area around each link. These areas correspond to the model of the space that could be obtained by ultrasonic-sensors attached to both sides of each link. Each fuzzy unit produces an output relative to the joint error and the closeness to an obstacle. The application of this approach is presented for 2dof and 3dof planar manipulators in static environments.

The work by (Althoefer, 1997) presents a briefly discussion of the application of this approach where a 3dof planar manipulator deals with a point obstacle whose trajectory is known.

In (Zavlangas & Tzafestas, 2000), a modified version of this approach is presented to consider the specific case of the implementation to an Adept 1 industrial manipulator in a 3D space, the Adept 1 architecture corresponds to that of a SCARA manipulator. The modification consists of the way the nearness to an obstacle is determined. In this case, the area surrounding the links is said to be a cylindrical volume surrounding each link. Again, as in the original work presented by Althoefer, the start and goal of the manipulator are both specified in the joint space of the robot as a set of given configurations. This implementation, due the nature of the SCARA configuration, can be simplified to just analysing a 2dof planar manipulator.

In (Mbede et al., 2000), a similar approach is presented where an harmonic potential field is used to drive a 2dof planar manipulator to its desired goal, which has been described in the joint space. The cases of static obstacles and that of a trajectory-known point obstacle are presented and successful results are reported. Later, in (Mbede & Wei, 2001), a neuro-fuzzy implementation of this approach is used to solve the same proposed cases.

3. Fuzzy-GA trajectory planner

The fuzzy-GA trajectory planner (FuGATP) approach presented in this section employs a simple-GA based trajectory planner (simple-GABTP) to initially drive the manipulators towards their goals. Individual fuzzy units provide a correction in the displacement of each articulation in the event that a link is approaching an obstacle. The goals are specified in workspace coordinates to take full advantage of the dexterity of the redundant manipulators.

The manipulators are modelled as obstacles in the workspace using the artificial potential field (APF) method. Since this is a local approach, the APF is only calculated for the near vicinity of the links of the manipulator to save time in the modelling process. Because each manipulator is considered by the other as a mobile obstacle where the position and orientation on their links are constantly changing, there would be a major computational problem if the APF for the whole of the shared workspace were to be calculated each time one of the manipulators has moved as would be necessary in a global approach.

The simple-GABTP algorithm, illustrated in figure 6, carries out an initial estimation of the motion of the manipulators without considering the presence of any obstacles. This calculation is performed by finding the best set of joint displacements within a pre-established range of $\Delta\theta$'s per unit of time until the goal is reached, without solving the inverse kinematics of the manipulator or dealing with singularities in the case of redundant manipulators.

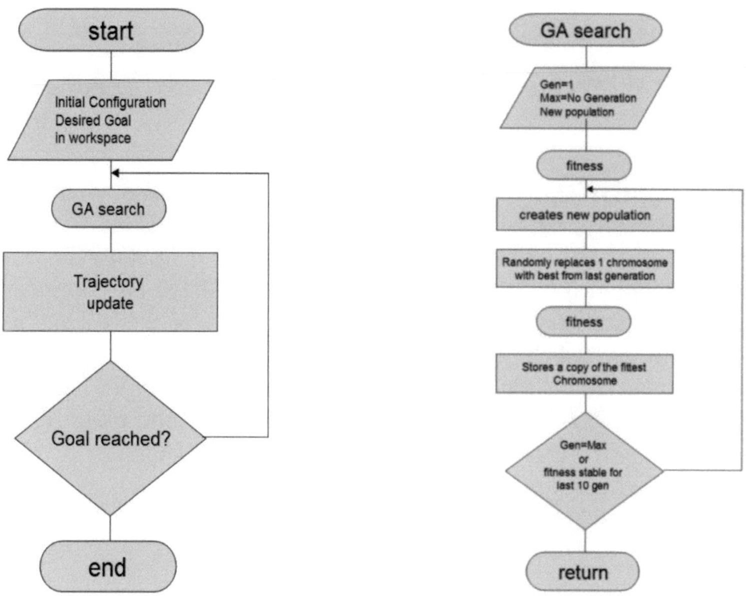

(a) (b)

Figure 6. Simple GABT algorithm. (a) General outline of the planner; (b) GA search process

In order to favour fully-extended motion of the manipulator to minimise unnecessary displacements of the links when moving in a space free from obstacles, the fitness function for the GA search considers the degree of extension of the manipulator, equation 4.

$$f = \frac{1}{e^{\frac{error}{R}}} \tag{4}$$

Where:

$error = \left\| (x_g, y_g) - (x, y) \right\|$

$R = \left\| (x_o, y_o) - (x, y) \right\|$

$(x_o, y_o) = Manipulator's\ base\ coordinates$

$(x_g, y_g) = Goal\ coordinates$

$(x, y) = DKM\left((\theta_1 + \Delta\theta_1), (\theta_2 + \Delta\theta_2), \cdots, (\theta_n + \Delta\theta_n)\right)$

$DKM =$ Direct Kinematic Model as a function of $\theta's$

Each chromosome of the population considered by the GA is composed of substrings of equal length corresponding to a particular joint variation. Figure 7 shows how these strings are decoded from the binary representation into joint increments/decrements which are then applied to the DKM of the manipulators to calculate the new error due to these variations. In this representation, m determines the minimum increment and n is the number of joints in the manipulator. The algorithm returns the set of $\Delta\theta$'s that best reduces the error, updates the configuration of the manipulator and continues until the goals are reached.

[X X X X X . . . X ¦ X X X X X . . . X ¦ . . . ¦X X X X X . . . X]

$\pm\ 2^1\ 2^0\ 2^{-1}\ 2^{-2}\ \ldots\ 2^{-m}\quad \pm\ 2^1\ 2^0\ 2^{-1}\ 2^{-2}\ \ldots\ 2^{-m}\quad \cdots\quad \pm\ 2^1\ 2^0\ 2^{-1}\ 2^{-2}\ \ldots\ 2^{-m}$

$\pm\ \Delta\theta_1 \qquad \pm\ \Delta\theta_2 \qquad \pm\ \Delta\theta_n$

Figure 7. Chromosome structure

Once the simple-GABTP algorithm has determined an initial set of configurations that best reduce the Cartesian error between the current end-effector positions of the manipulator, the fuzzy units provide a correction based on the magnitude of the AFP that modifies the original displacement of the articulations given by the GABTP. This correction ranges from slowing the articulation to completely change the direction of the original motion given by the GABTP. The fuzzy units determine the best direction to correct the original motion of an articulation by determining the direction of the motion that implies an increment on the magnitude of the APF at the particular instant consider by the planner. This results in a signed correction to affect the initial estimation given by the GABTP.

The direct output from the FuGABTP algorithm is given by:

$$[\Delta\theta_1\ \ \Delta\theta_2 \cdots \Delta\theta_n]_{FuGABTP} = [\Delta\theta_1\ \ \Delta\theta_2 \cdots \Delta\theta_n]_{GABTP} + [\Delta\theta_1\ \ \Delta\theta_2 \cdots \Delta\theta_n]_{FuzzyCorrection} \tag{5}$$

The diagram shown in figure 8 illustrates the general outline of the trajectory planner.

The fuzzy correction is achieved by individual Mamdami-type fuzzy units that provide a correction value per articulation in order to avoid collision. Each of the units has three inputs: obstacle direction, APF magnitude and error; and a single output named correction, as illustrated in Figure 9.

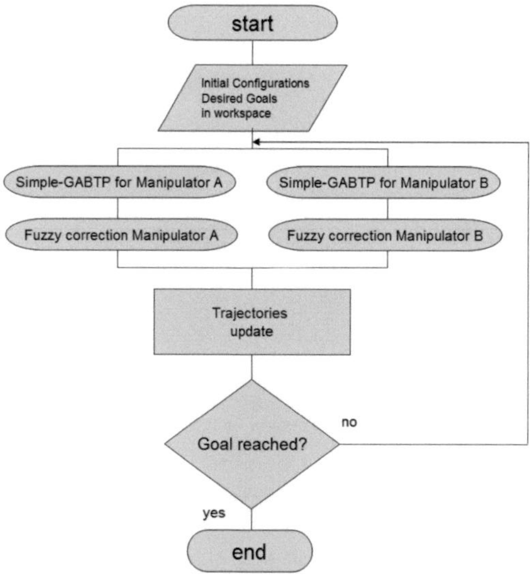

Figure 8. General outline of Fuzzy-GA trajectory planner

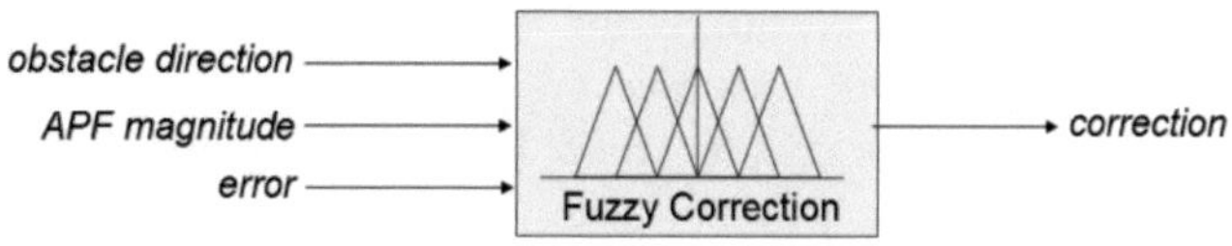

Figure 9. Fuzzy correction unit

The obstacle direction can either be right if the magnitude of the APF increases by moving the considered link around its axis of movement in a positive direction, or left if the magnitude of the APF increases in the opposite direction.

The magnitude of the APF indicates how close a link is from a nearby obstacle, and the error is the Cartesian error between the actual position of the end-effector and the goal. Finally, the correction output indicates the positive or negative correction needed to ensure that no collision takes place. The range of each variable, or universe of disclosure D, is partitioned into fuzzy sets $\mu_1 \dots \mu_p$. Each of the sets $\mu_{\tilde{p}}$, $\tilde{p}$ = 1,..., p represents a mapping $\mu_p(d) : D \to [0,1]$ by which the degree of membership of d within a fuzzy set is associated with a number in the interval [0,1].

The functions employed to represent the fuzzy sets have been chosen to be asymmetrical triangular functions as they are easily computed. The degree of membership of each input is calculated by :

$$\mu_{\tilde{p}}(d) = \begin{cases} \min\left(\dfrac{(d - ml_{\tilde{p}})}{mc_{\tilde{p}} - ml_{\tilde{p}}}, 0\right) & \text{if } d \le mc_{\tilde{p}} \\ \min\left(\dfrac{(d - mr_{\tilde{p}})}{mc_{\tilde{p}} - mr_{\tilde{p}}}, 0\right) & \text{if } d > mc_{\tilde{p}} \end{cases} \tag{6}$$

where:

$ml_{\tilde{p}}$ $mr_{\tilde{p}}$ = x-coordinates of the left and right zero crossing, and

$mc_{\tilde{p}}$ = x-coordinate where the fuzzy set becomes 1

For the values that lie either at the left or right of each input range, the triangular functions are continued as constant values of magnitude 1 as:

$$\mu_1(d) = \begin{cases} 1 & \text{if } d \le mc_1 \\ \min\left(\dfrac{(d - mr_1)}{mc_1 - mr_1}, 0\right) & \text{if } d > mc_1 \end{cases} \tag{7}$$

and

$$\mu_p(d) = \begin{cases} \min\left(\dfrac{(d - ml_p)}{mc_p - ml_p}, 0\right) & \text{if } d \le mc_p \\ 1 & \text{if } d > mc_p \end{cases} \tag{8}$$

Figures 10 and 11 illustrate the partition of the fuzzy sets that describe each of the fuzzy variables.

The defuzzification of the data into a crisp output is accomplished by combining the results of the inference process and then computing the "fuzzy centroid" of the area. The weighted strengths of each output member function are multiplied by their respective output membership function centre points and summed. Finally, this area is divided by the sum of the weighted member function strengths and the result is taken as the crisp output.

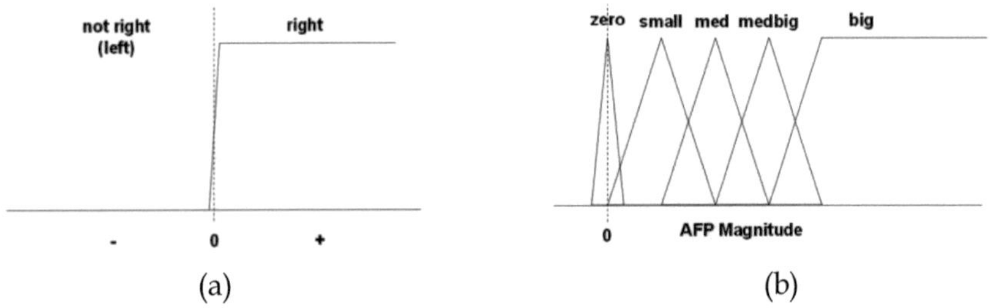

Figure 10. Fuzzy sets (a) Obstacle direction, (b) APF Magnitude

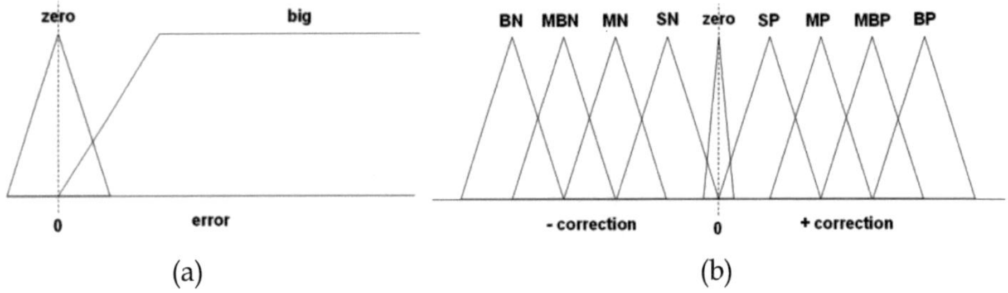

Figure 11. Fuzzy sets (a) Error, (b) Output correction

The correction output of each fuzzy unit is given as the product of the intersection of the following fuzzy rules:

If (obstacle is left) and (APF Magnitude is small) then (Correction is SN)
If (obstacle is left) and (APF Magnitude is med) then (Correction is MN)
If (obstacle is left) and (APF Magnitude is medbig) then (Correction is MBN)
If (obstacle is left) and (APF Magnitude is big) then (Correction is BN)
If (obstacle is left) and (APF Magnitude is zero) then (Correction is zero)
If (obstacle is right) and (APF Magnitude is small) then (Correction is SP)
If (obstacle is right) and (APF Magnitude is med) then (Correction is MP)
If (obstacle is right) and (APF Magnitude is medbig) then (Correction is MBP)
If (obstacle is right) and (APF Magnitude is big) then (Correction is BP)
If (obstacle is right) and (APF Magnitude is zero) then (Correction is zero)
If (obstacle is left) and (error is zero) then (Correction is zero)
If (obstacle is right) and (error is zero) then (Correction is zero)

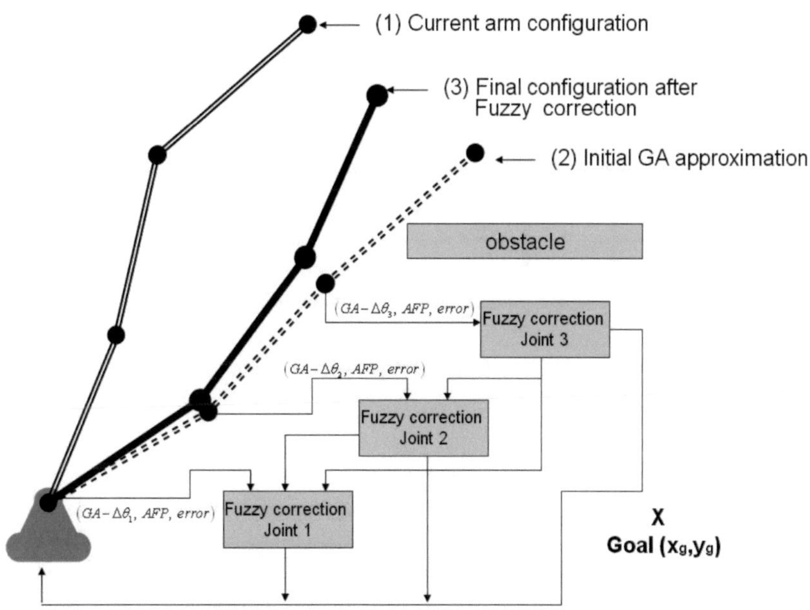

Figure 12. Fuzzy-GA trajectory planner (FuGATP) approach, schematic representation (Merchán-Cruz & Morris, 2006)

The final fuzzy correction for each joint is also affected by the correction of the distal links since the motion of the distal links is not only given by the action of a single articulation. The overall approach is illustrated for a 3dof manipulator in Figure 12 where (1) shows the manipulator at the current configuration, (2) indicates the initial GA estimation that simply reduces the Cartesian error between the current position of the end-effector and its goal, and finally, (3) shows the final position of the manipulator after the fuzzy units have corrected the final $\Delta\theta$'s for the manipulator. Since the $\Delta\theta$'s are obtained by the planner, the velocities and accelerations necessary to calculate the necessary torque of each articulation are easily calculated as a function of time.

3.1 Simulations for two manipulators sharing a common workspace

In this section the described approach is applied to solve the motion planning problem for the 3dof & 7dof two-manipulators systems. Initially the problem of static obstacles is dealt with by considering the manipulator B as a stationary obstacle. Finally, the approach is tested when both manipulators are required to move to reach their independent goals while considering the other manipulator as a mobile obstacle with unknown trajectory.

3.1.1 Static case

For the study of the static obstacle case, the manipulator A is required to reach a goal in the workspace from a given starting configuration while the manipulator B remains in a stationary position where it behaves as a static obstacle. The following cases are presented, as they test the performance of the algorithm in different circumstances and are considered to be representative of the common task that the manipulator performs in its workspace. Tables 1 and 2 summarise these cases for the 3dof and 7dof manipulators.

The example cases for the 3dof system corroborate the application of the proposed approach for a planar manipulator with low redundancy, while the example cases solved for the 7dof manipulator probe that the algorithm is also applicable for highly redundant manipulators. Table 3 summarises the results for the application of the Fuzzy-GA trajectory planner considering a static obstacle.

	Manipulator A			Goal (x,y)		Manipulator B			Goal (x,y)	
Case	θ_{1i}	θ_{2i}	θ_{3i}	x	y	θ_{1i}	θ_{2i}	θ_{3i}	x	y
I	90	0	0	0	-2	0	0	0	1.5	0
II	90	0	0	1.9	-0.5	0	0	0	1.5	0
III	-35	20	15	1.95	0.5	0	0	0	1.5	0

Table 1. Representative cases for the two-3dof manipulator system

	Manipulator A							Goal (x,y)	
Case	θ_{1i}	θ_{2i}	θ_{3i}	θ_{4i}	θ_{5i}	θ_{6i}	θ_{7i}	x	y
I	90	0	0	0	0	0	0	0	-4
II	90	0	0	0	0	0	0	2.75	-1.705
III	90	0	0	0	0	0	0	3.5	-1.1
IV	90	0	0	0	0	0	0	3.95	-0.45
V	85	-25	-25	-35	0	0	0	2.75	-1.705

	Manipulator B							Goal (x,y)	
Case	θ_{1i}	θ_{2i}	θ_{3i}	θ_{4i}	θ_{5i}	θ_{6i}	θ_{7i}	x	y
I	0	0	0	0	0	0	0	1.75	0
II	0	0	0	0	0	0	0	1.75	0
III	0	0	0	0	0	0	0	1.75	0
IV	0	0	0	0	0	0	0	1.75	0
V	0	0	0	0	0	0	0	1.75	0

Table 2. Representative cases for the two-7dof manipulator system

Figure 13 illustrates the described path and trajectory profiles for case III of the 7dof system. The figure shows how the manipulators move from the given start configuration to the end-effector goal in the workspace. Collision with the static manipulator is successfully avoided.

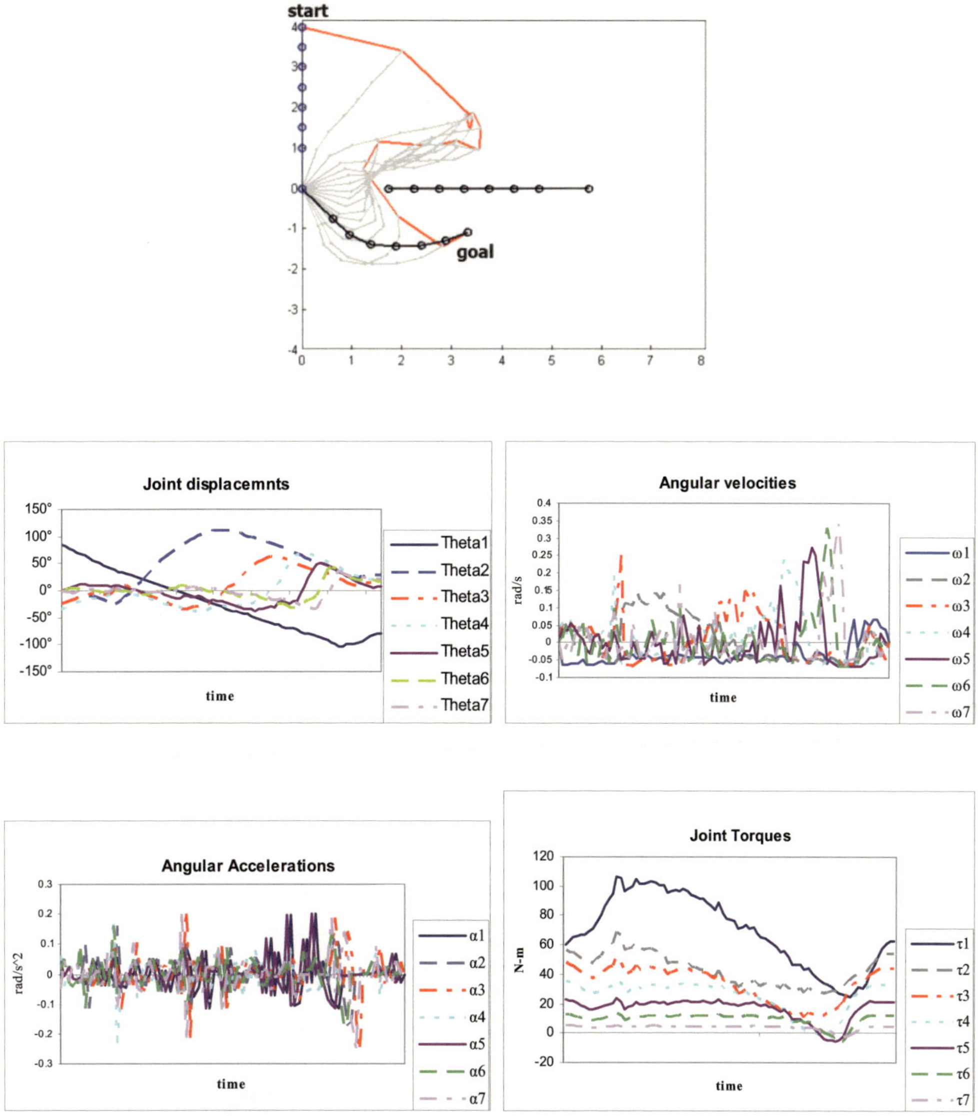

Figure 13. Described path and trajectory profiles for the 7dof manipulator case III, considering a static obstacle obtained with the FuGATP algorithm

The combination of the simple-GABTP algorithm with the fuzzy correction drives the manipulator towards the desired goal in the workspace while keeping the manipulator away from the obstacle by moving the link away from it as a result of the correction action determined by the fuzzy units.

	3dof		
Case	*Trajectory Segments*	*Time (s)*	*Avg per segment (s)*
I	63	9	0.14
II	61	17	0.28
III	50	7	0.14

	7dof		
Case	*Trajectory Segments*	*Time (s)*	*Avg per segment (s)*
I	182	36	0.20
II	177	34	0.19
III	165	31	0.19
IV	199	40	0.20
V	170	35	0.20

Table 3. Summary of results for the two-manipulator system considering one as a static obstacle

3.1.2 Dynamic case

In this section, the FuGATP algorithm is applied to solve the trajectories for two manipulators sharing a common workspace where both manipulators have the same priority when moving towards individual goals. As in the previous section, the examples presented here are just a selection to illustrate the applicability of the algorithm when one manipulator has to consider the other as a mobile obstacle of unknown trajectory. Table 4 summarises these representative cases where each one tests the algorithm under different circumstances.

	Manipulator A							Goal (x,y)	
Case	θ_{1i}	θ_{2i}	θ_{3i}	θ_{4i}	θ_{5i}	θ_{6i}	θ_{7i}	x	y
I	-15	15	0	0	0	0	0	3.8	1
II	-90	0	0	0	0	0	0	0.5	3.5
III	90	0	0	0	0	0	0	3.8	-0.5

	Manipulator B							Goal (x,y)	
Case	θ_{1i}	θ_{2i}	θ_{3i}	θ_{4i}	θ_{5i}	θ_{6i}	θ_{7i}	x	y
I	-15	15	0	0	0	0	0	1.95	-1
II	-90	0	0	0	0	0	0	5.25	-3.5
III	90	0	0	0	0	0	0	2	0.5

Table 4. Representative cases for the two-7dof manipulator system with individual goals

When individual goals have been specified and these require that both manipulators move, the algorithm starts solving in parallel the trajectory for both manipulators as described in

the algorithm outline (Figure 8) by considering the current individual configuration of each manipulator. The simple-GABTP built in the algorithm provides an initial approximation which is corrected by the fuzzy units in the event that this proposed configuration lays within the vicinity of an obstacle, in this case the links of the other manipulator. The components of the algorithm remain the same is with the static obstacle case, with the only difference that the manipulator B is now required to move.
The case of two 7dof planar manipulators is shown in Figure 14, where the manipulators are required to reach the goals from the starting configuration according to case I defined in Table 4. In this example, the applicability of the present approach for a highly redundant system is proven.

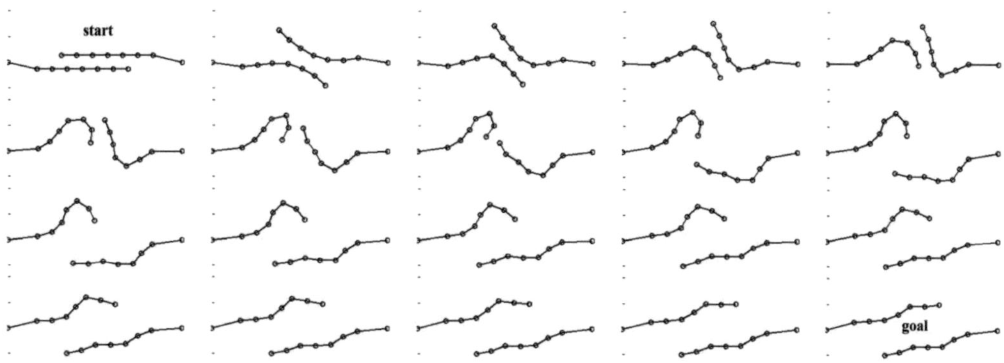

Figure 14. Detailed sequence for the two 7dof manipulators with individual goals Case I

As seen in Figure 14, the FuGATP maintains the manipulators moving towards their goals as a result of the simple-GATP which optimises the configuration of the manipulators to reduce the error to the desired goals. Solving in parallel for both manipulators, as described in Figure 8, produces individual trajectories that do not require from further scheduling as required by traditional decoupled methods. Given that both manipulators have the same priority, both adapt their trajectories when necessary to avoid collision. This avoids over compensating the motion of a single given manipulator as happens with conventional hierarchical approaches. Table 5 summarises the results for the representative cases where each one tests the algorithm under different circumstances.

	7dof		
Case	***Trajectory Segments***	***Time (s)***	***Avg per segment (s)***
I	71	14	0.20
III	128	26	0.20
III	138	27	0.20

Table 5. Summary of results for dynamic cases (two-7dof manipulator system)

4. Conclusion

This chapter has presented a discussion on the application of soft-computing techniques for the trajectory planning of multi-robot manipulator systems. The application of such techniques derived on an approach based on a combination of fuzzy logic and genetic algorithms to solve in parallel the trajectory planning of two manipulators sharing a common workspace where both manipulators have to adjust their trajectories in the event of a possible collision.

The planner combines a simple-GABTP with fuzzy correction units. The simple-GABTP solves the trajectory planning problem as an initial estimation without considering the influence of any obstacles. Once this stage is reached, fuzzy units assigned to each articulation verify that the manipulator is not facing a possible collision based on the magnitude of the APF exerted by the manipulators when considered as obstacles. If the initial output by the GABTP takes a link or links of the manipulator near the influence of the modelled APF, the fuzzy units evaluate a correction value for the $\Delta\theta$ that corresponds to the link and articulation involved in the possible collision and modifies not only the $\Delta\theta$ for that articulation, but also modifies the $\Delta\theta$'s for the more anterior articulations in case where a distal articulation is involved.

The approach presented here has shown a robust and stable performance throughout the different simulated cases. It produced suitable trajectories that are easily obtained since the direct output from the FuGABTP algorithm is a set of $\Delta\theta$'s after each iteration. When a time dimension is added, these provide not only positional information but also the angular velocities and accelerations necessary to describe the trajectory profiles of each manipulator and allow the calculation of the necessary torques to be applied at each joint to produce the obtained trajectory.

Since the planning is done in parallel for both manipulators, no further planning or scheduling of their motions are necessary as is required with decoupled approaches since the trajectories are free from any collision. The implementation of the simple-GABTP in the algorithm solves the problem of over compensation associated with the fuzzy units by maintaining the direction of the manipulators towards the desired goal at all times, keeping the links of the manipulator from over-reacting when approaching an obstacle.

Finally, the proposed algorithm was applied to solve the trajectory planning problem of two manipulators sharing a common workspace with individual goals. In this case each manipulator considers the other as a mobile obstacle of unknown trajectory. The considered cases were successfully solved, obtaining suitable trajectories for both manipulators, obtaining an average of 0.20 seconds per iteration for the 7dof systems in the dynamic cases.

In conclusion, the approach discussed in this chapter could be applicable for real-time application since the compilation of the algorithms that conform the suggested approach into executable files could reduce even more the iteration time.

5. Acknowledgments

The authors would like to thank the National Council for Science and Technology (CONACYT) Mexico, and the Instituto Politécnico Nacional, for the means and support to carry this investigation through the projects SEP-CONACYT 2005/49701 and the corresponding SIP projects.

6. References

Althoefer, K. (1996). *Neuro-fuzzy motion planning for robotic manipulators*. Department of Electronic and Electrical Engineering. London, UK, King's College London, Unuversity of London: 178.

Althoefer, K. (1997). *Neuro-fuzzy motion planning for robotic manipulators*. Department of Electronic and Electrical Engineering. London, UK, King's College London, Unuversity of London: 178.

Althoefer, K. & D. A. Fraser (1996). Fuzzy obstacle avoidance for robotic manipulators. *Neural Network World* 6(2): 133-142.

Althoefer, K., B. Krekelberg, et al. (2001). Reinforcement learningin a rule-based navigator for robotic manipulators. *Neurocomputing* 37(1): 51-70.

Althoefer, K., L. D. Seneviratne, et al. (1998). Fuzzy navigation for robotic manipulators. International Journal of Uncertainty, *Fuzziness and Knowledge-based Systems* 6(2): 179-188.

Ata, A. A. (2007). Optimal Trajectory Planning of Manipulators: A Review. *Journal of Engineering Science and Technology* 2(1): 32-54.

Barraquand, J. & J. C. Latombe (1991). Robot motion planning: a distributed representation approach. *International Journal of Robotics Research* 10(6): 628-649.

Bessiere, P., J. Ahuactzin, et al. (1993). The 'Adriane's Clew' algorithm: Global planning with local methods. *Proceedings of the International Conference on Intelligent Robots and System*, IEEE: 1373-1380.

Brooks, R. A. (1986). A roboust layered control system for a mobile robot *IEEE Journal of Robotics and Automation* 2(1): 14-23.

Connolly, C. I. (1994). *Harmonic Functions as a basis for motor control and planning*. Department of Computer Science. Amherst, MA, USA., University of Massachusetts.

Connolly, C. I., J. B. Burns, et al. (1990). Path planning using Laplace's equation. *Proc. of the 1990 IEEE International Conference on Robotics and Automation*: 2102-2106.

Connolly, C. I. & R. A. Grupen (1993). On the application of harmonic functions to robotics. *Journal of Robotc Systems* 10(7): 931-946.

Coste-Maniere, E., L. Adhami, et al. (2003). Optimal Planning of Robotically Assisted Heart Surgery: Transfer Precision in the Operating Room. *8th International Symposium on Experimental Robotics*: 721-732.

Chang, C., M. J. Chung, et al. (1994). Collision avoidance of two general manipilators by minimum delay time. *IEEE Transactions on Systems, Man and Cybernetics* 24(3): 517-522.

Chen, M. & A. M. S. Zalzala (1997). "A genetic approach to motion planning of redundant mobile manipulator systems considering safety and configuration." Journal of Robotc Systems 14(7): 529-544.

Chen, P. C. & Y. K. Hwang (1998). SANDROS: A dynamic graph search algorithm for motion planning. *IEEE Transactions on Robotics and Automation* 14(3): 392-403.

Davidor, Y. (1991). Genetic algorithms and robotics: a heuristic strategy for optimization, *World Scientific*.

Doyle, A. B. (1995). *Algorithms and computational techniques for robot path planning*. Bangor, The University of Wales.

Doyle, A. B. & D. I. Jones (1996). Robot path planning with genetic algorithms. *Proc. of 2nd Portuguese Conference on Automatic Control* (CONTROLO): 312-318.

Eldershaw, C. (2001). *Heuristic algorithms for motion planning*. Oxford, UK, The University of Oxford: 149.

Fortune, S., G. Wilfong, et al. (1987). Coordinated motion of two robot arms. *International Conference on Robotics and Automation*.

Fraichard, T. & C. Laugier (1992). Kinodynamic Planning in a Structured and Time-Varying 2D Workspace. *Proc. of the 1992 International Conference on Robotics and Automation:* 1500-1505.

Freund, E. & H. Hoyer (1985). Collision avoidance in multi-robot systems. Robotics Research: *Proc. of The Second International Symposium*. H. Hanafusa and H. Inoue. Cambridge, MA USA, MIT Press: 135-146.

Gill, M. A. C. & A. Y. Zomaya (1998). Obstacle avoidance in multi-robot systems, *World Scientific.*

Goldberg, D. E. (1983). Computer-aided gas pipeline operation using genetic algorithms and rule learning (*Doctoral dissertation*), University of Michigan.

Holland, J. H. (1975). *Adaptation in natural and artificial systems*, University of Michigan Press.

Hwang, Y. K. & N. Ahuja (1992). Gross motion planning - A survey. *ACM Computing Surveys* 24(3): 219-291.

Isto, P. (2003). *Adaptive probabilistic raodmap construction with multi-heristic local planing*. Department of Computer Science and Engineering. Espoo, Finland, Helsinki University of Technology: 62.

Khatib, O. (1986). Real-time obstacle avoidance for manipulators and mobile robots. *International Journal of Robotics Research* 5(1): 90-98.

Kim, J. & P. K. Kolsha (1992). Real time obstacle avoidance using harmonic potential functions. *IEEE Transactions on Robotics and Automation* 8(3): 338-349.

Koditschek, D. E. (1991). The control of natural motion in mechanical systems. *Journal of Dynamic Systems, Measurement, and Control* 113: 547-551.

Koditschek, D. E. (1991). Some applications of natural motion. *Journal of Dynamic Systems, Measurement, and Control* 113: 552-557.

Kubota, N., T. Arakawa, et al. (1997). Trajectory generation for redundant manipulator using virus evolutionary genetic algorithm. *Proc. of International Conference on Robotics and Automation* (ICRA). Albuquerque, NM, USA., IEEE: 205-210.

Latombe, J. C. (1991). *Robot Motion Planning*. Boston, MA, USA, Kluwer Academic.

Lee, B. H. & C. S. G. Lee (1987). Collision free motion planning of two robots. *IEEE Transactions on Systems, Man and Cybernetics* 17(1): 21-32.

Lee, D. (1996). *The map-building abd exploration strategies of a simple sonar-equiped mobile robot: An experomental, quantitative evaluation*. UK, Cambridge University Press.

Liu, Y. (2003). Interactive reach planning for animated characters using hardware acceleration. *Computer and Information Science*. Pittsburg, Univesity of Pennsylvania: 95.

Lozano-Peréz, T. (1983). Spatial planning: A configuration space approach. *IEEE Transactions on Computing* C-32(2): 108-119.

Lozano-Peréz, T. & M. A. Wesley (1979). An algorithm for planning collision-free paths among polyhedral obstacles. *Communications of the ACM* 22(10): 560-570.

Mamdani, E. H. (1974). Application of fuzzy algorithms for control of a simple dynamic plant. *Proc. IEE* 121(12): 1585-1588.

Masehian, E. & D. Sedighizadeh (2007). Classic and Heuristic Approaches in Robot Motion Planning - A Chronological Review. *Proceedings of World Academy of Science, Engineering and Technology* 23: 101-106.

Mbede, J. B., X. Huang, et al. (2000). Fuzzy motion planning among dynamic obstacles using artificial potential field for robot manipulators. *Robotics and Autonomous Systems* 32: 61-72.

Mbede, J. B. & W. Wei (2001). Fuzzy and recurrent neural network motion control among dynamic obstacles for robot manipulators. *Journal of Intelligent and Robotic Systems* 30: 155-177.

Merchán-Cruz, E. A., L. H. Hernández-Gómez, et al. (2007). Exploiting Monotony on a Genetic Algorithm Based Trajectory Planner (GABTP) for Robot Manipulators. 16th International *Conference on Applied Simulation and Modelling,* Palma de Mallorca, Spain, IASTED.

Merchán-Cruz, E. A. & A. S. Morris (2004). GA based trajectory planner for robot manipulators sharing common workspace. *Proceedings of the IASTED International conference on Applied Simulation and Modelling*, Rodos, Greece.

Merchán-Cruz, E. A. & A. S. Morris (2006). Fuzzy-GA-based trajectory planner for robot manipulators sharing a common workspace. *IEEE Transactions on Robotics* 22(4): 613-624.

Nearchou, A. C. & N. A. Aspragathos (1997). A genetic path planning algorithm for redundant articulated robots. *Robotica* 15: 213-224.

Parker, J. K., A. R. Khoogar, et al. (1989). Inverse kinematics of redundant robots using genetic algorithms. *International Conference on Robotics and Automation*, Scottsdale, AZ.

Rana, A. (1996). *Soft computing algorithms for motion planning of robotic manipulators.* Sheffield, University of Sheffield.

Rana, A. & A. M. S. Zalzala (1996). An evolutionary planner for near time-optimal collision-free motion of multi-arm robotic manipulators. *Proc. of International Conference on Control.* Exeter, UK. 1: 29-35.

Ridao, M. A., J. Riquelme, et al. (2001). An evolutionary and local search algorithm for motion planning of two manipulators. *Journal of Robotic Systems* 18(8): 463-476.

Rimon, E. & D. E. Koditschek (1992). Exact robot navigation using artificial potential functions. *IEEE Transactions on Robotics and Automation* 8(5): 501-518.

Schwartz, J. T. & M. Sharir (1983). On the 'piano movers' problem II: General techniques for computing topological properties of real algebraic manifolds. *Advances in Applied Mathematics* 4: 298-351.

Sugihara, K. & J. Smith (1997). Genetic algorithms for adaptive motion planning of autonomus mobile robots. *Technical Report*. Manoa, University of Hawaii.

Tombropoulos, R. Z., J. R. Adler, et al. (1999). CARABEAMER: A treatment planner for a robotic radiosurgical system with general kinematics. *Medical Image Analysis* 3(3): 237-264.

Vadakkepat, P., C. T. Kay, et al. (2000). Evolutionary artificial potential fields and their application in real time robot path planning. *Proc. of the 2000 Congress of Evolutionary Computation* (CEC2000), San Diego, California.

Volpe, R. & P. K. Khosla (1990). Manipulator control with superquadric artificial potential functions: Theory and experiments. *IEEE Transactions on Systems, Man and Cybernetics* 20(6): 1423-1436.

Wang, Q. & A. M. S. Zalzala (1996). Genetic control of near time-optimal motion for an industrial robot arm. *Proc. of International Conference on Robotics and Automation* (ICRA). Minneapolis, MI, USA, IEEE: 2592-2597.

Wise, D. K. & A. Bowyer (2000). A survey of global configuration-space mapping techniques for a single robot in a static environment. *International Journal of Robotics Research*: 762-779.

Zadeh, L. (1997). What is soft computing? www.ece.nus.edu.sg/stfpage/elepv/softc_def.html 2003.

Zadeh, L. A. (1965). Fuzzy Sets. *Information and Control* 8(3): 338-353.

Zavlangas, P. & S. G. Tzafestas (2000). Industrial robot navigation and obstacle avoidance using fuzzy logic. *Journal of Intelligent and Robotic Systems* 27: 85-97.

21

Robot Control Using On-Line Modification of Reference Trajectories

Javier Moreno-Valenzuela
Centro de Investigación y Desarrollo de Tecnología Digital del IPN
Mexico

1. Introduction

In practice, the robot motion is specified according to physical constraints, such as limited torque input. Thus, the computation of the desired trajectory is constrained to attain the physical limit of actuator saturations. See, e.g., (Pfeiffer & Johanni, 1987), for a review of trajectory planning algorithms considering the robot model parameters and torque saturation. Besides, trajectories can also be planned irrespectively of the estimated robot model, i.e., simply by using constraints of position, velocity and acceleration at each time instant, see, e.g., (Cao et al., 1997; Macfarlane & Croft, 2003).
Once that the reference trajectory is specified, the task execution is achieved in real time by using a trajectory tracking controller. However, if parametric errors in the estimated model are presented, and considering that the desired trajectories require to use the maximum torque, no margin to suppress the tracking error will be available. As consequence, the manipulator deviates from the desired trajectory and poor execution of the task is obtained. On-line time-scaling of trajectories has been studied in the literature as an alternative to solve the problem of trajectory tracking control considering constrained torques and model uncertainties.
A technique for time-scaling of off-line planned trajectories is introduced by Hollerbach (1984). The method provides a way to determine whether a planned trajectory is dynamically realizable given actuator torque limits, and a mode to bring it to one realizable. However, this method assumes that the robot dynamics is perfectly known and robustness issues were not considered. It is noteworthy that this approach has been extended to the cases of multiple robots in cooperative tasks (Moon & Ahmad, 1991) and robot manipulators with elastic joints (De Luca & Farina, 2002). In order to tackle the drawback of the assumption that the robot model in exactly known, Dahl and Nielsen (1990) proposed a control algorithm that result in the tracking of a time-scaled trajectory obtained from a specified geometric path and a modified on-line velocity profile. The method considers an internal loop that limits the slope of the path velocity when the torque input is saturated. Other solutions have been proposed in, e.g., (Arai et al., 1994; Eom et al., 2001; Niu & Tomizuka, 2001).
In this chapter, an algorithm for tracking control of manipulators under the practical situation of limited torques and model uncertainties is introduced. The proposed approach consists of using a trajectory tracking controller and an algorithm to obtain on-line time-

scaled reference trajectories. This is achieved through the control of a time-scaling factor, which is motivated by the ideas introduced by Hollerbach (1984). The new method does not require the specification of a velocity profile, as in (Dahl & Nielsen, 1990; Dahl, 1992; Arai et al., 1994). Instead, we formulate the problem departing from the specification of a desired path and a desired timing law. In addition, experiments in a two degrees-of-freedom direct-drive robot show the practical feasibility of the proposed methodology.

2. Robot model, desired motion description and control problem formulation

2.1 Robot model

The dynamics in joint space of a serial-chain n-link robot manipulator, considering the presence of friction at the robot joints, can be written as (Sciavicco & Siciliano, 2000)

$$M(q)\ddot{q} + C(q,\dot{q})\dot{q} + g(q) + f(\dot{q}) = \tau, \tag{1}$$

where q is the $n \times 1$ vector of joint displacements, $\dot{q}$ is the $n \times 1$ vector of joint velocities, τ is the $n \times 1$ vector of applied torque inputs, $M(q)$ is the $n \times n$ symmetric positive definite manipulator inertia matrix, $C(q,\dot{q})\dot{q}$ is the $n \times 1$ vector of centripetal and Coriolis torques, $g(q)$ is the $n \times 1$ vector of gravitational torques, and $f(\dot{q})$ is the vector of forces and torques due to the viscous and Coulomb friction, which depends on joint velocities. Besides, the i th element of the friction vector $f(\dot{q})$ satisfies,

$$|f_i(\dot{q})| \le f_{Ci} + f_{vi} |\dot{q}_i|,$$

where f_{Ci} and f_{vi} are positive constants, and $i = 1,\dots,n$.
Let Ψ denotes the torque space, defined as

$$\Psi = \{\tau \in R^n : -\tau_i^{max} < \tau_i < \tau_i^{max}\},\ i = 1,\dots,n, \tag{2}$$

with $\tau_i^{max} > 0$ the maximum torque input for the i th joint. It is assumed that

$$f_{Ci} + |g_i(q)| < \tau_i^{max} i = 1,\dots,n. \tag{3}$$

where $g_i(q)$ is the i th element of the vector $g(q)$. Assumption (3) ensures that any joint position q will be reachable.

2.2 Desired motion description and control problem formulation

The desired motions of a robot arm can be formulated in terms of a desired geometric path in the joint space, and an associated timing law. A desired path is given as a function

$$q_d(s) : [s_0, s_f] \to R^n, \tag{4}$$

where $s_f > s_0 > 0$, and s is called path parameter. It is assumed that the desired path attains

$$\frac{d}{ds} q_d(s) \neq 0.$$

This assumption is required in the problem design of time optimal trajectories (see e.g. (Pfeiffer & Johanni, 1987)) and also will be necessary in the results exposed in this chaper. The motion specification is completed by specifying a timing law (Hollerbach, 1984):

$$s(t):[t_0,t_f]\rightarrow[s_0,s_f], \tag{5}$$

which is a strictly increasing function. The timing law (5) is the path parameter as a time function. As practical matter, it is possible to assume that the path parameter $s(t)$ is a piecewise twice differentiable function. The first derivative $\dot{s}(t)$ is called *path velocity* and the second derivative $\ddot{s}(t)$ is called *path acceleration* (Dahl, 1990).
A nominal trajectory is obtained from the path $q_d(s)$ and the timing law, i.e.,

$$q_r(t)=q_d(s(t)). \tag{6}$$

2.3 Control problem formulation

Let us suppose that only estimations of the robot model (1) are available, namely $\hat{M}(q)$ for the inertia matrix, $\hat{C}(q,\dot{q})$ for the centripetal and Coriolis matrix, $\hat{g}(q)$ for the vector of gravitational forces, and $\hat{f}(\dot{q})$ for the vector of friction forces.
The control problem is to design an algorithm so that forward motion along the path $q_d(\sigma(t))$, with $\sigma(t)$ a strictly increasing function, is obtained as precise as possible, i.e.,

$$\| q_d(\sigma(t))-q(t)\| \leq \varepsilon,\ \forall\, t\geq t_0, \tag{7}$$

where ε is a positive constant, and considering that the input torque τ must be into the admissible torque space, i.e., $\tau\in\Psi$.
It is important to remark that in our approach is not necessary that the specified nominal trajectories (6), which are encoded indirectly by the desired path $q_d(s)$ and the timing law $s(t)$, evaluated for the estimated robot model produce torque into the admissible torque space Ψ in (2). In other words, it is not necessary the assumption

$$\hat{M}(q_r)\ddot{q}_r+\hat{C}(q_r,\dot{q}_r)\dot{q}_r+\hat{g}(q_r)+\hat{f}(\dot{q}_r)=\tau\in\Psi.$$

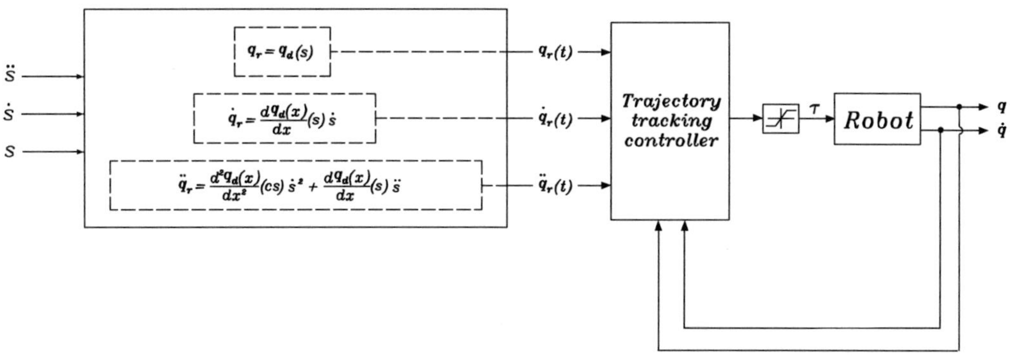

Figure 1. Block diagram of the classical approach of tracking control of nominal trajectories

This is because the proposed algorithm generate on-line new trajectories that produce admissible control inputs.

3. Tracking control of nominal trajectories

Trajectory tracking control is the classical approach to solve the motions tasks specified by the nominal trajectory (6). Since we have assumed the partial knowledge of the robot model, a controller that guarantees robust tracking of the nominal trajectory $q_r(t) = q_d(s(t))$ must be used. Many algorithms have been proposed with this aim, between them we have the following computed torque-based controller (Lewis et al., 1993)

$$\tau = \hat{M}(q)\left[\frac{d^2}{dt^2}q_d(s) + K_v\dot{e}_p + K_p e_p\right] + \hat{C}(q,\dot{q})\dot{q} + \hat{g}(q) + \hat{f}(\dot{q}), \tag{8}$$

where

$$\begin{aligned} e_p(t) &= q_d(s(t)) - q(t), \\ \dot{e}_p(t) &= \dot{q}_d(s(t)) - \dot{q}(t), \end{aligned}$$

$$\frac{d^2}{dt^2}q_d(s) = \frac{d^2 q_d(x)}{dx^2}(s)\dot{s}^2 + \frac{dq_d(x)}{dx}(s)\ddot{s},$$

and K_p, K_v are $n \times n$ symmetric positive definite matrices.

Considering unconstrained torque input, the control law (8) guarantees that the signals $e_p(t)$ and $\dot{e}_p(t)$ are uniformly ultimately bounded (see e.g. (Lewis et al., 1993). Figure 1 shows a block diagram of implementation of the trajectory tracking controller (8), having as inputs the path parameter $s(t)$, the path velocity $\dot{s}(t)$, and the path acceleration $\ddot{s}(t)$. With respect to the performance requirement (7), in the case of the classical trajectory tracking control, we have that $\sigma(t) = s(t)$.

Notwithstanding, the torque capability of the robot actuators is limited, and frequently desired trajectories are planned using this fact. Thus, it does not exist room for extra torque of the feedback control action for compensating the model disturbances; hence a poor tracking performance (7) is presented.

4. Tracking control of on-line time-scaled reference trajectories

Let us define a new path parameter σ in the following way

$$\sigma(t) = c(t)s(t), \tag{9}$$

where $c(t)$ is a scalar function called *time-scaling factor,* and $s(t)$ is the path parameter given as a time function in an explicit way. The desired trajectory is obtained using the new path parameter (1) into the path (4) as a function of time:

$$\bar{q}_r(t) = q_d(\sigma(t)) = q_d(c(t)s(t)). \tag{10}$$

The signal $c(t)$ in (9) is a time-scaling factor of the nominal trajectory $\overline{q}_r(t)$ in (10). The introduction of a time-scaling factor $c(t)$ is also used in the non robust algorithm proposed in (Hollerbach, 1984). Let us notice that the nominal (desired) value of the time scaling factor $c(t)$ is one. It is easy to see that when time-scaling is not present, the time evolution of $\sigma(t)$ is identical to $s(t)$, and $\overline{q}_r(t) = q_r(t)$, where $q_r(t)$ is given by (6). An important point in the definition of the time-scaling factor $c(t)$ is that if $c(t) > 1$ the movement is sped up and if $c(t) < 1$ the movement is slowed down. Thus, movement speed can be dynamically changed to compensate errors in the resulting tracking performance of trajectory (10).
The time derivative of $\sigma(t)$ in (9) is given as follows

$$\dot{\sigma} = \dot{c}s + c\dot{s}.$$

We call to the signal $\dot{c}(t)$ *time-scaling velocity* of the nominal trajectory. Further, the path acceleration $\ddot{\sigma}$ is given by

$$\ddot{\sigma} = \ddot{c}s + \gamma, \tag{11}$$

where

$$\gamma = 2\dot{c}\dot{s} + c\ddot{s}. \tag{12}$$

The signal $\ddot{c}(t)$ in equation (11) is called *time-scaling acceleration*.
Let us consider the trajectory tracking controller given by

$$\tau = \hat{M}(q)\left[\frac{d^2}{dt^2}q_d(cs) + K_v\dot{\overline{e}}_p + K_p\overline{e}_p\right] + \hat{C}(q,\dot{q})\dot{q} + \hat{g}(q) + \hat{f}(\dot{q}), \tag{13}$$

where

$$\begin{aligned} \overline{e}_p(t) &= \overline{q}_d(c(t)s(t)) - q(t), \\ \dot{\overline{e}}_p(t) &= \dot{\overline{q}}_r(c(t)s(t)) - \dot{q}(t), \end{aligned}$$

$$\frac{d^2}{dt^2}q_d(cs) = \frac{d^2q_d(x)}{dx^2}(cs)\underbrace{[\dot{c}s + c\dot{s}]^2}_{\dot{\sigma}^2} + \frac{dq_d(x)}{dx}(cs)\underbrace{[\ddot{c}s + \gamma]}_{\ddot{\sigma}},$$

and K_p, K_v are $n \times n$ symmetric positive definite matrices. Thus, under the control scheme (13) we have that performance requirement (7) is given with $\sigma(t) = c(t)s(t)$.
It is worth noting that controller (13) can be written in a parametric form by factoring the time scaling acceleration $\ddot{c}$, i.e.

$$\tau = \beta_1(c,s,q)\ddot{c} + \beta_2(c,\dot{c},s,\dot{s},q,\dot{q}), \tag{14}$$

where

$$\beta_1(c,s,q) = \hat{M}(q)\frac{dq_d(x)}{dx}(cs)s, \tag{15}$$

$$\begin{aligned}\beta_2(\sigma,\dot{\sigma},q,\dot{q}) &= \hat{M}(q)\frac{dq_d(x)}{dx}(cs)\gamma \\ &+\hat{M}(q)\left[\frac{d^2q_d(x)}{dx^2}(cs)[\dot{c}s+c\dot{s}]^2+\right. \\ &\left.+K_v\dot{e}_p+K_pe_p\right] \\ &+\hat{C}(q,\dot{q})\dot{q}+\hat{g}(q)+\hat{f}(\dot{q}),\end{aligned} \tag{16}$$

and γ is defined in (12).

The path acceleration $\ddot{\sigma}$ in equation (11) is expressed in terms of the time-scaling factor $c(t)$, the parameter $s(t)$, and successive time derivatives. In this way, instead of constraining the path acceleration $\ddot{\sigma}(t)$, as done in (Dahl & Nielsen, 1990; Dahl, 1992), for enforcing the control torque into the admissible limits, it is possible to constrain the scaling acceleration $\ddot{c}(t)$ using the limits of torque input for each joint i, that is to say

$$\ddot{c}_i^{min} \le \ddot{c} \le \ddot{c}_i^{max}.$$

The signals $\ddot{c}_i^{min}$ and $\ddot{c}_i^{max}$ are computed in the following way:

$$\ddot{c}_i^{max} = \begin{cases} [\tau_i^{max} - \beta_{2i}] / \beta_{1i}, & \forall \beta_{1i} > 0, \\ [-\tau_i^{max} - \beta_{2i}] / \beta_{1i}, & \forall \beta_{1i} < 0, \\ \infty, & \forall \beta_{1i} = 0, \end{cases}$$

$$\ddot{c}_i^{min} = \begin{cases} [-\tau_i^{max} - \beta_{2i}] / \beta_{1i}, & \forall \beta_{1i} > 0, \\ [\tau_i^{max} - \beta_{2i}] / \beta_{1i}, & \forall \beta_{1i} < 0, \\ -\infty, & \forall \beta_{1i} = 0, \end{cases}$$

where β_{1i} and β_{2i} are given explicitly in equations (15), and (16), respectively.

Let us remark that for computing the values of $\ddot{c}_i^{min}$ and $\ddot{c}_i^{max}$, the assumption that $s_0 > 0$ is required to avoid that $\ddot{c}_i^{min}$ and $\ddot{c}_i^{max}$ become undetermined.

The limits of the time scaling acceleration $\ddot{c}$, which depends on s, $\dot{s}$, $\ddot{s}$, c, $\dot{c}$, and measured signals $q_i, \dot{q}_i$, are given by

$$\ddot{c}^{max} = \min_i \left\{ \ddot{c}_i^{max} \right\}, \tag{17}$$

$$\ddot{c}^{min} = \max_i \left\{ \ddot{c}_i^{min} \right\}, \tag{18}$$

The limits (17)-(18) provide a way of modifying the reference trajectory (10), via alteration of $\ddot{c}(t)$, so that the torque limits are satisfied. If the resulting nominal reference trajectory $\overline{q}_r(c(t)s(t))$ is inadmissible in the sense of unacceptable torques, then it will be modified by limiting the time scaling acceleration $\ddot{c}(t)$ to satisfy $\tau_i \in \Psi$.

Note that when the time-scaling acceleration $\ddot{c}(t)$ is modified, the value of the time-scaling factor $c(t)$ will change. Thus, considering the limits (17) and (18), we must design an internal feedback to drive the time-scaling factor $c(t)$ in a proper way. The proposed internal feedback is given by

$$\frac{d}{dt}c = \dot{c}, \tag{19}$$

$$\frac{d}{dt}\dot{c} = sat(u;\ddot{c}^{min},\ddot{c}^{max}), \tag{20}$$

with the initial conditions $c(t_0)=1$, and $\dot{c}(t_0)=0$, the saturation function

$$sat(u;\ddot{c}^{min},\ddot{c}^{max}) = \begin{cases} u & \forall \quad \ddot{c}^{min} \le u \le \ddot{c}^{max}, \\ \ddot{c}^{min} & \forall \quad u < \ddot{c}^{min}, \\ \ddot{c}^{max} & \forall \quad u > \ddot{c}^{max}, \end{cases}$$

and u properly designed. The input torques are kept within the admissible limits by using the internal feedback (19)-(20), which provides a way to scale in time the reference trajectories $q_d(c(t)s(t))$. We adopt the idea of specifying a time varying desired time-scaling factor. Let us consider

$$u = \ddot{r} + k_{vc}\dot{\tilde{c}} + k_{pc}\tilde{c}, \tag{21}$$

where k_{vc} and k_{pc} strictly positive constants,

$$\tilde{c}(t) = r(t) - c(t), \tag{22}$$

and $r(t)$ is the desired time-scaling reference. The control law (21) is used in the internal feedback (19)-(20).

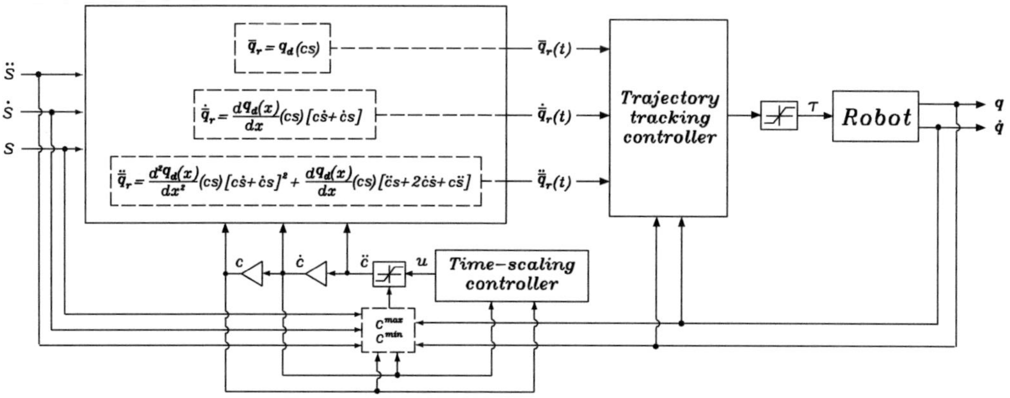

Figure 2. Block diagram of proposed approach of tracking control of on-line time-scaled references trajectories

If the generated control torque is admissible, then the equation (20) satisfies $\ddot{c}=u$, u defined in (21). Therefore we can write

$$\ddot{\tilde{c}} + k_{vc}\dot{\tilde{c}} + k_{pc}\tilde{c} = 0.$$

Because k_{vc} and k_{pc} are strictly positive constants, $\tilde{c}(t) \to 0$ as $t \to \infty$ in an exponential form. We define the desired time-scaling factor $r(t)$ in accord to the following equation

$$\dot{r}(t) = -k_r \tilde{c}(t), \quad r(t_0) = 1, \tag{23}$$

where k_r is a strictly positive constant, and $\tilde{c}$ is defined in (22). The value of $\ddot{r}(t)$ in (21) is computed differentiating (23) with respect to time.

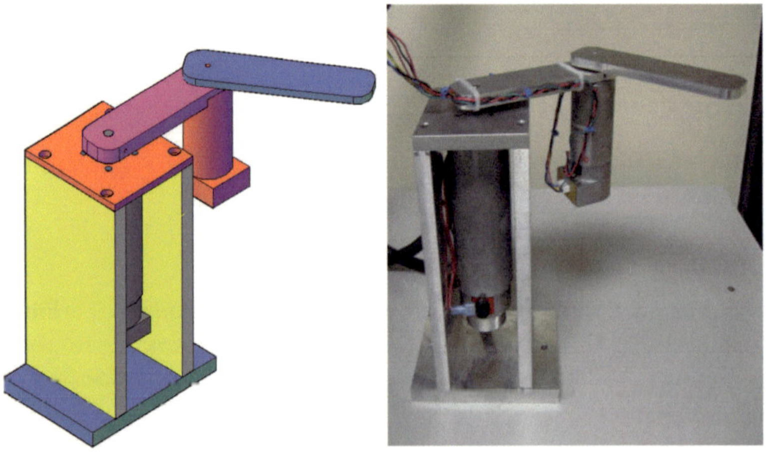

Figure 3. Experimental robot arm

The proposed dynamics for $r(t)$ in (23) obeys the error between $c(t)$ and $r(t)$. Thus, when torque saturation is detected the time-scaling factor $c(t)$ is modified (see equations (19)-(20)) to make the reference trajectory $q_d(c(t)s(t))$ feasible, i.e. to make the input torque admissible. The desired time-scaling factor $r(t)$ can be interpreted as an *estimation* of the actual time-scaling factor $c(t)$. Figure 2 depicts a block diagram of the controller and time-scaling algorithm (14), (19) and (20).

It is worthwhile to note that the proposed time-scaling method is independent of the definition of path parameter $s(t)$, Thus, several definitions of the path parameter $s(t)$ considering different characteristics in its time evolution can be used along the lines of the proposed time-scaling algorithm. See, e.g, (Macfarlane & Croft, 2003; Cao et al., 1997), for planning algorithms of the timing law $s(t)$.

5. Experimental results

A planar two degrees-of-freedom direct-drive arm has been built at the CITEDI-IPN Research Center. The system is composed by two DC Pittman motors operated in current mode with two Advanced Motion servo amplifiers. A Sensoray 626 I/O card is used to read encoder signals (with quadrature included), while control commands are transferred through the D/A channels. The control system is running in real-time with a sampling rate of 1 kHz on a PC over *Windows XP* operating system and *Matlab Real-Time Windows Target*. Figure 3 shows the experimental robot arm.

5.1 Practical considerations

The torque delivered by DC motor actuators is given by

$$\tau = Kv,$$

where $K = diag\{k_1,...,k_n\}$ is a positive definite matrix that contains the voltage to torque conversion constants of the motors, and $v \in R^n$ is a vector that contains the applied voltage. In this situation, the proposed control law (13) can be implemented as

$$v = \hat{K}^{-1}\left[\hat{M}(q)\left[\frac{d^2}{dt^2}q_d(cs) + K_v\dot{\bar{e}}_p + K_p\bar{e}_p\right] + \hat{C}(q,\dot{q})\dot{q} + \hat{g}(q) + \hat{f}(\dot{q})\right],$$

where $\hat{K} = diag\{\hat{k}_1,...,\hat{k}_n\}$ is an estimation of K. Note that torque limits Ψ in (2) can be indirectly specified by voltage limits

$$\Upsilon = \{v \in R^n : -v_i^{max} < \tau_i < v_i^{max}\},\ i = 1,\ldots,n,$$

with $v_i^{max} > 0$ the maximum voltage input for the i th joint. In order to obtain the real-time time-scaling factor $c(t)$, the parameterization (15) –(16) and limits (17)-(18) should be used with v_i^{max} instead of τ_i^{max}.

For simplicity, we have selected

$$\hat{K} = diag\{1,1\}\ [Nm / Volt].$$

A constant matrix for the estimation of the inertia matrix is proposed, i.e.,

$$\hat{M} = diag\{0.2, 0.01\}\ [Nm \cdot sec^2 / rad], \tag{24}$$

while $\hat{C}(q,\dot{q}) = 0$ for the centripetal and Coriolis matrix, $\hat{g}(q) = 0$ for the vector of gravitational forces, and $\hat{f}(\dot{q}) = 0$ for the friction forces vector. The voltage limits

$$v_1^{max} = 5\ [\text{Volt}] \text{ and } v_2^{max} = 0.5\ [\text{Volt}]. \tag{24}$$

were specified. We have assumed that with the voltage limits (24), the assumption (3) is attained. In other words, we considered that there is enough voltage room to compensate the disturbances due to the friction $f(\dot{q})$ present in the robot model (1). Note that the vector of gravitational forces $g(q)$ is zero, since the joints of the experimental robot arm move in the horizontal plane.

The requested task is to drive the arm in such a way that its joint position $q(t)$ traces a desired path with prescribed velocity in the joint configuration space. The path should be traced with nominal constant tangent velocity. The desired path is

$$q_d(s) = \begin{bmatrix} r_0 \cos(v_0 s) \\ r_0 \sin(v_0 s) \end{bmatrix}, \tag{25}$$

where r_0=1 [rad] and v_0=2.5 [rad/sec]. The timing law, i.e., the path parameter $s(t)$ as function of time, is specified as follows

$$s(t) = \begin{cases} \frac{v_0}{2} t^2 + 0.001, \text{ for } t < 1, \\ v_0 [t - 0.5], \text{ for } t \geq 1, \end{cases}$$

which is piecewise twice differentiable. In the simulations, the initial configuration of the arm is $q_1(0) = 1$ [rad], $q_2(0) = 0$ [rad], and $\dot{q}_1(0) = \dot{q}_1(0) = 0$ [rad/sec]. It is noteworthy that $s_0 = s(0) > 0$ is satisfied.

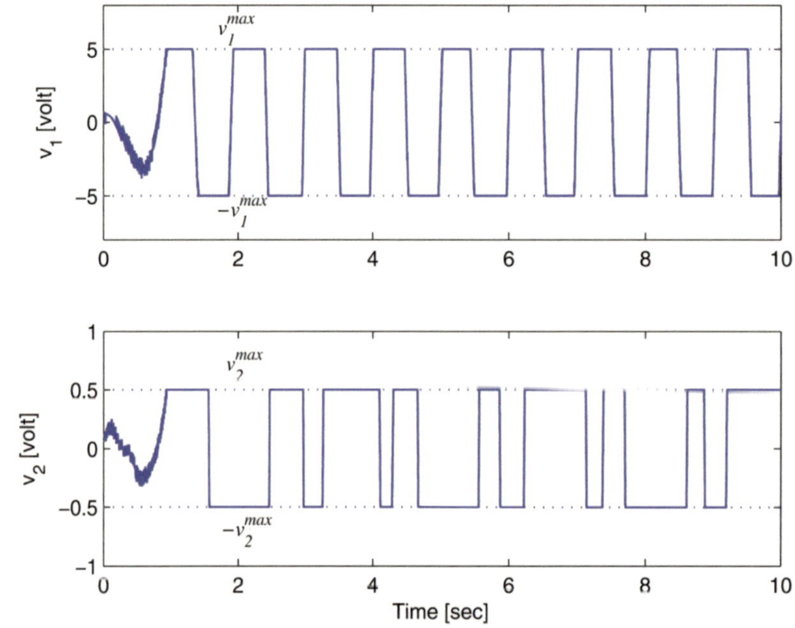

Figure 4.Tracking control of nominal trajectories: Applied voltage

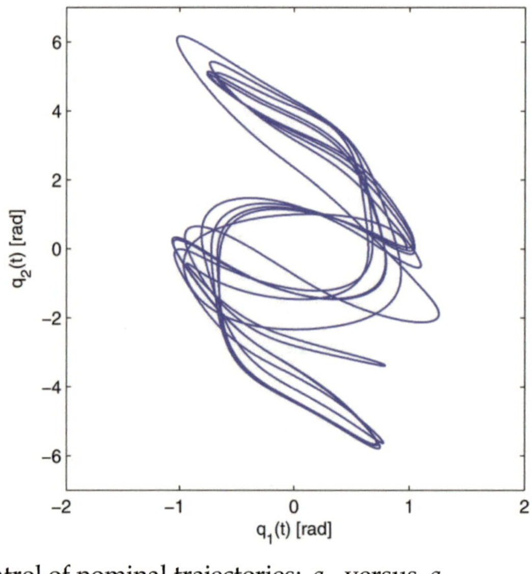

Figure 5. Tracking control of nominal trajectories: q_2 versus q_1

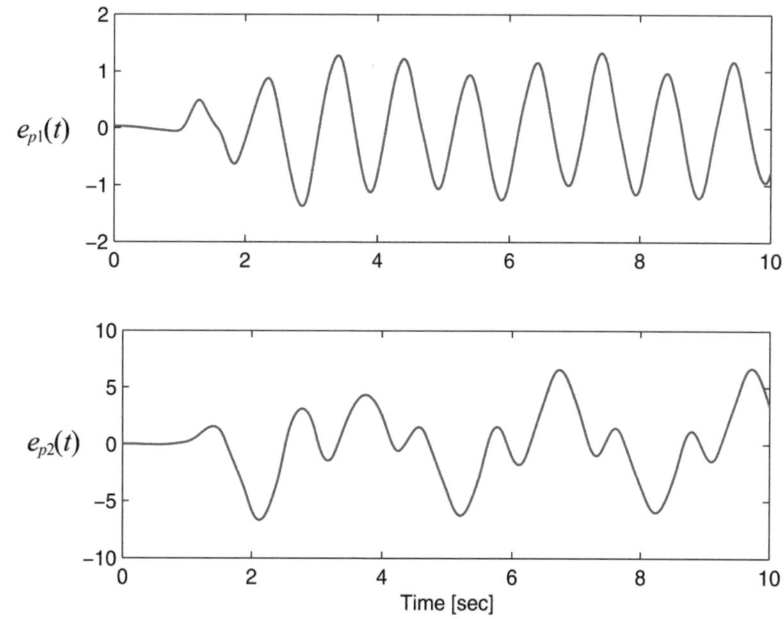

Figure 6. Tracking control of nominal trajectories: Tracking errors

5.2 Tracking control of nominal trajectories

In order to compare the performance of the proposed algorithm, we carried out a simulation with the trajectory tracking controller given by (8). Thus, considering the error signal

$$e_p(t) = q_r(t) - q(t),$$

where the desired nominal trajectory $q_r(t)$ is defined in (6), the resulting trajectory tracking controller is given by

$$v = \hat{M}\left[\ddot{q}_r + K_v\dot{e}_p + K_p e_p\right], \tag{26}$$

which results from the above specifications. The used gains were

$$K_p \;=\; diag\{100, 400\}\,[1/sec^2], \tag{27}$$

$$K_v \;=\; diag\{10, 20\}\,[1/sec]. \tag{28}$$

The experimental results are presented in Figure 4 that shows the applied voltage, in Figure 5 that depicts the traced path in q_1 - q_2 coordinates, and in Figure 6 that describes the time evolution of the tracking errors.

We note in Figure 4 that both control voltages are saturated at the same time intervals. This is undesirable, because it produces deviations from the specified desired path, as it can be shown in Figure 5. On the other hand, tracking errors $e_{p1}(t)$ and $e_{p2}(t)$ are large, as illustrated in Figure 6. Specifically, we have considered $t \geq 2$ [sec] for the steady state regimen, thus $\max_{\forall t \geq 2}\{e_{p1}(t)\} = 1.38$ [rad] and $\max_{\forall t \geq 2}\{e_{p2}(t)\} = 6.5$ [rad].

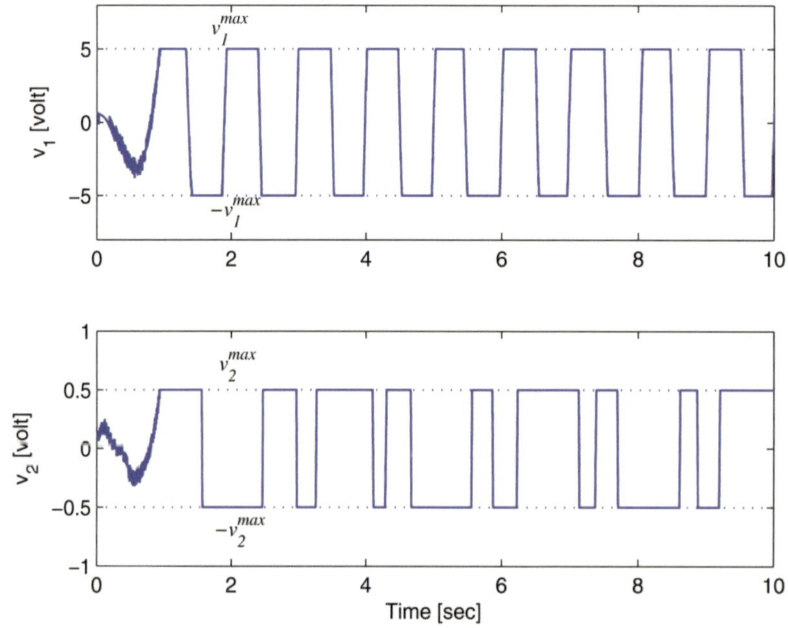

Figure 7. Tracking control of on-line time-scaled references trajectories: Applied voltage

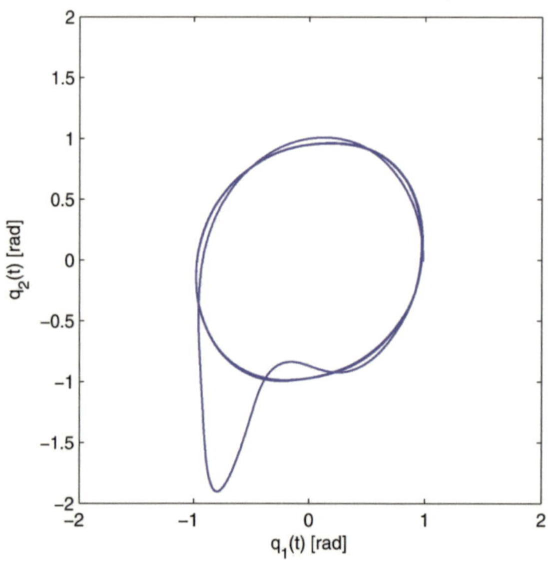

Figure 8. Tracking control of on-line time-scaled references trajectories: q_2 versus q_1

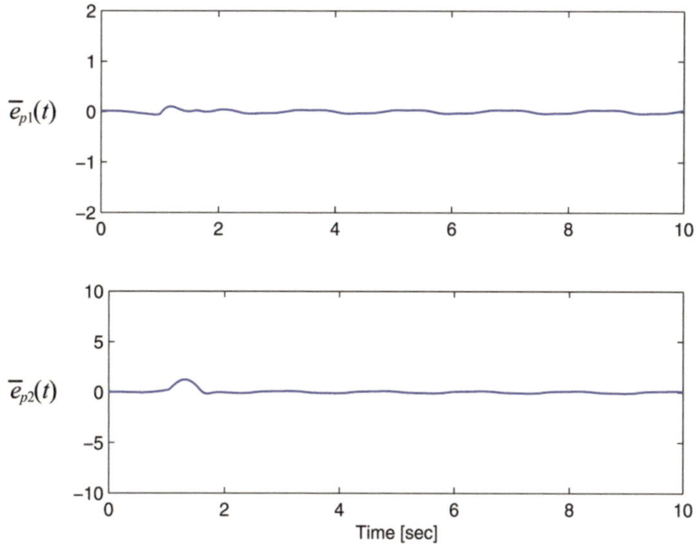

Figure 9. Tracking control of on-line time-scaled references trajectories: Tracking errors

5.3 Tracking control of on-line time-scaled reference trajectories

By taking into account the practical considerations in Section 5.2, we implemented in real-time the proposed scheme of tracking control of on-line time-scaled reference trajectories (14), (19), (20), and (21)-(23). The control gains (27)-(28) were used in the controller (13), the gains k_{pc} =50 and k_{vc} =14, in the internal feedback (21), and $k_r = 50$ in the system (23).

Figure 7 shows the applied voltage, Figure 8 illustrates the traced path in q_1 - q_2 coordinates, Figure 9 depicts the tracking errors, and finally Figure 10 shows the time evolution of the time-scaling fact

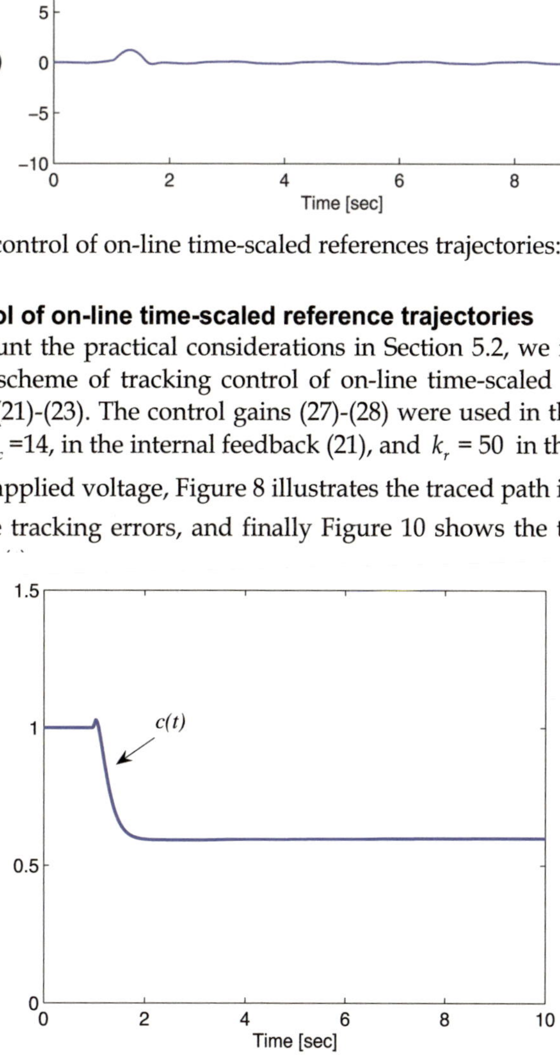

Figure 10. Tracking control of on-line time-scaled references trajectories: Time-scaling factor

It is possible to observe that the applied voltage remains within the admissible limits, and simultaneous saturation does not occur. Moreover, tracking errors are drastically smaller than

the tracking errors obtained with tracking control of nominal trajectories (26). See Figure 6 and Figure 9 to compare the performance of the tracking errors $e_p(t)$ with respect to the tracking errors $\bar{e}_p(t)$. Particularly, $\max_{\forall t\geq 2}\{\bar{e}_{p1}(t)\} = 0.36$ [rad] and $\max_{\forall t\geq 2}\{\bar{e}_{p2}(t)\} = 0.1$ [rad], which are drastically smaller than values obtained for the classical trajecory tracking controller.
Finally, Figure 10 shows that the time-scaling factor $c(t)$ tends to 0.6, thus the tracking accuracy is improved because the reference trajectory $q_d(c(t)s(t))$ is slowed down.

6. Conclusion

An approach for trajectory tracking control of manipulators subject to constrained torques has been proposed. A secondary loop to control the time-scaling of the reference trajectory is used, then the torque limits are respected during the real-time operation of the robot. The proposed algorithm does not require the specification of a velocity profile, as required in (Dahl & Nielson, 1990; Dahl, 1992; Arai et al. 1994), but instead a timing law should be specified.

7. References

Arai, H., Tanie, K. & Tachi, S. (1994). Path tracking control of a manipulator considering torque saturation. *IEEE Transactions on Industrial Electronics*, Vol. 41, No. 1, pp. 25-31.

Cao, B., Dodds, G.I. & Irwin, G.W. (1997). Constrained time-efficient and smooth cubic spline trajectory generation for industrial robots. *IEE Proc.-Control Theory and Applications*, Vol. 44, No. 5, pp. 467-475.

Dahl, O. & Nielsen, L. (1990). Torque-limited path following by on-line trajectory time scaling. *IEEE Trans. Rob. Aut.* Vol. 6, No. 5, pp. 554-561.

Dahl, O. (1992). *Path Constrained Robot Control*, Doctoral Dissertation, Department of Automatic Control, Lund Institute of Technology.

De Luca, A. & Farina, R. (2002). Dynamic scaling of trajectories for robots with elastic joints., *Proc. of the 2002 IEEE International Conference on Robotics and Automation*, Washington, May, pp. 2436-2442.

Eom, K.S., Suh, I.H. & Chung, W.K. (2001). Disturbance observer based path tracking of robot manipulator considering torque saturation. *Mechatronics*, Vol. 11, pp. 325-343.

Hollerbach, J.M. (1984). Dynamic scaling of manipulator trajectories. *Journal of Dynamic Systems, Measurement and Control*, Vol. 106, No. 1, pp. 102-106.

Lewis, F.L., Abdallah, C.T. & Dawson, D.M. (1993). *Control of Robot Manipulators*, Macmillan Publishing Company, New York.

Macfarlane, S. & Croft, E.A. (2003). Jerk bounded manipulator trajectory planning: Design for real-time applications. *IEEE Trans. Rob. Aut.*, Vol. 19, No. 1, pp. 42-52.

Moon, S.B. & Ahmad, S. (1991). Time scaling of cooperative multirobot trajectories. *IEEE Trans. Systems Man and Cybernetics*, Vol. 21, No. 4, pp. 900-908.

Niu, W. & Tomizuka, M. (2001). A new approach of coordinated motion control subjected to actuator saturation. *Journal of Dynamic Systems, Measurement and Control*, Vol. 123, pp. 496-504.

Pfeiffer, F. & Johanni, R. (1987). A concept for manipulator trajectory planning. *IEEE Journal of Robotics and Automation*, Vol. RA-3, No. 2, pp. 115-123.

Sciavicco, L. & Siciliano, B. (2000). *Modeling and Control of Robot Manipulators*, Springer, London.

22

Motion Behavior of Null Space in Redundant Robotic Manipulators

Tsuyoshi Shibata and Toshiyuki Murakami
Keio University, Department of System Design Engineering
Japan

1. Introduction

As well known, robot manipulator is utilized in many industrial fields. However manipulator motion has a single function at most and is so limited because of low degree-of-freedom motion. To improve this issue, it is necessary for the robot manipulator to have redundant degree-of-freedom motion. The motion spaces of redundant manipulator are divided into work space motion and null space motion. A key technology of the redundant manipulator is dexterous use of null space motion. Then, the important issue is how to design null space controller for the dexterous motion control. However the control strategy of null space motion based on the stability analysis has not been established yet. Focusing on this issue, this chapter shows PID controller based on stability analysis considering passivity in null space motion of redundant manipulator. The PID controller which is passive controller plays a role to compensate disturbance of null space motion without deteriorating asymptotic stability. On the other hand, the work space control is stably achieved by a work space observer based PD control.

In this chapter, work space controller and null space controller are considered separately. In the proposed controller design, the work space observer that compensates the work space disturbance affecting position response of end-effector is one of important key technologies [1]. Furthermore it brings null space motion without calculating kinematic transformation. Then the null space input can be determined arbitrary in joint space. In this chapter, PID control is utilized to synthesize the null space input. The null space motion is also affected by unknown disturbance even if the work space observer compensates the work space disturbance completely. In particular, null space disturbance denotes an interactive effect among null space responses. The large interaction affects the stability of not only the null space but also the work space motion. To increase both the stability margin and robustness of null space motion, this chapter introduces PID controller considering passivity [2]. The validity of the proposed approach is verified by simulations of 4-link redundant manipulator.

2. Kinematics and Dynamics of Redundant Manipulator

2.1 Dynamics Modeling of Redundant Manipulator

This chapter assumes n-link robot manipulator with all revolute-type joints. When the end-effector is free to move, the motion of the manipulator is governed by the Lagrange equation. It is described in terms of joint coordinated $\boldsymbol{q} = (q_1, \cdots, q_n)^T$, that is, the joint angle vector. Then the motion equation of the robot manipulator

$$R(\boldsymbol{q})\ddot{\boldsymbol{q}} + \frac{1}{2}\dot{R}(\boldsymbol{q})\dot{\boldsymbol{q}} + \mathcal{S}(\boldsymbol{q}, \dot{\boldsymbol{q}})\dot{\boldsymbol{q}} + \boldsymbol{g}(\boldsymbol{q}) = \boldsymbol{\tau} \tag{1}$$

can be obtained by the Lagrange equation. Here $R(\boldsymbol{q})$ is the inertial matrix and symmetric and positive definite. $(\frac{1}{2}\dot{R}(\boldsymbol{q})\dot{\boldsymbol{q}} + \mathcal{S}(\boldsymbol{q}, \dot{\boldsymbol{q}})\dot{\boldsymbol{q}})$ denotes the Coriolis and centrifugal torques. $\boldsymbol{\tau} \in R^n$ is the vector of input torques generated at joint actuators. $\mathcal{S}(\boldsymbol{q}, \dot{\boldsymbol{q}})$ is skew-symmetric, since it is expressed in

$$\mathcal{S}(\boldsymbol{q}, \dot{\boldsymbol{q}})\dot{\boldsymbol{q}} = \frac{1}{2}\left[\dot{R}(\boldsymbol{q})\dot{\boldsymbol{q}} - \frac{\partial}{\partial \boldsymbol{q}}\left\{\dot{\boldsymbol{q}}^T R(\boldsymbol{q})\dot{\boldsymbol{q}}\right\}^T\right]. \tag{2}$$

$\mathcal{S}(\boldsymbol{q}, \dot{\boldsymbol{q}})$ satisfies

$$\dot{\boldsymbol{q}}^T \mathcal{S}(\boldsymbol{q}, \dot{\boldsymbol{q}})\dot{\boldsymbol{q}} = 0. \tag{3}$$

$\boldsymbol{g}(\boldsymbol{q}) = \frac{\partial U(\boldsymbol{q})}{\partial \boldsymbol{q}}$ is the gravity force vector, where $U(\boldsymbol{q})$ denotes a potential energy. In (1), robot manipulator is passive between input $\boldsymbol{\tau}$ and output $\dot{\boldsymbol{q}}$. This passivity relation plays an important role in building a bridge between the energy conservation law in physics and the operational input-output relation in systems theory [2].

2.2 Kinematic Modeling of Redundant Manipulator

In n-link manipulator, the kinematic relation can be written by (4)-(5). $\boldsymbol{x} \in R^m$ $(n > m)$ is a position vector in work space motion.

$$\dot{\boldsymbol{x}} = J_{aco}(\boldsymbol{q})\dot{\boldsymbol{q}} \tag{4}$$

$$\ddot{\boldsymbol{x}} = J_{aco}(\boldsymbol{q})\ddot{\boldsymbol{q}} + \dot{J}_{aco}(\boldsymbol{q})\dot{\boldsymbol{q}} \tag{5}$$

Here $J_{aco} \in R^{m \times n}$ is the Jacobian matrix, and it is shown in (6).

$$J_{aco}(\boldsymbol{q}) = \frac{\partial \boldsymbol{f}(\boldsymbol{q})}{\partial \boldsymbol{q}} \tag{6}$$

The inverse kinematics of (5) is given as

$$\ddot{\boldsymbol{q}} = J_{aco}^{+}(\ddot{\boldsymbol{x}} - \dot{J}_{aco}\dot{\boldsymbol{q}}) + (I - J_{aco}^{+}J_{aco})\ddot{\boldsymbol{\phi}}. \tag{7}$$

J_{aco}^{+} is the pseudo inverse matrix of Jacobian, which is defined by

$$J_{aco}^{+} = J_{aco}^{T}(J_{aco}J_{aco}^{T})^{-1}. \tag{8}$$

In the right-hand side of (7), $\ddot{\boldsymbol{\phi}}$ is the input vector of null space. Using (7), the acceleration references of work space are resolved into joint space of redundant manipulator.

2.3 Acceleration Reference for Null Space Motion

From the kinematic relation, an equivalent motion equation is derived in the work space [1]. In the proposed approach, the joint acceleration reference is given by

$$\ddot{\boldsymbol{q}}^{ref} = J_{aco}^{+}\ddot{\boldsymbol{x}}^{ref}. \tag{9}$$

The superscript ^{ref} denotes motion command in the work space. To consider the null space motion based on the joint acceleration reference, arbitrary reference of the joint acceleration $\ddot{\boldsymbol{q}}_{null}^{ref}$ space motion is added to (9). Then, the joint acceleration reference $\ddot{\boldsymbol{q}}^{ref}$ is redefined as follows.

$$\ddot{\boldsymbol{q}}^{ref} = J_{aco}^{+}\ddot{\boldsymbol{x}}^{ref} + \ddot{\boldsymbol{q}}_{null}^{ref} \tag{10}$$

Using (5) and (10), the equivalent acceleration error $\ddot{\boldsymbol{e}}_{work}$ in the work space is given as follows. Here, in (5), joint acceleration controller ($\ddot{\boldsymbol{q}} = \ddot{\boldsymbol{q}}^{ref}$) is assumed by joint disturbance observer as shown in Fig. 2.

$$\begin{aligned} \ddot{\boldsymbol{e}}_{work} &= \ddot{\boldsymbol{x}}^{ref} - \ddot{\boldsymbol{x}} \\ &= -J_{aco}\ddot{\boldsymbol{q}}_{null}^{ref} - \dot{J}_{aco}\dot{\boldsymbol{q}} \end{aligned} \tag{11}$$

When the acceleration error $\ddot{\boldsymbol{e}}_{work}$ is calculated or estimated as $\widehat{\ddot{\boldsymbol{x}}}^{dis} = \ddot{\boldsymbol{e}}_{work}$, it may be fed back to compensate the acceleration error in the work space. Then, (10) is rewritten as follows. Practically, $\ddot{\boldsymbol{e}}_{work}$ is estimated by a work space observer shown in Fig. 2.

$$\ddot{\boldsymbol{q}}^{ref} = J_{aco}^{+}(\ddot{\boldsymbol{x}}^{ref} + \widehat{\ddot{\boldsymbol{x}}}^{dis}) + \ddot{\boldsymbol{q}}_{null}^{ref} \tag{12}$$

Assuming that $\widehat{\ddot{\boldsymbol{x}}}^{dis} = \ddot{\boldsymbol{e}}_{work}$ and substituting (11) into (12),

$$\ddot{\boldsymbol{q}}^{ref} = J_{aco}^{+}(\ddot{\boldsymbol{x}}^{ref} - \dot{J}_{aco}\dot{\boldsymbol{q}}) + (I - J_{aco}^{+}J_{aco})\ddot{\boldsymbol{q}}_{null}^{ref} \tag{13}$$

is obtained. (13) shows that the feedback of the estimated acceleration error of the work space brings the null space motion without calculating the matrix $(I - J_{aco}^{+}J_{aco})$.
When $J_{aco}^{+}(\ddot{\boldsymbol{x}}^{ref} + \widehat{\ddot{\boldsymbol{x}}}^{dis})$ is defined as $\ddot{\boldsymbol{q}}_{work}^{ref}$, (12) is expressed in

$$\ddot{\boldsymbol{q}}^{ref} = \ddot{\boldsymbol{q}}_{work}^{ref} + \ddot{\boldsymbol{q}}_{null}^{ref}. \tag{14}$$

(14) shows that work space controller and null space controller are separetely designed to determine each acceleration reference. To achieve this, the work space observer is employed in the work space to estimate the acceleration error given by (11) [1].

3. Design of Work Space Controller and Null Space Controller

This section describes design procedure of work space and null space controllers. In the work space, the work space observer based PD control is utilized. The work space observer suppresses work space disturbance $\widehat{\ddot{\boldsymbol{x}}}^{dis}$. Hence the work space constitutes robust acceleration control system. As described before, the work space and the null space motion are determined separately according to the final acceleration reference of joint space in (13). Also the disturbance of null space is considered independently without affecting the work space disturbance. In the proposed approach, PID control is used as null space controller to suppress the null space disturbance. Here this paper proves that disturbance of null space is compensated by PID controller. Fig.1 shows a whole block diagram of the proposed method.

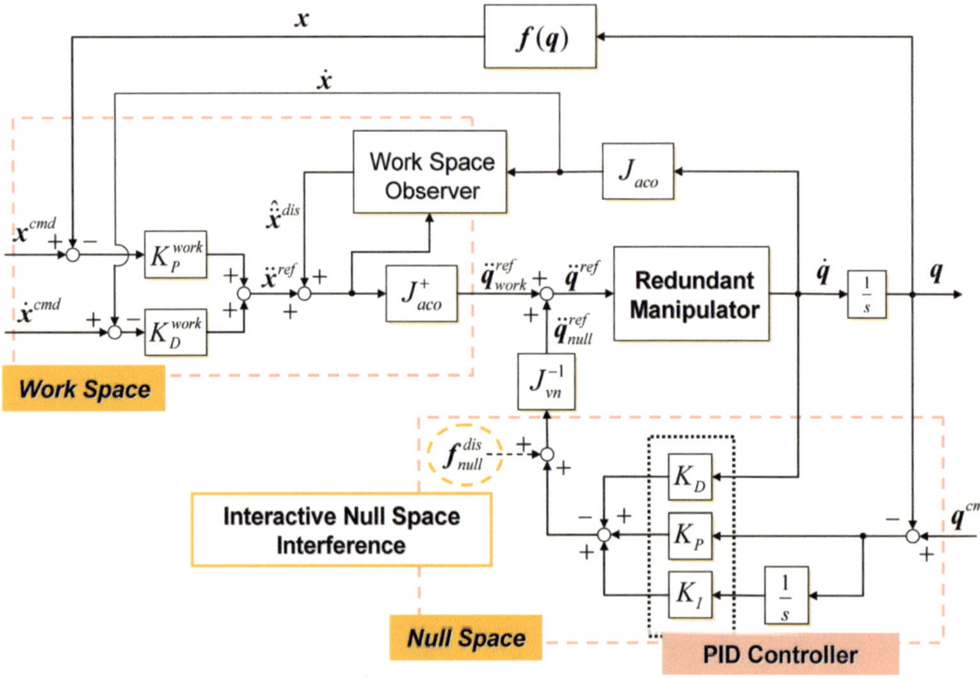

Figure 1. A whole block diagram of proposed method

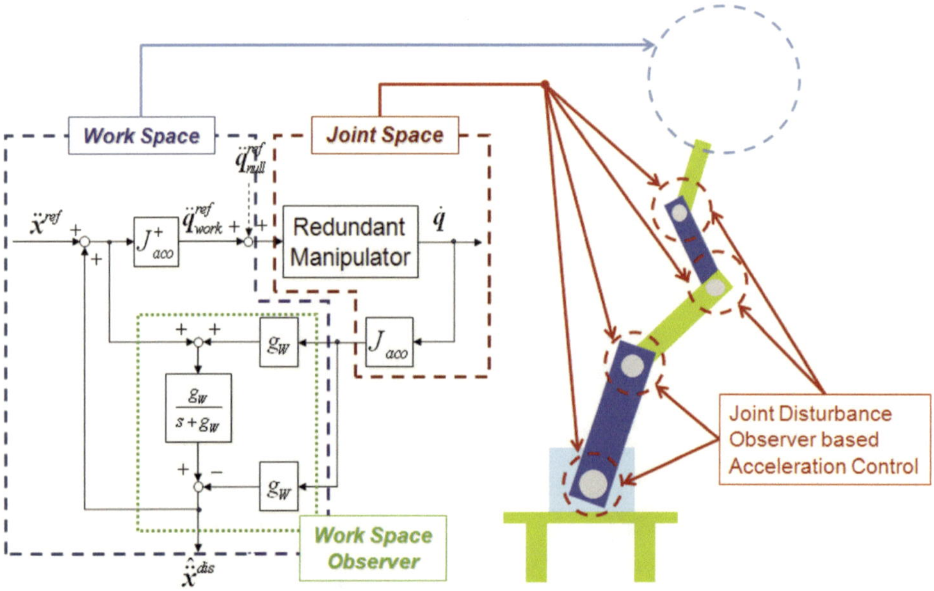

Figure 2. Work space observer

3.1 Work Space Controller

In the work space controller, the acceleration reference is given by

$$\ddot{\boldsymbol{x}}^{ref} = K_P^{work}(\boldsymbol{x}^{cmd} - \boldsymbol{x}) + K_D^{work}(\dot{\boldsymbol{x}}^{cmd} - \dot{\boldsymbol{x}}) \tag{15}$$

where $\boldsymbol{x}^{cmd} \in R^m$ denotes a position command vector. K_P^{work}, $K_D^{work} \in R^{m\times m}$ are positive definite diagonal matrices. Furthermore the work space observer is employed, and the estimated disturbance acceleration $\hat{\ddot{\boldsymbol{x}}}^{dis}$ is fed back as shown in Fig.2 [1].

3.2 Null Space Controller

Passivity of PID Controller

In this section, passivity of PID controller is shown. This is given by H. K. Khalil [4] and B. Brogliato *et al.* [5]. The general formulation of PID controller is given in

$$\boldsymbol{\tau} = K_P\boldsymbol{e} + K_I\int_0^t \boldsymbol{e}(\tau)d\tau + K_D\dot{\boldsymbol{e}}, \tag{16}$$

where $\boldsymbol{e} \in R^n$ and $\boldsymbol{\tau} \in R^n$ are error and input respectively. $K_P,\ K_I,\ K_D \in$ are $R^{n\times n}$ parameters and they are positive definite diagonal matrices. In this paper, only a servo or a regulator problem is considered for null space stability, and a tracking problem is not treated. Then, the error $\boldsymbol{e}$. is written in

$$\boldsymbol{e} = \boldsymbol{q}^{cmd} - \boldsymbol{q}, \tag{17}$$

where $\boldsymbol{q}^{cmd} \in R^n$ denotes a command vector of the joint angle. Here, $\boldsymbol{q}^{cmd} = \boldsymbol{0}$ is regulator problem. (16) is rewritten in

$$\boldsymbol{\tau} = K_P(\boldsymbol{q}^{cmd} - \boldsymbol{q}) + K_I\int_0^t(\boldsymbol{q}^{cmd} - \boldsymbol{q}(\tau))d\tau - K_D\dot{\boldsymbol{q}}. \tag{18}$$

Here a state variable vector $\boldsymbol{z}$ is additionally introduced. The integral term of (18) is replaced with

$$\int_0^t(\boldsymbol{q}^{cmd} - \boldsymbol{q}(\tau))d\tau = \boldsymbol{z}. \tag{19}$$

Hence, (18) is expressed in

$$\boldsymbol{\tau} = K_P(\boldsymbol{q}^{cmd} - \boldsymbol{q}) + K_I\boldsymbol{z} - K_D\dot{\boldsymbol{q}}. \tag{20}$$

Then, the storage function of PID controller (20) is expressed in (21).

$$S_c(\boldsymbol{z},\boldsymbol{q}) = \frac{1}{2}\boldsymbol{z}^T K_I\boldsymbol{z} + \frac{1}{2}(\boldsymbol{q}^{cmd} - \boldsymbol{q})^T K_D(\boldsymbol{q}^{cmd} - \boldsymbol{q}) \tag{21}$$

(21) is the positive definite function on $(\boldsymbol{z},\boldsymbol{q})$. Here the storage function (21) is technical function to prove passivity. The time differentiation of (21) brings

$$\dot{S}_c = (\boldsymbol{q}^{cmd} - \boldsymbol{q})^T\boldsymbol{\tau} - (\boldsymbol{q}^{cmd} - \boldsymbol{q})^T K_P(\boldsymbol{q}^{cmd} - \boldsymbol{q}). \tag{22}$$

This equation means that the PID controller given by (20) is passive between input $(\boldsymbol{q}^{cmd} - \boldsymbol{q} = \boldsymbol{e})$ and $\boldsymbol{\tau}$.

Closed-loop Stability Analysis

In the proposed strategy, the PID parameters are decided so that closed-loop system is asymptotically stable. A proof of asymptotical stabilization is considered based on S. Arimoto [2, 3] and M. W. Spong *et al.* [8]. Here the proposed method is applied to the motion equation of joint space. From (1) and (20), the following equation is obtained.

$$\begin{aligned} &R(\boldsymbol{q})\ddot{\boldsymbol{q}} + \frac{1}{2}\dot{R}(\boldsymbol{q})\dot{\boldsymbol{q}} + S(\boldsymbol{q},\dot{\boldsymbol{q}})\dot{\boldsymbol{q}} + \boldsymbol{g}(\boldsymbol{q}) \\ &- K_P(\boldsymbol{q}^{cmd} - \boldsymbol{q}) - K_I\boldsymbol{z} + K_D\dot{\boldsymbol{q}} = \boldsymbol{0} \end{aligned} \tag{23}$$

Here an equilibrium point of (23) is $(\boldsymbol{z},\boldsymbol{q},\dot{\boldsymbol{q}}) = (K_I^{-1}\boldsymbol{g}(\boldsymbol{q}^{cmd}),\boldsymbol{q}^{cmd},\boldsymbol{0})$ and it is proved that the equilibrium point $(\boldsymbol{z},\boldsymbol{q},\dot{\boldsymbol{q}}) = (K_I^{-1}\boldsymbol{g}(\boldsymbol{q}^{cmd}),\boldsymbol{q}^{cmd},\boldsymbol{0})$ of the closed-loop system (23) is asymptotically stable. Here the following assumptions are satisfied.

[Assumption 1]
PID parameters $K_P,\ K_I,\ K_D$ are positive definite diagonal matrices in (20).
Also they are set to high gains.

[Assumption 2]
It is assumed that K_P is high gain so that it satisfies (24).

$$(\boldsymbol{q}^{cmd} - \boldsymbol{q})^T K_P(\boldsymbol{q}^{cmd} - \boldsymbol{q}) \geq 16(\boldsymbol{q}^{cmd} - \boldsymbol{q})^T \boldsymbol{g}(\boldsymbol{q}) \tag{24}$$

[Assumption 3]
A constant value α is set to a small gain. Also K_D is adjusted to satisfy (25).

$$\alpha K_P - K_I \geq \frac{1}{8}\alpha K_P \tag{25}$$

[Assumption 4]
α is set to a small gain. Also K_D is chosen to satisfy (26).

$$K_D - \alpha R(\boldsymbol{q}) \geq \frac{1}{4}K_D \tag{26}$$

[Proposition 1]
If the PID parameters $K_P,\ K_I,\ K_D$ are chosen to satisfy Assumption 1-4 with some $\alpha > 0$, then the equilibrium point $(\boldsymbol{z},\boldsymbol{q},\dot{\boldsymbol{q}}) = (K_I^{-1}\boldsymbol{g}(\boldsymbol{q}^{cmd}),\boldsymbol{q}^{cmd},\boldsymbol{0})$ of the closed- loop system (23) is asymptotically stable.

(proof) Carrying out the stability analysis, this paper proposes the Lyapunov function candidate $S(\boldsymbol{z},\boldsymbol{q},\dot{\boldsymbol{q}})$ given by (27).

$$\begin{aligned} S(\boldsymbol{z},\boldsymbol{q},\dot{\boldsymbol{q}}) = &\frac{1}{2}\alpha\boldsymbol{z}^T K_I\boldsymbol{z} + \frac{1}{2}\alpha(\boldsymbol{q}^{cmd} - \boldsymbol{q})^T K_D(\boldsymbol{q}^{cmd} - \boldsymbol{q}) \\ &+\boldsymbol{z}^T K_I(\boldsymbol{q}^{cmd} - \boldsymbol{q}) + \frac{1}{2}\dot{\boldsymbol{q}}^T R(\boldsymbol{q})\dot{\boldsymbol{q}} + U(\boldsymbol{q}) \\ &+\frac{1}{2}(\boldsymbol{q}^{cmd} - \boldsymbol{q})^T K_P(\boldsymbol{q}^{cmd} - \boldsymbol{q}) \\ &-\alpha(\boldsymbol{q}^{cmd} - \boldsymbol{q})^T R(\boldsymbol{q})\dot{\boldsymbol{q}} \end{aligned} \tag{27}$$

The Lyapunov function candidate, that is, (27) consists of the total energy function of the closed-loop system given by (23). $\frac{1}{2}\boldsymbol{z}^T K_I\boldsymbol{z} + \frac{1}{2}(\boldsymbol{q}^{cmd} - \boldsymbol{q})^T K_D(\boldsymbol{q}^{cmd} - \boldsymbol{q})$ denotes the storage function of the PID controller shown in (20). Also $\frac{1}{2}\dot{\boldsymbol{q}}^T R(\boldsymbol{q})\dot{\boldsymbol{q}} + U(\boldsymbol{q})$ is the energy of the robot manipulator given by (1). Passivity of robot manipulator can be proved easily, when the time differentiation of $\frac{1}{2}\dot{\boldsymbol{q}}^T R(\boldsymbol{q})\dot{\boldsymbol{q}} + U(\boldsymbol{q})$ is considered by using (1). As a result, passivity of PID controller and robot manipulator is available. Then, if Assumption 1-4 are satisfied, (27) is semi-positive definite. Here the stability theorem of semi-positive definite Lyapunov function [9] is utilized. The time differentiation of (27) along (23) is calculated as (28), by using (3).

$$\begin{aligned}\dot{S}(\boldsymbol{z},\boldsymbol{q},\dot{\boldsymbol{q}}) \;=\; & -(\boldsymbol{q}^{cmd}-\boldsymbol{q})^T\{\alpha K_P - K_I\}(\boldsymbol{q}^{cmd}-\boldsymbol{q}) \\ & -\dot{\boldsymbol{q}}^T\{K_D - \alpha R(\boldsymbol{q})\}\dot{\boldsymbol{q}} \\ & +\alpha(\boldsymbol{q}^{cmd}-\boldsymbol{q})^T\left\{-\frac{1}{2}\dot{R}(\boldsymbol{q}) + \mathcal{S}(\boldsymbol{q},\dot{\boldsymbol{q}})\right\}\dot{\boldsymbol{q}} \\ & +\alpha(\boldsymbol{q}^{cmd}-\boldsymbol{q})^T\boldsymbol{g}(\boldsymbol{q})\end{aligned} \tag{28}$$

Considering Assumption 1-4, (28) is rewritten as

$$\begin{aligned}\dot{S}(\boldsymbol{z},\boldsymbol{q},\dot{\boldsymbol{q}}) \;\leq\; & -\frac{\alpha}{16}(\boldsymbol{q}^{cmd}-\boldsymbol{q})^T K_P(\boldsymbol{q}^{cmd}-\boldsymbol{q}) - \frac{1}{4}\dot{\boldsymbol{q}}^T K_D\dot{\boldsymbol{q}} \\ & +\alpha(\boldsymbol{q}^{cmd}-\boldsymbol{q})^T\left\{-\frac{1}{2}\dot{R}(\boldsymbol{q}) + \mathcal{S}(\boldsymbol{q},\dot{\boldsymbol{q}})\right\}\dot{\boldsymbol{q}}.\end{aligned} \tag{29}$$

Since $R(\boldsymbol{q})$ is constant or a trigonometric function, each entry of $\dot{R}(\boldsymbol{q})$ and $\mathcal{S}(\boldsymbol{q},\dot{\boldsymbol{q}})$ decrease as $\dot{\boldsymbol{q}}$ approaches $\mathbf{0}$. Hence, when $\dot{\boldsymbol{q}}$ approaches $\mathbf{0}$, (28) is satisfying (30).

$$\dot{S}(\boldsymbol{z},\boldsymbol{q},\dot{\boldsymbol{q}}) \leq -\frac{\alpha}{16}(\boldsymbol{q}^{cmd}-\boldsymbol{q})^T K_P(\boldsymbol{q}^{cmd}-\boldsymbol{q}) - \frac{1}{4}\dot{\boldsymbol{q}}^T K_D\dot{\boldsymbol{q}} \tag{30}$$

Finally, invoking the LaSalle's theorem [6], the equilibrium point $(\boldsymbol{z},\boldsymbol{q},\dot{\boldsymbol{q}}) = (K_I^{-1}\boldsymbol{g}(\boldsymbol{q}^{cmd}), \boldsymbol{q}^{cmd}, \mathbf{0})$ of the closed-loop system (23) is asymptotically stable, when dynamic range of manipulator motion is less than natural frequency of mechanical system.
As shown in the above analysis, Assumption 1-4 are designed by PID controller. Because the Lyapunov function candidate (27) consists of the storage function of PID controller. Assumption 2 and 4 show that PID controller can compensate nonlinear terms. Here, in this case, PID controller guarantees local asymptotic stability. However, as redundant degree-of-freedom increases, it becomes easy to decide configuration satisfying Assumption 1. Therefore position control by PID controller is effective use in redundant systems.

Null Space Controller

The null space controller is given as follows.

$$\ddot{\boldsymbol{q}}_{null}^{ref} = J_{vn}^{-1}\left\{K_P(\boldsymbol{q}^{cmd}-\boldsymbol{q}) + K_I\boldsymbol{z} - K_D\dot{\boldsymbol{q}}\right\} \tag{31}$$

where $J_{vn} = I \in R^{n\times n}$ denotes an equivalent inertial matrix. In practical implementation, J_{vn} is gain to adjust acceleration control of each joint depending on a torque limit of joint. (31) shows that PID controller (20) is used as null space controller $\ddot{\boldsymbol{q}}_{null}^{ref}$. PID parameters are decided so that the closed-loop system (23) is asymptotically stable. Here (31) satisfies

Assumptions 1-4 and the null space disturbance can be suppressed by tuning PID parameters.

4. Simulations

In this section, simulations are carried out to confirm the proposed strategy. The kinematic model of 4-link redundant manipulator is shown in Fig.3. Manipulator parameters are summarized in Table 1.

Link	1	2	3	4
Length [m]	0.2650	0.2450	0.1950	0.1785
Mass [kg]	2.7	2.0	1.0	0.4
Reduction ratio	1/100	1/100	1/100	1/100
Resolution of encoder [pulse/rev]	1024	1000	1000	1000
Moment of inertia [kgm²]	0.019	0.0038	0.081	0.043
Torque constant [Nm/A]	20.6	0.213	5.76	4.91

Table 1. Manipulator parameters

The joint angle vector and position vector are defined as $\boldsymbol{q} = (q_1, q_2, q_3, q_4)^T$ and $\boldsymbol{x} = (x, y)^T$ respectively.

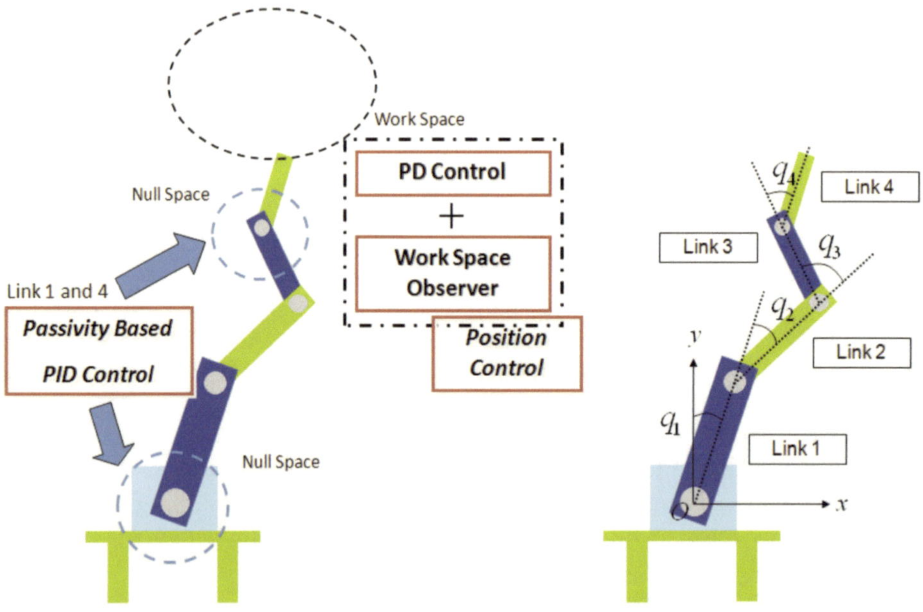

Figure 3. Model of experimental 4-link redundant manipulator

In the simulation, a comparison study of the proposed approach and null space observer is implemented. As described before, the proposed approach is position control in work space and PID control in null space. The compared approach is position control in work space and PD control using null space observer shown in Fig.4. Null space observer is designed based on disturbance observer [7]. Although a structure of null space observer based approach seems PID control, by point of feedback of acceleration disturbance $\hat{\ddot{q}}^{dis}$, it is different from proposed method. The initial configuration of manipulator is as follows.

$$\boldsymbol{q}(0) = \begin{bmatrix} \frac{\pi}{3} & -\frac{\pi}{6} & \frac{\pi}{6} & \frac{\pi}{6} \end{bmatrix}^T$$

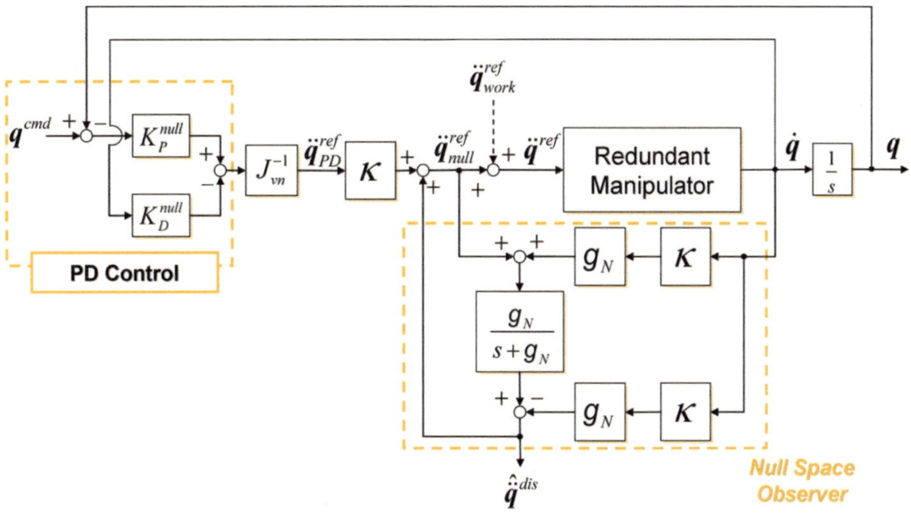

Figure 4. Null space observer based approach

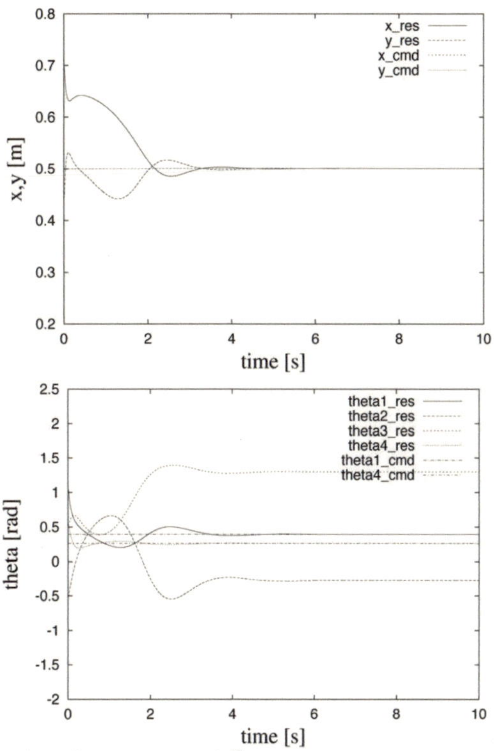

Figure 5. Simulation results of position and θ responses in PID control

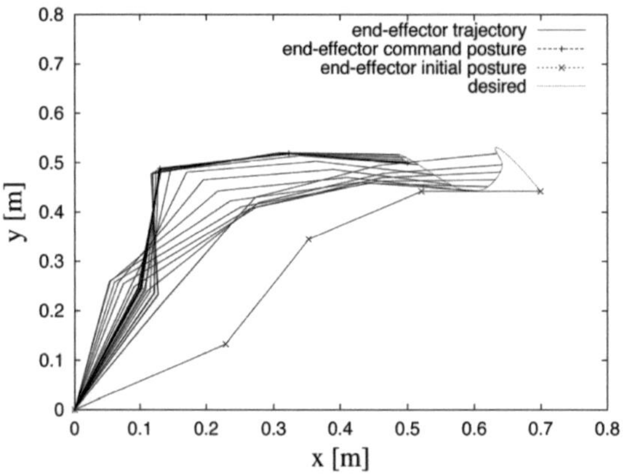

Figure 6. Simulation of manipulator trajectory in PID control

The position commands of end-effector are shown in

$$\boldsymbol{x}^{cmd} = \begin{bmatrix} 0.5 & 0.5 \end{bmatrix}^T, \ \dot{\boldsymbol{x}}^{cmd} = \begin{bmatrix} 0.0 & 0.0 \end{bmatrix}^T.$$

The controller of work space uses the PD controller (15). K_P^{work} and K_D^{work} are given in

$$K_P^{work} = \begin{bmatrix} 400 & 0 \\ 0 & 400 \end{bmatrix}, \ K_D^{work} = \begin{bmatrix} 60 & 0 \\ 0 & 60 \end{bmatrix}.$$

These parameters are decided by a natural frequency and a damping coefficient.
The gain of work space observer g_W is 50rad/s.
PID controller given by (31) is employed in joint 1 and 4.

$$q_1^{cmd} = \frac{\pi}{8}, \ q_4^{cmd} = \frac{\pi}{12}$$

In joint 2 and 3, velocity feedback control called a null space dumping is implemented for stability improvement of null space motion. Then, the command angle is given as a time function so that the joint angles converge to them in 5.0s.

4.1 PID control

PID parameters are as follows.

$$K_P = \text{diag}(1000, 0, 0, 1000), \ K_I = \text{diag}(2500, 0, 0, 2500),$$
$$K_D = \text{diag}(60, 60, 60, 60)$$

First of all, the higher gain K_I is selected according to system noise. Then other parameters are adjusted by using α. Therefore PID parameters satisfy Assumptions 1-4, sufficiently. Fig.5 and 6 show position and θ response.

4.2 Null Space Observer Based Approach

PD controller of null space is shown in

$$\ddot{\boldsymbol{q}}_{PD}^{ref} = J_{vn}^{-1}\left\{K_P^{null}(\boldsymbol{q}^{cmd} - \boldsymbol{q}) - K_D^{null}\dot{\boldsymbol{q}}\right\} \tag{32}$$

where $J_{vn} = I \in R^{n \times n}$ denotes the inertia matrix. PD controller given by (31) is set to joint 1 and 4. In joint 2 and 3, the null space dumping is implemented for stability improvement of null space motion. PD parameters are as follows.

$$K_P^{null} = \text{diag}(400, 0, 0, 400),\ K_D^{null} = \text{diag}(60, 60, 60, 60)$$

They are decided so that the first precision improves. The null space observer gain g_N is 50rad/s. Fig.7 and 8 show position and θ response.

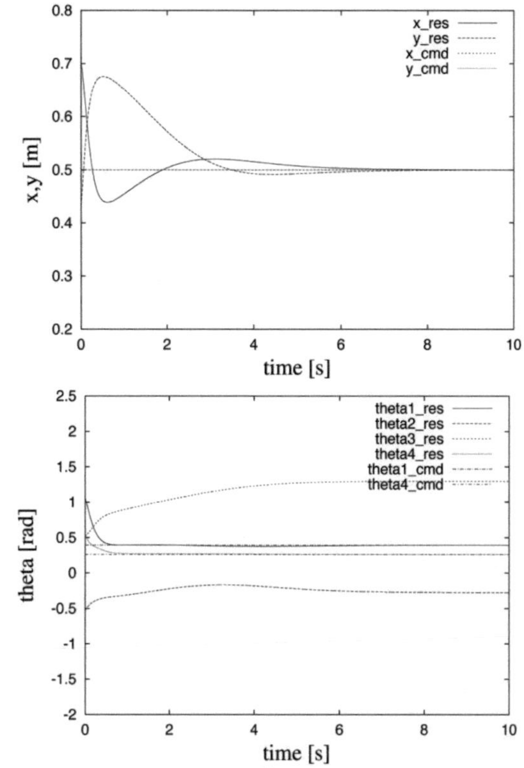

Figure 7. Simulation Results of Position and θ Responses in Null Space Observer Based Approach

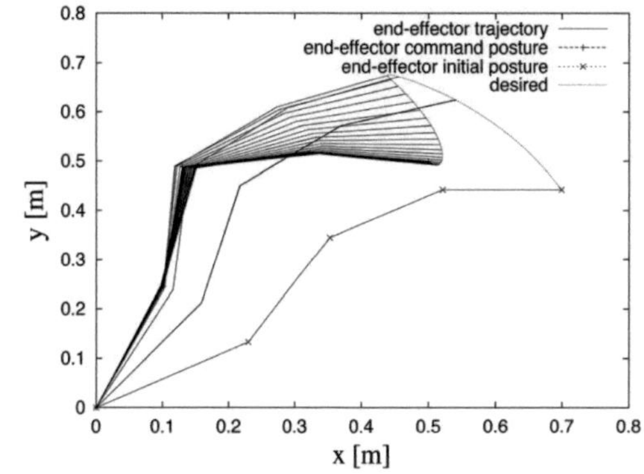

Figure 8. Simulation of Manipulator Trajectory in Null Space Observer Based Approach

4.3 Results and Discussions

In simulation results, the position and θ responses coincided with each command. It is realized that the PID control and null space observer based approach are the same stabilizing ability. Hence, the proposed simple PID controller has an equivalent performance of null space observer.

5. Conclusions

This chapter shows the control design of null space motion by PID controller. When the work space observer is emplyed in work space controller, work space and null space motion are determined independently. Then the PD based work space controller makes work space motion stable, but global stability of null space motion is not always guranteed. To improve the stability and the robustness of null space motion, PID controller considering passivity is useful and the design strategy of PID controller is established for the motion control of null space in this chapter. Several simulations of 4-link manipulator are implemented to confirm the validity of the proposed method. Finally, the feasibility of control design of null space by "classical" PID control and practical/effective use of null space motion are shown.

6. References

T. Murakami, K. Kahlen, and R. W. De Doncker, Robust motion control based on projection plane in redundant manipulator, *IEEE Trans. Ind. Electron.*, vol.49, no.l, pp.248-255, 2002. [1]

S. Arimoto, Control theory of non-linear mechanical systems, *Oxford Science Publications*, 1996. [2]

S. Arimoto, Fundamental problems of robot control: Part I, Innovations in the realm of robot servo-loops, *Robotica*, vol.13, pp.10—27, 1995. [3]

H. K. Khalil, Nonlinear systems, 3rd ed., *NJ:Prentice-Hall*, 2002. [4]

B. Brogliato, R. Lozano, B. Maschke, and O. Egeland, Dissipative systems analysis and control —theory and applications—, *Springer*, 2006. [5]

J. P. LaSalle and S. Lefschetz, Stability by Lyapunov's direct method, *Academic Press*, 1961. [6]

K. Ohnishi, M. Shibata, and T. Murakami, Motion control for advanced mechatronics, *IEEE/ASME Trans. Mech.*, vol.1, no.l, pp.56-67, 1996. [7]

M. W.Spong and M. V. Vidyasagar, *Robot dynamics and control*, John Wiley & Sons, 1989. [8]

R. Sepulchre, M. Jankovic, and P. Kokotovic, *Constructive Nonlinear Control*, Springer, 1997. [9]

23

Paddle Juggling by Robot Manipulator with Visual Servo

Akira Nakashima, Yoshiyasu Sugiyama and Yoshikazu Hayakawa
Nagoya University
Japan

1. Introduction

Juggling is a typical example to represent dexterous tasks of humans. There are many kinds of juggling, e.g., the ball juggling and club juggling in Fig. 1. Other kinds are the toss juggling, bouncing juggling, wall juggling, wall juggling, box juggling, devil sticks, contact juggling and so on. These days, readers can watch movies of these juggling in the website youtube and must be surprised at and admire many amazing dexterous juggling.
The paddle juggling is not major juggling and may be most easy juggling in all kinds of juggling. The paddle juggling means that one hits a ping-pong ball iteratively in the inverse direction of the gravity by a racket as shown in Fig. 2. Most people can easily hit a ball a few times as the left figure since it is easy to hit the ball iteratively. However, most people finally miss out on the juggling as the right figure since it is difficult to control the hitting position. Therefore, realizing the paddle juggling by a robot manipulator is challenging and can lead to solution of human dexterity.
As mentioned previously, the paddle juggling of a ball by a robot manipulator is composed of three parts: the first part is to iterate hitting the ball, the second part is to regulate the incident angle of the ball to the racket, and the third part is to regulate the hitting position and the height of the hit ball. M. Buehler et. al. (Buehler, Koditschek, and Kindlmann 1994) proposed the mirror algorithms for the paddle juggling of one or two balls by a robot having one degree of freedom in two dimensional space, where the robot motion was symmetry of the ball motion with respect to a horizontal plane. This method achieved hitting the ball iteratively. R. Mori et. al. (R. Mori and Miyazaki 2005) proposed a method for the paddle juggling of a ball in three dimensional space by a racket attached to a mobile robot, where the trajectory of the mobile robot was determined based on the elevation angle of the ball. This method achieved hitting the ball iteratively and regulating the incident angle. These algorithms are simple and effective for hitting the ball iteratively. However, their method does not control the hitting position and the height of the ball. On the other hand, S. Schaal et. al. (Schaal and Atkeson 1993) proposed a open loop algorithm for one-dimensional paddle juggling of one ball. S Majima et. al. (Majima and Chou 2005) proposed a method for the paddle juggling of one ball with the receding horizon control base on the impulse information of the hit ball. Their methods achieved regulating the height of the ball. However, the control problem is only considered in one dimensional space. This study propose a method to achieve paddle juggling of one ball by a racket attached to a robot

manipulator with two visual camera sensors. The method achieve hitting the ball iteratively and regulating the incident angle and the hitting position of the ball. Furthermore, the method has the robustness against unexpected disturbances for the ball. The proposed method is composed of juggling preservation problem and ball regulation problem. The juggling maintenance problem means going on hitting the ball iteratively. The ball regulation problem means regulating the position and orientation of the hit ball. The juggling preservation problem is achieved by the tracking control of the racket position for a symmetry trajectory of the ball with respect to a horizontal plane. This is given by simply extending the mirror algorithm proposed in (Buehler, Koditschek, and Kindlmann 1994). The ball regulation problem is achieved by controlling the racket orientation, which is determined based on a discrete transition equation of the ball motion. The effectiveness of the proposed method is shown by an experimental result.

Figure 1. Examples of Juggling: Ball Juggling and Club Juggling

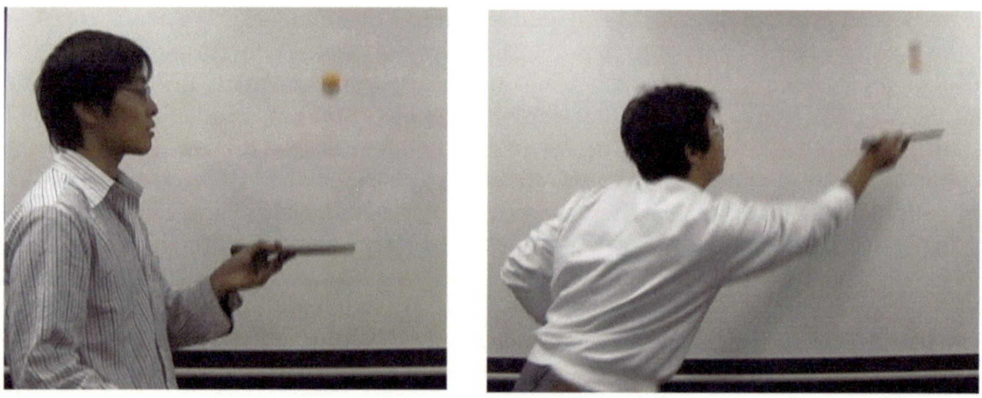

Figure 2. Paddle Juggling: Stable and Unstable Juggling

Section 2 shows the experimental equipments. In Section 3, the system configuration and the models of the manipulator and the ball motion are shown. As for the preliminary of the paddle juggling, the manipulator dynamics are linearized in Section 4. In Section 5, the control designs for the juggling preservation and the ball regulation problems are shown.

An experimental result is shown in Section 6 and the conclusion in this paper and our further Manipulator, works are described in Section 7.

2. Experiment System

We use two CCD cameras and a robot manipulator as shown in Fig. 3 for realizing the paddle juggling. With these equipments, the experiment system is constructed as Fig. 4 composed of a robot manipulator, DOS/V PC as a controller, two CCD cameras, an image processing implement, two image displays and a ping-pong ball. The robot manipulator has 6 joints J_i (i = 1, ... ,6), to which AC servo motors with encoders are attached. The motors are driven by the robot drivers with voltage inputs corresponding the motor torques from the PC through the D/A board. The joint angles are calculated by counting the encoder pluses with the counter board. On the other hand, the image processing implement can detect pixel and calculate the center point of the detected pixel area in two assigned squared areas with thresholds of the brightness. Furthermore, the assigned squared areas can track the calculated center points such that the points are within the areas. The tracked center points are sent to the PC as the image futures. The image data and the image futures can be watched by the two image displays. The cameras are located as the image plane corresponding to the CCD camera include the robot manipulator. Since the ball is very smaller than the robot, the image pixel area of the ball in the image plane has the almost same size as the ball. Therefore, we put the tracking squared areas to the detected pixel of the ball and calculate the position of the ball from the tracked image futures.
The robot is made by DENSO, Inc.. The physical parameters of the robot are shown in Fig 5. J_i and W_i (i = 1, ... ,6) denote the ith joint and the center of mass of the ith link respectively. The values of W_{1r}, W_2, W_3, W_4, W_{56} are 6.61, 6.60, 3.80, 2.00 and 2.18 [kg] respectively. The rated torques are 0.25[N-m] (J_1-J_3) and 0.095[N-m] (J_4-J_6). The reduction gear ratios are 241.6, 536, 280, 200, 249.6 and 139.5. The racket is a square plate made from aluminum. The sides, the thickness and the mass are 150, 3[mm] and 0.20[kg] respectively. The ball is a ping-pong ball with the radius and the mass being 35[mm] and 2.50 x 10^{-3}[kg].
The image processing implement is QuickMAG (OKK, Inc.) and the CCD cameras are CS8320 (Tokyo Electronic Equipment, Inc.). The pixel and the focal length of the each camera are 768(H) x 498(W) and 4.8[mm] respectively. The sampling period is 120[Hz]. The PC for the QuickMAG is the FMV-DESK POWER S III 167 (Fujitsu, Inc.). The QuickMAG can track boxed areas to two targets simultaneously and the boxes can be resized easily.

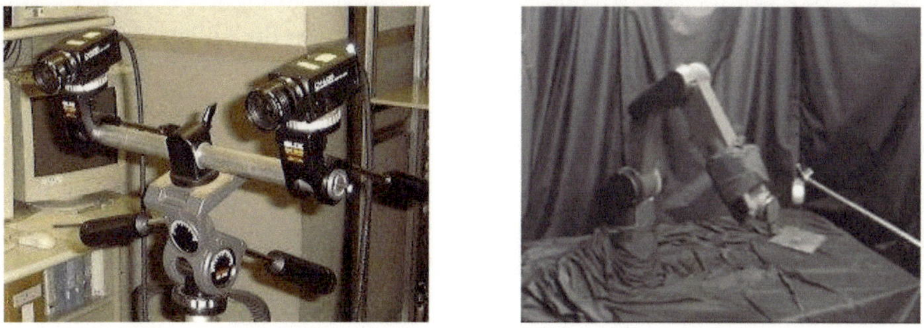

Figure 3. Two CCD cameras and a Robot

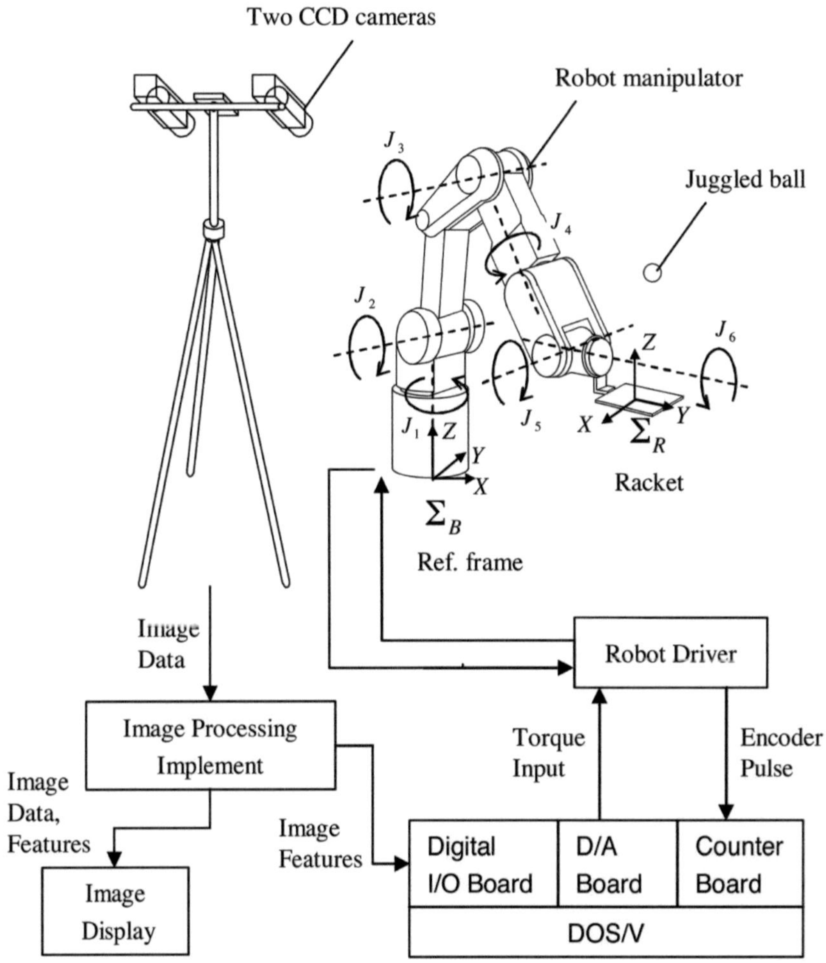

Figure 4. Padde Juggling of a Ball by Camera-Robot System

The PC is the DELL OptiPlex Gxi (CPU:Celeron IGHz, Memory: 256MB). The control method is programmed in C language with Borland C++ 5.5. The D/A board is the DA12-8 (PC) with 12 bit resolution and 9 ch (PC) (CONTEC, Inc.). The counter board is the CNT24-4 (PC) with up-down count in 4 channels (CONTEC, Inc.). The digital board is the IBX-2752C with the 32 number of I/O (Interface, Inc.).

3. Modeling

3.1 System Configuration

In this paper, we consider control of paddle juggling of one ball by a racket attached to a robot manipulator with two-eyes camera as shown in Fig. 4. In this paper, the juggled ball is supposed to be smaller and lighter than the racket. The reference frame Eg is attached at the base of the manipulator. The position and orientation of the racket are represented by the frame Σ_R, which is attached at the center of the racket. The camera is calibrated with respect to Σ_B. Therefore, the position of the juggled ball can be measured as the center of the ball which the two cameras system detects. (B. K. Ghosh and Tarn 1999). It is obvious that the orientation of the racket about z-axis with respect to Eg does not effect on the ball motion at the time when the racket hits the ball. Therefore we need only 5 degrees of freedom of the manipulator. In the following discussion, the fourth joint J_4 is assumed to be fixed by an appropriate control.

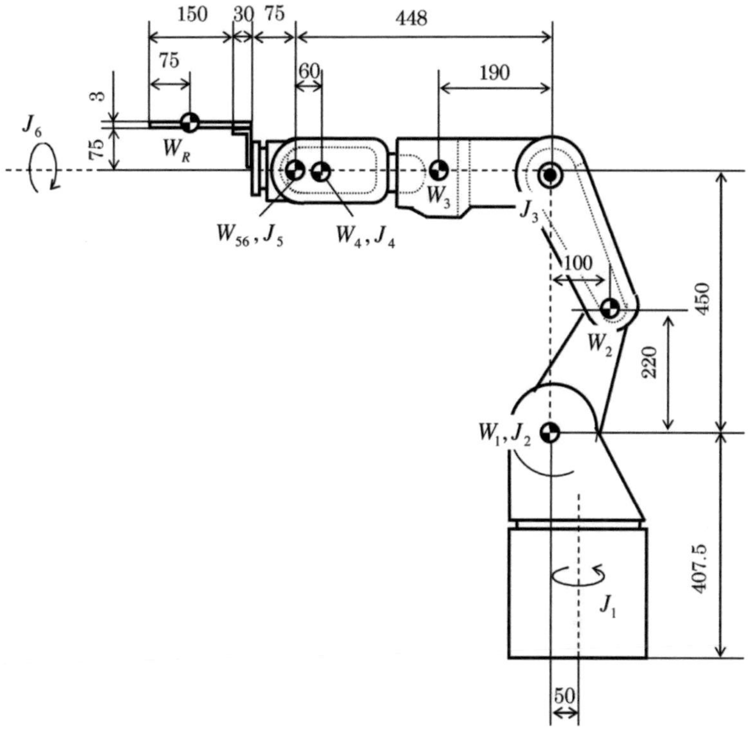

Figure 5. The physical parameters of the manipulator

3.2 Dynamical Equation of Manipulator

The dynamical equation of the manipulator is given by

$$M(q)\ddot{q} + C(q, \dot{q})\dot{q} + N(q) = \tau, \tag{1}$$

where $q \in \mathbb{R}^5$ and $\tau \in \mathbb{R}^5$ describe the joint angles and the input torques respectively, and $M \in \mathbb{R}^{5\times5}$, $C \in \mathbb{R}^{5\times5}$ and $N \in \mathbb{R}^5$ are the inertia matrix, the coriolis matrix and the gravity term respectively. For the following discussion, we introduce the following coordinate transformations:

$$^{B}p_R := h_1(q) \in \mathbb{R}^3, \ ^{B}\theta_R := h_2(q) \in \mathbb{R}^2, \tag{2}$$

where $^{B}p_R = [\ ^{B}p_{Rx}\ ^{B}p_{Ry}\ ^{B}p_{Rz}\ ]^{\mathrm{T}} \in \mathbb{R}^{3_3}$ is the position of Σ_R with respect to Σ_B, $[\ ^{B}\theta_{Rx}\ ^{B}\theta_{Ry}\ ]^{\mathrm{T}} \in \mathbb{R}^2$ is the orientation of Σ_R; with respect to x- and y- axes of Σ_B. The left superscript B of the vector denotes that the vectors are expressed with respect to the frame Σ_B. This notation is utilized for another frames in the following. The functions $\boldsymbol{h}_1(\boldsymbol{q})$ and $\boldsymbol{h}_1(\boldsymbol{q})$ describe the relationships between the position/orientation of the racket and the joint angles respectively. Differentiating (2) with respect to time t and getting together the resultant equations yield the velocity relationship

$$\dot{\boldsymbol{x}}_R = \boldsymbol{J}_R(\boldsymbol{q})\dot{\boldsymbol{q}}, \tag{3}$$

where

$$\boldsymbol{J}_R := [\,(\partial \boldsymbol{h}_1/\partial \boldsymbol{q})^{\mathrm{T}}\ (\partial \boldsymbol{h}_2/\partial \boldsymbol{q})^{\mathrm{T}}\,]^{\mathrm{T}} \in \mathbb{R}^{5\times 5}$$
$$\boldsymbol{x}_R := [\,{}^B\boldsymbol{p}_R^{\mathrm{T}}\ {}^B\boldsymbol{\theta}_R^{\mathrm{T}}\,] \in \mathbb{R}^5.$$

Applying the relationship (3) for the dynamical equation (1) yields the dynamical equation with respect to the position and orientation of the racket

$$\boldsymbol{M}_R(\boldsymbol{q})\ddot{\boldsymbol{x}}_R + \boldsymbol{C}_R(\boldsymbol{q},\dot{\boldsymbol{q}})\dot{\boldsymbol{x}}_R + \boldsymbol{N}_R(\boldsymbol{q}) = \boldsymbol{J}_R(\boldsymbol{q})^{-\mathrm{T}}\boldsymbol{\tau}, \tag{4}$$

where

$$\boldsymbol{M}_R := \boldsymbol{J}_R^{-\mathrm{T}}\boldsymbol{M}\boldsymbol{J}_R^{-1} \in \mathbb{R}^{5\times 5}$$
$$\boldsymbol{C}_R := \boldsymbol{J}_R^{-\mathrm{T}}\boldsymbol{C}\boldsymbol{J}_R^{-1} + \boldsymbol{J}_R^{-\mathrm{T}}\boldsymbol{M}\tfrac{d}{dt}\boldsymbol{J}_R^{-1} \in \mathbb{R}^{5\times 5}$$
$$\boldsymbol{N}_R := \boldsymbol{J}_R^{-\mathrm{T}}\boldsymbol{N} \in \mathbb{R}^5.$$

3.3 Equation of Motion and Rebound Phenomenon of Ball

For the modeling of the equation of motion and the rebound phenomenon of the ball, we make the following assumptions:

[Assumption 1] There does not exist the air resistance for the ball.

[Assumption 2] The rebound phenomenon between the ball and the racket is the perfectly elasticity. The surfaces of the ball and the racket are uniform, therefore, the restitution coefficients are isotopic without the incident direction.

[Assumption 3] The mass of the racket is bigger than the mass of the ball such that the racket velocity does not change due to the rebound phenomenon.

Define the ball position by ${}^B\boldsymbol{p}_b := [\,{}^Bp_{bx}\ {}^Bp_{by}\ {}^Bp_{bz}\,]^{\mathrm{T}} \in \mathbb{R}^3$ From Assumption 1, the equation of motion of the ball is given by

$${}^B\dot{p}_{bx} = v_{b_{xy}}\cos\theta,\ {}^B\dot{p}_{by} = v_{b_{xy}}\sin\theta, \tag{5}$$

$${}^B\ddot{p}_{bz} = -mg, \tag{6}$$

where $v_{b_{xy}}$ and θ are the velocity and the direction angle in the (x,y) plane respectively, m[kg] is the mass of the ball and g[m/s^2] is the gravitational constant. Note that (5) represents the ball motion in the (x, y) plane, which is the uniform motion, and $v_{b_{xy}}$ and θ are determined by the motion and the orientation of the racket at the hitting. From Assumption 2 and 3, the mathematical model of the rebound phenomenon is given by

$${}^B\dot{\boldsymbol{p}}_b(t_r+dt) = \boldsymbol{R}_{BR}\boldsymbol{E}\boldsymbol{R}_{BR}^{\mathrm{T}}({}^B\dot{\boldsymbol{p}}_b(t_r) - {}^B\dot{\boldsymbol{p}}_R(t_r)) + {}^B\dot{\boldsymbol{p}}_R(t_r), \tag{7}$$

where t_r is the start time of the collision between the ball and the racket, dt is the time interval of the collision, $\boldsymbol{R}_{BR} \in \mathbb{R}^{3\times3}$ is the rotation matrix from Σ_R to Σ_B and $\boldsymbol{E} := \mathrm{diag}(e_{xy}, e_{xy}, -e_z) \in \mathbb{R}^{3\times3}$ represents the restitution coefficients of the x, y and z directions with respect to Σ_R.

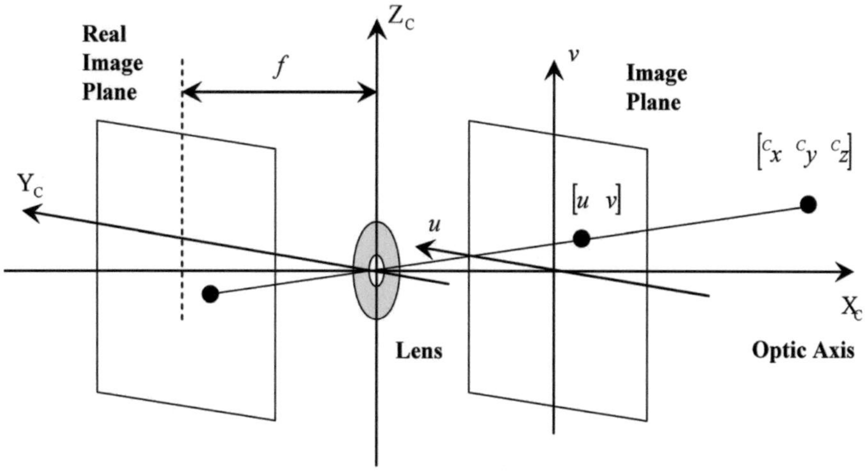

Figure 6. Perspective Transformation Model of Lens

3.4 Perspective Transformation Model of Cameras

Figure 6 shows the perspective transformation of a lens as the pin-hole camera. The camera frame Σ_C is attached the center of the lens and a point p is denoted by a position vector ${}^C\boldsymbol{p} := [{}^Cx\ {}^Cy\ {}^Cz]^{\mathrm{T}} \in \mathbb{R}^3$ expressed in Σ_C. The point p is projected on the real image plane with the focal length f in the inverse direction of the X_C axis. For simplicity, the image feature vector $\boldsymbol{\xi} := [u\ v]^{\mathrm{T}} \in \mathbb{R}^2$ is defined as the coordinates of the projected point on the image plane with the focal length f in the direction of the X_C axis. From the geometric view point, the relationship between the image feature $\boldsymbol{\xi}$ and the position ${}^C\boldsymbol{p}$ is given by

$$\begin{bmatrix} u \\ v \end{bmatrix} = \begin{bmatrix} \dfrac{f}{\alpha_u} \dfrac{{}^Cy}{{}^Cx} \\ \dfrac{f}{\alpha_v} \dfrac{{}^Cz}{{}^Cx} \end{bmatrix}, \tag{8}$$

where α_u and α_v [m/pixel] are the lengths of the unit pixel in the u and v directions. These camera parameters are f = 8.5 x 10~3[m], α_u = 8.4 x 10~7 and α_v = 9.8 x 10^7 [m/pixel].

4. Linearizing Compensator for Manipulator

As for the preliminary to control the paddle juggling, we linearize the manipulator dynamics (4) by the following linearizing compensator:

$$\boldsymbol{\tau} = \boldsymbol{J}_R^{\mathrm{T}}(\boldsymbol{M}_R \boldsymbol{u}_R + \boldsymbol{C}_R \dot{\boldsymbol{x}}_R + \boldsymbol{N}_R), \tag{9}$$

where $\boldsymbol{u}_R := [\ \boldsymbol{u}_{Rp}^{\mathrm{T}}\ \ \boldsymbol{u}_{R\theta}^{\mathrm{T}}\ ]^{\mathrm{T}} \in \mathbb{R}^5$ is the new input for $\ddot{\boldsymbol{x}}_R$. Substituting (9) into (4) results in

$$M_R(\ddot{x}_R - u_R) = 0.$$

Since $M_R = J_R^{-T} M J_R^{-1}$ is the positive definite matrix because the inertia matrix M is positive definite, M_R always has the inverse matrix. Therefore, we get the linearized equations given by

$$^B\ddot{p}_R = u_{Rp}, \tag{10}$$

$$^B\ddot{\theta}_R = u_{R\theta}. \tag{11}$$

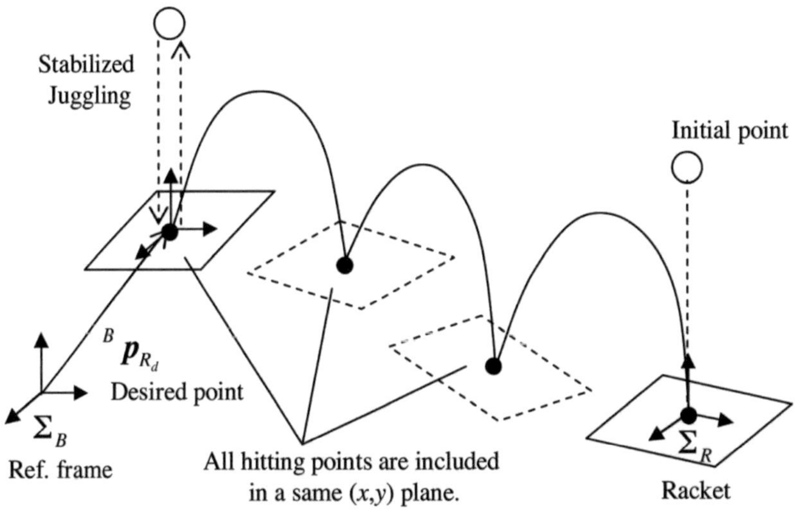

Figure 7. The concept for Padde Juggling of one Ball

The tracking control of $^B p_R$ and $^B\theta_R$ can be easily realized by PD controller for u_{Rp} and $u_{R\theta}$, for example. In the following discussion, let us consider the desired values of $^B p_R$ and $^B\theta_R$ to realize the paddle juggling of the ball.

5. Control Design for Paddle Juggling

5.1 Control Objectives

Figure 7 illustrates the control purpose. The initial point of the ball is above the racket and the ball is freely released. The control purpose is to achieve the paddle juggling of the ball at the desired hitting point $^B p_{R_d}$ by hitting the ball with the racket iteratively. This is described by the following specific control problems:

1. [Juggling Preservation Problem]
   The juggling maintenance problem means going on hitting the ball iteratively. The control of the ball position is not included in this problem.
2. [Ball Regulation Problem]
   The ball regulation problem means regulating the hitting position of the hit ball. The regulation of the ball height is not included in this problem.

In the following sections, the control designs for these control problems are shown.

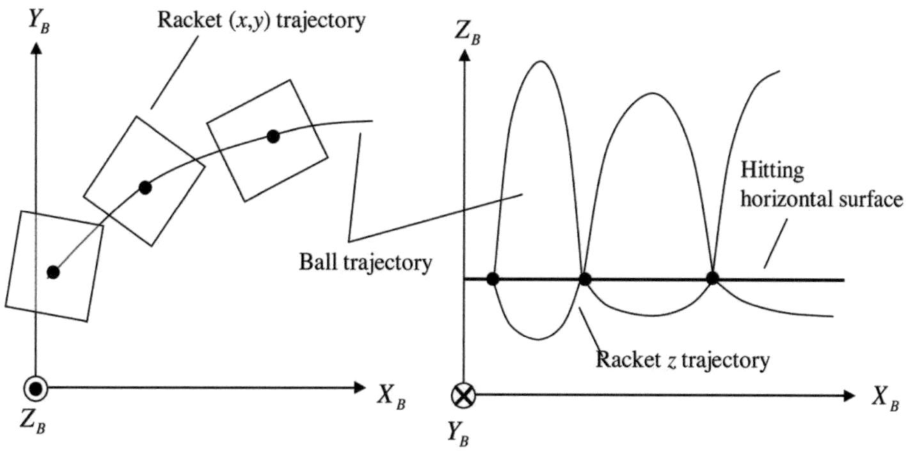

(a) Trajectory of (*x*,*y*) (b) Smaller opposite trajectory of *z*

Figure 8. Control Scheme for the Racket Position Control

5.2 Control Design for Juggling Preservation Problem

The control design for the juggling preservation problem with controlling the racket position ${}^{B}\boldsymbol{p}_{R}$ is shown. We firstly consider that the ball is always hit by the racket in a same (*x*,*y*) plane as shown in Fig. 7. This plane is called the hitting horizontal plane in this paper. More specific determination of the racket position is shown in Fig. 8. As in the left figure (a), the *x* and *y* coordinates of the racket are forced to follow the *x* and *y* coordinates of the ball. On the other hand, as in the right figure (b), the *z* coordinate of the racket is forced to follow the symmetric trajectory of the *z* coordinate of the ball with respect to the hitting horizontal plane. The desired value of the racket position ${}^{B}\boldsymbol{p}_{R_d}$ to satisfy the mentioned in the above is given by

$$ {}^{B}\boldsymbol{p}_{R_d} := \begin{bmatrix} {}^{B}p_{b_x} \\ {}^{B}p_{b_y} \\ {}^{B}p_h - k_h({}^{B}p_{b_z} - {}^{B}p_h) \end{bmatrix}, \tag{12} $$

where ${}^{B}p_h$ is the height of the hitting horizontal plane and $0< k_h <1$ the constant to make the *z* coordinate of the racket small appropriately. Note that effects on the height of the hitting ball.

Due to determining the desired value of the racket position ${}^{B}\boldsymbol{p}_{R}$ by (12), there always exists the racket under the ball and the ball is hit in the same horizontal plane automatically. This determination of the racket position is based on the mirror algorithm (Buehler, Koditschek, and Kindlmann 1994).

5.3 Control Design for Ball Regulation Problem

For preliminary, we introduce a frame based on the states of the ball at the hitting. In Fig. 9 (a), the origin of Σ_H is the hitting position. The Y_H- axis is defined as the direction of the ball motion in the (*x*,*y*)-plane at the hitting, and the Z_H- axis is defined as the same direction of Z_B of Σ_B. The X_H- axis is defined such that Σ_H is the right-handed frame. θ_H and θ_R denote the angles of Y-axes of Σ_H and Σ_R respectively. The relationship between these two frames is represented by the rotation matrix

$$\boldsymbol{R}_{RH} := \begin{bmatrix} \cos(\theta_R - \theta_H) & \sin(\theta_R - \theta_H) \\ -\sin(\theta_R - \theta_H) & \cos(\theta_R - \theta_H) \end{bmatrix}, \tag{13}$$

where

$$\theta_H := \sin^{-1}\left(\frac{\dot{p}_{b_y}}{\sqrt{\dot{p}_{b_x}^2 + \dot{p}_{b_y}^2}}\right). \tag{14}$$

We consider the racket orientation ${}^H\boldsymbol{\theta}_R := [\ {}^H\theta_{R_x}\ \ {}^H\theta_{R_y}\ ]^{\mathrm{T}}$ with respect to Σ_H as shown in Fig. 9 (b).

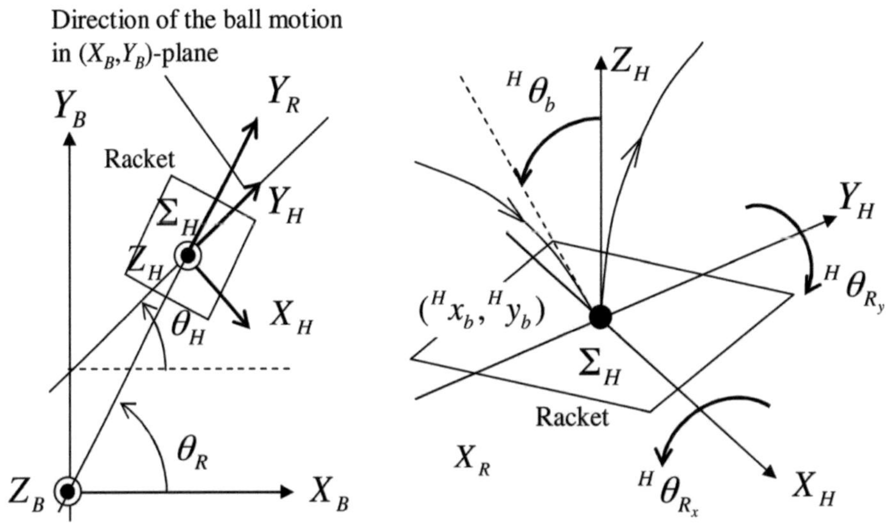

(a) Frame Σ_H defined at hitting (b) Racket orientation w.r.t. Σ_H

Figure 9. Racket Orientation w. r. t. Σ_H Denned by Ball Sates at Hitting

In the frame Σ_H defined above, let us consider the ball regulation problem by the racket orientation ${}^H\boldsymbol{\theta}_R$. For the derivation simplicity of the transition equation of the hit ball, the translational velocity of the racket is assumed to be zero. The control variables are the incidence angle ${}^H\theta_b[i]$ from the Z_H- axis and the position $({}^Hx_b[i], {}^Hy_b[i])$ at the hitting point as shown in Fig. 9 (b). The constant i (i = 0,1,2, ...) represents the number of the hitting. Note that ${}^H\theta_{R_x}$ effects on both $({}^Hy_b, {}^H\theta_b)$, and ${}^H\theta_{R_y}$ effects on only ${}^Hx_b$. Therefore, the ball transition by the hitting can be separated to the Y_H- and X_H- directions illustrated in Fig. 10. The figure (a) shows the transition of the incident angle ${}^H\theta_b$ and the position ${}^Hy_b$, and the figure (b) shows the transition of the position ${}^Hx_b$. From (5)-(7), the relationships between $({}^Hx_b[i], {}^Hy_b[i], {}^H\theta_b[i])$ and $({}^Hx_b[i+1], {}^Hy_b[i+1], {}^H\theta_b[i+1])$ are given by

$${}^Hx_b[i+1] \ = \ {}^Hx_b[i] + \frac{(v_{b_0}\cos{}^H\theta_b[i])^2 \sin 4\,{}^H\theta_{R_y}[i]}{g} \tag{15}$$

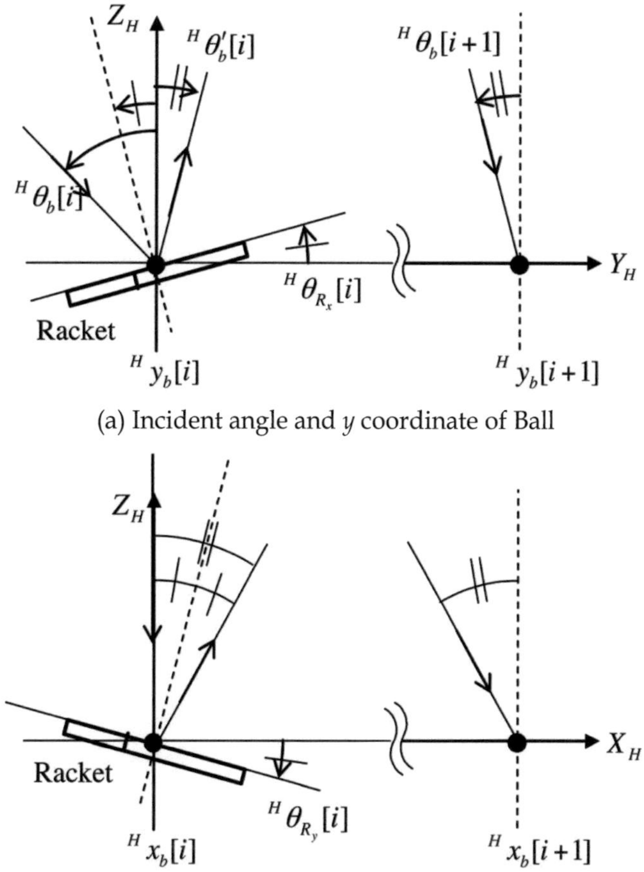

(a) Incident angle and y coordinate of Ball

(b) x coordinate of Ball

Figure 10. The ball state transition Due to the hitting

$$^Hy_b[i+1] = {}^Hy_b[i] - \frac{v_{b_0}^2 \sin 2^H\theta_b'[i]}{g} \tag{16}$$

$$^H\theta_b[i+1] = -^H\theta_b'[i], \tag{17}$$

where v_{b_0} is the magnitude of the velocity of the ball just before the hit and

$$^H\theta_b'[i] = -^H\theta_b[i] + 2^H\theta_{R_x}[i] \tag{18}$$

is the reflection angle of the ball after the hit. Since the initial $^H\theta_b[0] = 0$ from the control purpose and the desired value $^H\theta_{b_d} = 0$, the reflection angle $^H\theta_b'[i]$ can be small enough to assume that cos $^H\theta_b'[i] \simeq 1$ and sin $^H\theta_b'[i] \simeq \theta_b'[i]$. Due to this approximation, we get the linear discrete transition equation of the ball given by

$$\begin{bmatrix} {}^H x_b \\ {}^H y_b \\ {}^H \theta_b \end{bmatrix}[i+1] = \begin{bmatrix} 1 & 0 & 0 \\ 0 & 1 & \frac{2v_{b0}^2}{g} \\ 0 & 0 & 1 \end{bmatrix} \begin{bmatrix} {}^H x_b \\ {}^H y_b \\ {}^H \theta_b \end{bmatrix}[i] + \begin{bmatrix} 0 & \frac{4v_{b0}^2}{g} \\ -\frac{4v_{b0}^2}{g} & 0 \\ -2 & 0 \end{bmatrix} \begin{bmatrix} {}^H \theta_{R_x} \\ {}^H \theta_{R_y} \end{bmatrix}[i]. \tag{19}$$

Note that the sampling times between the states $[i]$ and $[i+1]$ ($i = 0,1,2, \ldots$) do not equal to each other. From the transition equation (19), the controllers for ${}^H\theta_{R_x}[i]$ and ${}^H\theta_{R_y}[i]$ to stabilize $({}^H x_b[i], {}^H y_b[i], {}^H\theta_b[i])$ are easily derived as

$$\begin{cases} {}^H\theta_{R_x}[i] = -k_{py}({}^H y_b[i] - {}^H y_{b_d}[i]) - k_\theta {}^H\theta_b[i] \\ {}^H\theta_{R_y}[i] = -k_{px}({}^H x_b[i] - {}^H x_{b_d}[i]) \end{cases}, \tag{20}$$

where ${}^H x_{b_d}[i]$ and ${}^H y_{b_d}[i]$ are the desired values expressed in $\Sigma_H[i]$, and k_{px},k_{py} and $k_\theta > 0$ are the control gains to be determined such that the eigenvalues of the closed system is smaller thanl. Note that it is impossible to get ${}^H x_b[i]$, ${}^H y_b[i]$, ${}^H\theta_b[i]$, ${}^H x_{b_d}[i]$ and ${}^H y_{b_d}[i]$ because these variables are the states of the ball at the hitting. Therefore, we have to use a prediction for these variables as the following section.

5.4 Prediction of Ball State

As mentioned in Section 5.3, from (20), we need the ball velocity just before the ith hit $v_{bx}[i]$, $v_{by}[i]$ and $v_{bz}[i]$ to calculate the incident angle ${}^H\theta_b[i]$. In this subsection, a simple prediction method of these variables is proposed.

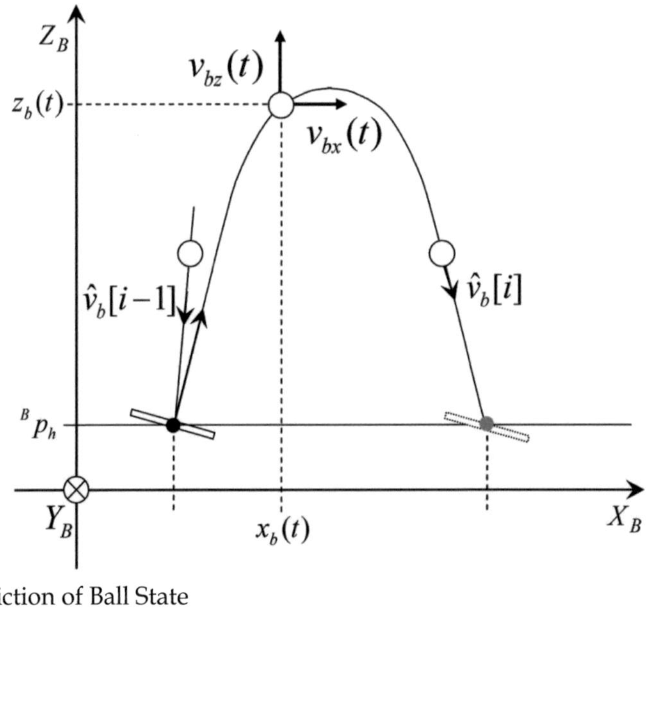

Figure 11. Prediction of Ball State

As shown in Fig. 11, ($x_b(t)$, $y_b(t)$, $z_b(t)$) are the ball position and ($v_{bx}(t)$,$v_{by}(t)$,$v_{bz}(t)$) are the ball velocity at time t between the $i-$ 1th and ith hits. $y_{bx}(t)$ and $v_{by}(t)$ are omitted for simplicity. Suppose the position and velocity of the ball can be detected by the visual sensor. From Assumption 1, the velocity ($v_{bx}[i]$, $v_{by}[i]$) at the ith hit are the same as ($v_{bx}(t)$,$v_{bz}(t)$). For the velocity $\hat{v}_{bz}[i]$, the predicted z velocity $v_{bz}[i]$ can be obtained from $v_{bz}(t)$ and $z_b(t)$ before the ith hit based on the energy conservation law:

$$\hat{v}_{bz}[i] = \sqrt{g(z_b(t) - p_h) + v_{bz}^2(t)}. \tag{21}$$

From this predicted velocity $\hat{v}_{bz}[i]$, the incident angle ${}^H\theta_b[i]$ is calculated as

$${}^H\hat{\theta}_b[i] = \tan^{-1}\frac{\sqrt{v_{bx}^2[i] + v_{by}^2[i]}}{v_{bz}[i]}. \tag{22}$$

Note that it is sufficient to replace ${}^Hx_b[i]$ and ${}^Hy_b[i]$ to ${}^Hx_b(t)$ and ${}^Hy_b(t)$ in the racket angle controller (20) because ${}^Hx_b(t)$ and ${}^Hy_b(t)$ equal ${}^Hx_b[i]$ and ${}^Hy_b[i]$ when the racket hits the ball. Therefore, it is not necessary to predict ${}^Hx_b[i]$ and ${}^Hy_b[i]$.

6. Experimental Result

The effectiveness of the proposed method is shown by a experimental result. In the control design, we need the variables ${}^Hx_b[i]$, ${}^Hy_b[i]$, ${}^H\theta_b[i]$, ${}^Hx_{b_d}[i]$ and ${}^Hy_{b_d}[i]$, which are calculated by the position and the velocity of the ball at the hit. As for a simple prediction of these variables, we use the position and velocity of the ball being the height 0.015[m] from the hitting horizontal plane as those of the ball at the hitting. The initial state of the robot manipulator is set such that the racket is horizontal as shown in Fig. 4. The initial position of the ball is above the racket. The desired value of the hitting position is set to ${}^B\boldsymbol{p}_H = [0.35\ -0.65\ 0.35]^T$ [m]. The control gains are set to $k_{p_x} = k_{p_y} = 20 \times \frac{\pi}{180}$ [rad/m] and k_θ = 1.25. The control gain k_h effecting on the ball height is set to

$$k_h = \begin{cases} 0.275 & (E_b \le \bar{E}_b) \\ 0.275 - \frac{0.75}{g}(E_b - \bar{E}_b) & (E_b > \bar{E}_b) \end{cases},$$

where

$$E_b := ({}^Bp_{b_z} - {}^Bp_h)g + \frac{1}{2}\|{}^B\boldsymbol{p}_b\|^2,\ \bar{E}_b := {}^B\bar{p}_{br}g.$$

E_b, is the total energy of the unit mass of the ball and ${}^B\bar{p}_{br}$ is the threshold value for the ball height. In this experiment, ${}^B\bar{p}_{br}$ is set to ${}^B\bar{p}_{br}$ = 0.070 [m].

The time interval to make the experiment is set to 100[s]. The experimental result is shown in Fig. 12. The left figure shows the trajectories of the ball and racket positions in the time interval 0-20[s]. The solid and the dashed trajectories represent the ball and the racket respectively. The solid horizontal lines are the desired hitting value. We can confirm that the x and y coordinates of the racket follow those of the ball and the z coordinate of the racket is symmetric to the ball with respect to the horizontal surface ${}^Bp_h$ = 0.35[m]. This result shows the effectiveness of the juggling preservation by controlling the racket position. On the other hand, the right figure shows the ball trajectory in the (x,y) plane. The white circle is the initial position and the black circle is the desired hitting point. The big circle having the radius 0.05[m] is depicted with the center begin the desired hitting position. We can confirm that the most of the ball trajectory is included in the small circle. This result shows the effectiveness of the ball regulation problem by controlling the racket orientation. However, the average x and y positions have the errors 1.4[cm] and 3.6[cm] from the desired hitting position. As for this reason, the calculation errors and the uncertainty of the robot model can effect on the errors of the hitting position. On the other hand, the spread of the ball can be caused by the air resistance.

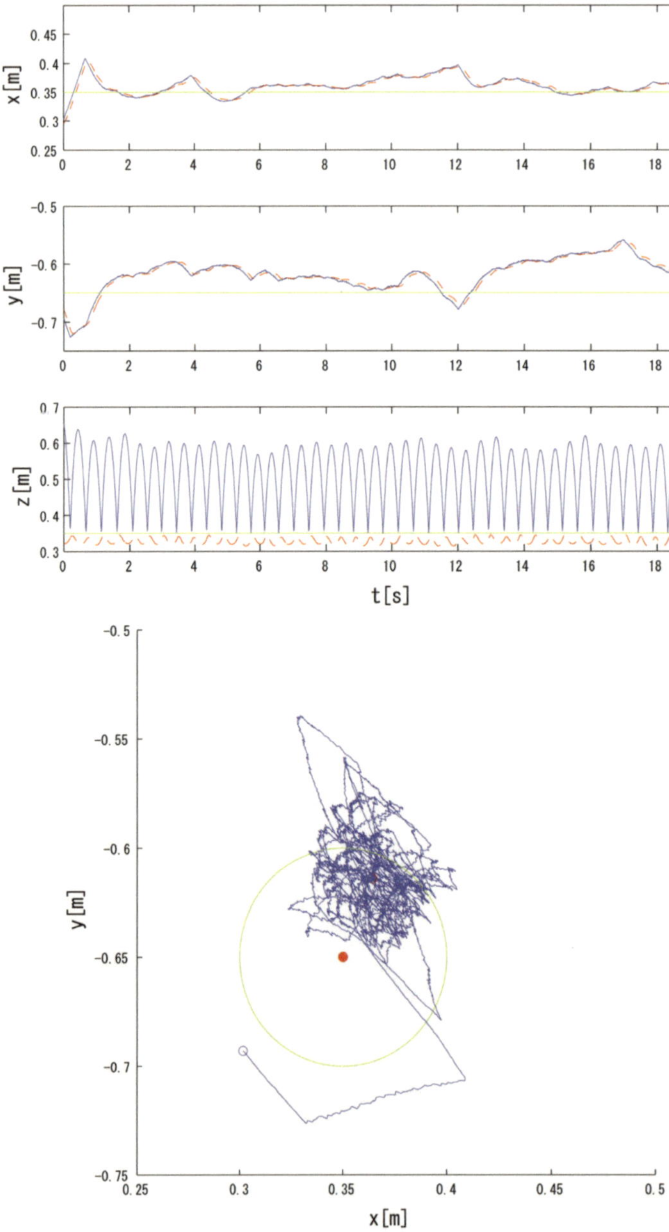

Figure 12. Experimental result

Next, the experiment with the disturbances for the ball is shown in Fig 13. The conditions for this experiment are same of the previous experiment. Not that the left figure is depicted in the time interval 0-100[s]. As for the disturbances, the ball was picked at random by a stick. We can confirm that the racket follows the ball and the ball juggling is preserved. This result shows the effectiveness of the juggling preservation problem and the robustness of the method. Readers can see the movie to represent the robustness in the website (Nakashima).

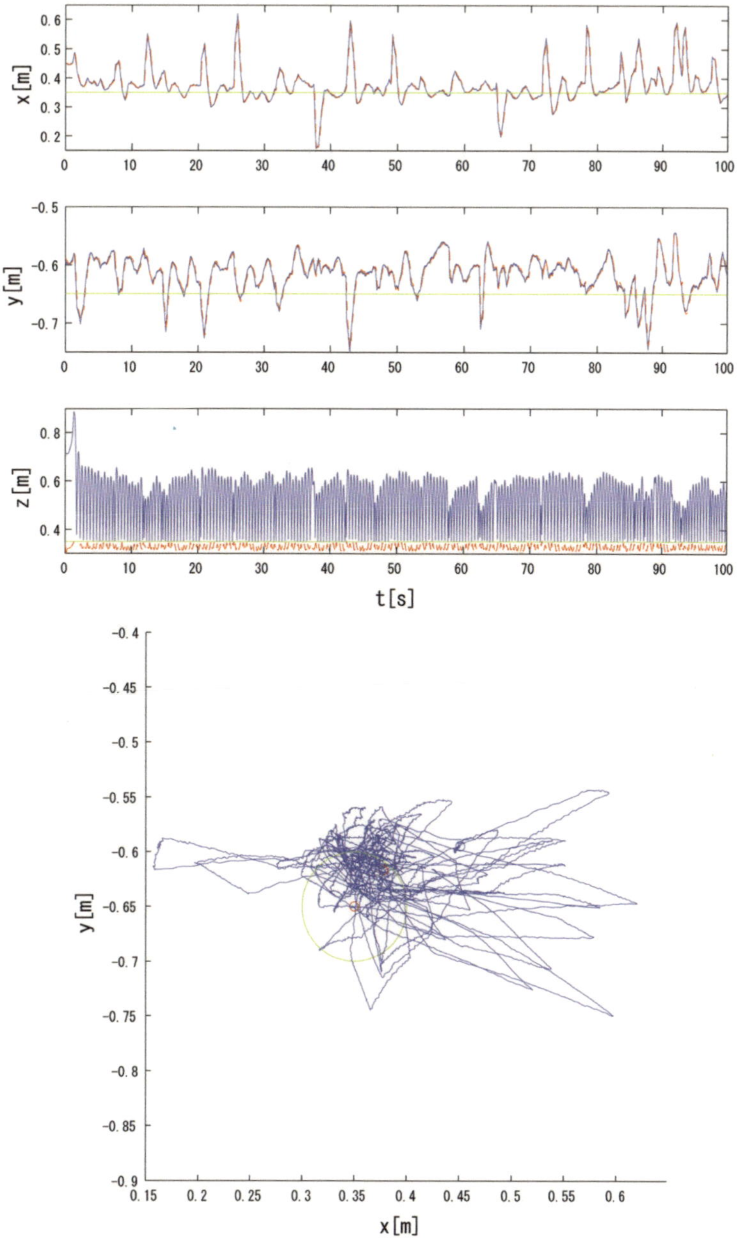

Figure 13. Experimental result with the disturbances for the ball

7. Conclusion

We realized the paddle juggling by a robot manipulator in this research. The key points are the following:

1. The method can hit the ball iteratively without the prediction of the fall point of the ball.
2. The method regulate the hitting position based on the ball state with the only prediction of the velocity in the z direction.

An example of another approach to the paddle juggling is to determine the racket position, angle and velocity based on the predicted ball trajectory in real time. This approach may be available while this needs more information about the ball than our proposed method. On the other hand, our method always tracks the ball in (x,y) plane and predicts only the z velocity. This contributes on the simplicity and the robustness. Furthermore, our method realizes the hitting with only synchronizing the ball in z direction. This contributes on no determination of the racket velocity for hitting the ball. To sum up, the simplicity and synchronous are significant to realize the paddle juggling. We hope that this may indicate that human do juggling based on the same strategy.

8. References

B. K. Ghosh, N. Xi, and T. J. Tarn. 1999. Control in Robotics and Automation: Sensor-Based Integration. Academic Press.

Buehler, M., D. E. Koditschek, and P. J. Kindlmann. 1994. Planning and Control of Robotic Juggling and Catching Tasks. *Int. J. Robot. Res.* 13 (2): 101-118.

Majima, S., and K. Chou. 2005. A Receding Horizon Control for Lifting Ping-Pong Ball. *IFAC World Congress*. Th-A04-TP/3.

Nakashima, A. http://www.haya.nuem.nagoya-u.ac.jp/~akira/data65s.wmv.

R. Mori, K. Hashimoto, E Takagi, and E Miyazaki. 2005. Examination of Ball Lifting Task using a Mobile Robot. *Proc. IEEE/RSJ Inter. Conf. Int. Robot. Sys.* 369-374.

Schaal, S., and C. G. Atkeson. 1993. Open Loop Control Strategies for Robot Juggling. *Proc. IEEE Inter. Conf. on Robot. Automat.* 913-918.

24

An Industrial Robot as Part of an Automatic System for Geometric Reverse Engineering

Mohamed Rahayem
Örebro University
Sweden

1. Introduction

In many areas of engineering, medical sciences or art there is a strong demand to create appropriate computer representations of existing objects from large sets of measured 3D data points. This process is called Geometric Reverse Engineering (GRE).
GRE is an important topic in computer vision but due to the advances in high accuracy laser scanning technology, GRE is now also an important topic in the geometric modelling community where it is used to generate accurate and topologically consistent CAD models for manufacturing and prototyping. Commercial GRE systems are available but still developing. There are many research groups dealing with the GRE problem and its applications in universities all over the world, e.g., Geometric Modelling Laboratory at the Hungarian Academy of Sciences and Geometric Computing & Computer Vision at Cardiff university.
The CAD technology research group at Örebro University in Sweden started its GRE research in 2001. Its work is focused on problems related to automatic GRE of unknown objects. A laboratory environment has been created with a laser profile scanner mounted on an industrial robot controlled and integrated by software based on the open source CAD system Varkon. The basic principles and the system were described by (Larsson & Kjellander, 2004). They give the details of motion control and data capturing in (Larsson & Kjellander, 2006) and present an algorithm for automatic path planning in (Larsson & Kjellander, 2007).
Automatic GRE requires an automatic measuring system and the CAD research group has chosen to use the industrial robot as the carrier of the laser scanner. This has many advantages but a problem with the approach is that the accuracy of the robot is relatively low as compared to the accuracy of the laser scanner. It is therefore of certain interest to investigate how different parts of the system will influence the final accuracy of GRE operations.
The author knowing that the scanner is more accurate than the robot (Rahayem et al., 2007) it is also of interest to investigate if and how 2D profile data from the scanner can be used to increase the accuracy of GRE operations. Section 3 presents the measuring system. A detailed investigation of the accuracy of the system is given in (Rahayem et al., 2007). Section 4 includes an introduction of a typical GRE operation (segmentation). The accuracy of planar segmentation based on 3D point clouds is investigated and published in (Rahayem

et al., 2008). Finally, this chapter will present the implementation of a planar segmentation algorithm supported by experimental results.

2. Geometric Reverse Engineering

In the development of new products, computer aided design (CAD) systems are often used to model the geometry of the objects to be manufactured. Geometric Reverse Engineering GRE is the reverse process where the objects already exist and CAD models are created by interpreting geometric data measured directly from the surfaces of the objects. One application of GRE is to create a three-dimensional (3D) CAD model of some object *X* and use the model to manufacture new objects that are partly or completely copies of *X*. An introduction to GRE which is often referred to is a paper by (Vàrady et al., 1997), which divides the GRE process into four steps. The flow chart in Fig. 1 describes the sequence of these steps. The forthcoming sections will explain each step in more detail. See (Farin et al., 2002) for more details about GRE.

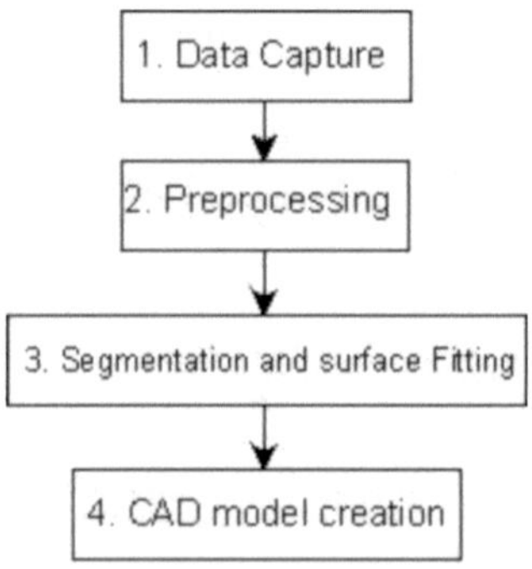

Figure 1. Basic steps of GRE

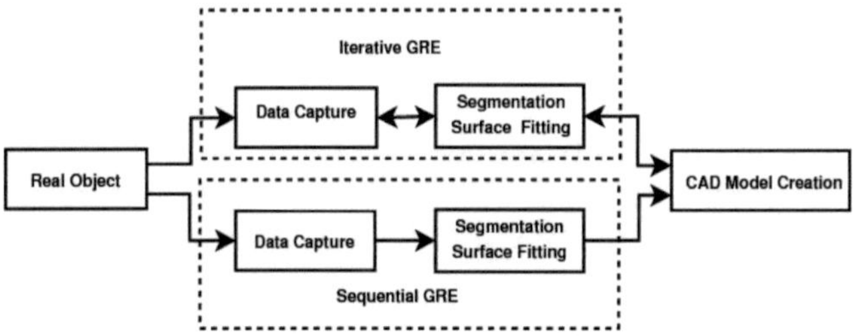

Figure 2. Sequential and iterative GRE

2.1 Data Capture

There are many different methods for acquiring shape data. The most popular methods are described in detail in (Vàrady et al., 1997). Essentially, each method uses some mechanism for interacting with the surface or the shape of the measured object. There are *non-contact methods*, where light, sound or magnetic fields are used with or without camera, and *tactile methods* where the surface is touched by using mechanical probes at the end of an arm. In each case an appropriate analysis must be performed to determine positions of points on the object's surface from physical readings obtained. In some laser range finders, for example a camera calibration is performed to determine the distance between the point and the center of the sensor. From a practical point of view, *tactile methods* are more robust (i.e. less noisy, more accurate, more repeatable), but they are also the slowest method for data acquisition. Coordinate Measuring Machines (CMM) can be programmed to follow paths along a surface and collect very accurate and almost *noise-free* data. In addition, step 1 can be done *manually* or *automatically*. Manual measuring enables full control over the measurement process and an experienced operator can optimize the point cloud to produce the best possible result. Automatic measuring can produce good results if optimized for a specific class of objects with similar shape. For unknown objects, practically automatic measuring has not reached the state where it can compete with the manual measuring as far as the quality of the point cloud is concerned. One reason for this is that automated processes do not adapt the density of the point cloud to the shape of the object. Planar regions need fewer points than sharp corners or edges to be interpreted correctly in steps 2.2 and 3.3. For optical measurement systems, like laser scanners, it is also important that the angle between the laser source and the camera in relation to the normal of the surface is within certain limits. If they are not, the accuracy will drop dramatically. A flexible system is described in (Chan et al., 2000), where a CMM is used in combination with a profile scanner. In (Callieri et al., 2004), a system based on a laser range camera mounted on the arm of an industrial robot in combination with a turntable, is presented. In this case, the robot moves the camera to view the object from different positions. In (Seokbae et al., 2002) a motorized rotary table with two degrees of freedom and a scanner mounted on a CNC machine with four degrees of freedom are used. There are many practical problems with acquiring usable data, for instance:

- **Calibration** is an essential part of setting up the position and orientation of the measuring device. If not done correctly, the calibration process may introduce systematic errors. Laser scanner accuracy also depends on the resolution and the frame rate of the video system used.
- **Accessibility** means the issue of scanning data that is not easily acquired due to the shape or topology of the part. This usually requires multiple scans and/or a flexible device for sensor orientation, for instance a robot with many degrees of freedom. Still, it may be impossible to acquire data through holes, etc.
- **Occlusion** means the blocking of the scanning medium due to shadowing or obstruction. This is primarily a problem with optical scanners.
- **Noise and incomplete data** is difficult to eliminate. Noise can be introduced in a multitude of ways, from extraneous vibrations, specular reflections, etc. There are many different filtering approaches that can be used. An important question is whether to eliminate the noise before, after or during the model building stages. Noise filtering can also destroy the sharpness of the data, sharp edges may disappear and be replaced by smooth blends.
- **Surface finish**, i.e., the smoothness and material coating can dramatically affect the data acquisition process. Both contact and non-contact data acquisition methods will produce more noise with a rough surface than a smooth one. See (Vàrady et al., 1997) for more details about data acquisition and its practical problems.

2.2 Pre-processing

As indicated in section 2 the main purpose of GRE is to convert a *discrete* set of data into an explicit CAD model. The pre-processing step takes care of the acquired data and prepares it for segmentation and surface fitting. This may involve such things as *merging of multiple point clouds, point reduction, triangulation, mesh smoothing, noise filtering , etc.* The discrete data set typically consists of (x,y,z) coordinate values of measured data points. The points may be *organized* in a pattern or *unorganized.* In the last case, the points come from random or manual point wise sampling, in the former case the measurements may have taken place along known scan paths, which results in a sequence of scan lines or profiles, see (Larsson & Kjellander, 2006). Another important issue is the *neighborhood information,* for example a regular mesh implicitly gives connectivity except at step discontinuities. In other words, a topology is created that connects neighbouring points with each other, usually into a triangular or polygonal mesh. The measurement result can then be displayed as a shaded faceted surface, see (Roth & Wibowoo, 1997) and (Kuo & yan, 2005).

2.3 Segmentation and surface fitting

This section will briefly introduce segmentation and surface fitting. Later on, section 4 gives more detials and describes different segmentation methods. In reality, the surfaces of the object can be divided into sub surfaces, which meet along sharp or smooth edges. Some of them will be simple surfaces such as planes or cylinders, while others will need to be represented by more general free-form surfaces. The tasks to be solved at this stage of shape reconstruction are:

- **Segmentation**, each point set will be divided into subsets, one for each natural surface so that each subset contains just those points sampled from a particular natural surface.
- **Classification** to decide what type of surface each subset of points belongs to (e. g. planar, cylindrical, etc).
- **Fitting** to find the surface of the given type which is the best fit to the points in the given subset. See (Chivate & Jablokow, 1995); (Fisher, 2004) ; (Benkö et al., 2004) for more information about surface fitting for GRE.

Essentially, two different approaches of segmentation could be introduced, namely *Edge based* and *Face* or *Region based* methods see (Woo et al., 2002) and (Petitjean, 2002).

2.4 CAD model creation

The previous section described various alternatives for extracting geometric information from a dense set of measured data points. Due to the diversity of surface representations and applicable algorithms, the input for the model creation can be in various formats at different levels of geometrical and topological completeness. For this reason, to present a uniform approach for the final model creation would be very difficult see (Vàrady et al., 1997). (Benkö et al., 2001) describes a method to create an explicit CAD model in the form of a Boundary Representation (B-rep).

2.5 Sequential and iterative GRE

The above steps 2.1, 2.2, 2.3 and 2.4 are usually sequential and automatic but may involve different software products and threshold values and usually need to be adjusted manually to yield good results. To completely automate the entire GRE process steps 2.1 through 2.4 need to be integrated into a single system. Such a system would need a numerically controlled mechanical device for orientation of the sensor and an automatic planning algorithm to control the movements, see (Larsson & Kjellander, 2004) and (Larsson & Kjellander, 2006).

An iterative approach could then be applied where the system during step 2.3 first makes a coarse measurement of the object and then iteratively more precise measurements based on the result of previous iterations. During segmentation and fitting it would also be able to increase the quality of the result by directing the measurement system to acquire complimentary data iteratively, see fig. 2. An example of an automatic GRE system is the system used to achieve the work in this chapter. This system is based on an industrial robot with a turntable, a laser profile scanner and the Varkon CAD software. The setup and installation of the system is described in (Larsson & Kjellander, 2004). In addition section 3 will give an introduction to the system and its specification. An industrial robot with a turntable is a very flexible device. It is fast and robust and relatively cheap. This is true also for a laser profile scanner. The combination is thus well suited for industrial applications where automatic GRE of unknown objects is needed in real time. A future goal is to implement all four steps of the GRE process into this system. The accuracy of a CAD model created by GRE depends of course on the basic accuracy of the measurement system used, but also on how the system is used and how the result is processed. It is important to state that the basic accuracy of an industrial robot is relatively low. The accuracy of a profile scanner is at least 10 times better, see (Rahayem et al., 2007). With a fully integrated GRE system, where the GRE software has access not only to 3D point data but also to 2D profile data and the GRE software can control the orientation of the scanner relative to the measured object's surface, it is a reasonable to believe that it may be possible to compensate to some extent for the low accuracy of the robot.

3. A system for automatic GRE

The automatic GRE system consists of a laser profile scanner mounted on an industrial robot with a turntable. The robot movement and scanning process is controlled trough the GRE software, which in turn is reachable via a communication interface over TCP/IP. The robot is an industrial robot of type ABB IRB 140 with S4C control system, i.e., a very common robot system. To increase the flexibility, a turntable is included in the setup. Fig. 3 shows the system. The system setup gives the potential of moving the scanner along any arbitrary scan path with a suitable and continuously varying orientation of scanner head. This ability is required to be able to perform an optimal scan, but it also increases the complexity of the scan path generation. The essential parts of the system are described in more detail below.

Figure 3. The robot measures an object

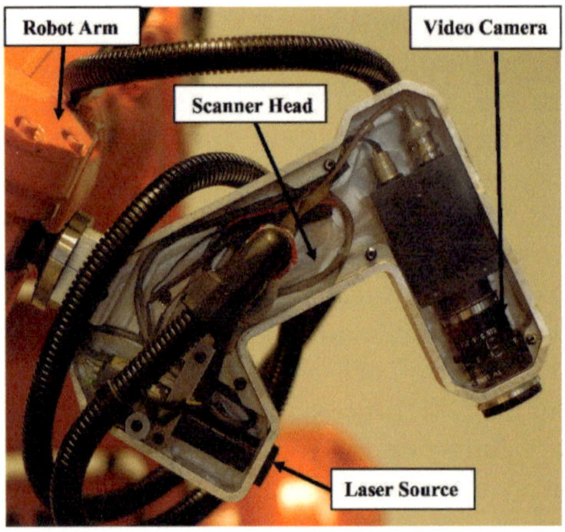

Figure 4. The scanner head

3.1 The robot and turntable

The robot arm is a standard ABB IRB 140 with six rotational joints, each with a resolution of 0.01^o. The robot manufacturer offers an optional test protocol for an individual robot arm called *Absolute accuracy*. According to the test protocol of our robot arm it can report its current position within ± *0.4 mm* everywhere within its working area under full load. The robot arm is controlled by a S4C controller which also controls the turntable. The turntable has one rotational joint. Repeating accuracy, according to the manufacturer, at equal load and radius 500 mm is ± *0.1 mm*. This corresponds to an angle of 0.01^o which is the same accuracy as the joints of the robot arm. See (ABB user's guide) and (ABB rapid reference manual) for more details on the robot. See (Rahayem et al., 2007) for more details on accuracy analysis. While not yet verified, the system could achieve even better accuracy for the following reasons:

- The calibration and its verification was performed at relatively high speed, while the part of GRE measuring that demands the highest accuracy will be performed at low speed.
- The weight of the scanner head is relatively low, which together with low scanning speed implies limited influence of errors introduced by the impact of dynamic forces on the mechanical structure.
- A consequence of using a turntable in combination with the robot is that a limited part of the robot's working range will be used. Certainly the error introduced by the robot's first axis will be less than what is registered in the verification of the calibration. Another possibility to realize this system would have been to use a CMM in combination with laser profile scanners with interfaces suited for that purpose. This would give a higher accuracy, but the use of the robot gives some other interesting properties:
  1. The robot used as a translation device and a measurement system is relatively cheap compared to other solutions that give the same flexibility.
  2. The robot is robust and well suited for an industrial environment, which makes this solution interesting for tasks where inspection at site of production is desirable.
  3. The system has potential for use in combination with tasks already robotized.

3.2 The laser profile scanner

The laser profile scanner consists of a line laser and a Sony XC-ST30CE CCD camera mounted in a scanner head which is manufactured in the Mechanical Engineering laboratory at Örebro University. The camera is connected to a frame grabber in a PC that performs the image analysis with software developed by Namatis AB Company in Karlskoga, Sweden. An analysis of the scanner head (camera and laser source), sources of errors and scanner head accuracy has been done by a series of experiments and shows that the accuracy of the scanner head is at least *10 times* better than the robot's accuracy (Rahayem et al., 2007) and (Rahayem et al., 2008). Fig. 4 shows the scanner head.

3.1.1 Accuracy of the scanner head

In (Rahayem et al., 2007), the authors proved that an accuracy of *0.05 mm* or better is possible when fitting lines to laser profiles. The authors also show how intersecting lines from the same camera picture can be used to measure distances with high accuracy. In a new series of experiments (Rahayem et al., 2008) have investigated the accuracy in measuring the radius of a circle. An object with cylindrical shape was measured and the camera captured pictures with the scanner head orthogonal with the cylinder axis. The cylinder will then appear as a circular arc in the scan window. The authors used a steel cylinder with $R=10.055mm$ measured with a Mitutoyo micrometer (0-25mm/0.001mm) and the experiment was repeated *100 times* with D increasing in steps of *1 mm* thus covering the scan window. To make it possible to distinguish between systematic and random errors each of the *100 steps* was repeated $n=10$ *times*, and in each of these the scanner head was moved *0.05 mm* in a direction collinear with the cylinder axis to filter out the effect of dust, varying paint thickness or similar effects. The total number of pictures analyzed is thus *1000*. For each distance D a least squares fit of a circle is done to each of the N pictures and the systematic and random errors were calculated using Esq. (1) and (2). The result was plotted in fig. 5 and 6.

$$E_s = R - \frac{\sum_{i=1}^{N} R_i}{N} \tag{1}$$

$$E_r = \frac{\sum_{i=1}^{N} abs\,(R_i + E_s) - R}{N} \tag{2}$$

E_s and E_r are the systematic and random radius errors. R and R_i are the true and measured radius and N the number of profiles for each D. The maximum size of the random error is less than *0.02 mm* for reasonable values of D. For more detail about accuracy analysis see (Rahayem et al., 2007) and (Rahayem et al., 2008).

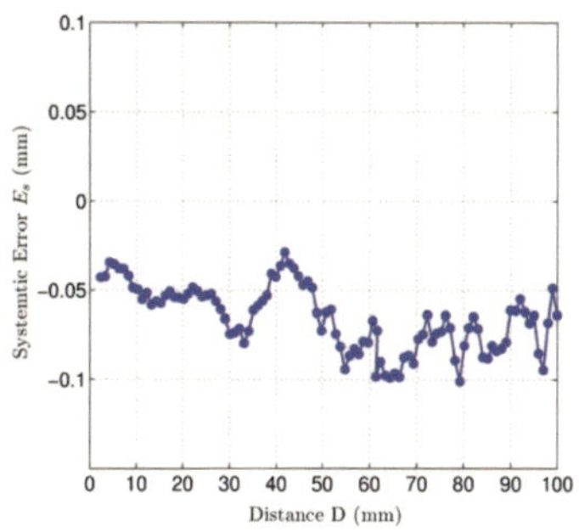

Figure 5. Systematic error in radius

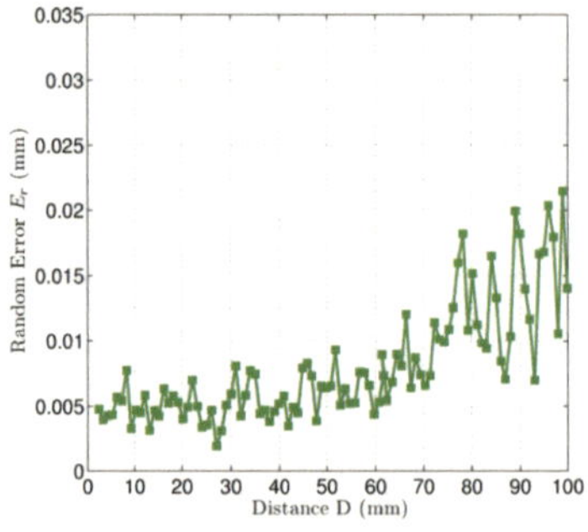

Figure 6. Random error in radius

3.3 The CAD system

A central part of our experimental setup is the Varkon CAD system. The CAD system is used for the main procedure handling, data representation, control of used hardware, decision making, simulation, verification of planned robot movements and the GRE process. The robot controller and the scanner PC are connected through TCP/IP with the GRE computer where the Varkon CAD system is responsible for their integration, see fig.7. Varkon started as a commercial product more than 20 years ago but is now developed by the research group at Örebro University as an open source project on Source Forge, see (Varkon). Having access to the C sources of a 3D CAD system with geometry, visualization, user interface etc, is a great advantage in the development of an automatic GRE process where data capturing, preprocessing, segmentation and surface fitting needs to be integrated. In addition, it gives a possibility to add new functions and procedures. Varkon includes a high level geometrically oriented modeling language *MBS*, which is used for parts of the GRE system that are not time critical but also to develop prototypes for testing before final implementation in the C sources.

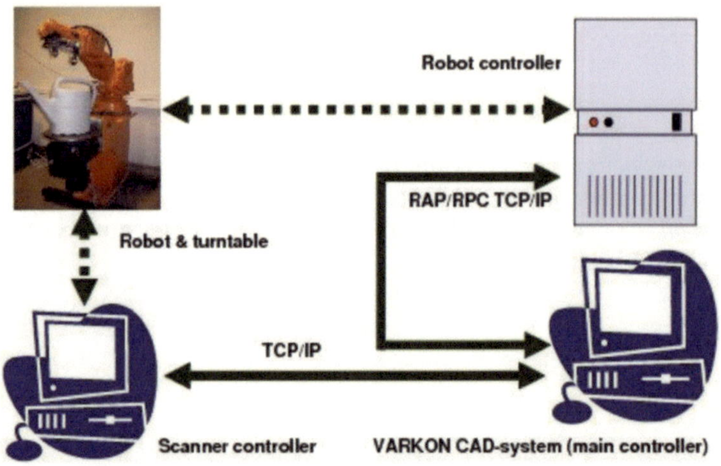

Figure 7. The sub-systems in combination

In general, GRE process as described earlier in section 2 above is purely sequential. A person operating a manual system may however decide to stop the process in step 2.2 or 2.3 and go back to step 2.1 in order to improve the point cloud. A fully automatic GRE system should

behave similar to a human operator. This means that the software used in steps 2.2 and 2.3 must be able to control the measurement process in step 2.1.
In a GRE system which is a fully automatic the goal of the first iteration may only be to establish the overall size of the object, i.e., its *bounding box*. Next iteration would narrow in and investigate the object in more detail. The result of each iteration can be used to plan the next iteration that will produce a better result. This idea leads to dividing the automatic GRE procedure into three different modules or steps, which will be performed in the following order:

- **Size scan** – to retrieve the object bounding box.
- **Shape scan** - to retrieve the approximate shape of the object.
- **GRE scan** - to retrieve the final result by means of integration with the GRE process.

Before proceeding to these steps in more detail I will give a little introduction to how path planning, motion control and data capturing procedures are implemented in the system.

3.4 Path planning

One of the key issues of an autonomous measuring system is the matter of path planning. The path planning process has several goals:

- Avoid collision.
- Optimize scanning direction and orientation.
- Deal with surface occlusion.

The process must also include a reliable self-terminating condition, which allows the process to stop when perceptible improvement of the CAD model is no longer possible. (Pito & Bajcsy, 1995) describe a system with a simple planning capability that combines a fixed scanner with a turntable. The planning of the scan process in such a system is a question of determining the Next Best View (NBV) in terms of turntable angles. A more flexible system is achieved by Combining a laser scanner with a CMM, see (Chan et al., 2000) and (Milroy et al., 1996). In automated path planning it is advantageous to distinguish between objects of *known* shape and objects of *unknown* shape, i.e., no CAD model exists in forehand. Several methods for automated planning of laser scanning by means of an existing CAD model are described in the literature, see for example (Xi & Shu, 1999); (Lee & Park, 2001); (Seokbae et al., 2002). These methods are not directly applicable to the system as the system is dealing with unknown objects. For further reading in the topic of view planning see (Scott et al., 2003), which is a comprehensive overview of view planning for automated three dimensional object reconstructions. In this chapter, the author uses manual path planning in order to develop automatic segmentation algorithms. In future work, the segmentation algorithms will be merged with the automatic planning. In manual mode the user manually defines the following geometrical data which is needed to define each scan path:

- Position curve.
- View direction curve.
- Window turning curve (optional).
- Tool center point z-offset (TCP z-offset).

Fig. 8 shows a curved scan path modeled as a set of curves.
The system then automatically processes them and the result after the robot has finished moving is Varkon *MESH* geometric entity for each scan path. In automatic mode, the system automatically creates the curves needed for each scan path. This is done using a process where the system first scans the object to establish its bounding box and then switches to an

algorithm that creates a MESH representation suitable for segmentation and fitting. This algorithm is published by (Larsson & Kjellander, 2006).

3.5 Motion control

To control the robot a concept of a scan path is developed, which is defined by geometrical data mentioned in the previous section, see fig. 8. This makes it possible to translate the scanner along a space curve at the same time as it rotates. It is therefore possible to orient the scanner so that, the distance and angle relative to the object is optimal with respect to accuracy, see (Rahayem et al., 2007). A full 3D orientation can also avoid occlusion and minimize the number of re-orientations needed to scan an object of complex shape. A full description of motion control is described in (Larsson & Kjellander, 2006).

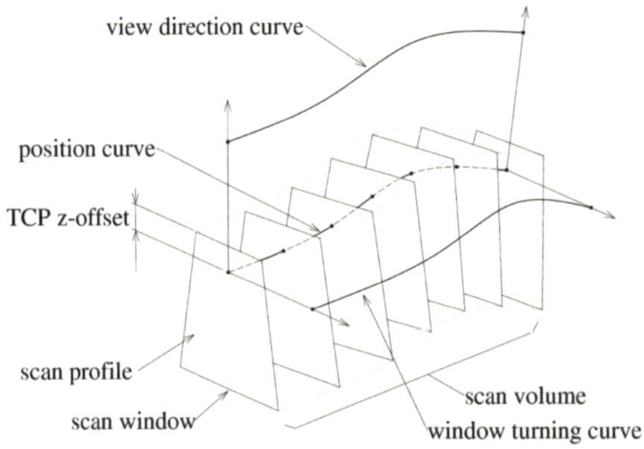

Figure 8. Curved scan path defined by three curves

3.6 Data capture and registration

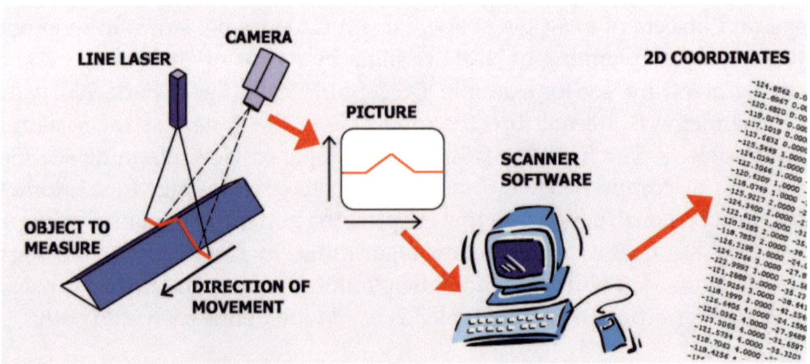

Figure 9. From laser profile to 2D coordinates

After the user has defined the geometrical data which defines a scan path, the Varkon CAD system calculates a series of robot poses and turntable angles and sends them to the robot. While the robot is moving it collects actual robot poses at regular intervals together with a time stamp for each actual pose. Similarly, the scanner software collects scan profiles at regular intervals, also with time stamps, see fig. 9. When the robot reaches the end of the scan path all data are transferred to the Varkon CAD system, where an actual robot pose is calculated for each scan profile by interpolation based on the time stamps. For each pixel in a profile its corresponding 3D coordinates can now be computed and all points are then connected into a triangulated mesh and stored as a Varkon MESH entity. Additional information like camera and laser source centers and TCP positions and orientations are stored in the mesh data structure to be used later into the 2D pre-processing and segmentation. The details of motion control and data capturing were published in (Larsson & Kjellander, 2006). Fig. 7 shows how the different parts of the system combine together and how they communicate.

3.7 The automatic GRE procedure

As mentioned above, our automatic GRE process is divided into three modules: size, shape and GRE scan module. These modules use all techniques described in 3.4, 3.5 and 3.6, and also common supporting modules such as the robot simulation which is used to verify the planned scanning paths. Each of the three modules performs the same principal internal iteration:

- Plan next scanning path from previous collected information (the three modules use different methods for planning).
- Verify robot movement with respect to the robot's working range, collision etc.
- Send desired robot movements to the robot.
- Retrieve scanner profile data and corresponding robot poses.
- Register collected data in an intermediate model (differs between the modules).
- Determine if self terminating condition is reached. If so, the iteration will stop and the process will continue in the next module until the final result is achieved.

The current state of the system is that the size scan and the shape scan are implemented. The principles of the size, shape and GRE modules are:

- **Size Scan module**. The aim of this module is to determine the object extents, i.e., its bounding box. It starts with the assumption that the size of the object is equal to the working range of the robot. It then narrows in on the object in a series of predefined scans until it finds the surface of the object and thus its bounding box. To save time, the user can manually enter an approximate bounding box as a start value.
- **Shape Scan module**. The implementation of this step is described in detail in (Larsson & Kjellander, 2007). It is influenced by a planning method based on an Orthogonal Cross Section (OCS) network published by (Milroy et al., 1996).
- **GRE Scan module**. This module is under implementation. The segmentation algorithms presented in this chapter will be used in that work. The final goal is to automatically segment all data and create an explicit CAD model.

4. Segmentation

Segmentation is a wide and complex domain, both in terms of problem formulation and resolution techniques. For human operators, it is fairly easy to identify regions of a surface that are simple surfaces like planes, spheres, cylinders or cones, while it is more difficult for a computer. As mentioned in section 2.3, the segmentation task breaks the dense measured point set into subsets, each one containing just those points sampled from a particular simple surface. During the segmentation process two tasks will be done in order to get the final segmented data. These tasks are *Classification* and *Fitting*. It should be clearly noted that these tasks cannot in practice be carried out in the sequential order given above, see (Vàrady et al., 1997).

4.1 Segmentation background

Dividing a range image or a triangular mesh into regions according to shape change detection has been a long-standing research problem. The majority of point data segmentation approaches can be classified into three categories. In (Woo et al., 2002) the authors defined the three categories as follows:

4.1.1 Edge-based approaches

The edge-detection methods attempt to detect discontinuities in the surfaces that form the closed boundaries of components in the point data. (Fan et al., 1987) used local surface curvature properties to identify significant boundaries in the data range. (Chen & Liu, 1997) segmented the CMM data by slicing and fitting them by two-dimensional NURBS curves. The boundary points were detected by calculating the maximum curvature of the NURBS curve. (Milory et al., 1997) used a semi-automatic edge-based approach for orthogonal cross-section (OCS) models. (Yang & Lee, 1999) identified edge points as the curvature extremes by estimating the surface curvature. (Demarsin et al., 2007) presented an algorithm to extract closed sharp feature lines, which is necessary to create such a closed curve network.

4.1.2 Region-based approaches

An alternative to edge-based segmentation is to detect continuous surfaces that have homogeneity or similar geometrical properties. (Hoffman & Jian, 1987) segmented the range image into many surface patches and classified these patches as planar, convex or concave shapes based on non-parametric statistical test. (Besl & Jian, 1988) developed a segmentation method based on variable order surface fitting. A robust region growing algorithm and its improvement was published by (Sacchi et al., 1999); (Sacchi et al., 2000).

4.1.3 Hybrid approaches

Hybrid segmentation approaches have been developed, where the edge and region-based approaches are combined. The method proposed by (Yokoya et al., 1997) divided a three dimensional measurement data set into surface primitives using bi-quadratic surface fitting. The segmented data were homogeneous in differential geometric properties and did not contain discontinuities. The Gaussian and mean curvatures were computed and used to perform the initial region based segmentation. Then after employing two additional edge-based segmentations from the partial derivatives and depth values, the final segmentation result was applied to the initial segmented data. (Checchin et al., 2007) used a hybrid approach that combined edge detection based on the surface normal and region growing to merge over segmented regions. (Zhao & Zhang, 1997) employed a hybrid method based on triangulation and region grouping that uses edges, critical points and surface normal. Most researches have tried to develop segmentation methods by exactly fitting curves or surfaces to find edge points or curves. These surface or curve fitting tasks take a long time and, furthermore it is difficult to extract the exact edge points because the scan data are made up of discrete points and edge points are not always included in the scan data. A good general overview and surveys of segmentation are provided by (Besl & Jian, 1988); (Petitjean, 2002); (Woo et al., 2002); (Shamir, 2007). Comparing the edge-based and region-based approaches makes the following observations:

- **Edge-based approaches** suffer from the following problems. Sensor data particularly from laser scanners are often unreliable near sharp edges, because of specular reflections there. The number of points that have to be used to segment the data is small, i.e., only points in the vicinity of the edges are used, which means that information from most of the data is not used to assist in reliable segmentation. In turn, this means a relatively high sensitivity to occasional spurious data points. Finding smooth edges which are tangent continuous or even higher continuity is very unreliable, as computation of derivatives from noisy point data is error-prone. On the other hand, if smoothing is applied to the data first to reduce the errors, this distorts the estimates of the required derivatives. Thus sharp edges are replaced by blends of small radius which may complicate the edge-finding process, also the positions of features may be moved by noise filtering.

- **Region-based approaches** have the following advantages; they work on a large number of points, in principle using all available data. Deciding which points belong to which surface is a natural by-product of such approaches, whereas with edge-based approaches it may not be entirely clear to which surface a given point belongs even after we have found a set of edges. Typically region-based approaches also provide *the best-fit surface* to the points as a final result.

Overall, Authors of (Vàrady et al., 1997); (Fisher et al., 1997); (Robertson et al., 1999); (Sacchi et al., 1999); (Rahayem et al., 2008) believe that *region-based approaches* rather than *edge-based approaches* are preferable. In fact, segmentation and surface fitting are like the *chicken and egg problem*. If the surface to be fitted is known, it could immediately be determined which sample points belonged to it.

It is worth mentioning that it is possible to distinguish between *bottom-up* and *top-down* segmentation methods. Assume that *a region-based approach* was adopted to segment data points. The class of *bottom-up* methods initially start from *seed points*. Small initial neighbourhoods of points around them, which are deemed to consistently belong to a single surface, are constructed. Local differential geometric or other techniques are then used to add further points which are classified as belonging to the same surface. The growing will stop when there are no more consistent points in the vicinity of the current regions.

On the other hand, the *top-down* methods start with the premise that *all* the points belong to a single surface, and then test this hypothesis for validity. If the points are in agreement, the method is done, otherwise the points are subdivided into two (or more) new sets, and the single surface hypothesis is applied recursively to these subsets to satisfy the hypothesis. Most approaches of segmentation seem to have taken the bottom-up approach, (Sapidis & Besl, 1995). While the top-down approach has been used successfully for image segmentation; its use for surface segmentation is less common.

A problem with the *bottom-up approaches* is to choose good seed points from which to start growing the nominated surface. This can be difficult and time consuming.

A problem with the *top-down approaches* is choosing where and how to subdivide the selected surface.

4.2 Planar segmentation based on 3D point could

Based on the segmentation described in sections 2.3 and 4.1, the author has implemented a *bottom-up* and *region-based* planar segmentation approach in the Varkon CAD system by using the algorithm described in (Sacchi et al., 1999) with a better region growing criterion. The segmentation algorithm includes the following steps:

1. **Triangulation** by joining points in neighbouring laser profiles (laser strips) into a triangular mesh. This is relatively easy since the points from the profile scanner are ordered sequentially within each profile and the profiles are ordered sequentially in the direction the robot is moving. The triangulation algorithm is described in (Larsson & Kjellander, 2006).
2. **Curvature estimation.** The curvature of a surface can be calculated by analytic methods which use derivatives, but this can not be applied to digitized (discrete) data directly and requires the fitting of a smooth surface to some of the data points. (Flynn & Jain, 1989) proposed an algorithm for estimating the curvature between two points on a surface which uses the surface normal change between the points. For more details about estimating curvature of surfaces represented by triangular meshes see (Gatzke, 2006). In order to estimate the curvature for every triangle in the mesh, for any pair of triangles which share an edge one can find the curvature of the sphere passing through the four vertices involved. If they are coplanar the curvature is zero. In order to compensate for the effect of varying triangle size, *compensated triangle normal* is used as follows:
   - Calculate the normal for each vertex, which is called *interpolated normal*, equal to the weighted average of the normals for all triangles meeting at this vertex. The weighting factor used for each normal is the area of its triangle.
   - Calculate the *compensated normal* for a triangle as the weighted average of the three interpolated normals at the vertices of the triangle, using as weighting factor for each vertex the sum of the areas of the triangles meeting at that vertex.
   - Calculate in a similar way the *compensated centre* of each triangle as the weighted average of the vertices using the same weighting factor as in the previous step.
   - For a pair of triangles with compensated centres C_1 and C_2 and $N_1 \otimes N_2$ is the cross product of the compensated normals, the estimated curvature is:

$$K_{1,2} = \frac{\| N_1 \otimes N_2 \|}{\| C_1 - C_2 \|} \quad (3)$$

   For a given triangle surrounded by three other triangles, three curvature values are estimated. In similar way another three curvature values will be estimated by pairing the compensated normals with the interpolated normals at each of the three vertices in turn.
   - The triangle curvature is equal to the mean of the maximum and minimum of the six curvature estimates obtained for that triangle.
3. **Find the seed** by searching the triangular mesh to find the triangle with lowest curvature. This triangle will be considered as *a seed.*
4. **Region Growing** adds connected triangles to the region as long as their normal vectors are reasonably parallel to the normal vector of the seed triangle. This is done by calculating a cone angle between the triangle normals using the following formula:

$$\alpha_{1,2} = \cos^{-1}(N_1 \bullet N_2) \quad (4)$$

Where, $N_1 \bullet N_2$ is the dot product of the compensated triangle normals of the two neighbouring triangles respectively.

5. **Fit a plane** to the current region. Repeat the steps 3 and 4 until all triangles in the mesh have been processed, for each segmented region. Fit a plane using Principle Components Analysis, see (Lengyel, 2002).

Fig. 10 shows the result of this algorithm.
The difference between the algorithm described in this section and Sacchi algorithm described in (Sacchi et al., 1999) is that Sacchi allowed a triangle to be added, if its vertices lie within the given tolerance of the plane associated with the region while the algorithm described here allows a triangle to be added if the cone angle between its compensated normal and the seed's normal lie within a given tolerance. This makes the algorithm faster than the Sacchi algorithm since it uses already calculated data for the growing process instead of calculating new data. For more details about this algorithm refer to (Rahayem et al. 2008).

The test object Mesh before segmentation Mesh after Segmentation
Figure 10. Planar segmentation based on point cloud algorithm

5. Conclusion and future work

The industrial robot equipped with a laser profile scanner is a desirable alternative in applications where high speed, robustness and flexibility combined with low cost is important. The accuracy of the industrial robot is relatively low, but if the GRE system has access to camera data or profiles, basic GRE operations like the fitting of lines can be achieved with relatively high accuracy. This can be used to measure for example distance or radius within a single camera picture. Experiments that show this are published in (Rahayem et al., 2007); (Rahayem et al., 2008).
The author also investigated the problem of planar segmentation and implemented a traditional segmentation algorithm section 4.2 based on 3D point clouds. From the investigations described above it is possible to conclude that the relatively low accuracy of an industrial robot to some extents can be compensated if the GRE software has access to data directly from the scanner. This is normally not the situation for current commercial solutions but is easy to realize if the GRE software is integrated with the measuring hardware, as in our laboratory system. It is natural to continue the work with segmentation of conic surfaces. Cones, cylinders, and spheres are common shapes in manufacturing. It is

therefore interesting to investigate if 2D profile data can be used in the GRE process also for these shapes. The theory of projective geometry can be used to establish the shape of a conic curve projected on a plane. A straight line projected on a conic surface is the inverse problem and it would be interesting to investigate if this property could be used for segmentation of conic surfaces. 2D conic segmentation and fitting is a well known problem, but the author has not yet seen combined methods that use 2D and 3D data to segment and fit conic surfaces. Another area of interest is to investigate to what extent the accuracy of the system would be improved by adding a second measurement phase (iteration) which is based on the result of the first segmentation. This is straightforward with the described system as the GRE software is integrated with the measurement hardware and can control the measurement process. The system would then use the result from the first segmentation to plan new scan paths where the distance and orientation of the scanner head would be optimized for each segmented region. I have not seen any work published that describes a system with this capability.

6. References

ABB, Absolute accuracy user's guide. 3HAC 16062-1/BW OS 4.0/Rev00

ABB, Rapid reference manual. 3HAC 77751.

Besl, P. & Jain, R. (1988). Segmentation through variable order surface fitting. *IEEE transactions on pattern analysis and machine intelligence,* Vol. 10, No. 2 pp. 167-192, ISSN 0162-8828

Benkö, P.; Martin, R. & Vàrady, T.; (2002). Algorithms for reverse engineering boundary representation models. *Computer Aided Design,* Vol. 33, pp. 839-851, ISSN 0010-4485

Benkö, P.; Kos, G.; Vàrady, T.; Andor, L. & Martin, R. (2002). Constrained fitting in reverse engineering. *Computer Aided Geometric Design,* Vol. 19, pp. 173-205, ISSN 0167-8396

Callieri, M. ; Fasano, A. ; Impoco G. ; Cignoni, P. ; Scopigno R. ; Parrini G. ; & Biagini, G. (2004). Roboscan : an automatic system for accurate and unattended 3D scanning, *Proceedings of 2nd international symposium on 3D data processing, visualization and transmission,* 341-350, ISBN 0769522238, Greece, pp. 805-812, 6-9 September, IEEE

Chivate, P. & Jablokow, A. (1995). Review of surface representations and fitting for reverse engineering. *Computers integrated manufacturing systems,* Vol. 8, No. 3,pp. 193-204, ISSN 0951-5240

Chen, Y. & Liu, C. (1997). Robust segmentation of CMM dada based on NURBS. *International journal of advanced manufacturing technology,* Vol. 13, pp. 530-534, ISSN 1433-3015

Chan, V.; Bradley, C. & Vickers, G. (2000). A multi-sensor approach for rapid digitization and data segmentation in reverse engineering. *Journal of manufacturing science and engineering,* Vol. 122, pp. 725-733, ISSN 1087-1357

Checchin, P. ; Trassoudaine, L.& Alizon, J. (1997). Segmentation of range images into planar regions, *proceedings of IEEE conference on recent advances in 3-D digital imaging and modelling,* pp. 156-163, ISBN 0-8186-7943-3, Ottawa, Canada

Demarsin, K.; Vanderstraeten, D.; Volodine, T. & Roose, D. (2007). Detection of closed sharp edges in point clouds using normal estimation and graph theory. *Computer Aided Design,* Vol. 39, pp. 276-283, ISSN 0010-4485

Flynn, P. & Jain, A. (1989). On reliable curvature estimation. *Proceedings of IEEE computer vision and pattern recognition,* pp. 110-116, ISSN 0-8186-1952-x

Fan, T.; Medioni, G. & Nevatia, R. (2004). Segmented description of 3D-data surfaces. *IEEE Transactions on robotics and automation,* Vol. 6, pp. 530-534, ISSN 1042-296X

Farin, G. ; Hoschek J. & Kim, M. (2002). *Handbook of computer aided geometric design,* Elsevier, ISBN 0-444-51104-0, Amsterdam, The Netherlands

Fisher, R.; Fizgibbon, A. & Eggert D. (1997). Extracting surface patches from complete range descriptions. *proceedings of IEEE conference on recent advances in 3-D digital imaging and modelling,* pp. 148-154, ISBN 0-8186-7943-3, Ottawa, Canada

Fisher, R. (2004). Applying knowledge to reverse engineering problems. *Computer Aided Design,* Vol. 36, pp. 501-510, ISSN 0010-4485

Gatzke, T. (2006). Estimating curvature on triangular meshes. *International journal of shape modeling,* Vol. 12, pp. 1-28, ISSN 1793-639X

Hoffman, R. & Jain, A. (1987). Segmentation and classification of range images. *IEEE transactions on pattern analysis and machine intelligence,* Vol. 9, pp. 608-620, ISSN 0162-8828

Kuo, K. & Yan, H. (2005). A Delaunay-based region growing approach to surface reconstruction from unorganized points. *Computer Aided Design,* Vol. 37, pp. 825-835, ISSN 0010-4485

Lengyel, E. (2002). *Mathematics for 3D game programming and computer graphics,* Charles river media, 1-58450-277-0, Massachusetts, USA

Larsson, S. & Kjellander, J. (2004). An industrial robot and a laser scanner as a flexible solution towards an automatic system for reverse engineering of unknown objects, *Proceedings of ESDA04 – 7th Biannual conference on engineering systems design and analysis,* pp.341-350, ISBN 0791841731, Manchester, 19-22 July, American society of mechanical engineers, New York

Larsson, S. & Kjellander, J. (2006). Motion control and data capturing for laser scanning with an industrial robot. *Robotics and autonomous systems,* Vol. 54, pp. 453-460, ISSN 0921-8890

Larsson, S. & Kjellander, J. (2007). Path planning for laser scanning with an industrial robot, *Robotics and autonomous systems,* in press, ISSN 0921-8890

Lee, K.; Park, H. & Son, S. (2001). A framework for laser scans planning of freeform surfaces. *The international journal of advanced manufacturing technology,* Vol. 17, ISSN 1433-3015

Milroy, M.; Bradley, C. & Vickers, G. (1996). Automated laser scanning based on orthogonal cross sections. *Machines vision and applications,* Vol. 9, pp. 106-118, ISSN 0932-8092

Milroy, M.; Bradley, C. & Vickers, G. (1997). Segmentation of a warp around model quadratic surface approximation. *Computer Aided Design,* Vol. 29, pp. 299-320, ISSN 0010-4485

Petitjean, S. (2002). A survey of methods for recovering quadrics in triangle meshes. *ACM computing surveys,* Vol. 34, pp. 211-262, ISSN 0360-0300

Pito, R. & Bajcsy R. (1995) A solution to the next best view problem for automated CAD model acquisition of free-form objects using range cameras. *Technical report, RASP laboratory,department of computer and information science, university of Pennsylvania.* URL:ftp://ftp.cis.upenn.edu/pub/pito/papers/nbv.ps.gz

Robertson, C. ; Fisher, R. ;Werghi, N. & Ashbrook, A. (1999). An improved algorithm to extract surfaces from complete range descriptions, *proceedings of world manufacturing conference,* pp. 592-598

Rahayem, M. ; Larsson, S.& Kjellander, J. (2007). Accuracy analysis of 3D measurement system based on a laser profile scanner mounted on industrial robot with a turntable, *proceedings of ETFA07- 12th IEEE conference on emerging technologies and factory automation,* pp. 380-383, ISBN 978-1-4244-0826-9, Patras, Greece

Rahayem, M. ; Larsson, S.& Kjellander, J. (2008). Geometric Reverse Engineering using a laser profile scanner mounted on an industrial robot, *proceedings of the 6th international conference of DAAAM Baltic industrial engineering,* pp. 147-154, ISBN 978-1-9985-59-783-5, Tallinn, Estonia

Roth, G. & Wibowoo, E. (1997). An efficient volumetric method for building closed triangular meshes from 3D images and point data. *Proceedings of the conference on computer graphics interface , ,* pp. 173-180, ISBN 0-9695338-6-1, Kelowna, Canada

Sapidis, S. & Besl, J. (1995). Direct construction of polynomial surfaces from dense range images through region growing. *ACM Transactions on Graphics,* Vol. 14, No. 2, pp. 171-200, ISSN 0730-0301

Sacchi, R. ; Poliakoff, J.& Thomas, P. (1999). Curvature estimation for segmentation of triangulated surfaces, *proceedings of the 2nd IEEE conference on 3-D digital imaging and modelling,* pp. 536-543, ISBN 0-7695-0062-5, Ottawa, Canada

Sacchi, R. ; Poliakoff, J.& Thomas, P. (2000). Improved extraction of planar segments scanned surfaces, *proceedings of IEEE international conference on information visualization,* pp. 325-330, ISBN 0-7695-0743-3, London, UK

Scott, W.; Roth, G. & Rivest, J. (2003). A view planning for automated three dimensional objects reconstruction and inspection. *ACM computing surveys,* Vol. 35, No. 1 pp. 64-96, ISSN 0360-0300

Seokbae, S.; Hyunpung, P. & Lee, K. (2002). Automated laser scanning system for reverse engineering and inspection, *International journal of machine tools & manufacture,* Vol. 42, pp. 889-897, ISSN 0890-6955

Shamir, A. (2007). A survey on mesh segmentation techniques, *Computer graphics forum,* in press, ISSN 1467-8659

Varkon homepage, url: http://varkon.sourceforge.net

Vàrady, T.; Martin, R., & Cox, J. (1997). Reverse Engineering of geometric models–an introduction. *Computer Aided Design,* Vol. 29, No. 4, pp. 255-268, ISSN 0010-4485

Woo, H.; Kang, E.; Wang, S. & Lee, K. (2002). A new segmentation method for point cloud data. *International journal of machine tools & manufacture,* Vol. 42, pp. 167-178, ISSN 0890-6955

Xi, F. & Shu, C. (1999). CAD- based path planning for 3D line laser scanning. *Computer Aided Design,* Vol. 31, pp. 473-479, ISSN 0010-4485

Yokoya, N. & Levine, M. (1997). Range image segmentation based on differential geometry: hybrid approach. *IEEE Transactions on pattern analysis and machine intelligence,* Vol. 11, pp. 643-649, ISSN 0162-8828

Yang, M. & Lee, E. (1999). Segmentation of a measured point data using a parametric quadric surface approximation. *Computer Aided Geometric Design,* Vol. 31, pp. 449-457, ISSN 0167-8396

Zhao, D. & Zhang, X. (1997). Range data based object surface segmentation via edges and critical points. *IEEE Transactions in Image processing,* Vol. 6, pp. 826-830, ISSN 1057-7149

25

Sensing Planning of Calibration Measurements for Intelligent Robots

Mikko Sallinen and Tapio Heikkilä
VTT Technical Research Centre of Finland
Finland

1. Introduction

1.1 Overview of the problem

Improvement of the accuracy and performance of robot systems implies both external sensors and intelligence in the robot controller. Sensors enable a robot to observe its environment and, using its intelligence, a robot can process the observed data and make decisions and changes to control its movements and other operations. The term *intelligent robotics*, or *sensor-based robotics*, is used for an approach of this kind. Such a robot system includes a manipulator (arm), a controller, internal and external sensors and software for controlling the whole system. The principal motions of the robot are controlled using a closed loop control system. For this to be successful, the bandwidth of the internal sensors has to be much greater than that of the actuators of the joints. Usually the external sensors are still much less accurate than the internal sensors of the robot. The types of sensors that the robot uses for observing its environment include vision, laser-range, ultrasonic or touch sensors. The availability, resolution and quality of data varies between different sensors, and it is important when designing a robot system to consider what its requirements will be. The combining of information from several measurements or sensors is called *sensor fusion*.

Industrial robots have high repeatable accuracy but they suffer from high absolute accuracy (Mooring et. al. 1991). To improve absolute accuracy, the kinematic parameters, typically Denavit - Hartenberg (DH) parameters or related variants can be calibrated more efficiently, or the robot can be equipped with external sensors to observe the robot's environment and provide feedback information to correct robot motions. With improved kinematic calibration the robot's global absolute accuracy is improved. While using external sensors the local absolute accuracy is brought to the accuracy level of the external sensors. The latter can also be called workcell calibration. For kinematic calibration several methods have been developed to fulfil the requirements of several applications. There are two main approaches for calibration (Gatla et. al. 2007): open loop and closed loop. Open loop methods use special equipment such as coordinate measuring machines or laser sensors to measure position and orientation of the robot end-effector. These methods are relatively expensive and time-consuming. The best accuracy will be achieved when using these machines as Visual Servoing tools where they guide the end-effector of the robot on-line (Blank et. al. 2007). Closed loop methods use robot joint measurements and end-effector state to form closed loop equations for calculating the calibration parameters. The state of the end effector can be

physically constrained, e.g., to follow a plane, or it can be measured with a sensor. These methods are more flexible but are more sensitive to quality of set of samples.
External sensors, like vision and laser sensors have been used extensively for a long time; the first visual guidance was demonstrated already in the 70's and 80's, e.g., carrying out simple assembly tasks by visual feedback in a look-and-move manner (Shirai & Inoue, 1971). Later on even heavy duty machinery have been equipped with multiple sensors to automatically carry out simple material handling applications (Vähä et. al. 1994). The challenge in using external sensor is to utilize the information from external sensors as efficiently as possible.
Simulation and off-line programming offer a flexible approach for using a robot system efficiently. Nowadays product design is based on CAD models, which are used also for simulation and off-line programming purposes. When the robot is working, new robot paths and programs can be designed and generated with off-line programming tools, but there is still a gap between the simulation model and an actual robot system, even if the dynamic properties of the robot are modelled in the simulation model. This gap can be bridged with calibration methods, like using sensor observations from the environment so that motions are corrected according to the sensor information. This kind of interaction improves the flexibility of the robot system and makes it more cost-effective in small lot sizes as well.
Sensing planning is becoming an important part of a flexible robot system. It has been shown, that even for simple objects the spatial relation between the measured object and the observing sensor can have a substantial impact on the final locating accuracy (Järviluoma & Heikkilä 1995). Sensing planning as its best includes a method for selecting optimal set of target features for measurements. The approach presented in this chapter, i.e., the purpose of the sensing planning is to generate optimal measurements for the robot, using accuracy or low level of spatial uncertainties as the optimality criterion. These measurements are needed e.g. in the calibration of the robot work cell. First implementations of such planning systems into industrial applications are now becoming a reality (Sallinen et. al. 2006).
This chapter presents a synthesis method for sensing planning based on minimization of a posteriori error covariance matrix and eigenvalues in it. Minimization means here manipulation of the terms in the Jacobian and related weight matrices to achieve low level of spatial uncertainties. Sensing planning is supported by CAD models from which planning algorithms are composed depending on the forms of the surfaces. The chapter includes an example of sensing planning for a range sensor - taken from industry - to illustrate the principles, results and impacts.

1.2 State-of-the-art

Methods for sensing planning presented in the literature can be divided into two main types: generate-and-test and synthesis (Tarabanis et. al. 1995). In addition to these, there are also other sensing planning types, including expert systems and sensor simulation systems (Tarabanis et. al. 1995). The quality of the measured data is very important in cases where only a sparse set of samples is measured using, e.g., a point-by-point sampling system or when the available data is very noisy. Third case for careful planning is for situations where there is only very limited time to carry out measurements such as real-time systems. Examples of systems yielding a sparse set of data include the Coordinate Measuring Machine (CMM) (Prieto et. al. 2001), which obtains only a few samples, or a robot system with a tactile sensor. Also, compared with vision systems, the amount of data achieved

using a point laser rangefinder or an ultrasound sensor is much smaller, unless they are scanning sensors. Several real-time systems have to process measurement data very fast and there quality of the data improves reliability significantly.
Parameters that a sensing planning system produces can vary (see figure 1). In the case of a vision system they can be the set of target features, the measurement pose or poses and optical settings of the sensor, and in some cases the pose of an illuminator (Heikkilä et. al. 1988, Tarabanis et. al. 1995). The method presented here focuses on calculating the pose of the sensor. The optical settings include those for internal parameters including visibility, field of view, focus, magnification or pixel resolution and perspective distortion. The illumination parameters include illuminability, the dynamic range of the sensor and contrast (Tarabanis et. al. 1995).

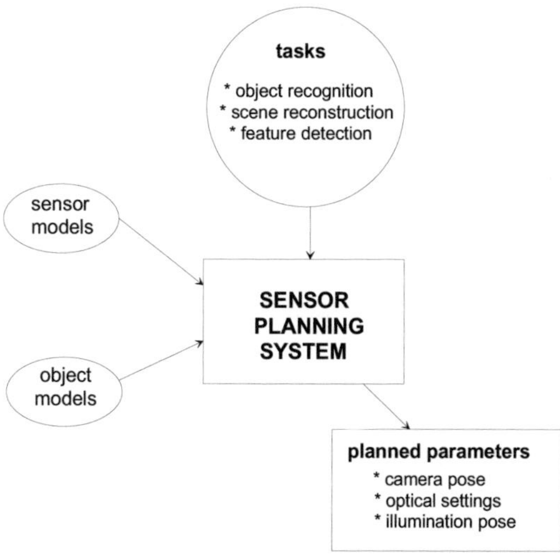

Figure 1. Sensing planning for computer vision (adapted from Tarabanis et. al. 1995)

In the object recognition that precedes the planning phase in general sensing planning systems, the object information is extracted from CAD models. The required parameters or other geometric information for sensing planning will be selected automatically or manually from the CAD models. This selection is based on Verification Vision Approach assuming that shape and form of the objects is known beforehand (Shirai 1987).
In the following chapters, pose estimation methods and related sensing planning for work object localization are described. In addition, some further analysis is done for the criteria used in the sensing planning. Finally, an example for work object localization with automatic sensing planning is described, followed by a discussion and conclusions.

2. Work object localization

In the robot-based manufacturing work cells, the work objects are fixed in the robot's working environment with fastening devices, like fixtures or jigs. These devices can be calibrated separately into the robot coordinate system. However, the attachment of the work object may not be accurate and even the main dimensions of the work object may be inaccurate, especially when considering cast work objects in foundries (Sallinen et. al. 2001). Geometric relationships with coordinate frames and transformations of the measuring system are illustrated in figure 2. In this case the sensor is attached to the robot TCP and the work object is presented in robot coordinate frame.

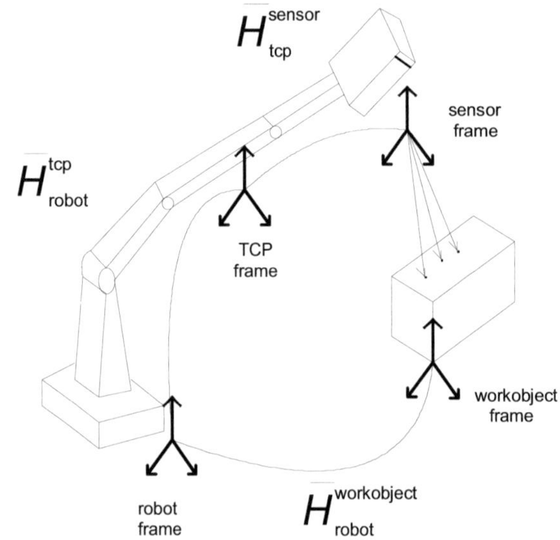

Figure 2. Coordinate frames and transformations for the work object localization

2.1 Estimation of the work object model parameters

In the work object pose estimation, the 3D point in sensor frame is transferred first into the robot wrist frame, then to the robot base frame and finally to the work object frame, where an error function is calculated for fitting the measurements to the object model. The idea is to fit the points measured from the object surface into reference model of the work object. All the points are transformed to the work object coordinate frame, see figure 2. As a reference information a CAD model of the work object is used.

We define the work object surface parameters by surface normal $\bar{n}$ and the shortest distance from the work object coordinate origin ***d***. To minimize the distance between the reference model and measured points, we need an error function. The error function for a point in the surface of the work object is now defined as

$$\bar{e}_{PtoS} = \bar{n} \cdot \bar{p} - d \tag{1}$$

where

$\bar{e}_{PtoS}$ is the error function from point to surface

$\bar{p}$ is the measured point

$\bar{n}$ is the surface normal vector

d is the shortest distance from surface to the work object origin

We define the pose of the work object as $\bar{m}_{workobject} = [x \quad y \quad z \quad \phi_x \quad \phi_y \quad \phi_z]$, which includes three translation and three rotation parameter (xyz euler angles). The corrections for estimated parameters are defined as additional transformations following the nominal one. The pose parameter (translations and rotations) values are updated in an iterative manner using linear models for the error functions. Linearization of the equation (1) gives the following partial derivatives for the work object pose parameters using the chain rule:

$$\frac{\partial \bar{e}_{PtoS}}{\partial \bar{m}_{workobject}} = \frac{\partial \bar{e}_{PtoS}}{\partial \bar{p}_P} \frac{\partial \bar{p}_P}{\partial \bar{m}_{workobject}} \tag{2}$$

Equation (2) can be written in matrix form as

$$\bar{J}_m^e = \frac{\partial \bar{p}_{PtoS}}{\partial \bar{m}_{workobject}} = \begin{bmatrix} n_x & n_y & n_z \end{bmatrix} \begin{bmatrix} 1 & 0 & 0 & 0 & p_{z,c} & -p_{y,c} \\ 0 & 1 & 0 & -p_{z,c} & 0 & p_{x,c} \\ 0 & 0 & 1 & p_{y,c} & -p_{x,c} & 0 \end{bmatrix} \tag{3}$$

where

$\bar{J}_m^e$ is the Jacobian matrix

n_x, n_y and n_z are the components of the surface normal vector.

$p_{x,c}$, $p_{y,c}$ and $p_{z,c}$ are the coordinates of the point in reference model.

Each sample (a point) on the surface produces a row into the compound work object localization Jacobian matrix. The Jacobian matrix, equation (3), is then used for the computation of correction increments for the work object localization, i.e., for the pose parameters:

$$\Delta \bar{m}_{workobject} = -\left(\overline{J_m^e}^T \bar{Q}^{-1} \overline{J_m^e} \right) \overline{J_m^e}^T \bar{Q}^{-1} \bar{e}_{PtoS} \tag{4}$$

where

$\Delta \bar{m}_{workobject}$ is an incremental correction transformation fort he work object pose,

$\bar{Q}$ is error covariance of the measurements $\bar{e}_{PtoS}$,

$\bar{J}_m^e$ are jacobians, equation (3)

$\bar{e}_{PtoS}$ is the error vector containing the error values, cf. equation (1)

In each iteration step we get a correction vector $\Delta \bar{m}_{workobject}$ and the estimated parameters are updated using the correction vector $\Delta \bar{m}_{workobject}$. Computation of corrections using equation 4 is repeated until corrections become close to zero.

2.2 Weight matrices for the workobject localization

We are calculating the matrices to weight more certain measurements more than uncertain ones and call here these as weight matrices. In this case the error covariances become somewhat more complicated. Work object localization includes errors coming from the robot joint control and range sensor measurements. In addition to this, work object localization is affected by the error of the sensor hand-eye calibration.

The Jacobian for the measurements can be written as follows:

$$\bar{J}_h^e = \frac{\partial \bar{e}_{PtoS}}{\partial \bar{h}} = \begin{bmatrix} \frac{\partial \bar{e}_{PtoS}}{\partial \bar{h}_{TCP}} & \frac{\partial \bar{e}_{PtoS}}{\partial \bar{h}_{S,TCP}} \end{bmatrix} \tag{5}$$

where

$\frac{\partial \bar{e}_{PtoS}}{\partial \bar{h}_{TCP}}$ is the partial derivative of the error function with robot TCP parameters

$\frac{\partial \bar{e}_{PtoS}}{\partial \bar{h}_{S,TCP}}$ is the partial derivative of the error function with sensor in robot TCP

The error function that is used to derive equation (5) is the one presented in equation (1). The Jacobian of the robot pose parameters $\partial \bar{e} / \partial \bar{h}_{TCP}$ can be written then as follows:

$$\bar{J}^{e}_{TCP} = \frac{\partial \bar{e}}{\partial \bar{p}_s} \frac{\partial \bar{p}_s}{\partial \bar{h}_{TCP}} \tag{6}$$

For the Jacobian for the parameters of the sensor hand eye calibration we get

$$\bar{J}^{e}_{hs,TCP} = \frac{\partial \bar{e}}{\partial \bar{h}_{S,TCP}} = \frac{\partial \bar{e}}{\partial \bar{p}} \frac{\partial \bar{p}}{\partial \bar{h}_{S,TCP}} \tag{7}$$

where

$$\frac{\partial \bar{e}}{\partial \bar{h}_{S,TCP}} = \begin{bmatrix} n_x & n_y & n_z \end{bmatrix} \bar{V}_P^T \bar{V}_R \begin{bmatrix} 1 & 0 & 0 & 0 & p_Z & -p_Y \\ 0 & 1 & 0 & -p_Z & 0 & p_X \\ 0 & 0 & 1 & p_Y & -p_X & 0 \end{bmatrix} \tag{8}$$

The covariance for the actual measurement is then an extension of the one in the sensor calibration:

$$\bar{R} = \begin{bmatrix} \bar{R}_{\Delta TCP,W} & \bar{0} \\ \bar{0} & \bar{R}_{\Delta S,TCP} \end{bmatrix} \tag{9}$$

where

$\bar{R}_{\Delta TCP,W}$ is the noise matrix of the robot TCP

$\bar{R}_{\Delta S,TCP}$ is the noise matrix between sensor origin and robot TCP

The covariance matrix $\bar{Q}$ for the measurements $\bar{J}^{e}_{m}$ is then

$$\bar{Q} = \bar{J}^{e}_{h} \bar{R} \bar{J}^{e\,T}_{h} \tag{10}$$

which is finally used in the estimator, see equation 4.

In the case of the work object localization, the estimated parameters $\bar{m}_{e,w-o}$ are $\bar{m}_{e,w-o} = \begin{bmatrix} x & y & z & \phi_x & \phi_y & \phi_z \end{bmatrix}$. The form of the partial derivative is close to equation (5), but due to a different error function the error derivative for the error covariance of the measuring calibration is written as

$$\frac{\partial \bar{e}_{PtoS}}{\partial \bar{h}_{S,TCP}} = \begin{bmatrix} n_x & n_y & n_z \end{bmatrix} \bar{V}_P^T \bar{V}_R \begin{bmatrix} 1 & 0 & 0 & 0 & p_Z & -p_Y \\ 0 & 1 & 0 & -p_Z & 0 & p_X \\ 0 & 0 & 1 & p_Y & -p_X & 0 \end{bmatrix} \tag{11}$$

where

$\lfloor n_x \quad n_y \quad n_z \rfloor$ is the partial derivative of the error function with respect to the point in the sensor frame, i.e. $\partial \bar{e} / \partial \bar{p}_s$.

The effect of the sensor calibration error, the partial derivatives of the error function with respect to parameters of the sensor calibration can be written as follows:

$$\frac{\partial \bar{e}_{PtoS}}{\partial \bar{h}_{S,TCP}} = \lfloor n_x \quad n_y \quad n_z \rfloor \bar{V}_P^T \bar{V}_R \bar{V}_T \begin{bmatrix} 1 & 0 & 0 & 0 & p_Z & -p_Y \\ 0 & 1 & 0 & -p_Z & 0 & p_X \\ 0 & 0 & 1 & p_Y & -p_X & 0 \end{bmatrix} \tag{12}$$

The equation (12) has been used to calculate weight matrices for the calculation of covariances in the case of work object location estimation. In addition to that, the robot joint uncertainty (noise in the joints propagated in to TCP pose uncertainty) has been taken into consideration.

3. Planning of the measurements

This chapter presents a new synthesis method for sensing planning based on minimization of a posteriori error covariance matrix and eigenvalues of it. The method is based on multiplication of the Jacobian matrix by the weight matrix, and minimization means here manipulation of the terms in the Jacobian and weight matrices to achieve a low level of spatial uncertainties. The minimization of error covariance matrix and effect of signal-to-noise ratio is illustrated in simulations, where the location of the measurement points in the surface of the work object affects to the pose uncertainties. This is because composition of equations is made intuitively from the equations composed for parameter estimation. The Jacobian and weight matrices may belong to any phase of the calibration of the robot system, and planning is carried out for each phase separately. The work is relying on the work done by Nakamura & Xu in (Nakamura & Xu 1989) and mathematical background for the work is in principal component analysis (PCA) introduced in (Hotelling 1933).

3.1 Error covariance matrix and SNR

The earlier reported methods for sensing planning using a posteriori error covariance matrices as evaluation criteria are suitable for many cases, but they have following shortcomings: Borgli & Caglioti (Borghi & Caglioti 1998) used the method only in a 2D case and based it on the calculation of range distances between a robot and a surface. This kind of criteria formation is not possible in hand-eye calibration, for instance, because the solution space is more complicated. Nilsson et. al. (Nilsson et. al. 1996) also used geometric distances and the respective uncertainties as criteria for sensing planning. These methods were designed for mobile robots and are more suitable for purposes in which there are usually 3 degrees of freedom and the minimization criteria can be composed based on simple geometric features of the environment. Condition number is also a criteria for evaluating uncertainties. Geometrical interpretation of condition number is a ratio of largest and smallest dimension of the uncertainty ellipsoid. This has been studied in (Motta & McMaster 1997) and (Zhao & Wang 2007). In most of the cases this gives a relatively good solution but if the criteria will be optimized, it leads to a situation where uncertainty ellipsoid is flat. It means uncertainties are very small in one direction and very large in other direction respectively. Nahvi presents a new observability index called Noise Amplification Index which is a ratio of maximum singular value to condition number of the measurement matrix (Nahvi & Hollenbach 1996). We use this criteria also for evaluating our sensing planning method later in chapter 6. First versions of sensing planning for Hand-eye calibration presented here were presented in (Sallinen & Heikkilä 2003).

The planning is focused on selecting reference features in the measurement pose. The poses are assumed to be given, and optimization considers selecting the features that will be measured. The goal is to minimize the uncertainties in an estimated pose and the selection of measurements that especially affect the rotation uncertainties is considered. As will be illustrated later, the estimation of translation parameters is not so sensitive to quality of measurements as the estimation of the orientation parameters of the pose. Here the minimization and maximization of matrices means selecting of values for parameters in matrix rows and columns. The goal of the selection is to achieve low level of uncertainties in terms of eigenvalues of the error covariance matrix. To simplify the evaluation of uncertainty criteria, multiplication and sum of all eigenvalues is used which is illustrated in simulations further in this chapter.

2.2 Computation of the covariance

The method presented here is based on minimizing a posteriori error covariance of the estimated parameters when using Bayesian-form modelling of the spatial uncertainties. This means that the sensor has to be placed in such poses that the noise in the estimated parameters will be minimized. In addition to noise, there are a lot of unmodelled error sources which cannot be adjusted.

The error covariance matrix $\vec{P}$ of the estimated parameters can be calculated as follows:

$$\vec{P} = (\vec{J}^T \vec{Q}^{-1} \vec{J})^{-1} \tag{13}$$

where

$\vec{J}$ is the Jacobian matrix for estimating the model parameters

$\vec{Q}$ is the weight matrix of the measurements

The error covariance matrix $\vec{P}$ consists of the product of the Jacobian matrix $\vec{J}$ and the weight matrix $\vec{Q}$. According to equation (13), the minimization leads to a situation in which $\vec{J}$ has to be maximized and $\vec{Q}$ has to be minimized. These matrices share the same parameters, however, so that minimization of $\vec{P}$ is not so straightforward. It may be carried out in two ways: by measuring optimal points for localization, i.e. each row in the Jacobian matrix will give a large amount of information and there are only a few rows, or else the number of rows in the Jacobian matrix can be increased but each row will give only a little information.

The control parameters of the sensing planning system presented here are parameters of the Jacobian matrix $\vec{J}$ and weight matrix $\vec{Q}$. These may include (x,y,z) point values of a calibration plate frame in hand-eye calibration, range of measurement in the sensor frame, point values on the work object surface or the orientation of the TCP of the robot, for example.

3.3 Computation of the criteria

When planning the set of measurements, the parameters to be estimated have to be considered first. The partial derivatives of the error function with respect to the estimated parameters comprise the parameters that affect the spatial uncertainties of the estimated parameters. The error function is the same as that used in parameter estimation, e.g. estimation of the pose of the work object. Estimation of rotation parameters requires at least a pair of points, which leads to a situation in which the distance between these points is maximized in order to achieve a high signal to noise ratio (SNR) and this is formulated as a max problem. Based on these parameters, the set of samples will be generated by selecting the maximum values of the composed parameters, taking into consideration the constraints of the parameter space. Thus the equation to be maximized for generating a set of samples based on a Jacobian matrix is the following:

$$\vec{\Psi}_J = \partial\vec{e}/\partial\vec{m} \tag{14}$$

where

$\vec{e}$ is the error function

$\vec{m}$ are the estimated parameters

The boundary values for (14) are usually set by the dimensions of the parameter space. It means the control parameters have to be manipulated to maximize the distance between the measurement points, i.e. sequential lines of Jacobian matrix. Signs of control parameters do not need to be considered because covariance matrix P includes multiplication of Jacobian matrix $\vec{J}^T\vec{J}$ and the result is always positive. The number of solution spaces depends on the amount of parameters in a Jacobian row. In the example above, there are two parameters and it will give two different point pairs: $(x \approx 0, y \approx c)$ and $(x \approx c, y \approx 0)$.

In general case, the Jacobian matrix $\partial\vec{e} / \partial\vec{m}$ determines the control parameters and may include position and orientation parameters from the world frame or work object frame as well as the sensor range or other internal or external parameters of the sensor. The composed position parameters from the partial derivative $\partial\vec{e} / \partial\vec{m}$ are maximized by selecting their largest possible values.

The idea is following the ideas presented by Lowe (Lowe 1985) where a small rotation around different axes can be projected as a small translation in a perpendicular axis. Using this principle, the set of samples is generated by maximizing the distances between the points.

2.4 Effect of the weight matrix

The weight matrix $\vec{Q}$ also has to be studied when considering the optimal positions and orientation. The weight matrix consists of partial derivatives of the error function with respect to the measurements $\vec{K}$ and their noise matrix, as follows:

$$\vec{Q} = \vec{K}\vec{R}\vec{K}^T \tag{15}$$

where

$\vec{K}$ is a matrix consisting of partial derivatives of the error function with respect to the input parameters

$\vec{R}$ is a matrix consisting of the noise affecting the input parameters

If the noise matrix $\vec{R}$ is assumed to be constant, then the goal will be to minimize the weight matrix $\vec{K}$, which can be written in partial derivative form:

$$\vec{K} = \partial e / \partial \vec{m}_{input} \tag{16}$$

where

e is the error function

$\vec{m}_{input}$ is the vector containing the measurement parameters

As in the case of computation of the criteria, the error function depends on the type of surface form and the measurement parameters of the robot system. E.g., the plane surface is a part of cube or plane -formed calibration plane for hand-eye calibration. The partial derivatives for the weight matrix can be written:

$$\vec{\Psi}_K = \partial e / \partial \vec{m}_{input} \tag{17}$$

Optimizing both equations (14) and (17) is not so simple, due to the interdependence of the parameters between the equations. Of the two, equation (14) has the larger effect on the estimation process, due to the duplicate terms in the Jacobian matrix $\vec{J}$ compared with the weight matrix $\vec{Q}$ in (13). The effect of equation (15) becomes noticeable when the noise of the input parameters is not linear.

Minimization of weight matrix $\vec{Q}$ is similar with maximizing the Jacobian matrix $\vec{J}$. In the case of weight matrix, also the point pairs of sequential rows are selected and distance between them are minimized. The weight matrix includes multiplication of matrices $\vec{K}$ such that sign does not have to be considered $(\vec{K}\vec{K}^T)$.

When forming the total measurement matrix $\vec{\Psi}$, the information obtained from both the Jacobian measurement matrix $\vec{\Psi}_J$ and the weight measurement matrix $\vec{\Psi}_K$ has to be considered.

4. Properties of the planning criterion

Here we verify the planning criterion by simulating measurements and evaluating results by two different methods. The selection of measurement points in sensing planning i.e. maximizing the distances between the points is shown here to illustrate the effect of given criteria for spatial uncertainties of the localized cube. The result of the sensing planning is verified by distributing the measurement points around the three sides of the work object and analyzing the spatial uncertainties of a corner point of the localized cube. The points included equal noise. Throughout this simulation, the criteria of selection of the measurement points are maximum distances from each others. The criteria used for evaluating the uncertainties were product of eigenvalues of error covariance matrix, sum of eigenvalues of error covariance matrix and volume of the uncertainty ellipsoid. In all the three different cases, two pairs of points from each surface was measured. The placing of the points to the surfaces was done similarly in each side: all the possible location variations of measurement points were calculated. The parameter that uncertainties were related to was a distance between measurement points on each surface. The criterion for selecting the pair of points is two distances that give the largest total distance. These two largest distances were summed together and total distance becomes then a sum of distances between two point pairs in each surface.

Figure 3 illustrates the uncertainties of a point in the calibrated cube frame when the measurement points are distributed around the three sides of the cube. In all the figures 3-5 the highest uncertainties are cut off to improve the illustration of the curves. From the figure 3 it can be seen that when the distance between the points on the surface of each side increases, the uncertainties of pose of the cube decreases. The trend of the curve seems to be that uncertainties are decreased dramatically first and after that, decrease is more stable.

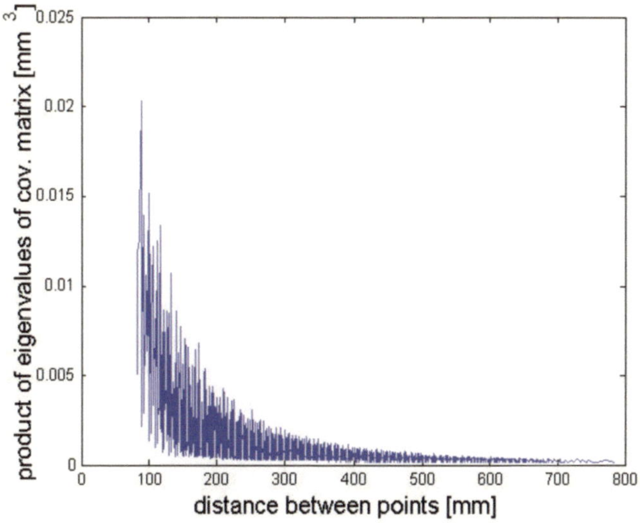

Figure 3. Product of eigenvalues of error covariance matrix vs. distance between the measurement points

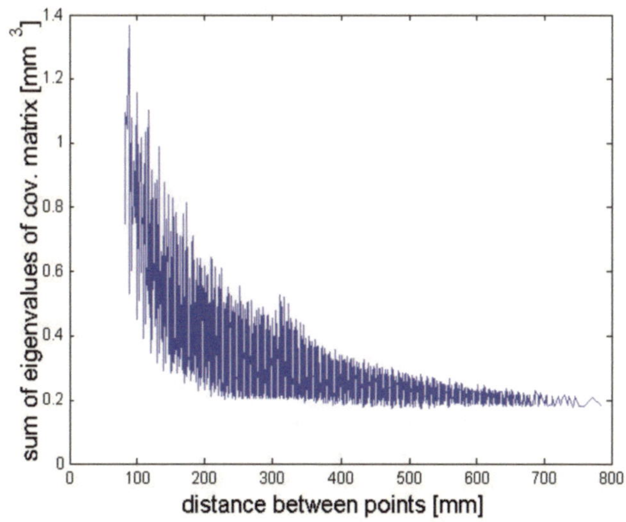

Figure 4. Sum of eigenvalues of error covariance matrix vs. distance between the measurement points

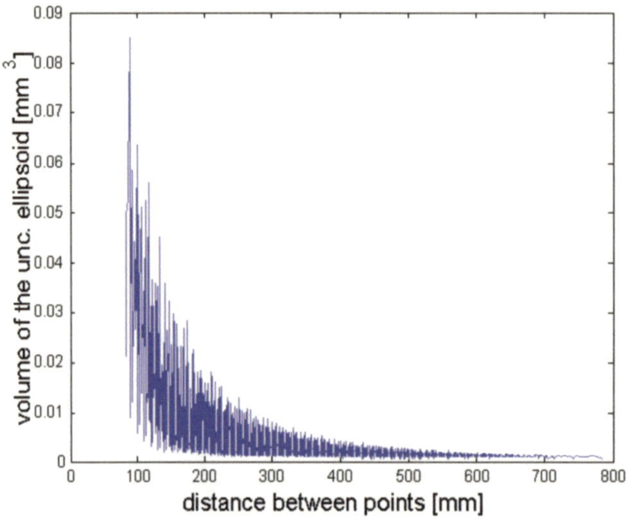

Figure 5. Volume of the uncertainty vs. distance between the measurement points

In the figure 4, sum of the eigenvalues of the error covariance matrix with respect to distance between the points is presented. As in the case of product of eigenvalues, the trend is that uncertainties are decreasing, though here not such fast.

In the last case, figure 5, volume of the uncertainty ellipsoid is illustrated for different distances between the points. Curve is close to product of eigenvalues, i.e. the curve is decreasing rather fast.

All the three different cases illustrated the same behaviour between uncertainties of the localized cube and distance between the measurement points. Here one criteria of uncertainties in simulations has been selected to simplify the interpretation of results, i.e. product on sum of eigenvalues, not single eigenvalues. This kind of method is suitable because usually the goal is to minimize all uncertainties of the system, not only within one direction.

The principle of using distances between measurement points can be generalized into other forms of set of points. The set of points may form a feature the distances of which will be considered. Here an example of localizing of a cube using three points forming a triangle in each surface is illustrated. The goal is to decrease uncertainties and the distance of points is calculated as a circle of triangle formed by the three measurement points. The uncertainties are illustrated in a same way as in case of two point pairs, product and sum of eigenvalues and the volume of the uncertainty ellipsoid.

However, if the number of points increases, the complexity of computing distances becomes higher and one has to be careful when calculating the distances.

The problem involved in planning the reference features illustrated in this chapter is that planning is based on the assumption that the feature contains very few or no spatial uncertainties. This means that the requirement for successful measurement is that the deviation between the nominal location of the pose of the work object, for example, and its actual location should be smaller than in the selection of details for the reference features. If the deviation is larger, there is a possibility that the robot will measure false features or even ones lying beyond the work object. The proposed planning method is illustrated in an example of work object localization.

5. Example: Planning of the measurements for workobject localization

We will consider here equations for planning the measurements required to localize a cubic work object. Of the six plane surfaces constituting the cube, three are measured for localization of the pose. The sensing planning starts out from the parameters to be estimated, i.e. the pose of the work object.

The pose of the work object is determined using three translation values and three orientation values:

$$\vec{m}_{cube} = \begin{bmatrix} x & y & z & \phi_x & \phi_y & \phi_z \end{bmatrix} \tag{18}$$

The measurement points are planned for each of the three plane surfaces, in order to minimize uncertainties in all six pose parameters. Each surface is determined using translation from the origin to the surface and two orientation values of the surface. The surfaces are termed here the x-plane, y-plane and z-plane, referring to normal directions in the world coordinate frame. Error functions for localizing the cube in the world frame can then be written for the sides as follows: the error function for the measurement point on the x-plane surface is $e_x = x$, for the measurement point on the y-plane surface $e_y = y$, and for the measurement point on the z-plane surface $e_z = z$. These error functions are used to compose the parameters for sensing planning. The Jacobian matrix for estimating the pose parameters of the work object for the measurements $1 \ldots n$ for three different surfaces can be written as (19).

The goal of the planning is to find measurements points (x,y,z) for each pair of rows in (19) such that they will give an optimum or close to optimum amount of information for parameter estimation.

$$\frac{\partial \vec{e}}{\partial \vec{m}_{cube}} = \begin{bmatrix} 1 & 0 & 0 & 0 & -z_{x,1} & y_{x,1} \\ \vdots & \vdots & \vdots & \vdots & \vdots & \vdots \\ 1 & 0 & 0 & 0 & -z_{x,n} & y_{x,n} \\ 0 & 1 & 0 & -z_{y,1} & 0 & x_{y,1} \\ \vdots & \vdots & \vdots & \vdots & \vdots & \vdots \\ 0 & 1 & 0 & -z_{y,n} & 0 & x_{y,n} \\ 0 & 0 & 1 & -y_{z,1} & x_{z,1} & 0 \\ \vdots & \vdots & \vdots & \vdots & \vdots & \vdots \\ 0 & 0 & 1 & -y_{z,n} & x_{z,n} & 0 \end{bmatrix} \tag{19}$$

where

$(x_1 \ldots x_n, y_1 \ldots y_n, z_1 \ldots z_n)$ are the measurements of the parameters (x,y,z) between $1 \ldots n$ for x, y and z surfaces

n is the number of measurements

To find the parameters to be maximized in each pair of rows in Jacobian matrix, the partial derivative of the error function with respect to the parameters to be estimated for the plane surface in the direction x can be written as (20). The whole pose includes 6 pose parameters, but each surface (here the x-plane) gives information only on three parameters and also information on these parameters for the Jacobian matrix, i.e. one translation and two orientation parameters.

$$\frac{\partial \vec{e}_x}{\partial \vec{m}_{cube,x}} = \begin{bmatrix} \frac{\partial e}{\partial x} & \frac{\partial e}{\partial \phi_y} & \frac{\partial e}{\partial \phi_z} \end{bmatrix} \tag{20}$$

The partial derivative of the error function with respect to the estimated pose parameter x is 1. The control parameters are then obtained from partial derivatives of the error function with respect to the orientation parameters, equation (18), when the constraints of the model are $0 \ldots n$ and $n+1 \ldots n+n$, as follows:

$$\vec{\Psi}_{i=0}^{i=n} = \max\{(y_1, z_1), (y_2, z_2) \ldots (y_{n-1}, z_{n-1}), (y_n, z_n)\}_{cube,x} \tag{21}$$

$$\vec{\Psi}_{i=0}^{i=n} = \max\{(x_1, z_1), (x_2, z_2) \ldots (x_{n-1}, z_{n-1}), (x_n, z_n)\}_{cube,y} \tag{22}$$

$$\vec{\Psi}_{i=0}^{i=n} = \max\{(x_1, y_1), (x_2, y_2) \ldots (x_{n-1}, y_{n-1}), (x_n, y_n)\}_{cube,z} \tag{23}$$

The minimization and maximization (21), (22) and (23) all contain the same parameters, so that the result of planning is that measurements will be minimized and maximized in these directions and the uncertainties in the pose of the work object will be minimized.

The criteria to be considered are the weight matrix of the error covariance matrix and computation of the parameters which affect the uncertainties in the estimated parameters, i.e. matrix $\vec{\Psi}_K$. This means that the weight matrix $\vec{K}$ has to be minimized, which is done by minimizing its parameters in a pair of rows. When locating the work object, as in the case of calculating the matrix $\vec{\Psi}_J$, each surface has to be considered separately, because weights for the surfaces are different. The input to the system is a measurement point p: $p = \lfloor p_x \quad p_y \quad p_z \rfloor$.

The weight matrix is composed by calculating the partial derivatives of an error function $e_x = x$ with respect to the input parameters. For the plane in the direction x, the weight matrix can be computed as follows:

$$\frac{\partial e_x}{\partial p} = [1 \quad 0 \quad 0] \tag{24}$$

The weight matrices for the plane surfaces in the directions y and z can be written in a similar way, where the error functions are $e_y = y$ and $e_z = z$. As illustrated above, the weight matrices are constant in all directions. It can be written for $1 \ldots n$ measurements for all in one matrix:

$$\vec{K}_{cube} = \begin{bmatrix} 1 & 0 & 0 \\ 0 & 1 & 0 \\ 0 & 0 & 1 \\ \vdots & \vdots & \vdots \\ 1_n & 0 & 0 \\ 0 & 1_n & 0 \\ 0 & 0 & 1_n \end{bmatrix} \tag{25}$$

The above example means that only the translation parameters of the measurement affect the uncertainties in the estimated parameters, and the measurement poses can be planned based on information from the Jacobian matrix $\vec{\Psi}_J$. This is because only the uncertainty attached to the measurement points in the work object coordinate frame was considered. The output of the sensing planning algorithm is a total measurement matrix $\vec{\Psi}$ which is a combination of the measurement matrix $\vec{\Psi}_J$ and the measurement matrix $\vec{\Psi}_K$.
The total number of features composed using sensing planning is 12. Depending on the requirements of the spatial uncertainties of the location of the work object, the measurement points are constantly distributed around these features.

6. Experimental tests in laboratory

6.1 Calculation of a pose of a cube

To test the sensing planning algorithm when estimating full 6 DOF pose, we run a test where we measured points from three sides of a cube. The marks were positioned around the cube using the PUMA 650 industrial robot with 6 degrees of freedom and all the points was generated off-line. The camera system we used was CCD stereo camera system, developed by Tomita et. al. More information about the camera system, see (Tomita et. al. 1998) and for test case (Sallinen 2002).

6.2 Results from the actual tests

The different set of samples was generated in a following way: planned set of samples means the sample set generated using the sensing planning method proposed in this chapter. Random set of samples is a set of points randomly selected from the surfaces of the cube and pattern is constantly selected points from three side of the cube. The noise level and the number of points (8 in each side) is same in all the set of samples. The table 1 illustrates the results from the pose estimation of a cube.
Results in the table 1 illustrates the goodness of the selection of the set of samples when using the proposed sensing planning method. In the table, eigenvalues of the pattern set is close to planned, but still worse than that. The remarkable improvement between pattern and planned compared with random and planned is the orientation of the eigenvectors. When using planning algorithm, the eigenvectors points almost towards to coordinate axis of the estimated cube. In the case of pattern eigenvectors are a little bit rotated and in the case of random, they point random directions. When the eigenvectors point to direction of axis, the eigenvalues are usually close to each other. This means the uncertainty ellipsoid has a form of a sphere.

Set of sample	Eigenvalue λ_1 [mm]	Eigenvalue λ_2 [mm]	Eigenvalue λ_3 [mm]	Eigenvectors
Planned	0.0381	0.0439	0.0339	0.9651 0.1782 0.1919 0.1577 -0.9805 0.1173 -0.2090 0.0830 0.9744
Random	0.0510	0.0402	0.0722	-0.7292 0.6107 -0.3086 0.6607 0.7457 -0.0856 -0.1778 0.2663 0.9473
Pattern	0.0493	0.0408	0.0453	-0.9369 0.3494 0.0133 0.2193 0.6168 -0.7560 0.2723 0.7054 0.6545

Table 1. Results from estimation of a cube

Compared to our previous studies, the optimal location of set of samples is more important than the noise level or the number of samples. When we have a good set of samples, the eigenvectors are not rotated and in that way eigenvalues are easier to control (e.g. changing the noise level or increasing the number of samples). In addition to eigenvalues, we calculated condition number, noise amplification indexes and volumes of the uncertainty ellipsoid for cube estimation. For more information about Noise Amplification Index, see (Nahvi & Hollenbach 1996). Table 2 illustrates the results:

Set of sample	Condition Number	Noise Amp. Index	Vol. Of Uncertainty Ellipsoid
Planned	565	$0.470*10^6$	$2.3760*10^{-4}$
Random	1721	$1.424*10^6$	$6.1957*10^{-4}$
Pattern	813	$0.594*10^6$	$3.8159*10^{-4}$

Table 2. Condition number, Noise amplification indexes and volume of uncertainty ellipsoid for different set of samples

Results in the table above show the efficiency of the proposed sensing planning method. As in the case of eigenvalues, pattern -form set of samples is better than random but not such good as planned. According to results, even the scale between different evaluation criteria seem to be close to each other.

7. Industrial application

The calibration and sensing planning algorithms were implemented in an industrial application to test the operation in foundry environment. The application consists of planning the localization measurements and localization of the work object. A touch sensor was selected as the measuring device because it is working well in harsh environments and is easy to use. Disadvantages include speed of measuring and sparse amount of measured data.

The target for sensor feedback was to localize a sand mould for robotized milling. It was done by six measurements made on its front surface. The measuring system was built in a demonstration robot cell at the foundry, Finland, see figure 6.

Figure 6. Actual robot system

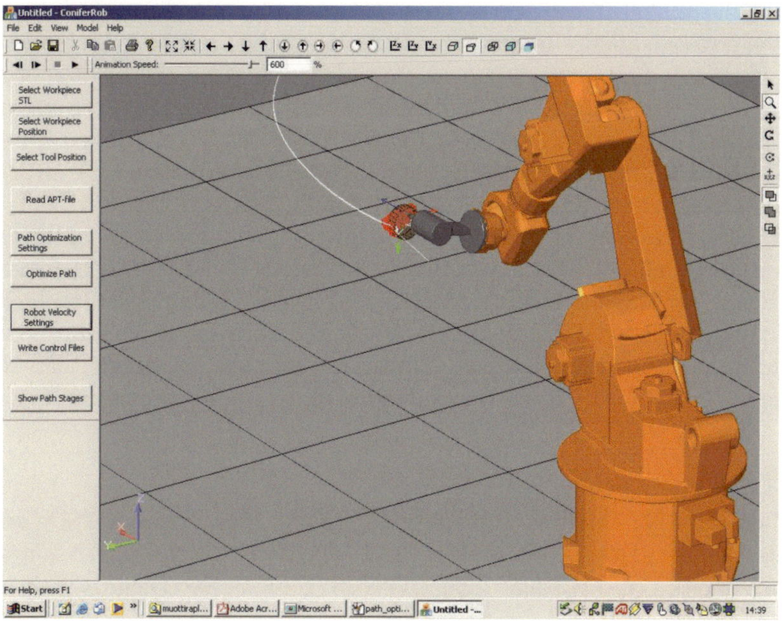

Figure 7. Sensing planning in the off-line programming software

The measuring system was implemented into a commercial robot off-line programming system which includes an easy-to-use graphical interface to facilitate use, see Figure 7. Software is distributed by Simtech Systems Ltd (Simtech 2008).
The planning and measuring system operated well in the foundry application. When measuring took time, it was very useful for operator to get verification that there are enough measurement points. Calibration was also successful in every case during the test period.

8. Discussion

Goal of the work was to achieve a flexible robot-based manufacturing work cell. A lot of work can be done off-line before executing the job with the actual robot. This preparatory work involves planning the measurements, simulating them and computing the related spatial uncertainties. The final goal for sensing planning is analysis of the results and a decision to increase the number of measurements or begin executing the job with the actual robot. The planning system uses initial CAD information obtained from the work cell, including the robot model and work object models, but is also able to adjust itself to changing conditions which are not all programmed beforehand. This adjustment is based on sensor observations on the environment and can be carried out on-line. The results of this chapter represent a system which is operating in the way described here. The system has been verified with point simulations, laboratory tests with a robot and a camera system and finally in industrial application. All phases show the goodness of the developed sensing planning method.

9. Conclusions

In this chapter we have presented a novel method for sensing planning which relies on modelling of the spatial uncertainties using covariance and trying to minimize criterion related to the error covariance matrix. The planning method can be used for generating an optimal or close to optimal measurements for robot and work cell calibration. When the behaviour of the spatial uncertainties in parameter estimation is known, the error covariance matrix can be used as a planning criterion and the results can be interpreted either as values or geometrically. The selection of set of points is based on minimizing the error covariance matrix and it leads to a situation where distances between measurement points are maximized. The reliability of the criteria is illustrated with simulations where pose uncertainties compared with distance between the measurement points. There is not such a close connection between uncertainty analysis and sensing planning in the literature as is proposed in this chapter, and as a result the new method of sensing planning is very flexible and easy to implement in conjunction with other planning operations than work object localization, e.g. hand-eye calibration or estimation of the model parameters for a surface.

10. References

Blank S., Shen Y., Xi N., Zhang C. & Wejinya U., (2007). High Precision PSD Guided Robot Localization: Design, Mapping, and Position Control. *Proceedings of IEEE International Conference Intelligent Robots and Systems*, San Diego, CA, USA, Oct 29 – Nov 2.

Borghi G. & Caglioti V., (1998) Minimum Uncertainty Explorations in the Self-Localization of Mobile Robots, *IEEE Transactions on Robotics and Automation*. Vol. 14, No. 6, pp. 902-911.

Edan Y., Rogozin D., Flash T. & Miles G. (2000). Robotic Melon Harvesting. *IEEE Transactions on Robotics and Automation*. Vol 16, No 6, Dec 2000. Pp. 831-834.

Gatla C., Lumia R., Wood J. & Starr G. (2007). An Automated Method to Calibrate Industrial Robots Using a Virtual Closed Kinemaatic Chain. *IEEE Transactions on Robotics*, Vol. 23, No 6.

Heikkilä T., Matsushita T. & Sato T., (1988). Planning of Visual Feedback with Robot-Sensor Co-operation. *Proceedings of the IEEE International Workshop on Intelligent Robots and Systems*, Tokio 1988, pp. 763 - 768.

Hotelling, (1933). Analysis of complex statistical variables into principal components. *Journal of Educational Psychics*, Vol. 24 pp. 417-441.

Järviluoma M. & Heikkilä T., (1995). An Ultrasonic Multi-Sensor System for Object Locating Applications. *Mechatronics*, Vol. 5, pp 433 - 440.

Lowe D., (1985). *Perceptual organisation and Visual Recognition*. Kluwer Academic, Hingham, U.K., 1985. 162 p.

Mooring B. W., Roth Z. S. & Driels M. S., (1991). *Fundamental of Manipulator Calibration*. John Wiley & Sons, inc. 324p.

Motta J.M. & McMaster R.S,, (1997). Improving Robot Calibration Results Using Modeling Optimisation. *IEEE Catalog Number* 97TH8280.

Nahvi A. & Hollenbach J., (1996). The Noise Amplification Index for Optimal Pose Selection in Robot Calibration. *Proceedings of IEEE International Conference on Robotics and Automation*, pp. 647-654.

Nakamura Y. & Xu Y., (1989). Geometical Fusion Method for Multi-Sensor Robotic Systems. *IEEE International Conference on Robotics and Automation*. Vol. 2, pp. 668-673.

Nilsson B., Nygårds J. & Wernersson Å., (1996). On 2D Posture Estimation with Large Initial Orientation Uncertainty and Range Dependant Measurement Noise. *ICARV'96*, Signapore, Dec 1996.

Prieto E., Redarce T., Boulanger P. & Lepage R., (2001) Tolerance control with high resolution 3D measurements. *IEEE International Conference on 3-D Digital Imaging and Modeling*. Page 339-346.

Sallinen M., (2002). Sensing Planning for Reducing the Workobject Localization Uncertainties, *Proceedings of the International Conference on Machine Automation* (ICMA2002), 11-13.9.2002, Tampere, Finland, pp. 329 – 337.

Sallinen M., Heikkilä T. & Sirviö M., (2001). A Robotic Deburring System of Foundry Castings Based on Flexible Workobject Localization, *Proceedings of SPIE Intelligent Robots and Computer Vision XX: Algorithms, Techniques, and Active Vision. Newton*, USA, 29 - 31 Oct. 2001, USA. Vol. 4572 (2001), pp. 476 – 487.

Sallinen M., Heikkilä T. & Sirviö M. (2006). Planning of sensory feedback in industrial robot workcells. *Proceedings of the IEEE International Conference on Robotics and Automation*. pp. 675-680.

Shirai Y., (1987). *Three-Dimensional Computer Vision*. Springler-Verlag. ISBN 3-540-15119-2, 296p.

Shirai Y. & Inoue H., (1971). Guiding a Robot by Visual Feedback. in Assembling Tasks. *Pattern Recognition* (5), No. 2, pp. 99 – 108.

Simtech (2008). Simtech Systems, www.easysimulation.com. Accessed June 2008.

Tarabanis K., A., Allen P. & Tsai R., (1995). A Survey of Sensor Planning in Computer Vision. *IEEE Transactions on Robotics and Automation*, Vol. 11, No. 1, Feb.9. pp.2429-2432

Tomita F., Yoshimi S., Ueshiba T., Kawai Y., Sumi Y., Matsushita T., Ichimura N., Sugimoto R. & Ishiyama Y., (1998). R&D of Versatile 3D Vision System VVV., *Proc. IEEE International Conference on System, Man and Cybernetics*, pp.4510-4516, San Diego, USA

Zhao H. & Wang S. (2007). Optimal Selection of Measurement Configurations for Robot Calibration with Solis&Wets Algorithm. *IEEE International Conference on Control and Automation.*

Vähä P., Heikkilä T., Röning J., & Okkonen J., (1994). Machine of the Future: An Intelligent Paper Roll Manipulator. *Mechatronics*, Vol. 4, No. 8 , pp. 861 - 877.

26

Robot Programming in Machining Operations

Bjørn Solvang[1], Gabor Sziebig[1] and Peter Korondi[2]
[1]*Narvik University College,* [2]*Budapest University of Technology and Economics*
[1]*Norway,* [2]*Hungary*

1. Abstract

Driven by the need for higher -flexibility and -speed during initial programming and path adjustments, new robot programming methodologies quickly arises. The traditional "Teach" and "Offline" programming methodologies have distinct weaknesses in machining applications. "Teach" programming is time consuming when used on freeform surfaces or in areas with limited visibility /accessibility. Also, "Teach" programming only relates to the real work-piece and there is no connection to the ideal CAD model of the work–piece. Vice versa during offline programming there is no knowledge about the real work-piece, only the ideal CAD model is used. To be able to relate to both the real- and ideal- model of the work-piece is especially important in machining operations where the difference between the models often represents the necessary material removal (Solvang et al., 2008 a).

In this chapter an introduction to a programming methodology especially targeted for machining applications is given. This methodology use a single camera combined with image processing algorithms like edge- and colour- detection, combines information of the real and ideal work-piece and represents a human friendly and effective approach for robot programming in machining operations.

2. Introduction

Industrial robots are used as transporting devices (material handling of work pieces between machines) or in some kind of additive- (e.g. assembly, welding, gluing, painting etc.) or subtractive- manufacturing process (e.g. milling, cutting, grinding, de-burring, polishing etc.). Also the industrial robot controller has good capability of I/O communication and often acts as cell controller in a typical set-up of a flexible manufacturing cell or system (Solvang et al., 2008 b)

Until now, driven by the need from the car manufactures, material handling and welding has been the most focused area of industrial robot development. Zhang et al. (2005) reports that material handling and the additive processes of welding counted for 80% of the industrial robot based applications in 2003, but foresights an extension into subtractive manufacturing applications. Brodgård (2007) also predict that a lightweight type robot, capable of both subtractive and additive manufacturing, will have an impact on the car industry and the small and medium sized enterprises (SMEs).

Large European research programs, like the framework programme 6 (FP 6) has put a major focus on the integration of the robot into the SME sector (SMErobot, 2005) and also in the newer research program FP 7: Cognitive Systems, Interaction, Robotics (CORDIS Information and Communication Technologies, 2008) there is focus onto the integration of the intelligent robot system with the human being.
In general there exist several strong incentives to open new areas for the usage of the industrial robot. However, many challenges in this development have been identified such as: lack of stiffness in the robot arm (Zhang et al., 2005) and difficulties in integration of various sensors due to vendor specific and their "normally closed" robot controllers.
In the case of subtractive manufacturing, contact forces must be taken into account. Forces must be measured and provided as feedback to the robot control system. Such force measurement usually involves sensors attached to the robot end-effector. At least older robot controllers do not support such possibility of sensor fusion within the control architecture. Several academic researches have been focusing in this field where some choose to entirely replace the original controller like Garcia et al. (2004) while others again include their own subsystem in-between the control loop, like Blomdell et al. (2005) and Thomessen & Lien (2000).
To address problems related to sensor integration in robotic applications several general architectures, named middleware's, are being developed for easy system integration. In Table 1. some developers of robot middlewares are identified. There are several differences between these units, among others; how many robots are supported; plug and play discovery; etc. Each of theses middleware solutions are composed from modularized components and have a hierarchical setup (Solvang et al., 2008 b). Brodgård (2007) states that one prerequisite for successful introduction of the industrial robot to the SMEs, and their rapidly changing demands, are modularity in the assembly of the robot arm (task dependant assembly of the mechanical arm) but also modularity of the software elements in the controller. As of today the robot manufactures seem to show more interest in sensor integration, perhaps driven by the competition from the middleware community as well as the possible new market opportunities represented by the huge SME sector.
Another challenge related to the introduction of the industrial robot into lighter subtractive machining operations is an effective human-robot communication methodology. Traditionally, man-machine interaction was based on online programming techniques with the classical teach pendant. Later we have seen a development into virtual reality based offline programming methodologies. Actually today, a wide range of offline software tools is used to imitate, simulate and control real manufacturing systems. However, also these new methodologies lack capability when it comes to effective human-machine communication. Thomessen et al. (2004) reports that the programming time of grinding robot is 400 times the program execution time. Kalpakjian & Scmid (2006) states that a manual de-burring operation can add up to 10% on the manufacturing cost. In general manual grinding, de-burring and polishing are heavy and in many cases unhealthy work-tasks where the workers need to wear protective equipment such as goggles, gloves and earmuffs (Thomessen et al., 1999).

Name	*Middleware technology*	*Relevant contributors*
ASEBA (Magnenat et al., 2007)	CAN	EPFL
CLARAty (Nayar & Nesnas, 2007)	Multi-level mobility abstraction	NASA
Microsoft RS (Jackson, 2007)	Web-Services	Microsoft
Miro (Weitzenfeld et al., 2003)	CORBA	University of California
Orca (Ozaki & Hashimoto, 2004)	ICE	KTH Stockholm
OrIN (Mizukawa et al., 2002)	DCON, SOAP, XML	JARA
Open-R (Lopes & Lima, 2008)	Aperios OS	Sony
Orocos (Bruyninkx et al., 2003)	RealTime Toolkit	Katholieke Universiteit Leuven
Player (Kranz et al., 2006)	Client / Server architecture	Multiple
RT-Middleware (Ando et al., 2005)	CORBA	AIST
YARP (Metta et al., 2006)	Client / Server architecture	MIT
UPnP (Veiga et al., 2007)	HTTP, SOAP, XML	University of Coimbra
Urbi (Baillie, 2005)	Client / Server architecture	Gostai

Table 1. Middleware architectures, based on Solvang et al. (2008 b)

Thus, there is a need to develop new methodologies for programming of industrial robots, especially associated to lighter machining operations like grinding/de-burring and polishing

In this chapter, focus will be kept on the human friendly and effective communication between the human operator and the robot system.

The organization of this chapter is as follows: Section 3 gives an overview of the presented programming methodology while section 4 presents details of some system components. Section 5 concludes and gives recommendations for further work.

3. Programming of the industrial robot in lighter machining operations: A conceptual methodology

Based on the challenges introduced above, the authors have developed a conceptual programming methodology, especially targeted for lighter machining operations.

The concept of the methodology is captured in Fig.1. The key-issue of the proposed methodology is to capture the knowledge of a skilled operator and make a semi- automatic knowledge transfer to a robot system. The expertise of the worker is still needed and appreciated. The man- machine interaction is carried out in a human friendly way with a minimum time spent on robot programming.

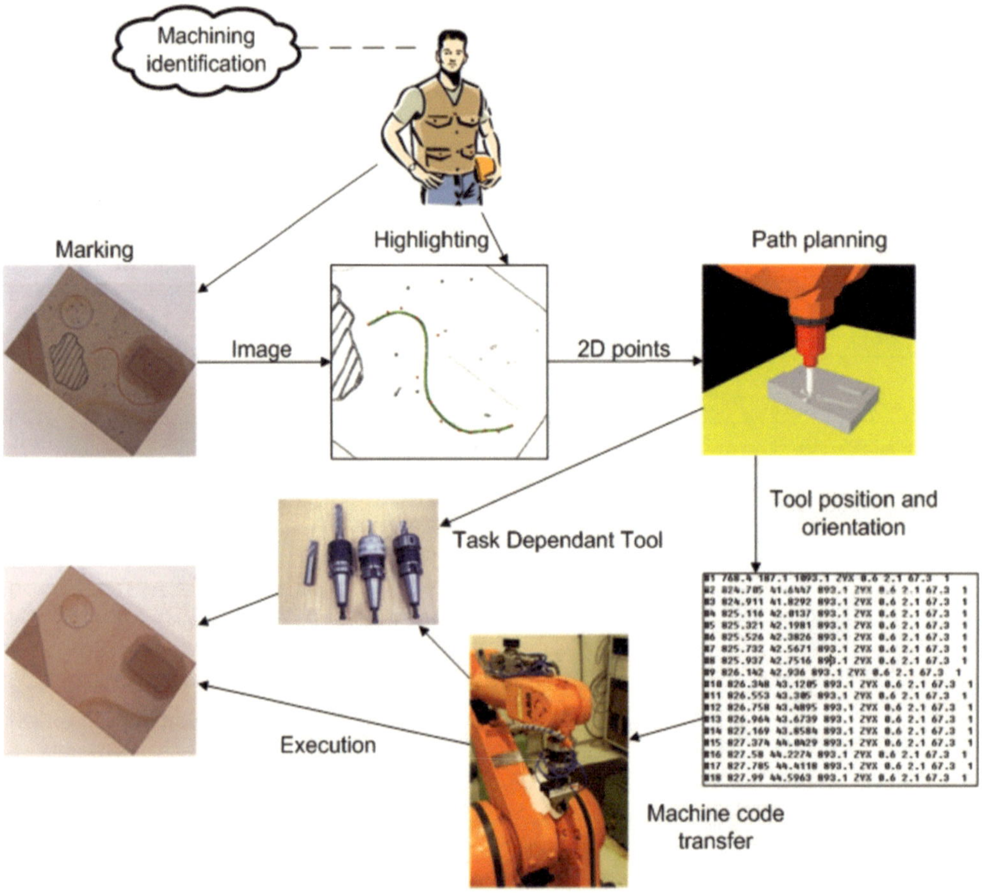

Figure 1. Conceptual overview

The concept is the following: a manufactured work-piece is inspected by an operator, who decides if there is any more machining necessary. The operator indicates the machining tasks by drawing (Marking) different shapes on the surface of the work-piece. Different colour mean different machining operation (e.g. green = polishing). The size of the machining is depending on the marking technique (e.g. curve = one path (if tool size equals curve's width)). After the marking a photo is taken. The marked surface is converted first to 2D points (with the help of an operator and image processing techniques) and second to robot position and orientation data. The operator is mainly involved in the error highlighting process. Finally the "error-free", cleaned work-piece is manufactured by the robot.

The methodology steps are formalized in the following sequence:

1. Work-piece inspection
2. Machining selection
3. Image processing (see Section 4.1)
4. Image to cutting points conversion (see Section 4.2)

5. Tool orientation establishment (see Section 4.3)
6. Locating the work-piece (see Section 4.4)
7. Simulation
8. Execution of path

The steps in details are the following:

First of all the work-piece error identification should be carried out. This is done by a human operator, who can spot the error very easy and can identify that there are irregularities at a certain area, but cannot at the same time state the magnitude and exact location of the error.

Next the operator should try to determine how such an error will impact on the functionality or the aesthetics of the work-piece and state when ether this is a critical error in such aspects. In fact, these operator messages can be written directly onto the work-piece by using standard marker pens. At a later stage, in the image processing module, colour differentiation is used to pick out messages from the operator.

Further the operator should determine if the error is of type point, line or region. Also in this case different colour pens can be used to draw lines, to mark regions, to scatter point clouds directly onto the surface.

After the error identification and classification the operator should select an appropriate machine tool for the material removal. This of course requires an experienced operator which is trained to evaluate error sources, its significance and the available machine tools to correct the error. Cutting depths are, at this stage, unknown to the operator and represents a challenge. Increasing depth means higher cutting forces and is a challenge for the less stiff industrial robot, compared with the conventional NC-machine. Zhang et al. (2005) states that the NC machines typically are 50 times stiffer than the industrial serial robots. Unfortunately, what is gained in flexibility and large working area is paid off by a decreased stiffness of the robot. However, in case of doubt, high accuracy verification of cutting depths can be undertaken by measuring the work-piece in a coordinate measuring machine (CMM). Also, at a later stage in the programming methodology there is a possibility to check for cutting depths by setting a few teach points on top of the machining allowance along the machining path. Anyway in lighter subtractive machine processes forces are small, and could as well be monitored by a force sensor attached to the robot end-effector.

Assuming selection of a robot as machine tool, the next step in the procedure is image retrieval and processing. A single camera system is used to capture the area of error and the possible written messages from the operator. A key issue is to capture some known geometrical object on the picture which can be used for sizing the picture and establishes a positional relationship between the error and the known geometry. The final goal of the image processing is to give positional coordinates of the error source related and stored in a work-piece reference frame. Details of the image processing module will be presented further in section 4.1

The image processing module present the machining path as 2D coordinates with reference to a work-piece coordinate system. In the next module this path must transferred into 3D coordinates by determination of the depth coordinate. In an offline programming environment the 2D path is transferred to the surface of a CAD model of the work-piece by an automatic "hit and withdrawal" procedure. After recovering the depth coordinate the 3D cutting points (CPs) are known. This procedure is further described in section 4.2

The best cutting conditions, in a given cutting point, is met when a selected tool is aligned with the surface at certain angles. To enable such tool alignment, information of the surface

inclination must be determined. Such procedure starts with finding the surface coordinate system in each cutting point by seeking the surface inclination in two perpendicular directions. Tool orientation is further described in section 4.3.
Out in the real manufacturing area, the relationship between the work-piece coordinate system and the industrial robot must be established. Here some already existing (vendor specific) methodology can be utilised. However, in section 4.4 a general approach can be found.
Before execution, path simulations could be undertaken in an offline programming environment in order to seek for singularities, out of reach problems or collisions.
The generated path previously stored in the work-piece coordinate system can be transferred to robot base coordinates and executed. As mentioned above, cutting depth calculations may be undertaken in order to verify that the operation is within certain limits of the machinery.

4. Some system components

In this paragraph several of the important modules of the programming methodology is presented in an overview manner, seeking to give valuable ideas and references for the reader's more than detailed technical information of each of the building blocks.

4.1 Vision and image processing module

In order to help the operator during path planning, different type of image processing techniques are used. After a short introduction of these techniques, the module structure will be shown.
Usually an image is represented as a matrix of pixels in the computer's memory.

$$I = \begin{bmatrix} C_{1,1} & C_{1,2} & \cdot & \cdot & \cdot & \cdot & C_{1,n} \\ C_{2,1} & \cdot & & & & & \cdot \\ \cdot & & \cdot & & & & \cdot \\ \cdot & & & & & & \\ \cdot & & & & \cdot & & \cdot \\ \cdot & & & & & \cdot & \cdot \\ C_{m,1} & \cdot & \cdot & & \cdot & \cdot & C_{m,n} \end{bmatrix} \quad (1)$$

where $C_{m,n} \in ((0\ldots255),(0\ldots255),(0\ldots255))$ represents the pixel's colour. The three values give the decomposition of the colour in the three primary colours: red, green and blue. Almost every colour that is visible to human can be represented like this. For example white is coded as (255, 255, 255) and black as (0, 0, 0). This representation allows an image to contain 16.8 million of colours. With this decomposition a pixel can be stored in three bytes. This is also known as RGB encoding, which is common in image processing.
As the mathematical representation of an image is a matrix, matrix operations and functions are defined on an image.
Image processing can be viewed as a function, where the input image is I_1, and I_2 is the resulting image after processing.

$$I_2 = F(I_1) \tag{2}$$

F is a filter or transformation. Applying a filter changes the information content of the image.

Many types of filters exist. Some of them are linear and others are non-linear. Range is from basic filters (Colour extraction, Greyscale converter, Resize, Brightness, Rotation, Blending functions) to Matrix convolution filters (Edge detectors, Sharpen filters).

The basis of the convolution filters comes from signal processing (Smith, 2002). When you have a filter you can compute its response ($y(t)$) to an entering signal ($x(t)$), by convolving $x(t)$ and the response of the filter to a delta impulse ($h(t)$).

Continuous time (Smith, 2002):

$$y(t) = h(t) \times x(t) = \int_{-\infty}^{\infty} h(\alpha) \cdot x(t-\alpha) d\alpha = \int_{-\infty}^{\infty} h(t-\alpha) \cdot x(\alpha) d\alpha \tag{3}$$

Discrete time (Smith, 2002):

$$y[k] = h[k] \times x[k] = \sum_{i=-\infty}^{\infty} h[i] \cdot x[k-i] = \sum_{i=-\infty}^{\infty} h[k-i] \cdot x[i] \tag{4}$$

where $\times$ sign is the convolution integral. The same way that we can do for one-dimensional convolution, it can be easily adapted to image convolutions. To get the result image the original image has to be convolved with the image representing the impulse response of the image filter.

Two-dimensional formula (Smith, 2002):

$$y[r,c] = \frac{1}{\sum_{i,j} h[i,j]} \cdot \sum_{j=0}^{M-1} \sum_{i=0}^{M-1} h[j,i] \cdot x[r-j, c-i] \tag{5}$$

where y is the output image, x is the input image, h is the filter and width and height is M. For demonstration of the computation of (5) see Fig. 2., which shows an example of computing the colour value (0..255) of the output image's one pixel ($y[i,j]$).

So in general, the convolution filtering loops through all the pixel values of the input image and computes the new pixel value (output image) based on the matrix filter and the neighbouring pixels.

Let observe a sample work-piece in Fig. 3. The image processing will be executed on this picture, which shows a work-piece and on the surface of it, some operator instructions. The different colours and shapes (point, line, curve, and region) describe the machining type (e.g. green = polishing) and the shape defines the machining path.

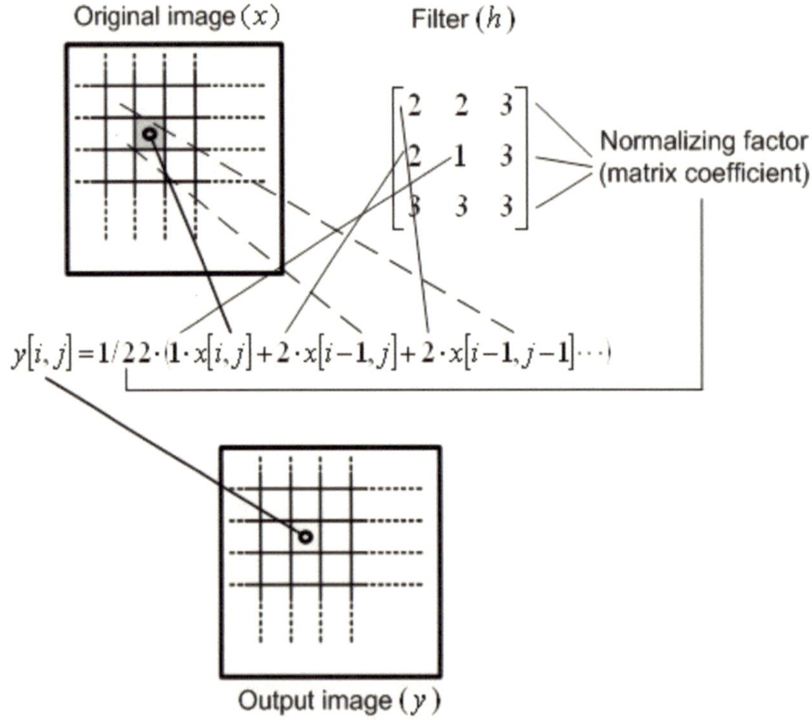

Figure 2. Demonstration of matrix convolution

Figure 3. Work-piece example

As mentioned before the best result filters are the matrix convolution filters. Edge detection is used to make transitions more visible, which results high accuracy in work-piece coordinate system establishment (as seen in Fig. 4. (a)) and colour filtering is used to highlight the regions of the error (Fig. 4. (b)).

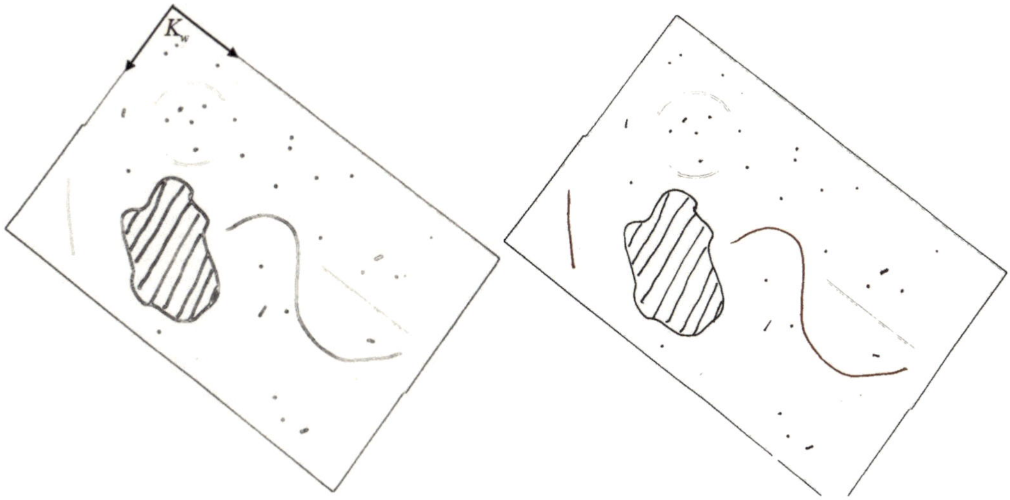

(a) (b)

Figure 4. (a) Work-piece coordinate system establishment. (b) Colour filtering

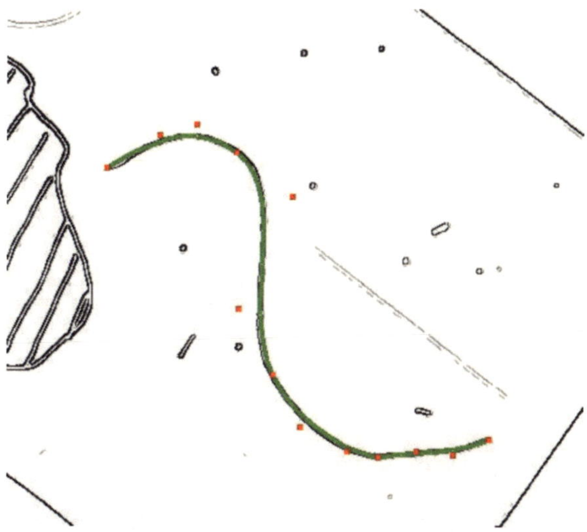

Figure 5. Curve highlighting example

The used Edge detector is the Canny edge detector. The canny edge detection is known as the optimal edge detector (Canny, 1986). With low error rate and low multiple responses (An edge is detected only once). Canny edge detection is built up from several steps for the best results. The steps contain smoothing filter, searching for edge strength (gradient of image), finding edge direction and eliminating streaks. The detailed step descriptions can be found in (Canny, 1986). Here only the filter matrix and gradient formula is presented, as proposed in (Canny, 1986):

$$G_x = \begin{bmatrix} -1 & 0 & 1 \\ -2 & 0 & 2 \\ -1 & 0 & 1 \end{bmatrix} \quad G_y = \begin{bmatrix} 1 & 2 & 1 \\ 0 & 0 & 0 \\ -1 & -2 & -1 \end{bmatrix} \quad |G| = |G_x| + |G_y| \tag{6}$$

This performs a 2D spatial gradient measurement on the image. Two filter matrices are used for the function, one estimating the gradient in x direction and the other estimating it in the y direction (6).
The operator executes the following tasks to select machining path:

- Scale picture according to a known geometric object
- Establish a work-piece coordinate system K_w
- Select machining type, based on the colour
- Highlight the 2D path for machining
- Save the 2D path for further processing with reference to K_w

Path highlighting consists of very basic steps. The operator selects what type of shape she/he wants to create and points on the picture to the desired place. The shape will be visible just right after the creation. The points and curves can be modified by "dragging" the points (marked as red squares) on the picture. A case of curve can be observed in Fig 5.
From the highlighting the machining path is generated autonomously. In case of a line or a curve the path is constructed from points, which meets the pre-defined accuracy. The region is split up into lines in the same way as the computer aided manufacturing software's (CAM) do.
The result of the module is a text file with the 2D coordinates of the machining and the machining types. This file is processed further in the next sections.

4.2 From image to 3D cutting points

As the result of the previous section is only 2D (x and y) coordinate, the depth coordinate (z) must also be defined. This is achieved by a standard commercial available simulation program, where the industrial robot maps the surface of the work-piece. The mapping process (as indicated in Fig. 6.) is a "hit and withdrawal" procedure: the robot moves along the existing 2D path and in every point of the path the robotic tool tries to collide with the work-piece surface. If there is a collision the z coordinate is stored and a cutting position p_w^n (index w= work -piece reference coordinate system and index n= cutting point number) is established (Sziebig, 2007).

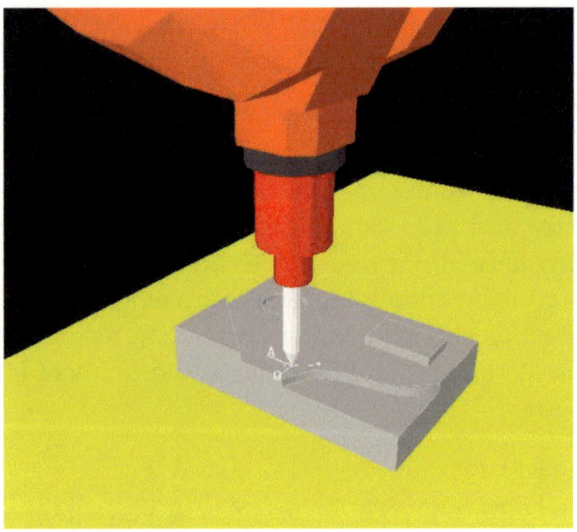

Figure 6. Demonstration of the "hit and withdrawal" procedure

4.3 Surface inclination and alignment of the cutting tool (3D to 6D)

From the previous paragraph, a set of cutting positions p_w^n were identified. However, a 3D positional description of the path is not enough.
To achieve the best and most effective cutting conditions, with a certain shaped and sized cutting tool, the tool must be aligned with the surface of the work piece at certain angles.
To enable such a tool alignment a surface coordinate system must be determined in each cutting point. The authors have previously developed an automatic procedure for finding the surface coordinate system (Solvang et al., 2008 a). According to this procedure the feed direction is determined as the normalized line X_n between two consecutive cutting points p_w^n and p_w^{n+1}.

$$X_n = \frac{p_w^{n+1} - p_w^n}{\left|p_w^{n+1} - p_w^n\right|} \tag{7}$$

Surface inclination Y_n in a direction perpendicular to the feed direction is also determined by an automatic collision detection procedure using a robot simulation tool to determine two points p_w^{n1} and p_w^{n2} perpendicular to the X_n line. This procedure consists of 6 steps, as shown in Fig.7.:

(0. Robot tool-axis Z_v ($|Z_v| = 1$) already aligned with current work- piece reference axis, Z_w ($|Z_w| = 1$)

1. Rotate robot tool axis Z_v around $Z_v \times X_n$ so that $Z_v \perp X_n$
2. Step along the robot tool- axis Z_v, away from the cutting point (p_w^n)
3. Step aside in a direction $Z_v \times X_n$
4. Move to collide with the surface and store position as p_w^{n1} ;

Relocate above the cutting point (p_w^n); according to step 2.

5. Step aside in a direction $-(Z_v \times X_n)$
6. Move to collide with the surface and store position as p_w^{n2} .

$$Y_n = \frac{p_w^{n2} - p_w^{n1}}{\left|p_w^{n2} - p_w^{n1}\right|} \tag{8}$$

The surface normal Z_n in the cutting point is found as

$$Z_n = X_n \times Y_n \tag{9}$$

In each cutting point p_w^n the directional cosines X_n, Y_n, Z_n forms a surface coordinate system K_n. To collect all parameters a (4×4) transformation matrix is created. The matrix (10) represents the transformation between the cutting point coordinate system K_n and the work piece current reference coordinate system K_w.

$$T_w^n = \begin{bmatrix} X_n & 0 \\ Y_n & 0 \\ Z_n & 0 \\ p_w^n & 1 \end{bmatrix} \tag{10}$$

Tool alignment angles are often referred to as the "lead" and "tilt" angles. The lead angle (β) is the angle between the surface normal Z_n and the tool axis Z_v in the feeding direction while the tilt angle (α) is the angle between the surface normal Z_n and the tool axis Z_v in a direction perpendicular to the feeding direction (Köwerich, 2002). Also a third tool orientation angle (γ), around the tool axis itself, enables usage of a certain side of the cutting tool, or to collect and direct cutting sparks in a given direction.
In case of the existence of a lead, tilt or a tool orientation angle the transformation matrix in (10) is modified according to:

$$T_w^n = \begin{bmatrix} \cos\gamma & \sin\gamma & 0 & 0 \\ -\sin\gamma & \cos\gamma & 0 & 0 \\ 0 & 0 & 1 & 0 \\ 0 & 0 & 0 & 1 \end{bmatrix} \cdot \begin{bmatrix} \cos\beta & 0 & -\sin\beta & 0 \\ 0 & 1 & 0 & 0 \\ \sin\beta & 0 & \cos\beta & 0 \\ 0 & 0 & 0 & 1 \end{bmatrix} \cdot \begin{bmatrix} 1 & 0 & 0 & 0 \\ 0 & \cos\alpha & \sin\alpha & 0 \\ 0 & -\sin\alpha & \cos\alpha & 0 \\ 0 & 0 & 0 & 1 \end{bmatrix} \cdot T_w^n \tag{11}$$

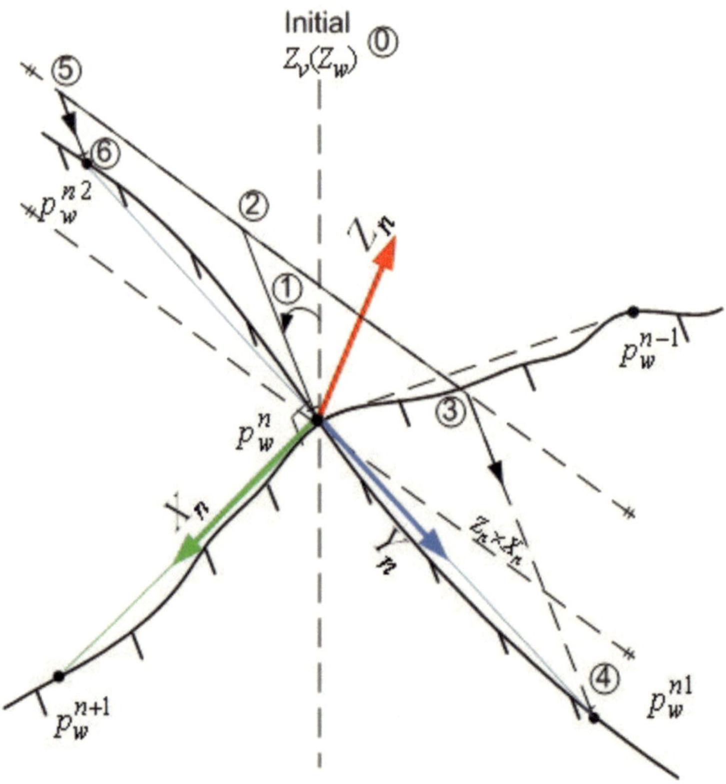

Figure 7. Steps in procedure to determine surface inclination

Fig. 8. shows the angular relationships described above.

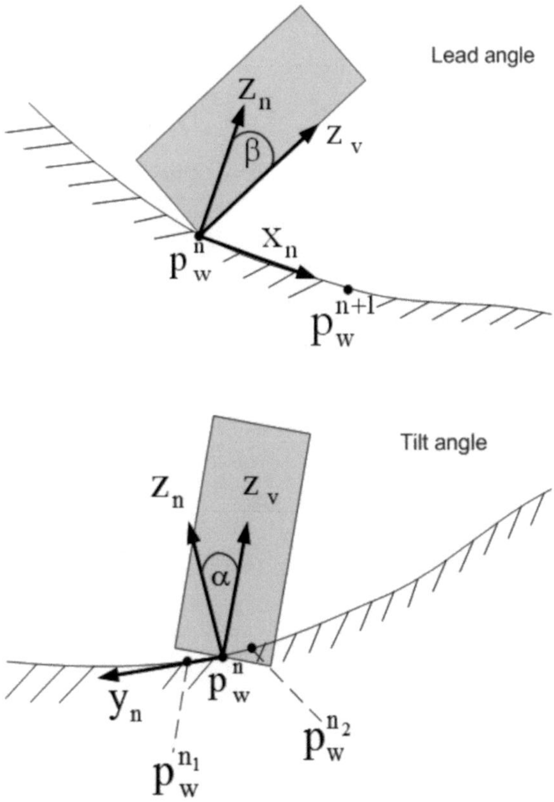

Figure 8. Angular relationships (based on Solvang et al., 2008 a)

4.4 Locating the work-piece

The robot path generated in the previous section is stored with reference to the work-piece coordinate system K_w. By such an arrangement the generated path is portable together with the work-piece. Next, before machining can start the path must be re-stored with reference to the robot base coordinate system K_0.

Typically, industrial robot systems have their own set-up procedure to determine such coordinate system relations but in many cases these are based on a minimum number of measurements and focus on a rapid simplicity more than accuracy. However, in robot machining operations the accuracy in reproducing the path heavily depends on the set-up procedure of these coordinate relationships. By adapting the typical coordinate set-up procedures found in coordinate measurement machines (CMMs) accuracy issues are well undertaken.

For the most significant (largest geometries) substitute (ideal) geometries are created based on robot point measurements.

The creation of substitute geometries is done by letting an operator identify what kind of ideal geometry should be created (plane, cylinder, sphere, cone etc.) Then, based on a minimum number of measurement points, such geometry is created so that the squared sum of the distance l_i to each measurement point is minimised according to the Gaussian Least Square Methodology (Martinsen,1992).

$$\sum_{i=1}^{m} l_i^{\,2} \rightarrow \min \tag{12}$$

These substitute geometries are given a vector description with a position and a directional (if applicable) vector.

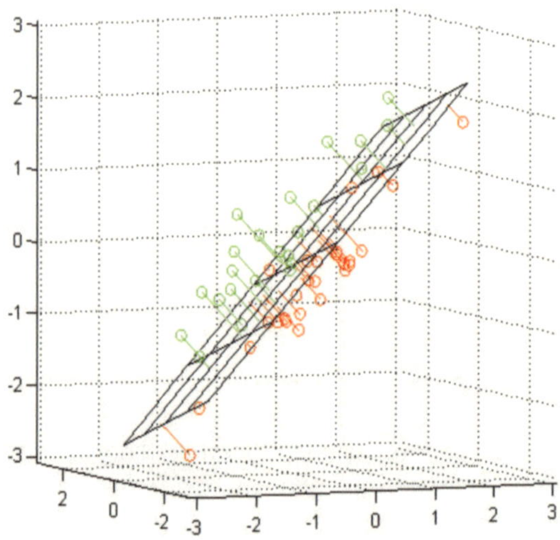

Figure 9. Probing and defining a plane

When establishing K_w, typically a directional vector from the most widespread geometry is selected as one of the directional cosines X_w, Y_w, or Z_w. The next axis could be defined by the intersection line from the crossing of two substitute geometries while the final third is found with the cross product of the two first directional cosines. The origin of K_w is typically located in the intersection of three substitute geometries. Fig. 9. shows this methodology for a given work-piece.

The directional cosines X_w, Y_w, or Z_w and the origin p_0^w is collected in the transformation matrix T_0^w, as introduced in the previous paragraph. Finally, the T_0^w transformation matrix is multiplied with the T_w^n transformation to create the total transformation matrix T_0^n:

$$T_0^n = T_w^n \cdot T_0^w \tag{13}$$

(13) hold information on the position and orientation of the machining path in each cutting point, related to robot base coordinate system K_0. According to (10) position data is extracted from the matrix as the last row in the matrix and orientation is given by the by the directional cosines. The nine parameter directional cosines can be reduced to a minimum of three angular parameters dependant on the choice of the orientation convention. In Craig, (2005) details of most used orientation conventions and transformation are found.

Finally, the robot position and orientation in each cutting point are saved as a robot coordinate file (vendor dependant).

Before execution of the machining process, simulations can be undertaken to search for singularities, out of reach or collisions.

5. Conclusion and future work

In this chapter a conceptual methodology for robot programming in lighter subtractive machining operations have been presented, seeking to give the reader an overview of the challenges and possible solutions for effective communication between the human and the robot system. Some key-modules have been elaborated, in order to let the readers in on more technical details necessary when developing their own systems.

Some important features of the proposed methodology are:

- User communication is human friendly and rapid
- Information of the real work-piece is combined with the ideal CAD-model
- Lines, regions ,dots and messages can be drawn directly onto the work-piece
- Uses only one camera
- Semi -automatic robot path generation

The authors have carried out some initial laboratory tests with the various modules around the proposed methodology (Solvang et al., 2008 a), (Solvang et al., 2007) and (Sziebig, 2007). These results are very promising, first of all operator communication is improved and the operator is presented only those tasks he can master intuitively. When it comes to accuracy measurements results shows that the selection of camera is important.

A high resolution camera (2856*2142) produced twice as good results as a low resolution web camera (640*480). To measure the complete error chain, a set of experiments was carried out with an ABB irb 2000 robot. For these tests the low resolution web camera were used to capture a hand drawn line on a sculptured surface After the image processing the robot was instructed to follow the generated path.. The deviation between the hand drawn and the robot drawn answer was approximate 1 mm, the same accuracy as for the web camera.

The initial test shows promising results for the methodology. Tests were carried as path generations only, without any machining action undertaken. The next step would be to look at accuracy tests under stress from the machining process. For these test we need to integrate our equipment with a robot system, preferably equipped with a force sensor at the end-effector.

8. References

Ando, N.; Suehiro, T.; Kitagaki, K. & Kotoku, T. (2005). RT-middleware: distributed component middleware for RT (robot technology), *Proceedings of IEEE/RSJ International Conference on Intelligent Robots and Systems (IROS 2005)*, pp. 3933- 3938, ISBN 0-7803-8912-3, Aug. 2005.

Baillie, J.C. (2005). URBI: towards a universal robotic low-level programming language, *Proceedings of IEEE/RSJ International Conference on Intelligent Robots and Systems (IROS 2005)*, pp. 820-825, ISBN 0-7803-8912-3, Aug. 2005.

Blomdell, A.; Bolmsjo, G.; Brogårdh, T.; Cederberg, P.; Isaksson, M.; Johansson, R.; Haage, M.; Nilsson, K.; Olsson, M.; Olsson, T.; Robertsson, A. & Wang, J. (2005). Extending an industrial robot controller: implementation and applications of a fast open sensor interface. *IEEE Robotics & Automation Magazine*, Vol. 12, No. 3, Sep. 2005, pp. 85-94, ISSN 1070-9932

Brogårdh T. (2007). Present and future robot control development—An industrial perspective. *Annual Reviews in Control*, Vol. 31, No. 1, 2007, pp. 69-79, ISSN 1367-5788

Bruyninckx, H.; Soetens, P. & Koninckx B. (2003). The real-time motion control core of the Orocos project, *Proceedings of IEEE International Conference on Robotics and Automation (ICRA '03)*, pp. 2766-2771, ISBN: 0-7803-7736-2, Sep. 2003.

Canny, J. (1986). A computational approach to edge detection. *IEEE Transactions on Pattern Analysis and Machine Intelligence*, Vol. 8, No. 6, Nov. 1986, pp. 679–698, ISSN 0162-8828

CORDIS (2008). *European Union Seventh Framework Program (FP7) to improve the competitiveness of European industry*, http://cordis.europa.eu/fp7/ict/, 2008-2012.

Craig, J.J. (2005). *Introduction to robotics, Mechanics and control*, Pearson Education, ISBN 0-13-123629-6, USA

Garcia, J.G.; Robertsson, A.; Ortega, J.G. & Johansson, R. (2004) Sensor fusion of force and acceleration for robot force control, *Proceedings of IEEE/RSJ International Conference on Intelligent Robots and Systems (IROS 2004)*, pp. 3009-3014, ISBN 0-7803-8463-6, Sep. 2004.

Jackson, J. (2007). Microsoft robotics studio: A technical introduction. *IEEE Robotics & Automation Magazine*, Vol. 14, No. 4, Dec. 2007, pp. 82-87 ISSN: 1070-9932

Köwerich, S. (2002). *5 akse maskinering: En state-of-the-art studie*, SINTEF report STF 38 A97247, ISBN 82-14-00695-3, Norway

Kranz, M.; Rusu, R.B.; Maldonado, A.; Beetz, M. & Schmidt, A. (2006). A Player/Stage System for Context-Aware Intelligent Environments, *Proceedings of the System Support for Ubiquitous Computing Workshop (UbiSys 2006) at 8th International Conference of Ubiquitous Computing (UbiComp 2006)*, pp. 17-21, ISBN 3-5403-9634-9, Sep. 2006.

Lopes, N. & Lima P. (2008). OpenSDK - An Open-source Implementation of OPEN-R, *Proceedings of 7th International Joint Conference on Autonomous Agents and Multi-Agent Systems (AAMAS 2008)*, ISBN 978-0-9817-3813-0, May. 2008.

Magnenat, S.; Longchamp, V. & Mondada, F. (2007). ASEBA, an event-based middleware for distributed robot control, *Proceedings of Workshops DVD of IEEE/RSJ International Conference on Intelligent Robots and Systems (IROS 2007)*, ISBN 978-1-4244-0912-9, Oct. 2007.

Martinsen, K. (1992). *Kompendium i måleteknikk og vektorielle toleranser*, Norges Tekniske Høgskole, Norway

Metta, G.; Fitzpatrick, P. & Natale, L. (2006). YARP: Yet Another Robot Platform. *International Journal on Advanced Robotics Systems*, Vol. 3, No. 1, Mar. 2006, pp. 43-48, ISSN 1729-8806

Mizukawa, M.; Matsuka, H.; Koyama, T.; Inukai, T.; Noda, A.; Tezuka, H.; Noguchi, Y. & Otera, N. (2002). ORiN: open robot interface for the network - the standard and unified network interface for industrial robot applications, *Proceedings of the 41st SICE Annual Conference (SICE 2002)*, pp. 925- 928, ISBN 0-7803-7631-55-7, Aug. 2002.

Nayar, H. D. & Nesnas, I. A. (2007). Measures and Procedures: Lessons Learned from the CLARAty Development at NASA/JPL, *Proceedings of Workshops DVD of IEEE/RSJ International Conference on Intelligent Robots and Systems (IROS 2007)*, ISBN 978-1-4244-0912-9, Oct. 2007.

Ozaki, F. & Hashimoto, H. (2004). Open Robot Controller Architecture (ORCA), the Robot Platform. *Toshiba Review*, Vol. 59, No. 9, 2004, pp. 20-24, ISSN:0372-0462

SMErobot (2005). *European Union Sixth Framework Program (FP6) proposal number 011838*, http://www.smerobot.org/, 2005-2009.

Smith, S. (2002). *Digital Signal Processing: A Practical Guide for Engineers and Scientists*, Elsevier, ISBN 0-7506-7444-X, USA

Solvang, B.; Korondi, P.; Sziebig, G. & Ando, N. (2007). SAPIR: Supervised and Adaptive Programming of Industrial Robots, *Proceedings of 11th International Conference on Intelligent Engineering Systems (INES 2007)*, pp. 281-286, ISBN 1-4244-1147-5, Jun. 2007.

Solvang, B.; Sziebig, G. & Korondi, P. (2008 a). Vision Based Robot Programming, *Proceedings of IEEE International Conference on Networking, Sensing and Control (ICNSC 2008)*, pp. 949-954, ISBN 978-1-4244-1685-1, Apr. 2008.

Solvang, B.; Sziebig, G. & Korondi, P. (2008 b). Multilevel Control of Flexible Manufacturing Systems, *Proceedings of IEEE Conference on Human System Interactions* (HSI'08), pp. 785–790, ISBN 1-4244-1543-8, May. 2008.

Sziebig, G. (2007). Interactive vision-based robot path planning, *Master of Science thesis*, Budapest University of Technology and Economics, 82 pp., May 2007.

Thomessen, T.; Lien, T. K. & Solvang, B. (1999). Robot control system for heavy grinding applications, *Proceedings of 30th International Symposium on Robotics*, pp. 33–38, Oct. 1999.

Thomessen, T. & Lien T. K. (2000). Robot control system for safe and rapid programming of grinding applications. *Industrial Robot: An International Journal*, Vol. 27, No. 6, 2002, pp. 437-444, ISSN 0143-991X

Thomessen, T.; Sannæs, P. K. & Lien, T. K. (2004). Intuitive Robot Programming, *Proceedings of 35th International Symposium on Robotics*, ISBN 0-7695-0751-6, Mar. 2004.

Veiga, G.; Pires J.N. & Nilsson, K. (2007). On the use of service oriented software platforms for industrial robotic cells, *Proceedings of Intelligent Manufacturing Systems (IMS2007)*, May. 2007.

Weitzenfeld, A.; Gutierrez-Nolasco, S. & Venkatasubramanian, N. (2003). MIRO: An Embedded Distributed Architecture for Biologically inspired Mobile Robots, *Proceedings of 11th IEEE International Conference on Advanced Robotics (ICAR'03)*, ISBN 972-96889-9-0, Jun. 2003.

Zhang, H.; Wang, J.; Zhang, G.; Gan, Z.; Pan, Z.; Cui, H. & Zhu, Z. (2005). Machining with flexible manipulator: toward improving robotic machining performance, *Proceedings of IEEE/ASME International Conference on Advanced Intelligent Mechatronics*, pp. 1127-1132, ISBN 0-7803-9047-4, July 2005.

27

Dynamic Visual Servoing with an Uncalibrated Eye-in-hand Camera[1]

Hesheng Wang and Yun-Hui Liu
The Chinese University of Hong Kong
China

1. Introduction

Visual servoing is an approach to control motion of a robot manipulator using visual feedback signals from a vision system. Though the first systems date back to the late 1970s and early 1980s, it is not until the middle 1990s that there is a sharp increase in publications and working systems, due to the availability of fast and affordable vision processing systems (Hutchinson et al, 1996).
There are many different ways of classifying the reported results: based on number of cameras used, generated motion command (2D, 3D), camera configuration, scene interpretation, underlying vision algorithms. We will touch upon these issues briefly in the following sections.

1.1 Image-based and Position-based Visual Servoing

Visual servo robot control overcomes the difficulties of uncertain models and unknown environments. Existing methods can be classified into two basic schemes, namely position-based visual servo control (Fig. 1) and image-based visual servo control (Fig. 2). In both classes of methods, object feature points are mapped onto the camera image plane, and measurements of these points are used for robot control. A combination of the two schemes is called hybrid visual servoing.
A position-based approach first uses an algorithm to estimate the 3-D position and orientation of the robot manipulator or the feature points from the images and then feeds the estimated position/orientation back to the robot controller. The main advantage of position-based visual servoing is that it controls the camera trajectory in the Cartesian space, which allows it to easily combine the visual positioning task with obstacles avoidance and singularities avoidance. Position-based methods for visual servoing seem to be the most generic approach to the problems, as they support arbitrary relative position with respect to the object. The major disadvantage of position-based methods is that the 3D positions of the feature points must be estimated. In position-based visual servoing, feedback is computed using estimated quantities that are a function of the system calibration parameters. Hence, in some situations, position-based control can become extremely sensitive to calibration error. Since 3-D position/orientation estimation from images is subject to big noises,

[1] This work is partially supported by Hong Kong RGC under the grant 414406 and 414707, and the NSFC under the projects 60334010 and 60475029. Y.-H. Liu is also with Joint Center for Intelligent Sensing and Systems, National University of Defense Technology, Hunan, China.

position-based methods are weak to disturbances and measurement noises. Therefore, the position-based visual servoing is usually not adopted for servoing tasks.

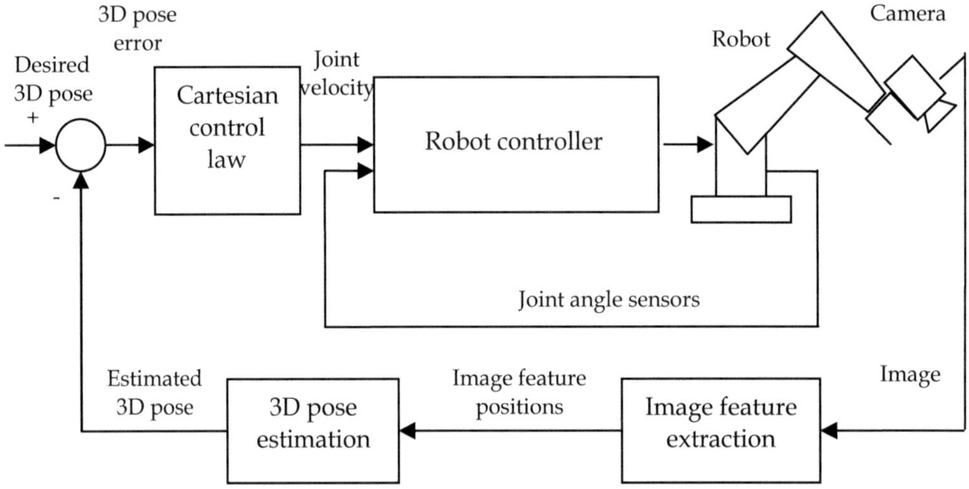

Figure 1. Position-based visual servoing

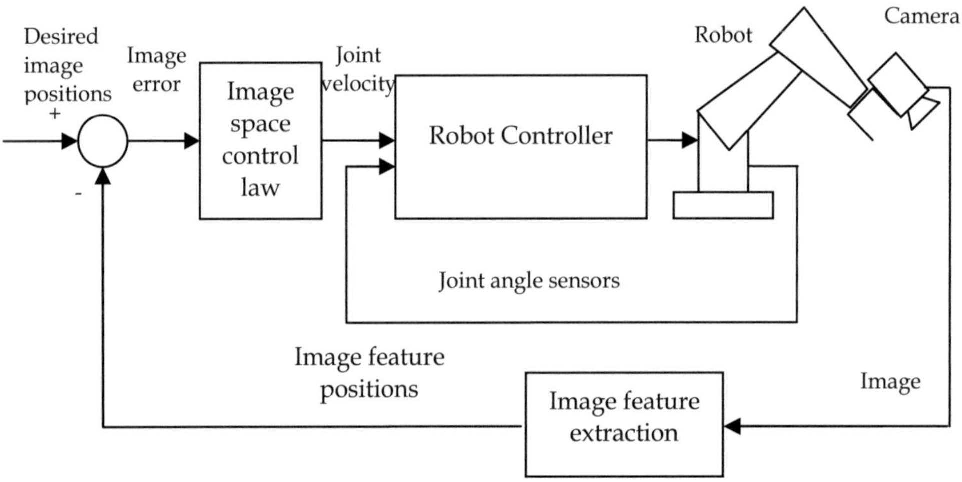

Figure 2. Image-based visual servoing

An image-based approach selects a set of feature points on the robot or the target object and directly employs their projection signals on the image plane of the camera in robot control. The general approach used in the image-based visual control methods is to control the robot motion in order to move the image plane features to desired positions. This usually involves the calculation of an image Jacobian or a composite Jacobian, the product of the image and robot Jacobian. A composite Jacobian relates differential changes in joint angles to differential changes in image features. The image-based control has the input command described directly in the feature space; it is then easy to generate the input trajectory. Since the feedback signals are projection errors on the image plane, image-based controllers are considered more robust to disturbances than position-based methods. Coarse calibration only affects the rate of convergence of the control law in the sense that a longer time is needed to reach the desired position. One disadvantage of image-based methods compared to position-based methods is the presence of singularities in the feature mapping function, which reflect themselves as unstable points in the inverse Jacobian control law. The estimation of the image Jacobian requires knowledge of the camera intrinsic and extrinsic parameters. Extrinsic parameters also represent a rigid mapping between the scene or some reference frame and the camera frame. If one camera is used during the servoing process, the depth information needed to update the image Jacobian is lost. Therefore, many of the existing systems usually rely on a constant Jacobian that is computed for the desired camera/end-effector pose. This is one of the drawbacks of this approach, since the convergence is ensured only around the desired position. This problem may be solved by adaptive estimation of the depth.

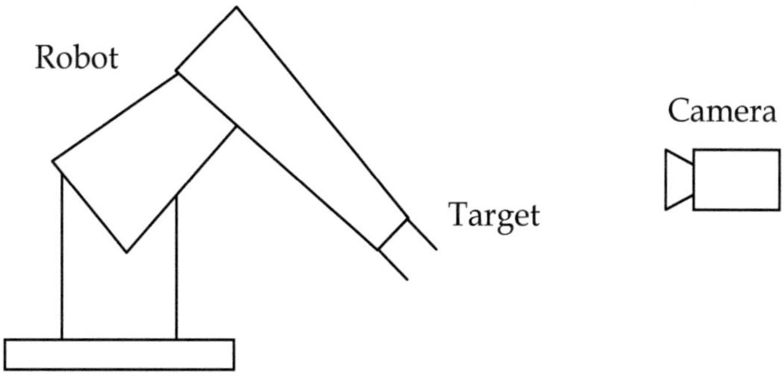

Figure 3. Fixed camera configuration

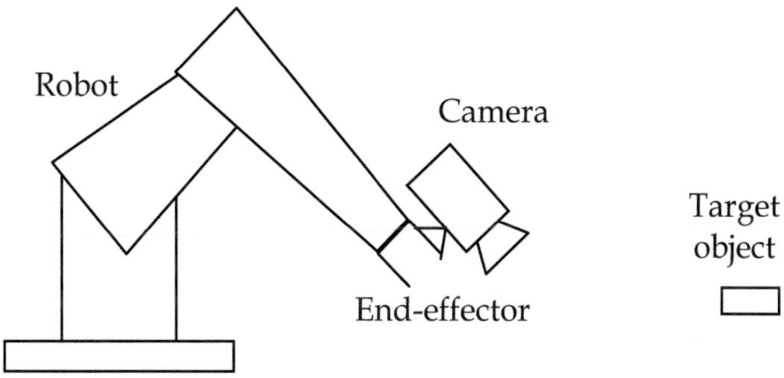

Figure 4. Eye-in-hand camera configuration

1.2 Fixed Camera and Eye-in-hand Configurations

There are two possible configurations, namely fixed camera (eye-to-hand) configuration (Fig. 3) and eye-in-hand configuration (Fig. 4), to set up a vision system for visual servo control of robot manipulators.

In a fixed camera setup, targets are mounted on a robot end-effector and the camera is fixed at a position near the manipulator. In this case, the camera does not move with the robot manipulator and its objective is to monitor motion of the robot.

In an eye-in-hand setup, the camera is mounted at the end-effector of the robot manipulator so that it moves with the manipulator. The camera is to measure information of objects in the surrounding environment. The objective of this approach is to move the manipulator in such a way that the projection of either a moving or a static object be always at a desired location in the image captured by the camera

While each configuration has its own advantages and drawbacks, they both can be found in real applications. Eye-in-hand systems can be widely found in research laboratories and are being used in tele-operated inspection systems in hazardous environments such as nuclear factories. Fixed camera setups are used in vision-based pick-and-place at manufacturing lines, robots assisted by networked cameras, etc.

1.3 Kinematics-based and Dynamic Visual Servoing

Existing methods can also be classified into kinematics-based (Fig. 5) and dynamic methods (Fig. 6). Kinematics-based methods do not consider the nonlinear dynamics of the robot manipulator and design the controller based on the kinematics only. Kinematics-based methods decouple the designs of motion controller of manipulators and the visual controller under the assumption that the manipulator can perfectly perform the position or velocity control required in visual servoing. They change visual servoing into a problem of designing the velocity or position of the end-effector of the robot using the visual feedback signals. It is well known that the nonlinear robot forces have significant impact on robot motion, especially when the robot manipulator moves at high speed. Neglecting them not only decays the control accuracy but also results in instability. In a rigorous sense, the stability is not guaranteed for all kinematics-based controllers. Kinematics-based controllers are suitable for slow robot motion only.

Dynamic methods design the joint input directly, instead of the velocity of the end-effector, using the visual feedback and include the nonlinear robot dynamics in the control loop. The nonlinear centrifugal and Coriolis forces are compensated for in the control loop. Dynamic controllers guarantee the stability and are suitable for both slow and fast robot motions. Compared to kinematics-based methods, work on dynamic visual servoing is relatively limited, though the importance has been recognized for a long time by many researchers. This is mainly due to the difficulties in incorporating the nonlinear forces in controller design and the existence of the nonlinear scaling factors corresponding to the reciprocal of the depths of the feature points in the perspective projection. Since it is difficult to measure or estimate on-line the depths of the feature points, most dynamic controllers are subject to planar motions of robot manipulators only.

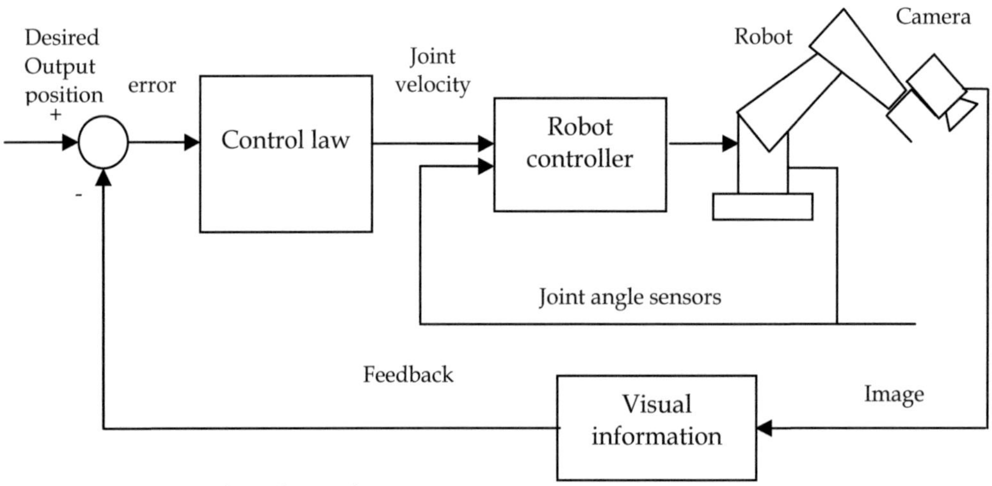

Figure 5. Kinematic-based visual servoing

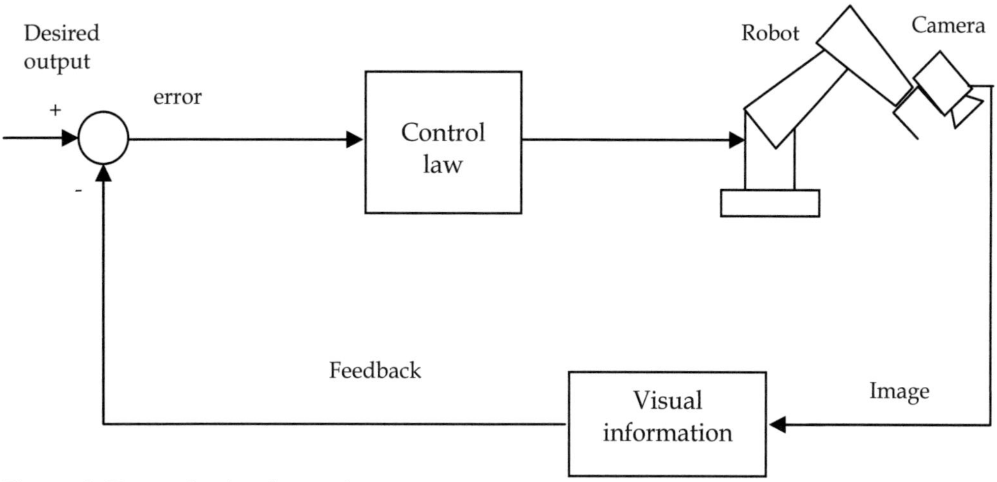

Figure 6. Dynamic visual servoing

1.4 Related Work of Visual Servoing with Uncalibrated Eye-in-hand Camera

Image-based eye-in-hand visual servoing is a problem of controlling the projections of image features to desired positions on the image plane of a camera mounted on a robot manipulator by controlling motion of the manipulator. Compared to a camera fixed in the workspace, an eye-in-hand camera enables the manipulator to view the workspace more flexibly. To implement a visual servo controller, an important step is to calibrate the intrinsic and extrinsic parameters of the camera. It is well known that camera calibration is costly and tedious. A survey of camera self-calibration could be referring to (Hemayed, 2003). To avoid camera calibration, many efforts have been made to uncalibrated visual seroving for which methods can be classified into kinematics-based controllers and dynamic controllers.

In kinematics-based methods, the typical idea for uncalibrated visual servoing is to estimate the image Jaocbian or the depths of the features on-line. (Papanikolopoulos et al, 1995) developed an on-line estimation method to determine the depth information of the feature point. (Malis et al, 1999; 2004) proposed 2-1/2-D visual servoing to deal with uncalibrated camera intrinsic parameters. (Yoshimi & Allan, 1995) proposed an estimator of the image Jacobian for a peg-in-hole alignment task. (Hosoda & Asada, 1994) designed an on-line algorithm to estimate the image Jacobian. (Ruf et al, 1997) proposed a position-based visual servoing by adaptive kinematic prediction. (Pomares et al, 2007) proposed adaptive visual servoing by simultaneous camera calibration. Since the design of visual servo controller does not take the nonlinear robot dynamics into account, the stability of the overall system is not guaranteed even if the visual controller is stable. This is point out by many researchers such as (Bishop & Spong, 1997; Cheah et al, 2003; Deng et al, 2002; Hsu and Aquino, 1999; Kelly et al, 1999; Kim et al, 1995; Dixon, 2007; Shen et al, 2003).

Dynamic methods design the joint input directly, instead of the velocity of the end-effector, using the visual feedback and include the nonlinear robot dynamics in the control loop. (Carelli et al, 2000) proposed an adaptive controller to estimate robot parameters for the eye-in-hand setup. (Kelly et al, 2000) developed an adaptive controller for visual servoing of planar manipulators. The controller developed by (Hashimoto et al, 2002; Nagahama et al 2002) incorporated the nonlinear robot dynamics in controller design and employed a motion estimator to estimate motion of an object in a plane. In our early work, we proposed a new adaptive controller for uncalibrated visual servoing of 3-D manipulator using a fixed camera (Liu et al, 2005, 2006; Wang & Liu, 2006; Wang et al, 2007). The underlying idea in our controller is the development of the concept of the depth-independent interaction matrix (or image Jacobian) matrix to avoid the depth that appears in the denominator of the perspective projection equation.

1.5 Outline of This Chapter

A brief review of uncalibrated visual servoing is presented in this Chapter. Some prerequisite knowledge including camera parameters, perspective projection models, coordinate transformations, geometry relationship are discussed in detail. Then we focus on one typical problem in visual servoing: visual servoing with uncalibrated eye-in-hand camera.

This paper extends our controller developed before (Liu et al, 2006) to image-based visual servoing of point features using an uncalibrated eye-in-hand camera. In an eye-in-hand setup in addition to the camera parameters, the 3-D coordinates of the features are not known either. Therefore, we need to estimate both the camera parameters and the 3-D coordinates on-line. The basic idea in the controller design is to use the depth-independent interaction (or image Jacobian) matrix so that the depth appearing in the denominator of the perspective projection equation can be avoided. To simultaneously estimate the camera parameters and the unknown 3-D coordinates of the features, we combine the Slotine-Li method with an on-line estimator designed on the basis of the idea of structure from motion in computer vision. The estimator uses the image sequence, captured during motion of the manipulator, to estimate the 3-D structure and the perspective projection matrix on-line by minimizing the Frobenius norm of the estimated projection errors. Based on the nonlinear dynamics of the manipulator, we have proved by the Lyapunov theory that the image errors will be convergent to zero and the unknown camera parameters and 3-D coordinates of the features are convergent to the real values up to a scale. Experiments have been conducted to validate the proposed methods.

This Chapter is organized as follows. Section 2 will review the kinematics and the dynamics. Section 3 will present adaptive image-based visual servoing with unknown target positions. In Section 4, we will discuss adaptive visual servoing with unknown camera parameters and target positions. Section 5 presents the experimental results and Section 6 concludes the major results in this Chapter.

2. Camera and Robot Model

2.1 Problem Definition

Consider an eye-in-hand set-up (Fig. 7), in which a vision system is mounted on the end-effector to monitor a set of target points. Assume that the target points are fixed ones but their coordinates are not known. Suppose that the camera is a pin-hole camera with perspective projection (Fig. 8). Furthermore, we will consider two cases by assuming that the intrinsic parameters of the camera and the extrinsic parameters, i.e. the homogeneous transform matrix between the camera and the end-effector, are known or unknown.

Problem 1: Given desired positions of the target points on the image plane, design a proper joint input for the manipulator such that the projections of the target points on the image plane is asymptotically convergent to the desired positions. Assuming that only the target 3-D position are unknown.

Problem 2: Given desired positions of the target points on the image plane, design a proper joint input for the manipulator such that the projections of the target points on the image plane is asymptotically convergent to the desired positions. Assuming that both the 3-D coordinates of the target points and the intrinsic and extrinsic parameters of the camera are unknown.

To simplify the discussion, we will assume that the target always remains in the field of view

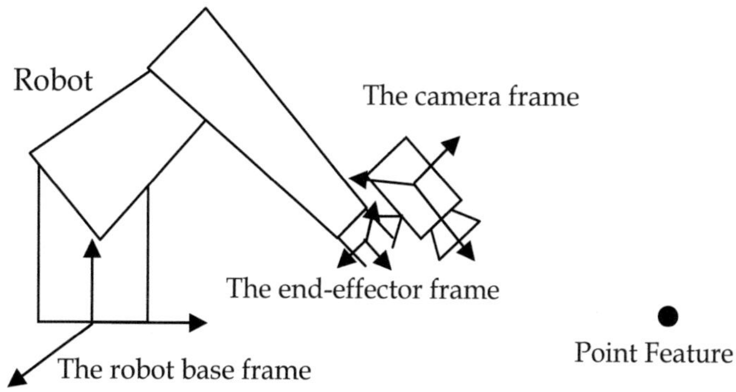

Figure 7. An eye-in-hand setup for visual servoing

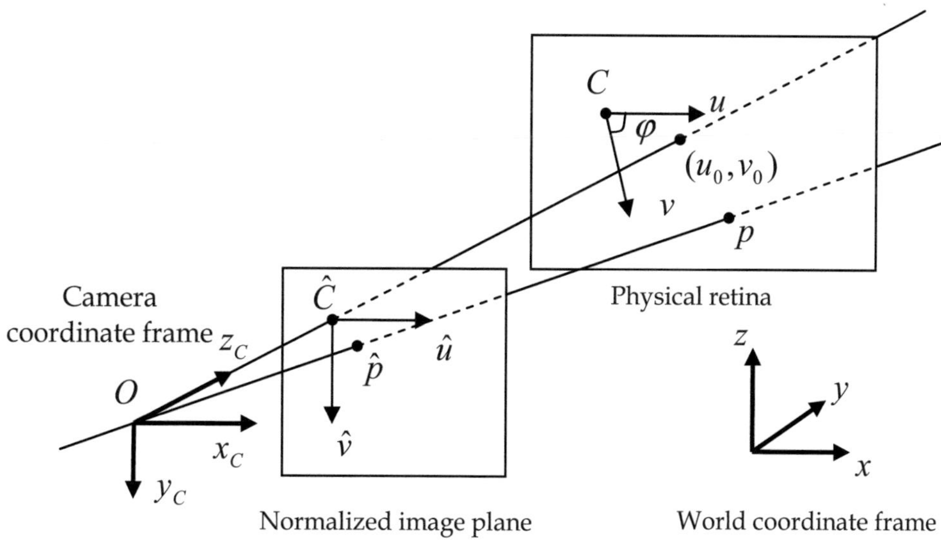

Figure 8. Perspective projection

2.2 Kinematics

In Fig. 7, three coordinate frames, namely the robot base frame, the end-effector frame, and the camera frame, have been set up to represent motion of the manipulator. Denote the joint angle of the manipulator by $\mathbf{q}(t)$, and the homogenous coordinates of the target w.r.t. the robot base and the camera frames by $\mathbf{x}$ and ${}^{c}\mathbf{x}$, respectively. Note that the $\mathbf{x}$ is a constant vector for the fixed target point and $\mathbf{x} = \begin{pmatrix} {}^{b}\mathbf{x} & 1 \end{pmatrix}^T$. Denote the homogenous transform matrix of the end-effector with respect to the base frame by $\mathbf{T_e(q)}$, which can be calculated from the kinematics of the manipulator. Denote the homogeneous transformation matrix of the camera frame with respect to the end-effector frame by $\mathbf{T_c}$, which represents the camera extrinsic parameters. From the forward kinematics of the manipulator, we have

$$ {}^{\mathbf{c}}\mathbf{x} = \mathbf{T_c^{-1} T_e^{-1}(q) x} \tag{1} $$

where

$$ \mathbf{T}_e^{-1}(t) = \begin{pmatrix} \mathbf{R}(t) & \xi(t) \\ 0 & 1 \end{pmatrix} \tag{2} $$

where $\mathbf{R}(t)$ and $\xi(t)$ are the rotation matrix and the translation vector from robot base frame to end-effector frame.

Let $\mathbf{y} = (u, v, 1)^T$ express the homogenous coordinates of the projection of the target point on the image plane. Under the perspective projection model,

$$ \mathbf{y} = \frac{1}{{}^{c}z} \mathbf{\Omega}\, {}^{\mathbf{c}}\mathbf{x} = \frac{1}{{}^{c}z} \mathbf{\Omega T_c^{-1} T_e^{-1}(q) x} \tag{3} $$

where $\mathbf{\Omega}$ is determined by the camera intrinsic parameters:

$$ \mathbf{\Omega} = \begin{pmatrix} \alpha & -\alpha \cot\varphi & u_0 & 0 \\ 0 & \dfrac{\gamma}{\sin\varphi} & v_0 & 0 \\ 0 & 0 & 1 & 0 \end{pmatrix} \tag{4} $$

where α and γ are the scalar factors of the u and v axes of the image plane. φ represents the angle between the two axes. ${}^{c}z$ is the depth of the feature point with respect to the camera frame. Denote by $\mathbf{M}$ the product of $\mathbf{\Omega}$ and $\mathbf{T_c^{-1}}$, i.e.,

$$ \mathbf{M} = \begin{pmatrix} \alpha \mathbf{r_1^T} - \alpha \cot\varphi \mathbf{r_2^T} + u_0 \mathbf{r_3^T} & \alpha p_x - \alpha \cot\varphi p_y + u_0 p_z \\ \dfrac{\gamma}{\sin\varphi} \mathbf{r_2^T} + v_0 \mathbf{r_3^T} & \dfrac{\gamma}{\sin\varphi} p_y + v_0 p_z \\ \mathbf{r_3^T} & p_z \end{pmatrix} \tag{5} $$

where $\mathbf{r_i^T}$ denotes the i-th row vector of the rotation matrix of $\mathbf{T_c^{-1}}$, and $(p_x, p_y, p_z)^T$ is its translational vector. The matrix $\mathbf{M}$ is called *perspective projection matrix,* which depends on the intrinsic and extrinsic parameters only. Eq. (3) can be rewritten as

$$ \mathbf{y} = \frac{1}{{}^{c}z} \mathbf{M T_e^{-1}(q) x} \tag{6} $$

The depth of the feature point is given by

$$ {}^{c}z = \mathbf{m}_3^{\mathrm{T}} \mathbf{T}_{\mathrm{e}}^{-1}(\mathbf{q})\mathbf{x} \tag{7}$$

Where $\mathbf{m}_3^{\mathrm{T}}$ denotes the third row vector of the perspective projection matrix **M**.

$$ {}^{c}\dot{z} = \underbrace{\frac{\partial(\mathbf{m}_3^{\mathrm{T}} \mathbf{T}_{\mathrm{e}}^{-1}(\mathbf{q})\mathbf{x})}{\partial \mathbf{q}}}_{\mathbf{n}^{\mathrm{T}}(\mathbf{q})} \dot{\mathbf{q}} = \mathbf{n}^{\mathrm{T}}(\mathbf{q})\dot{\mathbf{q}} \tag{8}$$

It is important to note:

Property 1: *A necessary and sufficient condition for matrix* **M** *to be a perspective projection matrix is that it has a full rank.*

By differentiating eq. (6), we obtain the following relation:

$$ \dot{\mathbf{y}} = \frac{1}{{}^{c}z}(\mathbf{M}\dot{\mathbf{T}}_{\mathrm{e}}^{-1}(\mathbf{q}) - \mathbf{y}\,{}^{\mathrm{c}}\dot{z}) = \frac{1}{{}^{c}z}\mathbf{A}(\mathbf{q})\dot{\mathbf{q}} \tag{9}$$

where the matrix **A(q)** is a matrix determined by

$$ \mathbf{A}(\mathbf{q}) = \underbrace{\begin{pmatrix} \mathbf{m}_1^{\mathrm{T}} - u\mathbf{m}_3^{\mathrm{T}} \\ \mathbf{m}_2^{\mathrm{T}} - v\mathbf{m}_3^{\mathrm{T}} \\ \mathbf{0} \end{pmatrix}}_{\mathbf{D}} \frac{\partial(\mathbf{T}_{\mathrm{e}}^{-1}(\mathbf{q})\mathbf{x})}{\partial \mathbf{q}} \tag{10}$$

It should be noted that the image Jacobian matrix is the matrix $\frac{1}{{}^{c}z}\mathbf{A}(\mathbf{q})$. Matrix **A(q)** differs from the image Jacobian matrix by the scale factor. Here we call **A(q)** *Depth-independent interaction matrix*. Note that in the matrix **A(q)** the coordinates vector **x** is a constant vector. The products of the components of the perspective project matrix **M** and the components of the vector **x** are in the linear form in the Depth-independent interaction matrix.

Proposition 1: Assume that the Jacobian matrix $\mathbf{J}(\mathbf{q}(t))$ of the manipulator is not singular. For any vector ${}^{b}\mathbf{x}$, the matrix $\frac{\partial(\mathbf{T}_{\mathrm{e}}^{-1}(t)\mathbf{x})}{\partial \mathbf{q}}$ has a rank of 3.

Proof: Substituting (10) into the matrix (9), we obtain

$$ \begin{aligned} \frac{\partial(\mathbf{T}_{\mathrm{e}}^{-1}(t)\mathbf{x})}{\partial \mathbf{q}} &= \frac{\partial\left(\mathbf{R}(t)^{b}\mathbf{x} + \xi(t)\right)}{\partial \mathbf{q}} \\ &= \left(sk\{\mathbf{R}(t)^{b}\mathbf{x} + \xi(t)\} \quad \mathbf{I}_{3\times3}\right)\begin{pmatrix} \mathbf{R}(t) & 0 \\ 0 & \mathbf{R}(t) \end{pmatrix}\mathbf{J}(\mathbf{q}(t)) \end{aligned} \tag{11}$$

where sk is a matrix operator and $sk(\mathbf{x})$ with vector $\mathbf{x} = [x_1 \quad x_2 \quad x_3]^T$ can be written as a matrix form

$$ sk(\mathbf{x}) = \begin{pmatrix} 0 & -x_3 & x_2 \\ x_3 & 0 & -x_1 \\ -x_2 & x_1 & 0 \end{pmatrix} \tag{12}$$

■

Obviously, the matrix in (11) has a rank of 3.

Proposition 2: *The Dept- independent interaction matrix* $\mathbf{A(q)}$ *has a rank of 2 if the matrix* $\mathbf{M}$ *is a perspective projection matrix.*

Proof: This proposition can be proved if we can show that the matrix **D** has a rank of 2. Assume that the rank of the matrix **D** is smaller than 2. Then, there exist nonzero scalars λ_1 and λ_2 such that

$$\lambda_1(\mathbf{m}_1 - u\mathbf{m}_3) + \lambda_2(\mathbf{m}_2 - v\mathbf{m}_3) = 0 \tag{13}$$

This equation can be written as

$$\lambda_1\mathbf{m}_1 + \lambda_2\mathbf{m}_2 - (\lambda_1 u + \lambda_2 v)\mathbf{m}_3 = 0 \tag{14}$$

If $(\lambda_1 u + \lambda_2 v)$ is not zero, eq. (14) means that the row vectors of **M** are linearly dependent. If the coefficient is zero, the first two row vectors are linearly dependent. However, the row vectors must be linearly independent because the rank of **M** is 3. Therefore, the rank of the matrix **D** is 2. ■

Property 2: *For any vector* $\boldsymbol{\rho}$ *, the product* $\mathbf{A(q)}\boldsymbol{\rho}$ *can be written as a linear form of the unknown parameters ,i.e.*

$$\mathbf{A(q)}\boldsymbol{\rho} = \mathbf{Q}(\boldsymbol{\rho}, \mathbf{q}, \mathbf{y})\boldsymbol{\theta} \tag{15}$$

where $\mathbf{Q}(\boldsymbol{\rho}, \mathbf{q}, \mathbf{y})$ *does not depend on the parameters representing the products of the camera parameters and the world coordinates of the target point.*

2.3 Robot Dynamics

The dynamic equation of a robot manipulator has the form:

$$\mathbf{H(q)}\ddot{\mathbf{q}} + (\frac{1}{2}\dot{\mathbf{H}}(\mathbf{q}) + \mathbf{C}(\mathbf{q}, \dot{\mathbf{q}}))\dot{\mathbf{q}} + \mathbf{G(q)} = \boldsymbol{\tau} \tag{16}$$

where $\mathbf{H(q)}$ is the positive-definite and symmetric inertia matrix. $\mathbf{C}(\mathbf{q}, \dot{\mathbf{q}})$ is a skew-symmetric matrix. The term $\mathbf{G(q)}$ represents the gravitational force, and $\boldsymbol{\tau}$ is the joint input of the robot manipulator. The first term on the left side of (16) is the inertial force, and the second term represents the Colioris and centrifugal forces.

3. Adaptive Image-based Visual Servoing with Unknown Target Positions

Our objective is to develop a controller that is able to estimate the unknown parameters on-line while controlling motion of the manipulator. This problem differs from traditional adaptive control utilizing state feedback errors. The feedback signals here are the image errors, which are outputs of the system. Fortunately, the state information of the manipulator can be also obtained by the internal sensors, so the problem is slightly different from output adaptive control problems in nonlinear control theory. We propose to employ a hybrid feedback scheme that feeds back both the position and velocity (state) of the manipulator and the image errors (output).

In this section, we assume that the world coordinates of the target point are unknown. Therefore the unknown parameters $\boldsymbol{\theta}$ here is only the three coordinate components of the target position $\mathbf{x}$.

3.1 Controller Design

Denote the desired position of the target point on the image plane by $\mathbf{y_d}$. The image error is given by:

$$\Delta\mathbf{y} = \mathbf{y} - \mathbf{y_d} \tag{17}$$

Denote an estimation of the unknown parameters $\boldsymbol{\theta}$ by $\hat{\boldsymbol{\theta}}$. Using the estimation, we propose the following controller:

$$\boldsymbol{\tau} = \mathbf{G(q)} - \mathbf{K}_1\dot{\mathbf{q}} - (\hat{\mathbf{A}}^T(\mathbf{q}) + \frac{1}{2}\hat{\mathbf{n}}(\mathbf{q})\Delta\mathbf{y}^T)\mathbf{B}\Delta\mathbf{y} \tag{18}$$

The first term is to cancel the gravitational force. The second term is a velocity feedback. The last term represents the visual feedback. $\hat{\mathbf{A}}(\mathbf{q})$ is the estimated Depth-independent interaction matrix calculated by the estimated parameters. $\hat{\mathbf{n}}(\mathbf{q})$ is an estimation of vector $\mathbf{n(q)}$, $\mathbf{K}_1$ and $\mathbf{B}$ are positive definite gain matrices. It is important to note that $\frac{1}{{}^c z}$ does not appear in the controller. The quadratic form of $\Delta\mathbf{y}$ is to compensate for the effect of the scale factor. Using the Depth-independent interaction matrix and including the quadratic term differentiates our controller from other existing ones. By substituting (18) into (16), we obtain:

$$\begin{aligned} &\mathbf{H(q)}\ddot{\mathbf{q}} + (\frac{1}{2}\dot{\mathbf{H}}(\mathbf{q}) + \mathbf{C}(\mathbf{q},\dot{\mathbf{q}}))\dot{\mathbf{q}} = -\mathbf{K}_1\dot{\mathbf{q}} \\ &-(\mathbf{A}^T(\mathbf{q}) + \frac{1}{2}\mathbf{n}(\mathbf{q})\Delta\mathbf{y}^T)\mathbf{B}\Delta\mathbf{y} - [(\hat{\mathbf{A}}^T(\mathbf{q}) - \mathbf{A}^T(\mathbf{q})) + \frac{1}{2}(\hat{\mathbf{n}}(\mathbf{q}) - \mathbf{n})\Delta\mathbf{y}^T]\mathbf{B}\Delta\mathbf{y} \end{aligned} \tag{19}$$

From the Property 2, the last term can be represented as a linear form of the estimation errors of the parameters as follows:

$$-[(\hat{\mathbf{A}}^T(\mathbf{q}) - \mathbf{A}^T(\mathbf{q})) + \frac{1}{2}(\hat{\mathbf{n}}(\mathbf{q}) - \mathbf{n})\Delta\mathbf{y}^T]\mathbf{B}\Delta\mathbf{y} = \mathbf{Y}(\mathbf{q},\mathbf{y})\Delta\boldsymbol{\theta} \tag{20}$$

where $\Delta\boldsymbol{\theta} = \hat{\boldsymbol{\theta}} - \boldsymbol{\theta}$ is the estimation error and the regressor $\mathbf{Y(q,y)}$ does not depend on the unknown parameters.

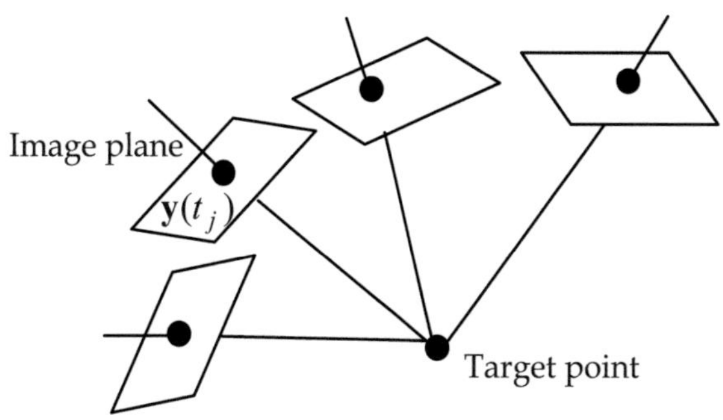

Figure 9. Projections of the target point on image planes

3.2 Estimation of the Parameters

With the motion of the robot manipulator, the camera moves and captures the images of the target point from different view points (Fig. 9). What we need to estimate is the 3-D coordinates of the target point. The problem is similar to the projective structure from motion problem in computer vision. However, it should be pointed out that there are two differences here: 1) only the initial parameters need to be estimated and the remaining can be obtained from motion of the manipulator and 2) the estimation must be conducted on-line. In the projective structure from motion, the basic idea is to find a solution of matrix **M** and **x** that minimizes the Frobenius norm given by

$$\mathbf{E} = \sum_{j=1}^{l} \left\| {}^{c}z\mathbf{y}(t_j) - \mathbf{M}\mathbf{T}_{\mathbf{e}}^{-1}(\mathbf{q}(t_j))\mathbf{x} \right\|^2 \tag{21}$$

where l denotes the number of images captured by the cameras at difference configurations of the manipulator. Here t_j represents the time instant when the j-th image was captured. $\mathbf{y}(t_j) = (u(t_j), v(t_j), 1)^T$ is image coordinates of the target point on the j-th image and $\mathbf{q}(t_j)$ represents the corresponding joint angles. Note that the l images can be selected from the trajectory of camera. By substituting the estimated parameters in the Frobenius norm, we note that

$$\begin{aligned} & {}^{c}\hat{z}(t_j)\mathbf{y}(t_j) - \mathbf{M}\mathbf{T}_{\mathbf{e}}^{-1}\hat{\mathbf{x}} \\ & = \mathbf{y}(t_j)({}^{c}\hat{z}(t_j) - {}^{c}z(t_j)) - (\mathbf{M}\mathbf{T}_{\mathbf{e}}^{-1}\hat{\mathbf{x}} - \mathbf{M}\mathbf{T}_{\mathbf{e}}^{-1}\mathbf{x}) \\ & = \mathbf{W}(\mathbf{x}(t_j), \mathbf{y}(t_j))\Delta\boldsymbol{\theta} \end{aligned} \tag{22}$$

From the result in computer vision, we have
Proposition 3: *If 2 images are selected from the trajectory of the camera, the equation*

$$\mathbf{W}(\mathbf{x}(t_j), \mathbf{y}(t_j))\Delta\boldsymbol{\theta} = 0, \quad \forall j = 1,2 \tag{23}$$

implies the estimated parameters can be estimated to the real values up to a scale..
Based on the discussion above, we propose the following adaptive rule:

$$\frac{d}{dt}\hat{\boldsymbol{\theta}} = -\boldsymbol{\Gamma}^{-1}\{\mathbf{Y}^{\mathbf{T}}(\mathbf{q},\mathbf{y})\dot{\mathbf{q}} + \sum_{j=1}^{l} \mathbf{W}^{\mathbf{T}}(\mathbf{x}(t_j), \mathbf{y}(t_j))\mathbf{K}_3\mathbf{W}(\mathbf{x}(t_j), \mathbf{y}(t_j))\Delta\boldsymbol{\theta}\} \tag{24}$$

where $\boldsymbol{\Gamma}$ and $\mathbf{K}_3$ are positive-definite and diagonal gain matrices. To simplify the notation, let

$$\mathbf{W}_{\mathbf{j}} = \mathbf{W}(\mathbf{x}(t_j), \mathbf{y}(t_j)) \tag{25}$$

Remark 1: *The adaptive algorithm (24) differs from Slotine and Li's algorithm (Slotine &Li, 1987) due to the existence of the last term and the potential force, which ensures the convergence of the estimated parameters to real values.*

3.3 Stability Analysis

This section analyses the stability of the proposed controller.

Theorem 1: *Under the control of the controller (18) and the adaptive algorithm (24) for parameters estimation, the image error of the target point is convergent to zero, i.e.*

$$\lim_{t\to\infty} \Delta\mathbf{y} = 0 \tag{26}$$

Furthermore, if a sufficient number of images are used in the adaptive algorithm, the estimated parameters are convergent to the real ones up to a scale.

Proof: Introduce the following positive function:

$$V(t) = \frac{1}{2}\{\dot{\mathbf{q}}^{\mathbf{T}}\mathbf{H}(\mathbf{q})\dot{\mathbf{q}} + {}^{c}z\Delta\mathbf{y}^{\mathbf{T}}\mathbf{B}\Delta\mathbf{y} + \Delta\boldsymbol{\theta}^{\mathbf{T}}\boldsymbol{\Gamma}\Delta\boldsymbol{\theta}\} \tag{27}$$

Notice that ${}^{c}z > 0$. Multiplying the $\dot{\mathbf{q}}^{\mathbf{T}}$ from the left to (19) results in

$$\dot{\mathbf{q}}^{\mathbf{T}}\mathbf{H}(\mathbf{q})\ddot{\mathbf{q}} + \frac{1}{2}\dot{\mathbf{q}}^{\mathbf{T}}\dot{\mathbf{H}}(\mathbf{q})\dot{\mathbf{q}} = -\dot{\mathbf{q}}^{\mathbf{T}}\mathbf{K}_1\dot{\mathbf{q}} - \dot{\mathbf{q}}^{\mathbf{T}}\mathbf{A}^{\mathbf{T}}(\mathbf{q})\mathbf{B}\Delta\mathbf{y} - \frac{1}{2}\dot{\mathbf{q}}^{\mathbf{T}}\mathbf{n}(\mathbf{q})(\Delta\mathbf{y}^{\mathbf{T}}\mathbf{B}\Delta\mathbf{y}) + \dot{\mathbf{q}}^{\mathbf{T}}\mathbf{Y}(\mathbf{q},\mathbf{y})\Delta\boldsymbol{\theta} \tag{28}$$

From equation (9), we have

$$\dot{\mathbf{q}}^{\mathbf{T}}\mathbf{A}^{\mathbf{T}}(\mathbf{q}) = {}^{c}z\dot{\mathbf{y}}^{\mathbf{T}} = {}^{c}z\Delta\dot{\mathbf{y}}^{\mathbf{T}} \tag{29}$$

Multiplying the $\Delta\boldsymbol{\theta}^{\mathbf{T}}$ from the left to (24), we obtain

$$\Delta\boldsymbol{\theta}^{\mathbf{T}}\boldsymbol{\Gamma}\Delta\dot{\boldsymbol{\theta}} = -\Delta\boldsymbol{\theta}^{\mathbf{T}}\mathbf{Y}^{\mathbf{T}}(\mathbf{q},\mathbf{y})\dot{\mathbf{q}} - \sum_{j=1}^{l}\Delta\boldsymbol{\theta}^{\mathbf{T}}\mathbf{W}_{\mathbf{j}}^{\mathbf{T}}\mathbf{K}_3\mathbf{W}_{\mathbf{j}}\Delta\boldsymbol{\theta} \tag{30}$$

Differentiating the function $V(t)$ results in

$$\dot{V}(t) = \dot{\mathbf{q}}^{\mathbf{T}}(\mathbf{H}(\mathbf{q})\ddot{\mathbf{q}} + \frac{1}{2}\dot{\mathbf{H}}(\mathbf{q}))\dot{\mathbf{q}} + \Delta\dot{\boldsymbol{\theta}}^{\mathbf{T}}\boldsymbol{\Gamma}\Delta\boldsymbol{\theta} + {}^{c}z\Delta\dot{\mathbf{y}}^{\mathbf{T}}\mathbf{B}\Delta\mathbf{y} + \frac{1}{2}{}^{c}\dot{z}\Delta\mathbf{y}^{\mathbf{T}}\mathbf{B}\Delta\mathbf{y} \tag{31}$$

Note that

$${}^{c}\dot{z} = \mathbf{n}^{\mathbf{T}}\dot{\mathbf{q}} \tag{32}$$

By combining the equations (28)-(32), we have

$$\dot{V}(t) = -\dot{\mathbf{q}}^{\mathbf{T}}\mathbf{K}_1\dot{\mathbf{q}} - \sum_{j=1}^{l}\Delta\boldsymbol{\theta}^{\mathbf{T}}\mathbf{W}_{\mathbf{j}}^{\mathbf{T}}\mathbf{K}_3\mathbf{W}_{\mathbf{j}}\Delta\boldsymbol{\theta} \tag{33}$$

obviously we have $\dot{V}(t) \le 0$, and hence $\dot{\mathbf{q}}$, $\Delta\mathbf{y}$ and $\Delta\boldsymbol{\theta}$ are all bounded. From (19) and (24), we know $\ddot{\mathbf{q}}$ and $\dot{\hat{\boldsymbol{\theta}}}$ are bounded respectively. Differentiating the function $\dot{V}(t)$ resulting in

$$\ddot{V}(t) = -\ddot{\mathbf{q}}^{\mathbf{T}}\mathbf{K}_1\dot{\mathbf{q}} - \sum_{j=1}^{l}(\Delta\dot{\boldsymbol{\theta}}^{\mathbf{T}}\mathbf{W}_{\mathbf{j}}^{\mathbf{T}} + \Delta\boldsymbol{\theta}^{\mathbf{T}}\dot{\mathbf{W}}_{\mathbf{j}}^{\mathbf{T}})\mathbf{K}_3\mathbf{W}_{\mathbf{j}}\Delta\boldsymbol{\theta} \tag{34}$$

Then, we can conclude that $\ddot{V}(t)$ is bounded. Consequently, from barbalat's lemma, we have

$$\begin{aligned}&\lim_{t\to\infty}\dot{\mathbf{q}} = 0\\ &\lim_{t\to\infty}\dot{\hat{\boldsymbol{\theta}}} = 0\\ &\lim_{t\to\infty}\mathbf{W}(x(t_j), y(t_j))\Delta\boldsymbol{\theta} = 0\end{aligned} \tag{35}$$

The convergence of $\mathbf{W}_j\Delta\boldsymbol{\theta}$ implies that the estimated matrix $\hat{\boldsymbol{\theta}}$ is convergent to the real values up to a scale. In order to prove the convergence of the image error, consider the invariant set of the system when $\dot{V}(t)=0$. From the close loop dynamics (19), at the invariant set,

$$(\hat{\mathbf{A}}^{\mathrm{T}}+\frac{1}{2}\hat{\mathbf{n}}\Delta\mathbf{y}^{\mathrm{T}})\mathbf{B}\Delta\mathbf{y}=0 \tag{36}$$

It is reasonable to assume that the manipulator is not at the singular configurations so that $\mathbf{J(q)}$ is of full rank. Note that

$$\hat{\mathbf{A}}^{\mathrm{T}}+\frac{1}{2}\hat{\mathbf{n}}\Delta\mathbf{y}^{\mathrm{T}}=\left[\frac{\partial(\mathbf{T}_{\mathbf{e}}^{-1}(\mathbf{q})\hat{\mathbf{x}})}{\partial\mathbf{q}}\right]^T\underbrace{\begin{pmatrix}\mathbf{m}_1^{\mathrm{T}}+(0.5\Delta u-u)\mathbf{m}_3^{\mathrm{T}}\\ \mathbf{m}_2^{\mathrm{T}}+(0.5\Delta v-v)\mathbf{m}_3^{\mathrm{T}}\\ 0\end{pmatrix}^T}_{\mathbf{E}} \tag{37}$$

Since it is guaranteed that $\hat{\mathbf{x}}\neq 0$ at the invariant set, the matrix $\frac{\partial(\mathbf{T}_{\mathbf{e}}^{-1}(\mathbf{q})\hat{\mathbf{x}})}{\partial\mathbf{q}}$ is not singular from Proposition 1. Then we proposed

Proposition 4: *The matrix* $\hat{\mathbf{A}}^{\mathrm{T}}+\frac{1}{2}\hat{\mathbf{n}}\Delta\mathbf{y}^{\mathrm{T}}$ *has a rank of 2 if the matrix* $\hat{\mathbf{M}}$ has a rank of 3.

This proposition can be proved if we can show that the matrix **E** has a rank of 2. the proof is similar to Proposition 2.

Therefore, from (36), it is obvious that at the invariant set the position error on the image plane must be zero, e.g.

$$\Delta\mathbf{y}=0 \tag{38}$$

From the Barbalat's lemma, we can conclude the convergence of the position error of the target point projections on the image plane to zero and convergence of the estimated parameters to the real values

4. Adaptive Image-based Visual Servoing with Unknown Camera Parameters and Target Positions

In this section, we assume that in addition to the 3-D coordinates of the target points, the intrinsic and extrinsic parameters of the camera are also unknown. Denote all the possible products of the components of matrix **M** and vector **x** by a vector $\boldsymbol{\theta}_l$, which represents all the unknown parameters including the camera intrinsic and extrinsic parameters and the unknown world coordinates of the target point. It should be pointed out that there are 39 unknown parameters in total. Here we fixed one parameters $\hat{p}_z=1$ and define the left 38 parameters as uniform parameters. Here we define $\hat{\boldsymbol{\theta}}_l=\frac{\hat{m}_{ij}\hat{x}_k}{\hat{p}_z}$ and $\boldsymbol{\theta}_l=\frac{m_{ij}x_k}{p_z}$, where

$$\forall\, l=1,\ldots,38;\ i,j=1,2,3,\ k=1,2,3,4 \text{ or } i=1,2,\ j,k=4.$$

4.1 Controller Design

Using the estimation, we propose the following controller similar to the previous section:

$$\boldsymbol{\tau} = \mathbf{G}(\mathbf{q}) - \mathbf{K}_1\dot{\mathbf{q}} - (\hat{\mathbf{A}}^{\mathrm{T}}(\mathbf{q}) + \frac{1}{2}\hat{\mathbf{n}}(\mathbf{q})\Delta\mathbf{y}^{\mathrm{T}})\mathbf{B}\Delta\mathbf{y} \tag{39}$$

By substituting (39) into (16), we obtain:

$$\begin{aligned}&\mathbf{H}(\mathbf{q})\ddot{\mathbf{q}} + (\frac{1}{2}\dot{\mathbf{H}}(\mathbf{q}) + \mathbf{C}(\mathbf{q},\dot{\mathbf{q}}))\dot{\mathbf{q}} = -\mathbf{K}_1\dot{\mathbf{q}}\\ &-(\frac{\mathbf{A}^{\mathrm{T}}(\mathbf{q})}{p_z} + \frac{1}{2}\frac{\mathbf{n}(\mathbf{q})\Delta\mathbf{y}^{\mathrm{T}}}{p_z})\mathbf{B}\Delta\mathbf{y} - [(\hat{\mathbf{A}}(\mathbf{q}) - \frac{\mathbf{A}^{\mathrm{T}}(\mathbf{q})}{p_z}) + \frac{1}{2}(\hat{\mathbf{n}}(\mathbf{q}) - \frac{\mathbf{n}}{p_z})\Delta\mathbf{y}^{\mathrm{T}}]\mathbf{B}\Delta\mathbf{y}\end{aligned} \tag{40}$$

From the Property 2, the last term can be represented as a linear form of the estimation errors of the parameters as follows:

$$-[(\hat{\mathbf{A}}(\mathbf{q}) - \frac{\mathbf{A}^{\mathrm{T}}(\mathbf{q})}{p_z}) + \frac{1}{2}(\hat{\mathbf{n}}(\mathbf{q}) - \frac{\mathbf{n}}{p_z})\Delta\mathbf{y}^{\mathrm{T}}]\mathbf{B}\Delta\mathbf{y} = \mathbf{Y}(\mathbf{q},\mathbf{y})\Delta\boldsymbol{\theta}_l \tag{41}$$

where $\Delta\boldsymbol{\theta}_l = \hat{\boldsymbol{\theta}}_l - \boldsymbol{\theta}_l$ is the estimation error and the regressor $\mathbf{Y}(\mathbf{q},\mathbf{y})$ does not depend on the unknown parameters.

4.2 Estimation of the Parameters

The idea is same as previous section, we minimizes the Frobenius norm, Since there are 38 unknowns in $\boldsymbol{\theta}_l$, 19 images are necessary for estimating the parameters. By substituting the estimated parameters in the Frobenius norm, we note that

$$\begin{aligned}&{}^c\hat{z}(t_j)\mathbf{y}(t_j) - \hat{\mathbf{M}}\mathbf{T}_{\mathbf{e}}^{-1}\hat{\mathbf{x}}\\ &= \mathbf{y}(t_j)({}^c\hat{z}(t_j) - \frac{{}^c z(t_j)}{p_z}) - (\hat{\mathbf{M}}\mathbf{T}_{\mathbf{e}}^{-1}\hat{\mathbf{x}} - \frac{\mathbf{M}\mathbf{T}_{\mathbf{e}}^{-1}\mathbf{x}}{p_z})\\ &= \mathbf{W}(\mathbf{x}(t_j),\mathbf{y}(t_j))\Delta\boldsymbol{\theta}_l\end{aligned} \tag{42}$$

From the result in computer vision, we have

Proposition 5: *If a sufficient number of images are selected from the trajectory of the camera, then*

$$\mathbf{W}(\mathbf{x}(t_j),\mathbf{y}(t_j))\Delta\boldsymbol{\theta} = 0, \quad \forall j = 1,2,\ldots,l \tag{43}$$

implies one of the following two cases:

(1) the estimated parameters can be estimated to the real values up to a scale, i.e.

$$\hat{\boldsymbol{\theta}} = \lambda\boldsymbol{\theta} \tag{44}$$

By the definition of $\boldsymbol{\theta}_l$*, we can estimate the perspective projection matrix to the real one up to a scale.*

$$\hat{\mathbf{M}} = \lambda\mathbf{M} \tag{45}$$

then matrix $\hat{\mathbf{M}}$ *a rank of 3 from Property 1.*

(2) The estimated parameters are zero.

Proposition 6: *If* $\mathbf{W}(\mathbf{x}(t_j),\mathbf{y}(t_j))\Delta\boldsymbol{\theta}_l = 0$, *we can avoid the trivial solution of* $\hat{\boldsymbol{\theta}}_l = 0$.

Proof: By fixing the parameter $\hat{p}_z = 1$, if $\hat{\boldsymbol{\theta}}_l = 0$, we can obtain ${}^c\hat{z} = \hat{\mathbf{m}}_3^{\mathbf{T}}\mathbf{T}_{\mathbf{e}}^{-1}(\mathbf{q})\hat{\mathbf{x}} = 1$, which implies that $\mathbf{W}(\mathbf{x}(t_j),\mathbf{y}(t_j))\Delta\boldsymbol{\theta}_l = \mathbf{y}(t_j)$. It is obviously wrong since we know that $\mathbf{W}(\mathbf{x}(t_j),\mathbf{y}(t_j))\Delta\boldsymbol{\theta}_l = 0$. So we can avoid the trivial solution.

Remark 2: By Proposition 6 and (44), it is easy to guaranteed that $\hat{\mathbf{x}} \neq 0$.
The adaptive rule is simlar as (24)

$$\frac{d}{dt}\hat{\boldsymbol{\theta}}_l = -\boldsymbol{\Gamma}^{-1}\{\mathbf{Y}^{\mathbf{T}}(\mathbf{q},\mathbf{y})\dot{\mathbf{q}} + \sum_{j=1}^{l}\mathbf{W}^{\mathbf{T}}(\mathbf{x}(t_j),\mathbf{y}(t_j))\mathbf{K}_3\mathbf{W}(\mathbf{x}(t_j),\mathbf{y}(t_j))\Delta\boldsymbol{\theta}_l\} \tag{46}$$

4.3 Stability Analysis

This section analyses the stability of the proposed controller.
Theorem 2: *Under the control of the controller (39) and the adaptive algorithm (46) for parameters estimation, the image error of the target point is convergent to zero, i.e.*

$$\lim_{t\to\infty}\Delta\mathbf{y} = 0 \tag{47}$$

Furthermore, the estimated parameters are convergent to the real values up to a scale:
Proof: Introduce the following positive function:

$$V(t) = \frac{1}{2}\{\dot{\mathbf{q}}^{\mathbf{T}}\mathbf{H}(\mathbf{q})\dot{\mathbf{q}} + \frac{{}^c z}{p_z}\Delta\mathbf{y}^{\mathbf{T}}\mathbf{B}\Delta\mathbf{y} + \Delta\boldsymbol{\theta}^{\mathbf{T}}\boldsymbol{\Gamma}\Delta\boldsymbol{\theta}\} \tag{48}$$

Notice that $p_z > 0$ and ${}^c z > 0$. Multiplying the $\dot{\mathbf{q}}^{\mathbf{T}}$ from the left to (19) results in

$$\begin{aligned}\dot{\mathbf{q}}^{\mathbf{T}}\mathbf{H}(\mathbf{q})\ddot{\mathbf{q}} + \frac{1}{2}\dot{\mathbf{q}}^{\mathbf{T}}\dot{\mathbf{H}}(\mathbf{q})\dot{\mathbf{q}} &= -\dot{\mathbf{q}}^{\mathbf{T}}\mathbf{K}_1\dot{\mathbf{q}} - \frac{1}{p_z}\dot{\mathbf{q}}^{\mathbf{T}}\mathbf{A}^{\mathbf{T}}(\mathbf{q})\mathbf{B}\Delta\mathbf{y} \\ &- \frac{1}{2p_z}\dot{\mathbf{q}}^{\mathbf{T}}\mathbf{n}(\mathbf{q})(\Delta\mathbf{y}^{\mathbf{T}}\mathbf{B}\Delta\mathbf{y}) + \dot{\mathbf{q}}^{\mathbf{T}}\mathbf{Y}(\mathbf{q},\mathbf{y})\Delta\boldsymbol{\theta}_l\end{aligned} \tag{49}$$

Differentiating the function *V(t)* results in

$$\dot{V}(t) = \dot{\mathbf{q}}^{\mathbf{T}}(\mathbf{H}(\mathbf{q})\ddot{\mathbf{q}} + \frac{1}{2}\dot{\mathbf{H}}(\mathbf{q}))\dot{\mathbf{q}} + \Delta\dot{\boldsymbol{\theta}}^{\mathbf{T}}\boldsymbol{\Gamma}\Delta\boldsymbol{\theta} + \frac{{}^c z}{p_z}\Delta\dot{\mathbf{y}}^{\mathbf{T}}\mathbf{B}\Delta\mathbf{y} + \frac{1}{2p_z}{}^c\dot{z}\Delta\mathbf{y}^{\mathbf{T}}\mathbf{B}\Delta\mathbf{y} \tag{50}$$

By proper simplification, we have

$$\dot{V}(t) = -\dot{\mathbf{q}}^{\mathbf{T}}\mathbf{K}_1\dot{\mathbf{q}} - \sum_{j=1}^{l}\Delta\boldsymbol{\theta}^{\mathbf{T}}\mathbf{W}_{\mathbf{j}}^{\mathbf{T}}\mathbf{K}_3\mathbf{W}_{\mathbf{j}}\Delta\boldsymbol{\theta}_l \tag{51}$$

obviously we have $\dot{V}(t) \leq 0$, and hence $\dot{\mathbf{q}}$, $\Delta\mathbf{y}$ and $\Delta\boldsymbol{\theta}_l$ are all bounded. From (19) and (46), we know $\ddot{\mathbf{q}}$ and $\dot{\hat{\boldsymbol{\theta}}}_l$ are bounded respectively. Differentiating the function $\dot{V}(t)$ resulting in

$$\ddot{V}(t) = -\ddot{\mathbf{q}}^{\mathbf{T}}\mathbf{K}_1\dot{\mathbf{q}} - \sum_{j=1}^{l}(\Delta\dot{\boldsymbol{\theta}}^{\mathbf{T}}\mathbf{W}_{\mathbf{j}}^{\mathbf{T}} + \Delta\boldsymbol{\theta}^{\mathbf{T}}\dot{\mathbf{W}}_{\mathbf{j}}^{\mathbf{T}})\mathbf{K}_3\mathbf{W}_{\mathbf{j}}\Delta\boldsymbol{\theta}_l \tag{52}$$

Similar as Theorem 1, From the Barbalat's lemma, we can conclude the convergence of the position error of the target point projections on the image plane to zero and convergence of the estimated parameters to the real values.

4.4 Extension to Multiple Feature Points

This controller can be directly extended to visual seroving using a number S of feature points. In control of multiple feature points, the dimension of the depth independent interaction matrix will increase, and hence the major difference will be the difference in computation time. A controller similar to that in (39) can be designed by

$$\boldsymbol{\tau} = \mathbf{G}(\mathbf{q}(t)) - \mathbf{K}_1\dot{\mathbf{q}}(t) - \sum_{i=1}^{S}(\hat{\mathbf{A}}_i^T(t) + \frac{1}{2}\hat{\mathbf{n}}_i(t)\Delta\mathbf{y}_i^T(t))\mathbf{B}\Delta\mathbf{y}_i(t) \tag{53}$$

where $\hat{\mathbf{A}}_i(t)$ and $\hat{\mathbf{n}}_i(t)$ have a similar form to Eq. (10) and Eq. (7) for point i, respectively. $\Delta\mathbf{y}_i(t)$ is the image position error for i-th feature point. The image errors are convergent to zero when the number n of degrees of freedom of the manipulator is larger than or equal to 2S.

When n=6 and $S \geq 3$, the convergence of image errors can also be guaranteed. It is well known that three non-collinear projections of fixed feature points on the image plane can uniquely define the 3-D position of the camera, and hence the projections of all other fixed feature points are uniquely determined. Therefore, if the projections of three feature points whose projections are convergent to their desired positions on the image plane, so are the projection of all other feature points. We can conclude the convergence of the image errors when the manipulator has equal or more degrees of freedom than 2S, or when the manipulator has 6 degrees of freedom.

5. Experiments

We have implemented the controller in a 3 DOF robot manipulator in the Chinese University of Hong Kong. Fig. 10 shows the experiment setup system. The moment inertia about its vertical axis of the first link of the manipulator is 0.005kgm^2, the masses of the second and third links are 0.167 kg, and 0.1 kg, respectively. The lengths of the second and third links are, 0.145m and 0.1285m, respectively. The joints of the manipulator are driven by Maxon brushed DC motors. The powers of the motors are 20 watt, 10 watt, and 10watt, respectively. The gear ratios at the three joints are 480:49, 12:1, and 384:49, respectively. High-precision Encoders are used to measure the joint angles. The joint velocities are obtained by differentiating the joint angles. A frame processor MATROX PULSAR installed in a PC with Intel Pentium II CPU acquires the video signal. This PC processes the image and extracts the image features.

5.1 Control of One Feature Point with Unknown Target Positions

We first set initial positions and record the image position of the point and then move the end-effector to another position and record that image position as the desired ones (Fig. 11). The control gains are $\mathbf{K}_1 = 30$, $\mathbf{B} = 0.00006$, $\mathbf{K}_3 = 0.001$, $\boldsymbol{\Gamma} = 5000$. The transformation matrix of the base frame respect to the vision frame is $\mathbf{T} = \begin{bmatrix} 0 & 0 & 1 & 0 \\ 0 & -1 & 0 & 0 \\ 1 & 0 & 0 & 1 \\ 0 & 0 & 0 & 1 \end{bmatrix}$, the initial estimated target position is $\mathbf{x} = \begin{bmatrix} 0.85 & -0.15 & 0.05 \end{bmatrix}^T$. The real camera intrinsic parameters are $a_u = 871$, $a_v = 882$, $u_0 = 381$, $v_0 = 278$.

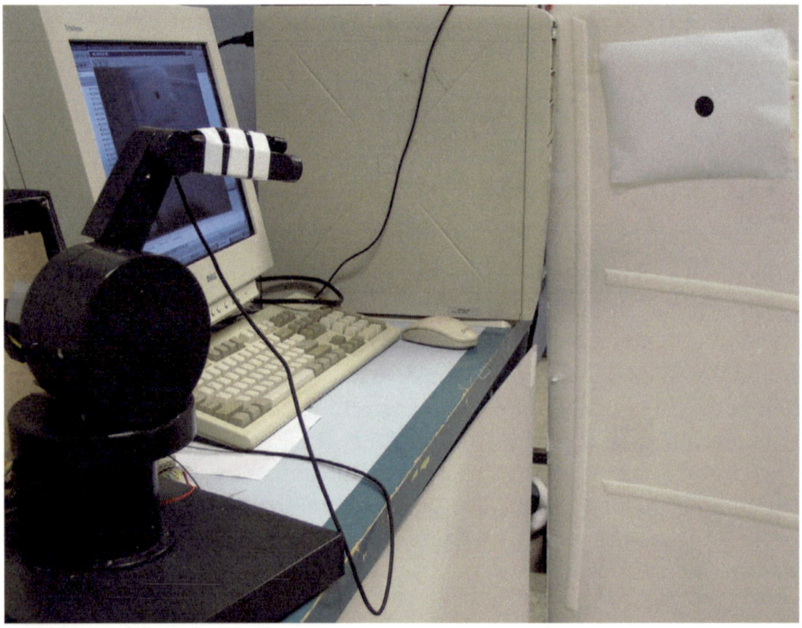

Figure 10. The experiment setup system

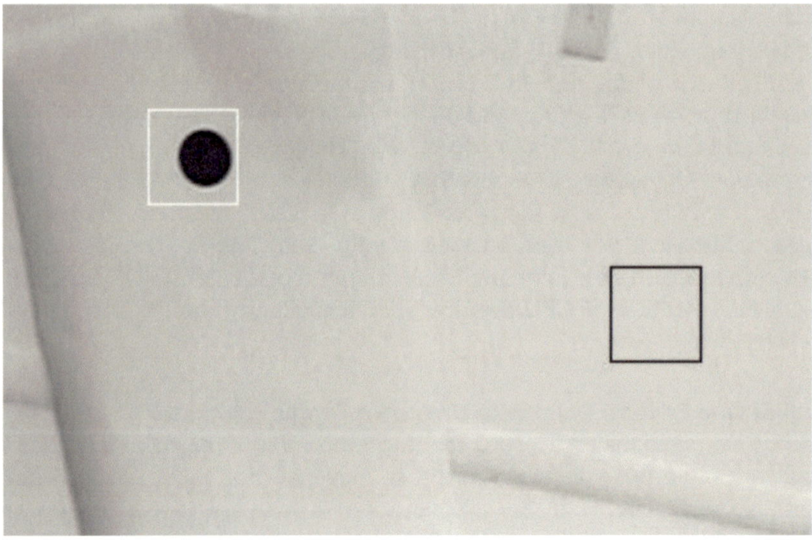

Figure 11. The initial and desired (black square) positions

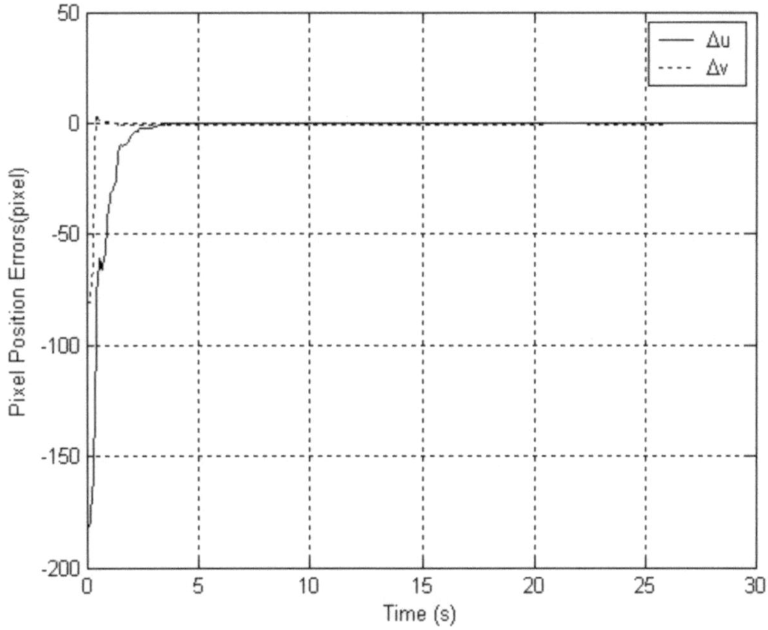

Figure 12. Position errors of the image feature

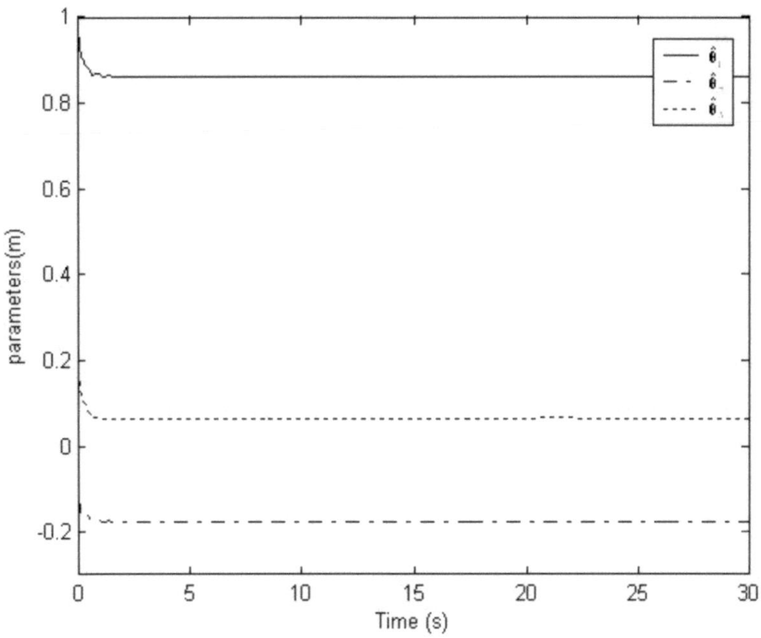

Figure 13. The profile of the estimated parameters

As shown in Fig. 12, the image feature points asymptotically converge to the desired ones. The results confirmed the convergence of the image error to zero under control of the proposed method. Fig. 13 plots the profiles of the estimated parameters, which are not convergent to the true values. The sampling time in the experiment is 40ms.

5.2 Control of One Feature Point with Unknown Camera Parameters and Target Positions

The initial and final positions are same as Fig. 10. The control gains are the same as previous experiments. The initial estimated transformation matrix of the base frame respect to the vision frame is $\hat{\mathbf{T}} = \begin{bmatrix} -0.3 & 0 & 0.95 & 0 \\ 0 & -1 & 0 & 0 \\ 0.95 & 0 & 0.3 & 1 \\ 0 & 0 & 0 & 1 \end{bmatrix}$, The initial estimated target position is $\hat{\mathbf{x}} = [1 \quad -0.1 \quad 0.2]^T$ m. The estimated intrinsic parameters are $\hat{a}_u = 1000$, $\hat{a}_v = 1000$, $\hat{u}_0 = 300$, $\hat{v}_0 = 300$. The image errors of the feature points on the image plane are demonstrated in Fig. 14. Fig. 15 plots the profiles of the estimated parameters.

This experiment showed that it is possible to achieve the convergence without camera parameters and target position information, by using the adaptive rule.

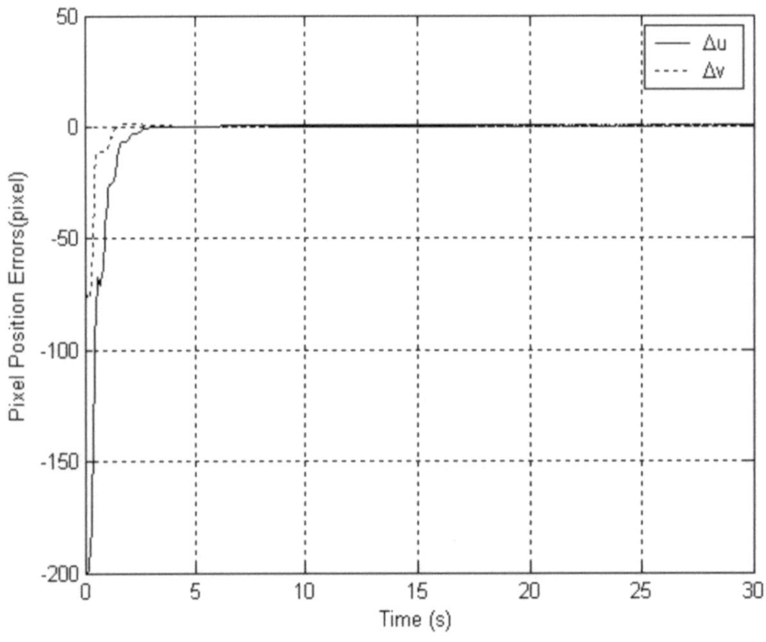

Figure 14. Position errors of the image feature

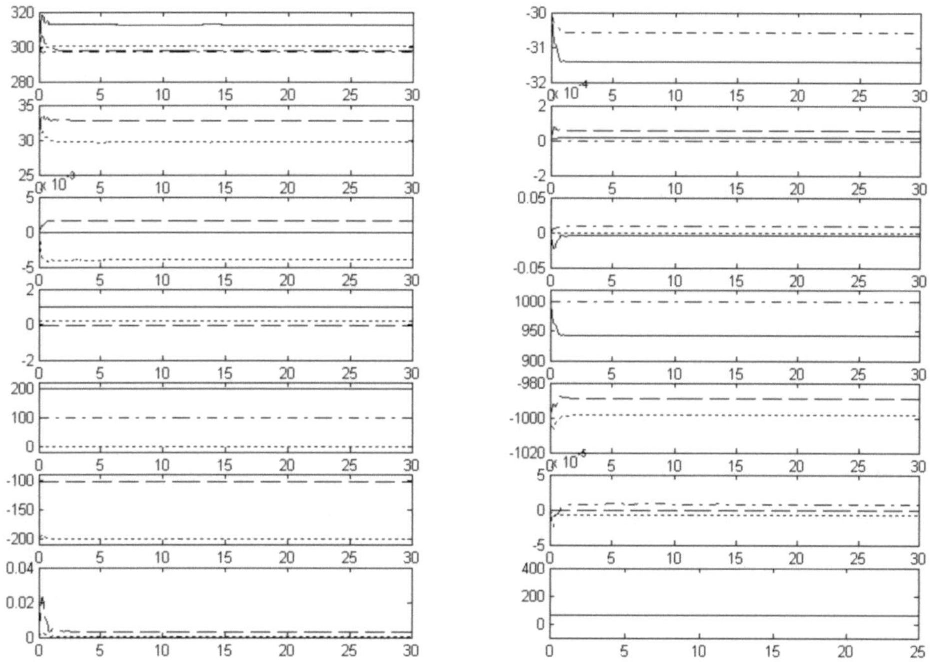

Figure 15. The profile of the estimated parameters

5.3 Control of Three Feature Points

In the third experiment, we control three feature points whose coordinates with respect to the end-effector frames are $\hat{\mathbf{x}}_1 = [1 \quad -0.1 \quad 0.2]^T$ m, $\hat{\mathbf{x}}_2 = [1 \quad -0.15 \quad 0.25]^T$ m, and $\hat{\mathbf{x}}_3 = [1 \quad -0.15 \quad 0.15]^T$ m, respectively.

The initial and desired positions of the feature points are shown in Fig. 16. The image errors of the feature points on the image plane are demonstrated in Fig. 17. The experimental results confirmed that the image errors of the feature points are convergent to zero. The residual image errors are within one pixel. In this experiment, we employed three current positions of the feature points in the adaptive rule. The control gains used the experiments are $\mathbf{K}_1 = 18$, $\mathbf{B} = 0.000015$, $\mathbf{K}_3 = 0.001$, $\mathbf{\Gamma} = 10000$. The true values and initial estimations of the camera intrinsic and extrinsic parameters are as the same as those in previous experiments.

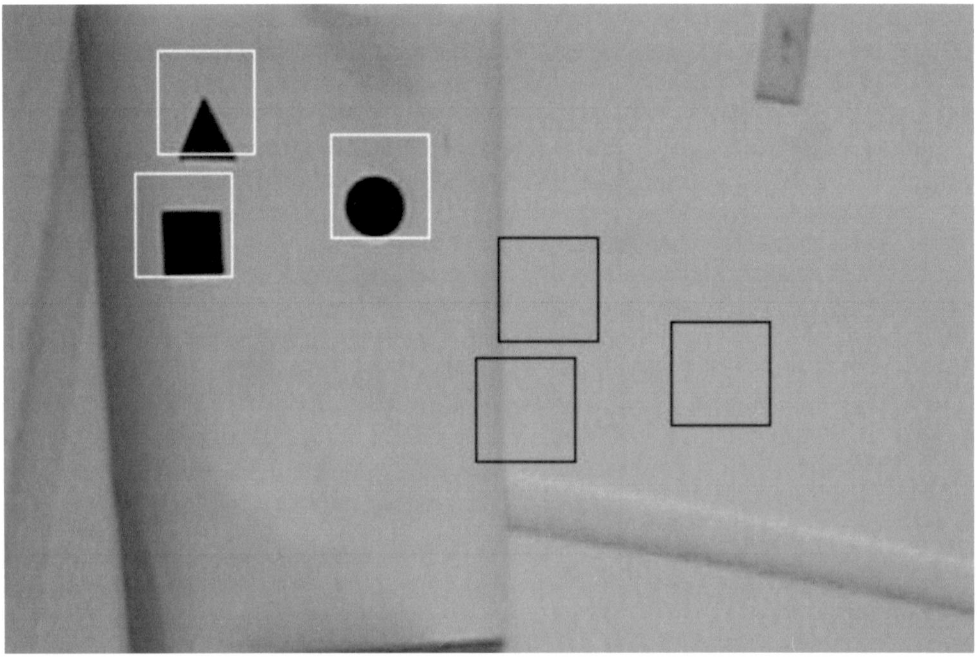

Figure 16. The initial and desired (black square) positions

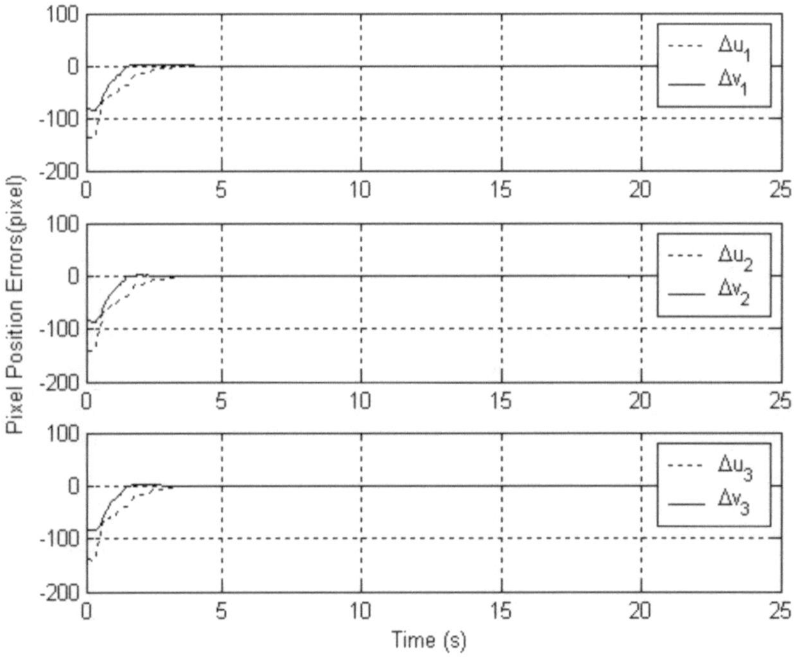

Figure 17. Position errors of the image features

6. Conclusion

This Chapter presents a new adaptive controller for dynamic image-based visual servoing of a robot manipulator uncalibrated eye-in-hand visual feedback. To cope with nonlinear dependence of the image Jacobian on the unknown parameters, this controller employs a matrix called depth-independent interaction matrix which does not depend on the scale factors determined by the depths of target points. A new adaptive algorithm has been developed to estimate the unknown intrinsic and extrinsic parameters and the unknown 3-D coordinates of the target points. With a full consideration of dynamic responses of the robot manipulator, we employed the Lyapunov method to prove the convergence of the postion errors to zero and the convergence of the estimated parameters to the real values. Experimental results illustrate the performance of the proposed methods.

7. References

Bishop. B. & Spong. M.W. (1997). Adaptive Calibration and Control of 2D Monocular Visual Servo System, *Proc. of IFAC Symp. Robot Control*, pp. 525-530.

Carelli. R.; Nasisi. O. & Kuchen. B. (1994). Adaptive robot control with visual feedback, *Proc. of the American Control Conf.*, pp. 1757-1760.

Cheah. C C.; Hirano. M.; Kawamura. S. & Arimoto. S. (2003). Approximate Jacobian control for robots with uncertain kinematics and dynamics, *IEEE Transactions on Robotics and Automation*, vol. 19, no. 4, pp. 692 - 702.

Deng. L.; Janabi-Sharifi. F. & Wilson. W. J. (2002). Stability and robustness of visual servoing methods, *Prof. of IEEE Int. Conf. on Robotics and Automation*, pp. 1604 - 1609.

Dixon. W. E. (2007). Adaptive regulation of amplitude limited robot manipulators with uncertain kinematics and dynamics," *IEEE Trans. on Automatic Control*, vol. 52, no. 3, pp. 488 --493.

Hashitmoto. K.; Nagahama. K. & Noritsugu. T. (2002). A mode switching estimator for visual seroving, *Proc. of IEEE Int. Conf. on Robotics and Automation*, pp. 1610- 1615.

Hemayed. E. E. (2003). A survey of camera self-calibration, *Proc. IEEE Conf. on Advanced Video and Signal Based Surveillance*, pp. 351-357.

Hosada. K. & Asada. M. (1994). Versatile Visual Servoing without knowledge of True Jacobain, *Proc. of IEEE/RSJ Int. Conf. on Intelligent Robots and Systems*, pp.186-191.

Hsu. L. & Aquino. P. L. S. (1999). Adaptive visual tracking with uncertain manipulator dynamics and uncalibrated camera, *Proc. of the 38th IEEE Int. Conf. on Decision and Control*, pp. 1248-1253.

Hutchinson. S.; Hager. G. D. & Corke. P. I. (1996). A tutorial on visual servo control, *IEEE Trans. on Robotics and Automation*, vol. 12, no. 5, pp. 651-670.

Kelly. R.; Reyes. F.; Moreno. J. & Hutchinson. S. (1999). A two-loops direct visual control of direct-drive planar robots with moving target, *Proc. of IEEE Int. Conf. on Robotics and Automation*, pp. 599-604.

Kelly. R.; Carelli. R.; Nasisi. O.; Kuchen. B. & Reyes. F. (2000). Stable Visual Servoing of Camera-in-Hand Robotic Systems, *IEEE/ASME Trans. on Mechatronics*, Vol. 5, No.1, pp.39-48.

Liu. Y. H.; Wang. H. & Lam. K. (2005). Dynamic visual servoing of robots in uncalibrated environments, *Proc. of IEEE Int. Conf. on Robotics and Automation*, pp. 3142-3147.

Liu. Y. H.; Wang. H.; Wang. C. & Lam. K. (2006). Uncalibrated Visual Servoing of Robots Using a Depth-Independent Image Jacobian Matrix, *IEEE Trans. on Robotics*, Vol. 22, No. 4.

Liu. Y. H.; Wang. H. & Zhou. D. (2006). Dynamic Tracking of Manipulators Using Visual Feedback from an Uncalibrated Fixed Camera, *Proc. of IEEE Int. Conf. on Robotics and Automation*, pp.4124-4129.

Malis. E.; Chaumette. F. & Boudet S. (1999). 2-1/2-D Visual Servoing, *IEEE Transaction on Robotics and Automation*, vol. 15, No. 2.

Malis E. (2004). Visual servoing invariant to changes in camera-intrisic parameters, *IEEE.Trans. on Robotics and Automation*, vol. 20, no.1, pp. 72-81.

Nagahama. K.; Hashimoto. K.; Norisugu. T. & Takaiawa. M. (2002). Visual servoing based on object motion estimation, *Proc. IEEE Int. Conf. on Robotics and Automation*, pp. 245-250.

Papanikolopoulos. N. P.; Nelson. B. J. & Khosla. P. K. (1995). Six degree-of-freedom hand/eye visual tracking with uncertain parameters, *IEEE Trans. on Robotics and Automation*, vol. 11, no. 5, pp. 725-732.

Pomares. J.; Chaumette. F. & Torres. F. (2007). Adaptive Visual Servoing by Simultaneous Camera Calibration, *Proc. of IEEE Int. Conf. on Robotics and Automation*, pp.2811-2816.

Ruf. A.; Tonko. M.; Horaud. R. & Nagel. H.-H. (1997) Visual tracking of an end effector by adaptive kinematic prediction, *Proc. of IEEE/RSJ Int. Conf. on Intelligent Robots and Systems*, pp. 893-898.

Shen. Y.; Song. D.; Liu. Y. H. & Li. K. (2003). Asymptotic trajectory tracking of manipulators using uncalibrated visual feedback," *IEEE/ASME Trans. on Mechatronics*, vo. 8, no. 1, pp. 87-98.

Slotine. J. J. & Li. W. (1987), On the adaptive control of robot manipulators, *Int. J. Robotics Research*, Vol. 6, pp. 49-59.

Wang H. & Liu. Y. H. (2006). Uncalibrated Visual Tracking Control without Visual Velocity, *Proc. of IEEE Int. Conf. on Robotics and Automation*, pp.2738-2743.

Wang. H. & Liu. Y. H. (2006). Uncalibrated Visual Tracking Control without Visual Velocity, *Proc. of IEEE Int. Conf. on Robotics and Automation*, pp.2738-2743.

Wang. H.; Liu. Y. H. & Zhou. D. (2007). Dynamic visual tracking for manipulators using an uncalibrated fxied camera, *IEEE Trans. on Robotics*, vol. 23, no. 3, pp. 610-617.

Yoshimi. B. H. & Allen. P. K. (1995). Alignment using an uncalibrated camera system, *IEEE.Trans. on Robotics and Automation*, vol. 11, no.4, pp. 516-521.

28

Vision-Guided Robot Control for 3D Object Recognition and Manipulation

S. Q. Xie, E. Haemmerle, Y. Cheng and P. Gamage
Mechatronics Engineering Group, Department of Mechanical Engineering,
The University of Auckland
New Zealand

1. Introduction

Vision-guided robotics has been one of the major research areas in the mechatronics community in recent years. The aim is to emulate the visual system of humans and allow intelligent machines to be developed. With higher intelligence, complex tasks that require the capability of human vision can be performed and replaced by machines. The applications of visually guided systems are many, from automatic manufacturing (Krar and Gill 2003), product inspection (Abdullah, Guan et al. 2004; Brosnan and Sun 2004), counting and measuring (Billingsley and Dunn 2005) to medical surgery (Burschka, Li et al. 2004; Yaniv and Joskowicz 2005; Graham, Xie et al. 2007). They are often found in tasks that demand high accuracy and consistent quality which are hard to achieve with manual labour. Tedious, repetitive and dangerous tasks, which are not suited for humans, are now performed by robots. Using visual feedback to control a robot has shown distinctive advantages over traditional methods, and is commonly termed visual servoing (Hutchinson et al. 1996). Visual features such as points, lines, and regions can be used, for example, to enable the alignment of a manipulator with an object. Hence, vision is a part of a robot control system providing feedback about the state of the interacting object.

The development of new methods and algorithms for object tracking and robot control has gained particular interest in industry recently since the world has stepped into the century of automation. Research has been focused primarily on two intertwined aspects: tracking and control. Tracking provides a continuous estimation and update of features during robot/object motion. Based on this sensory input, a control sequence is generated for the robot. More recently, the area has attracted significant attention as computational resources have made real-time deployment of vision-guided robot control possible. However, there are still many issues to be resolved in areas such as camera calibration, image processing, coordinate transformation, as well as real time control of robot for complicated tasks.

This chapter presents a vision-guided robot control system that is capable of recognising and manipulating general 2D and 3D objects using an industrial charge-coupled device camera. Object recognition algorithms are developed to recognize 2D and 3D objects of different geometry. The objects are then reconstructed and integrated with the robot controller to enable a fully vision-guided robot control system. The focus of the chapter is placed on new methods and technologies for extracting image information and controlling a

serial robot. They are developed to recognize an object to be manipulated by matching image features to a geometrical model of the object and compute its position and orientation (pose) relative to the robot coordinate system. This absolute pose and cartesian-space information is used to move the robot to the desired pose relative to the object. To estimate the pose of the object, the model of the object needs to be established. To control the robot based on visual information extracted in the camera frame, the camera has to be calibrated with respect to the robot. In addition, the robot direct and inverse kinematic models have to be established to convert cartesian-space robot positions into joint-space configurations. The robot can then execute the task by performing the movements in joint-space.
The overall system structure used in this research is illustrated in Figure 1. It consists of a number of hardware units, including the vision system, the CCD video camera, the robot controller and the ASEA robot. The ASEA robot used in this research is a very flexible device which has five degrees of freedoms (DOF). Such a robot has been widely used in industry automation and medical applications. Each of the units carries out a unique set of inter-related functions.

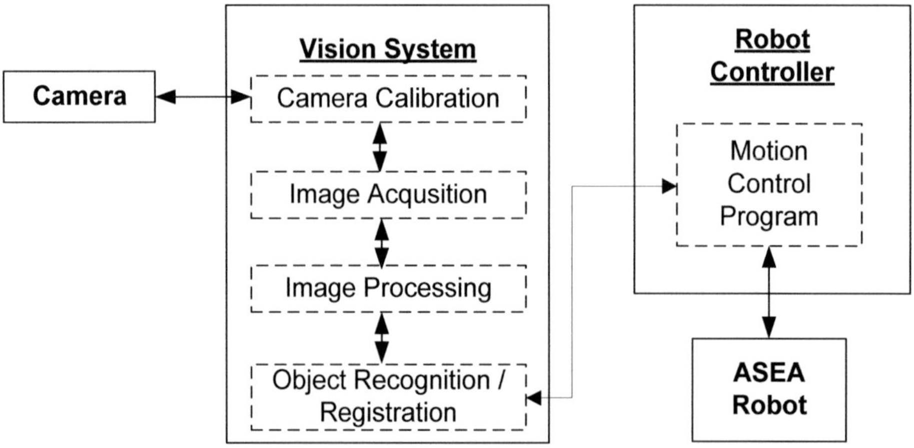

Figure 1. The overall system structure

The vision system includes the required hardware and software components for collecting useful information for the object recognition process. The object recognition process undergoes five main stages: (1) camera calibration; (2) image acquisition; (3) image and information processing; (4) object recognition; and (5) results output to the motion control program.

2. Literature Review

2D object recognition has been well researched, developed and successfully applied in many applications in industry. 3D object recognition however, is relatively new. The main issue involved in 3D recognition is the large amount of information which needs to be dealt with. 3D recognition systems have an infinite number of possible viewpoints, making it difficult to match information of the object obtained by the sensors to the database (Wong, Rong et al. 1998).
Research has been carried out to develop algorithms in image segmentation and registration to successfully perform object recognition and tracking. Image segmentation, defined as the

separation of the image into regions, is the first step leading to image analysis and interpretation. The goal is to separate the image into regions that are meaningful for the specific task. Segmentation techniques utilised can be classified into one of five groups (Fu and Mui 1981), threshold based, edge based, region based, classification (or clustering) based and deformable model based. Image registration techniques can be divided into two types of approaches, area-based and feature-based (ZitovZitova and Flusser 2003). Area-based methods compare two images by directly comparing the pixel intensities of different regions in the image, while feature-based methods first extract a set of features (points, lines, or regions) from the images and then compare the features. Area-based methods are often a good approach when there are no distinct features in the images, rendering feature-based methods useless as they cannot extract useful information from the images for registration. Feature-based methods on the other hand, are often faster since less information is used for comparison. Also, feature-based methods are more robust against viewpoint changes (Denavit and Hartenberg 1995; Hartley and Zisserman 2003), which is often experienced in vision-based robot control systems.

Object recognition approaches can be divided into two categories. The first approach utilises appearance features of objects such as the colour and intensity. The second approach utilises features extracted from the object and only matches the features of the object of interest with the features in the database. An advantage of the feature-based approaches is their ability to recognise objects in the presence of lighting, translation, rotation and scale changes (Brunelli and Poggio 1993; ZitovZitova and Flusser 2003). This is the type of approach used by the PatMax algorithm in our proposed system.

With the advancement of robots and vision systems, object recognition which utilises robots is no longer limited to manufacturing environments. Wong et al. (Wong, Rong et al. 1998) developed a system which uses spatial and topological features to automatically recognise 3D objects. A hypothesis-based approach is used for the recognition of 3D objects. The system does not take into account the possibility that an image may not have a corresponding model in the database since the best matching score is always used to determine the correct match, and thus is prone to false matches if the object in an image is not present in the database. Büker et al. (BBuker, Drue et al. 2001) presented a system where an industrial robot system was used for the autonomous disassembly of used cars, in particular, the wheels. The system utilised a combination of contour, grey values and knowledge-based recognition techniques. Principal component analysis (PCA) was used to accurately locate the nuts of the wheels which were used for localisation purposes. A coarse-to-fine approach was used to improve the performance of the system. The vision system was integrated with a force torque sensor, a task planning module and an unscrewing tool for the nuts to form the complete disassembly system.

Researchers have also used methods which are invariant to scale, rotation, translation and partially invariant to affine transformation to simplify the recognition task. These methods allow the object to be placed in an arbitrary pose. Jeong et al. (Jeong, Chung et al. 2005) proposed a method for robot localisation and spatial context recognition. The Harris detector (Harris and Stephens 1988) and the pyramid Lucas-Kanade optical flow methods were used to localise the robot end-effector. To recognise spatial context, the Harris detector and scale invariant feature transform (SIFT) descriptor (Lowe 2004) were employed. Peña-Cabrera et al. (Pena-Cabrera, Lopez-Juarez et al. 2005) presented a system to improve the performance of industrial robots working in unstructured environments. An artificial neural network (ANN) was used to train and recognise objects in the manufacturing cell. The object recognition process utilises image histogram and image moments which are fed into the ANN to determine what the object is. In Abdullah et al.'s work (Abdullah, Bharmal et al. 2005), a robot vision system was successfully used for sorting meat patties. A modified Hough transform (Hough 1962) was used to detect the centroid of the meat patties, which was used to guide the robot to pick up individual meat patties. The image processing was embedded in a field programmable gate array (FPGA) for online processing.

Even though components such as camera calibration, image segmentation, image registration and robotic kinematics have been extensively researched, they exhibit shortcomings when used in highly dynamic environments. With camera calibration in real time self-calibration is a vital requirement of the system while with image segmentation and registration it is crucial that the new algorithms will be able to operate under the presence of changes in lighting conditions, scale and with blurred images, e.g. due to robotic vibrations. Hence, new robust image processing algorithms will need to be developed. Similarly there are several approaches available to solve the inverse kinematics problem, mainly Newton-Raphson and Neural Network algorithms. However, they are hindered by accuracy and time inefficiency respectively. Thus it is vital to develop a solution that is able to provide both high accuracy and a high degree of time efficiency at the same time.

3. Methods

The methods developed in our group are mainly new image processing techniques and robot control methods in order to develop a fully vision-guided robot control system. This consists of camera calibration, image segmentation, image registration, object recognition, as well as the forward and inverse kinematics for robot control. The focus is placed on improving the robustness and accuracy of the robot system.

3.1 Camera Calibration

Camera calibration is the process of determining the intrinsic parameters and/or the extrinsic parameters of the camera. This is a crucial process for: (1) determining the location of the object or scene; (2) determining the location and orientation of the camera; and (3) 3D reconstruction of the object or scene (Tsai 1987).

A camera has two sets of parameters, intrinsic parameters which describe the internal properties of the camera, and extrinsic parameters which describe the location and orientation of the camera with respect to some coordinate system. The camera model utilised is a pinhole camera model and is used to relate 3D world points to 2D image projections. While this is not a perfect model of cameras used in machine vision systems, it gives a very good approximation and when lens distortion is taken into account, the model is sufficient for the most common machine vision applications. A pinhole camera is modelled by:

$$x = K[R \mid t]X \tag{1}$$

Where X is the 3D point coordinates and x the image projection of X. (R,t) are the extrinsic parameters where R is the 3×3 rotation matrix and t the 3×1 translation vector. K is the camera intrinsic matrix, or simply the camera matrix, and it describes the intrinsic parameters of cameras:

$$K = \begin{bmatrix} f_x & 0 & c_x \\ 0 & f_y & c_y \\ 0 & 0 & 1 \end{bmatrix} \quad (2)$$

where (f_x, f_y) are the focal lengths and (c_x, c_y) the coordinates of the principal point along the major axes x and y, respectively. The principal point is the point at which the light that passes through the image is perpendicular to the image, and this is often, but not always, at the centre of an image.

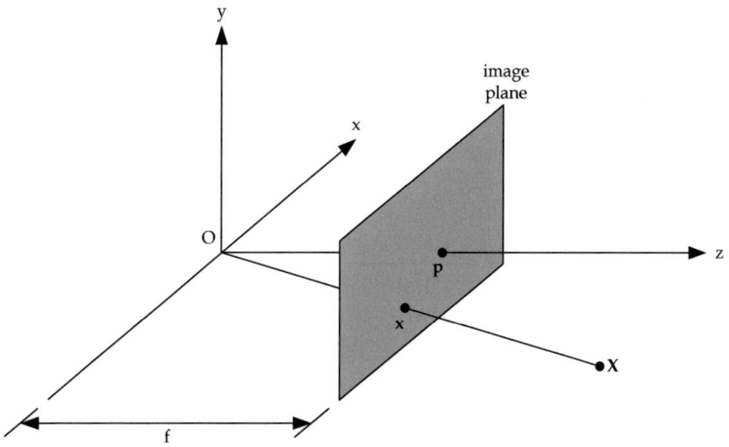

Figure 2. Pinhole camera showing: camera centre (O), principal point (p), focal length (f), 3D point (X) and its image projection (x)

As mentioned previously, lens distortion needs to be taken into account for pinhole camera models. Two types of distortion exist, radial and tangential. An infinite series is required to model the two types of distortions, however, it has been shown that tangential distortions can often be ignored, in particular for machine vision application. It is often best to limit the number of terms for the distortion coefficient for radial distortion for stability reasons (Tsai 1987). Below is an example of how to model lens distortion, taking into account both tangential and radial distortions, using two distortion coefficients for both:

$$\tilde{x} = x + x\left[k_1 r^2 + k_2 r^4\right] + \left[2p_1 xy + p_2\left(r^2 + 2x^2\right)\right] \quad (3)$$

$$\tilde{y} = y + y\left[k_1 r^2 + k_2 r^4\right] + \left[2p_1 xy + p_2\left(r^2 + 2y^2\right)\right] \quad (4)$$

$$r = \sqrt{x^2 + y^2} \quad (5)$$

where (x, y) and $(\tilde{x}, \tilde{y})$ are the ideal (without distortion) and real (distorted) image physical coordinates, respectively. r is the distorted radius.

3.2 Object Segmentation

Image segmentation is the first step leading to image analysis. In evaluation of the five methodologies (Fu and Mui 1981), the deformable model based segmentation holds a distinct advantage over the others when dealing with image-guided robotic systems. Many visually-guided systems use feature based approaches for image registration. Such systems may have difficulty handling cases in which object features become occluded or deformation alters the feature beyond recognition. For instance, systems that define a feature as a template of pixels can fail when a feature rotates relative to the template used to match it. To overcome these difficulties, the vision system presented in this chapter incorporates contour tracking techniques as an added input along with the feature based registration to provide better information to the vision-guided robot control system. Hence, when a contour corresponding to the object boundary is extracted from the image, it provides information about the object location in the environment. If prior information about the set of objects that may appear in the environment is available to the system, the contour is used to recognise the object or to determine its distance from the camera. If additional, prior information about object shape and size will be combined with the contour information, the system could be extended to respond to object rotations and changes in depth.

3.2.1 Active Contours Segmentation

The active contours segmentation methodology implemented is a variation of the one proposed by (Kass, Witkin et al. 1988).

The implemented methodology begins by defining a contour parameterized by arc length s as

$$C(s) \equiv \{(x(s), y(s)) : 0 \leq s \leq L\} : \Re \rightarrow \Omega \qquad (6)$$

where L denotes the length of the contour C, and Ω denotes the entire domain of an image $I(x, y)$. The corresponding expression in a discrete domain approximates the continuous expression as

$$C(s) \approx C(n) = \{(x(s), y(s)) : 0 \leq n \leq N, s = 0 + n\Delta s\} \qquad (7)$$

where $L = N\Delta s$. An energy function $E(C)$ can be defined on the contour such as

$$E(C) = E_{\text{int}} + E_{\text{ext}} \qquad (8)$$

where E_{int} and E_{ext} respectively denote the internal energy and external energy functions. The internal energy function determines the regularity and smooth shape, of the contour. The implemented choice for the internal energy is a quadratic functional given by

$$E_{\text{int}} = \sum_{n=0}^{N} \left(a|C'(n)|^2 + \beta|C''(n)|^2 \right) \Delta s \qquad (9)$$

Here a controls the tension of the contour, and β controls the rigidity of the contour. The external energy term determines the criteria of contour evolution depending on the image $I(x, y)$, and can be defined as

$$E_{ext} = \sum_{n=0}^{N} E_{img}(c(n)) \Delta s \qquad (10)$$

where E_{img} denotes a scalar function defined on the image plane, so the local minimum of E_{img} attracts the active contour to the edges. An implemented edge attraction function is a function of image gradient, given by

$$E_{\text{img}}(x,y) = \frac{1}{\lambda \left| \nabla G_{\sigma} \times I(x,y) \right|} \tag{11}$$

where G_σ denotes a Gaussian smoothing filter with the standard deviation σ, and λ is a suitably chosen constant. Solving the problem of active contours is to find the contour C that minimises the total energy term E with the given set of weights α, β and λ. The contour points residing on the image plane are defined in the initial stage, and then the next position of those snake points are determined by the local minimum E. The connected form of those points are considered as the contour to proceed with. Figure 3 shows an example of the above method in operation over a series of iterations.

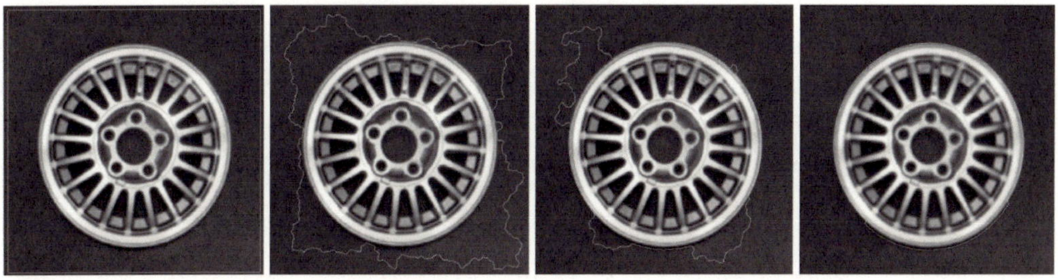

(a) (b) (c) (d)

Figure 3. (a) Original level set contour, (b) Level set contour after 20 iterations, (c) Level set contour after 50 iterations, (d) Level set contour after 70 iterations

3.3 Image Registration and Tracking

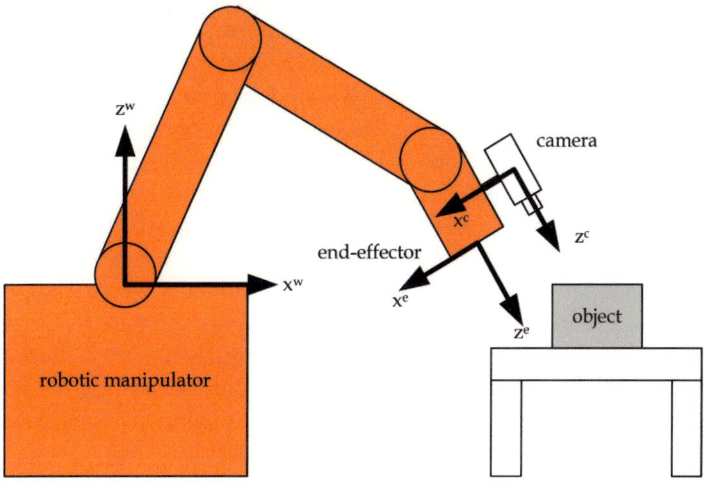

Figure 4. Eye-in-hand vision system

An important role of machine vision in robotic manipulators is the ability to provide feedback to the robot controller by analysing the environment around the manipulator within the workspace. Thus a vision-guided robotic system is more suitable for tasks such as grasping or aligning objects in the workspace compared to conventional feedback controls, for example force feedback controllers. The main challenge in vision-based robot control systems is to extract a set of robust and useful feature points or regions and using these features to control the motion of the robotic manipulator in real-time.
To accurately provide feedback information from the vision system, the camera first needs to be aligned with the robotic manipulator. Figure 4 shows an eye-in-hand configuration (Spong, Hutchinson et al. 2006), which has the camera mounted on the end-effector. The 'eye-in-hand' setup is often the preferred configuration in machine vision applications since it provides a better view of the object. The base of the robotic manipulator contains the world coordinate system (x^w, y^w, z^w), which is used as the reference for all the other coordinate systems. The end-effector's coordinate system (x^e, y^e, z^e) has a known transformation with respect to the world coordinate. The camera's coordinate system is aligned to the end-effector by a number of methods, one of which is based on the geometric relationship between the origin of the end-effector and the camera coordinate. By knowing the geometric relationship of the end-effector and the camera, the rigid transformation or homogeneous transformation of the two components can be defined:

$$P^{camera} = H^{camera}_{end\,effector} P^{end\,effector} \tag{12}$$

where P^{camera} is the homogeneous representation of a 3D point in the camera coordinate,

$$P^{camera} = \begin{bmatrix} p^{camera} \\ 1 \end{bmatrix} \tag{13}$$

Using the homogeneous transformation, the information obtained by the vision system can be easily converted to suit the robotic manipulator's needs. In order for the vision system to track an object in the workspace, either continuous or discontinuous images can be captured. Continuous images (videos) provide more information about the workspace using techniques such as optical flow (Lucas and Kanade 1981), however require significantly more computation power in some applications, which is undesirable. Discontinuous images provide less information but can be more difficult to track an object, especially if both the object and the robotic manipulator are moving. However, the main advantage is that they require less computation as images are not being constantly compared and objects tracked. One method for tracking an object is by using image registration techniques, which aims at identifying the same features of an object from different images taken at different times and viewpoints.
Image registration techniques can be divided into two types of approaches, area-based and feature-based (Zitova and Flusser 2003). Area-based methods compare two images by directly comparing the pixel intensities of different regions in the image, while feature-based methods first extract a set of features (points, lines, or regions) from the images, the features are then compared. Area-based methods are often a good approach when there are no distinct features in the images, rendering feature-based methods useless as they cannot extract useful information from the images for registration. Feature-based methods on the other hand, are often faster since less information is used for comparison. Also, feature-based methods are

more robust against viewpoint changes (Denavit and Hartenberg 1955; Hartley and Zisserman 2003), which is often experienced in vision-based robot control systems.

3.3.1 Feature-Based Registration Algorithms

To successfully register two images taken from different time frames, a number of feature-based image registration methods, in particular, those based on the local descriptor method have been studied. In order for the algorithms to be effective in registering images of an object over a period of time, two types of rigid transformations have been analysed: (1) change of scale and (2) change of viewpoint. A change of scale implies that the camera has moved towards or away from the camera, hence the size of the object in the image has changed. A change in viewpoint implies that the camera has moved around in the environment and that the object is being viewed from a different location. Four algorithms were compared, namely: Scaled-Invariant Feature Transform (SIFT) (Lowe 2004), Principal Component Analysis SIFT (PCA-SIFT) (Ke and Sukthankar 2004), Gradient Location and Orientation Histogram (GLOH) (Mikolajczyk and Schmid 2004), and Speeded Up Robust Features (SURF) (Bay, Tuytelaars et al. 2006). The scale of the image is defined as the ratio of the size of the scene in an image with respect to a reference image:

$$s = \frac{h_{I,s}}{h_{I,r}} \tag{14}$$

where $h_{I,r}$ is the height of the scene in the reference image, and $h_{I,s}$ is the height of the scene in the image of scale s. From this equation it is clear that the scale is proportional to the height of the scene in the image. This is best illustrated in Figure 5, where Figure 5(a) shows the image used as the reference and has a scale of one by definition, and Figure 5(b) shows a new image with scale s.

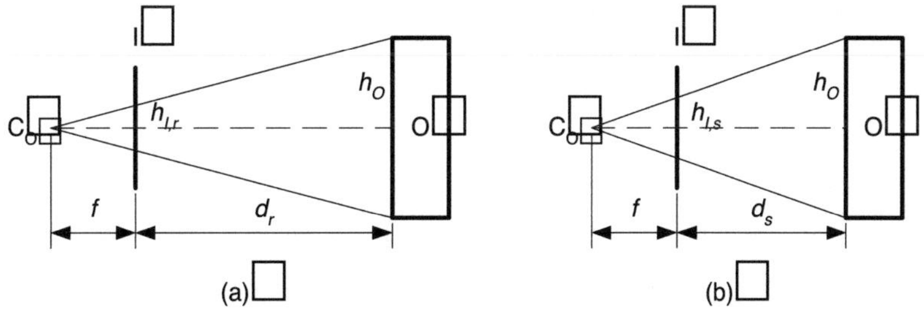

Figure 5. Change of scale of objects in images

3.4 Improved Registration and Tracking Using the Colour Information of Images

3.4.1 Colour model

One area of improvement over existing methods is the use of the information available from the images. Methods such as SIFT or SURF make use of greyscale images, however many images which need to be registered are in colour. By reducing a colour image to a greyscale one, a large proportion of information are effectively lost in the conversion process. To overcome this issue, it is proposed that colour images are used as inputs for the registration process, thus providing more unique information for the formation of descriptors, increasing the uniqueness of descriptors and enhancing the robustness of the registration process.

While it is possible to simply use the RGB values of the images, it is often not the desirable approach, since factors such as the viewing orientation and location of the light source affect these values. Many colour invariant models exist in an attempt to rectify this issue, and the $m_1m_2m_3$ model (Weijer and Schmid 2006) is utilised here. The advantage of the chosen model is that it does not require *a priori* information about the scene or object, as the model is illumination invariant, and is based on the ratio of surface albedos rather than the surface albedo itself. The model is a three component model, and can be defined as:

$$m_1 = \frac{R^{x_1} G^{x_2}}{R^{x_2} G^{x_1}}, \quad m_2 = \frac{R^{x_1} B^{x_2}}{R^{x_2} B^{x_1}}, \quad m_3 = \frac{G^{x_1} B^{x_2}}{G^{x_2} B^{x_1}} \tag{15}$$

where $\mathbf{x}_1, \mathbf{x}_2$ are the neighbouring pixels. Without loss of generality, m_1 is used to derive the results, by taking the logarithm on both sides:

$$\begin{aligned} \log\left(m_1\left(R^{\mathbf{x}_1} R^{\mathbf{x}_2} G^{\mathbf{x}_1} G^{\mathbf{x}_2}\right)\right) &= \log\left(\frac{R^{\mathbf{x}_1} G^{\mathbf{x}_2}}{R^{\mathbf{x}_2} G^{\mathbf{x}_1}}\right) \\ &= \log R^{\mathbf{x}_1} + \log G^{\mathbf{x}_2} - \log R^{\mathbf{x}_2} - \log G^{\mathbf{x}_1} \\ &= \log\left(\frac{R^{\mathbf{x}_1}}{G^{\mathbf{x}_1}}\right) - \log\left(\frac{R^{\mathbf{x}_2}}{G^{\mathbf{x}_2}}\right) \end{aligned} \tag{16}$$

From this, it can be shown that the colour ratios can be represented as the difference of the two neighbouring pixels:

$$d_{m_1}(\mathbf{x}_1, \mathbf{x}_2) = \left(log\left(\frac{R}{G}\right)\right)^{\mathbf{x}_1} - \left(log\left(\frac{R}{G}\right)\right)^{\mathbf{x}_2} \tag{17}$$

3.4.2 Dimension Reduction

It should be noted that one main issue which arises from using the colour information of images is the increase in the amount of data which needs to be dealt with. By computing descriptors for the model described above, the dimension of each descriptor is increased three-fold. For example, in the case of the SURF descriptor, the dimension of each descriptor increases from 64 to 192. While this increase in dimension often aids in improving the robustness, a significant drop in computational speed is noted. To overcome this issue, Principal Component Analysis (PCA) (Pearson 1991) is applied. PCA is a statistical method for vector space transform, often used to reduce data to lower dimensions for analysis. The original data is transformed as follows:

$$Y = UX \tag{18}$$

where Y is the new data, based on the original data X, and the eigenvectors of the covariant matrix of X, U. The covariance matrix can be computed by first mean-shifting the data, then estimating:

$$\text{cov}(X) = \frac{1}{n-1} \sum_{i=1}^{n} \left(X_i - \overline{X}\right)\left(X_i - \overline{X}\right)^T \tag{19}$$

where, $\overline{X} = \frac{1}{n} \sum_{i=1}^{n} X_i$ is the sample mean, and n is the number of data entries. Approximately 20,000 image patches have been used for estimating the covariance matrix. Here, the SURF descriptor has been used to generate a set of descriptors for each image patch.

3.5 Object Recognition

The algorithm for object recognition is one of the most important parts of the developed vision system. This module is used to make decisions for the object recognition process and command different parts of the system to function.

3.5.1 Algorithm for 2D Object Recognition

Recognising 2D objects is a relatively simple task, since only the top view of the objects need to be considered. One simple algorithm is described below. Firstly, the program commands the VisionServer, part of the vision system, to take an image of the object from the top and match it against the images of all the models in the database. The VisionServer will then compute a score for each model to indicate how close each model matches with the object, with 1.0 being the highest and 0.0 being the lowest. Preceding this, the program can identify what the object is by comparing the scores of the various models.
One simple methodology is to compare the scores with a threshold value. The model is reported as the object only if the score is above the threshold value. Sometimes there will be more than one model getting a score that is bigger than the threshold value due to the similarity of the models in the database. In this case, the model with the highest score will be reported as the object. On the other hand, if none of the scores is above the threshold value, the program will report an error. After the object is recognised, the next 2D object can be processed using the same procedure. Since this algorithm is simple and only one image needs to be checked against other trained images in the system at each time, problems such as long computational time and view planning, which are common in 3D object recognition systems, do not exist.

3.5.2 Algorithm for 3D Object Recognition

Three-dimensional object recognition is much more complex compared to 2D cases. This is because a 3D object can have many different faces and sitting bases. The appearance of the object can be very different from one viewpoint to another. Moreover, an object recognition system often needs to identify more than one object. If the camera is only fixed in one place to take images, it may not have enough information to identify an object if it looks similar to more than one model in the database. It may also report the wrong model name when cluster or occlusion exists. Therefore, taking multiple images from different angles is a common approach for 3D object recognition systems.
Consider the case where there are several 3D objects to be recognised and there is no restriction on the number of sitting bases each object can have. One approach to solving this problem is to obtain and train several views of each object that can be seen by the camera and store the results in the system. In run-time, if any one of the images in the database looks similar to the object being recognised, the system will report the name of the model that this image belongs to. Otherwise, it will go to another viewpoint to take a new image and repeat the same process until it finds a trained image that looks similar to the new view of the run-time object. Sometimes, there may be two or more images from different models which appear similar to the view of the object due to the similarity of the models. In this case, the system will need to take another image from a new position and repeat the process again until there is only one image in the database which matches the run-time image. Figure 6 shows the flowchart of this algorithm. This algorithm is easy to implement but it has many drawbacks. One of the major problems is the long computational time needed. If

there is only one model in the database, this algorithm can perform reasonably fast. However, this is not a practical setup. If there are more models to be recognised or if the models have more complex shapes and more faces, there will be a lot more trained images in the database. For example, if there are n models to be identified and each has m faces, there will then be m×n trained images in the database of the system. Since there is no restriction on the number of sitting bases each object can have, the system has no idea which view of the object will face the camera. As a result, every time an image is taken, the system needs to compare the image with all the trained images in the database. This process is extremely time consuming and occupies a lot of computer resource. This problem is magnified when there are similar models in the database or when the system fails to find a trained image which looks similar to the run-time image because, as shown in the flowchart in Figure 6, the system will simply repeat the same process until it finds one model that matches. Another problem in this algorithm is if an object that has not been trained is put in the system for recognising, then the system will simply go into an endless loop and never finish since it can never find a model that matches the image of the new object. These disadvantages make it impractical to be used in real life applications and therefore, an improved method is derived.

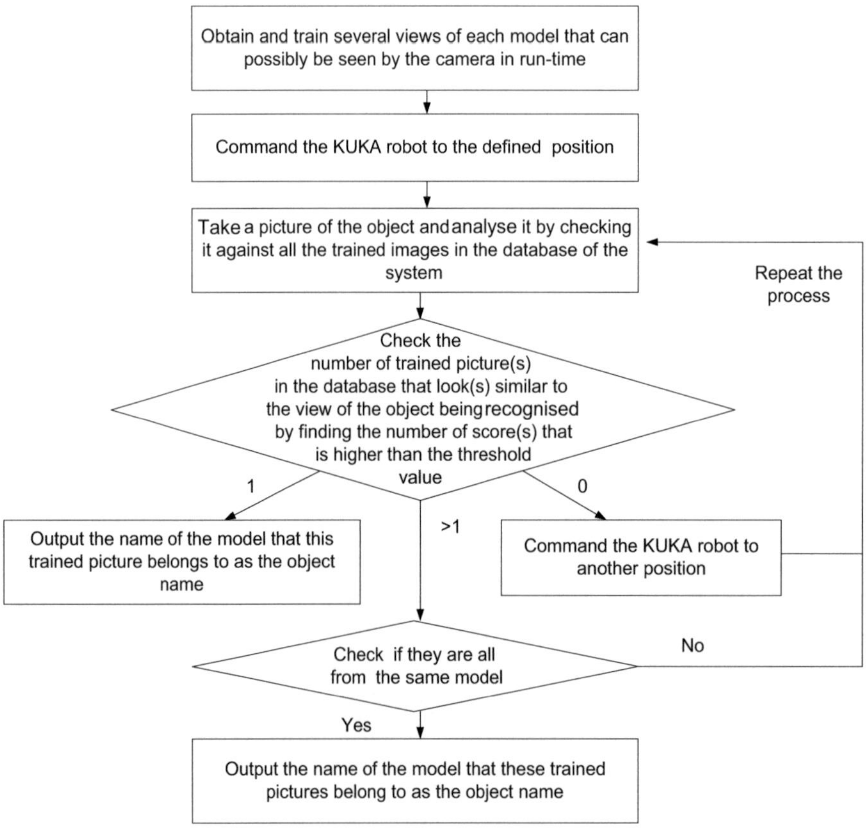

Figure 6. Flowchart of a simple algorithm for 3D object recognition

3.5.3 Improved Algorithm for 3D Object Recognition

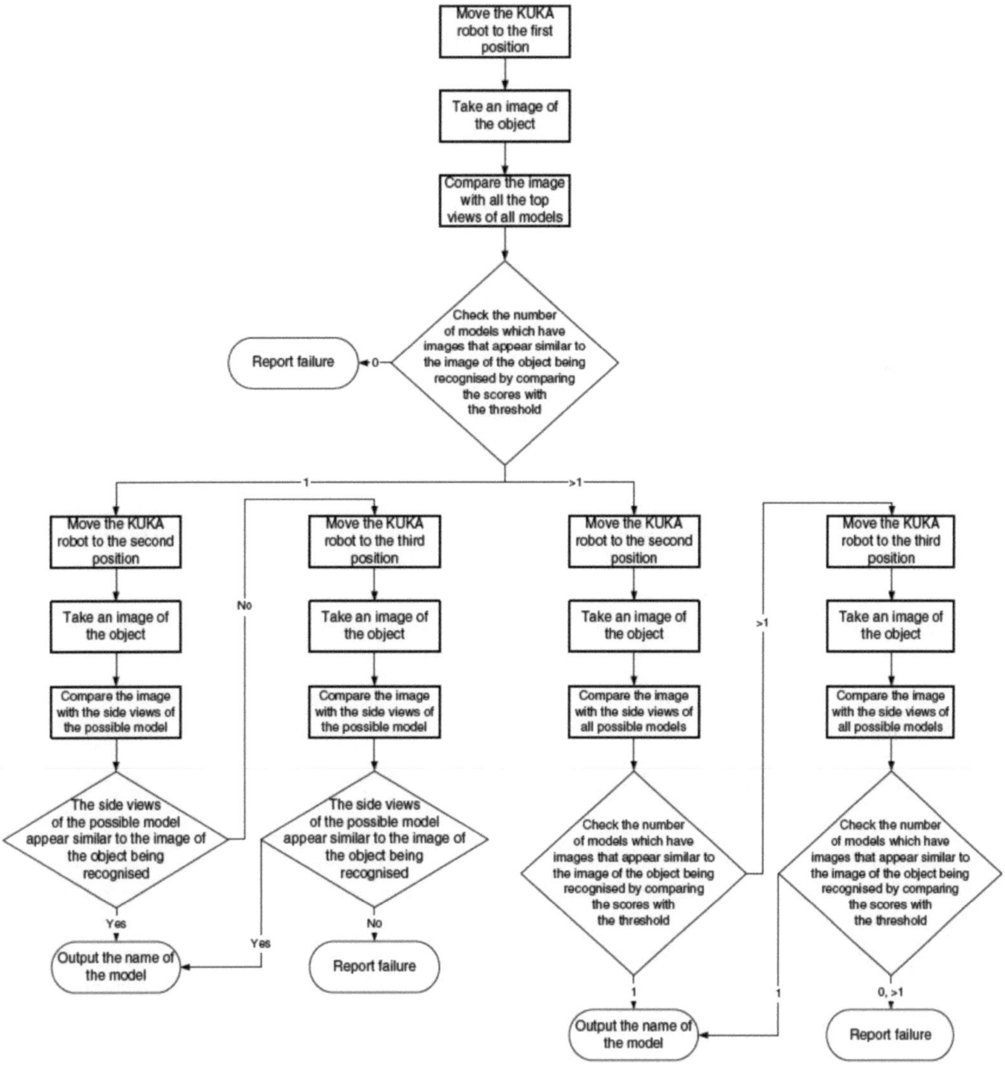

Figure 7. Flowchart of an improved algorithm for 3D recognition

To improve the above algorithm, one method which is used is to restrict all the 3D objects to one sitting base only. By placing this restriction, any 3D object will always have the same face showing up no matter what pose it is placed under the camera. This means that every time a new object recognition process starts, the robot can always go to the top (position 1) to take an image of the object and compare it with the top views of all other models in the database. After the images have been compared, the system will have an idea of which model the object being processed is likely to be by comparing the scores. It outputs what object it is based on the top view. If the system is uncertain what the object is, it goes to the side positions (either position 2 or 3) to check for the object's side views to further verify whether or not it is the one that the system initially determines before the system outputs the result. By allowing the system to output the result based on the top view, it can make the process performs faster, however, it also decreases the accuracy. On the other hand, if the system checks for at least two views of the object before it outputs the result, the system will perform slower, but will have a much higher accuracy. There is a trade-off between speed and accuracy and this decision should be made based on the situation. In this work, it was chosen to check for two views in for 3D object recognition to enhance the accuracy of the system.

Using the proposed approach, if the run-time object looks similar to two or more models from the top, the ASEA robot will analyse the side view of the run-time object from position 2 by checking it against all the side views of all the possible models. If the system can identify which model the object is, it will stop and output the result, otherwise, it will go to position 3 and repeat the same process. If the system still cannot determine what this object is, it will output a failure message and stop the object recognition process at this point. This is to prevent the system from running for a long period of time to handle complex or new objects.

This improved algorithm is fast and accurate compared with the simple 3D object recognition algorithm described in Figure 6. For every object recognition process, only a small portion of the trained images in the database needs to be looked at. This minimises the time required and the workload of the vision computer. In addition, when there is not sufficient information to identify the object due to the presence of similar models in the database, the system will automatically go to a new position to take another image of the object and perform the analysis again. If the system still cannot identify the object after three runs, it will stop and report a failure. This algorithm can also be used in cases where objects have two or more sitting bases. This is achieved by treating the object sitting on the other base as a separate object.

3.6 ASEA Robot Modelling and Simulation

The control of the ASEA robot directly affects the performance of the system. The processing time is one of the key factors to determining the performance. For example, if the robot moves too slowly, it will reduce its efficiency. However if the robot moves too fast, vibrations will affect the quality of the images taken. Therefore, the control of the ASEA robot has to be carefully considered.

One of the problems is to find a method to control the end-effector of the robot manipulator to follow a pre-defined trajectory. Normally a desired pose of the end-effector with respect to the base is given as well as the torque required to move a particular object. For serial robots, inverse kinematics algorithm is used to determine the desired rotational angle of each joint, given the desire position of end-effector. The corresponding torque of each motor needs be computed as well using inverse kinematics. In the kinematics model, the torques provided by the motors are the inputs into the plant, and the outputs are the resulting rotational angles and the angular velocities of the links. The operation is accomplished by actuating the five joint motors, by sensing the joint angles and angular velocities, and by comparing the desired angular positions with the actual ones. Figure 8 below shows a simulation model established to validate the control and kinematics of the ASEA robot.

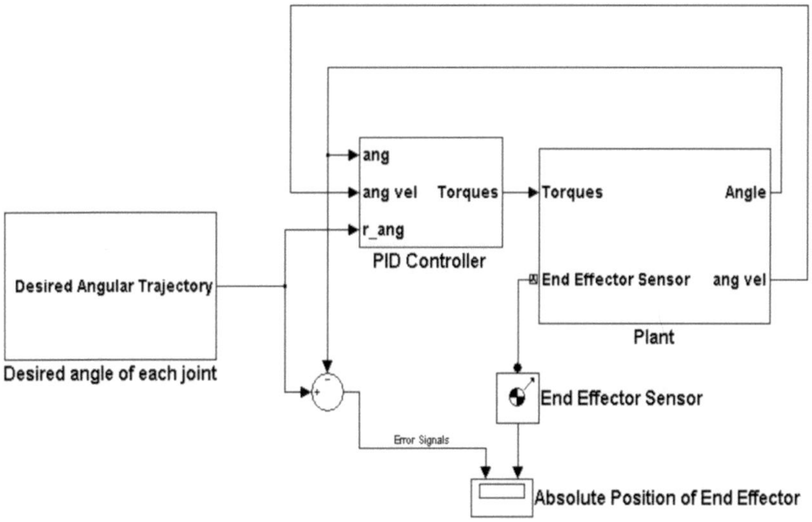

Figure 8. Overall control system modelled in Simulink

3.6.1 Kinematics Analysis

The ASEA robot manipulator has five serial links, which allows the robot to move with five degrees of freedom (DOF). The dimension of the robot manipulator is shown in Table 1. L1-L5 represent the five links that make up the serial robot while q1-q5 represent the five joints that form the serial structure. For the serial robot, a new inverse kinematics algorithm is developed using a combination of the numerical and analytic methods. The forward kinematics was implemented in combination with inverse kinematics in order to minimise the error of inverse kinematics solution. The basic principle is that the solution of inverse kinematics is fed to forward kinematics so that the desired position derived from calculated joint angles are computed. The difference between real desired position and calculated desired position is then calculated, which is added to inverse kinematics function. By repeating the loop a few times, the solution of inverse kinematics should be optimal.

Link	Length(mm)	Joint	Type of Rotation	Angle of Rotation(deg)
L1	730	q1	axis	360
L2	690	q2	vector	82.5
L3	670	q3	vector	63
L4	73	q4	vector	208.5
L5	20	q5	axis	331.5

Table 1. The ASEA robot parameters

3.6.2 Forward Kinematics

The end-effector position and orientation can be solved link by link from the base to the end-effector. If x, y, z, u and w are defined as instantaneous cartesian coordinates then the end-effector orientation can be solved using the following relations,

$$
\begin{aligned}
x &= \cos(q_1) \times (L_2 \times \cos(q_2) + L_3 \times \cos(q_2 + q_3) + (L_4 + L_5) \times \cos(q_2 + q_3 + q_4)) \\
y &= \sin(q_1) \times (L_2 \times \cos(q_2) + L_3 \times \cos(q_2 + q_3) + (L_4 + L_5) \times \cos(q_2 + q_3 + q_4)) \\
z &= L_2 \times \sin(q_2) + L_3 \times \sin(q_2 + q_3) + L_1 + (L_4 + L_5) \times \sin(q_2 + q_3 + q_4)) \\
u &= \cos(q_2 + q_3 + q_4) \\
w &= \sin(q_2 + q_3 + q_4)
\end{aligned}
\tag{20}
$$

3.6.3 Inverse Kinematics: Sugeno Fuzzy Inference System (SFIS)

To achieve a mapping between the cartesian variables and joint variables of the ASEA robot, a zero-order Sugeno Fuzzy Inference System (SFIS) was developed. The Sugeno system is based on the concepts of fuzzy set theory, fuzzy if-then rules, and fuzzy reasoning. Essentially the fuzzy system has three main components: a rule-base, which contains fuzzy if-then rules; a database defining input and output fuzzy membership functions (MF) used in the rule-base and an inferencing mechanism to perform fuzzy reasoning. While carrying out the inverse kinematic analysis it is required to find a set of joint variables ($q_1, q_2, \ldots q_5$) to achieve the specified end effector pose of the robot. The cartesian variables of the end effector are chosen to be fuzzy antecedent variables and each variable is represented by two gaussian fuzzy MF with linguistic terms such as Low and High. Similarly, constant functions are intuitively chosen for the output joint variables and a rule-base containing a total of 32 rules is generated based on the following rule (Jamwal and Shan 2005):

$$R = \prod_{j=1}^{n} m_j \tag{21}$$

where R represents the total number of rules; m_j is the number of linguistic terms of the jth linguistic variable and n is the total number of input variables. Thus in the present case the total number of rules becomes 2^5 or 32. A Sugeno inferencing (Jang, Sun et al. 1997) is selected due to its advantages over the Mamdani method. The rule-base defines the relationship between inputs and outputs and a typical Sugeno fuzzy rule has the following structure:

$$If \quad \begin{matrix} x \\ y \\ z \\ u \\ w \end{matrix} \quad are \quad \left[Low \ High \right] \quad Then \quad \begin{matrix} q_1 \\ q_2 \\ q_3 \\ q_4 \\ q_5 \end{matrix} \quad = \quad \begin{matrix} r_1 \\ r_2 \\ r_3 \\ r_4 \\ r_5 \end{matrix} \tag{22}$$

where x, y, z, u and w are instantaneous input cartesian coordinates and r_1, r_2, r_3, r_4 and r_5 are the crisp output values for the joint variables of ASEA robot. The fuzzy system so developed is not accurate (Jang, Sun et al. 1997) until the fuzzy parameters of antecedent and consequent variables are adjusted or tuned.

3.6.3.1 Consequent MF tuning

Once a database containing a set of inputs and outputs is available from the forward kinematic analysis, the consequent MFs are tuned to produce the desired model accuracy. To achieve this, a gradient-descent method is used to minimise an error function. Outputs from the SFIS are compared from the corresponding values available from forward kinematic analysis to find the error. The learning rule for the consequent crisp MF (r_i's) can be expressed as below:

$$r_i(t+1) = r_i(t) \quad \alpha \frac{\partial E}{\partial r_i} \qquad (23)$$

where E is Sum of Squared Errors (SSE) and α is the learning rate which decides the quantum of change in the parameters after each iteration.

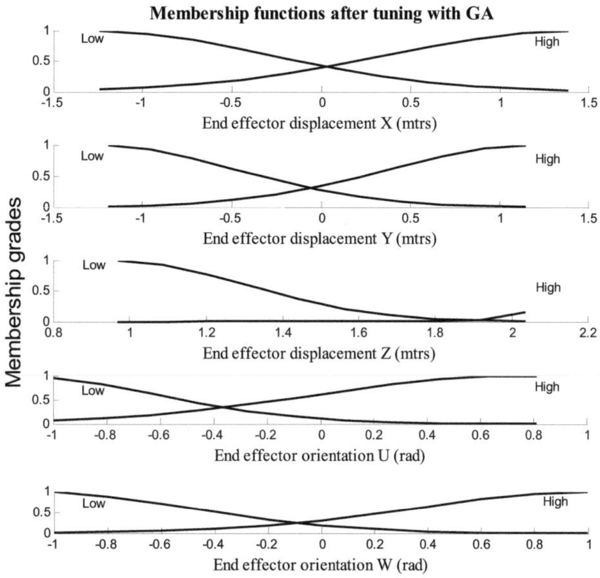

Figure 9. Antecedent membership functions after tuning with genetic algorithms

3.6.3.2 Antecedent MF tuning

Initially the antecedent variables (cartesian coordinates) have been fuzzified and two gaussian MF linguistically termed as Low (L) and High (H) are placed equally dividing their respective universe of discourses. The position of the point of minimum fuzziness or the mean of the gaussian MF and its standard deviation are the important parameters, which influence accurate inferencing. There exists an optimal set of mean and standard deviation of all the membership functions for which the fuzzy model shall be given a maximum accuracy. In the present application these values of MF are varied within ±33% of their values. For example, mean of the 'Low' membership function of end effector displacement 'X' has been varied between -1.7373 and -0.8686 metres and the standard deviation is varied between 0.66 and 1.33. Thus for five input variables consisting two MF each, there are 20 points of mean and standard deviation which require tuning. In the present work genetic algorithms (GA) have been used (Jamwal and Shan 2005) for this tuning and to represent 20 tuning points a binary string of 200 bits is used, wherein 10 bits are devoted to each tuning point. Initial population of 20 binary coded strings of 200 bits was generated using Knuth's random number generator. The roulette wheel selection method (Jang, Sun et al. 1997) is used for selection of binary strings in the mating pool from the initial population and the probabilities of crossover and mutation operators were kept as 0.95 and 0.01. A Matlab code has been written to implement a genetic algorithm to tune the fuzzy system.

3.6.4 Simulation Model

A simulation model is established using SimMechanics to validate the developed control and kinematic algorithms. The model is shown in Figure 10. The model directly communicates with the control algorithms developed in Simulink, hence the control algorithms are firstly tested before they are implemented. The simulation model also provides visual feedback on how the robot tracks predefined trajectories.

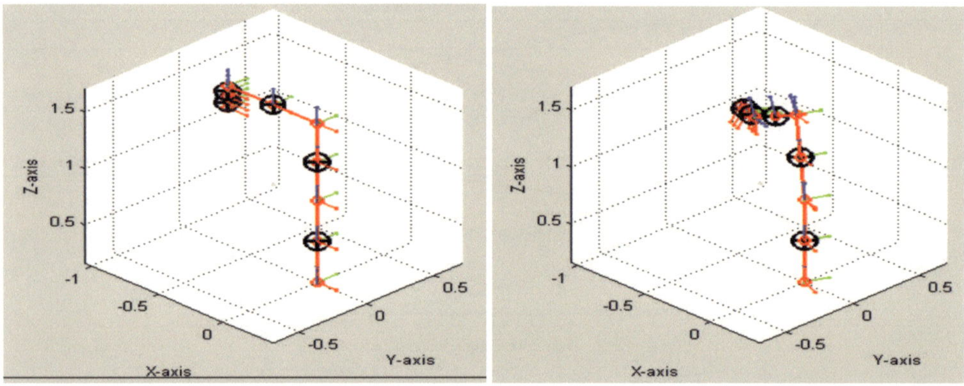

Figure 10. Simulation results of the simple integrated control system – starting point of the robot (left) and final position of the robot (right)

4. Results

In this section the results that have been achieved to date are presented. These are from the vision system, modelling and simulation of the robot, and the test results of the vision-guided robot.

4.1 Experimental Setup

The experimental setup is shown in Figure 11. The experiment consists of an ASEA robot and a small scale assembly line setup as shown. The ASEA robot is shown in its default position and carries a CCD camera on its end-effector for visual feedback. The assembly line runs at a constant speed, and carries a number of objects (car wheels). The experiment is set up to automatically inspect car wheels.

In the robot control system such as this, it is important to keep track of the camera poses for taking images. An appropriate viewpoint planning algorithm is required to decide the camera viewpoint to get the next best view of an object, if the first image does not provide sufficient information to identify what the object is. This prevents the camera from taking images from an inadequate viewpoint. In this research, the ASEA robot was only allowed to go to one of three pre-defined viewpoints to capture images. Three fixed viewpoints exist, one on top and two on the sides of the object. There are several software packages used in this system. Controlling the operations of the whole process, enabling communication between different parts and carrying out object recognition tasks are the main roles of the developed software system. The software used to facilitate the construction of the system include the VisionPro software package, LabVIEW Programming Language, and the ASEA Robot Programming Language. VisionPro is a very comprehensive software package for machine vision applications. It contains a collection of Microsoft's Component Object Model (COM) components and ActiveX controls that allow the development of vision applications in Visual Basic and Visual C++ programming environments with great flexibility.

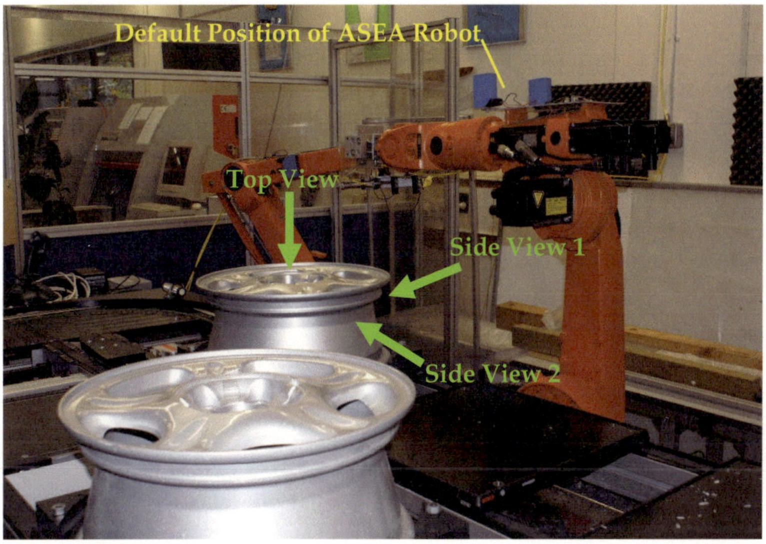

Figure 11. Experiment setup

4.2 Results: Feature Identification

The performance of the local descriptors are evaluated by using the criterion described in (Mikolajczyk and Schmid) which is based on the number of correct matches and number of false matches obtained for any image pair. A plot of the recall versus the 1-precision is computed for each image pair. In addition, the recall value for each image pair is computed and this is plotted against the viewpoint changes. Recall is a measure of how well the local descriptors performed, based on the ratio of the number of correct matches and the total number of corresponding regions, determined with the overlap error (Kadir, Zisserman et al. 2004):

$$\text{recall} = \frac{\#\text{ of correct matches}}{\#\text{ of correspondences}} \tag{24}$$

An overlap error of 50% is used. The 1-precision is a measure of the accuracy of the local descriptors:

$$1 - \text{precision} = \frac{\#\text{ of false matches}}{\#\text{ of matches}} \tag{25}$$

The performance of the descriptors for the two different transformation types are discussed below.

4.2.1 Scale Changes

The recall versus 1-precision plot for scale changes is shown in Figure 12(a). This curve is near straight, indicating that the performance of the descriptors do not change with the change in scale (Mikolajczyk and Schmid). Figure 12(b) shows the recall plot for scale changes. The first section of the plot shows the performance is not affected by the change in scale. The performance however decreases rapidly for scale values greater than two. Further study of the local descriptors suggests that the drop in the recall value is due to the loss of information in the images due to the images being taken from further distances away from the objects.

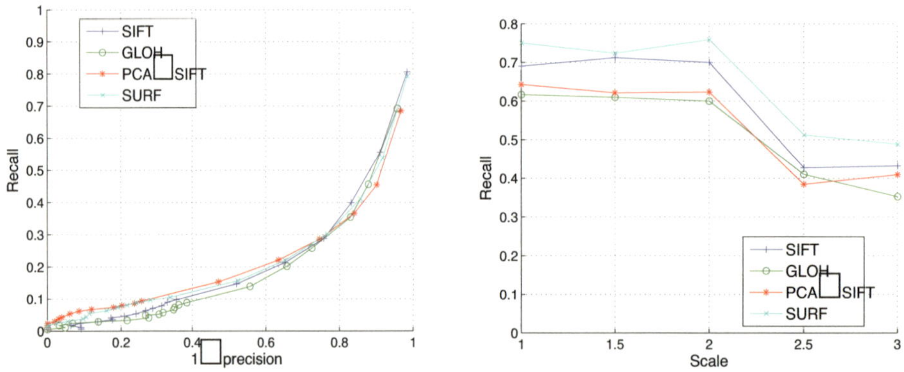

Figure 12. Recall v.s. 1-precision for a scale change of two (a) and recall for different scales (b)

The good performance for scale change is in agreement with previous research (Mikolajczyk and Schmid 2004) which indicates that the local descriptors are relatively robust to the above-mentioned transformations regardless of the type of scene used. As the descriptors are normalised for scale, good results for the scale changes are expected and this is true for scales of up to two for the carved flute. The poor performance of the techniques for scales of higher values is possibly caused by the removal of the background in all images and the scene of interest not covering the entire reference image. Therefore less information from the images is available compared to the evaluation studies in previous works (Mikolajczyk and Schmid 2004).

4.2.2 Viewpoint Changes

Viewpoint changes often pose the greatest challenge for image registration and this is reflected in the results obtained. Figure 13 shows the recall plotted against the 1-precision for different viewpoint changes. A slowly increasing curve in Figure 13(a) indicates that the performance of the local descriptors is influenced by the degradation of images, in particular viewpoint changes. The near-horizontal curve in Figure 13(b) indicates that the performance of the local descriptors is limited by the similarity of the structures of the scenes, and the descriptors cannot distinguish between the structures (Mikolajczyk and Schmid 2004).

Figure 14 shows the best recall values for the different viewpoint changes. The best recall value is defined as the value which has the maximum number of correct matches for the given number of correspondences, obtained by using different threshold values for identifying a pair of descriptors. As can be shown, the recall values degrade rapidly as the viewpoint angle increases, indicating the poor performance of the descriptors for dealing with viewpoint changes when using complex 3D scenes or scenes without distinctive features. Note that the recall value for a zero viewpoint angle is computed from two images of the scene taken from very close, but not identical, viewpoints due to the errors which exist in the turntable. The reason for a relatively low recall value despite the small difference in viewpoints is due to the way descriptors are matched. For low matching threshold values, not all the descriptors from the same region in the scene can be matched correctly, however for high threshold values mismatches occur, which decreases the recall value.

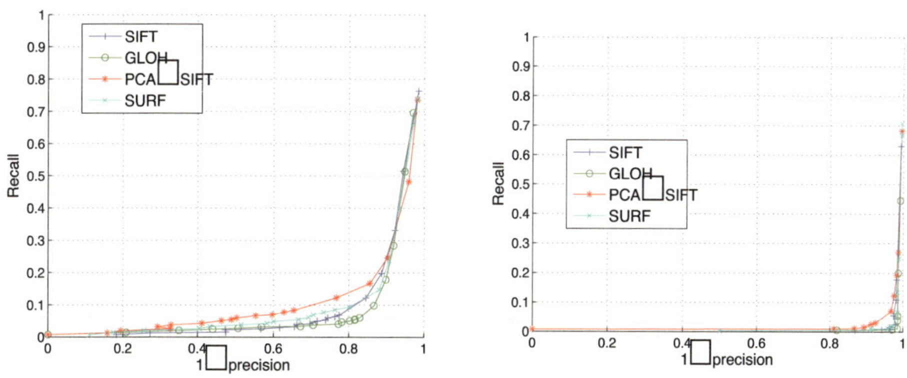

Figure 13. Recall v.s. 1-precision for a 5° viewpoint change (a), and 10° viewpoint change (b)

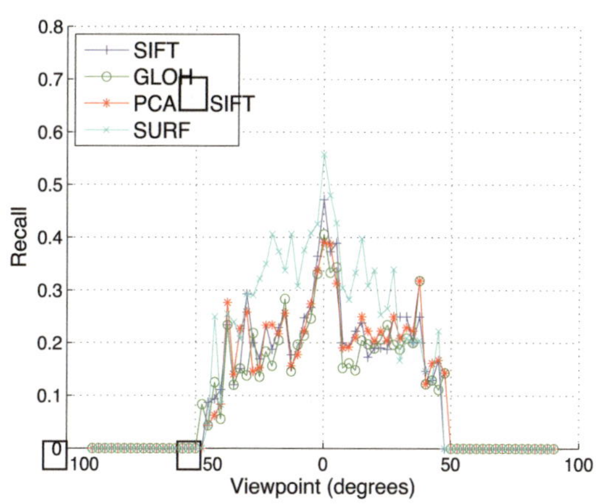

Figure 14. Recall values for different viewpoint changes

4.3 Results: Improved Feature Identification Using Colour Information of Images

The performances of several of the local descriptors were once again evaluated to quantify the benefit of utilising colour information during the feature identification process. The top two images shown in Figure 15 were taken from the active contours segmented tyre images of Figure 3.

(a) (b) (c) (d)

Figure 15. Images used for comparison of the performance of the improved feature identification approach. (a) is the original control image while (b), (c) and (d) have been rotated, translated and scaled respectively. The results of the registration process between (a) and (b) are given below

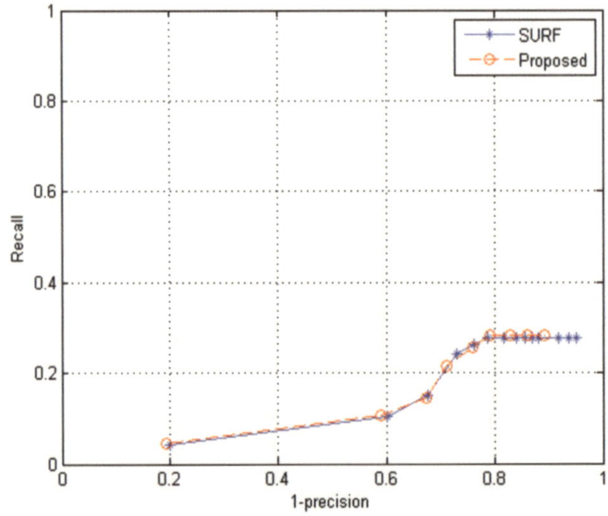

Figure 16. Results for the improved approach. The graphs show the recall against 1-precision values for the tyre images

Once again the recall versus 1-precision plot is used, which were discussed previously. As it can be shown in Figure 16 the improved descriptor performs slightly better than the SURF descriptors, indicating the robustness of the algorithm against changes in the image conditions (for example, illumination). This is due to the lack of sufficient colour information in the test images. Significant improvements are observed when colour images are used.

4.4 Results: SFIS

The inverse kinematics SFIS system was tuned using genetic algorithm for 50 cycles and the average sum of squared errors in prediction of joint variables was dropped from 2.339 to

0.000193 for a set of 100 random end-effector poses. The SFIS, after tuning of antecedent and consequent variables is now accurate enough and is available for implementation in SIMULINK for simulations and online use. In order to check the system accuracy, 50 random sets of Cartesian variables were given to the system as inputs and the results obtained were plotted in Figure 17. It can be observed that the sum of squared errors in joint variables is very small and falls in the range of $0-2\times10^{-3}$ radians. Further, the inference system so developed can be used for online control application, since it is time efficient and can provide results in less than 0.8 milliseconds.

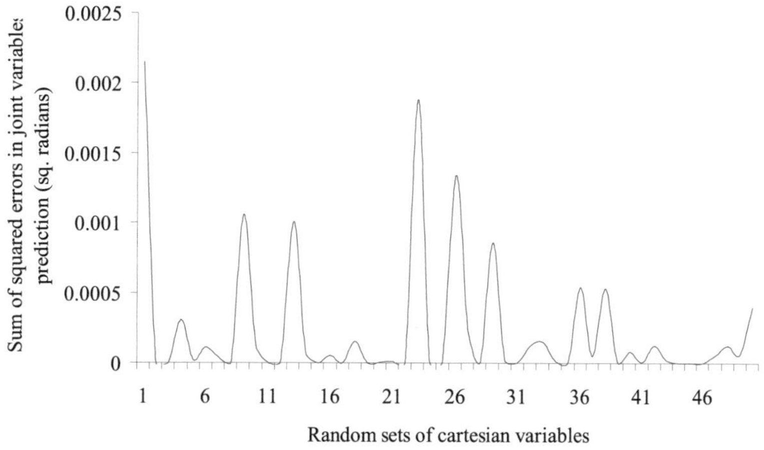

Figure 17. Sum of squared errors in joint variables for random sets of cartesian variables

5. Conclusions

Research into a fully automated vision-guided robot for identifying, visualising and manipulating 3D objects with complicated shapes is still undergoing major development world wide. The current trend is toward the development of more robust, intelligent and flexible vision-guided robot systems to operate in highly dynamic environments.

The theoretical basis of image plane dynamics and robust image-based robot systems capable of manipulating moving objects still need further research. Research carried out in our group has been focused on the development of more robust image processing methods and adaptive control of robot. Our research has been focused on manufacturing automation and medical surgery (Graham, Xie et al. 2007), with the main issues being the visual feature tracking, object recognition and adaptive control of the robot that requires both position and force feedback.

The developed vision-guided robot system is able to recognise general 2D and 3D objects using a CCD camera and the VisionPro software package. Object recognition algorithms were developed to improve the performance of the system and a user interface has been successfully developed to allow easy operation by users. Intensive testing was performed to verify the performance of the system and the results show that the system is able to recognise 3D objects and the restructured model is precise for controlling the robot. The kinematics model of the robot has also been validated together with the robot control algorithms for the control of the robot.

6. Future Research

Future work includes a number of improvements which will enhance the robustness and efficiency of the system. The first is the use of affine invariant methods for 3D object recognition. This method utilises local descriptors for feature matching and it has been shown to have higher success and accuracy rates for complex objects, cluster and occlusion problems (Bay, Tuytelaars et al. 2006). An automatic threshold selection process can also be implemented. This will eliminate the need to manually select a threshold for determining if an object is correctly recognised or not.

The current threshold value was determined empirically and is not guaranteed to work in every possible case. The system needs to be optimised to reduce the processing time required for recognising objects. This requires the development of more efficient image processing algorithms that are also robust and accurate.

Much work still needs to be done into the processing of images in real time, especially when objects become more complicated geometrically. A more robust algorithm for camera self-calibration that works for any robot set-up is also under development in our group.

The control of a serial robot still needs to be investigated for manipulating a wider range of 3D objects. New hybrid algorithms need to be developed for applications that require multi-objective control tasks, e.g. trajectory and force. Efficient algorithms also need to be developed for the inverse kinematics of a serial robot. In this regard, we will further improve the Sugeno Fuzzy Inference System for the inverse kinematics by optimising the number of fuzzy rules. The rule-base of the fuzzy system is equally important and has larger effect on time efficiency of the system.

7. References

Abdullah, M. Z., M. A. Bharmal, et al. (2005). High speed robot vision system with flexible end effector for handling and sorting of meat patties. *9th International Conference on Mechatronics Technology.*

Abdullah, M. Z., L. C. Guan, et al. (2004). The applications of computer vision system and tomographic radar imaging for assessing physical properties of food. *Journal of Food Engineering* 61: 125-135.

Bay, H., T. Tuytelaars, et al. (2006). SURF: Speeded up robust features. *Lecture Notes in Computer Science* 3951 NCS: 404-417.

Billingsley, J. and M. Dunn (2005). Unusual vision - machine vision applications at the NCEA. *Sensor Review* 25(3): 202-208.

Brosnan, T. and D.-W. Sun (2004). Improving quality inspection of food products by computer vision - A review. *Journal of Food Engineering* 61: 3-16.

Brunelli, R. and T. Poggio (1993). Face recognition: features versus templates. *IEEE Transactions on Pattern Analysis and Machine Intelligence* 15(10): 1042-1052.

Buker, U., S. Drue, et al. (2001). Vision-based control of an autonomous disassembly station. *Robotics and Autonomous Systems* 35(3-4): 179-189.

Burschka, D., M. Li, et al. (2004). Scale-invariant registration of monocular endoscopic images to CT-scans for sinus surgery. *Lecture Notes in Computer Science* 3217(1): 413-421.

Denavit, J. and R. S. Hartenberg (1955). A kinematic notation for lower pair mechanisms. *Journal of Applied Mechanics* 23: 215-221.

Fu, K. S. and J. K. Mui (1981). A survey on image segmentation. *Pattern Recognition* 13: 3-16.

Graham, A. E., S. Q. Xie, et al. (2007). Robotic Long Bone Fracture Reduction. Medical Robotics. V. Bozovic, *i-Tech Education and Publishing*: 85-102

Harris, C. and M. Stephens (1988). A combined corner and edge detector. *Proceedings of the 4th Alvey Vision Conference*: 147-151.

Hartley, R. and A. Zisserman (2003). *Multiple View Geometry in Computer Vision*, Cambridge University Press.

Hough, P. (1962, December). *Method and means for recognizing complex patterns.*

Jamwal, P. K. and H. S. Shan (2005). Prediction of Performance Characteristics of the ECM Process Using GA Optimized Fuzzy Logic Controller. *Proceedings of Indian International Conference on Artificial Intelligence.*

Jang, J. S. R., C. T. Sun, et al. (1997). *Neuro-Fuzzy and Soft computing.* NJ, Prentice Hall.

Jeong, S., J. Chung, et al. (2005). Design of a simultaneous mobile robot localization and spatial context recognition system. *Lecture Notes in Computer Science* 3683 NAI: 945-952.

Kadir, T., A. Zisserman, et al. (2004). An affine invariant salient region detector. *Lecture Notes in Computer Science* 3021: 228-241.

Kass, M., A. Witkin, et al. (1988). Snakes, active contour model. *International Journal of Computer Vision*: 321–331.

Ke, Y. and R. Sukthankar (2004). PCA-SIFT: A more distinctive representation for local image descriptors. *Proceedings of the IEEE Computer Society Conference on Computer Vision and Pattern Recognition* 2: 506-513.

Krar, S. and A. Gill (2003). *Exploring Advanced Manufacturing Technologies.* 200 Madison Avenue, New York., Industrial Press Inc.

Lowe, D. G. (2004). Distinctive image features from scale-invariant keypoints. *International Journal of Computer Vision* 60(2): 91-110.

Lucas, B. D. and T. Kanade (1981). An iterative image registration technique with an application to stereo vision. *Proceedings of Imaging understanding workshop,*: 121-130.

Mikolajczyk, K. and C. Schmid (2004). Comparison of affine-invariant local detectors and descriptors. *Proceedings of the 12th European Signal Processing Conference*: 1729-1732.

Mikolajczyk, K. and C. Schmid (2005). A performance evaluation of local descriptors. *Proceedings of the IEEE Computer Society Conference on Computer Vision and Pattern Recognition* 2: 257-263.

Pearson, K. (1991). On lines and planes of closest fit to systems of points in space. *Philosophical Magazine* 2(6): 559-572.

Pena-Cabrera, M., I. Lopez-Juarez, et al. (2005). Machine vision approach for robotic assembly. *Assembly Automation* 25(3): 204-216.

Spong, M. W., S. Hutchinson, et al. (2006). *Robot Modeling and Control*, John Wiley & Sons, Inc.

Tsai, R. Y. (1987). A versatile camera calibration technique for high-accuracy 3D machine vision metrology using off-the-shelf TV cameras and lenses. *IEEE Journal of Robotics and Automation* RA-3(4): 323-344.

Weijer, J. V. D. and C. Schmid (2006). Coloring Local Feature Extraction. *Lecture Notes in Computer Science* 3952: 334-348.

Wong, A. K. C., L. Rong, et al. (1998). Robotic vision: 3D object recognition and pose determination. *IEEE International Conference on Intelligent Robots and Systems* 2: 1202-1209.

Yaniv, Z. and L. Joskowicz (2005). Precise Robot-Assisted Guide Positioning for Distal Locking of Intramedullary Nails. *IEEE Transactions on Medical Imaging* 24(5): 624-625.

Zitova, B. and J. Flusser (2003). Image registration methods: A survey. *Image and Vision Computing* 21(11): 977-1000.